Modeling and Simulation Techniques in Structural Engineering

Pijush Samui
National Institute of Technology Patna, India

Subrata Chakraborty
Indian Institute of Engineering Science and Technology (IIEST), Shibpur, India

Dookie Kim
Kunsan National University, South Korea

A volume in the Advances in Civil and Industrial Engineering (ACIE) Book Series

IGI GLOBAL
DISSEMINATOR OF KNOWLEDGE

www.igi-global.com

Published in the United States of America by
IGI Global
Engineering Science Reference (an imprint of IGI Global)
701 E. Chocolate Avenue
Hershey PA, USA 17033
Tel: 717-533-8845
Fax: 717-533-8661
E-mail: cust@igi-global.com
Web site: http://www.igi-global.com

Library of Congress Cataloging-in-Publication Data

Names: Samui, Pijush, 1978- editor. | Chakraborty, Subrata, 1969- editor. | Kim, Dookie, 1969- editor.
Title: Modeling and simulation techniques in structural engineering / Pijush Samui, Subrata Chakraborty, and Dookie Kim, editors.
Description: Hershey, PA : Engineering Science Reference, [2017] | Includes bibliographical references and index.
Identifiers: LCCN 2016017352| ISBN 9781522505884 (hardcover) | ISBN 9781522505891 (ebook)
Subjects: LCSH: Structural engineering--Mathematics. | Structural analysis (Engineering)--Data processing. | Structural frames--Simulation methods. | Buildings--Computer simulation.
Classification: LCC TA643 .M57 2017 | DDC 624.101/1--dc23 LC record available at https://lccn.loc.gov/2016017352

This book is published in the IGI Global book series Advances in Civil and Industrial Engineering (ACIE) (ISSN: 2326-6139; eISSN: 2326-6155)

British Cataloguing in Publication Data
A Cataloguing in Publication record for this book is available from the British Library.

All work contributed to this book is new, previously-unpublished material. The views expressed in this book are those of the authors, but not necessarily of the publisher.

For electronic access to this publication, please contact: eresources@igi-global.com.

Advances in Civil and Industrial Engineering (ACIE) Book Series

ISSN: 2326-6139
EISSN: 2326-6155

Mission

Private and public sector infrastructures begin to age, or require change in the face of developing technologies, the fields of civil and industrial engineering have become increasingly important as a method to mitigate and manage these changes. As governments and the public at large begin to grapple with climate change and growing populations, civil engineering has become more interdisciplinary and the need for publications that discuss the rapid changes and advancements in the field have become more in-demand. Additionally, private corporations and companies are facing similar changes and challenges, with the pressure for new and innovative methods being placed on those involved in industrial engineering.

The **Advances in Civil and Industrial Engineering (ACIE) Book Series** aims to present research and methodology that will provide solutions and discussions to meet such needs. The latest methodologies, applications, tools, and analysis will be published through the books included in **ACIE** in order to keep the available research in civil and industrial engineering as current and timely as possible.

Coverage

- Operations Research
- Materials Management
- Production Planning and Control
- Engineering Economics
- Structural Engineering
- Productivity
- Quality Engineering
- Coastal Engineering
- Urban Engineering
- Optimization Techniques

IGI Global is currently accepting manuscripts for publication within this series. To submit a proposal for a volume in this series, please contact our Acquisition Editors at Acquisitions@igi-global.com or visit: http://www.igi-global.com/publish/.

The Advances in Civil and Industrial Engineering (ACIE) Book Series (ISSN 2326-6139) is published by IGI Global, 701 E. Chocolate Avenue, Hershey, PA 17033-1240, USA, www.igi-global.com. This series is composed of titles available for purchase individually; each title is edited to be contextually exclusive from any other title within the series. For pricing and ordering information please visit http://www.igi-global.com/book-series/advances-civil-industrial-engineering/73673. Postmaster: Send all address changes to above address.

Titles in this Series

For a list of additional titles in this series, please visit: www.igi-global.com

Computational Modeling of Masonry Structures Using the Discrete Element Method
Vasilis Sarhosis (Newcastle University, UK) Katalin Bagi (Budapest University of Technology and Economics, Hungary) José V. Lemos (National Laboratory for Civil Engineering, Portugal) and Gabriele Milani (Technical University in Milan, Italy)
Engineering Science Reference • copyright 2016 • 505pp • H/C (ISBN: 9781522502319) • US $210.00 (our price)

Handbook of Research on Emerging Innovations in Rail Transportation Engineering
B. Umesh Rai (Chennai Metro Rail Limited, India)
Engineering Science Reference • copyright 2016 • 664pp • H/C (ISBN: 9781522500841) • US $235.00 (our price)

Advanced Research on Nanotechnology for Civil Engineering Applications
Anwar Khitab (Mirpur University of Science and Technology, Pakistan) and Waqas Anwar (Mirpur University of Science and Technology, Pakistan)
Engineering Science Reference • copyright 2016 • 339pp • H/C (ISBN: 9781522503446) • US $195.00 (our price)

Emerging Challenges and Opportunities of High Speed Rail Development on Business and Society
Raj Selladurai (Indiana University Northwest, USA) Peggy Daniels Lee (Kelley School of Business, Indiana University, USA) and George VandeWerken (Providence Bank, USA)
Engineering Science Reference • copyright 2016 • 289pp • H/C (ISBN: 9781522501022) • US $210.00 (our price)

Advanced Manufacturing Techniques Using Laser Material Processing
Esther Titilayo Akinlabi (Univeristy of Johannesburg, South Africa) Rasheedat Modupe Mahamood (University of Johannesburg, South Africa & University of Ilorin, Nigeria) and Stephen Akinwale Akinlabi (University of Johannesburg, South Africa)
Engineering Science Reference • copyright 2016 • 288pp • H/C (ISBN: 9781522503293) • US $175.00 (our price)

Handbook of Research on Applied E-Learning in Engineering and Architecture Education
David Fonseca (La Salle Campus Barcelona, Universitat Ramon Llull, Spain) and Ernest Redondo (Universitat Politècnica de Catalunya, BarcelonaTech, Spain)
Engineering Science Reference • copyright 2016 • 569pp • H/C (ISBN: 9781466688032) • US $310.00 (our price)

Emerging Design Solutions in Structural Health Monitoring Systems
Diego Alexander Tibaduiza Burgos (Universidad Santo Tomás, Colombia) Luis Eduardo Mujica (Universitat Politecnica de Catalunya, Spain) and Jose Rodellar (Universitat Politecnica de Catalunya, Spain)
Engineering Science Reference • copyright 2015 • 337pp • H/C (ISBN: 9781466684904) • US $235.00 (our price)

IGI GLOBAL
DISSEMINATOR OF KNOWLEDGE
www.igi-global.com

701 E. Chocolate Ave., Hershey, PA 17033
Order online at www.igi-global.com or call 717-533-8845 x100
To place a standing order for titles released in this series, contact: cust@igi-global.com
Mon-Fri 8:00 am - 5:00 pm (est) or fax 24 hours a day 717-533-8661

Dedicated to Amitava Pal

Table of Contents

Detailed Table of Contents

The close relationship between noise and vibration is reviewed and analyzed for the suppression of noise and vibration in structures. The suppression of noise and vibration by acoustic means are addressed. For the first, an analysis is carried out by representing noise by monopoles and higher harmonics, and to devise a straight-forward method to counter their influence by selective secondary acoustic source. The second problem is analyzed using a methodology developed earlier for the computational scheme for the calculation of the acoustic disturbance to the aeroelasticity of structures. The generic approach of the latter consists of three parts. The first is the formulation of the acoustic wave propagation governed by the Helmholtz equation by using boundary element approach, to allow the calculation of the acoustic pressure on the acoustic-structure boundaries. The structural dynamic problem is formulated using finite elements. The third part involves the calculation of the unsteady aerodynamics loading on the structure using generic unsteady aerodynamics computational method.

Smart structures are complex systems situated in even more complex and large scale urban environments. This chapter opens the field of agent based modelling and simulation (ABMS) to civil engineers. ABMS offers a wide range of tools for implementing simulation models of systems with high degrees of interconnectivity and a large number of component subsystems. The ease of use for specialized engineers and the capabilities of integration with existent technologies and infrastructures, make agent based models a very attractive way to incorporate the social system in the design process of buildings. Moreover, ABMS allows for the testing and validation of structure wide control and automation systems. This chapter presents past and current efforts of using agent based modelling for smart structures, as well as the main challenges brought by this new interdisciplinary research domain.

This chapter presents in one hand the definition of the pushover procedure based on Eurocode 8, and displacement based and force base elements which will be used in this study, and the application of the pushover procedure based to Eurocode 8 to nonlinear framed structure model and comparison with time history analysis. As test-bed application, we consider a two-story reinforced concrete building in Bonefro, Italy. It is representative of typical residential building construction in Italy in the 1970's and 1980's. The aim of this chapter is to compare 2D model for both modal and uniform load distributions with time history analysis. In another hand an Application to two-story RC Frame using force base elements without soil and with taking into account soil in order to have an idea of the soil modelisation effect with the structure in the results

Determination of transverse mechanical properties from ring specimens machined from tubular components is not straightforward due to presence of combined tension and bending stresses. Zircaloy tubes as used in nuclear reactors are manufactured through a complicated process of pilgering and heat-treatment and hence, the properties need to be determined in the as-manufactured condition. In this work, the authors perform ring-tensile tests on specimens of Zircaloy pressure tubes of Indian pressurized heavy water reactor in order to carry out integrity assessment of these tubes. As the loading condition in this test imposes both membrane and bending stresses in the cross-section of the ring, 3-D finite element analysis of the test setup was carried out in order to determine material stress-strain curve using an iterative technique. The effect of the design of the loading mandrel on the experimental stress-strain data has been investigated in detail. To validate the methodology, miniature tensile specimens have been tested and the data has been compared to those of ring specimens.

Wind turbines are slender flexible structures susceptible to strong wind fluctuations. The flexible wind turbine structure, when subjected to strong dynamic forces, it leads to an ideal condition for induced vibrations and resonance problems. Hence studying the dynamic response of these critical structures using the computational and experimental procedures becomes of utmost importance. This chapter reviews the theories used for the dynamic analysis of a modern day offshore wind turbine structure and applies these theories in analyzing realistic situations for offshore turbines under wave and wind action. The first half of the chapter gives a broad overview on the concepts of structural dynamics of wind turbine structures with illustrative examples that will enable the user to understand the methodology used to analyze these structures. The latter half of the chapter deals with the computational aspect of the analysis and focuses on the use of finite element software ANSYS 14 to model these critical structures.

A comparative study of various metamodelling approaches namely the least squares method (LSM), moving least squares method (MLSM) and artificial neural network (ANN) based response surface method (RSM) are presented to demonstrate the effectiveness to approximate the nonlinear dynamic response of structure required for efficient seismic reliability analysis (SRA) of structures. The seismic response approximation by the LSM, MLSM and ANN based RSMs are explained with a brief note on the important issue of ground motion bin generation. The procedure adopted herein for SRA is based on the dual response surface approach. In doing so, the repetition of seismic intensity for SRA at different intensity levels is avoided by including this as one of the predictors in the seismic response prediction model. A nonlinear SDOF system has been taken up to elucidate the effectiveness of various metamodels in SRA.

Pipelines are being used to convey different sorts of fluids from hazardous and toxic substances to high pressure steam. Piping systems are subjected to various external loads leading to major failures with gross plastic deformation. Pipe bends are incorporated into piping systems not only to change the direction of flow but also to provide flexibility, hence they are considered to be critical components and its safe design under various loads becomes important. Earlier studies of pipe bends utilized analytical methods to determine the plastic loads. The evolution of FEM and the advancements in computational capabilities have enabled analysts to generate large number of data which is expensive and time consuming with experimental investigations. In this chapter, the major studies on pipe bends by various researchers are explored. Different studies on pipe bends namely stress analysis and the influence of geometric shape imperfections are also presented.

The exponential growth of computational power during the last few decades has enabled the finite element analysis of many real-life engineering systems which are too complex to be analytically solved in a closed form. In the traditional deterministic finite element analysis, system parameters such as mass, geometry and material properties are assumed to be known precisely and defined exactly. However, in practice most of the data used in the solution process of many practical engineering systems are either collected from experiments or acquired as empirical data from the past, which are usually ill defined, imprecise and uncertain in nature. This work presents a practical approach based on High Dimensional Model Representation (HDMR) for analyzing the response of structures with fuzzy parameters. The proposed methodology involves integrated finite element modelling, HDMR based surrogate model, and explicit fuzzy analysis procedures.

For investigation of fracture behavior and assessment of remaining life of critical thin-walled tubular components in industry, the transverse mechanical property and the axial fracture properties are essential. However, evaluation of these properties by machining suitable standard specimens from these components directly and subjecting them to standard tests is usually not feasible. In this chapter, the nonlocal version of the Rousselier's damage model has been used to predict the fracture resistance behavior of double-edged-notched-tensile specimens made from Zircaloy-4 material. Initially, the micro-mechanical parameters have been determined from the testing of ring-type specimens. Subsequently, these parameters were used in finite element analysis of the double-edged-notched-tensile specimen in order to predict the crack growth behavior and the crack path under applied displacement-controlled loading conditions. The fracture resistance behavior obtained in terms of J-R curve was also compared with the corresponding J-R curves of axially-cracked pin-loading-tension specimens.

Squat walls are widely used in structural engineering. The empirical equations currently used to calculate the peak shear strength of squat walls do not correlate well with the experimental results. Another limitation is the reliance on the use of many assumed intermediate parameters. This chapter explores the use of MARS and BPNN approaches to build predictive peak shear strength models of squat walls based on an extensive experimental database from the literature. First the MARS methodology and its associated procedures will be explained in detail. Analyses of the database are then carried out to verify the MARS's capacity in modelling the non-linear interactions between variables without making any specific assumptions. The performances of the built MARS and BPNN models are compared in terms of predictive accuracy, parameter relative importance, parametric analysis and model interpretability. Design charts are also proposed based on parametric studies using the developed models.

In this book chapter, a system identification method for modeling nonlinear behavior of smart buildings is discussed that has a significantly low computation time. To reduce the size of the training data used for the adaptive neuro-fuzzy inference system (ANFIS), principal component analysis (PCA) is used, i.e., PCA-based adaptive neuro-fuzzy inference system: PANFIS. The PANFIS model is evaluated on a seismically excited three-story building equipped with a magnetorheological (MR) damper. The PANFIS model is trained using an artificial earthquake that contains a variety of characteristics of earthquakes. The trained PANFIS model is tested using four different earthquakes. It was demonstrated that the proposed PANFIS model is effective in modeling nonlinear behavior of a smart building with significant reduction in computational loads.

Uncertainty quantification of civil engineering structures are often time consuming. Hence, one has to rely on equivalent models (EM) that represent the original functions with a specific level of accuracy. In this chapter, a fully equivalent operational model, referred as, Polynomial Correlated Function Expansion (PCFE) has been discussed. PCFE facilitates a systematic mapping between the input and output by expressing the output as a ranked order of component functions, with higher order component functions representing higher-order cooperative effect. The component functions are expressed in terms of extended bases and the unknown coefficients associated with the bases are determined by employing homotopy algorithm (HA). HA determines the unknown coefficients by minimizing the least-squared error and imposing an additional criteria defined as an objective function. Implementation of PCFE has been illustrated with two large scale problems. Results obtained have been compared with other popular techniques. For both the problems, PCFE outperforms popular EMs available in literature.

Chapter 13

M.K. Samal, Bhabha Atomic Research Centre, India

In this chapter, a mathematical model for rate of formation of chromium carbides near the grain boundary, which is a pre-cursor to chromium depletion and corresponding sensitization behavior in stainless steels, is presented. This model along with the diffusion equation for chromium in the grain has been used to obtain chromium depletion profiles at various time and temperature conditions. Finite difference method has been used to solve the above equations in the spherical co-ordinate system and the results of time-temperature-sensitization diagrams of four different types of alloys have been compared with those of experiment from literature. For the problem of low temperature sensitization and corresponding inter-granular corrosion in austenitic stainless steel, it is very difficult to carry out experiment at higher temperatures and justify its validity at lower operating temperatures by extrapolation. The development of predictive models is highly useful in order to design the structures for prevention of corrosion of the material in aggressive environments.

Chapter 14

Sharvil Alex Faroz, Indian Institute of Technology Bombay, India
Nikil Narayan Pujari, Indian Institute of Technology Bombay, India
Rohit Rastogi, Bhabha Atomic Research Centre, India
Siddhartha Ghosh, Indian Institute of Technology Bombay, India

Engineering phenomena described through models come with uncertainties arising out of the modelling assumptions and observational statistical discrepancies. A probabilistic framework helps in quantifying the uncertainties involved. Conventional statistics are not very useful for any inferring from a limited amount of data which is often the case for many real life engineering problem. Bayesian inference tackles this issue by integrating past experience with current information. This chapter introduces the basic concepts of Bayesian inference and illustrates through three problems in structural engineering, how it can be useful. The emphasis is in showing how Markov chain Monte Carlo (MCMC) simulation is used in Bayesian inference.

Chapter 15

Behrouz Behnam, Amirkabir University of Technology (Tehran Polytechnic), Iran

Post-earthquake fire (PEF) is one of the most complicated problems resulting from earthquake, presenting a serious risk to urban buildings. As most standards and codes ignore the possibility of PEF, buildings are too weak under PEF loads. This chapter is to investigate the effects of PEF loads on partially damaged RC buildings located in urban regions. To do that, a methodology named sequential analysis is introduced here via which the structural performance at various performance levels is evaluated under fire and PEF scenarios. Numerically, in order to simulate the earthquake loads, conventional pushover analysis is employed, with an explanation presented in the chapter to introduce the pushover analysis, its advantages and its limitations. To simulate the fire loads, standard fire curve (ISO 834) is used for simplicity.

Preface

The performance of structure is assessed through conducting experiment or analysis of mechanical model. Experimental techniques are time consuming and expensive. Engineering community always prefer model for determination of performance of structure over experimental technique. Thus, the development of analytical and numerical model is the major focus for understanding performance of structure. The rapid advance in information processing systems in recent decades directed engineering and research towards the development of various computational methods that can model engineering phenomena realistically. The accuracy of the different models will be different. Therefore, it is necessary to develop the best possible model for a particular problem in structural engineering. Therefore, the development of new model is an imperative task for better understanding of behavior of structure. This book is an attempt to present various computational techniques for application in structural engineering. It consists of 16 chapters, and aims to present the developments in advanced computational techniques and new research areas in the diverse field of structural engineering.

In Chapter 1, the close relationship between noise and vibration is reviewed and analyzed for the suppression of noise and vibration in structures. For the first, an analysis is carried out by representing noise by monopoles and higher harmonics, and to devise a straight-forward method to counter their influence by selective secondary acoustic source. The second problem is analyzed using a methodology developed earlier for the computational scheme for the calculation of the acoustic disturbance to the aeroelasticity of structures. The generic approach of the latter consists of three parts. The first is the formulation of the acoustic wave propagation governed by the Helmholtz equation by using boundary element approach, to allow the calculation of the acoustic pressure on the acoustic-structure boundaries. The structural dynamic problem is formulated using finite elements. The third part involves the calculation of the unsteady aerodynamics loading on the structure using generic unsteady aerodynamics computational method. Chapter 2 opens the field of agent based modelling and simulation (ABMS) to civil engineers. ABMS offers a wide range of tools for implementing simulation models of systems with high degrees of interconnectivity and a large number of component subsystems. The ease of use for specialized engineers and the capabilities of integration with existent technologies and infrastructures, make agent based models a very attractive way to incorporate the social system in the design process of buildings. Moreover, ABMS allows for the testing and validation of structure wide control and automation systems. Chapter 3 presents the past and current efforts of using agent based modelling for smart structures, as well as the main challenges brought by this new interdisciplinary research domain. The chapter presents in one hand the definition of the pushover procedure based on Eurocode 8, and displacement based and force base elements which will be used in this study, in another hand the application of the pushover procedure based to Eurocode 8 to nonlinear framed structure model and comparison with time history analysis. As test-bed application, a two-story reinforced concrete building in Bonefro, Italy is considered. It is representative of typical residential building construction in Italy in the 1970's and 1980's. In chapter 4, the authors present the ring-tensile tests on specimens of Zircaloy pressure tubes of Indian pressurized heavy water reactor in order to carry out integrity assessment of these tubes. Determination of transverse mechanical properties from ring specimens machined from tubular components is not straightforward due to presence of combined tension and bending stresses. Zircaloy tubes as used in nuclear reactors are manufactured through a complicated process of pilgering and heat-treatment and hence, the properties need to be determined in the as-manufactured condition. As the loading condition in this test imposes both membrane and bending stresses in the cross-section of the ring, 3-D finite element analysis of the test setup was carried out in order to determine material stress-strain curve using an iterative technique. The effect of the design of the loading mandrel on the experimental stress-strain data has been investigated in detail. To validate the methodology, miniature tensile specimens have been tested and the data has been compared to those of ring specimens. Chapter 5 reviews the theories used for dynamic analysis of a modern day offshore wind turbine structure and applies these theories in analyzing realistic situations for offshore turbines under wave and wind action. The first half of the chapter provides a broad overview on the concepts of structural dynamics of wind turbine structures with illustrative examples that will enable the user to understand the methodology used to analyze these structures. The latter half of the chapter deals with the computational aspect of the analysis and focuses on the use of finite element software ANSYS 14 to model these critical structures. In chapter 6, a comparative study of various metamodelling approaches namely the least squares method (LSM), moving least squares method (MLSM) and artificial neural network (ANN) based response surface method (RSM) are presented to demonstrate the effectiveness to approximate the nonlinear dynamic response of structure required for efficient seismic reliability analysis (SRA) of structures. The seismic response approximation by the LSM, MLSM and ANN based RSMs are explained with a brief note on the important issue of ground motion bin generation. The procedure adopted herein for SRA is based on a dual response surface approach. In doing so, the repetition of seismic intensity for SRA at different intensity levels is avoided by including this as one of the predictors in the seismic response prediction model. A nonlinear SDOF system has been taken up to elucidate the effectiveness of various metamodels in SRA. In chapter 7, the major studies on pipe bends by various researchers are explored thoroughly through extensive literature survey. Pipelines are being used to convey different sorts of fluids from hazardous and toxic substances to high pressure steam. Piping systems are subjected to various external loads leading to major failures with gross plastic deformation. Pipe bends are incorporated into piping systems not only to change the direction of flow but also to provide flexibility, hence they are considered to be critical components and its safe design under various loads becomes important. Earlier studies of pipe bends utilized analytical methods to determine the plastic loads. The evolution of FEM and the advancements in computational capabilities have enabled analysts to generate large number of data which is expensive and time consuming with experimental investigations. Different studies on pipe bends namely stress analysis, creep analysis, fatigue analysis, analysis with cracks and the influence of geometric shape imperfections on pipe bends are also presented. Chapter 8 presents a practical approach based on High Dimensional Model Representation (HDMR) for analyzing response of structures with fuzzy parameters. The exponential growth of computational power during the last few decades has enabled the finite element analysis of many real-life engineering systems which are too complex to be analytically solved in a closed form. In the traditional deterministic finite element analysis, system parameters such as mass, geometry and material properties are assumed to be known precisely and defined exactly.

However, in practice most of the data used in the solution process of many practical engineering systems are either collected from experiments or acquired as empirical data from the past, which are usually ill defined, imprecise and uncertain in nature. The proposed methodology involves integrated finite element modelling, HDMR based surrogate model, and explicit fuzzy analysis procedures. In chapter 9, the nonlocal version of the Rousselier's damage model has been used to predict the fracture resistance behavior of double-edged-notched-tensile specimens made from Zircaloy-4 material. For investigation of fracture behavior and assessment of remaining life of critical thin-walled tubular components in industry, the transverse mechanical property and the axial fracture properties are essential. However, evaluation of these properties by machining suitable standard specimens from these components directly and subjecting them to standard tests is usually not feasible. Initially, the micro-mechanical parameters have been determined from the testing of ring-type specimens. Subsequently, these parameters were used in finite element analysis of the double-edged-notched-tensile specimen in order to predict the crack growth behavior and the crack path under applied displacement-controlled loading conditions. The fracture resistance behavior obtained in terms of J-R curve was also compared with the corresponding J-R curves of axially-cracked pin-loading-tension specimens. Chapter 10 explores the use of MARS and BPNN approaches to build predictive peak shear strength models of squat walls based on an extensive experimental database from the literature. Squat walls are widely used in structural engineering. The empirical equations currently used to calculate the peak shear strength of squat walls do not correlate well with the experimental results. Another limitation is the reliance on the use of many assumed intermediate parameters. First the MARS methodology and its associated procedures will be explained in detail. Analyses of the database are then carried out to verify the MARS's capacity in modelling the non-linear interactions between variables without making any specific assumptions. The performances of the built MARS and BPNN models are compared in terms of predictive accuracy, parameter relative importance, parametric analysis and model interpretability. Design charts are also proposed based on parametric studies using the developed models. In chapter 11, a system identification method for modeling nonlinear behavior of smart buildings is discussed that has a significantly low computation time. To reduce the size of the training data used for the adaptive neuro-fuzzy inference system (ANFIS), principal component analysis (PCA) is used, i.e., PCA-based adaptive neuro-fuzzy inference system: PANFIS. The PANFIS model is evaluated on a seismically excited three-story building equipped with a magnetorheological (MR) damper. The PANFIS model is trained using an artificial earthquake that contains a variety of characteristics of earthquakes. The trained PANFIS model is tested using four different earthquakes. It was demonstrated that the proposed PANFIS model is effective in modeling nonlinear behavior of a smart building with significant reduction in computational loads. In chapter 12, a fully equivalent operational model, referred as, Polynomial Correlated Function Expansion (PCFE) has been discussed. Uncertainty quantification of civil engineering structures are often time consuming. Hence, one has to rely on equivalent models (EM) that represent the original functions with a specific level of accuracy PCFE facilitates a systematic mapping between the input and output by expressing the output as a ranked order of component functions, with higher order component functions representing higher-order cooperative effect. The component functions are expressed in terms of extended bases and the unknown coefficients associated with the bases are determined by employing homotopy algorithm (HA). HA determines the unknown coefficients by minimizing the least-squared error and imposing an additional criteria defined as an objective function. Implementation of PCFE has been illustrated with two large scale problems. Results obtained have been compared with other popular techniques. For both the problems, PCFE outperforms popular EMs available in literature. In chapter 13, a mathematical model for rate of formation of chromium carbides near the grain boundary, which is a pre-cursor to chromium depletion and corresponding sensitization behavior in stainless steels, has been presented. This model along with the diffusion equation for chromium in the grain has been used to obtain chromium depletion profiles at various time and temperature conditions. Finite difference method has been used to solve the above equations in the spherical co-ordinate system and the results of time-temperature-sensitization diagrams of four different types of alloys have been compared with those of experiment from literature. For the problem of low temperature sensitization and corresponding inter-granular corrosion in austenitic stainless steel, it is very difficult to carry out experiment at higher temperatures and justify its validity at lower operating temperatures by extrapolation. The development of predictive models is highly useful in order to design the structures for prevention of corrosion of the material in aggressive environments. Chapter 14 introduces the basic concepts of Bayesian inference and illustrated through three problems in structural engineering, how it can be useful. Engineering phenomena described through models come with uncertainties arising out of the modelling assumptions and observational statistical discrepancies. A probabilistic framework helps in quantifying the uncertainties involved. Conventional statistics are not very useful for any inferring from a limited amount of data which is often the case for many real life engineering problems. Bayesian inference tackles this issue by integrating past experience with current information. The emphasis is in showing how Markov chain Monte Carlo (MCMC) simulation is used in Bayesian inference. Chapter 15 is to investigate the effects of Post-earthquake fire (PEF) loads on partially damaged RC buildings located in urban regions. To do that, a methodology named sequential analysis is introduced here via which the structural performance at various performance levels is evaluated under fire and PEF scenarios. PEF is one of the most complicated problems resulting from earthquake, presenting a serious risk to urban buildings. As most standards and codes ignore the possibility of PEF, buildings are too weak under PEF loads. Numerically, in order to simulate the earthquake loads, conventional pushover analysis is employed, with an explanation presented in the chapter to introduce the pushover analysis, its advantages and its limitations. To simulate the fire loads, standard fire curve (ISO 834) is used for simplicity. Chapter 16 presents potential simulation approaches that are possible to utilize for SIA&C in relation to prioritization as well as evaluation of current status of a SIA&C organization. Structural integrity assessment and control (SIA & C) are vital for existing ageing as well as newly built offshore and onshore structures. The SIA & C becomes highly sensitive to interventions under a potential loss of structural integrity (SI) especially when there are inherent constraints present in carrying out engineering work in hazardous environments such as petroleum production and process facilities (P&PFs). The challenges have been further exacerbated by the constantly ageing onshore and offshore structures whilst it is the necessity of carrying out life extension at the verge of their design service lives. Local and international regulations demand the implementation of appropriate strengthening, modification and repair (SMR) plans when significant changes in the SI have been revealed.

All the chapters are written by more than fifteen experts in the subjects with a hope that the book will give impetus for application of new modeling techniques for assessing performance of structure. We are deeply indebted and wish to thank all the authors for their valuable contributions. We thank all reviewers and editorial board members for their helpful supports. A special thanks goes to the staff members of IGI Global. We would also like to acknowledge the support of our institutions during preparation of this edited book.

Chapter 1
Acoustic Method for the Suppression of Acoustic and Aerodynamically Induced Vibration on Structures

Harijono Djojodihardjo
Independent Researcher, Indonesia

ABSTRACT

The close relationship between noise and vibration is reviewed and analyzed for the suppression of noise and vibration in structures. The suppression of noise and vibration by acoustic means are addressed. For the first, an analysis is carried out by representing noise by monopoles and higher harmonics, and to devise a straight-forward method to counter their influence by selective secondary acoustic source. The second problem is analyzed using a methodology developed earlier for the computational scheme for the calculation of the acoustic disturbance to the aeroelasticity of structures. The generic approach of the latter consists of three parts. The first is the formulation of the acoustic wave propagation governed by the Helmholtz equation by using boundary element approach, to allow the calculation of the acoustic pressure on the acoustic-structure boundaries. The structural dynamic problem is formulated using finite elements. The third part involves the calculation of the unsteady aerodynamics loading on the structure using generic unsteady aerodynamics computational method.

INTRODUCTION

Noise and vibration occur in many engineering and domestic systems, are closely related and have received considerable attention. In industry, for example, engines, fans, transformers and compressors radiate noise. In cars, boats, trains and aircraft, for example, noise reduces comfort. With the progressive trends of the utilization of lightweight materials and more powerful and high performance systems, increased vibration and noise levels may prevail if these problems are not given due considerations. Structural-acoustic interaction problems have shown enhanced interest in the aerospace and automotive

DOI: 10.4018/978-1-5225-0588-4.ch001

industries. Acoustic excitation of the fuselage has been one of the main concerns during certain flight operations (see Figure 1). Under such conditions, the coupling of the acoustic and the structure systems and determining their combined random responses become extremely important.

Associated with man-machine system interaction, the primary concern with noise in the low frequency range is not only the potential risk of damage to the hearing, since low frequency noise is annoying and during periods of long exposure causes fatigue, discomfort and loss of concentration. Reduced concentration may also lead to an increased risk for accidents. The masking effect which low frequency noise has on speech also reduces speech intelligibility (Johansson, 2000; Lagö, 1996; Lagö, Johansson & Hellström, 1997), which is considered to be disturbing. Active noise control is a significant demand for human comfort and environmental requirements, including by the aviation authorities.

Noise problems may be solved using traditional passive approaches and/or approaches based on the concept of active noise control (Kuo & Morgan, 1996; Juang, 1993; Johansson et al, 1997; Kempton, 1976; Germain, 2000; Johansson, 2000; Pabst et al, 2009; Griffin, 2006; Kusni & Soenarko, 2001). The choice of technique is determined by the characteristics of the noise as well as of the application. Conventional passive approaches consist of absorbers and/or reflectors/barriers. Active noise control can be carried out using secondary noise source (Johansson, 2000; Kusni, Soenarko & Djojodihardjo, 2014). Similarly, for vibration, active vibration control can also be carried out using noise source.

Efforts for noise control have been carried out in many literatures, among others in the work by Kempton (1976), Germain (2000), Johansson (2000), Pabst et al (2009), Griffin (2006) and Kusni et al (2001), among others. Establishing very high frequency bound is a challenging problem in active noise reduction by using secondary source. In this regards, the distance of the secondary source with respect to the primary source is a factor that can seriously inhibit effective active noise reduction. Such problem may arise in the use of monopole secondary source. Such issue motivates the development and utilization of the method of multipole secondary source. The advantage of using the multipole secondary source is in the reduction of higher frequency noise and the number of secondary sources, which could lead further to a more compact active noise reduction system. Although the concept of the multipole secondary source is a widely known technique, the use of multipole secondary source for noise elimination has a limitation, since it is impossible to use the multipole secondary source of infinite order. To solve the problem, an optimization of secondary source strength will be performed so that the radiation of

Figure 1. Dominant cabin noise due to engine noise; other source of noise in boundary layer noise on the outer surface of the fuselage

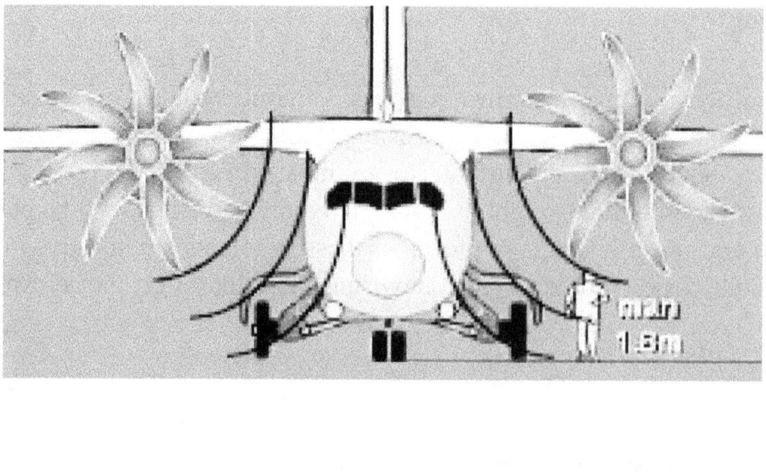

the sound power of the sound field can be minimized. Such effort will allow efficient sound reduction using limited multipole order. Such attempt has been made by Kusni (2001, 2002), Kusni and Soenarko (2001), and Kusni, Soenarko and Djojodihardjo (2014). The Concept of Sound Reduction with Multipole Source has been employed in the application of the Boundary Integral Method to sound radiation problems Seybert et al. (1984). In aerodynamics, the use of mutipoles has been utilized by Hess and Smith (1967), Djojodihardjo and Widnall (1969), as also comprehensively described by Ashley and Rodden (1972), among others. Utilization of monopole and multipole acoustic sources have also been discussed by Djojodihardjo (2009), Yu et al (2010) and Djojodihardjo (2013). Techniques on the use of multipole sources for description and cancellation of sound fields have been discussed by Nelson et al (1986), Ruijgrok (1993). Koopman et al (1989), Bolton and Beauvilain (1992), Beauvilain and Bolton (1993), Bolton et al (1995). These techniques have inspired one of the developments of the present work.

Other related works that should be considered in developing the present work include the work of Chan (2012), who carried out work focused on the control and suppression of vortex shedding of flow past bluff bodies. Two different computational fluid dynamic methods have been used in this research–the Semi-Implicit Method for Pressure Linked Equation that is widely employed in today's commercial incompressible flow solvers, and the high-order spectral difference method, recently developed for compressible flow solution. Experimental measurements have been performed to further validate the results. Chan proposed two very different suppression techniques, which are a passive control using a thin splitter plate positioned downstream of the bluff body, and an active control by way of counter rotating a cylinder pair. Chen (1985) investigated flow-induced vibration associated with the response of structures placed in or conveying fluid flow. His work covers those cases in which an interaction develops between fluid-dynamic forces and the inertia, damping or elastic forces in the structures. His extensive and systematic work covers flow-induced vibration of complex and diverse phenomena. Typical examples comprise subcritical vibration of nuclear fuel assemblies, galloping of transmission lines, flutter of pipes conveying fluid, and whirling of heat exchanger tube banks.

In view of the progress of research work on the aerodyamics and aeroacoustics of three-dimensional cavities at transonic speeds, Khanal (2010) carried out a numerical investigation of the Aerodynamic Noise Generation Mechanism in Transonic Cavity Flows with special attention to identify the flow structures responsible for resonant tones. The cavity length to depth (L=D) ratio values were chosen to focus the investigations mainly in the open cavity flow regime. The cavity geometries considered also included the cavities with stores. Hansen et al (2007) carried out optimization of active and semi-active noise and vibration control systems. Active and semi-active noise control system designs were considered to be multi-variable optimization processes. Legendre (2014) considers various effects of vortices on the propagation of acoustic waves, from simple convection effects to instabilities in the acoustic phenomena, including absorption, reflection and refraction effects. His work focuses on the effects of mean flow vorticity on the acoustic propagation.

In the present work, a unified analysis of the interaction between acoustic excitation and acoustic field at certain domain as well as structural vibration at certain boundary is carried out using fundamental approach to investigate the noise and/ or vibration alleviation. The analysis will start with the propagation of sound from a certain source, and eventually work out the interaction between the propagated sound wave to a noise source field at a certain domain or boundary, or the fluid-structure interaction between the propagated sound wave and the vibration of the structure. Accordingly, each problem is discussed and generic examples are worked out.

TYPES OF ACOUSTIC SOUND GENERATION

In dealing with noise control of fluid-borne sound, the source mechanisms should be identified, and to this end it can broadly be organized in three general classes based on their phenomenological origin, as exhibited in Table 1 (Chan, 2012).

In the first category, the sound phenomenon generally contains the most efficient source mechanisms in terms of the ratio of radiated power to supplied mechanical or electrical power. The most common example of sources in this category is the radiation from vibrating bodies where fluid volumes are displaced in an unsteady manner. There is, however, some qualifications necessary for this rule and that is since vibrating bodies simultaneously exert forces on the fluid. Fluctuations in the volume can also come about due to unsteady flow like in pulsating ventilation flows or due to injections of heat as in combustion processes where the temporally varying thermal expansions are the underlying fluctuations, and as one injects or pumps in air in a fluid space which is otherwise at rest.

In the second category unsteady forces are exerted on a fluid without any net volume displacement. This category incorporates mechanisms such as propeller noise, valve noise and the noise from rustling leaves. The presence of a rigid body in a moving fluid, however, constrains it locally and a small proportion of the flow energy is converted to acoustic by the forces internal to the fluid. The complexity of the fluid dynamic processes underlying this conversion makes it difficult to give simple qualitative descriptions (Sound Generation-Compendium, 2013).

The third category encompasses mechanisms where neither a net volume velocity nor a net force is produced. The most common example of this mechanism is the turbulent flow from a jet as for example the jet engine exhaust. For such a source mechanism, there is no interaction with solid surfaces. It stems from the shear stresses at the interface between fluid elements with differing velocities and which are also unsteady. The basic problem with such mechanisms is that they cannot be accommodated by the inhomogeneous linearized wave equation. Employing Lighthill's so-called "acoustic analogy" (Sound Generation-Compendium, 2013; Lighthill, 1952; Jacobsen and Juhl, 2011), however, the non-linear features of the mechanism can be incorporated as a source-term in the inhomogeneous wave equation. These three categories are associated with the three different ways in which kinetic energy can be converted into acoustic energy (Legendre, 2014). These are:

1. By forcing the mass in a fixed region of space to fluctuate, such as in a loud-speaker diaphragm embedded in a very large baffle.

Table 1. Acoustic source categories

Category	Phenomenon	Examples
1	Volume: • Fluctuating • Displacing • Injecting	Loudspeaker Exhaust pipes Combustion
2	Force or Pressure Fluctuating	Propeller Flow Obstructions
3	Shear Stress Fluctuating	Jet Engines Pressurized Air Nozzles

Adapted from Chan, 2012.

2. By forcing the momentum in a fixed region of space to fluctuate, or forcing the rates of mass flux across fixed surfaces to vary; both these occur when a solid object vibrates after being struck.
3. By forcing the rates of momentum flux across fixed surfaces to vary, as when sound is generated aerodynamically with no motion of solid boundaries.

In the development of aeroacoustics, Lighthill (1952) addressed the sound generated aerodynamically, that is, as a by-product of an airflow, as distinct from sound produced by the vibration of solids. Fluid turbulent flows tend to generate noise. This differs from sound produced by the vibration of solids. Lighthill initiated a theory, based on the equations of motion of a gas, for the purpose of estimating the sound radiated from a fluid flow, with rigid boundaries, which as a result of instability containing regular fluctuations or turbulence. The sound field is that which would be produced by a static distribution of acoustic quadrupoles whose instantaneous strength per unit volume is $\rho v_i v_j + p_{ij} - a_0^2 \rho \delta_{ij}$, where ρ is the density, v_i the velocity vector, p_{ij} the compressive stress tensor, and a_0 the velocity of sound outside the flow. This quadrupole strength density may be approximated in many cases as $\rho_0 v_i v_j$. The radiation field is deduced by means of retarded potential solutions. In it, the intensity depends crucially on the frequency as well as on the strength of the quadrupoles, and as a result increases in proportion to a high power, near the eighth, of a typical velocity U in the flow. Physically, the mechanism of conversion of energy from kinetic to acoustic is based on fluctuations in the flow of momentum across fixed surfaces. The efficiency is also increased, particularly for the sound emitted forwards, in the case of fluctuations convected at a not negligible Mach number. But the evolution of a general procedure for estimating the intensity of the sound produced in terms of the details of the fluid flow is one fundamental problem, because it is concerned with uncovering the mechanism of conversion of energy between two of its forms, namely, the kinetic energy of fluctuating shearing motions and the acoustic energy of fluctuating longitudinal motions.

THE WAVE AND HELMHOLTZ EQUATIONS FOR SOUND PROPAGATION AND ACOUSTIC SCATTERING

Assuming barotropic fluids, the usual assumptions for acoustic problems are that acoustic waves are perturbations of the medium (fluid) density ρ (r,t), pressure p (r,t), and mass velocity, v (r,t), where t is time (Duraiswami, 2006). It is also assumed that the medium is inviscid, and that perturbations are small, so that

$$\rho = \rho_0 + \rho', p = p_0 + p', \rho' << \rho_0, p' << p_0, \left|v'\right| << c \sim \sqrt{\frac{p_0}{\rho_0}} \tag{1}$$

Here the perturbations are near an initial spatially uniform state (ρ_0, p_0) of the fluid at rest ($v_0 = 0$) and are denoted by primes. The latter equation states that the mass velocity of the fluid is much smaller than the speed of sound c in that medium. In what follows, for convenience the prime symbols have been dropped for all the fluctuating sound pressure perturbation quantities.

The corresponding wave equation for three dimensions is given by (Ruijgrok, 1993; Lighthill, 1952; Dowling & Ffowcs Williams, 1983; Norton & Karzcub, 2003; Azarpeyvand, 2005):

$$\nabla^2 p - \frac{1}{c^2}\frac{\partial^2 p}{\partial t^2} = 0 \tag{2}$$

In order to satisfy Equation 2 the pressure must be position and time dependent, $p= p(x, y, z, t)$. Following the inhomogeneous wave equation, for example, in terms of sound pressure, becomes:

$$\nabla^2 p - \frac{1}{c^2} p = -\frac{\partial^2 m}{\partial t^2} + \nabla \cdot f \tag{3}$$

where the two terms on the right-hand side represent the acoustic sources. These terms vanish individually outside the source region. If, in addition, the possibility that the fluid is moving violently is to be accounted for such as in a jet exhaust, a convective part must be incorporated in the momentum equation and, following Lighthill's analogy (Johansson, 2000; Lagö, 1996; Lagö, Johansson and Hellström, 1997), a third source term is introduced such that the general form of the inhomogeneous wave equation for non-viscous fluids becomes,

$$\nabla^2 p - \frac{1}{c^2} p = -\frac{\partial^2 m}{\partial t^2} + \nabla \cdot f + Q \tag{4}$$

in which the so-called Lighthill-tensor is given by

$$Q = \sum_{i=1}^{3}\sum_{j=1}^{3} \frac{\partial^2 \left(\rho u_i u_j\right)}{\partial x_i \partial x_j} \tag{5}$$

For a monopole, the "mass acceleration" per unit volume generated at a point $\mathbf{x}_0$ in space can be written as,

$$\frac{\partial^2 m}{\partial t^2} == -\ddot{M}\delta\left(x - x_0\right) \tag{6}$$

The free-field Green's function is readily established by employing a unit strength monopole yielding,

$$g\left(\boldsymbol{x}|\boldsymbol{x}_0\right) == \frac{e^{-jkR}}{4\pi R} \tag{7}$$

where $\boldsymbol{x}$ is the observation point and $R=\boldsymbol{x}-\boldsymbol{x}_0$ the distance between this point and the source position.

PROPAGATION OF ACOUSTIC WAVE DUE TO A MONOPOLE IN FREE SPACE

Propagation of acoustic wave due to a monopole in free space can be modeled and demonstrated by utilizing a monopole source and expanding it using Taylor series (Kempton, 1976) into a series of multipole sources. If g(x) is an analytic function, it can be expanded using Taylor series as follows:

$$g(x+h)=\sum_{n=0}^{\infty}\frac{h^n}{n!}\frac{\partial^n}{\partial x^n}g(x) \tag{8}$$

Here h is an incremental increase of variable x, in which the formulation above will be convergent within the range $|x|>|h|$. Figure 2 exhibits the geometry and coordinate system used. Equation 8 also implies that the sound pressure field of a monopole source located at - h on the x-axis can be expressed as a multipole point field of infinite order.

The sound pressure generated at any point M in the O-xyz frame due to a monopole primary source located along the x-axis at –h can be expressed as,

$$P_{pm}(x,y,z)=\frac{j\omega\rho}{4\pi}Q_{pm}\frac{e^{-jkr_h}}{r_h} \tag{9}$$

Figure 2. Geometry and coordinate system

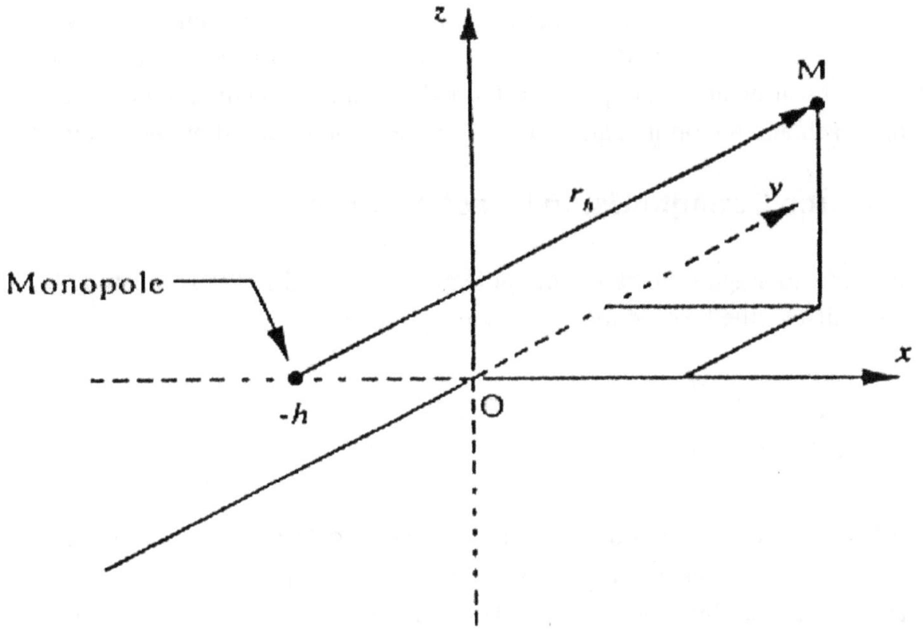

where Q_{pm} is the strength of the monopole primary source, $r_h = \sqrt{(x+h)^2 + y^2 + z^2}$, ρ is the air density, $\omega = 2\pi f$, where f is the prevailing frequency, and $k = \omega / c$, where c is the sound velocity, and $j = \sqrt{-1}$. Expanding the right hand side of Equation 9 in Taylor series, one obtains,

$$P_{pm}(x,y,z) = \frac{j\omega\rho}{4\pi} \sum_{n=0}^{\infty} Q_{pm} \frac{h^n}{n!} \frac{\partial^n}{\partial x^n} \left(\frac{e^{-jkr}}{r} \right) \tag{10}$$

where $r = \sqrt{x^2 + y^2 + z^2}$. For n = 0, $\left(\frac{j\omega\rho}{4\pi}\right) Q_{pm} \left(\frac{e^{-jkr}}{r}\right)$ is the sound pressure field formed by a monopole point located at the center of the coordinate system. Similarly, for n = 1, $\left(\frac{j\omega\rho}{4\pi}\right) Q_{pm} h \left(\frac{\partial}{\partial x}\right)\left(\frac{e^{-jkr}}{r}\right)$ represents the sound pressure field formed by a dipole point located at the origin of the coordinate system with the dipole axis along the x-axis. Hence the general form $\left(\frac{j\omega\rho}{4\pi}\right) Q_{pm} \left(\frac{h^n}{n!}\right)\left(\frac{\partial^n}{\partial x^n}\right)\left(\frac{e^{-jkr}}{r}\right)$ represents the n-th order longitudinal multipole field component with its axis along the x-axis and located at the origin of the coordinate system (consequently, the monopole source can be defined as a zeroth-order multipole source).

Equation 10 also implies that the sound pressure field produced by a monopole located at –h along the x-axis can be regarded as an infinite series of multipole points with each of them situated at the center of the coordinate system. Therefore, if a multipole characterized by Equation 10 is located at the center of the coordinate system, it can be regarded as a series of primary sources of sound. Accordingly, the problem is transformed into the determination of the strength of the secondary sources for sound pressure level minimization at certain point or defined region. The latter can be obtained in two ways: with a direct approach and the optimization of the source strength based on the minimal sound power.

Secondary Source Strength Using Direct Approach

For this purpose, consider again that the monopole primary sound field (as illustrated in Figure 3) can be expressed explicitly in the form of multipole source strength

$$P_{pm}(x,y,z) = \frac{j\omega\rho}{4\pi} \sum_{n=0}^{\infty} Q_{sn} \frac{\partial^n}{\partial x^n} \left(\frac{e^{-jkr}}{r} \right) \tag{11}$$

The multipole source strength in Equation 11, Q_{sn}, can be determined by equating Equation 11 with Equation 10 for known monopole primary source strength, Q_{pm}, and the distance of the secondary source, h, which in principle is a straight forward boundary value problem. Clearly the monopole in the series expansion has the same source strength as the primary monopole, i.e., $Q_{s0} = Q_{pm}$. The dipole secondary source has a source strength, $Q_{s1} = hQ_{pm}$, and, in general, the nth order multipole component has a source strength of $Q_{sn} = h^n \frac{Q_{pm}}{n!}$, and so forth. This approach is used to define secondary source strength, thus

Figure 3. A monopole source

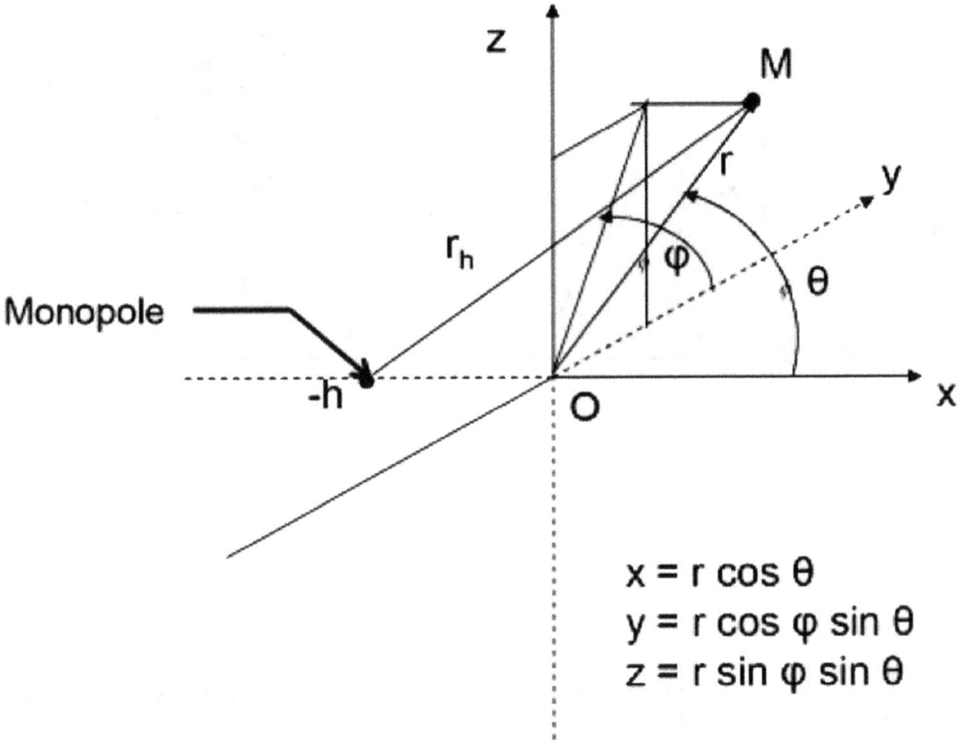

the attribute direct approach. Q_{sn}'s are the source strengths required to produce exactly a monopole primary sound pressure field when the Taylor series expansions of Equation 10 and 11 are carried out to an infinite number of terms. In practice, only a series number of terms will be utilized. Consequently, with a limited number of multipole components, Q_{sn} obtained with the direct approach may not result in an optimal monopole-equivalent primary field (as sketched in Figure 4). Therefore, an alternative procedure is suggested, as outlined below, to obtain a minimal composite sound strength by a combination of primary and secondary source radiation.

Secondary Source Strength Using Optimized Approach

The optimal secondary source strength can be established by resorting to the following procedure (Kusni, Soenarko and Djojodihardjo, 2014). Firstly, the sound power has to be determined. The sound power, W, radiated by an acoustic source, can be expressed as

$$W = \iint_S I \cdot ndS \qquad (12)$$

where I is an acoustic intensity vector; n is an outward normal vector to the surface S enclosing the source, and dS is an incremental element of that surface. Assuming harmonic motion, the average intensity is given by $I = \left(\frac{1}{2}\right) \mathrm{Re}(p * v)$, where p is the sound pressure, v is the velocity vector provided

Figure 4. A primary and a secondary source in a sound field alleviation setting

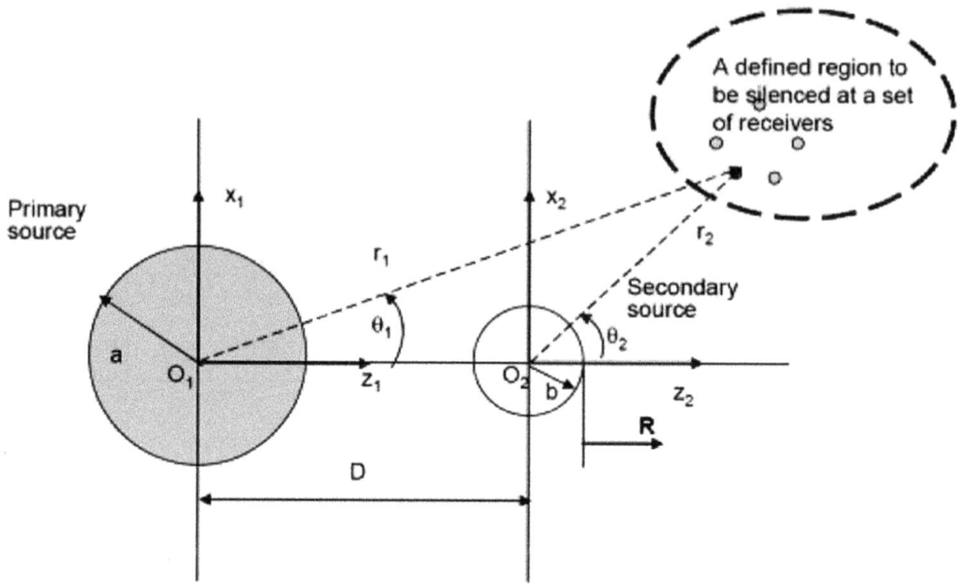

by Euler's equation, $v = -\left(\frac{1}{j\omega\rho}\right)\nabla P$, Re{.} expresses the real part of the argument, superscript * expresses a complex conjugate and ∇ is a gradient operator. Integration of Equation 40 over the surface of the sphere S yields the sound strength which can be expressed as

$$W = \int_0^{2\pi}\int_0^{2\pi} I_r r^2 \sin\theta d\theta d\varphi \tag{13}$$

Here $I_r = \left(\frac{1}{2}\right)\mathrm{Re}\left\{P * V_r\right\}$ is a radial component of the sound intensity, V_r is a radial component of the particle velocity and the angles θ and ϕ are shown in Figure 4.

For the various cases discussed in this work, the distribution of sound pressure is spherically symmetric to the x-axis, so that there is no pressure variation withϕ. Hence Equation 13 can be further reduced to

$$W = 2\pi r^2 \int_0^{\pi} I_r \sin\theta d\theta \tag{14}$$

Prior to the use of Equation 14 for calculating the sound power, it is desirable to utilize a combination of monopole primary source and multipole secondary source. To that end it is convenient to utilize spherical coordinate system. The sound field produced by a monopole primary source having a source strength Q_{pm} and located at – h along the x-axis can be expressed as

$$P_{pm}(r,\theta) = \frac{j\omega\rho}{4\pi} Q_{pm} \frac{e^{-jkr_h}}{r_h} \tag{15}$$

where r_h is distance of monopole primary source to the field point.

By reference to Figure 4, the following coordinate transformation can be carried out, by utilizing the relationship $r_h = \sqrt{r^2 + h^2 + 2rh\cos\theta}$. To illustrate the procedure, one may utilize the secondary source set to be comprised of secondary monopole, dipole, longitudinal quadrupole and longitudinal octupole sources, which are successively defined as follows;

$$P_{sm}(r,\theta) = \frac{j\omega\rho}{4\pi} Q_{sm} \frac{e^{-jkr}}{r} \tag{16}$$

$$P_{sd}(r,\theta) = \frac{\omega\rho}{4\pi} k\cos\theta \left(1 + \frac{1}{jkr}\right) Q_{sd} \frac{e^{-jkr}}{r} \tag{17}$$

$$P_{sq}(r,\theta) = -\frac{j\omega\rho}{4\pi} k^2 \left[\left(1 + \frac{3}{jkr} + \frac{3}{\left(jkr\right)^2}\right)\cos^2\theta - \left(\frac{1}{jkr} + \frac{1}{\left(jkr\right)^2}\right)\right] Q_{sq} \frac{e^{-jkr}}{r} \tag{18}$$

$$P_{so}(r,\theta) = -\frac{\omega\rho}{4\pi} k^3 \cos\theta \left[\begin{array}{l} \left(1 + \frac{6}{jkr} + \frac{15}{\left(jkr\right)^2} + \frac{15}{\left(jkr\right)^3}\right)\cos^2\theta \\ -\left(\frac{3}{jkr} + \frac{9}{\left(jkr\right)^2} + \frac{9}{\left(jkr\right)^3}\right) \end{array}\right] Q_{so} \frac{e^{-jkr}}{r} \tag{19}$$

The total sound power of primary and secondary source consisting of monopole, dipole, quadrupole, and octupole secondary sources can be obtained from Equations 15 to 19 (Kusni, Soenarko and Djojodihardjo, 2014; Djojodihardjo and Safari, 2006). Then the problem of noise alleviation in a defined region as illustrated in Figure 4 is to optimize the secondary source power set for minimal total sound power level in the defined region.

HELMHOLTZ INTEGRAL EQUATION FOR THE ACOUSTIC FIELD

For an exterior acoustic problem, as depicted in Figure 5, the problem domain V is the free space V_{ext} outside the closed surface. The problem domain V is the exterior region. V is considered enclosed in

Figure 5. Exterior problem for homogeneous Helmholtz equation

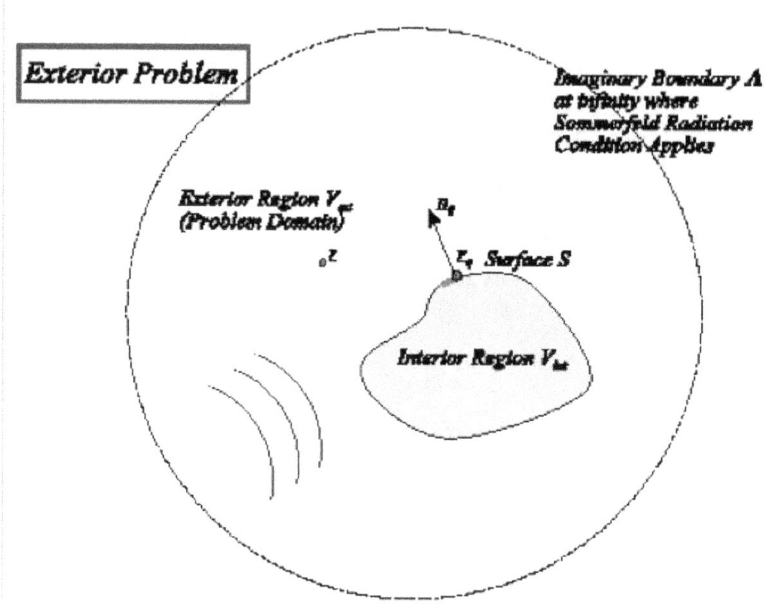

between the surface S and an imaginary surface Λ at a sufficiently large distance from the acoustic sources and the surface S such that the boundary condition on Λ satisfies Sommerfeld's acoustic radiation condition as the distance approaches infinity. Green's first and second theorem provide the basis for transforming the differential equations in the problem domain V and the boundary conditions on the surface S into an integral equation over the surface S.

For time-harmonic acoustic problems in fluid domains, the corresponding boundary integral equation is the Helmholtz integral equation (Dowling & Ffowcs Williams, 1983; Norton & Karzcub, 2003; Djojodihardjo & Safari, 2005; Chandler-Wilde & Langson, 2007; Wrobel, 2002).

$$cp(R) = \int_S \left(p(R)\frac{\partial g}{\partial \hat{n}_0} - g(|R - R_0|)\frac{\partial p}{\partial \hat{n}_0} \right) dS \tag{20}$$

where $\hat{n}_0$ is the surface unit normal vector, and the value of c depends on the location of R in the fluid domain as is well known, and where g the free-space Green's function. $\mathbf{R}_0$ denotes a point located on the boundary S, and from Equation 7, is given by

$$g(|R - R_0|) = \frac{e^{-ik|R-R_0|}}{4\pi|R - R_0|} \tag{21}$$

To solve Equation 20 with g given by Equation 21, one of the two physical properties, acoustic pressure and normal velocity, must be known at every point on the boundary surface. The specific normal impedance, which describes the relationship between pressure and normal velocity, can also serve as a boundary condition. At the infinite boundary, the Sommerfeld radiation condition in three dimensions can be written as (Dowling & Ffowcs Williams, 1983; Norton & Karzcub, 2003):

$$\lim_{|R-R_0|\to\infty} r\left(\frac{\partial g}{\partial r}+ikg\right)\Rightarrow 0 \text{ as } r\to\infty \tag{22}$$

which is satisfied by the fundamental solution.

Helmhotz Integral Equation

In previous work (Djojodihardjo & Safari, 2005; Djojodihardjo, 2007, 2008, 2013), taking advantage of the development in BEM-FEM formulation and computational procedure of Holström (2001), only the incident pressure is considered as the acoustic excitation. In the present work, scattering pressure is also considered in the problem formulation to give the total pressure which serves as an acoustic excitation on the structure. For scattering problems, Equation 20 only gives the solution to the scattered wave. The boundary conditions are however given in terms of the total wave, which is the sum of the incident and the scattered waves. It is necessary to modify the equation to include the incident wave. For exterior scattering problems, the modification can be carried out by adding to the scattered wave integral equations the result of applying the interior Helmholtz equations to the incident wave p_{inc}. Adding to the integral equation for the scattering wave in the exterior problem, we have

$$\int_S \left[p(R)\frac{\partial g(|R-R_0|)}{\partial n}-\frac{\partial p(R)}{\partial n}g(|R-R_0|)\right]dS=\begin{cases}p(R), & \mathrm{R}\in \mathrm{V}_{ext}\\ \frac{1}{2}p(R), & \mathrm{R}\in \mathrm{S}\\ 0, & \mathrm{R}\in \mathrm{V}_{int}\end{cases} \tag{23}$$

For total pressure p, which is $p = p_{inc} + p_{scatter}$ or on the simplified manner, we have

$$cp(R)-p_{inc}(R)=\int_S \left[p(R)\frac{\partial G(R-R_0)}{\partial \hat{n}_0}-\frac{\partial p(r_q)}{\partial \hat{n}_0}G(R-R_0)\right]dS \tag{24}$$

where

$$c=\begin{cases}1 & ,\ \mathrm{R}\in \mathrm{V}_{ext}\\ 1/2 & ,\ \mathrm{R}\in \mathrm{S}\\ \Omega/4\pi & ,\ \mathrm{R}\in \mathrm{S}\ \text{(non smooth surface)}\\ 0 & ,\ \mathrm{R}\in \mathrm{V}_{int}\end{cases} \tag{25}$$

BOUNDARY ELEMENT DISCRETIZATION OF THE HELMHOLTZ INTEGRAL EQUATION FOR THE ACOUSTIC FIELD

For the sake of completeness, much of the author's previous work (Djojodihardjo, 2007, 2008 and 2013) are reproduced in the present development. The first step in discretizing the Helmholtz equation is dividing the continuous system into a system of discretized element with N number of elements. Following the standard procedure in defining the elements on the boundary surface S, the discretized boundary integral equation became,

$$cp - p_{inc} - \sum_{j=1}^{N} \int_{S_j} p \frac{\partial g}{\partial \hat{n}} dS = -\sum_{j=1}^{N} \int_{S_j} g \frac{\partial p}{\partial \hat{n}} dS \tag{26}$$

where i indicates field point, j source point and S_j surface element j. For simplicity, introduce $\frac{\partial g}{\partial n} \equiv \bar{g}$. Then Equation 26 can be written as

$$cp_i - p_{inc} - \sum_{j=1}^{N} \int_{S} p\bar{g} dS = i\rho_0 \omega \sum_{j=1}^{N} \int_{S} gv dS \tag{27}$$

For constant element, p and v in Equation 27 are assumed to be constant over each element, and therefore they can be taken out of the integrals, and denoted by p_j and v_j, respectively, for any surface element j. By placing the node of the surface elements on the smooth part of the surface, c = ½. Defining $\bar{H}_{ij} = \int_{S_j} \bar{g} dS$ and $G_{ij} = \int_{S_j} g dS$, then Equation 27 becomes,

$$\frac{1}{2} p_i - p_{inc} - \sum_{j=1}^{N} \bar{H}_{ij} p_j = i\rho_0 \omega \sum_{j=1}^{N} G_{ij} v_j \tag{28}$$

Introducing

$$H_{ij} = \frac{1}{2} \delta_{ij} - \bar{H}_{ij} \tag{29}$$

where δ_{ij} is the Kronecker's delta, Equation 29 reduces to

$$\sum_{j=1}^{N} H_{ij} p_j = i\rho_0 \omega \sum_{j=1}^{N} G_{ij} v_j + p_{inc} \tag{30}$$

The discretized equation forms a set of simultaneous linear equation, where the point r is chosen to be the nodal points of the boundary elements on the surface S (Djojodihardjo and Safari, 2006; Chandler-Wilde. and Langson, 2007; Wrobel, 2002). In matrix form:

$$\left[\mathbf{H}\right]\left\{p\right\} = i\rho_0\omega\left[\mathbf{G}\right]\left\{v\right\} + \left\{p_{inc}\right\} \tag{31}$$

where, $\mathbf{H}$ and $\mathbf{G}$ are two NxN matrices of influence coefficients, while p and v are vectors of dimension N representing total pressure and normal velocity on the boundary elements. This matrix equation can be solved if the boundary condition $\frac{\partial p}{\partial n}$ and the incident field p_{inc} are known.

Boundary Elements

Holström (2001) has outlined the use of constant boundary elements and quadrilateral (four-node) boundary element following the above procedure. In this particular work, quadrilateral boundary element has been utilized throughout, and in addition, special care is taken on those at the poles for spherical acoustic source (Azarpeyvand, 2005). Monopole Green's free-space fundamental solution has been utilized. Hence

$$\mathbf{G}_{ij} = \int_{S_j} g dS = \int_{S_j} g\left(\left|R_j - R_i\right|\right) dS = \int_{S_j} \frac{e^{-ik\left|R_j - R_i\right|}}{4\pi\left|R_j - R_i\right|} \tag{32}$$

In Cartesian coordinate system, this equation can be expressed as

$$\mathbf{G}_{ij} = \int_{S_j} \frac{e^{-ik\sqrt{\left(x_j - x_i\right)^2 + \left(y_j - y_i\right)^2 + \left(z_j - z_i\right)^2}}}{4\pi\sqrt{\left(x_j - x_i\right)^2 + \left(y_j - y_i\right)^2 + \left(z_j - z_i\right)^2}} dS \tag{33}$$

where R_j is the midpoint of element j and R_i is the location of the node i. For i=j, and to avoid singularity in calculating the value of the Green's function in the integral, four Gauss points are used on element j, instead of the midpoint. These Gauss points in the isoparametric system are defined as (Wrobel, 2002; Djojodihardjo and Safari, 2005; Djojodihardjo, 2007, 2008; Holström, 2001):

$$\begin{aligned} \left(\xi_1, \eta_1\right) &= \frac{1}{\sqrt{3}}\left(1, -1\right) \quad \left(\xi_2, \eta_2\right) = \frac{1}{\sqrt{3}}\left(1, 1\right) \\ \left(\xi_3, \eta_3\right) &= \frac{1}{\sqrt{3}}\left(-1, 1\right) \quad \left(\xi_4, \eta_4\right) = \frac{1}{\sqrt{3}}\left(-1, -1\right) \end{aligned} \tag{34}$$

and G_{ii} is then chosen to be

$$\mathbf{G}_{ij}\Big|_{i=j} = \mathbf{G}_{ii} = \left.\frac{\mathbf{G}_{ii}\left(\xi_1, \eta_1\right) + \mathbf{G}_{ii}\left(\xi_2, \eta_2\right) + \mathbf{G}_{ii}\left(\xi_3, \eta_3\right) + \mathbf{G}_{ii}\left(\xi_4, \eta_4\right)}{4}\right|_{R_i = R_j} \tag{35}$$

To calculate $\bar{H}_{ij} = \int_{S_j} \bar{g} dS$, the derivative $\bar{g} = \frac{\partial g}{\partial \hat{n}}$ has to be evaluated

$$\bar{H}_{ij} = \int_{S_j} \bar{g} dS = \int_{S_j} \frac{\partial g}{\partial \hat{n}} dS = \int_{S_j} \left(\nabla g\right)^T \hat{n} dS \tag{36}$$

where

$$\hat{n} = \begin{bmatrix} n_x \\ n_y \\ n_z \end{bmatrix} \tag{37}$$

and

$$\nabla g = \begin{bmatrix} -\dfrac{x e^{-ik\sqrt{(x_i-x_j)^2+(y_i-y_j)^2+(z_i-z_j)^2}}}{4\pi\sqrt{(x_i-x_j)^2+(y_i-y_j)^2+(z_i-z_j)^2}}\left(ik+\dfrac{1}{\sqrt{(x_i-x_j)^2+(y_i-y_j)^2+(z_i-z_j)^2}}\right) \\ -\dfrac{y e^{-ik\sqrt{(x_i-x_j)^2+(y_i-y_j)^2+(z_i-z_j)^2}}}{4\pi\sqrt{(x_i-x_j)^2+(y_i-y_j)^2+(z_i-z_j)^2}}\left(ik+\dfrac{1}{\sqrt{(x_i-x_j)^2+(y_i-y_j)^2+(z_i-z_j)^2}}\right) \\ -\dfrac{z e^{-ik\sqrt{(x_i-x_j)^2+(y_i-y_j)^2+(z_i-z_j)^2}}}{4\pi\sqrt{(x_i-x_j)^2+(y_i-y_j)^2+(z_i-z_j)^2}}\left(ik+\dfrac{1}{\sqrt{(x_i-x_j)^2+(y_i-y_j)^2+(z_i-z_j)^2}}\right) \end{bmatrix} \tag{38}$$

when i ≠j, $H_{ij} = \bar{H}_{ij}$. When i = j, the $\bar{H}_{ii}$ terms are identically zero since the normal $\hat{n}$ and the element coordinates are perpendicular to each other:

$$H_{ij} = \int_{S_i} \frac{\partial g}{\partial \hat{n}} dS = \int_{S_i} \frac{\partial g}{\partial r}\frac{\partial r}{\partial \hat{n}} dS \equiv 0 \tag{39}$$

and therefore H_{ii} = ½ following the definition in Equation 29.

For a four-node iso-parametric quadrilateral elements, the pressure p and the normal velocity **v** at any position on the element can be defined by their nodal values and linear shape functions, i.e.,

$$v(\xi,\eta) = N_1 v_1 + N_2 v_2 + N_3 v_3 + N_4 v_4 = \begin{bmatrix} N_1 & N_2 & N_3 & N_4 \end{bmatrix} \begin{bmatrix} v_1 \\ v_2 \\ v_3 \\ v_4 \end{bmatrix} \tag{40}$$

$$p\left(\xi.\eta\right) = N_1p_1 + N_2p_2 + N_3p_3 + N_4p_4 = \begin{bmatrix} N_1 & N_2 & N_3 & N_4 \end{bmatrix} \begin{bmatrix} p_1 \\ p_2 \\ p_3 \\ p_4 \end{bmatrix} \tag{41}$$

where the shape functions are, in the element coordinate system,

$$\begin{aligned} N_1 &= \frac{1}{4}\left(\xi-1\right)\left(\eta-1\right) \quad N_2 = \frac{1}{4}\left(\xi+1\right)\left(\eta-1\right) \\ N_3 &= \frac{1}{4}\left(\xi+1\right)\left(\eta+1\right) \quad N_4 = \frac{1}{4}\left(\xi-1\right)\left(\eta+1\right) \end{aligned} \tag{42}$$

The four node quadrilteral element can have any arbitrary orientation in the three-dimensional space.

VIBRATION SUPPRESSION THROUGH FLUID-STRUCTURE COUPLING

As elaborated in (Wrobel, 2002; Djojodihardjo, 2013, 2015), following Weaver and Johnston (1987), the BE region is treated as a super finite element and its stiffness matrix is computed and assembled into the global stiffness matrix, and identified as the coupling to finite elements. The state of affairs is schematically depicted in Figure 8(a). The utilization of FEM on the structural domain leads to a system of simultaneous equations which relate the displacements at all the nodes to the nodal forces. In the BEM, on the other hand, a relationship between nodal displacements and nodal tractions is established. Representing the elastic structure by FE model, the structural dynamic equation of motion is given by Bisplinghoff, Ashley and Halfman (1955)

$$\left[\mathbf{M}\right]\left\{\ddot{x}\right\} + \left[\mathbf{C}\right]\left\{\dot{x}\right\} + \left[\mathbf{K}\right]\left\{x\right\} = \left\{\mathbf{F}\right\} \tag{43}$$

where **M**, **C** and **K** are structural mass, damping and stiffness, respectively, which are expressed as matrices in a FE model, while **F** is the given external forcing function vector, and $\{x\}$ is the structural displacement vector. The incorporation of the self-excited aerodynamic effects to the structural dynamics equation can be written as (Weaver & Johnston, 1987; Bisplinghoff, Ashley & Halfman, 1955; Rodden & Johnson, 1994; Zona Technology, Inc, 2004):

$$\left[\mathbf{M}\right]\left\{\ddot{x}\right\} + \left[\mathbf{C}\right]\left\{\dot{x}\right\} + \left[\mathbf{K}\right]\left\{x\right\} - q_{\infty}\left[\mathbf{A}\left(ik\right)\right]\left\{x\right\} = \left\{0\right\} \tag{44}$$

where **A**(*ik*) is an aerodynamic influence coefficient after applying aero-structure coupling from the control points of aerodynamic boxes to the structural finite element grid points as elaborated in (Djojodihardjo & Safari, 2005; Djojodihardjo, 2007).

Taking into account the acoustic pressure p on the structure at the fluid-structure interface as a separate excitation force, the acoustic-structure problem can be obtained from Equation 44 by introducing a fluid-structure coupling term given by $[\mathbf{L}]\{p\}$. It follows that

$$[\mathbf{M}]\{\ddot{x}\}+[\mathbf{C}]\{\dot{x}\}+[\mathbf{K}]\{x\}-q_{\infty}[\mathbf{A}(ik)]\{x\}+[\mathbf{L}]\{p\}=\{\mathbf{F}\} \tag{45}$$

where $\mathbf{L}$ is a coupling matrix of size $M \times N$ in the BEM/FEM coupling thus formulated. M is the number of FE degrees of freedom and N is the number of BE nodes on the coupled boundary. For the BE part of the surface at the fluid-structure interface a, Equation 30 can be utilized.

The global coupling matrix $\mathbf{L}$ transforms the acoustic fluid pressure acting on the nodes of boundary elements on the entire fluid-structure interface surface a, to nodal forces on the finite elements of the structure. Hence $\mathbf{L}$ consists of n assembled local transformation matrices L_e, given by

$$L_e = \int_{S_e} N_F^T n N_B \, dS \tag{46}$$

in which N_F is the shape function matrix for the finite element and N_B is the shape function matrix for the boundary element. It can be shown that:

$$N_F = \begin{bmatrix} 1 & 0 & 0 & 0 & 0 \\ 0 & 1 & 0 & 0 & 0 \\ 0 & 0 & 1 & 0 & 0 \end{bmatrix} [N_i] \tag{47}$$

The rotational parts in N_F are neglected since these are considered to be small in comparison with the translational ones in the BE-FE coupling, consistent with the assumptions in structural dynamics as, for example, stipulated in (Bisplinghoff, Ashley & Halfman, 1955; Rodden & Johnson, 1994; Zona Technology, Inc, 2004; Beer & Watson, 1992).

For the normal fluid velocities and the normal translational displacements on the shell elements at the fluid-structure coupling interface, a relationship, which takes into account the velocity continuity over the coinciding nodes, should be established. This relationship is given by

$$\mathrm{v} = i\omega(\mathbf{T}.x) \tag{48}$$

Similar to $\mathbf{L}$, $\mathbf{T}(n \times m)$ is also a global coupling matrix that connects the normal velocity of a BE node with the translational displacements of FE nodes obtained by taking the transpose of the boundary surface normal vector n (Beer and Watson, 1992). The local transformation vector T_e can then be written as:

$$T_e = n^T \tag{49}$$

The presence of an acoustic source can further be depicted by Figure 5 and Figure 6(b). Three regions are considered, i.e. *a, b* and c; region *a* is the fluid-structure interface region, where FEM mesh and BEM mesh coincide and region *b* and *c* is the region where all of the boundary conditions (pressure or velocity) are known.

For the coupled FEM-BEM regions, BEM equation can now be written as:

$$\begin{bmatrix} \mathbf{H}_{11} & \mathbf{H}_{12} & \mathbf{H}_{13} \\ \mathbf{H}_{21} & \mathbf{H}_{22} & \mathbf{H}_{23} \\ \mathbf{H}_{31} & \mathbf{H}_{32} & \mathbf{H}_{33} \end{bmatrix} \begin{Bmatrix} p_a \\ p_b \\ p_c \end{Bmatrix} = i\rho_0\omega \begin{bmatrix} \mathbf{G}_{11} & \mathbf{G}_{12} & \mathbf{G}_{13} \\ \mathbf{G}_{21} & \mathbf{G}_{22} & \mathbf{G}_{23} \\ \mathbf{G}_{31} & \mathbf{G}_{32} & \mathbf{G}_{33} \end{bmatrix} \begin{Bmatrix} v_a \\ v_b \\ v_c \end{Bmatrix} + \begin{Bmatrix} p_{inc_a} \\ p_{inc_b} \\ p_{inc_c} \end{Bmatrix} \tag{50}$$

Considering $\mathbf{v}_a = i\omega\left(\mathbf{T}.x\right)$, BEM equation can be modified as:

$$\begin{bmatrix} \mathbf{H}_{11} & \mathbf{H}_{12} & \mathbf{H}_{13} \\ \mathbf{H}_{21} & \mathbf{H}_{22} & \mathbf{H}_{23} \\ \mathbf{H}_{31} & \mathbf{H}_{32} & \mathbf{H}_{33} \end{bmatrix} \begin{Bmatrix} p_a \\ p_b \\ p_c \end{Bmatrix} = i\rho_0\omega \begin{bmatrix} \mathbf{G}_{11} & \mathbf{G}_{12} & \mathbf{G}_{13} \\ \mathbf{G}_{21} & \mathbf{G}_{22} & \mathbf{G}_{23} \\ \mathbf{G}_{31} & \mathbf{G}_{32} & \mathbf{G}_{33} \end{bmatrix} \begin{Bmatrix} i\omega\left[\mathbf{T}\right]x \\ v_b \\ v_c \end{Bmatrix} + \begin{Bmatrix} p_{inc_a} \\ p_{inc_b} \\ p_{inc_c} \end{Bmatrix} \tag{51}$$

or:

$$\begin{aligned} \mathbf{H}_{11}p_a + \mathbf{H}_{12}p_b + \mathbf{H}_{13}p_c &= -\rho_0\omega^2\mathbf{G}_{11}\mathbf{T}x + i\rho_0\omega\mathbf{G}_{12}v_b + i\rho_0\omega\mathbf{G}_{13}v_c + \left\{p_{inc_a}\right\} \\ \mathbf{H}_{21}p_a + \mathbf{H}_{22}p_b + \mathbf{H}_{23}p_c &= -\rho_0\omega^2\mathbf{G}_{21}\mathbf{T}x + i\rho_0\omega\mathbf{G}_{22}v_b + i\rho_0\omega\mathbf{G}_{23}v_c + \left\{p_{inc_b}\right\} \\ \mathbf{H}_{31}p_a + \mathbf{H}_{32}p_b + \mathbf{H}_{33}p_c &= -\rho_0\omega^2\mathbf{G}_{31}\mathbf{T}x + i\rho_0\omega\mathbf{G}_{32}v_b + i\rho_0\omega\mathbf{G}_{33}v_c + \left\{p_{inc_c}\right\} \end{aligned} \tag{52}$$

Figure 6. (a). Schematic of fluid-structure interaction domain (b) schematic of FE-BE problem representing quarter space problem domains for half wing

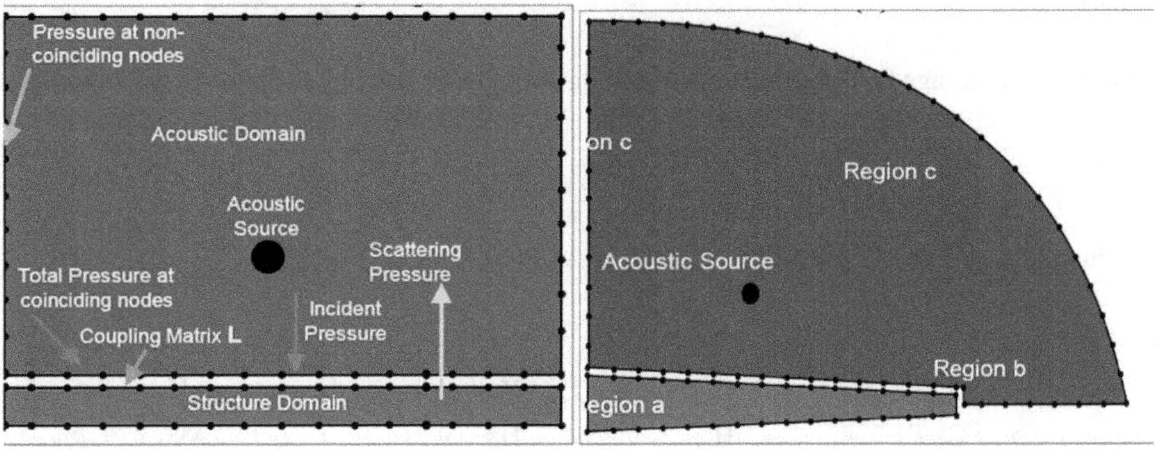

If the pressure boundary condition on b (p_b), velocity boundary condition on c (v_c), and the incident pressure on a, b and c are known, by taking to the left side all the unknown the above equation can be written as:

$$\begin{aligned} \rho_0\omega^2\mathbf{G}_{11}\mathbf{T}x + \mathbf{H}_{11}p_a - i\rho_0\omega\mathbf{G}_{12}v_b + \mathbf{H}_{13}p_c &= -\mathbf{H}_{12}p_b + i\rho_0\omega\mathbf{G}_{13}v_c + \left\{p_{inc_a}\right\} \\ \rho_0\omega^2\mathbf{G}_{21}\mathbf{T}x + \mathbf{H}_{21}p_a - i\rho_0\omega\mathbf{G}_{22}v_b + \mathbf{H}_{23}p_c &= -\mathbf{H}_{22}p_b + i\rho_0\omega\mathbf{G}_{23}v_c + \left\{p_{inc_b}\right\} \\ \rho_0\omega^2\mathbf{G}_{31}\mathbf{T}x + \mathbf{H}_{31}p_a - i\rho_0\omega\mathbf{G}_{32}v_b + \mathbf{H}_{33}p_c &= -\mathbf{H}_{32}p_b + i\rho_0\omega\mathbf{G}_{33}v_c + \left\{p_{inc_c}\right\} \end{aligned} \quad (53)$$

Since the pressure p on FEM equation lies in region a, Equation 45 can be written as

$$[\mathbf{M}]\{\ddot{x}\} + [\mathbf{C}]\{\dot{x}\} + [\mathbf{K}]\{x\} - q_\infty[\mathbf{A}(ik)]\{x\} + [\mathbf{L}]\{p_a\} = \{\mathbf{F}\} \quad (54)$$

where p_a is the total acoustic pressure resulting from the application of acoustic disturbance force to the structure, which consists of the incident acoustic pressure p_{inc} and scattering acoustic pressure p_{sc}. The terms $\mathbf{H}_{ij}$ and $\mathbf{G}_{ij}$ results from the partition of the acoustic domain and their related influence coefficients, as stipulated by and consequence of the zoning implied by Figure 6; the detail is elaborated in (Djojodihardjo & Safari, 2006) and (Djojodihardjo, 2013).

The scattering acoustic pressure will be dependent on the dynamic response of the structure due to the incident acoustic pressure. Following the general practice in structural dynamics, solutions of Equation 54 are sought by considering synchronous motion with harmonic frequencyω. Correspondingly, Equation 54 reduces to:

$$[\mathbf{K} + i\omega\mathbf{C} - \omega^2\mathbf{M}]\{\bar{x}\} - q_\infty[\mathbf{A}(ik)]\{x\} + [\mathbf{L}]\{\bar{p}_a\} = \{\bar{\mathbf{F}}\} \quad (55)$$

where

$$x = \bar{x}e^{i\omega t};\; p_a = \bar{p}_a e^{i\omega t} \quad (56)$$

or, dropping the bar sign for convenience, but keeping the meaning in mind, Equation 55 can be written as

$$[\mathbf{K} + i\omega\mathbf{C} - \omega^2\mathbf{M}]\{x\} - q_\infty[\mathbf{A}(ik)]\{x\} + [\mathbf{L}]\{p_a\} = \{\mathbf{F}\} \quad (57)$$

Combining Equation 53 and 57, the coupled BEM-FEM equation can then be written as:

$$\begin{bmatrix} [\mathbf{K} + i\omega\mathbf{C} - \omega^2\mathbf{M}] - q_\infty[\mathbf{A}(ik)]\{x\} & \mathbf{L} & 0 & 0 \\ \rho_0\omega^2\mathbf{G}_{11}T & \mathbf{H}_{11} & -i\rho_0\omega\mathbf{G}_{12} & \mathbf{H}_{13} \\ \rho_0\omega^2\mathbf{G}_{21}T & \mathbf{H}_{21} & -i\rho_0\omega\mathbf{G}_{22} & \mathbf{H}_{23} \\ \rho_0\omega^2\mathbf{G}_{31}T & \mathbf{H}_{31} & -i\rho_0\omega\mathbf{G}_{32} & \mathbf{H}_{33} \end{bmatrix} \begin{Bmatrix} x \\ p_a \\ v_b \\ p_c \end{Bmatrix} = \begin{Bmatrix} \mathbf{F} \\ -\mathbf{H}_{12}p_b + i\rho_0\omega\mathbf{G}_{13}v_c + p_{inc_a} \\ -\mathbf{H}_{22}p_b + i\rho_0\omega\mathbf{G}_{23}v_c + p_{inc_b} \\ -\mathbf{H}_{32}p_b + i\rho_0\omega\mathbf{G}_{33}v_c + p_{inc_c} \end{Bmatrix}$$

(58)

This equation forms the basis for the treatment of the fluid-structure interaction in a unified fashion. The solution vector consisting of the displacement vector of the structure and total acoustic pressure on the boundaries of the acoustic domain, including the acoustic-structure interface, can be obtained by solving Equation 58 as a dynamic response problem.

TREATMENT OF ACOUSTIC-ARODYNAMIC COUPLING: ACOUSTIC-AERODYNAMIC ANALOGY

At this point, the solution approach philosophy is in order. Analogous to the treatment of dynamic aeroelastic stability problem of structure, in which the aerodynamic effects can be distinguished into motion independent (self-excited) and motion induced aerodynamic forces, the effect of acoustic pressure disturbance to the aeroelastic structure (acousto-aero-elastic problem) can be viewed to consist of structural motion independent incident acoustic pressure (excitation acoustic pressure) and structural motion dependent acoustic pressure, which is known as the scattering pressure. However, the scattering acoustic pressure is also dependent on the incident acoustic pressure. The consequence of such treatment has been adopted in the above section and will further be implemented in the subsequent development.

For aeroelastic calculation purposes, further treatment to simplify Equation 58 will be carried out. Since the pressure boundary condition on b ($p_b = 0$) and velocity boundary condition on c ($v_c = 0$), Equation 53 can be written as:

$$\rho_0\omega^2\mathbf{G}_{11}\mathbf{T}x + \mathbf{H}_{11}p_a - i\rho_0\omega\mathbf{G}_{12}v_b + \mathbf{H}_{13}p_c = p_{inc_a} \tag{59a}$$

$$\rho_0\omega^2\mathbf{G}_{21}\mathbf{T}x + \mathbf{H}_{21}p_a - i\rho_0\omega\mathbf{G}_{22}v_b + \mathbf{H}_{23}p_c = p_{inc_b} \tag{59b}$$

$$\rho_0\omega^2\mathbf{G}_{31}\mathbf{T}x + \mathbf{H}_{31}p_a - i\rho_0\omega\mathbf{G}_{32}v_b + \mathbf{H}_{33}p_c = p_{inc_c} \tag{59c}$$

Since $\mathbf{G}_{22}$ and $\mathbf{H}_{33}$ is square matrix, Equation 59b and 59c can be written as

$$v_b = -\frac{i}{\rho_0\omega}\left[\mathbf{G}_{22}\right]^{-1}\left(\rho_0\omega^2\mathbf{G}_{21}\mathbf{T}x + \mathbf{H}_{21}p_a + \mathbf{H}_{23}p_c - p_{inc_b}\right) \tag{60a}$$

$$p_c = -\left[\mathbf{H}_{33}\right]^{-1}\left(\rho_0\omega^2\mathbf{G}_{31}\mathbf{T}x + \mathbf{H}_{31}p_a - i\rho_0\omega\mathbf{G}_{32}v_b - p_{inc_c}\right) \tag{60b}$$

Substitution of Equation 60b into equation 59b yields

$$\rho_0 \omega^2 \mathbf{A}_{21} \mathbf{T}x + \mathbf{B}_{21} p_a - i\rho_0 \omega \mathbf{A}_{22} v_b + \mathbf{B}_{23} p_{inc_c} - p_{inc_b} = \{0\} \tag{61}$$

where

$$\begin{aligned}
\mathbf{A}_{21} &= \left(\mathbf{G}_{21} - \mathbf{H}_{23}\left[\mathbf{H}_{33}\right]^{-1}\mathbf{G}_{31}\right)\\
\mathbf{B}_{21} &= \left(\mathbf{H}_{21} - \mathbf{H}_{23}\left[\mathbf{H}_{33}\right]^{-1}\mathbf{H}_{31}\right)\\
\mathbf{B}_{23} &= \left(\mathbf{H}_{23}\left[\mathbf{H}_{33}\right]^{-1}\right)\\
\mathbf{A}_{22} &= \left(\mathbf{G}_{22} - \mathbf{H}_{23}\left[\mathbf{H}_{33}\right]^{-1}\mathbf{G}_{32}\right)
\end{aligned} \tag{62}$$

Since $\mathbf{A}_{22}$ is square matrix, Equation 61 can be written as:

$$v_b = -\frac{i}{\rho_0 \omega}\left[\mathbf{A}_{22}\right]^{-1}\left(\rho_0 \omega^2 \mathbf{A}_{21} \mathbf{T}x + \mathbf{B}_{21} p_a + \mathbf{B}_{23} p_{inc_c} - p_{inc_b}\right) \tag{63}$$

Substitution of Equation 60a into equation 59c yields

$$\rho_0 \omega^2 \mathbf{C}_{31} \mathbf{T}x + \mathbf{A}_{31} p_a + \mathbf{A}_{33} p_c + \mathbf{B}_{32} p_{inc_b} - p_{inc_c} = \{0\} \tag{64}$$

where

$$\begin{aligned}
\mathbf{C}_{31} &= \left(\mathbf{G}_{31} - \mathbf{G}_{32}\left[\mathbf{G}_{22}\right]^{-1}\mathbf{G}_{21}\right)\\
\mathbf{A}_{31} &= \left(\mathbf{H}_{31} - \mathbf{G}_{32}\left[\mathbf{G}_{22}\right]^{-1}\mathbf{H}_{21}\right)\\
\mathbf{A}_{33} &= \left(\mathbf{H}_{33} - \mathbf{G}_{32}\left[\mathbf{G}_{22}\right]^{-1}\mathbf{H}_{23}\right)\\
\mathbf{B}_{32} &= \left(\mathbf{G}_{32}\left[\mathbf{G}_{22}\right]^{-1}\right)
\end{aligned} \tag{65}$$

Since $\mathbf{A}_{33}$ is square matrix, Equation 64 can be written as:

$$p_c = -\left[\mathbf{A}_{33}\right]^{-1}\left(\rho_0 \omega^2 \mathbf{C}_{31} \mathbf{T}x + \mathbf{A}_{31} p_a + \mathbf{B}_{32} p_{inc_b} - p_{inc_c}\right) \tag{66}$$

Substitution of Equation 63 and 66 into Equation 60a

$$\rho_0 \omega^2 \mathbf{D}_{11} \mathbf{T}x + \mathbf{E}_{11} p_a + \mathbf{F}_{12} p_{inc_b} + \mathbf{F}_{13} p_{inc_c} = \left\{p_{inc_a}\right\} \tag{67}$$

where

$$\begin{aligned}
\mathbf{D}_{11} &= \left(\mathbf{G}_{11} - \mathbf{G}_{12}\left[\mathbf{A}_{22}\right]^{-1}\mathbf{A}_{21} - \mathbf{H}_{13}\left[\mathbf{A}_{33}\right]^{-1}\mathbf{C}_{31}\right) \\
\mathbf{E}_{11} &= \left(\mathbf{H}_{11} - \mathbf{G}_{12}\left[\mathbf{A}_{22}\right]^{-1}\mathbf{B}_{21} - \mathbf{H}_{13}\left[\mathbf{A}_{33}\right]^{-1}\mathbf{A}_{31}\right) \\
\mathbf{F}_{12} &= \mathbf{G}_{12}\left[\mathbf{A}_{22}\right]^{-1} - \mathbf{H}_{13}\left[\mathbf{A}_{33}\right]^{-1}\mathbf{B}_{32} \\
\mathbf{F}_{13} &= \mathbf{H}_{13}\left[\mathbf{A}_{33}\right]^{-1} - \mathbf{G}_{12}\left[\mathbf{A}_{22}\right]^{-1}\mathbf{B}_{23}
\end{aligned} \tag{68}$$

Since $\mathbf{E}_{11}$ is square matrix

$$p_a = -\left[\mathbf{E}_{11}\right]^{-1}\left(\rho_0\omega^2\mathbf{D}_{11}\mathbf{T}\left\{x\right\} - p_{inc_a} + \mathbf{F}_{12}p_{inc_b} + \mathbf{F}_{13}p_{inc_c}\right) \tag{69}$$

Matrix $\mathbf{E}_{11}$ and $\mathbf{D}_{11}$ are also a square matrix; hence finally by substituting Equation 69 into Equation 58 the BEM-FEM aero-acoustic-structure coupling can be obtained as:

$$\begin{aligned}
&\left[\mathbf{K} + i\omega\mathbf{C} - \omega^2\mathbf{M}\right]\left\{x\right\} - q_\infty\left[\mathbf{A}\left(ik\right)\right]\left\{x\right\} \\
&\qquad + \left[\mathbf{L}\right]\left(-\left[\mathbf{E}_{11}\right]^{-1}\left(\rho_0\omega^2\mathbf{D}_{11}\mathbf{T}\left\{x\right\} - p_{inc_a} + \mathbf{F}_{12}p_{inc_b} + \mathbf{F}_{13}p_{inc_c}\right)\right) = \left\{\mathbf{F}\right\}
\end{aligned} \tag{70}$$

Incident pressure on region *b* and *c* will not influence the stability problem associated with the structures, and may at this point be disregarded. Hence, without considering damping matrix **C**Equation 70 simplifies to:

$$\begin{aligned}
&\left[\mathbf{K} - \omega^2\mathbf{M}\right]\left\{x\right\} - q_\infty\left[\mathbf{A}\left(ik\right)\right]\left\{x\right\} \\
&\qquad - \rho_0\omega^2\left[\mathbf{L}\right]\left[\mathbf{E}_{11}\right]^{-1}\left[\mathbf{D}_{11}\right]\left[\mathbf{T}\right]\left\{x\right\} = -\left[\mathbf{L}\right]\left[\mathbf{E}_{11}\right]^{-1}\left\{p_{inc_a}\right\} + \left\{\mathbf{F}\right\}
\end{aligned} \tag{71a}$$

or

$$\left[\mathbf{K} - \omega^2\mathbf{M}\right]\left\{x\right\} - q_\infty\left[\mathbf{A}\left(ik\right)\right]\left\{x\right\} - \rho_0\omega^2\left[\mathbf{F}_{ac_{sc}}\left(k_w\right)\right]\left\{x\right\} = \left\{\mathbf{F}_{ac_{inc}}\left(k_w\right)\right\} + \left\{\mathbf{F}\right\} \tag{71b}$$

where

$$\begin{aligned}
&\left[\mathbf{F}_{ac_{sc}}\left(k_w\right)\right] = \left[\mathbf{L}\right]\left[\mathbf{E}_{11}\right]^{-1}\left[\mathbf{D}_{11}\right]\left[\mathbf{T}\right] \\
&\left\{\mathbf{F}_{ac_{inc}}\left(k_w\right)\right\} = -\left[\mathbf{L}\right]\left[\mathbf{E}_{11}\right]^{-1}\left\{p_{inc_a}\right\}
\end{aligned} \tag{72}$$

Equation 71a will not be solved directly since the size of the mass and stiffness matrices of the aircraft model are very large. Instead one uses the modal approach where the structural deformation $\{x\}$ is transformed to the generalized coordinate $\{q\}$ given by the following relation:

$$x = \Phi q \tag{73}$$

where Φ is the modal matrix whose columns contain the lower order natural mode. Pre-multiplying by Φ^T and converting dynamic pressure q_∞ into reduced frequency (k) as elaborated by Djojodihardjo and Safari (2006), Equation 71b can then be written as:

$$\Phi^T \left[\mathbf{K} - \omega^2 \left(\mathbf{M} + \frac{\rho}{2} \left(\frac{L}{k} \right)^2 \left[\mathbf{A}(ik) \right] + \rho_0 \left[\mathbf{F}_{ac_{sc}} \left(k_w \right) \right] \right) \right] \Phi \{q\} = \Phi^T \left\{ \mathbf{F}_{ac_{inc}} \left(k_w \right) \right\} + \Phi^T \left\{ \mathbf{F} \right\} \tag{74}$$

since all of the acoustic terms are functions of wave number (k_w), Equation 74 can be solved by utilizing iterative procedure. Incorporation of the scattering acoustic term along with the aerodynamic term in the second term of Equation 74 can be regarded as one manifestation of the acoustic-aerodynamic analogy followed in this approach.

Further method of approach for the solution of the acousto-aeroelastic problem is then dealt with. Following the same procedure as developed in earlier work (Djojodihardjo & Safari, 2005; Djojodihardjo, 2007, 2008), the acoustic excitation is incorporated by coupling it to the unsteady aerodynamic load in the flutter stability formulation. Linearity and principle of superposition has been assumed. Hence the acoustic loading can be superposed to the aerodynamic loading on the structure, and form the modified aeroelastic equation (acousto-aeroelastic equation) of the structural dynamic problem associated with acoustic and aerodynamic excitation. In the earlier work, tacit consideration is only given to the incident acoustic pressure as the acoustic excitation, without considering the acoustic scattering effects, and without considering **L**. The incident pressure p_{inc} was also assumed to belong to a certain class that allows its incorporation in the aerodynamics term. Thus Equation 74 was treated in a decoupled fashion. Such approach has given promising results.

In the present development, rigorous consideration has been devoted to the acoustic scattering problem. Two generic approaches to solve Equation 58 can be followed. The first is to solve Equation 58 as a stability equation in a "unified treatment," and the disturbance acoustic pressure already incorporates the total pressure, i.e. the incident plus the scattering acoustic pressure. The treatment of the incident acoustic pressure p_{inc} in Equation 71 follows similar approach adopted for the scattering acoustic pressure by tuning p_{inc} so that it behaves like the aerodynamic terms in the modal Equation 71. This is considered logical for the problem considered since the intention is to look at its enhancement effect to the aerodynamic one, thus only a class of p_{inc} will meet the eigenvalue requirements of Equation 71. This approach has been elaborated further by Djojodihardjo. The second generic approach is to solve Equation 58 as a dynamic response problem due to acoustic excitation. The left hand side of Equation 58 incorporates the scattering acoustic pressure term. Appropriate algebraic manipulation is carried out to allow modal approach of p_{inc} (Djojodihardjo, 2007, 2008, 2013, 2015).

PROOF OF CONCEPT GENERIC EXAMPLES: NOISE SUPPRESSION

As a first example, the reduction of noise pressure consisting of several single frequencies will be elaborated. Figure 7 is the graph of sound pressure in the time domain obtained by measuring the aircraft noise emitted by the propeller of Cessna 150.

The measurement is carried out while the aircraft is on ground. Figure 8 shows the response of noise frequency of the aircraft engine. Cessna 150 aircraft has a single engine and a two-bladed propeller, driven by a four-cylinder piston engine. The peaks in the frequency response shown in Figure 9 are related to the frequency due to the rotation of the propeller blades and the combustion cycles in the piston engine. The magnitude of the frequency as a result of the rotation of the propeller blade, f_1, is expressed by:

$$f_1 = B\frac{n_p}{60}, Hz \tag{75}$$

where B is the number of propeller blades and n_p is rotational frequency of propeller in rpm. For the aircraft considered, the number of blades B=2 and n_p = 2400 rpm.

Hence

$$f_1 = 2\frac{2400}{60} = 80\,Hz \tag{75a}$$

Figure 7. Results of droning sound reduction consisting of a single sound frequency
Reproduced from Kusni et al, 2014.

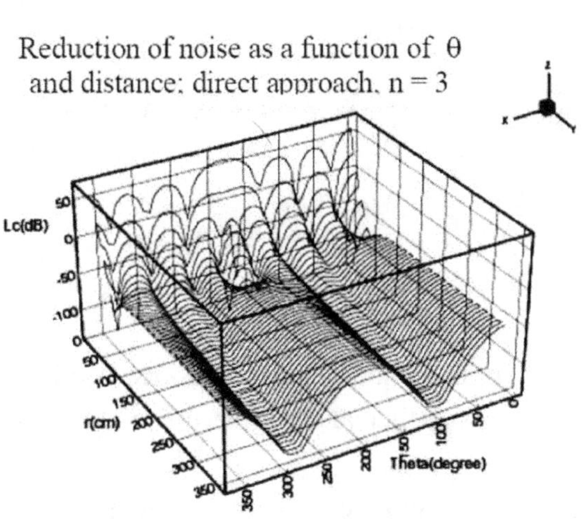

(a) Noise reduction as function of position (radius and theta)

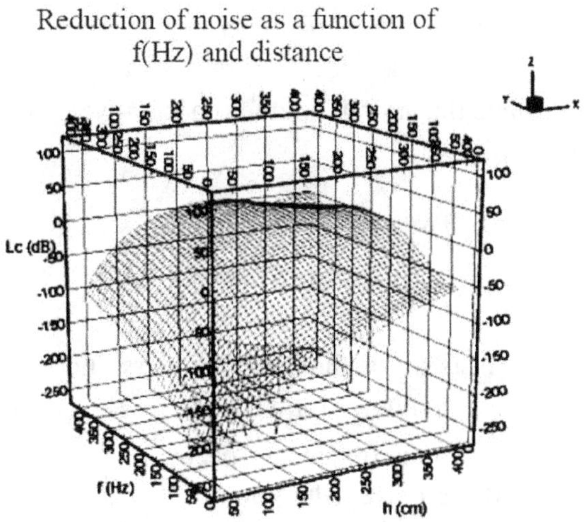

(b) Noise reduction versus frequency variation and distance between primary to secondary source

Figure 8. Frequency spectra of C-150 propeller noise
Reproduced from Kusni et al, 2014.

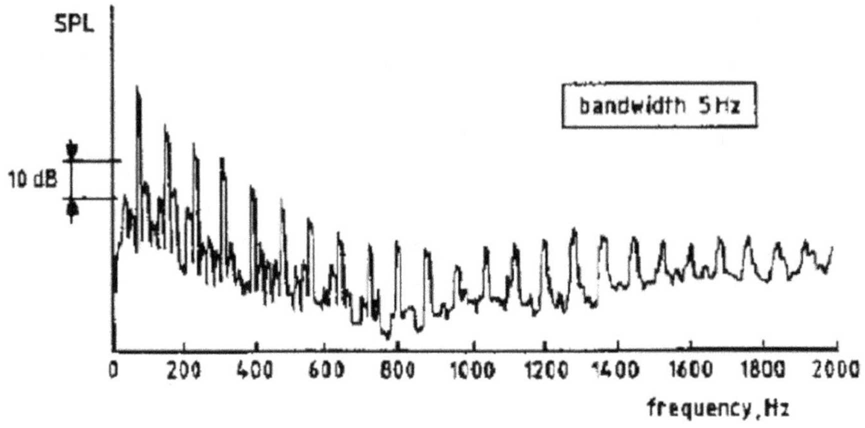

Figure 9. Time response diagram of Cessna C-150 noise; inset: a schematic picture of Cessna C-150

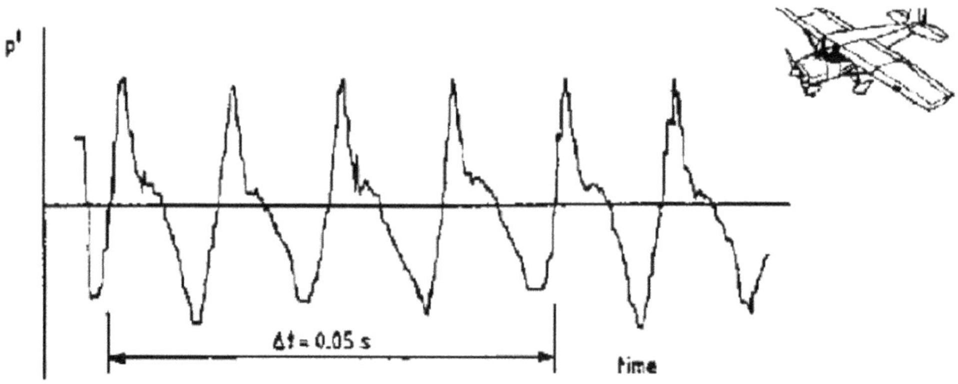

Since the aircraft engine is a four-cycle one, the frequency due to the combustion cycles, f_e, can be expressed as

$$f_e = N\frac{n/2}{60}, Hz \tag{75b}$$

where N is the number of cylinders and n is a velocity of engine rotation in rpm. Since N=4 and $n = n_p = 2400$ rpm, then $f_e = f_1$. Hence the contribution of propeller and engine noise cannot be separated in the frequency spectra.

Next the noise reduction of a Broadband White Noise will be elaborated. There are three kinds of variation carried out as related to this broadband noise reduction. The first is the variation of the distance between the primary and the secondary sources. Figure 10 shows that the farther the distance between the primary and secondary sources, the more difficult is the reduction of the broadband noise (1-20.000

Figure 10. The results of the C-150 aircraft propeller noise reduction
Reproduced from Kusni et al, 2014.

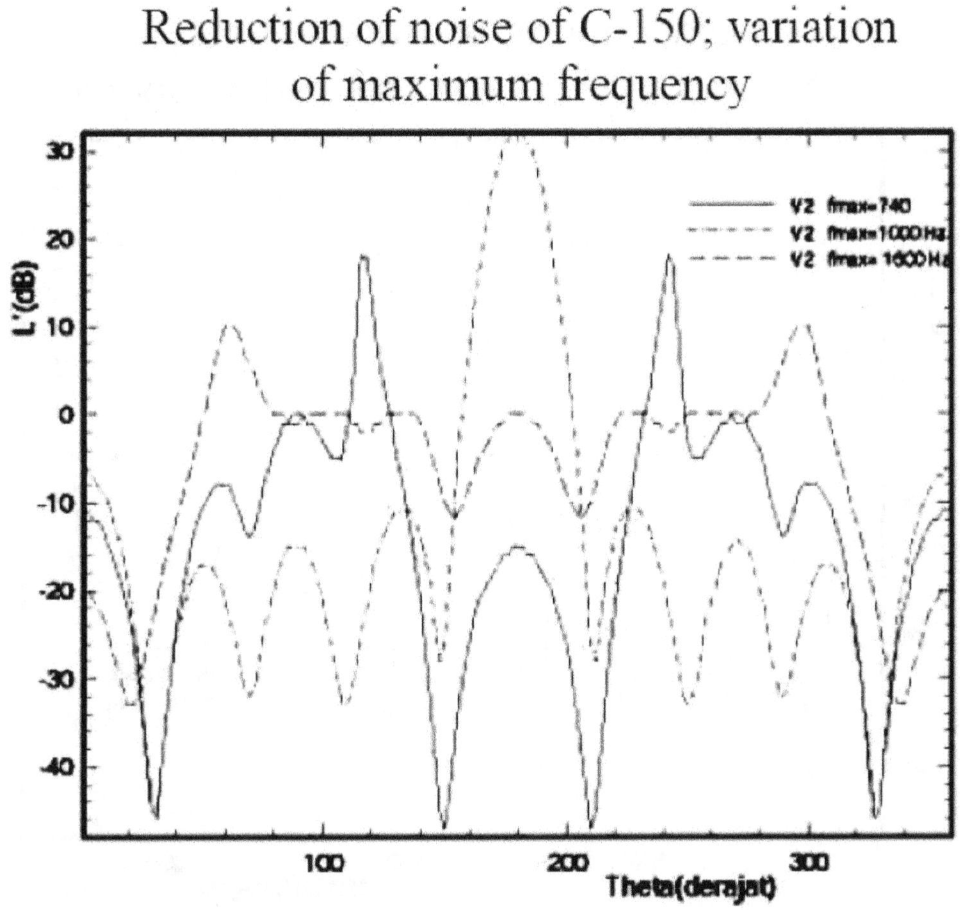

Hz). The second is the variation of the number of frequency contents. Figure 11 shows that the number of frequency content does not influence the magnitude of the noise reduction. The third case illustrates the capability of the noise reduction technique to identify the degree of white noise. As exhibited in Figures12 and 13, the results show that the broadband noise is not really a white noise, since the value of maximum frequency used is limited (below 20.000 Hz)

PROOF OF CONCEPT GENERIC EXAMPLES: FLUID EXCITED VIBRATION SUPPRESSION BY ACOUSTIC MEANS.

Numerical simulation of Coupled BEM-FEM is elaborated in what follows, with examples reproduced from (Djojodihardjo, 2013, 2015). The equivalent BAH wing structure (Bisplinghoff, Ashley & Halfman, 1955) and the surrounding boundary representing quarter space of the problem are discretized as shown in Figure 14. For computational convenience, the problem domain is divided into two parts. First the

Figure 11. The results of variation of distance of the primary to secondary sources
Reproduced from Kusni et al, 2014.

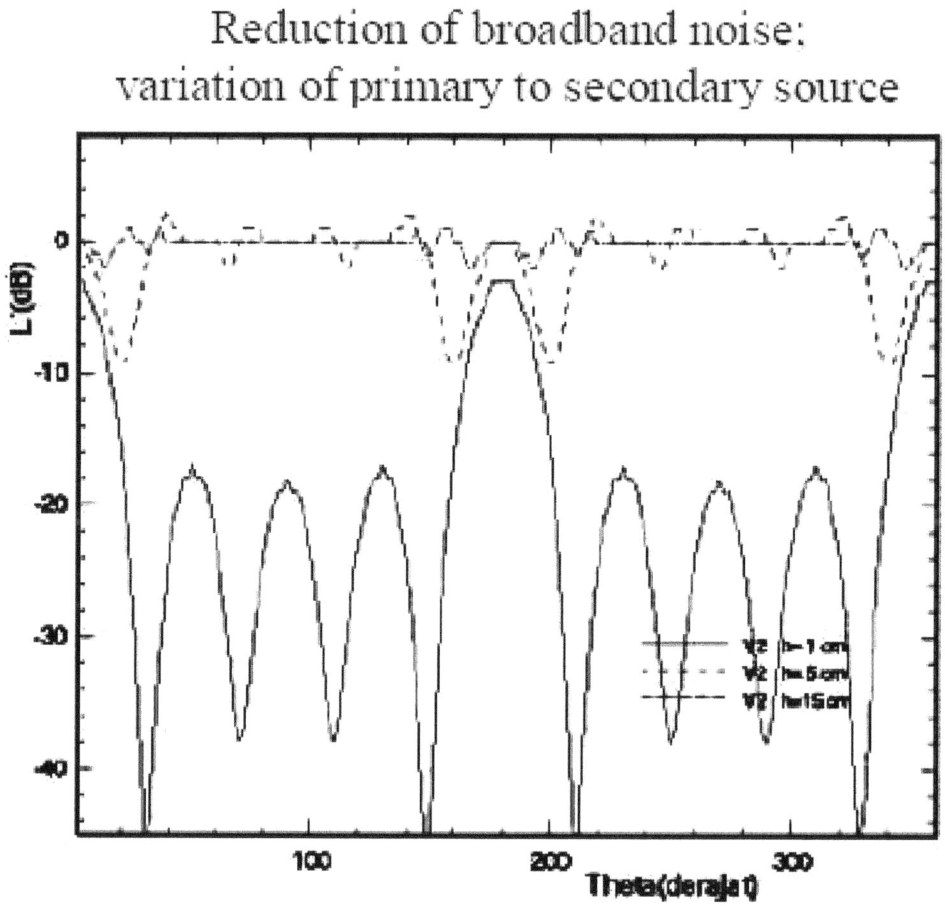

near field region is a quarter space with radius of two times the BAH wing span and is relatively more densely discretized; second, the intermediate to far field region is a quarter space with radius ten times the BAH wing span and is less densely discretized compared to the near field region, and is modeled with BEM only.

The BAH wing which is modeled as FEM and BEM is subjected to an excitation due to an acoustic monopole source; the acoustic medium is air with density $\rho = 1.225$ kg/m^3 and the sound velocity is $c = 340$ m/s. The monopole acoustic source is placed at the intersecting line of the half span and half chord planes of the BAH wing structure, and at about 0.1 m above the wing surface. The monopole source has the frequency of 10 Hz, radius of $a = 0.1$ m. The result of applying Equation 58 for $F = 0$ is presented as the incident, total and scattering pressures drawn as color-coded diagrams in Figure 15, which qualitatively exhibit the expected behavior. It could be added that in the example considered, CHIEF method has also been utilized in the BEM part to take care of the fictitious frequency problem, so that such phenomena can be eliminated in the BEM-FEM acoustic-structure coupling.

Figure 12. The reduction of broadband white noise results for case 2
Reproduced from Kusni et al, 2014.

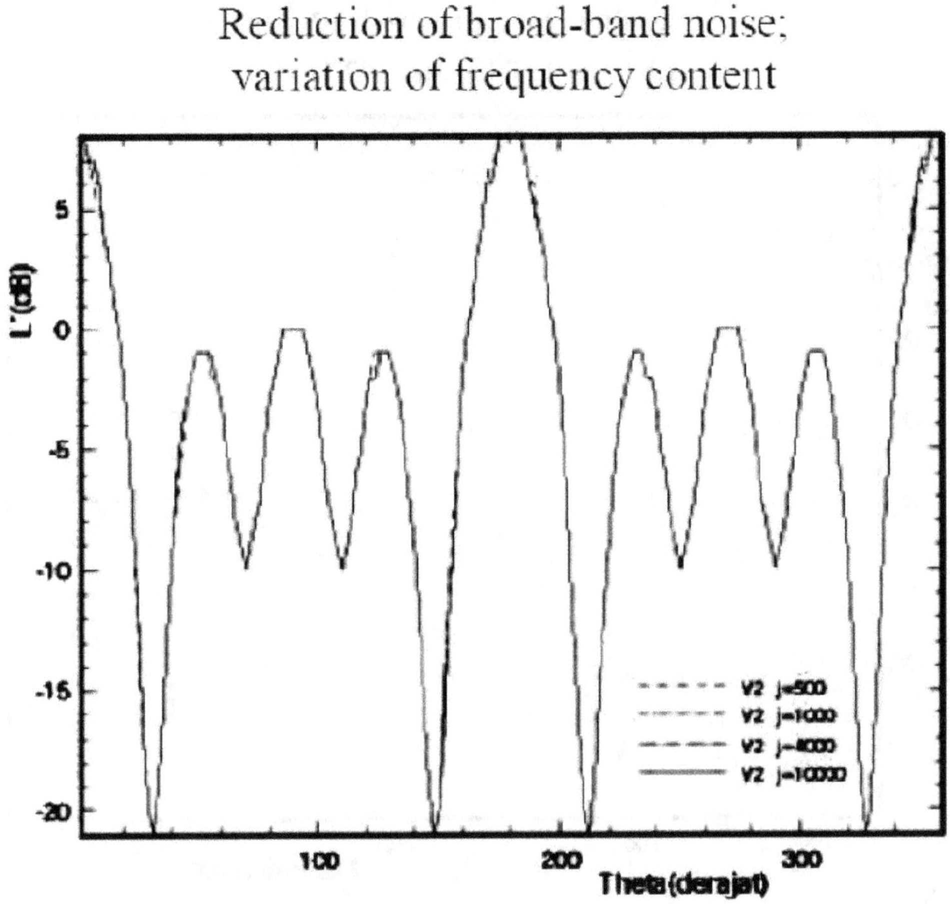

Flutter Calculation for Coupled Unsteady Aerodynamic and Acoustic Excitations

In earlier work (Djojodihardjo & Tendean, 2004; Djojodihardjo & Safari, 2005), the acoustic excitation is incorporated by coupling it to the unsteady aerodynamic load in the flutter stability calculation. Linearity and principle of superposition has been assumed. Hence the acoustic loading can be superposed to the aerodynamic loading on the structure, and form the modified aeroelastic equation (acousto-aeroelastic equation) of the structural dynamic problem associated with acoustic and aerodynamic excitation. In the earlier work, tacit consideration was only given to the incident acoustic pressure as the acoustic excitation, without considering the acoustic scattering effects, and without considering **L**.

The incident pressure p_{inc} was also assumed to belong to a certain class that allows its incorporation in the aerodynamics term. Thus Equation 58 was treated in a decoupled fashion; such approach has resulted in the delay of the inception of flutter. Figure 16 shows the unsteady aerodynamics pressure (C_p) and mode shape of the structure when flutter occurs.

Figure 13. The reduction of broadband white noise results for case 3
Reproduced from Kusni et al, 2014.

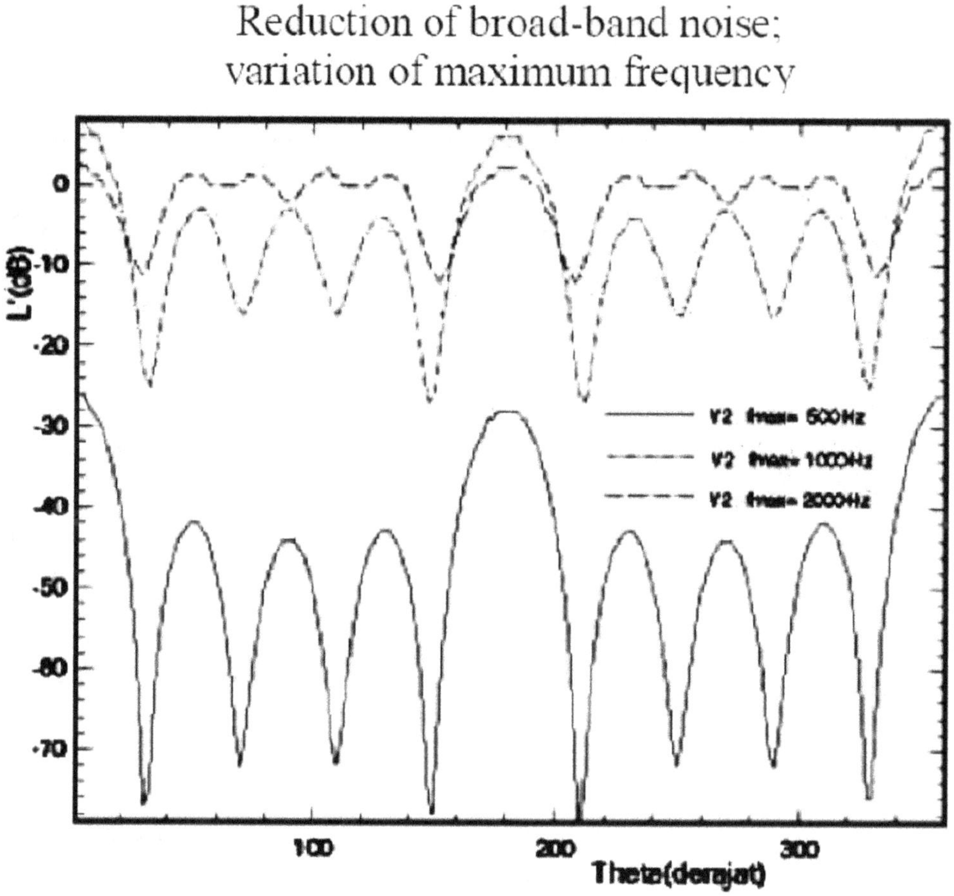

Figure 14. 3-D domain representing of BAH wing structure and its surrounding boundary

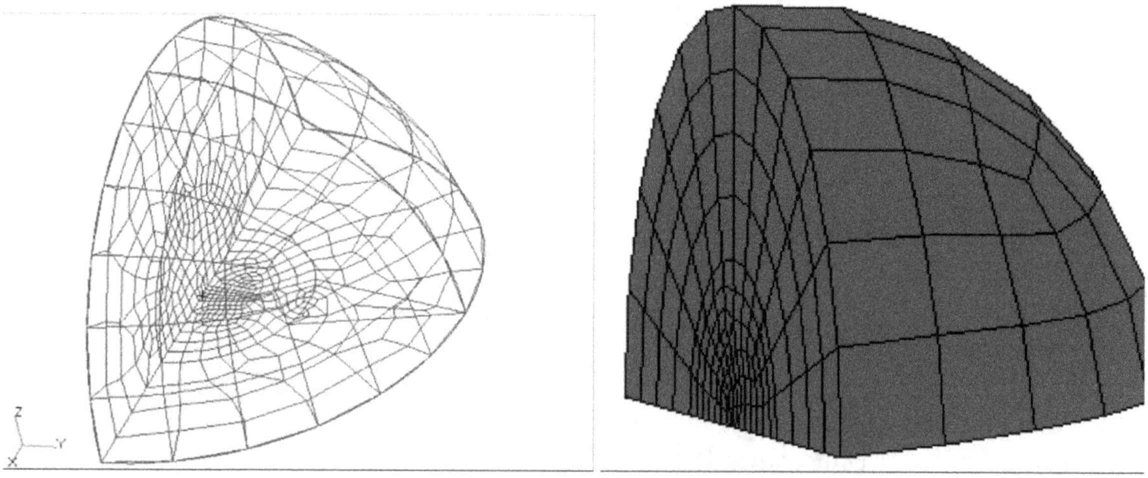

Figure 15. Pressure distribution on symmetric equivalent BAH wing: a) Incident pressure [dB] from monopole acoustic source as an acoustic excitation and b) deformation and total acoustic pressure response [dB]

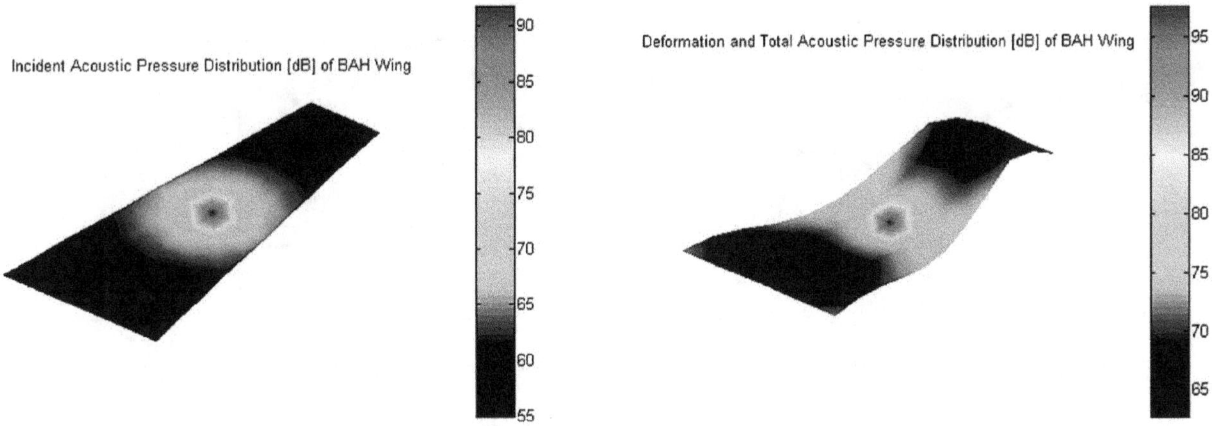

Figure 16. (a) Unsteady aerodynamics pressure distribution (C_p), and (b) mode shape of the wing structure when flutter occurs

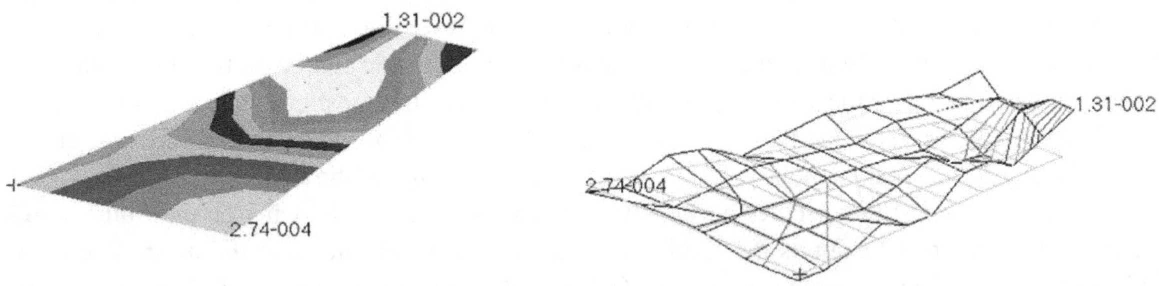

In the present development, rigorous consideration has been devoted to the acoustic scattering problem. By solving Equation 58 as stability equation in a "unified treatment" and the disturbance acoustic pressure already incorporates the total pressure, which has been "tuned" to behave like the aerodynamic terms in the modal Equation 74. The solution is exhibited as Figure17, which shows that the flutter inception is delayed at a higher speed also.

AERODYNAMIC AND ACOUSTIC COUPLING PARAMETRIC STUDY

The computational scheme for the distribution of acoustic pressure on the surface of the pulsating sphere, the total acoustic pressure for coupled BEM-FEM problem, and the influence of placing an acoustic monopole above a three dimensional wing (an equivalent BAH wing) to the flutter velocity, by using coupled BEM-FEM formulation for the acoustic incident pressure induced by the monopole source have been validated using NASTRAN® in (Djojodihardjo, 2008) and ZAERO® (Zona Technology, Inc, 2004).

Figure 17. Damping and frequency diagram for BAH wing calculated using V-g method written in MATLAB® for the acousto-aeroelastic problem (the total acoustic pressure already incorporates the scattering pressure)

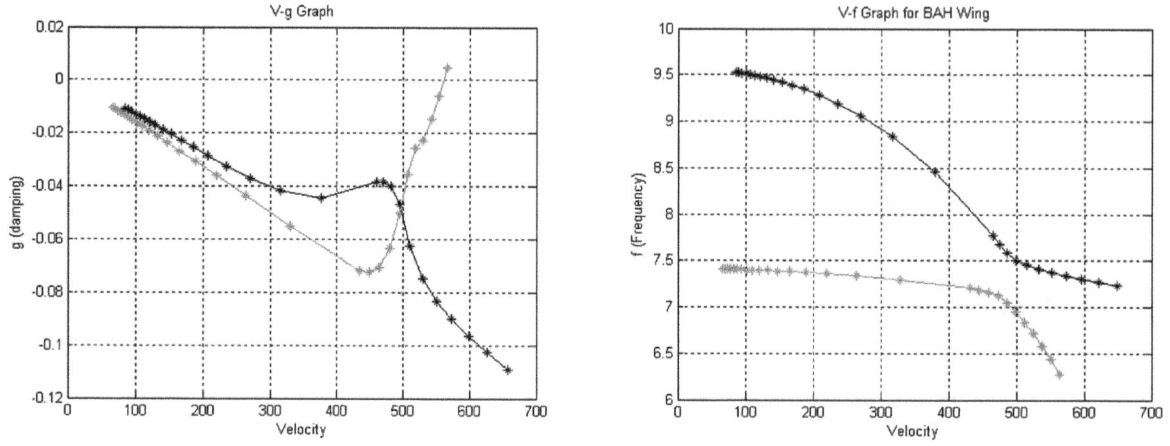

The calculation of the unsteady aerodynamic terms in Equation 74 is carried out using Doublet Point Method following Houbolt (1969) and Ueda and Dowell (1982), and developed into an in-house MATLAB® routine (Djojodihardjo and Safari, 2005). It is of interest to look into some simple applications to obtain the usefulness of the method. Along this line, several parametric studies are carried out. The first study looks into the influence of the intensity of the acoustic source on the flutter stability by varying its location above the wing. Figure 18(a) indicates that the most effective way in placing the acoustic monopole source is on the tip of the wing and Figure 18(b) indicates that the most effective way in placing the acoustic monopole source is on the trailing edge of the wing.

Next the influence of the distance between the acoustic source and the wing on the flutter stability is investigated. Figure 19 exhibits the results of such study and indicates that the most effective way in placing the acoustic monopole source is on the nearest distance from the wing. These results serve

Figure 18. The influence of acoustic monopole source intensity on flutter velocity as a function of monopole position: (a) at midchord along wing span, (b) at wing-tip section along the chord

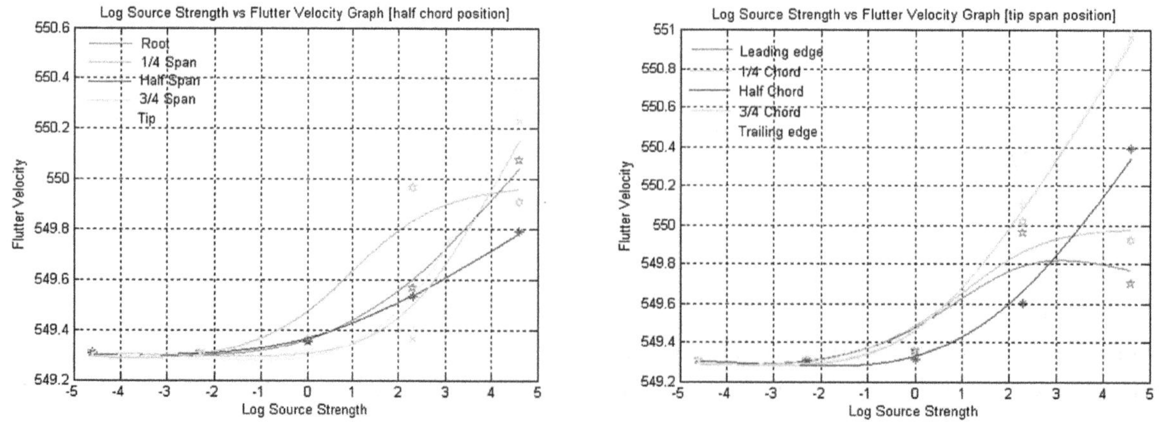

Figure 19. The influence of acoustic monopole source strength on flutter velocity as a function of the distance of the monopole source above the wing from the tip-chord point of the wing-tip section

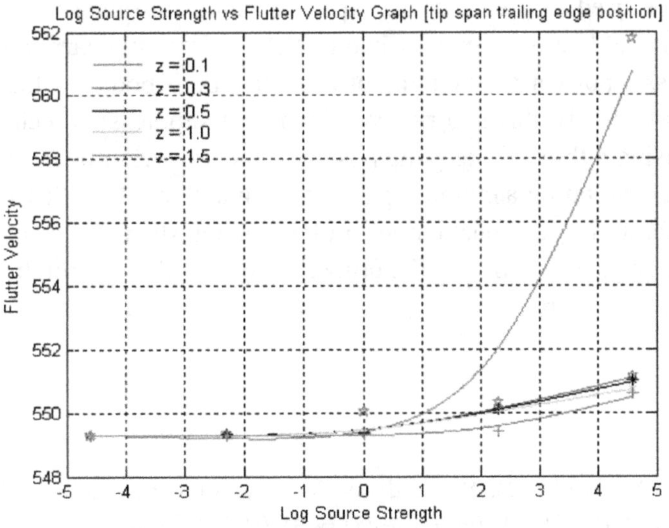

to indicate the logical trend of such problem, which will be useful for further practical applications. However, the favorable effect of the introduction of a monopole source closer to the wing should be accompanied by the increase of its strength.

CONCLUSION

The close relationship between noise and vibration has been elaborated and analyzed for the interest of the suppression or alleviation of noise and vibration occurring in many engineering and domestic systems. Two problems have been addressed, analyzed and studied; one is the suppression of noise and the second the suppression of vibration, both by acoustic means. For the first problem, an analysis is carried out on the representation of noise by monopoles and their higher harmonics, and to devise a straight-forward method to counter their influence in a certain domain by secondary acoustic source should be selective.

The second problem has been elaborated and analyzed by using a methodology developed earlier for the computational scheme for the calculation of the influence of the acoustic disturbance to the aeroelasticity of structures. The generic approach capitalizes on the formulation of the acoustic wave propagation governed by the Helmholtz equation by using boundary element approach, which then allows the calculation of the acoustic pressure on the acoustic-structure boundaries. The second part addresses the structural dynamic problem, which is formulated using finite element approach. The third part involves the calculation of the unsteady aerodynamics loading on the structure using generic unsteady aerodynamics computational method. A unified approach to allow effective solution of the problem formulated as Equation 87 is elaborated. Examples presented for both problems have shown the plausibility of the methods proposed, which serve as proof of concept.

In this conjunction, it is noted that NASA Marshall Space Flight Center rationale and methodology in designing launch vehicle/payload hardware is that these must be designed and built to withstand the exposure to various environments during fabrication, transportation, integration, launch, reentry, and landing. These environments include static, dynamic, thermal, electrical, corrosive, etc [58]. The method presented here could be used to predict such exposure before testing verification. In closing, some recent related methods could be mentioned. Chan (2012) has recently carried out research focused on the control and suppression of vortex shedding of flow past bluff bodies, which is also related to their generated noise suppression. To this end two very different suppression techniques were introduced, i.e. a passive control using a thin splitter plate positioned downstream of the bluff body and an active control by way of counter rotating a cylinder pair. Laksminanth et al [60] carried out a detailed study of the mechanics of vibration and acoustic noise in Permanent Magnet Synchronous Motors (PMSM) due to electromagnetic origins. To this end the utilization of wavelet controller in the area of noise suppression and filtering was explored.

REFERENCES

Ashley, H., & Rodden, W. P. (1972). Wing-Body Aerodynamic Interaction. *Annual Review of Fluid Mechanics*, *4*(1), 431–472. doi:10.1146/annurev.fl.04.010172.002243

Azarpeyvand, M. (2005). Active Noise Cancellation of a Spherical Multipole Source. *Acoustical Physics*, *51*(6), 6. doi:10.1134/1.2130891

Beauvilain, T. A., & Bolton, J. S. (1993). Cancellation of radiated sound fields by the use of multipole secondary sources. *Proceedings of the second Conference on Recent Advances in Active Control of Sound and Vibration* (Technomic, Lancaster, PA) (pp. 957-968).

Beer, G., & Watson, J. O. (1992). Introduction to Finite and Boundary Element Methods for Engineers. Chichester, England: John Wiley & Sons Ltd.

Bisplinghoff, R. L., Ashley, H., & Halfman, R. (1955). *Aeroelasticity*. Reading, Massachusetts: Addison Wesley.

Bolton, J. S., & Beauvilain, T. A. (1992). Multipole sources for cancellation of radiated sound fields. *The Journal of the Acoustical Society of America*, *91*(4), 2349. doi:10.1121/1.403449

Bolton, J. S., Gardner, B. K., & Beauvilain, T. A. (1995). Cancellation by the use of secondary multipoles. *The Journal of the Acoustical Society of America*, 1995, 98.

Chan, A. S. (2012). Control And Suppression Of Laminar Vortex Shedding Off Two-Dimensional Bluff Bodies [PhD thesis]. Stanford.

Chan, A. S. (2012). Control and Suppression of Laminar Vortex Shedding off Two-Dimensional Bluff Bodies [PhD thesis]. Stanford.

Chandler-Wilde, S., & Langson, S. (2007). Boundary element methods for acoustics. Retrieved from http://www.personal.reading.ac.uk/~sms03snc/fe_bem_notes_sncw.pdf

Chen, S. S. (1985). Flow-Induced Vibration Of Circular Cylindrical Structures. Argonne National Laboratory, The University of Chicago.

Djojodihardjo, H. (2007). BEM-FEM Acoustic-Structural Coupling for Spacecraft Structure Incorporating Treatment of Irregular Frequencies. *Proceedings of the58th International Astronautical Congress*, Hyderabad.

Djojodihardjo, H. (2008). Unified BE-FE Aerodynamic-Acoustic-Structure Coupling Scheme for Acoustic Effects on Aeroelastic Structures. *Proceedings of the 26th International Congress Of the Aeronautical Sciences*, Anchorage.

Djojodihardjo, H. (2009). Application of Fast Multipole Boundary Element Method for Unified BEM-FEM Acoustic-Structural Coupling. *Proceedings International Astronautical Congress*.

Djojodihardjo, H. (2013). *Unified Aerodynamic-Acoustic Formulation for Aero-Acoustic Structure Coupling. Journal of Mechanics Engineering and Automation*, *3*, 209–220.

Djojodihardjo, H. (2013). BEM-FEM Coupling for Acoustic Effects On Aeroelastic Stability Of Structures. *CMES*, *91*(3), 205–234.

Djojodihardjo, H. (2015). Vibro-Acoustic Analysis of the Acoustic-Structure Interaction of Flexible Structure Due to Acoustic Excitation. *Acta Astronautica*, *108*(March–April).

Djojodihardjo, H., & Safari, I. (2005). Further Development of the Computational Technique for Dynamics of Structure Subjected to Acoustic Excitation, paper IAC-05-C2.2.08. *Presented at the56th International Astronautical Congress*, Fukuoka Japan.

Djojodihardjo, H., & Safari, I. (2006). Unified Computational Scheme For Acoustic Aeroelastomechanic Interaction, paperIAC-06-C2.3.09. *Presented at the57th International Astronautical Congress*, Valencia, Spain.

Djojodihardjo, H & Tendean, E. (2004, October). A Computational Technique for the Dynamics of Structure Subject to Acoustic Excitation. *Proceedings of ICAS '04*, Vancouver.

Djojodihardjo, R. H., & Widnall, S. E. (1969). A Numerical Method for the Calculation of Nonlinear, Unsteady Lifting Potential Flow Problems. *AIAA Journal*, *7*(10), 2001–2009. doi:10.2514/3.5494

Djojodihardjo, H., & Tendean, F. (2001). Structure-Acoustic Analysis Using BEM/FEM; Implementation [Master's Thesis]. Structural Mechanics & Engineering Acoustics, LTH, Sweden.

Dowling, A. P., & Ffowcs Williams, J. E. (1983). Sound and Sources of Sound. Chichester: John Wiley & Sons.

Duraiswami, R. (2006). The Wave and Helmholtz Equations. *MD Medical Newsmagazine*.

Germain, P. (2000, May). *Active Control of Run-up Noise from Propeller Aircraft* [M.App.Sc. Thesis]. Uni. British Columbia.

Griffin, J. R. (2006). The Control Of Interior Cabin Noise Due To A Turbulent [MSc Thesis].

Hansen, C. H., Qiu, X.-J., Barrault, G., Howard, C. Q., Petersen, C. D., & Singh, S. (2007). Optimization Of Active And Semi-Active Noise and Vibration Control Systems. *Proceedings of the14th International Congress of Sound and Vibration*, Cairns, Australia.

Hess, J. L., & Smith, A. M. O. (1967). Calculation of Potential Flow about Arbitrary Bodies. *Progress in Aeronautical Sciences*, *8*, 1–138. doi:10.1016/0376-0421(67)90003-6

Houbolt, J.C. (1969). Some New Concepts in Oscillatory Lifting Surface Theory.

Jacobsen, F., & Juhl, P. (2011). *Sound Radiation, U.* Southern Denmark.

Johansson, S. (2000). Active Control of Propeller Induced Noise in Aircraft[Ronneby, Sweden]. *Department of Telecommunications and Signal Processing Blekinge Institute of Technology*, *SE–372*, 25.

Johansson, S. (2000, December). Active Control of Propeller Induced Noise in Aircraft - Algorithms & Methods [PhD Thesis]. Department of Telecommunications and Signal Processing, Blekinge Institute of Technology, Sweden.

Johansson S., Claesson I., Nordebo S. & Sjösten P. (1997, June). Evaluation of Multiple Reference Active Noise Control Algorithms on Dornier 328 Aircraft Data. *IEEE Trans. on Speech and Audio Processing*.

Juang, B. H. (1993). *Fundamentals of speech Recognition*. Prentice-Hall, Inc.

Kempton, J. (1976). The ambiguity of acoustic sources: A possibility for active noise control? *Journal of Sound and Vibration*, *48*(4), 475–483. doi:10.1016/0022-460X(76)90551-4

Khanal, B. (2010). A Numerical Investigation of the Aerodynamic Noise Generation Mechanism in Transonic Cavity Flows [PhD Thesis]. Cranfield.

Koopman, G. H., Song, L., & Fahnline, J. B. (1989). A method for computing acoustic fields based on the principle of wave superposition. *The Journal of the Acoustical Society of America*, *86*(6), 2433–2438. doi:10.1121/1.398450

Kuo, M., & Morgan, D. R. (1996). *Active Noise Control Systems. John Wiley & Sons, Inc.*

Kusni, M. (2002). The Active Noise Control Of Propeller Noise Using a Multipole Secondary Source.

Kusni, M, & Soenarko, B. (2001). The Active Noise Control of Free Sound Field Using A multipole Secondary Source. *J. Science and Math*, 2(2).

Kusni, M., & Soenarko, B. (2001). Dipole Wave Cancellation Using a Multipole Secondary Source. *Proc. of the Symposium on Wave Propagation, Institut Teknologi Bandung*.

Kusni, M., Soenarko, B., & Djojodihardjo, H. (2014). Application of a Multipole Secondary Source for Propeller Active Noise Control, Applied. *Mechanics of Materials*, *564*, 135–142. doi:10.4028/www.scientific.net/AMM.564.135

Lagö, Th. L. (1996). *Frequency Analysis of Helicopter Sound in the AS323 Super Puma* (Technical Report).

Lagö, Th. L., Johansson, S., & Hellström, P. A. (1997, February). Analysis of Helicopter Sound for the Development of a New Generation Active Headset. *Proceedings of the IMAC-XV Conference*, Orlando, Florida.

Legendre, C. (2014). On the interactions of sound waves and vortices [PhD Thesis]. Université Libre de Bruxelles.

Lighthill, M. J. (1952). On sound generated aerodynamically, General Theory, Proceedings of the Royal Society of London. *Series A, Mathematical and Physical Sciences*, 211(1107), 564-587.

NASA. (2013). Development of Vibroacoustic and Shock Design and Test Criteria. *NASA, MSFC-STD, 3676-A*(January), 10.

Nelson, P.A., Curtis, A.R.D. & Elliott, S.J. (1986). Optimal multipole source distribution for the active suppression and absorption of acoustic radiation. *Proc. Euromech Colloc.*

Norton, M.P. & Karzcub, D.G. (2003, October 16). Fundamentals of Noise and Vibration Analysis for Engineers. Cambridge University Press.

Pabst, O., Kletschkowski, Th., & Sachau, D. (2009, July 5-9). Combined Active Noise Control and Audio In A Light Jet. *Proceedings of the 16th Int.Congress on Sound and Vibration*, Krakow.

Rodden, W. P., & Johnson, E. H. (1994). MSC/NASTRAN Aeroelastic Analysis User's Manual, Version 68, pp. 237-257.

Ruijgrok, G. J. J. (1993). *Elements of Aviation Acoustics*. Delft University Press.

Seybert, A. F., Soenarko, B., Rizzo, F. J., & Shippy, D. J. (1984). Application of the BIE Method to Sound Radiation Problems Using an Isoparametric Element. *J. of Vib., Acoust. Stress and Reliab.*, *106*(3), 414–420. doi:10.1115/1.3269211

Sound Generation-Compendium, Sources Of Sound. (2013). Retrieved from www.akustik.tu-berlin.de

Ueda, T., & Dowell, E. H. (1982). A New Subsonic Method for Lifting Surfaces in Subsonic Flow. *AIAA Journal*, *20*(3), 348–355. doi:10.2514/3.7916

Weaver, W., & Johnston, P. R. (1987). *Structural Dynamics by Finite Elements*. Englewood Cliffs, New Jersey: Prentice-Hall, Inc.

Wrobel, L. C. (2002). The Boundary Element Methods, Vol.1, Applications in Thermo-Fluids and Acoustics. Chichester, West Sussex England: John Wiley & Sons.

Yu, K. H., Kadarman, A. H., & Djojodihardjo, H. (2010). Development and implementation of some BEM variants - A critical review. *Engineering Analysis with Boundary Elements*, *34*(10), 884–899. doi:10.1016/j.enganabound.2010.05.001

Zona Technology, Inc. (2004). ZAERO Theoretical Manual (13th ed.).

Chapter 2
Agent Based Modelling of Smart Structures:
The Challenges of a New Research Domain

Andreea Ion
University Politehnica of Bucharest, Romania

Monica Patrascu
University Politehnica of Bucharest, Romania

ABSTRACT

Smart structures are complex systems situated in even more complex and large scale urban environments. This chapter opens the field of agent based modelling and simulation (ABMS) to civil engineers. ABMS offers a wide range of tools for implementing simulation models of systems with high degrees of interconnectivity and a large number of component subsystems. The ease of use for specialized engineers and the capabilities of integration with existent technologies and infrastructures, make agent based models a very attractive way to incorporate the social system in the design process of buildings. Moreover, ABMS allows for the testing and validation of structure wide control and automation systems. This chapter presents past and current efforts of using agent based modelling for smart structures, as well as the main challenges brought by this new interdisciplinary research domain.

INTRODUCTION

Agent based modelling and simulation (ABMS) is a field that deals primarily with the computational techniques applied in studies of complex and large scale systems. ABMS is a powerful tool for simulating complex and dynamic behaviours of systems with important consideration regarding the interactions between entities in systems of systems.

The most common uses of ABMS are in social simulation and optimisation problems, such as traffic flow. Other applications refer to manufacturing and supply chains, economics with orientation towards financial markets, city infrastructure especially transportation and electric grids, biology and medicine for animals and behaviour of cells simulation.

DOI: 10.4018/978-1-5225-0588-4.ch002

The purpose of a structure is to shelter life, protect its inhabitants. But more than that, buildings host an array of activities vital to both the 21st century society and the growth of civilization. From the daily regular operations of the social structures accommodated within buildings, to the critical, time sensitive responses of control systems in order to protect human life, the current modelling directions should strive to include bio-social components when talking about structural performance. Moreover, the complexity of structural systems increases when taking into account the design of control systems, from environmentally friendly and energy saving automation systems, to robust control of seismic vibration, to efficient and dedicated safety protocols, and so on.

Additionally, special attention must be given to the unexpected disturbances of daily life, such as earthquakes, fires and other events. Specialized equipment (actuators, transducers, and controllers) has developed significantly, adding to the complexity of models necessary to design, control, monitor, and evaluate smart structures.

In order to integrate the requirements of civil engineering with the methodology of systems engineering, it is necessary that the models (and subsequent simulation environments) include the interactions between infrastructure and humans, control systems and social behaviour, structural performance and purpose of functionalities.

Thus, applying ABMS techniques in the design process of smart structures is the next step in the complementary vision of buildings that offer both protection and comfort, with focus on hosting daily life activities, in the larger context of a smart world.

In this chapter, a view of agent oriented programming is discussed, when applied to structural systems under the perspectives previously described, as the go-to tool when developing agent based models. The advantages of using ABMS depend on the problem that needs to be solved and a very important influence is given by the tools that the users choose. Some of these languages, environments, and frameworks are offering more accessible methods for the implementation of the standard concepts that describe the ABMS.

Moreover, the specificity of these computational modelling techniques make way for the subsequent development of simulation models, which are powerful tools in understanding complex systems, that offer integration of non-biological and biological systems within simulations, and that ultimately allow engineers to test and analyse what-if types of scenarios, safely observing and recording unexpected effects of various events that can take place within a building.

Thus, opening ABMS to the field of civil engineering can prove useful in long term, especially within the technological progress whirlwind of the current decade. The main objectives of this chapter are to introduce civil engineers to agent based computational techniques, to offer solutions for integrating the control systems component in the performance analysis of structures, and last, but not least, to open a new research area in agent based simulation models applied to structural engineering.

AGENTS AND MULTI-AGENT SYSTEMS

The Agent Based Modelling and Simulation Concept

When everything happens at once, wide and fast moving problems simply route around any central authority. (Kelly, 1995)

Along with the increase in urban complexity, technological advancement has brought forth more and more complex approaches to structural design and purpose. The development of the "smart" urban paradigm includes both micro and macro implementations. While the latter affect city-wide systems, like the energy grid, the former are endeavours more commonly known as smart home automation, intelligent building management, or variations thereof. The scale of these systems relative to the size of an average urban conglomerate does not reduce their inherent complexity. In an automated building, hundreds of concurrent processes must perform within nominal parameters, at all times, without interfering with one another (most of these systems are distributed or decentralized). Adding the human component in the mix, either as part of the governed processes or as disturbances to other processes, raises not only the number of systems belonging to a building at any one point, but also the degree of interactions between these entities.

The requirements for the automation solutions to be adaptive, combined with the requirements for safety and comfort of the biological component, while also taking into account the socio-economical purposes of civil structures, all these factors make smart buildings to be viewed as systems of systems, with a high degree of interconnectivity, parallelism, and distribution. The reductionist modelling approach does not do any justice to such entities. Thus, using modelling tools that are themselves complex is the most natural way to try and capture the essence of smart structures. This is the holistic perspective. Agents and multi-agent systems comprise one of the technologies congruent with it.

Why agents? Agents are quite versatile tools. In real world applications, their most enticing properties are their propensity to scalability when grouped into multi-agent systems (giving them the ability to manage large scale complex processes), their inherent parallelism, the distributed nature of multi-agent systems (for resources and expertise). Moreover, agents can form heterogeneous systems, allowing for customization based on purpose. Last, but not least, agents and multi-agent systems are viable for integration with legacy components and pre-existing technology. Therefore, it is not a stretch to use this technology for the modelling and simulation of complex structures.

To date, there is no universally accepted definition of the term agent (debates are ongoing in the field of computer science). What all definitions currently have in common is that an agent is an entity. Further than that, all manners of defining agents are based on their properties, which determine their applicability and field of use.

In ABMS, an agent is a persistent entity characterised by representable states, that interacts with the environment or with other agents, making changes in the environment or in the states of the other agents. To this respect, an agent based model (ABM) is a collection of agents and their states, and the rules that govern the agent-agent and agent-environment interactions.

As a system, an agent can have inputs (from other agents, from the environment), outputs (toward other agents or the environment), but it can also generate outputs that return to the agent self (feedback) in order to change the agent's internal state. An agent's internal state is governed by rules. When these rules can be changed by previous states, the agent also has the capability of learning. The behaviour of an agent is represented by the set of actions it performs (outputs) according to internal state and inputs.

The ABM itself can also have inputs, outputs, and states. However, from the ABM perspective, the states of the entire multi-agent system integrate their interactions, by using rules to describe these behaviours. There are plenty rule-based technologies nowadays, some of the most common including encapsulated rule bases (if-then-else structures similar to expert systems), multi-criteria decision making, machine learning, evolutionary computing, inference engines, etc.

An agent based model (ABM) is a computational/simulation model that is aligned with the generative experiment (Bertolotti, 2014). This type of model neither solves, nor estimates the behaviour of the system it represents, but instead it constructs ways to evaluate interconnections and interdependencies (agent to agent and agent to environment), interactive environments, behaviours, and the rules governing action sequences and information flows.

There are two types of ABM:

- **Growth Models:** In which a population of heterogeneous agents (that are set into a relevant environment and are allowed to interact based on simple local rules) organize in a bottom-up. manner
- **Predictive (What-If) Models:** In which a population of well known agents (that are set in various, and usually unexpected, environmental situations) is observed in terms of generated behaviours based on context.

It is interesting to note that, even though agent based models do not have equilibrium or stability conditions pre-specified, they still reach equilibrium states. These states are a result of the local behaviours of the entities within the ABM. It is this property that allows for agent based models to be fully parallelizable, completely distributed, able to emulate the complex behaviour of human groups. It makes ABM a useful tool in the simulation of social systems.

Agent Classifications

As previously stated, agents are most often described by their properties. Thus, agent classifications are currently organized by application field in order to differentiate between types of agents.

The Nwana Classification (Nwana, 1996) divides agents based on:

- **Capacity to Travel through a Network:** Static or mobile.
- **Cognitive Capacity:** Deliberative (able of symbolic reasoning based on which they partake in planning, negotiation, or coordination of other agents) or reactive (responding to environmental changes based on simple stimulus-response rules).
- **Properties of Autonomy, Learning, and Cooperation:**
    - **Collaborative Agents:** Cooperation and autonomy.
    - **Interface Agents:** Autonomy and learning.
    - **Learning-Collaborative Agents:** Learning and cooperation.
    - **Intelligent Agents:** All three properties.

A more detailed class-type division has been laid out by Maalal and Addou (2011):

- **Reactive Agents:** No intelligence; able to perceive the environment and act upon it; stimulus-response behavior.
- **Cognitive Agents:** Synthetic intelligence; symbolic representation of the world, knowledge, and objectives; able of planning their behaviour, learning, communication.
- **Intentional Agents:** Also known as BDI (beliefs-desires-intentions) agents; emulate the model of human intelligence and perspective of the world by using knowledge, beliefs, desires, intentions, choices, etc.

- **Rational Agents:** Capable of generating such behaviours that allow them to achieve their goals with the highest degree of success.
- **Adaptive Agents:** Capable of adapting to environmental changes; high degree of intelligence; capable of changing their own objectives and knowledge according to inputs.
- **Communicative Agents:** Capable of transmitting information to their surrounding entities.

The CitySCAPE classification (Patrascu & Dragoicea, 2014) is specifically constructed for smart buildings and smart cities (integrating both the control and modelling approaches), and it divides agents based on:

- **Level of Intelligence:**
    - **Reactive Agents:** Reflexive agents, either simple or with internal rules.
    - **Deliberative Agents:** Reasoning, intelligent inference systems, adaptive behaviours.
    - Composed reactive-deliberative agents.
- **Composition:**
    - **Singular:** These agents exist on their own in the system and usually perform specific tasks; they receive directives from other entities.
    - **Emergent:** A group of agents displaying a bottom-up organizational pattern.
- **Entity Type:**
    - **Embodied:** These agents have a physical body (hardware).
    - **Pure Software:** These agents exist only in the virtual world and interact only with each other through various communication systems.
- **Human Interaction:**
    - **With Strong Social Skills:** These agents are required to interact with humans; highly autonomous due to their non-deterministic environment.
    - **With Weak Social Skills:** These agents only interact with other agents.
- **Goal:**
    - **Control and Monitoring:** These agents perform all the actions required for the control of the devices throughout the building.
    - **Maintenance and Fault Diagnosis:** These agents perform all actions required for maintenance & fault diagnosis; their goal is intrinsically different than the one of the control and monitoring agents, due to the particularities of decisions and actions that these agents must take.
    - **Protection:** These agents are tasked with protection of human life, from hazard detection to hazard mitigation, including management of panic control devices and of the human component during emergency.
- **Interaction Protocols:**
    - **Communicative Agents:** Able to transmit information.
    - **Cooperative Agents:** Able to negotiate grouping into multi-agent systems.
- **Action Type:**
    - **Autonomous Agents:** Capable of making abstract decisions.
    - **Dependent Agents:** Acting or sensing agents capable of low level task execution.

Based on the way a multi-agent system is formed, there are three properties that can be highlighted:

- **Cooperation:** The ability of an agent to team up with other agents for a common goal.
- **Coordination:** The ability of agents to manage the interdependencies between their behaviours.
- **Negotiation:** The ability to reach a consensus regarding common interest problems.

Multi-Agent Systems Architectures

A multi-agent system architecture proposes a methodology that offers specifications regarding building the component agents, which involves the construction of a set of modules forming the agent, the manner in which the modules interact with each other, and how the input data and the internal state determine the future actions of the agent (rules). There are four major architecture types: deliberative, BDI, reactive, and hybrid, to which this chapter adds the cell-tissue configuration which is specifically created for the agent based modelling and simulation of urban systems.

Deliberative agents are systems with an explicit internal symbolic world model, a planning module, and an execution module (Figure 1). These agents make decisions through logic symbolic reasoning (Wooldridge & Jennings, 1995). For building such an agent, there are two issues that need to be taken into account: (a) transduction (transposing the real world into a precise symbolic description) and (b) representation and reasoning (designing the symbolic representation of information about the entities and complex processes of the real world and designing the reasoning mechanism that uses this information). In what concerns planning and execution, there are two main directions: hierarchical and non-linear planning. An implementation example of the deliberative architecture is Homer (Vere & Bickmore, 1990).

BDI (beliefs-desires-intentions) is an architecture based on practical reasoning (Wooldridge, 2009) that focuses on the role of beliefs, desires, and intentions (Figure 2). Practical reasoning is oriented toward actions, to the process of accomplishing what must be done and what can be done. The definition of the BDI architecture is inspired by the practical reasoning capabilities of human entities, and is comprised of two activities: deliberation (deciding what is desired, resulting in intentions), and means-ends reasoning (deciding how to act, resulting in planning). In this case, the beliefs are incorporating the agent's perception of the environment, desires are the effect of beliefs and can sometimes be inconsistent and

Figure 1. Deliberative agent

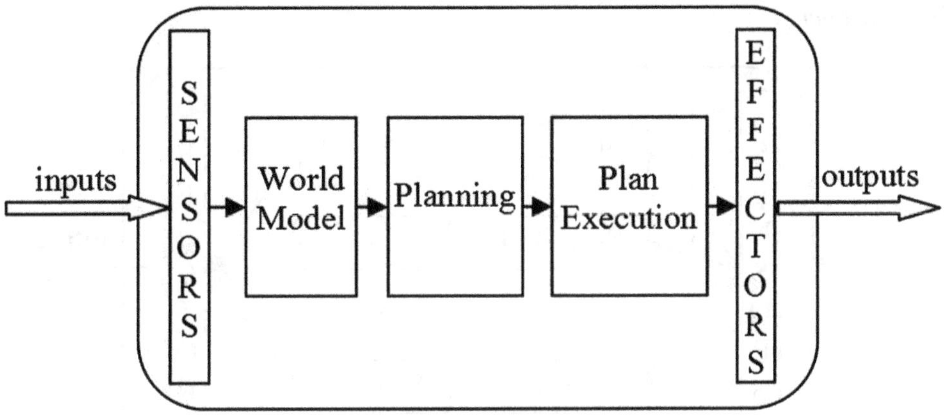

Figure 2. BDI agent

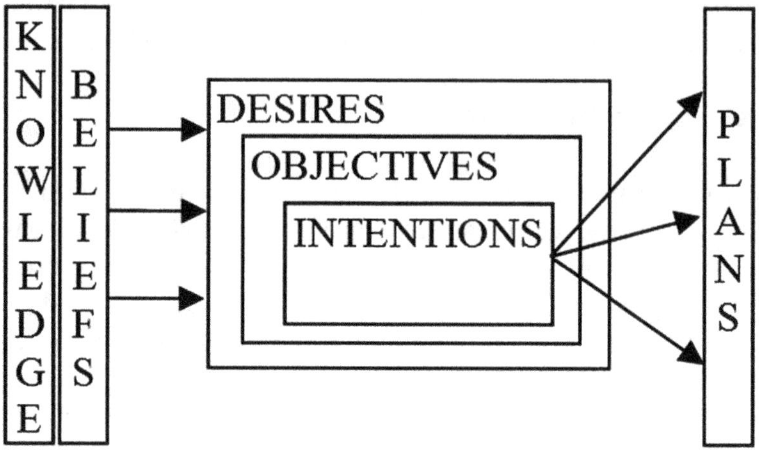

unrealistic, while intentions are objectives to be carried out (an objective becomes intention when the agent engages in achieving it, creating a plan of actions to be executed). An implementation example of the BDI architecture is IRMA (Bratman, Israel, & Pollack, 1988; Pollack & Riguette, 1990).

Reactive agents have a very simple internal world representation, but implement a tight connection between perception and action (Wooldridge, 2009). This approach is congruent with the hypothesis that intelligence is the product of the interaction between the agent and its environment. These agents implement reflexive (stimulus-action) types of behaviour (Figure 3). One of the most known reactive architectures is subsumption (Brooks, 1985) and is widely used nowadays for low level robot behaviours. Simple reactive agents are useful in the modelling and simulation of swarm processes, especially for biological systems in which emergence is observed.

Hybrid architectures combine the reactive and deliberative types by layering these two behaviours (Wooldridge, 2009), either (a) horizontally (in which each layer is directly connected to sensors and effectors), or (b) vertically (in which only one layer has access to inputs or outputs, respectively). Figure 4

Figure 3. Reactive agent

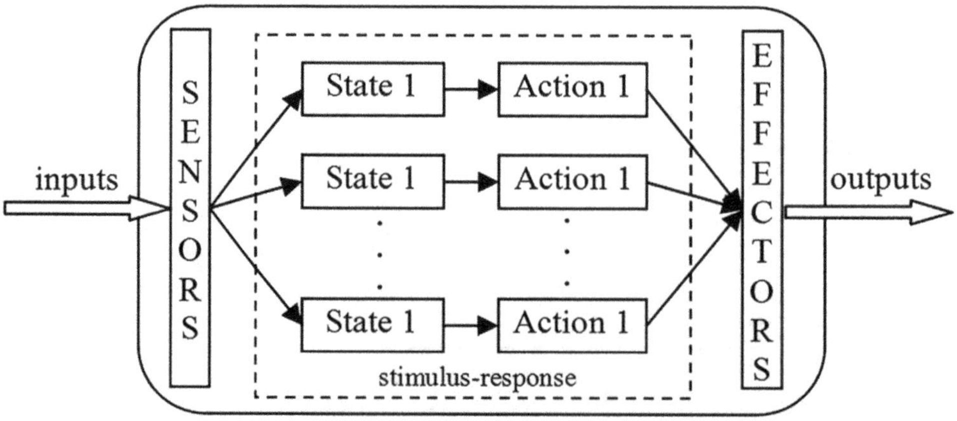

illustrates these two configurations. Thus, a hybrid agent is comprised of two subsystems, making use of symbolic representations and planning, while also being able to react reflexively to dynamic changes in the environment. A few interesting examples of hybrid agents are Touring Machines (Ferguson, 1992), the Procedural Reasoning System (Georgeff & Lansky, 1987), and InteRRAP (Muller & Pischel, 1993).

The cell-tissue configuration is an agent-MAS architecture ensemble, in which the component agents are named cells, and the multi-agent system formed is called a tissue. This concept is aligned with the CitySCAPE design framework (Patrascu, Dragoicea, & Ion, 2014). In this chapter, the authors define this configuration for use in ABMS, mainly for the modelling and simulation of intelligent building control and monitoring systems.

A cell agent for simulation is defined as a set of inference agents A_I, sensing agents A_S, and acting agents A_A, organized around a nucleus agent N:

$$\langle \cdot \rangle = \{A_I, A_S, A_A, N\} \tag{1}$$

This configuration allows for reflexive low level behaviour (like sensing and acting) to be integrated naturally with more complex processes, like decision-making using inference machines. This "agents within agents" structure offers scalability, the possibility of modelling increasingly complex building management systems as realistically as possible, and follows the intrinsic organization of feedback control systems. For instance, if an actuator for seismic vibration mitigation fails, using the cell structure, this event can be easily inserted into simulations, without re-designing the model or the interactions between components. In this chapter, the following terminology is used: an *objective* describes the purpose of a cell, a *goal* the purpose of a tissue, while a *task* describes the purpose of the cell component low level agents.

A nucleus agent:

$$N = \{U_{comm}, N_{alg}\} \tag{2}$$

is comprised of one communication unit U_{comm} and a (set of) algorithm(s) for aggregation and disintegration N_{alg} of cell agents based on necessities/objectives.

The nucleus is an agent with strong communication capabilities, since it manages the receiving of the objectives and the organization of the cell (aggregation or disintegration). The communication units of the nuclei U_{comm} deal with three types of information:

Figure 4. Hybrid configurations: (a) horizontal architecture; (b) vertical architecture

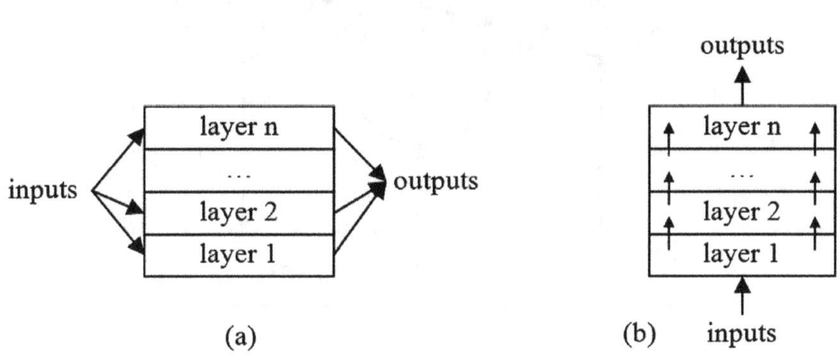

- *Objective information* can be received from units in other tissues that require certain actions to be performed by the tissue that incorporates the considered cell; for example: the command to clear a road to facilitate emergency vehicles access.
- *General information* needs to be relayed to other cells and is not necessarily pertinent to the current cell; for example: the request to clear a road in another part of the city - this information is thus analyzed and then relayed to other cells.
- *Reply information* confirms or denies the success with which the received objectives have been carried on; for example: the successful clearing of a cell controlled intersection, which will transmit the clear request to the next intersection and confirmation of green light or clear ahead to the intervention vehicles.

Figure 5 shows the main components of a cell agent: A_{Ii} is an inference agent running a control algorithm or a reasoning process, A_{Ai} are acting agents (effectors in general, actuators for control systems), A_{Si} are sensing agents (sensors or perception algorithms), N_i cell nucleus, d_i environmental data, y_i processed sensor data, u_i computed commands/decisions, a_i executed actions, obj_i cell objective of controlling a specific process received by the nucleus agent *a priori* (for the design of an aggregation mechanism and objective evaluation by the nucleus, please consult Patrascu, Dragoicea and Ion (2014)), tsk_i desired tasks according to objectives (for example the task of a sensing agent is to collect raw data, process it, and transmit it to the inference agent; whose task is in turn to elaborate control decisions based on the received data and transmit it to the acting agents; who in turn are tasked with executing the command).

A tissue is a set of n cells with the same goal and concurrent objectives:

$$\text{«}\cdot\text{»} = \{\langle\cdot\rangle_i \mid i=1,n\} \tag{3}$$

Figure 5. Cell agent configuration

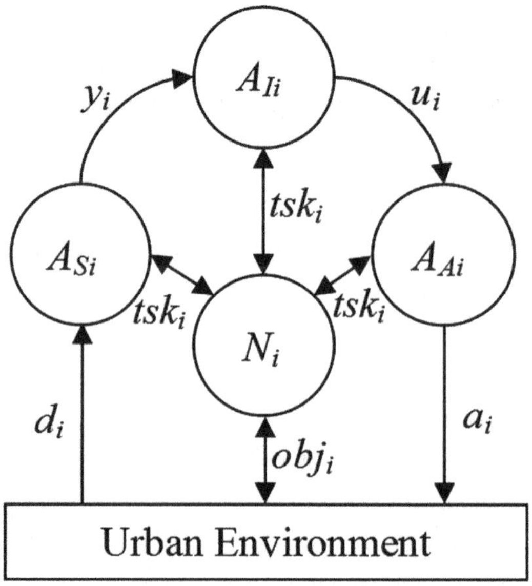

A tissue is a form of collaborative multi-agent system with emergent properties. The first effect of cell collaboration and creation of tissues is the interconnection of the components that form the system. This kind of interconnectivity depends upon a large exchange of data and information that is necessary to the components in order to fulfil their objectives and tasks. Through the emergent properties and the methodology of communication between cells, the overcharge of the communication network is avoided.

There are a few remarks to be made regarding the autonomy of the agents within the cell-tissue configuration:

- **Autonomy of Cells:**
  - Each cell acts individually to accomplish its objective.
  - When a disturbance is detected, the cell changes its internal behaviour without being externally prompted by a user or by another algorithm.
- **Autonomy of Tissues:**
  - Tissues can alternate their global behaviour (between emergency mode and nominal operation) without being prompted by other agents, tissues or users in the environment.
  - The internal state of the tissue is determined by the behaviour of each cell.

Naturally, all these architectures and configurations have their own advantages and disadvantages that might enhance or limit their performance and suitability based on application domain. Table 1 surmises a few of the advantages and limitations of the presented architectures.

LANGUAGES AND FRAMEWORKS

Recent research in Multi-Agent Systems (MASs) has led to the development of programming languages and tools (Tan, Chai, & Liu, 2011) appropriate for the practical implementation of such systems, thus the new programming paradigms are an essential requirement for the transfer of technology and have become a high interest topic in MAS development. In the past few years, a significant number of agent oriented programming (AOP) languages have already been developed, from declarative approaches, to imperative or hybrid variations (Bordini et al., 2006), most of which co-exist with supporting platforms for their semantic particularities (Kravari & Bassiliades, 2015).

Declarative AOP languages are partially characterized by their formal nature, firmly based on logic, of which more common are: FLUX (high level programming language with logic agents), MDPL (Multi-Dimensional Dynamic Logic Programming incorporates knowledge representation for the Minerva framework for reasoning mechanisms), DALI (a logic programming language with respect to rule based agents, either reactive or proactive), ReSpecT (includes logic programming with rule based semantic), CLAIM (Computational Language for Autonomous, Intelligent and Mobile Agents as part of the Himalaya framework: Hierarchical Intelligent Mobile Agents for building Large-scale and Adaptive systems based on Ambients).

The imperative approach in AOP is less common due to the use of particularized abstractions (entities or instruments that populate the agent environments) that are tightly tied to the actual design of agents. Even so, there are imperative AOPs out there, like JAL (JACK Agent Language), that allows the development of multi-agent systems. JAL is based on reactive planning for the BDI (Belief-Desire-Intention) agent architecture and is not a logic language, but more of an extension to the well-known Java.

Table 1. Advantages and limitations of agent architectures

	Advantages	Limitations
Deliberative	• Useful in dangerous environments • Allows transposing (optimal) solution finding procedures for general problem classes into programming languages	• Requiring constant updating of the internal model due to the dynamic nature of the world, most often in parallel with planning • The representation language is expressive enough to be used in any field, but limited for computational purposes • Planning is implemented through classical methods that search for optimal solutions, when a sub-optimal less costly one would work the same or better
BDI	• Intuitive, offers functional decomposition of modules • Easy to define formally	• High degree of difficulty in balancing mental activities, costly reasoning being necessary for launching intentions • Changes in the environment trigger restating the entire reasoning procedure • High computational cost
Reactive	• Simplicity of agents • Flexibility and adaptability, which make it ideal for non-deterministic environments • Avoiding complex planning and reasoning • Robustness for simple tasks	• The lack of a world model requires sufficient input information from the environment, which might not be available • Decisions are made based on local data, not taking into account a global view of the world • No long term planning • Limited applicability (basic robots, games, simulations) • High difficulty at implementing learning
Hybrid	• Horizontal architectures: each layer is an agent, offering simplicity and independence • Vertical architectures: low complexity	• Horizontal architectures: inter-layer competition can cause incoherence; the need for a mediator between layers increases complexity • Vertical architectures: less flexible and fault non-tolerant • Lack of a generic design methodology • High degree of specificity depending on the application for which the agents are designed • Difficult integration with formal theories
Cell-tissue	• Combines the advantages of deliberative and reactive architectures • The main nucleus procedures can be implemented using any programming language and can also be published as web services when computational resources are limited or shared • Modularity at a functional level • Distributed systems that do not require a central node • Reduced communication over the agent network in tissues • Adaptive and flexible, due to the ad-hoc and on demand creation of cells • Wide applicability	• If the controller has a high degree of intelligence, due to the dynamic nature of the world, the deliberation of the action may take more time • The high specificity of the details that need to be provided in the objective sent to the nucleus • Cooperation between different cells can have high costs regarding execution time

Some AOP languages combine the declarative and imperative approaches, making way for hybrids such as Jason (an extension of the AgentSpeak(L) logic programming language for Java), 3APL (An Abstract Agent Programming Language implementing cognitive agents), IMPACT, AF-APL (Agent Factory Agent Programming Language) or Jadex.

Further support for agent-oriented programming and application development is offered by mainstream versatile platforms and environments, from C# (a declarative, imperative, functional, generic, object oriented, component oriented language; used in combination with the Microsoft .NET Framework and Aplode-E3 for the development of MASs) and Python (an object oriented language, interpreted and interactive, which permits imperative, functional or procedural programming), to NetLogo (a programming language dedicated to multi-agent systems that offers an integrated modelling environment), GOAL (a language for programming rational agents), and Java (an imperative language, essential for creating cross-platform portable code, using the JADE framework for implementation of MASs, supporting asynchronous agent models, distributed systems, agent communication and so forth).

Regardless of the chosen programming language from the above mentioned ones or existing in literature, the complexity of implementing an ABMS increases along with the complexity of the problem. Taking in consideration that the concept of ABMS has wide applicability, the need for easier tools and languages that could be used by all kinds of engineers arises. In this respect, some simulator environments can be noted:

- **AMP Project**: Provides extensible frameworks and exemplary tools for representing, editing, generating, executing and visualizing agent-based models (ABMs) along with behavioural and functional features. AMP has two main themes that complement but don't depend on one another: modelling and execution platform (Eclipse, 2015).
- **Ascape:** A tool for developing and exploring general-purpose agent-based models. Models can be developed using less code than in other tools. It offers a broad array of modelling and visualization tools (Parker, 2001).
- **Breve:** An open-source software package which facilitates building 3D simulations of multi-agent systems and artificial life written in Python (Klein, 2008).
- **Entorama:** A tool for 3D multi-agent modelling and simulation specifically designed for simulation of decentralized systems; it is useful in modelling large-scale systems in which the engineer needs to create hundreds of agents, each with its own specific behaviour and observe the behaviour that emerges at the collective level. The goal of this project is to allow a wide variety of models while simplifying the model development process and eliminating programming overhead (Cace, 2006).
- **FLAME (Flexible Large-scale Agent-based Modeling Environment):** A very general system for building detailed agent-based models, and it generates highly efficient simulation software that can run on any computing platform (specially on High Performance parallel supercomputers-HPC) with any operating system; current available literature reveals this framework as the only one with this capability, although more are surely on their way. Because it is very general, the user has the possibility to set up large scale models with millions of agents (Chin et al., 2012).
- **MASON:** A fast discrete-event multi-agent simulation library core programmed in Java, designed to be the foundation for large custom-purpose Java simulations (Luke, Cioffi-Revilla, Panait, Sullivan, & Balan, 2005).
- **MASyV:** An open-source Multi-Agent System Visualization platform. A user can write a simple ABM simulation in C, creating agent images in Gimp, and MASyV provides a set of libraries of functions which the client can make use of to speak to the UI which takes care of displaying and moving the agents. It also comes with a few pre-made simulations which can save some time if the users are implementing something similar (e.g. ants eating and laying pheromone trails, infection spreading cell-to-cell or through the diffusion of virus with or without the presence of immune cells).
- **MetaABM:** A high-level architecture for designing, executing and systematically studying ABM models.

- **NetLogo:** A cross-platform multi-agent programmable modelling environment (Berryman & Angus, 2010; Wilensky, 1999).
- **Player/Stage:** A free software tool for robot and sensor applications (Vaughan, 2008).
- **Repast:** A free and open source agent-based modelling toolkit that simplifies model creation and use (North et al., 2013).
- **RoboCup-Rescue Project:** An initiative to promote research and development in this socially significant domain at various levels involving multi-agent team work coordination, physical robotic agents for search and rescue, information infrastructures, personal digital assistants, a standard simulator and decision support systems, evaluation benchmarks for rescue strategies and robotic systems that are all to be integrated into a comprehensive future system (RoboCup, 2006). Robocup-Rescue initiatives are divided into two main directions:
    - **The RoboCup-Rescue Simulation Project:** An open resource of research results. This project is itself divided into two main directions, namely the Virtual Robots and the Agents Simulation projects which target various challenges that exist at the single robot level and the multi-agent system level.
    - The challenges for robots in the Real Robots project range from mechatronics for advanced locomotion over perception and planning, up to providing full autonomy. The robots are evaluated in special test settings, the NIST rescue arenas.
- **SeSAm (Shell for Simulated Agent Systems):** A tool that provides a generic environment for modelling and experimenting with agent-based simulation. It is a system focused on providing a means for the easy construction of complex models, which include dynamic interdependencies or emergent behaviour (SeSAm Multiagent Simulation [SESAM], 2012).
- **SPARK (Simple Platform for Agent-Based Representation of Knowledge):** A cross-platform, free software for multi-scale agent-based modelling (ABM) (Solovyev et al., 2010).
- **StarLogo:** A programmable modelling environment for exploring the workings of decentralized systems (Klopfer, 2009).
- **TerraME:** A programming environment for spatial dynamical modelling. It supports cellular automata, agent-based models, and network models running in 2D cell spaces. It provides an interface to the TerraLib geographical database, allowing models direct access to geospatial data. Its modelling language has in-built functions that make it easier to develop multi-scale and multi-paradigm models for environmental applications (TerraME: Simulation and Modelling of Terrestrial Systems [TerraME], 2015).

If a distinction could be made between platforms, the criteria would be the adopted programming language. In this case, by far, the main programming language most models have adopted is Java. About 42% of the platforms employ Java as their primary programming language. Some of the toolkits that support Java programming are: ADK, AgentBuilder, AnyLogic, Ascape, Cougaar, DOMAR-J, jEcho, ECJ, FAMOJA, iGen, Jade, JAS, JASA, JCA-Sim, jES, JESS, Madkit, Mason, Moduleco, Omonia, RepastS, RepastJ, SimPack, SOAR6, Sugarscape, VSEit (Nikolai & Madey, 2009).

A few toolkits that support C++ programming are AgentBuilder, DeX, iGen, LSD, Madkit, MOOSE, Jade's Sim++, SimBioSys, SimPack, SOAR6. For C programming, the known toolkits are AgentBuilder, Echo, iGen, Madkit, MAML, SOAR6. The last section refers to the toolkits that support programming in Logo dialects and the most popular are NetLogo and StarLogo. Concerning the visual programming toolkits, some of them are MAML, PS-I, RepastS, SeSAm, SimPlusPlus, SME, Zeus, StarLogo TNG, AgentSheets.

Ultimately, the advantages of using ABMS depend on the nature of the problem at hand, being heavily influenced by the user's choice in tools. Some of these languages, environments, and frameworks might be more comprehensible to some engineers, while others might be more implementation friendly. In all, the selection of these tools needs to be carefully considered, in order to obtain relevant simulation models.

MODELLING SMART BUILDINGS AND CONTROL SYSTEMS

Types of Control Systems Found in Smart Structures

In order to understand the heterogeneity of automation systems throughout a structure, even one that cannot be categorized as smart, a brief review of the types of control systems found in buildings is included. This is, nevertheless, a far from exhaustive effort, as such systems are continuously being developed, tested, and implemented. The point of this section is to move past the way in which control systems are currently analyzed, which is independently of one another and of the human component. With ABMS, the integration of these systems in simulation is possible, and understanding their diverse nature constitutes one of the most important challenges smart building modelling currently faces.

For the modelling and simulation of different utilities (energy, water, temperature, etc.) inside a building, intelligent systems (expert and knowledge based systems, neural networks, fuzzy logic and combinations thereof) are the most popular. These are used in combination with the classic advanced method of process control like adaptive systems (self-tuning or with reference model), optimal and robust controllers, predictive (model predictive MPC or general predictive GPC control systems), multi-model techniques or MIMO (multiple-input-multiple-output) controllers.

In the work of Privara, Široký, Ferkl, and Cigler (2011), a model predictive controller (MPC) is implemented and used to control the temperature in a building. The proposed solution takes into account both the weather forecast and thermal model of a building, in order to minimize energy consumption. The model of the building, which is a MIMO system, is identified by means of subspace methods. During simulations, the authors observed that the identification data is not satisfactory and needs special treatment, but the implemented controller achieved savings of 17–24% compared to the building's previous controller.

Modelling a building reveals important issues that need to be solved. A review of such problems for building information modelling (BIM) is provided by Cerovsek (2011), along with some tools, standards, and comprehensive recommendations for interoperability, integration, model-based communication and collaboration in building projects. In order to improve the process of modelling a building, the author implemented a methodological framework CASB (Complex Adaptive System with BIM). It includes standardization, inter-operability, and intelligence connected to a five dimensional information space: process, product, project, resource, and knowledge. CASB can connect different BIM Models, its parts and external services, and adapt to a changing environment so that it has the capacity to change and learn from past activities. A combination of the BIM, 4D technology, construction simulation, and safety analysis as an integrated solution for information sharing and exchanging among architects, civil engineers, and construction managers is presented by Hu, Zhang, and Deng (2008). The improved time-dependent structure analysis method proposed in the paper performs as expected on a safety analysis system named SABIC. This research explores the integration of architecture design, structure calculation, and construction management.

Structural integrity under external disturbances also needs to be taken into account. A comparative study for seismic vibration control algorithms is presented by Patrascu, Dumitrache, and Patrut (2012). While the controlled object is nonlinear (MIMO system), often with multiple degrees of freedom, a high importance in control goes to actuator choice and design. The authors studied two types of actuators for seismic control, with five control algorithms and under various seismic disturbances. The analysis presented in this paper offers a basis for choosing control algorithms for seismic vibration control, according to the structural response component that is desired to be minimized. Moreover, seismic dampers themselves, as actuators, can be included in control loops that add to the systems' overall robustness. For example, a conventional PID controller is designed using genetic algorithms for a magnetorheological damper with non-linear behaviour in the work of Patrascu and Ion (2015).

Integrating Formal Methods

With the rapid development of high-rise buildings, the emergence of these complex architectural forms has brought many new problems to the seismic analysis of structures and seismic design. In the work of Ji, Zhang, Zhao, Yuan, and Liu (2012), the building seismic theory using the ANSYS general purpose finite element software for seismic analysis of the tall building structures is described and a finite element model is built. The dynamic characteristics of the entire structure are determined with modal analysis. The results obtained after response spectrum analysis determined the dynamic response of structures under seismic action and drawn the maximum internal force of the structure of each component, that can provide the basis for structural design. Linear static free vibration and nonlinear dynamic time domain analyses are performed by Islam, Hussain, Jameel, and Jumaat (2012) for both base isolated and non-isolated buildings under site specific bi-directional earthquakes. The study shows that for medium rise building construction, isolation can significantly reduce seismic response in soft to medium stiff soil. Results show that the building becomes more stable compared to the fixed base structure and the modelled non-linear system is suitable to cope with precise non-linearities, giving the building more flexibility.

These formal methods, widely used and highly reliable, present yet another challenge, as they should be integrated in future ABMS research. Choosing the most suitable agent architecture for the agent based approach can offer ways to make use of these valuable formal methods.

Agent Based Simulations and Models

There is growing interest in reducing building energy consumption through increased sensor data and increased computational support for building control. The goal of reduced energy is often coupled with the desire for improved occupant comfort. In the work of Klein et al., 2012, the authors present and implement a multi-agent comfort and energy system to model alternative management and control of building systems. The proposed solution also includes coordination for both building system devices and building occupants through direct changes to occupant meeting schedules using multi-objective Markov Decision Problems (MDP). The model is implemented and tested with input from a real-world building. The results are compared with the data from the real building and a significant decrease of energy consumption is obtained.

Considering the team work that is needed in construction, a different and interesting approach is presented by Kado, Ibrahim, Khaidzir, and Anwar (2014). The authors identified the parameters involved in building consultant processes. They used a thematic analysis approach to identify parameters for the subsequent development of an agent-based simulation model which simulates the interactions among the involved stakeholders. The model was validated with construction experts and with readings obtained from software reports.

An investigation regarding the usefulness of the agent concepts for the design of control systems for intelligent buildings is presented in the work of Davidsson and Boman (2000). Some factors that contributed to the solution choice were the close mapping that it offered between the entities of the application domain and the entities of the software, the concurrent nondeterministic nature of the activities inside the building. The obtained results showed that the approach is viable and that considerable energy savings are possible while at the same time providing added value for the customer.

Case Study: HVAC Control System Simulation

This case study presents an agent based simulation model implemented using JADE, featuring a Smart Building, in which an HVAC system is implemented (allowing control of temperature, humidity, CO_2 concentration, with a bonus of window glass break detection). This ABM uses the cell-tissue configuration. Thus, the control is performed by a cell agent (Figure 6), inside of which the inference module

Figure 6. HVAC control system cell with supervisor connection

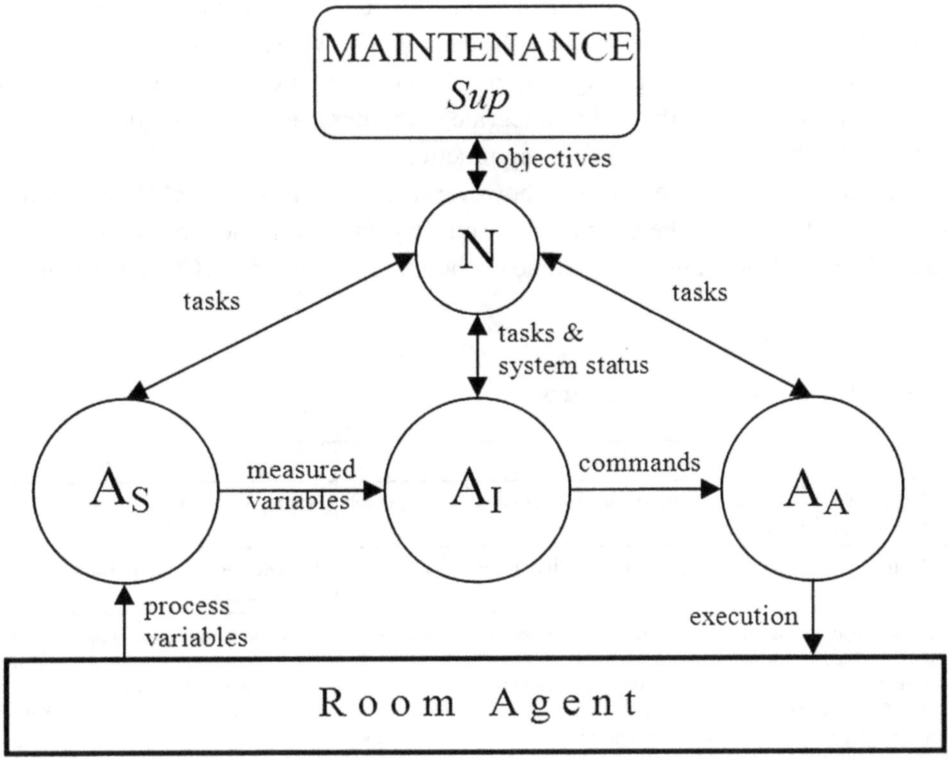

receives information from sensor agents for the considered process variables, applies proper control laws for each variable accordingly, and then transmits commands to the actuators of the system.

The role of the nucleus *N* in the cell agent is to monitor the completion of tasks inside the cell. Based on the analysis of sensor outputs, generated commands, and actuator performance, the inference agent can detect the situations in which the HVAC system is no longer controllable (for example, if the windows are broken, or there is sensor failure, etc.), in which case it will inform the nucleus and an urgent objective will be sent to a maintenance crew supervisor.

Figure 6 presents a conceptual view of this model, in which the RoomAgent represents the simulated environment, modelled as an interconnected set of first order behaviours, due to the slow dynamic of the process. The acting agents A_A are actuators like coolers, heaters, water dispersion mechanisms, fans and even human entities from the building personnel, should the need arise. The sensing agents A_S are a series of sensors that simulate measurement of the controlled variables throughout the system. In this implementation, the inference agent A_I contains a comprehensive rulebase designed to control the system and to analyse critical failures within the room. Table 2 offers a few examples of outputs throughout the system.

Further ABMS Challenges and Remarks

Based on technological progress, future structural analysis should be able to simulate the entire life of a building, from design, construction, maintenance, to deterioration. There are many challenges that start at this point toward improving the tools that structural engineers use nowadays. With all the progress made in the past years concerning computational techniques and operations, there are still many gaps in modelling structures, like large amounts of data needed to be processed and managed, modelling techniques that have execution times expressed in days, existing software and tools that are hard to maintain and update. From all the existing challenges, gaps, directions that could be researched, the ones that concern this chapter the most are those that integrate computer science techniques into computational models used in civil and mechanical engineering practice.

The studies reported in literature have not, however, focused on using ABMS techniques in designing activities such as the ones in the construction industry. However, there are many challenges ready to be tackled by the scientific community. Some of these are presented in (Chen & El-Metwally, 2011) and refer to:

Table 2. System behaviour in various contexts

Output	Interpretation
22:21:02.408: Heating system is increasing temperature in room two! Temperature is too low and humidity is too high!	Heating regulator at work in normal operation.
22:21:02.768: Cooling system is off in room two! Temperature is too low and humidity is ok!	The inference agent informing the nucleus of system status.
17:15:29.360: In room two, the maximum range (5 m/s) between air sensor 1 and air sensor 2 has been exceeded. Current range: 11.0 **17:15:29.363:** System's parameters cannot reach their optimal value! **17:15:29.379:** HVAC System has been shut down! **17:15:29.379:** team@192.168.0.100:1111/JADE is going to room two to check the air ducts and broken air sensors!	A critical malfunction in the ventilation ducts. The inference agent A_I informs the nucleus of system status. Human intervention teams are dispatched by the Maintenance Supervisor.

- Better use of parallelization; using current technology regarding shared and distributed memory, computer architectures will lead to the expansion of scope for civil and mechanical engineering problems that refer to computational simulation.
- Improved analytic formulations for handling the complexity of multi-physics simulations.
- Support for distributed computing architectures for distance computing, geographically disparate research teams, network and communication infrastructures.
- Data management tools; large-scale simulation techniques generating volumes of data that need to be managed and processed, especially techniques that are used in design-analysis cycles.
- Developing and maintaining tools for creating these structural models and simulations.

A remark can be made at this point: a complete life cycle of a building should be an interesting topic for researchers. For example, if an objective would be to simulate the life of a building, then new modelling techniques need to be developed and implemented to help civil and mechanical engineers to design a stable and smart structure, to observe component behaviours and to decrease the risk of casualties in case of natural disasters or emergency scenarios. The impact on society would be enormous if a simulation can be achieved on how urban areas behave during natural disasters, like earthquakes, tsunamis, or fires, and how people can decrease the time needed for evacuation during emergency situations, and where to build emergency centers in order to save and secure more people.

A direction in which the advancements in smart structures currently leads is the development of a new classes of smart materials (Anderson, Crowson, & Chandra, 1992). A smart material will have characteristics similar with the ones designed for smart structures, meaning embedded sensors, actuators, and control systems that are integrated into the structures, giving them the capabilities to respond to external or internal factors and to act in consequence. An implementation of this type will also need specialized solutions for power and maintenance management.

Another remark belongs here: based on these hypotheses it would be possible to model an entire building having a micro-structure or any component that helps the construction of the building modelled as an agent or a multi-agent system. In this way, the emergent behaviour of all components can be observed, and how different elements behave and influence the others. Even if the pieces of structures that are used nowadays in buildings do not include sensors and controllers, the first step in checking the improvements that could be brought by this development consists in modelling the solution and simulating its behaviour to certain emergency scenarios.

In many situations, the key ingredient in computing environments is the ability to visualize and interactively investigate alternative configurations in the overall process. As a result, the area of intelligent systems and artificial intelligence provides the foundation tools for designing self-adaptive mechanisms and structures. A basis for knowing which tools are available and which to use in generating shapes for buildings is presented by Reichl (2013). In the first part of the paper some procedural generation methods are described that are used to synthesize geometry, emphasizing those useful for building generation. The second part of the paper presents implementation and results of a developed plug-in, Faid, which is based on one of the techniques described in the first part, CGA shape grammars.

Some approaches considered a combination between ABMS and System Dynamics (SD) generating hybrid simulation models. Lättilä, Hilletoft h, and Lin (2010) present five different situations that allow combination of these methods. An interesting review of the domains where ABSM gained ground is presented by Macal and North (2010). This paper provides a brief introduction to ABMS, illustrates main concepts and foundations, discusses some recent applications across a variety of disciplines, and identifies methods and toolkits for developing agent models.

CONCLUSION

This chapter describes the nature and rationale of the agent based modeling and simulation (ABMS) concept. The main definitions of agents are presented, along with the most important properties that characterize them used in the implementation and development process of agent models. Based on the agents' capabilities to cooperate and organize, driven by the desire of achieving their tasks, multi-agent systems can be formed and their main basic architectures are presented, with advantages and limitations. Due to the multitude and variety of existing platforms for ABMS, a selection of the most important is made and described in order to facilitate ease of use for civil engineers. The chapter draws on applications of ABMS to illustrate the power and the usefulness of ABMS methodology. A case study of an HVAC control system model is also presented.

Therefore, ABMS is applicable to systems where there are decisions and behaviours that can be defined, where it is important that agents (or component elements) adapt and change, learn and engage in strategic collaboration, where it is important that agents form organizational structures, when scaling-up to superior levels of complexity is needed.

The main challenges of applying ABMS to smart structures consist in verification and validation of large scale models, implementation of representations and standards for code exchange between models, implementing specific properties along with the formal models of structures. Because of the various types of elements that can be contained by a structure, such an implementation could be unjustifiably time consuming. As presented in this chapter, there are various platforms for modelling and simulation of complex systems that do not require advanced programming skills, just a choice of the most suitable toolkit to be used in the modelling of the considered system.

Agent based models are an abstraction of reality, and there are different ways to simplify physical systems in a meaningful way, while the assumptions made during model development are visible in simulation. ABMS development for engineering practice enables a more thorough exploration of the design space at a much lower cost, it is an efficient tool to reduce product cost, it provides a faster turnaround of a product cycle, it is interdisciplinary, and it is suitable to engineering education for lifelong learning.

REFERENCES

Anderson, G. L., Crowson, A., & Chandra, J. (1992). Introduction to smart structures. In *Intelligent Structural Systems* (pp. 1–8). Springer Netherlands. doi:10.1007/978-94-017-1903-2_1

Berryman, M., & Angus, S. D. (2010). *Tutorials on Agent-based modelling with NetLogo and Network Analysis with Pajek* [Doctoral dissertation]. World Scientific Publishing Company.

Bertolotti, T. (2014). Generative and Demonstrative Experiments. In *Model-Based Reasoning in Science and Technology* (pp. 479–498). Springer Berlin Heidelberg. doi:10.1007/978-3-642-37428-9_27

Bordini, R. H., Braubach, L., Dastani, M., Seghrouchni, A. E. F., Gomez-Sanz, J. J., Leite, J., ... & Ricci, A. (2006). A survey of programming languages and platforms for multi-agent systems. *Informatica, 30*(1).

Branimir Cace. (2006). ENTORAMA multi-agent modeling and simulation tool. Retrieved from http://www.entorama.com/

Bratman, M. E., Israel, D., & Pollack, M. E. (1988). Plans and resource-bounded practical reasoning. *Computational Intelligence*, *4*(4), 349–355. doi:10.1111/j.1467-8640.1988.tb00284.x

Brooks, R. (1986). A robust layered control system for a mobile robot. *IEEE Journal of* Robotics and Automation, *2*(1), 14–23.

Cerovsek, T. (2011). A review and outlook for a 'Building Information Model'(BIM): A multi-standpoint framework for technological development. *Advanced Engineering Informatics*, *25*(2), 224–244. doi:10.1016/j.aei.2010.06.003

Chen, W. F., & El-Metwally, S.E.D.E. (2011). *Understanding Structural Engineering: From Theory to Practice*. CRC Press.

Chin, A. L., Worth, A. D., Greenough, A. C., Coakley, S., Holcombe, M., & Kiran, M. (2012). FLAME: An approach to the parallelisation of agent-based applications. *Work (Reading, Mass.)*, *501*, 63259.

Davidsson, P., & Boman, M. (2000). A multi-agent system for controlling intelligent buildings. *Proceedings of the Fourth International Conference on MultiAgent Systems '00* (pp. 377-378). IEEE. doi:10.1109/ICMAS.2000.858483

Eclipse. (2015). Agent Modeling Platform from the Eclipse Foundation. Retrieved from http://www.eclipse.org/amp/

Ferguson, I. A. (1992). *Touring Machines: An architecture for dynamic, rational, mobile agents* [Doctoral dissertation]. University of Cambridge.

Georgeff, M. P., & Lansky, A. L. (1987, July). *Reactive reasoning and planning* (Vol. 87). AAAI.

Hu, Z., Zhang, J., & Deng, Z. (2008). Construction process simulation and safety analysis based on building information model and 4D technology. *Tsinghua Science and Technology*, *13*(Suppl. 1), 266–272. doi:10.1016/S1007-0214(08)70160-3

Islam, A. S., Hussain, R. R., Jameel, M., & Jumaat, M. Z. (2012). Non-linear time domain analysis of base isolated multi-storey building under site specific bi-directional seismic loading. *Automation in Construction*, *22*, 554–566. doi:10.1016/j.autcon.2011.11.017

Ji, J., Zhang, W., Zhao, W., Yuan, C., & Liu, Y. (2012). Research of Seismic Testing and Dynamic Character of High-rise Building Structure Based on ANSYS. *JDCTA: International Journal of Digital Content Technology and its Applications*, 6(8), 63-71.

Kado, S. Z., Ibrahim, R., Khaidzir, M., & Anwar, K. (2014, May). Agent parameters for building consultant teamwork in the agent-based simulation model development. *Proceedings of the 2014 International Symposium on Technology Management and Emerging Technologies (ISTMET)* (pp. 212-217). IEEE.

Kelly, K. (1995). *Out of control: the new biology of machines, social systems and the economic world*. Basic Books.

Klein, J. (2008). Breve: a 3d Simulation Environment for Multi-Agent Simulations and Artificial Life. Retrieved from http://www.spiderland.org/

Klein, L., Kwak, J. Y., Kavulya, G., Jazizadeh, F., Becerik-Gerber, B., Varakantham, P., & Tambe, M. (2012). Coordinating occupant behavior for building energy and comfort management using multi-agent systems. *Automation in Construction*, *22*, 525–536. doi:10.1016/j.autcon.2011.11.012

Klopfer, E., Scheintaub, H., Huang, W., Wendel, D., & Roque, R. (2009). The simulation cycle: Combining games, simulations, engineering and science using StarLogo TNG. *E-Learning and Digital Media*, *6*(1), 71–96.

Kravari, K., & Bassiliades, N. (2015). A Survey of Agent Platforms. *Journal of Artificial Societies and Social Simulation*, *18*(1), 11. doi:10.18564/jasss.2661

Lättilä, L., Hilletofth, P., & Lin, B. (2010). Hybrid simulation models–when, why, how? *Expert Systems with Applications*, *37*(12), 7969–7975. doi:10.1016/j.eswa.2010.04.039

Luke, S., Cioffi-Revilla, C., Panait, L., Sullivan, K., & Balan, G. (2005). Mason: A multiagent simulation environment. *Simulation*, *81*(7), 517–527. doi:10.1177/0037549705058073

Maalal, S., & Addou, M. (2012). A new approach of designing Multi-Agent Systems. With a practical sample. *International Journal of Advanced Computer Science and Applications*, *2*(11), 148–157.

Macal, C. M., & North, M. J. (2010). Tutorial on agent-based modelling and simulation. *Journal of simulation*, 4(3), 151-162.

Müller, J. P., & Pischel, M. (2011). The agent architecture inteRRaP: Concept and application.

Nikolai, C., & Madey, G. (2009). Tools of the trade: A survey of various agent based modeling platforms. *Journal of Artificial Societies and Social Simulation*, *12*(2), 2.

North, M. J., Collier, N. T., Ozik, J., Tatara, E. R., Macal, C. M., Bragen, M., & Sydelko, P. (2013). Complex adaptive systems modeling with repast simphony. *Complex adaptive systems modeling*, *1*(1), 1-26.

Nwana, H. S. (1996). Software agents: An overview. *The Knowledge Engineering Review*, *11*(03), 205–244. doi:10.1017/S026988890000789X

Parker, M. T. (2001). What is Ascape and why should you care. *Journal of Artificial Societies and Social Simulation*, *4*(1), 5.

Pătraşcu, M., & Drăgoicea, M. (2014). Integrating agents and services for control and monitoring: managing emergencies in smart buildings. In *Service Orientation in Holonic and Multi-Agent Manufacturing and Robotics* (pp. 209–224). Springer International Publishing. doi:10.1007/978-3-319-04735-5_14

Patrascu, M., Dragoicea, M., & Ion, A. (2014, October). Emergent intelligence in agents: A scalable architecture for smart cities. *Proceedings of the 2014 18th International Conference on System Theory, Control and Computing (ICSTCC)* (pp. 181-186). IEEE. doi:10.1109/ICSTCC.2014.6982412

Patrascu, M., Dumitrache, I., & Patrut, P. (2012). A comparative study for advanced seismic vibration control algorithms. *UPB Sci. Bull., Series C, 74*(4), 3–16.

Patrascu, M., & Ion, A. (2015). *Seismic Vibration Control with Semi-Active Dampers*. Bucharest: MatrixRom.

Pollack, M. E., & Ringuette, M. (1990). *Introducing the Tileworld: Experimentally evaluating agent architectures* (pp. 183–189). Defense Technical Information Center.

Privara, S., Široký, J., Ferkl, L., & Cigler, J. (2011). Model predictive control of a building heating system: The first experience. *Energy and Building, 43*(2), 564–572. doi:10.1016/j.enbuild.2010.10.022

Reichl, B. P. (2013) *Procedural modeling of buildings* [Dissertation]. Masarikova University, Brno.

RoboCup Rescue League. (2006). RoboCup Rescue. Retrieved from http://www.robocuprescue.org/

SESAM. (2012). SeSAm Multiagent Simulation. Retrieved from http://130.243.124.21/sesam/index.php

Simulation and Modelling of Terrestrial Systems. (2015, June 22). *TerraME*. Retrieved from http://www.terrame.org/doku.php?id=start&rev=1434979167

Solovyev, A., Mikheev, M., Zhou, L., Dutta-Moscato, J., Ziraldo, C., An, G., & Mi, Q. et al. (2010, April). SPARK: a framework for multi-scale agent-based biomedical modeling.*Proceedings of the 2010 Spring Simulation Multiconference* (p. 3). Society for Computer Simulation International. doi:10.1145/1878537.1878541

Tan, W., Chai, Y., & Liu, Y. (2011). A message-driving formalism for modeling and simulation of multi-agent supply chain systems. *Journal of Systems Science and Systems Engineering, 20*(4), 385–399. doi:10.1007/s11518-011-5182-x

Vaughan, R. (2008). Massively multi-robot simulation in stage. *Swarm Intelligence, 2*(2-4), 189–208. doi:10.1007/s11721-008-0014-4

Vere, S., & Bickmore, T. (1990). A basic agent. *Computational Intelligence, 6*(1), 41–60. doi:10.1111/j.1467-8640.1990.tb00128.x

Wilensky, U. (1999). NetLogo. Retrieved from http://ccl.northwestern.edu/netlogo/

Wooldridge, M. (2009). *An introduction to multiagent systems*. John Wiley & Sons.

Wooldridge, M., & Jennings, N. R. (1995). Agent theories, architectures, and languages: a survey. In *Intelligent agents* (pp. 1–39). Springer Berlin Heidelberg. doi:10.1007/3-540-58855-8_1

KEY TERMS AND DEFINITIONS

Agent: A persistent systemic entity, capable of sensing its environment and acting upon it, governed by internal rules.

Agent Based Model: A computational or simulation generative model.

Control System: A set of control algorithms, actuators, sensors, and auxiliary elements that manages, commands, directs or regulates the behaviour of other devices or systems.

Model: An abstract representation of a physical entity or process.

Multi-Agent System: A collection of agents.

Simulation: The representation of the behaviour or characteristics of one system or process through the use of a model.

Smart Structure: Structures equipped with sensors, actuators, and control systems able to react to environmental changes, both to internal (e.g. fire) and external disturbances (e.g. seismic vibrations).

Chapter 3
Definition of Static Nonlinear Procedure and Flexibility-Based Model with Application on 2D Model for an Existing Structure and Comparing Results with Time History Analysis

Mourad Belgasmia
University of Setif, Algeria

ABSTRACT

This chapter presents in one hand the definition of the pushover procedure based on Eurocode 8, and displacement based and force base elements which will be used in this study, and the application of the pushover procedure based to Eurocode 8 to nonlinear framed structure model and comparison with time history analysis. As test-bed application, we consider a two-story reinforced concrete building in Bonefro, Italy. It is representative of typical residential building construction in Italy in the 1970's and 1980's. The aim of this chapter is to compare 2D model for both modal and uniform load distributions with time history analysis. In another hand an Application to two-story RC Frame using force base elements without soil and with taking into account soil in order to have an idea of the soil modelisation effect with the structure in the results

INTRODUCTION

Modern seismic design codes allow engineers to use either linear or nonlinear analyses to compute design forces and design displacements. In particular, Eurocode 8. (2003) contains four methods of analysis: simplified static analysis, modal analysis, nonlinear pushover analysis and nonlinear time-history analysis. These methods refer to the design and analysis of framed structures, mainly buildings and bridges.

DOI: 10.4018/978-1-5225-0588-4.ch003

The two nonlinear methods require advanced models and advanced nonlinear procedures in order to be fully applicable by design engineers. This chapter gives an overview of the steps followed to implement state-of-the-art nonlinear models in Z_Soil (Zimmermann et al., 2008) and presents a test-bed application to a reinforced concrete building.

Displacement based and force base elements are used in this study. The first is a classical two-node, displacement-based, Euler-Bernoulli frame element. The second is a two-node, force-based, Euler Bernoulli frame element. The main advantage of the second element is that it is "exact" within the relevant frame element theory. This implies that one element per frame member (beam or column) is used in preparing the frame mesh, thus leading to a reduction of the global number of degrees of freedom, see the work of Belgasmia and Moussaoui (2015). The complete theory for the force-based element can be found in Spacon (1994), Spacon et al. (1996) and Aydinoglu, (2004).

The nonlinear response of 2D model of an existing building is presented as a first full-size application. The building is a residential two-storey reinforced concrete building in Bonefro, Italy. It is representative of typical residential building construction in Italy in the 1970's and 1980's. The design spectrum for the building was obtained from EC8 using the local soil properties and the peak ground acceleration given by the new Italian seismic map.

NONLINEAR STATIC PUSHOVER ANALYSIS ACCORDING TO EUROCODE 8

The nonlinear pushover analysis consists of applying monotonically increasing constant shape lateral load distributions to the structure under consideration. The structure model can be either 2D or 3D. In particular, EC8 states that for buildings with plan regularity, 2D analysis of single plane frames can be performed, while for buildings with plan irregularity a complete 3D model is necessary. Given that the nonlinear methods are particularly interesting for existing buildings, which are rarely regular, a 3D model is required in most cases.

The nonlinear pushover procedure in EC8 follows the N2 method developed by Fajfar (1999, 2002). The N2 method was developed using a shear building model, i.e. a frame model with floors rigid in their planes. Furthermore, vertical displacement is typically neglected in the method and only the two horizontal ground motion components, *x* and *y*, are considered. Extension to the general case of a fully deformable frame is straightforward.

The N2 method consists of applying two load distributions to the frame:

1. A "modal" (or "triangular") pattern, that is a load shape proportional to the mass matrix multiplied by the first elastic mode shape, $P^1 = M\Phi_1$;
2. A "uniform" pattern, that is a mass proportional load shape, $P^2 = MR$.

where M is the mass matrix, Φ_1 is the first mode shape and R a vector of 1s corresponding to the degrees of freedom parallel to the application of the ground motion and 0s for all other degree of freedom (dofs). In the N2 method Φ_1 is normalized so that the top floor displacement is 1, i.e. $\Phi_{1,n} = 1$. The two load distributions are schematically shown in Figure 1. The applied lateral load distributions are increased and the response is plotted in terms of base shear V_b vs. top floor displacement D (for example center of mass of the top floor). This is the so-called pushover curve or capacity curve (also shown schematically in Figure 1).

The N2 procedure transforms the response of the MDOF system into the response of an equivalent SDOF system. This is necessary in order to compare the building capacity curve of Figure 1 with the demand, expressed in the design codes by the design spectra, which refer to SDOF systems. The transformation MDOF – SDOF, as developed by Fajfar (1999, 2002) is treated in more detail in the following section.

Equivalent SDOF Model and Capacity Diagram in Eurocode 8

As previously stated, Fajfar (1999, 2002) assumes that the building is a shear frame, i.e. the floors are rigid in their own plane. If the vertical displacements of the building are neglected, the floor displacements are given by the three degrees of freedom shown in Figure 2. The degrees of freedom are typically taken at the center of mass. Note that the beams can deform outside the floor plane, that is the nodes have out-of-floor plane rotational degrees of freedom.

The theoretical derivation of the transformation procedure is as follows. The equations of motion of a MDOF building subjected to base ground motion model is:

$$M\ddot{U} + F(U) = -MRa \qquad (1)$$

where damping is neglected, M is the mass matrix (assumed diagonal in the original derivation of the N2 method), U and F are vectors representing displacements and internal forces, respectively, R is the influence vector and *a* is the ground acceleration as function of time, i.e. $a = a(t)$. *a* is given in one direction only. In the linear elastic case $F = KU$ (where K is the structure stiffness matrix), in the nonlinear case F depends on the displacement history. For unidirectional ground motion, for example in the direction *x*, the influence vector R consists of 1s in correspondence to the dofs in the *x* direction, and 0s for all other dofs. For example, for the frame of Figure 4, considering only the 12 dofs of the 4 stories, $R^T = \{1 \;\; 1 \;\; 1\}$, with $1 = \{1 \;\; 0 \;\; 0\}$.

The first assumption (and approximation) of the N2 method is that the displacement **U** has a constant shape that does not change during the response to the ground motion:

Figure 1. Load distribution for pushover analysis according to EC8 and pushover response curves

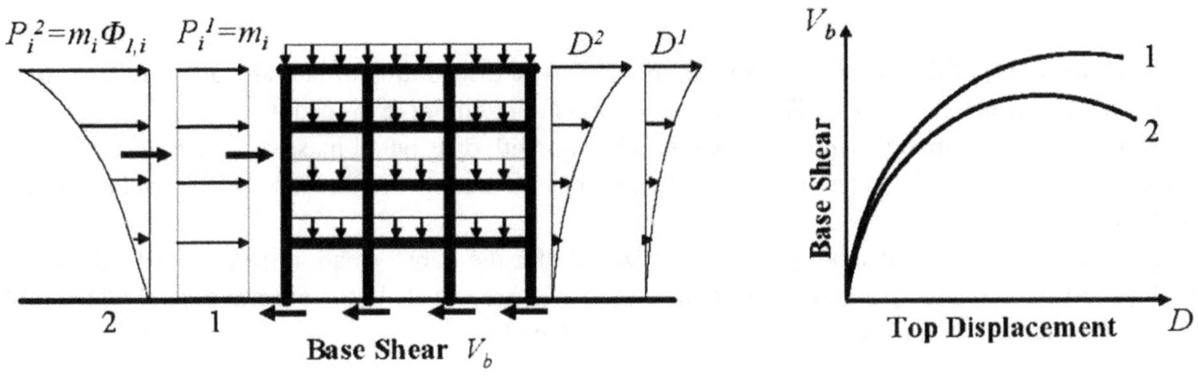

Figure 2. Rigid slab degrees of freedom in 3D shear frame (for simplicity, column axial deformability is neglected)

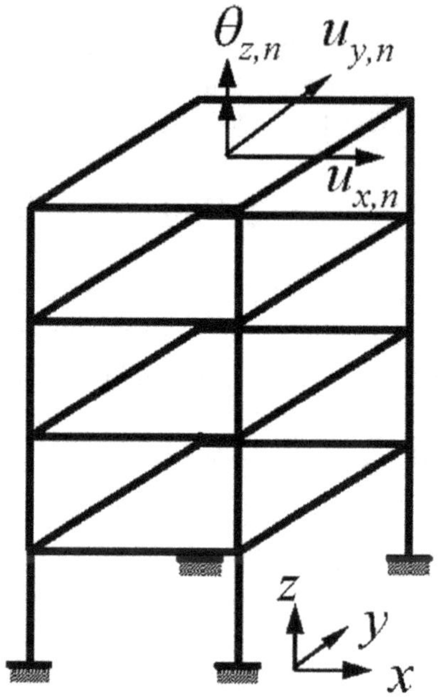

$$U = \Phi D_t \quad (2)$$

or $U(x,t) = \Phi(x) D_t(t)$, where D_t is the intensity at the pseudo-time t of the displacement shape Φ, where **x** indicates that the displacement shape depends on the degree of freedom location. For convenience, Φ is normalized in such a way that the top-story displacement is equal to 1, i.e. $\Phi_n = 1$. This way D_t gives the top floor displacement at time t.

In the pushover analysis of frames with rigid floors, lateral loads are applied at the centre of mass of each storey. The vector of the lateral loads P is

$$P = p\Psi = pM\Phi \quad (3)$$

The magnitude of the lateral load is p, i.e. $p = p(t)$. The distribution of lateral loads is related to the assumed displacement shape Φ. This is the second assumption of the procedure.

In more complex models with deformable slabs and with distributed masses at each node the load, the load vector **P** is applied to all degrees of freedoms with mass in the direction of the applied ground motion.

Note that the displacement shape Φ is needed only for the transformation from the MDOF system to the equivalent SDOF system of the nonlinear pushover procedure. In the general case of a 3D building, Φ has nonzero components in the six dofs of each node.

From Equation 3 it follows that in a shear frame the lateral force in the *i*-th storey is proportional to the component Φ_i of the assumed displacement Φ, weighted by the storey mass m_i. If the ground motion is applied in the *x* direction

$$P_i = p m_i \Phi_{x,i} \tag{4}$$

From statics it follows that

$$P = F \tag{5}$$

that is the internal forces F are equal to the pseudo-static external forces P.

By introducing Equations 2, 3 and 5 into Equation 1, and by pre-multiplying by Φ^T, we obtain

$$\Phi^T M \Phi \ddot{D}_t + \Phi^T M \Phi \, p = \Phi^T M \, R \, a \tag{6}$$

The left term of Equation 6 is then divided and multiplied by $\Phi^T M \ R$ to obtain

$$\underbrace{\Phi^T M \ R}_{m^*} \underbrace{\frac{\Phi^T M \ \Phi}{\Phi^T M \ R}}_{\frac{1}{\Gamma}} \ddot{D}_t + \underbrace{\frac{\Phi^T M \ \Phi}{\Phi^T M \ R}}_{\frac{1}{\Gamma}} \underbrace{\Phi^T M \ R p}_{V_b} = \underbrace{\Phi^T M \ R a}_{m^*} \tag{7}$$

where m^* is the mass of the SDOF equivalent to the MDOF building

$$m^* = \Phi^T MR \tag{8}$$

For a shear building and ground motion applied in the *x* direction:

$$m^* = \sum m_i \Phi_{x,i} \tag{9}$$

The constant Γ controls the transformation from MDOF to SDOF and back:

$$\Gamma = \frac{\Phi^T MR}{\Phi^T M\Phi} \tag{10}$$

For a shear building and ground motion in the *x* direction:

$$\Gamma = \frac{\sum m_i \Phi_{x,i}}{\sum m_i \Phi_{x,i}^2} \tag{11}$$

Γ is a factor that, for Φ equal to the mode shape of one the building modes, corresponds to the mode participation factor. In the development of the N2 method, Φ can be any reasonable deformed shape.

V_b is the base shear of the MODF building in the direction of the ground motion. From Equation 3 it follows that:

$$V = \Phi^T M\, R\, p \tag{12}$$

For a shear building and ground motion applied in the x direction:

$$V_x = p\sum m_i \Phi_{x,i} = \sum P_{x,i} \tag{13}$$

Equation 7 thus becomes the equation of the SDOF equivalent to the MODF building

$$m^* \ddot{D}^* + F^* = m^* a \tag{14}$$

where:

$$D^* = \frac{D_t}{\Gamma} \qquad F^* = \frac{V_b}{\Gamma} \tag{15}$$

The above derivation allows the transformation of the MDOF pushover capacity curves of Figure 2 into pushover curves for the equivalent SDOF system, as shown in Figure 3.

Figure 3. Capacity curve transformation from response of MDOF to equivalent SDOF

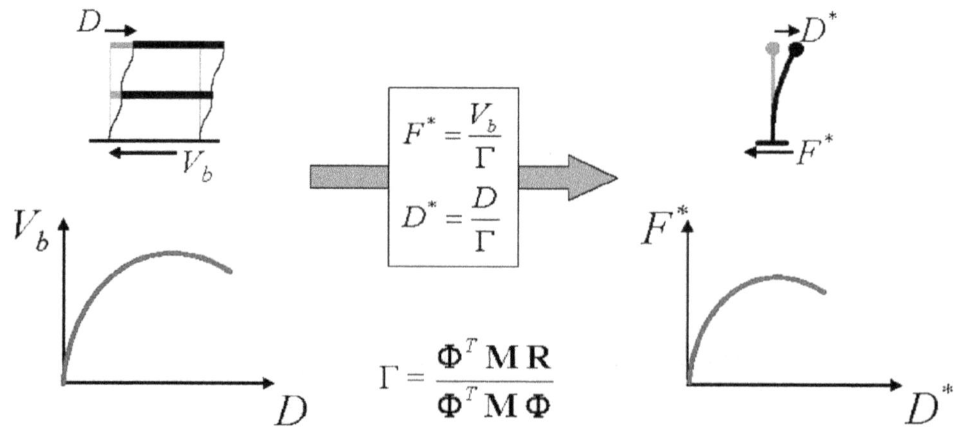

Both force and displacement axes are scaled by the same factor Γ. The stiffness of the system remains the same. Note that the transformation factor Γ depends on the shape of the assumed displacement shape Φ, and is thus different for different choices of Φ. As shown in Figure 3, two forms of Φ are suggested in EC8:

1. $\Phi = \Phi_I$, thus $\Gamma = \dfrac{\Phi_I^T M R}{\Phi_I^T M \Phi_I}$
2. $\Phi = R$, thus $\Gamma = 1$.

Linearization of the Capacity Curve and Comparison to Demand Spectrum

In order to compare the capacity curve to the demand curve given by the design spectrum, the nonlinear pushover curves of the SDOF are approximated by elastic-perfectly plastic (or bilinear) curves. According to Eurocode 8 (2003), this transformation can be based on the equal energy principle. A target displacement D_m^* is assumed, and equal energy is assumed between bilinear and nonlinear pushover curves. This simple procedure is illustrated in Figure 4.

The bilinearization of Figure 6 gives the yield force and the yield displacement $D_y^* = 2\left(D_m^* - \dfrac{E_m^*}{F_y^*}\right)$, which allow the initial elastic period to be computed as:

$$T^* = 2\pi\sqrt{\frac{m^* D_y^*}{F^*}} \tag{16}$$

Secondly, the capacity curve is transformed into capacity spectrum by normalizing the force F^* with respect to the SDOF weight $m^* g$. The resulting capacity spectrum s is shown in Figure 5

Figure 4. Bilinearization of the capacity curve of SDOF

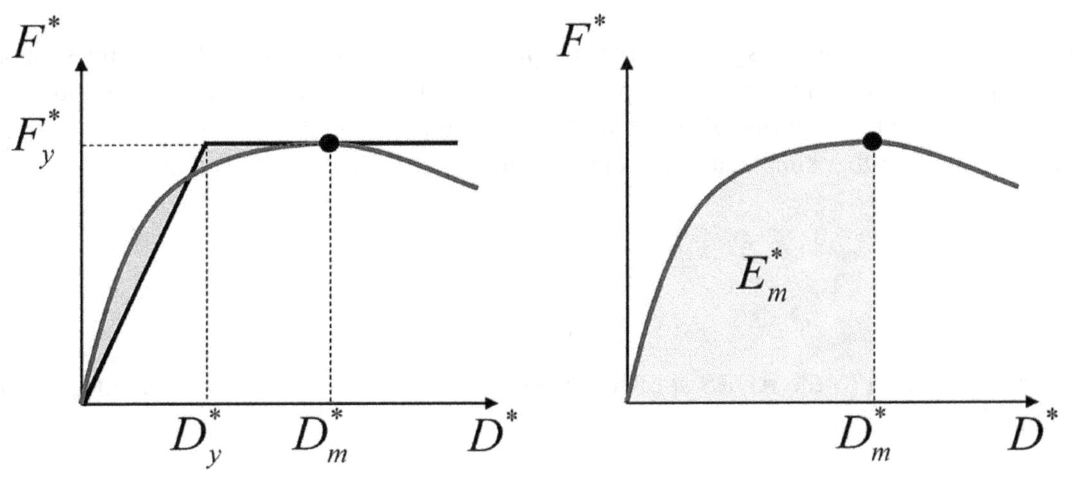

Figure 5. SDOF capacity spectrum

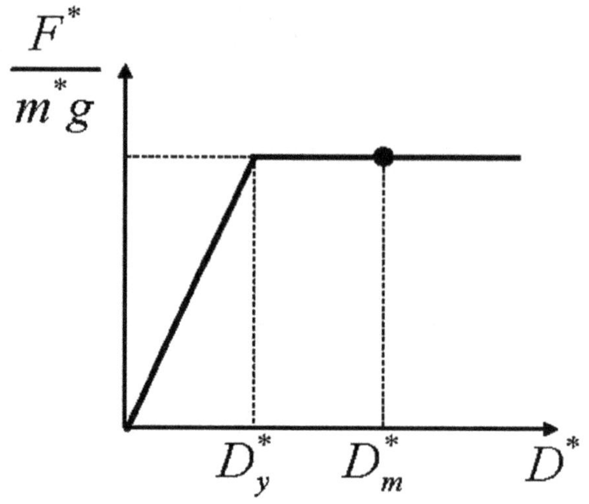

Figure 6. Transformation to ADRS linear spectrum

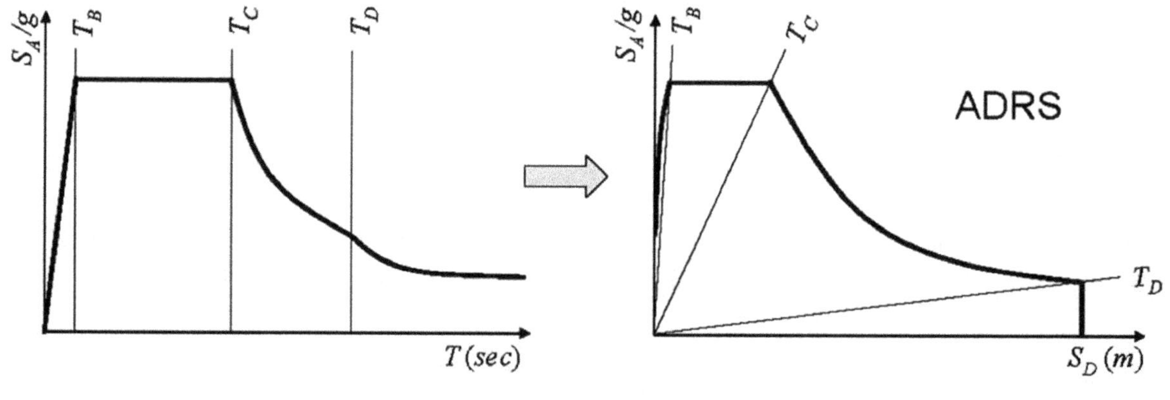

The demand on the building is given by the design spectrum provided by the design codes. In order to compare capacity and demand, the first step is to transform the format of the design spectrum from the classical Acceleration[1]*A* vs Period *T* format to the ADRS format, i.e. Acceleration *A* vs. Displacement *D*. The procedure is rather simple, as Acceleration and Displacement are related by

$$S_D = \left(\frac{T}{2\pi}\right)^2 S_A \tag{17}$$

The transformation to the ADRS spectrum is shown in Figure 6 Lines from the origin represent constant period.

The capacity spectrum of Figure 5 is now compared to the ADRS demand spectrum of Figure 6. The comparison is not immediate, because the capacity spectrum is nonlinear, while the ADRS spectrum given by the design codes is linear.

For a SDOF system with a bilinear plastic behavior, the acceleration spectrum S_A and the displacement spectrum S_D can be determined as Vidic (1994).

$$S_A = \frac{S_{Ae}}{R_\mu} \quad (18)$$

$$S_D = \frac{\mu}{R_\mu} S_{De} = \frac{\mu}{R_\mu} \frac{T^2}{4\pi^2} S_{Ae} = \mu \frac{T^2}{4\pi^2} S_{Ae} \quad (19)$$

where subscript *e* indicates elastic, μ is the ductility factor = maximum inelastic displacement/yield displacement, and R_μ is the reduction factor due to ductility[2]. The reduction factor R_μ can be found in different ways, some analytical, other approximated. In the simple version of the N2 method, the following approximated expressions are given:

$$R_\mu = \left(\mu - 1\right) \frac{T}{T_C} + 1 \; T \prec T_C \quad (20)$$

$$R_\mu = \mu \; T \succ T_C \quad (21)$$

where T_C is a characteristic period of the ground motion that depends on the soil type and is given by EC8. It typically corresponds to the transition from the constant acceleration range (short-period range) to the constant velocity range (medium-period range) in the response spectrum. The above equations suggest that in the short-period range the equal displacement principle is applied (elastic and inelastic SDOFs have the same maximum displacement), while in the medium- and long-period range the equal energy principle is applied. These principles are shown in Figure 7.

Equation 18, together with Equations 20 and 21 are used to obtain inelastic demand spectra of constant ductility, as shown in Figure 8. Note that this procedure is approximate, and that inelastic demand spectra can be determined by a rigorous (but more complex) procedure by using nonlinear dynamic analysis.

Using the procedures illustrated in the section, the seismic demand on the equivalent SDOF can be determined. The steps are schematically illustrated in Figure 9 for a bilinear oscillator with medium or long elastic period T*. Given the elastic demand spectrum and the bilinear capacity spectrum, from a theoretical point of view the target displacement is determined by finding the inelastic demand spectrum of ductility $\overline{\mu}$ that intersects the capacity spectrum in a point corresponding to a capacity ductility $\overline{\mu}$. In other words, the design point is given by the point with equal demand and capacity ductility.

In practice, this is done very easily using the approximate procedure given by Equations 18, 20 and 21:

Figure 7. Transformation elastic response – bilinear response: equal maximum displacement and equal energy assumptions

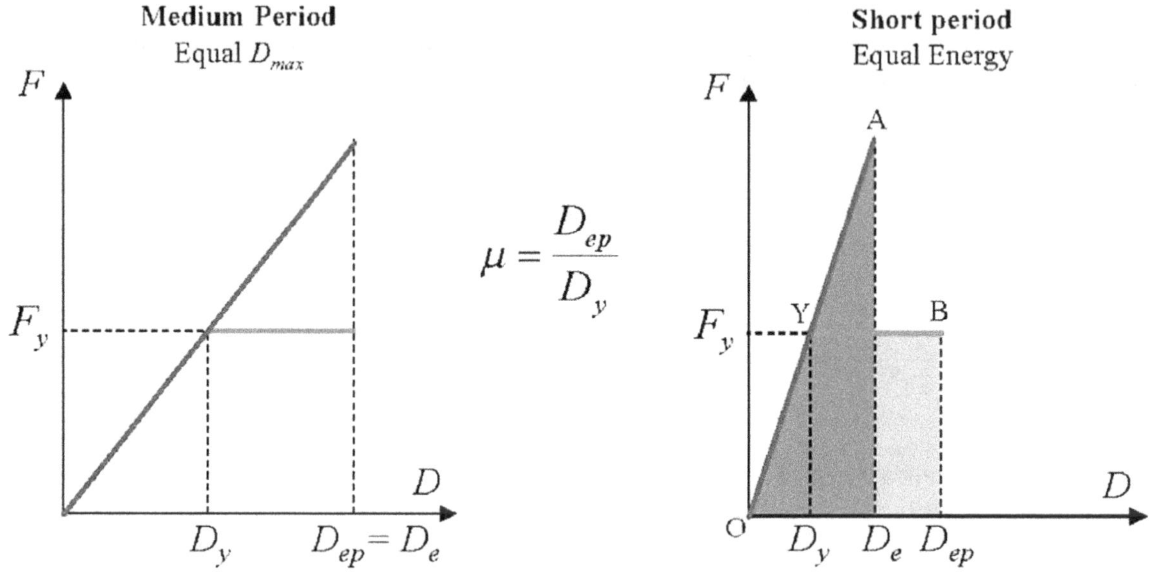

Figure 8. Demand spectra for constant ductilities in AD format (based on EC8 spectrum for Zone 1, Soil type A)

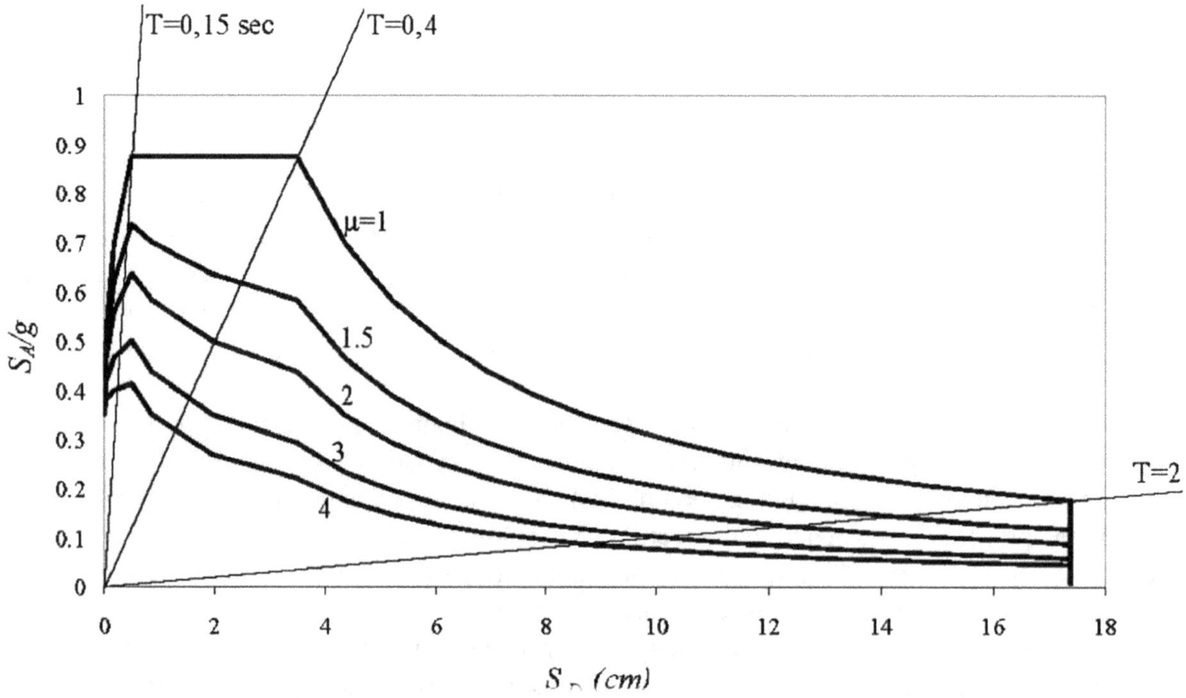

Figure 9. Capacity and Demand spectra for medium and long period T^*: determination of the target displacement for SDOF

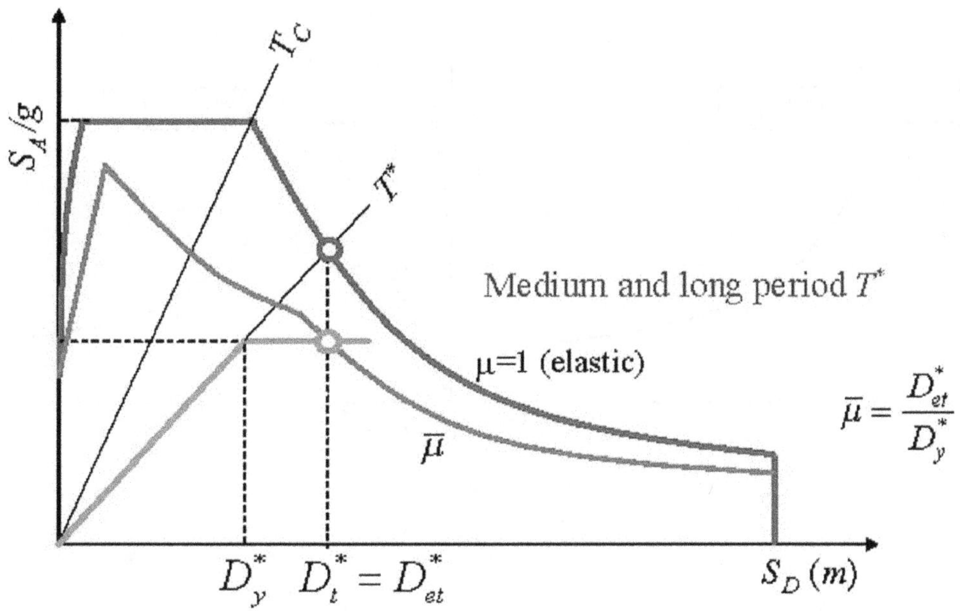

$T^* \prec T_C$ (Short periods)

$\frac{F^*_y}{m^*} \geq S_A(T^*) \Rightarrow$ the response remains linear elastic (left of Figure 10)

$D^*_t = D^*_{el}$

$\frac{F^*_y}{m^*} \prec S_A(T^*) \Rightarrow$ the response enters the nonlinear plateau (right of Figure 10)

$$D^*_t = \frac{D^*_{et}}{q_u}\left(1 + (q_u - 1)\frac{T_C}{T^*}\right) \geq D^*_{et}$$

where $q_u = \frac{S_{Ae}\left(T^*\right)}{F^*_y / m^*}$ is the reduction factor.

$T^* \geq T_C$ (Medium and long periods) (Figure 9).

*Figure 10. Capacity and demand spectra for short period T**

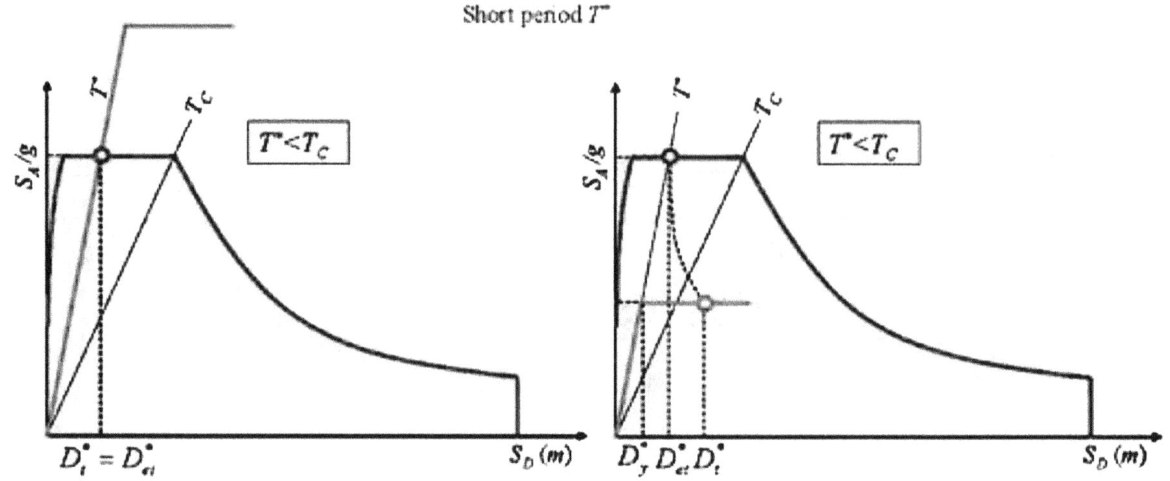

$$D_t^* = D_{et}^*$$

Note that the above procedure is based on the assumption of a tentative target displacement D_m^* (from which the equal energy principle is used to obtain the bilinear capacity curve in Figure 4). If the target displacement D_t^* is very different from the assumed value D_m^*, then the procedure must be repeated, setting for instance $D_m^* = D_t^*$. This is a simple iterative procedure that converges very rapidly.

The target displacement at the top of the building is obtained by inverting the transformation in Figure 3, i.e.

$$D_t = \Gamma D_t^* \tag{22}$$

One minor issue concerns the point at which the pushover curve is stopped. While it is not necessary to continue the pushover analysis to unreasonable values of the top floor displacement (this would mean long computational times and convergence problems at large displacement values), there is no general rule on when to stop the pushover curve. This means that it may happen that the pushover curve is stopped at a displacement level smaller that the computed target displacement. In this case, the pushover analysis must be repeated and it must be stopped at higher top displacement values. It is suggested to push the structure to top-displacements of the order of 2%-3% *h*, where *h* is the entire height of the building, for pushover analysis at the Ultimate and Collapse Limit States.

Finally, note that in general for each seismic input direction (*x* and *y*), four different pushover analysis should be performed. Besides considering two different load shapes (Figure 1) it is in general necessary to consider the forces applied both with the positive and the negative sign, as the irregularity of the building may lead to different responses in the positive and negative directions.

Flexibility Method

If equilibrium is considered in the deformed element configuration in Figure 11, the relation between nodal forces $\overline{Q}$ in the system without rigid body modes and internal forces f(x) is

$$f(x) = b\left[x, w(x)\right]\overline{Q} \tag{23}$$

where

$$b\left[x, w(x)\right] = \begin{bmatrix} 1 & 0 & 0 \\ -w(x) & \xi & \xi - 1 \end{bmatrix}, \xi = \frac{x}{L} \tag{24}$$

Is the matrix of displacement-dependent force interpolation functions. Since shear deformations are neglected, the shear force does not appear in Eq. (24), but it can be determined a posteriori from the equilibrium condition (see Figure 11)

$$V = -Hw + T \tag{25}$$

The weak form of the compatibility condition,

$$\int_0^L \delta F(x) D(x) dx \tag{26}$$

Leads to three compatibility equations for the frame element without rigid body modes. one for the axial displacement $\overline{q}_1$ and two for the end rotations $\overline{q}_2$ and $\overline{q}_3$ in Figure 11, the latter are identical with the linear case. The former becomes

$$\int_0^L \delta N \left[u'(x) + \frac{1}{2}\left[w'(x)\right]^2 - \varepsilon(x)\right] dx = 0 \tag{27}$$

Figure 11. Equilibrium in deformed configuration

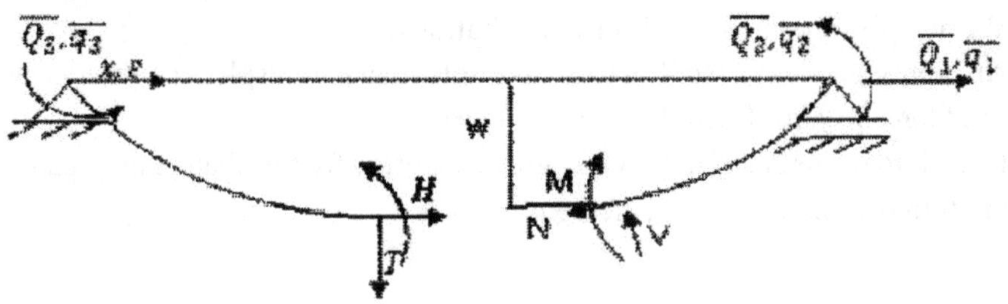

After integrating the preceding expression by parts and accounting for the boundary terms. The compatibility condition for $\overline{q}_1$ reduces to

$$\overline{q}_1 = \int_0^L \varepsilon(x)dx - \int_0^L \frac{1}{2}\chi(x)w(x)dx \tag{28}$$

Thus the complete set of governing equations, the flexibility-based is

$$\overline{q}_1 = \int_0^L b^{*t}\left[x, w(x)\right] D(x)dx \tag{29}$$

with

$$b^*\left[x, w(x)\right] = \begin{bmatrix} 1 & 0 & 0 \\ -\frac{w(\xi)}{2} & \xi & \xi - 1 \end{bmatrix} \tag{30}$$

In geometrically nonlinear see Valipour et al (2010) and in the flexibility-based analysis the evaluation of the flexibility matrix F requires special attention. This is because both *b(x)* and *b*(x)* depend on the transverse displacements *w(x)*. Starting from the governing equation in Equation 29 we obtain

$$F = \frac{\partial \overline{q}}{\partial \overline{Q}} = \int_0^L \left[b^{*t}(x) \frac{\partial D(x)}{\partial \overline{Q}} + \frac{\partial b^{*t}(x)}{\partial \overline{Q}} D(x) \right] dx$$

$$F = \frac{\partial \overline{q}}{\partial \overline{Q}} = \int_0^L \left[b^{*t}(x) \frac{\partial D(x)}{\partial f(x)} \frac{\partial f(x)}{\partial \overline{Q}} + \frac{\partial b^{*t}(x)}{\partial w(x)} D(x) \frac{\partial w(x)}{\partial \overline{Q}} \right] dx$$

$$F = \frac{\partial \overline{q}}{\partial \overline{Q}} = \int_0^L \left[b^{*t}(x) f\, b(x) + b^{*t}(x) f \frac{\partial b(x)}{\partial w(x)} \overline{Q} \frac{\partial w(x)}{\partial \overline{Q}} \right] dx + \int_0^L \frac{\partial b^{*t}(x)}{\partial w(x)} D(x) \frac{\partial w(x)}{\partial \overline{Q}} dx \tag{31}$$

where $f = K^{-1}$ is the section flexibility matrix. Since Eq. (31) is derived from Eq. (29), which is based on a variational principle, it consists of three contributions.

The first term represents the "direct" change in the element end displacements $\overline{q}$ caused by a change in the element end forces $\overline{Q}$ for fixed displacements *w(x)*.

The second and third terms are of indirect nature accounting for the change in $\overline{q}$ caused by the change in *w(x)*, which in turn originates from a change in $\overline{Q}$.

DYNAMICS IN Z_SOIL

Equations of Motion

For nonlinear dynamics we consider the following equation of motion

$$Ma + Cv + N\left(d\right) = F \tag{32}$$

with Rayleigh damping:

$$C = \alpha M + \beta K \tag{33}$$

where α and β are obtained by imposing damping rates at 2 different frequencies (i,j) using:

$$\left(\frac{1}{2}\right)\begin{bmatrix} \dfrac{1}{\omega_i} & \omega_j \\ \dfrac{1}{\omega_j} & \omega_j \end{bmatrix}\begin{Bmatrix} \alpha \\ \beta \end{Bmatrix} = \begin{Bmatrix} \xi_i \\ \xi_j \end{Bmatrix} \tag{34}$$

Remarks:

1. C will allow taking care of minor nonlinearities ignored in N(d)
2. K will be approximalies by the initial stiffness

Seimic Input

Seismic input will be given in terms of imposed displacements or accelerations of the ground, these input can be specified no matter if algorithms solve in terms of accelerations or displacements, or velocities.

Nonlinear Newmark Algorithm for Arbitrary Dynamic Input

Consider the system of equations:

$$M\,a + Cv + N\left(d\right) = F \tag{35}$$

which must be satisfied for any: $t_{n+1} = \left(n+1\right)\cdot \Delta t$ (36)

The algorithm consists of the following 3 equations:

$$\begin{aligned} &Ma_{n+1} + Cv_{n+1} + N\left(d_{n+1}\right) = F_{n+1} \\ &d_{n+1} = d_n + \Delta t v_n + \left(\Delta t^2 / 2\right)\left[\left(1-2\beta\right)a_n + 2\beta a_{n+1}\right] \\ &v_{n+1} = v_n + \Delta t\left[\left(1-\gamma\right)a_n + \gamma a_{n+1}\right] \end{aligned} \tag{37}$$

and we define the predictors

$$\begin{aligned} &\tilde{d}_{n+1} = d_n + \Delta t v_n + \left(\Delta t^2/2\right)\left(1-2\beta\right)a_n \\ &\tilde{v}_{n+1} = v_n + \Delta t\left(1-\gamma\right)a_n \end{aligned} \tag{38}$$

Newton-Raphson Nonlinear Implicit Newmark Algorithm

Static equivalent formulation:

Let, in the vicinity of $d_{n+1}{}^{i+1}$:

$$N\left(d^{i+1}{}_{n+1}\right) \cong N\left(d^{i}{}_{n}\right) + \frac{dN}{dd}\cdot\Delta d = N\left(d^{i}{}_{n}\right) + K_T\cdot\Delta d \tag{39}$$

hence:

$$\begin{cases} K_T\cdot\Delta d = F_{n+1} - N\left(d^{i}{}_{n+1}\right)\left(\text{avec } d_n = d^0_{n+1}\right) \\ d_{n+1}{}^{i+1} = d_{n+1}{}^{i} + \Delta d \end{cases} \tag{40}$$

The algorithm is iterative, convergence is reached if:

$$\left\|F_{n+1} - N\cdot\left(d^{i+1}{}_{n+1}\right)\right\| < TOLERANCE \tag{41}$$

REMARKS

The following algorithmic choices are possible

- **Update K at Each Step and Iteration:** Full Newton-Raphson.
- **Opportunistic Update:** Modified Newton-Raphson.
- **K_T=K_0:** Constant stiffness algorithm.

Implicit Nonlinear Newmark Algorithm in Terms of Δd Initialisation

$$d_n = d_0 \tag{42}$$

$$v_n = v_0 \tag{43}$$

$$a_n = M^{-1}\left\{F_{n+1} - N^*\left(d_0\right)\right\} \tag{44}$$

For each Δt compute predictors:

$$\tilde{d}_{n+1} = d_n + \Delta t v_n + \frac{\Delta t^2}{2}\left(1-2\beta\right)a_n \tag{45}$$

$$\tilde{v}_{n+1} = v_n + \Delta t\left(1-\gamma\right)a_n \tag{46}$$

$$d_{n+1}^{i=0} = \tilde{d}_{n+1} \tag{47}$$

$$v_{n+1}^{i=0} = \tilde{v}_{n+1} \tag{48}$$

$$a_{n+1}^{i=0} = 0 \tag{49}$$

Solve:

$$\Delta F_{n+1}^{i} = F_{n+1} - N^*\left(d_{n+1}^{i}\right) \tag{50}$$

$$K_T^{*i} \cdot \Delta d = \Delta F_{n+1}^{i} \tag{51}$$

with

$$N^*\left(d^i_{n+1}\right) = Ma^i_{n+1} + Cv^i_{n+1} + N\left(d^i_{n+1}\right) \tag{52}$$

and

$$K^{*i}_T = \left(\frac{dN^*}{dd}\right) = \frac{M}{\Delta t^2\beta} + \frac{\gamma}{\Delta t\beta}\cdot C^i_T + K^i_T \tag{53}$$

Compute correctors:

$$a^{i+1}_{n+1} = \left[d^{i+1}_{n+1} - \tilde{d}_{n+1}\right]/\Delta t^2\beta \tag{54}$$

$$v^{i+1}_{n+1} = \tilde{v}_{n+1} + \Delta t\gamma a^{i+1}_{n+1} \tag{55}$$

Next iteration:

i=i+1, if ΔF^i_{n+1} >TOLERANCE (56)

Next step

n=n+1 if t<Tmax

Nonlinear HHT-α Algorithm for Arbitrary Dynamic Input

The original linear form reads:

$$Ma_{n+1} + (1+\alpha)Cv_{n+1} - \alpha Cv_n + (1+\alpha)Kd_{n+1} - \alpha Kd = F_{n+\alpha} \tag{57}$$

with

$$F_{n+\alpha} = (1+\alpha)F_{n+1} - \alpha F_n \tag{58}$$

the corresponding nonlinear statement can be written as

$$Ma_{n+1} + Cv_{n+\alpha} + N\left(d_{n+\alpha}\right) = F_{n+\alpha} \tag{59}$$

$$F_{n+\alpha} = (1+\alpha)F_{n+1} - \alpha F_n \tag{60}$$

$$\begin{aligned} d_{n+1} &= d_n + \Delta t v_n + \left(\Delta t^2/2\right)\left[(1-2\beta)a_n + 2\beta a_{n+1}\right] \\ v_{n+1} &= v_n + \Delta t\left[(1-\gamma)a_n + \gamma a_{n+1}\right] \end{aligned} \tag{61}$$

and

$$\left[{}^*\right]_{n+\alpha} = (1+\alpha)\left[{}^*\right]_{n+1} - \alpha\left[{}^*\right]_n \tag{62}$$

which can be implemented as above:

Remarks:

1. Hilber proposes: $-0.3 < \alpha < 0$, $\gamma = (1-2\alpha)/2$, $\beta = [(1-\alpha)**2]/4$;
2. $\alpha = 0$ corresponds to Newmark's algorithm.

Equivalent 2D Model of Bonefro Building

It is possible to deduce the equivalent rectangular beam which have the same inertia than beam with T form so having b/d and h/t we can deduce b'/b see Figure 12 and the work of Belgasmia et al (2007).

In our case b/d = 6.67m & h/t = 2.5m thus b'/b = 0.336m so b'= 0.672m.

Figure 12, Figure 13, and Figure 14 represent the 2D frame stadies and 2D model using Z_soil software.

Concrete E=2.7e7 KN/m2, υ=0.16 ft=300 KN/m2 fc=3000KN/m2 Steel E=2.1e8 KN/m2, υ=0.2, ft = fc =500000 KN/m2 (see Table 1)

Figure 12. Equivalent rectangular beam

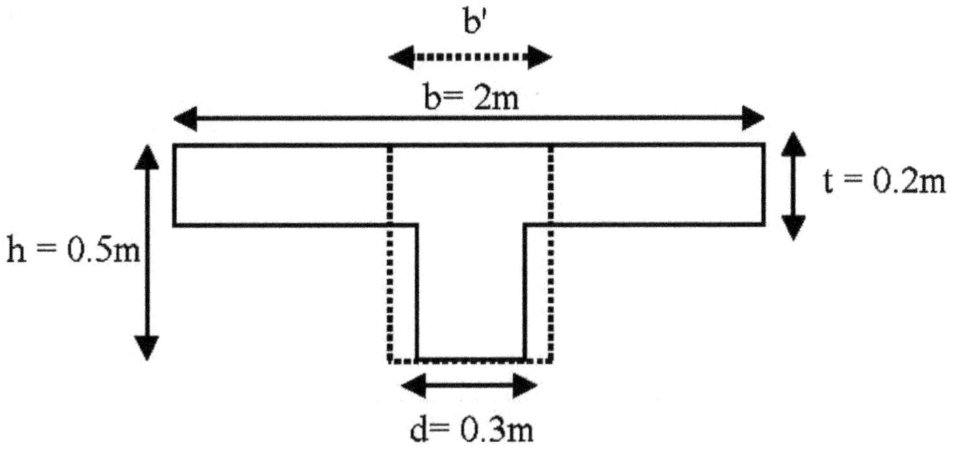

Figure 13. 2D frame stadies of bonefro building

Figure 14. 2D equivalent model of bonefro building using Z_soil software

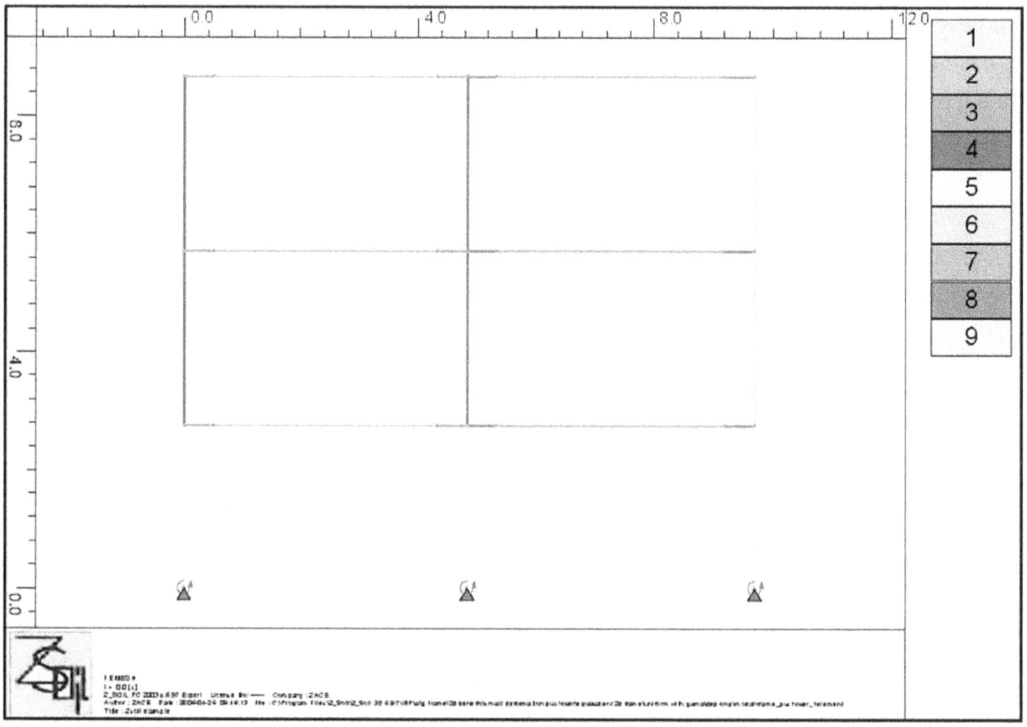

Table 1. Geometrical characteristics of equivalent 2D model

Elements and Characteristics	b(m)	h(m)	Reinforcement
Element 1	2.4	0.3	16Φ16 top; 16Φ16 bottom; 16 Φ 14 middle
Element 2	2.1	0.3	14Φ16 top; 14Φ16 bottom; 14 Φ 14 middle
Element 3	2.4	0.3	16Φ16 top; 16Φ16 bottom
Element 4	2.1	0.3	14Φ16 top; 14Φ16 bottom
Element 5	2.8	0.3	16Φ14 top; 24Φ14 bottom
Element 6	2.8	0.3	16Φ14 top; 20Φ14 bottom

Response of Bonefro Building to Ground Acceleration with Force-Based and Displacement-Based Elements

First, the response of a single 2D frame is studies. The side frame analyzed is shown in Figure 13. Its model with force-based elements in Z_Soil is shown in 14 see the work of Belgasmia et al (2007).

The response of the frame of to the El Centro accelerogram is shown in the following figures. Figure 15 shows the response obtained using force-based elements, while Figure 16 shows the response obtained

Figure 15. Response of 2D frame to El Centro accelerogram model with force-based elements: displacement history of three floors (d1 d2 d3)

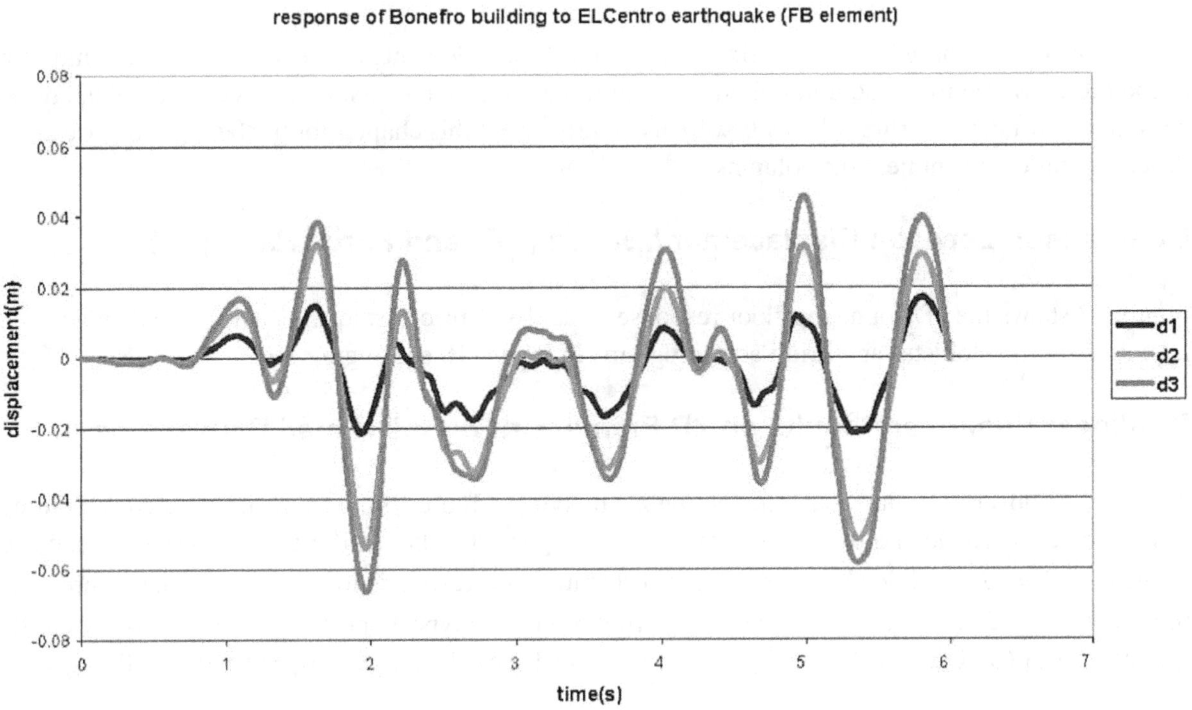

Figure 16. Response of 2D frame to El Centro accelerogram model with displacement-based elements displacement history of three floors (d1 d2 d3)

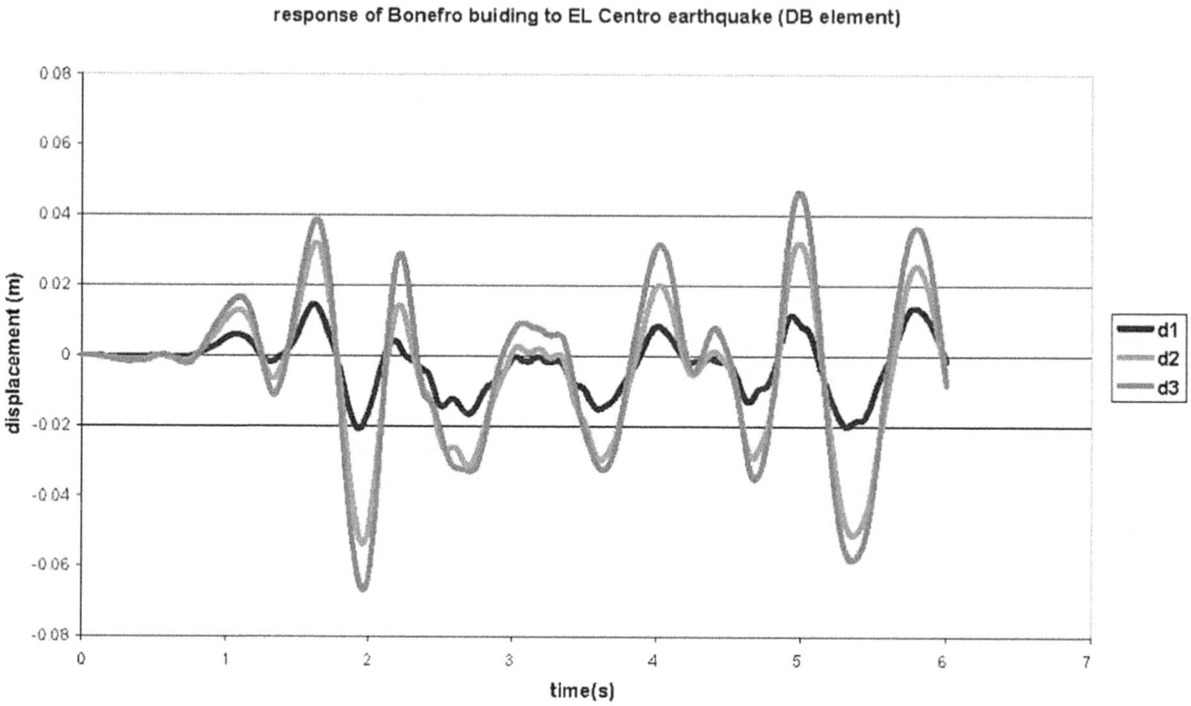

using displacement-based elements. Six displacement-based elements in every beam and column were needed to converge to the solution obtained with force-based elements. One force-based element per column was used, while three elements were used throughout this chapter for the beams, because of the different reinforcement near the columns.

Comparison between Displacement Based (DB) and Force Base (FB)

Figure 17 shows the 2D frame top floor response to the El Centro earthquake. The same displacement is plotted for the Hollister and Friuli accelerograms in Figure 18 and Figure 19.

Nonlinear Pushover Results on 2D Frame with Z_Soil (Small Displacements)

The EC8 pushover procedure is applied to the 2D model of Figure 13 the frame is modeled with force-based elements. The load distributions are shown in Figure 20. The "uniform" load distribution corresponds to $P = MR$, while the "modal" load distribution corresponds to $P = M\Phi_1$. The symbols are those of section 2.1. The elastic response spectrum refers to a type 1 spectrum of EC8 Part 1 §3.2.2.2, ground type A (S=1, T_B= 0,15 sec, T_C = 0,4 sec, T_D = 2 sec) 5% damping (η=1) and a_g = 0.15g.

Figure 17. Response of bonefro buiding to El Centro earthquake using displacement and force based element at top level (d3)

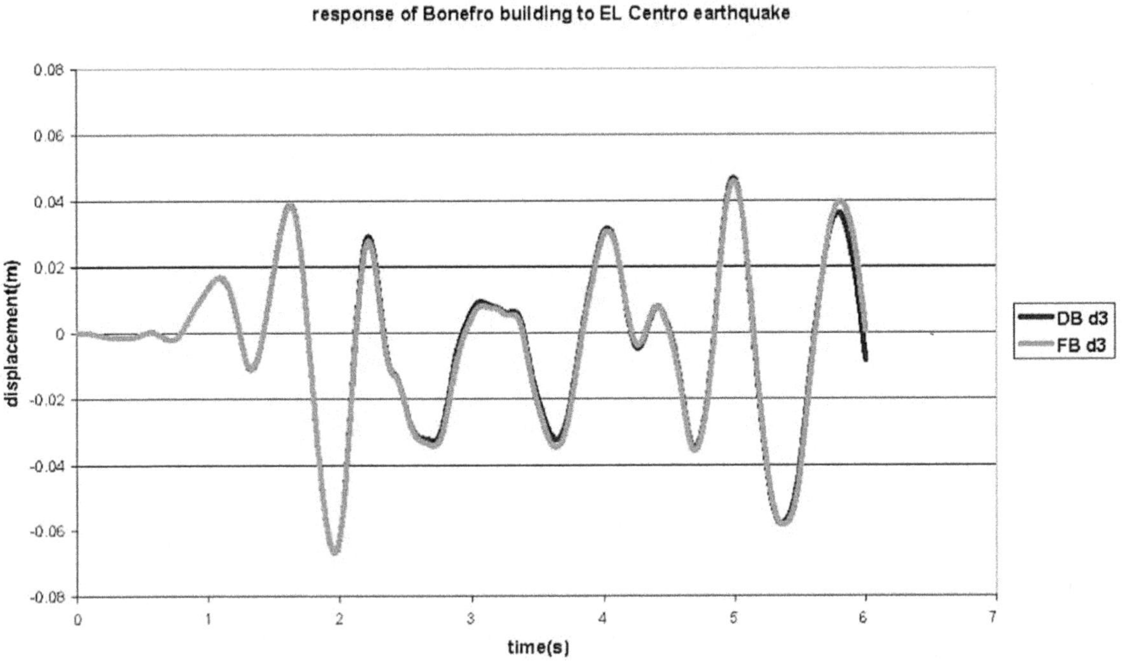

Figure 18. Response of bonefro buiding to Hollister earthquake using displacement and force based element at top level (d3)

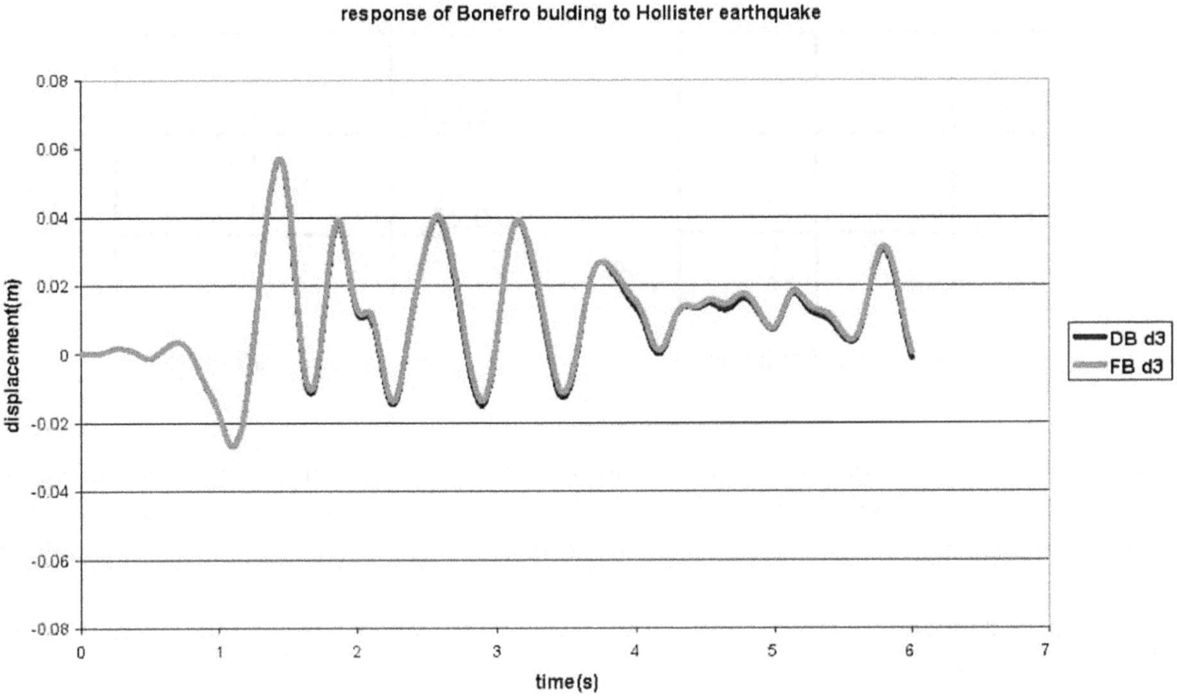

Figure 19. Response of bonefro buiding to Friuli earthquake using displacement and force -based element at top level (d3)

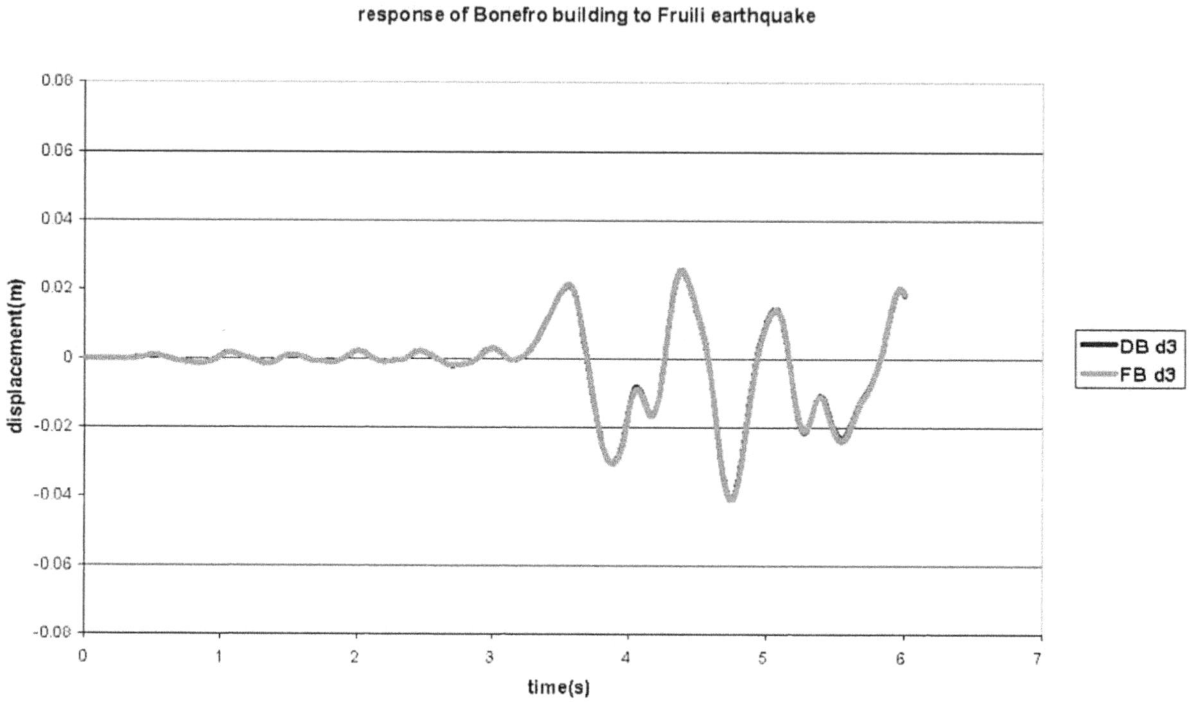

Figure 20. Load distributions used in the pushover analyses

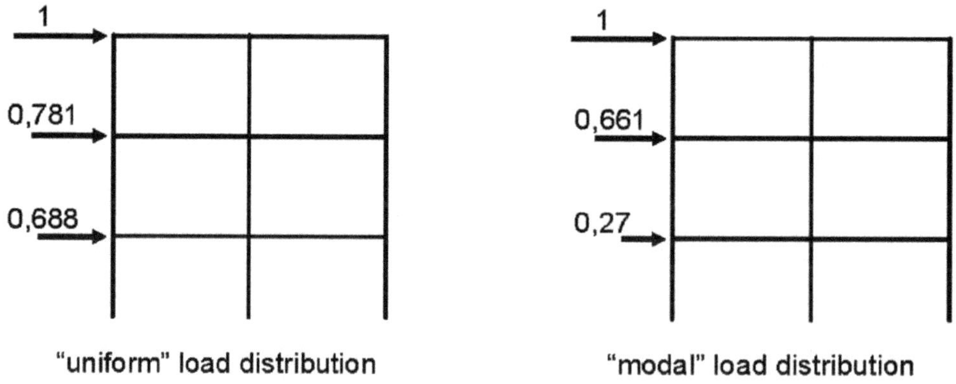

The nonlinear response of 2D model of an existing building is presented as a first application of the new capabilities in Z-soil see Belgasmia et al (2007). The reference frame for pushover analysis of the building in the x direction, load distribution defined earlier by $m_i\Phi_i$ (see §2.1.1.1) are shown in Figure 21.

Figure 22 and Figure 23 show the result of the pushover analyses for modal and uniform load distributions, respectively.

Figure 21. Two equivalent story building used for analysis with the loads distribution

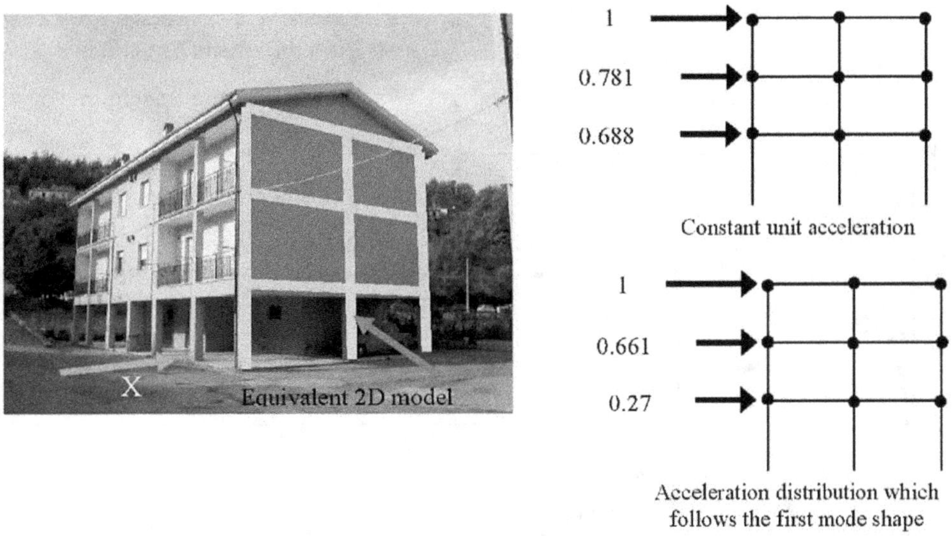

Figure 22. Determination of target displacement for modal load distribution (flexibility based)

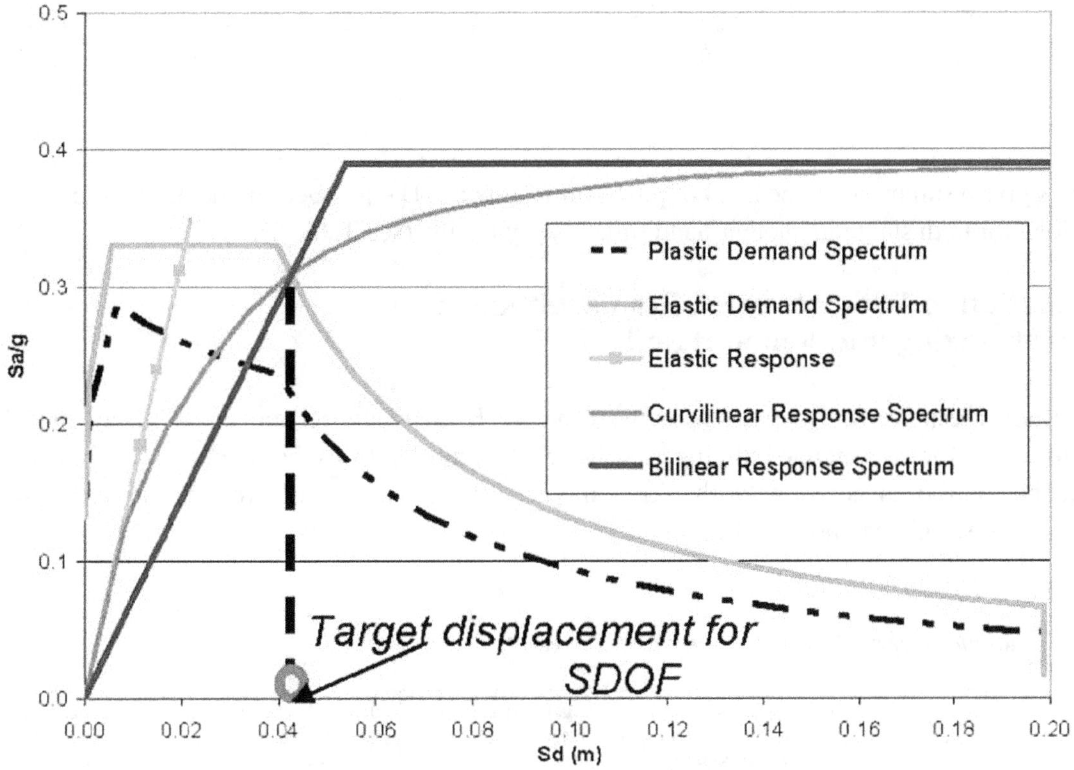

Figure 23. Determination of target displacement for uniform load distribution (flexibility based)

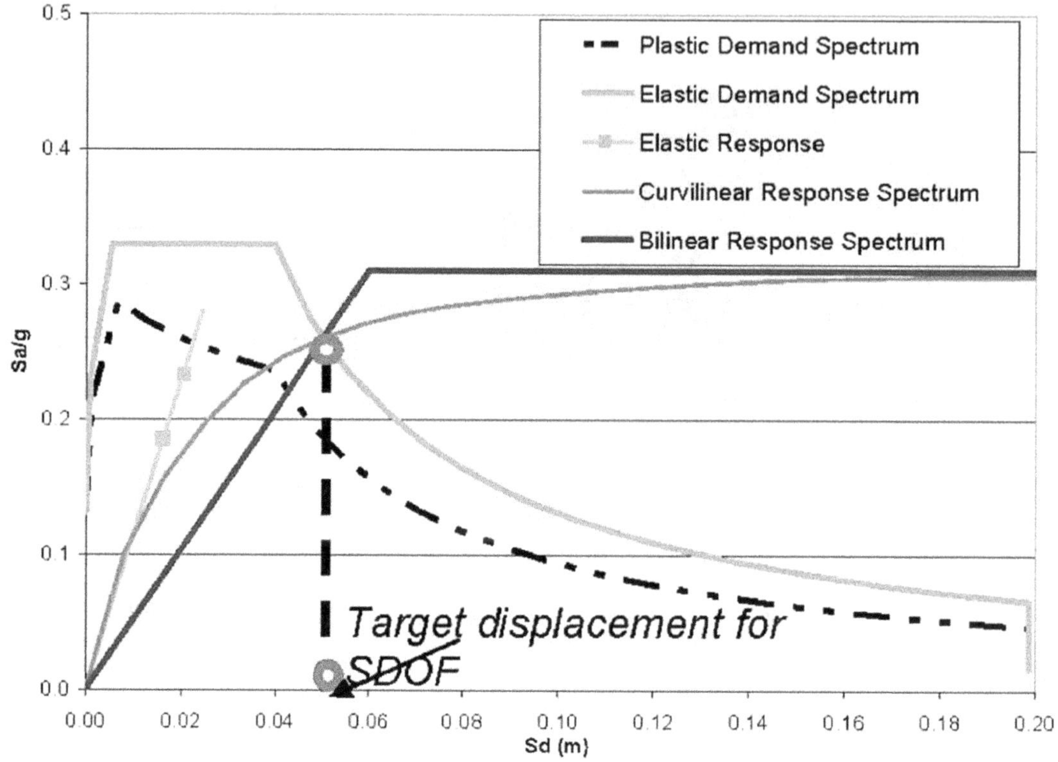

Let's give a summary in one hand of pushover results of 2D equivalent model under modal and uniform load for both single in another hand time history results (see Table 2).

Application to Two-Storey RC Frame without Soil and with Taking into Account Soil

The pushover response for two cases without soil and with taking into account soil of the 2D model of the two-storey reinforced concrete building see Figure 24 using flexibility-based model for superstructure.

The behavior of the structure with taking into account soil using pushover analysis are represented in Figure 25, see Command (2013).

Table 2. Summary of pushover and time history results

	Target Displacement under Uniform Load (SDOF)	**Target Displacement under Modal Load (SDOF)**	**Target Displacement Uniform Load (MDOF)**	**Target Displacement under Modal Load (MDOF)**
Pushover result of 2Dmodel	0.051m	0.042m	0.0612m	0.0504m
Time history result of 2D model			0.0625m	

Figure 24. Geometry and member properties of the typical frame structure analysed: (a) plan view, and (b) cross section A-A'

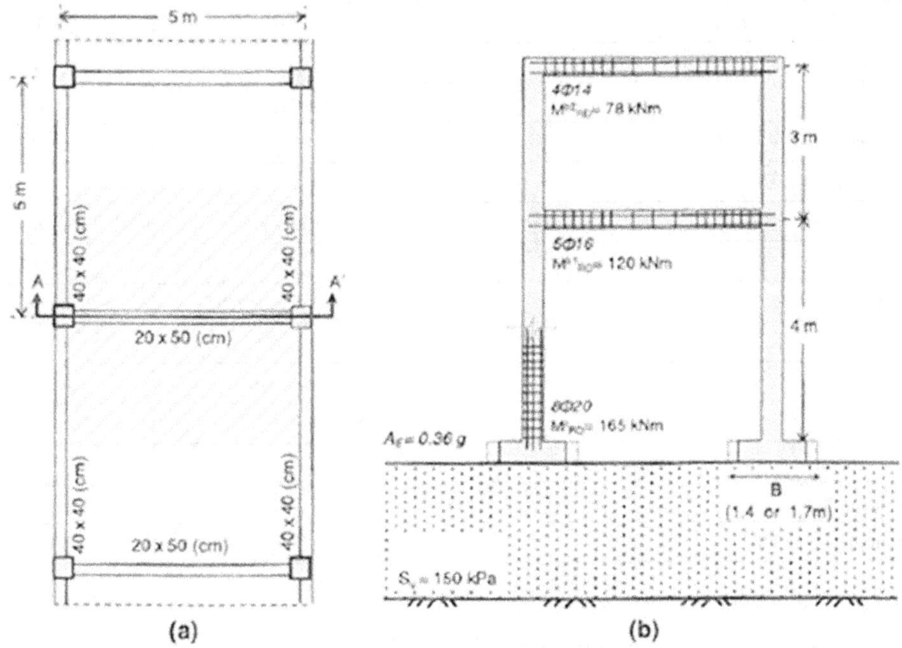

Figure 25. Displacement time history analysis results taking into account soil
Command., 2013.

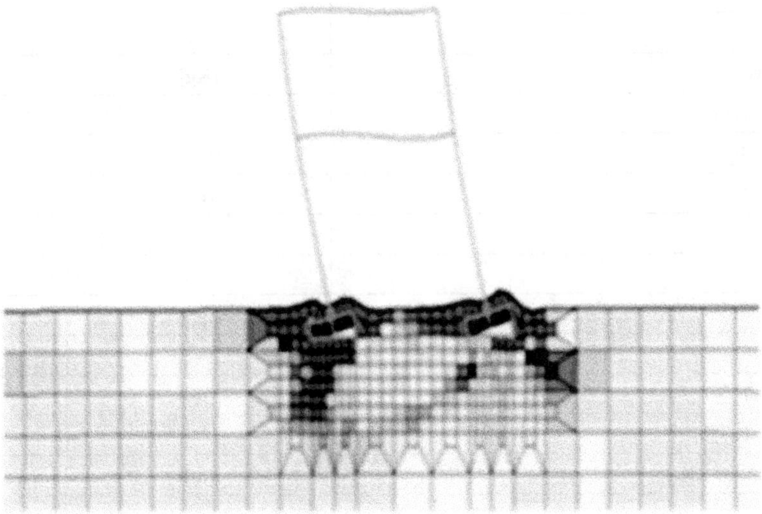

Tables 3 and 4 represent respectively, pushover analysis report for structure only without soil and pushover analysis report for structure in the case of taking into account the soil.

The multi degree of freedom displacement demand in the first and the second case are 6.5 and 9.65cm respectively. Table 5 summary all results of section 6, see Command (2013).

Table 3. Pushover analysis report for structure only

Pushover Analysis Report		
Item	**Unit**	**PSH1/Default**
MDOF Free vibr.period T	[s]	0.445066
MDOF Free vibr.period T*	[s]	0.837848363
SDOF equivalent mass M*	[Kg]	12369.6
Mass participation factor Gamma	-	1.20401
Bilinear yield force value Fy*	[KN]	48.9047294
Bilinear displ at yeild Dy*	[m]	0.07030178
Target displacement Dm*	[m]	0.166111577
SDOF displacement Dt*	[m]	0.053940741
Energy Em*	[KN*m]	6.404596979
Reduction factor qu	-	1
Demand ductility factor mi	-	3.079519728
Capacity ductility factor miC	-	2.362836024
MDOF displacement demande Dt	[m]	0.064945192

Table 4. Pushover analysis report for structure with taking into account soil

Pushover Analysis Report		
Item	**Unit**	**PSH1/Default**
MDOF Free vibr.period T	[s]	0.529679
MDOF Free vibr.period T*	[s]	1.1692452
SDOF equivalent mass M*	[Kg]	14036
Mass participation factor Gamma	-	1.28476
Bilinear yield force value Fy*	[KN]	19.58028002
Bilinear displ at yeild Dy*	[m]	0.048308883
Target displacement Dm*	[m]	0.075135711
SDOF displacement Dt*	[m]	0.075181918
Energy Em*	[KN*m]	0.998227533
Reduction factor qu	-	1.556275235
Demand ductility factor mi	-	1.556275235
Capacity ductility factor miC	-	1.555318745
MDOF displacement demande Dt	[m]	0.096590721

Table 5. Summary of results of section 6

	Pushover Analysis Results	
	Displacement Demand d_t (cm)	**Designed Moment $\|M\|max(d_t)$ (kNm)**
Structure only	6.5	94
Structure with soil	9.65	37

CONCLUSION

After satisfactory comparison of flexibility and displacement based formulation, only the first one was used in this study. The target displacements for 2D model of Bonefro building in the case of two loads distribution are around 0.051, and 0.042m, for uniform and modal load distribution respectively these results are for SDOF in order to have the displacement of MDOF we must multiply the results of SDOF by Γ which is equal to 1.2, thus the MDOF results are 0.0612, and 0.0504m, for 2D and the case of uniform and modal load we notice that in both cases represented in Figure 22 and Figure 23 the structure remains elastic, and 2D analysis indicates a larger target displacement for uniform loading vs modal loading (Figures 22 & 23) 0.051m against 0.042m.(reference point P structural displacements 0.061m vs 0.05m). The comparison between pushover and dynamics gives difference of 2,08% (dmax dynamics -dmax uniform from pushover/ dmax dynamics), and a difference of 19,36% (dmax dynamics -dmax uniform from pushover/ dmax dynamics).

Taking soil into account in the case of pushover analysis leads to more important value of displacement (about 50% and 32%) than taking structure only. For the designed moment the results are very interesting so when we take soil into account leads to lower values of moment (about 59% and 61%) than taking structure only which is a gain.

REFERENCES

Aydinoglu, M. N. (2004, June 28 – July 1). An Improved Pushover Procedure for Engineering Practice: Incremental Response Spectrum Analysis (IRSA) in Performance-based seismic design. In Krawinkler & Fajfar (Eds.), Concepts and implementation PEER report, Bled, Slovenia (pp. 345-356).

Belgasmia, M., & Mossaoui, S. (2015, February). Comparison between static nonlinear and time history analysis using Flexibility-based model for an existing structure and effect of taking into account soil using domain reduction method for single media. *KSCE Journal of Civil Engineering*.

Belgasmia, M., & Moussaoui, S. (2013). Comparison of Static Pushover Analysis in the case of Small and Large Deformation with Time History Analysis Using Flexibility-Based Model for an existing Structure. *International Journal of Current Engineering and Technologie*, *3*(2), 655–665.

Belgasmia, M., Spacone. E., Urbanski, A., & Zimmermann, Th. (2007, August). Seismic evaluation of construction: static pushover procedure and time history analysis for nonlinearframes. Swiss Federal Institute of Technology in Lausanne, Switzerland, LSC Internal Report.

Command, S. (2013). Pushover Analysis with Zsoil Taking Soil into Account. *Proceedings of the Numerics in Geotechnics and Structures Symposium, 28 years of Zsoil and structure PC*, Switzerland.

Eurocode 8. (2003). Design of Structures for Earthquake Resistance. European Committee for Standardization.

Fajfar, P. (1999). Capacity Spectrum Method Based on Inelastic Demand Spectra. *Earthquake Engineering & Structural Dynamics*, *28*(9), 979–993. doi:10.1002/(SICI)1096-9845(199909)28:9<979::AID-EQE850>3.0.CO;2-1

Fajfar, P. (2002). Structural Analysis in Earthquake Engineering – A Breakthrough of Simplified Non-Linear Methods.*Proc. 12th European Conference on Earthquake Engineering*.

Spacone, E., Filippou, F.C., & Taucer F.F. (1996). Fiber Beam-Column Model for Nonlinear Analysis of R/C Frames. Part I: Formulation, . *Earthquake Engineering & Structural Dynamics*, *25*(7), 711–742.

Spacone, E. (1994). Flexibility-Based Finite Element Models for the Nonlinear Static and Dynamic Analysis of Concrete Frame Structures [Ph.D. Dissertation]. Department of Civil Engineering, University of California, Berkeley.

Valipour, H. R., & Foster, S. J. (2010). A Total Secant Flexibility-Based Formulation for Frame Elements with Physical and Geometrical Nonlinearities. *Finite Elements in Analysis and Design*, *46*(3), 288–297. doi:10.1016/j.finel.2009.11.002

Vidic, T., Fajfar, P., & Fischinger, M. (1994). Consistent Inelastic Design Spectra: Strength and Displacement. *Earthquake Engineering & Structural Dynamics*, *23*(5), 502–521. doi:10.1002/eqe.4290230504

Zimmermann, Th., Truty, A., Urbanski, A., & Podles, K. (2008). *Z-Soil User Manual*. Switzerland: Zace Services.

ENDNOTES

1. The design spectra are actually given in terms of pseudo-acceleration rather than acceleration, but for low damping the two are basically identical, thus the "pseudo-" is typically dropped. It is worth recalling that the pseudo-acceleration is defined as.
2. Note that the ductility reduction factor R_μ is different from the reduction factor R (called behavior factor q in EC8). q takes into account both energy dissipation and the overstrength R_s, that is $R= R_\mu R_s$.

Chapter 4
Experiments on a Ring Tension Setup and FE Analysis to Evaluate Transverse Mechanical Properties of Tubular Components

M.K. Samal
Bhabha Atomic Research Centre, India

K.S. Balakrishnan
Bhabha Atomic Research Centre, India

ABSTRACT

Determination of transverse mechanical properties from ring specimens machined from tubular components is not straightforward due to presence of combined tension and bending stresses. Zircaloy tubes as used in nuclear reactors are manufactured through a complicated process of pilgering and heat-treatment and hence, the properties need to be determined in the as-manufactured condition. In this work, the authors perform ring-tensile tests on specimens of Zircaloy pressure tubes of Indian pressurized heavy water reactor in order to carry out integrity assessment of these tubes. As the loading condition in this test imposes both membrane and bending stresses in the cross-section of the ring, 3-D finite element analysis of the test setup was carried out in order to determine material stress-strain curve using an iterative technique. The effect of the design of the loading mandrel on the experimental stress-strain data has been investigated in detail. To validate the methodology, miniature tensile specimens have been tested and the data has been compared to those of ring specimens.

INTRODUCTION

Pressure tubes act as channels of heat transfer in the pressurized water type of nuclear reactors and contain the fuel bundles. These tubes are subjected to high pressure, moderate temperature and high fluence of neutrons during the reactor operation. During service, the mechanical and fracture properties of these

DOI: 10.4018/978-1-5225-0588-4.ch004

tubes degrade due to various material ageing phenomena including those of irradiation. In order to have better mechanical and corrosion properties in the irradiated water environment and to ensure minimum absorption cross-section for neutrons, Zirconium based alloys are usually used for manufacturing of these tubes (Nikulina, 2003, 2004). The manufacturing route of these pressure tubes from the ingots is also complicated with several stages of mechanical working and heat-treatment procedures. For example, the pressure tubes of current Indian pressurized heavy water reactors (PHWRs) are manufactured from quadruple-melted ingots of Zr-2.5Nb alloys (Deshmukh et al., 2010). Similarly, the fuel-clad tubes are also manufactured from different types of Zirconium alloys through a series of mechanical working and heat-treatment processes. Due to the long term exposure to aggressive environments in the reactor, significant degradation in properties of the fuel-clad tubes has also been reported by various researchers in the literature (Edsinger et al., 2000; Murty, 1999, 2002; Wang & Murty, 1998; Leclercq et al., 2008).

For design and safety analysis of the pressure tubes as well as of the fuel-clad tubes, it is essential to evaluate the mechanical properties in the as-manufactured as well as in the in-service conditions. Due to the complex mechanical working and heat-treatment conditions experienced by these tubes during their manufacture, the mechanical properties of the pressure tubes are highly anisotropic. It is not possible to evaluate the transverse mechanical properties from the standard tensile tests as standard specimens with sufficient dimensions cannot be machined from the pressure tubes in the circumferential direction (the radius of curvature is too small for this purpose). Similar problem is also faced by the engineers in the mechanical property evaluation of fuel-clad tubes (Arsene & Bai, 1996; Bae et al., 2008; Seok et al., 2006; Lee & Hong, 2002; Bertolino et al. 2003; Fukuda et al., 2003; Videm & Lunde, 1979; Lee et al., 2001; Grigoriev et al. 1995, 1996, 1997; Grigoriev & Jakobsson, 2000, 2005; Nakatsuka et al., 1991; Yoshitake et al., 2004; Hsu et al., 2001; Josefsson & Grigoriev, 1996; Sainte Catherine et al. 2006; Yagnik et al., 2008; Samal et al., 2010a,b, 2012, 2013, 2014, 2015; Balakrishnan et al., 2014; Rashid et al., 2000; Bertsch & Hoffelner, 2006; O' Brien & Ferguson, 1983; Martín-Rengel et al., 2009, 2012; Shah et al., 2011).

Several types of geometry of specimens and the loading-mandrels were designed by various researchers the world-over in order to determine the mechanical and fracture properties of these tubes satisfactorily. A pin-loading tension (PLT) test was designed by Grigoriev et al. (1995) to evaluate the fracture toughness of Zicaloy-2 fuel claddings. Axially notched ring specimens cut from the fuel cladding section have been tested in a way similar to that used to test compact-tension specimens. Hsu et al. (2001) have designed an X-shaped specimen and have used it to determine the fracture toughness of the thin-walled tubular materials. In this method, two curved axially-cracked semi-cylindrical specimens (machined from the fuel-clad tubes) were glued together by epoxy adhesive to make an X-shaped specimen so as to minimize the bending effects of the loading. A modified ring tension test was carried out by Jossefson and Grigoriev (1996) for fracture toughness testing of cladding where flattening of the ring is prevented by means of a central piece which is inserted into the ring. The central piece has almost the same dimension as that of the ring's internal diameter. An internal conical mandrel technique has been used in Sainte Catherine et al. (2006) for evaluation of fracture toughness of fuel-clad tubes. A compilation of different types of specimens used by various researchers is presented in Yagnik et al. (2008).

In the recent years, specimens have been machined from the fuel-clad tubes in the form of rings of small width and tested with special types of loading mandrels. In this work, a similar technique is followed for the specimens machined from the pressure tubes, which has not received much attention in literature. However, the results from these tests cannot be used directly to evaluate the material stress-strain curve. The complexities lie in:

1. Existence of both tensile and bending stresses in the loading cross-section due to the curvature of the specimen,
2. Effect of lateral pressure and friction drag exerted by loading mandrels on the specimen,
3. Geometry of the loading mandrel and the initial gap between the mandrels,
4. Orientation of gauge-length of reduced cross-section of the ring-specimen with respect to the loading axis and so on.

The aim of this work is to investigate the effects of the above-mentioned factors on the load-deformation response of the ring-tension test setup with a detailed 3-D finite element (FE) analysis and develop a suitable inverse FE analysis procedure so that the mechanical properties in the transverse direction of the pressure tubes can be satisfactorily evaluated from the test data. The mechanical properties in the axial direction of the pressure tubes can also be determined from the inverse FE analysis of tensile test data obtained from testing of straight specimens machined from the axial direction of these tubes.

In this work, experiments have been conducted on the ring-type specimens machined from the pressure tubes of two different types of materials loaded with different types of mandrels and the results of load-deformation response have been compared with those of FE simulation. It can be concluded that the finite element analysis procedure of this complex loading setup should be used to evaluate the accurate transverse mechanical properties in terms of the material true stress-strain curve.

This chapter is divided into ten sections. A brief description of the materials used in this work, design of the ring specimen and the test setup is provided in Sections 2-4. Results from ring tension and miniature specimens machined from Zircaloy-2 and Zr-2.5Nb pressure tubes are presented in Sections 5-6. The method for evaluation of engineering stress-strain data for ring tensile specimens directly from the experiment is presented in Section-7. The details of the finite element analysis procedure used in this work for simulation of deformation behavior of the test setup are given in Section-8. The results from FE analysis have been compared with those of experiments and the implications of FE analysis have been discussed in Section-9 followed by conclusions and scope of future research in Section-10.

MATERIAL AND SPECIMEN USED IN THE TEST

In this work, specimens in the form of circular rings have been machined from both Zircaloy-2 as well as Zr-2.5%Nb pressure tubes. Both of these materials have been used in the manufacture of pressure tubes of Indian PHWRs. Zircaloy-2 has been used in older type of PHWRs whereas Zr-2.5%Nb is used now-a-days due to the important property of lower hydrogen pick-up of these alloys from the flowing coolant (i.e., heavy water) inside the pressure tubes. The chemical composition of both of these alloys is given in Table 1 and 2 respectively.

Table 1. Chemical composition of Zircaloy-2 pressure tube material

Element	Sn	Fe	Cr	Ni	O	Zr
Wt. %	1.5	0.09	0.1	0.05	0.12	rest

Table 2. Chemical composition of Zr-2.5Nb pressure tube material

Element	Al	C	P	N	O	Fe	Ni	Cr	Nb	Si	Sn	Zr
Wt. %	0.003	0.01	0.001	0.0026	0.11	0.13	0.004	0.017	2.58	0.003	0.0056	rest

It can be noted that Zr-2.5Nb alloys contains 2.58 wt. % of Nb in addition to other alloying elements. However, the wt. % of Cr and Ni are significantly lower in comparison to those of Zircaloy-2. The major alloying elements in Zr-2.5Nb pressure tubes are niobium and oxygen. Niobium is a beta stabilizer and the presence of the same in the alloys leads to the formation of a two-phase microstructure consisting of hcp α and bcc β phases. The presence of niobium in zirconium is known to increase the tensile and creep-strength of the alloy significantly. Hence, Zr-Nb alloys show improved creep resistance as compared to the other Zirconium alloys (e.g., Zircaloy-2). This is also one of the considerations for using Zr-2.5Nb alloy instead of Zircaloy-2 as the pressure tube material in the current Indian PHWRs.

The fabrication of these pressure tubes consists of four major steps, viz., hot extrusion, two cold-pilgering stages with an intermediate annealing step (at 550 deg. C for 6 hours) and a final autoclaving stage (in 10 bar steam atmosphere at 400 deg. C for 36 hours). These mechanical working conditions also affect the texture of this material which is expressed in terms of Kearns' factor 'f'. The preferential orientation (in terms of Kearns' factor) of basal poles (i.e., the <c> directions) along transverse, radial and longitudinal directions are of the order of 0.58, 0.39 and 0.03 respectively (Deshmukh et al., 2010). Hence, the mechanical properties of these tubes are highly anisotropic. The ring-tensile specimens have been machined from the as-manufactured tubes and tested in a ring-tension setup. The pictures of the

Figure 1. (a) Ring-tension specimens with a gauge-length of reduced cross-section machined from Indian PHWR pressure tubes and tested to failure in the ring-tension setup; (b) solid model of the ring-tension specimen, the symmetric form of which was used in FE analysis

(a)

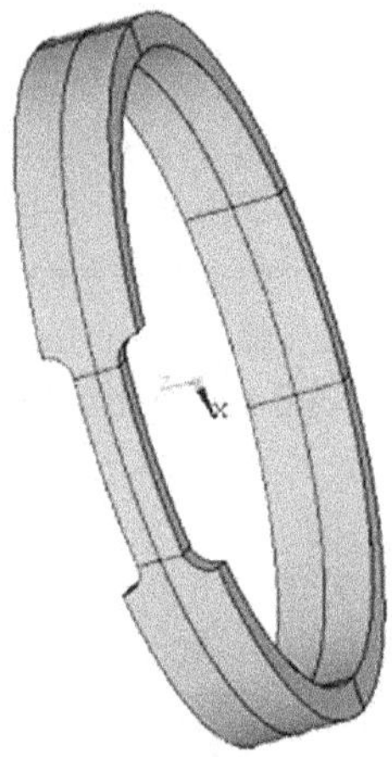

(b)

tested specimen are shown in Figure 1a. The 3-D solid model of the specimen used in the FE analysis is shown in Figure 1b.

The rings machined from the Zircaloy-2 pressure tube have internal diameter values of 83 mm and the value of thickness as 4.3 mm, whereas for the rings machined from Zr-2.5Nb pressure tubes, the corresponding dimensions are 83 mm and 3.6 mm respectively. A section of the ring specimen has a reduced cross-section (known as gauge-length section) in order to induce necking in the same region. The width of both of the ring specimens is 6 mm in the gauge-length section and 12.5 mm away from the gauge-length section. The specimen having a single reduced gauge-length in the periphery is also known as the single-gauge-length (SGL) specimen. The gauge length has a dimension of 25 mm in case of the Zircaloy-2 ring-specimen, whereas it is of 20 mm in case of the Zr-2.5Nb ring-specimen. The gauge-length can be oriented at different angles with respect to the loading axis of the testing machine. When the gauge-length is oriented at right angle to the loading axis, the arrangement is known as "12 'O' clock position" in this paper.

LOADING MANDREL AND THE TEST SETUP

The ring-tension setup consists of the ring-specimen and the loading mandrel. The loading mandrel can be of various designs. We have considered two types of loading mandrels in this work, i.e., 2-piece (Figure 2a) and 3-piece (Figure 3a) type of mandrel. The 2-piece mandrel consists of two semi-cylindrical mandrels with a gap between them and they load against the ring-specimen with the help of loading pins (Figure 2a). In case of the 3-piece mandrel, a central piece is inserted in between the top and the bottom loading mandrels in order to restraint the lateral movement of the ring-specimen during loading

Figure 2. Ring specimen with gage-length oriented at 12'O clock position in the test grips: (a) 2-piece mandrel, pins and the loading grip before assembling the parts; (b) 2-piece mandrel fitted with ring specimen and loaded in the grips

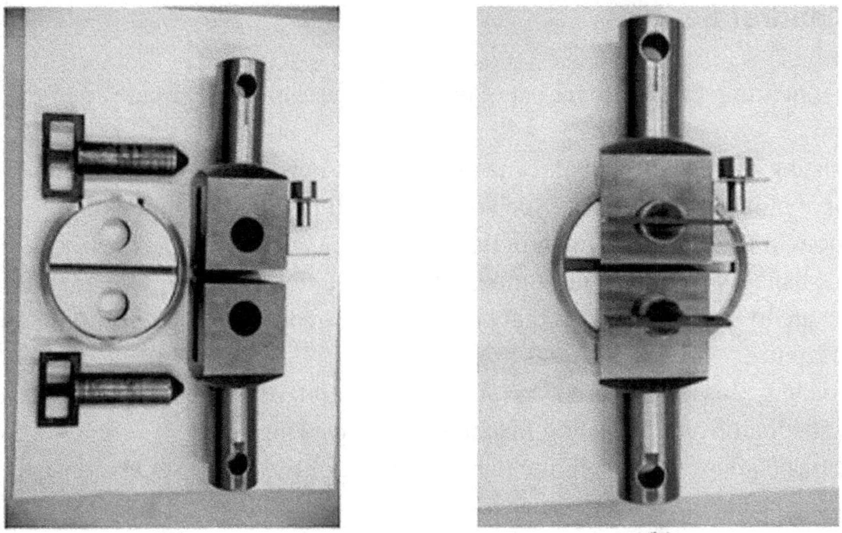

(a) (b)

Figure 3. Ring specimen with gage-length oriented at 12'O clock position in the test grips: (a) 3-piece mandrel and the SGL ring specimen before loading in the frame; (b) 3-piece mandrel fitted with ring specimen and loaded in the grips

(a)

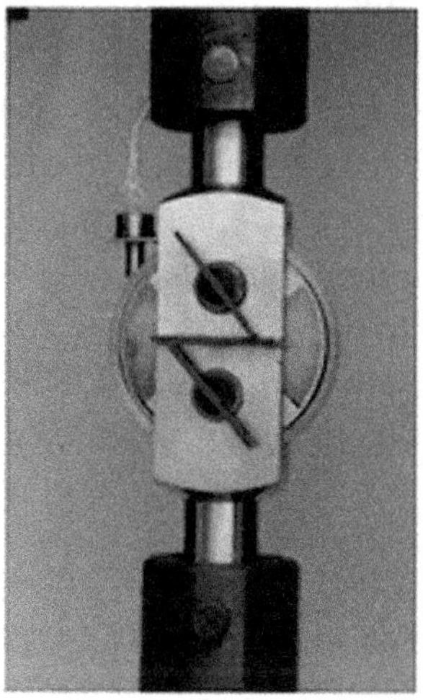

(b)

(Figure 3a). The loading arrangements with the 2-piece and 3-piece mandrels are shown in Figure 2b and Figure 3b respectively.

The types of test set up designed for conducting RTT were essentially of two kinds: (a) One having two- piece set of mandrels and (b) the other having three- piece set of mandrels. The two kinds of test set up differed with respect to the manner in which tensile loading was accomplished in the test ring. The basic difference in the two set up was with regard to the number of mandrels that would butt against the interior portions of the ring while the remaining parts were kept common to both the systems. One system consisted of two-piece set of mandrels while the other had three-piece set of mandrels.

Two-Piece Mandrel Set Up

The test system containing the two-piece set of mandrels was essentially made up of the following parts.

1. A set consisting of two semicircular mandrels which need to be positioned in close proximity to the interior contour of the test ring such that they will help in the application of tensile load to be brought about in the specimen in an indirect manner.
2. A set of Y-shaped yokes (two in number) that would act as the adjoining fixtures to which the mandrel set are to be attached such that both of them together would constitute the complete set of grips and fixtures.
3. A pair of loading pins (larger sized) to join the yokes and the mandrels. These permit the transmission of tensile load from the testing machine to the test ring.
4. A pair of female adaptors to facilitate the attachment of the entire set up consisting of (a), (b) and (c) to the crossheads of the testing machine.
5. A pair of connecting pins to join the female adaptors to the yokes.

6. A piece of Teflon tape to be inserted in the space between the circular mandrels and the test ring to minimize friction.

Figure 2a depicts a view of the ring tension test grips in which the two-piece set of mandrels has been placed. It also shows the various parts which constitute the test grips before assembling them while Figure 2b provides the assembled view of the grips & fixtures.

Three-Piece Mandrel Set Up

The three- piece set of grips was basically designed to take care of the unavoidable bending that would be experienced by the test ring when it is being subjected to pulling with the help of the two-piece type of grips. In this, a dog-bone piece was introduced in between the upper and lower mandrels such that it helped in bolstering the capacity of the entire set of mandrels to offer adequate support to the ring sample throughout its circumference as far as feasible. This is similar to the thinking originally suggested by Arsene and Bai (1996) who came out with the novel idea of introducing a dog-bone piece in between the mandrels to tackle the problem related to bending arising from the inadequate support in the case of nuclear fuel element cladding subjected to the application of tensile load.

Though a cue has been taken from that approach, substantial additional work has gone into applying this idea into practical terms in the present case. First of all, the size of the tube being handled in the present case is comparatively larger with regard to the diameter and wall thickness of nuclear fuel element cladding. Further, the type of specimens being handled is different and the friction conditions are also different. In view of these reasons it became absolutely essential to evaluate afresh about the impact of the new geometry on the results obtainable from the experiments being planned to carry out. The essential features of the grips are brought out in Figure 3.

The angle of the central dog-bone piece was appropriately designed such that the arc length produced at its periphery would be much more than that of the length corresponding to the reduced section (i.e. gauge length) of the test ring even after an extension of 15mm. This was done to ensure that complete support would be available for the gauge length portion of the ring tensile specimen throughout the entire testing process. The length of 15mm. would correspond to 60% of the gage length (i.e.25 mm,) of the ring tensile specimen to be used for the experiment. This is much more than the extension that would be needed to take care of the maximum expected value of ductility of zirconium alloy tubular material when they will be subjected to experimental evaluation.

Utmost care was taken in the design of the included angle of each pair of the mandrel set so that the gauge length (GL) portion would be fully supported throughout the experiment irrespective of the position at which the GL needs to be oriented with respect to the loading direction. This can be observed very clearly from Figure 3. The same yokes and adaptors used for the two-piece mandrel set were kept common to the three- piece set of mandrels as well since their basic purpose remained identical in both the cases. Thus the test system containing the three-piece set of mandrels was essentially made up of the following parts.

1. A set of mandrels having partly semi-circular shape which are required to be positioned at the upper and lower levels of the ring sample in close proximity to its interior contour.
2. A specially designed dog-bone mandrel piece located at the central region of the circular geometry of the system would to serve as the support for the other two load-bearing constituents of the man-

drel set. The angle at the edges of the dumbbell shape was so designed such that it would enable the central mandrel piece to have sufficiently large circular region that would be much more than the gauge length of the ring tensile sample. This was done particularly to ensure that full support to the GL region would exist even after it has extended by deformation.

3. The same Y-shaped yokes were kept common to both the systems as their purpose remained identically the same, i.e. to act as the fixtures for attaching the mandrel set.
4. The same pair of loading pins (larger sized) was kept common to both the systems for joining the yokes and the mandrels in order to allow the transmission of tensile load from the testing machine to the test ring.
5. The same pair of female adaptors was kept common to both the systems to facilitate the attachment of the entire set up consisting of (a), (b), (c) and (d) to the crossheads of the testing machine.
6. The same pair of connecting pins as earlier was used to join the female adaptors to the yokes.
7. The arch regions corresponding to the upper and lower mandrels also were made sufficiently large similar to the provision made at the edges of the dumbbell shape of the dog-bone mandrel piece. This particular provision was quite effective in ensuring that adequate support for the GL section of the test specimen would be available in the grips while conducting its tension test by keeping the GL oriented at 12'O clock or 3'O clock locations of the mandrel set as the case may be.
8. A piece of Teflon tape was inserted in the gap between the test ring and the mandrels to minimize friction.

DESIGN ASPECTS OF THE TEST SETUP

The basic problem with the ring tension test (RTT) methodology is that one does not have a prior understanding of how to obtain the correct and reliable data using such a specimen as it is not a well-accepted approach established by any standard testing agency like the ASTM, BS or ISI type of standard especially for metallic tubes. Further, one can conduct the experiment by positioning the gauge length region of the test sample at various locations (see Figure1) around the mandrels of the testing grips. It is apparent that these data sets would in turn very much vary depending on the way in which the reduced section is oriented with respect to the loading direction. Under such conditions, one will totally be in the dark regarding the veracity of the results generated. In this context it will then become mandatory to have a standard set of data derived through a method that is well-accepted and hence can be fully relied upon. It is in this respect that we find the exact utility of miniature plate tensile specimens sliced from the PT in the circumferential direction of the tube. In this work, it is therefore proposed to make use of the data derived from the miniature plate tensile specimens fabricated from the circumferential direction of the tube to act as the basis or standard against which the data obtainable from RTS could be compared and correlated with. Thus the data determined from the miniature plate tensile specimens would become the conventionally approved standard data for verifying and calibrating the output generated by using the RTS.

It is to be noted that the applicability of miniature plate tensile specimens to provide reliable and consistent data at the component level had been already experimentally established in Samal et al., (2013) by comparing the data generated on full thick conventional axial sheath tensile specimens and those of the axial miniature plate tensile specimens. The efficacy of miniature plate tensile specimens to provide reliable estimate of a tubular component was proven in the case of two types of PT materials, namely zircaloy-2 and Zr-2.5%Nb. The experimental evidence thus made it possible to conclude that the data generated on miniature plate tensile samples extracted from the circumferential direction of the tube can also be reliably taken as those corresponding to the same tube at the component level.

In the same vein it can be justifiably argued with equal conviction that the transferability of data at the component level especially for the transverse direction would become feasible when the data obtained from the RTT are standardized against those of the miniature plate tensile specimens. This will indeed be a real value addition as the search for a suitable methodology with an appropriate test specimen to be tested under a specific set of loading conditions would get automatically answered.

The procedure that was required to be employed was to generate various sets of data after placing the gauge length region of SGL type of specimen at certain locations around the mandrels of the test grips and then compare each one of them against that obtained from the miniature plate tensile specimen fabricated from the circumferential direction of the tube. This exercise would reveal the right location at which the gauge length region (GL) of an SGL type of specimen should be placed such that it would yield the correct and reliable data of the material being studied. However, adopting such an approach would be quite an involved and tedious work in view of the many possibilities of locating the GL at various angles with respect to the loading direction.

The above mentioned issues were distinctly solved through finite element method (FEM) as discussed in Samal et al. (2013). The FEM based analysis in indeed a virtual experiment in which the output can be obtained without actually conducting the test in reality. Hence it would save a lot of time, effort, material and resources in arriving at a possible answer with a reasonable level of accuracy. Thus it is quite a powerful tool that finds extensive application in difficult situations wherein an advanced idea regarding the possible outcome can be gauged without actually conducting the experiment as such. FEM based simulation analysis was carried out for both the materials using both types of grips having different sets of mandrels to apply the load onto the test specimen.

The FEM approach clearly showed that adopting the three-piece mandrel set and orienting the GL of the ring tensile specimen at 12'O clock location would provide a reliable estimate of the tensile property data for zircaloy-2 pressure tube material. It also showed that in the case of Zr-2.5%Nb PT the combination of two-piece mandrel set and orienting the GL of ring tensile specimen at 12'O clock location would give the correct set of data.

RESULTS FROM RING TENSION AND MINIATURE SPECIMENS MACHINED FROM ZIRCALOY-2 PRESSURE TUBES

A few typical load-displacement graphs obtained by conducting ring tension tests at room temperature using ring tensile type of specimens are brought out in Figures 4 to 6 pertaining to those obtained from the rings made out of the Zr-2 pressure tube. It can be observed that the maximum loads as obtained from the two tests with identical specimens are quite different (Figure 4 vs. Figure 6).

The test with the 2-piece mandrel shows a higher maximum load (at onset of necking) compared to that of the test with 3-piece mandrel. This is because of the effect of contact conditions existing at the interface of the mandrel and the ring-specimen. It is also evident from the typical load vs. displacement graphs shown from Figure 4 to Figure 6 that the test graphs corresponding to the RTT done with the GL positioned at 12'O clock using 3-piece mandrel shows better ductility and lower strength values compared to all other positions in the case of Zr-2 type of PT. After ruling out many identical positions it is apparent that there are two relevant locations or directions at which the gauge length (GL) of the test

Figure 4. Load-displacement graph of Zr-2 from ring tension test at room temperature with its GL positioned at 12' clock using 2-piece mandrel

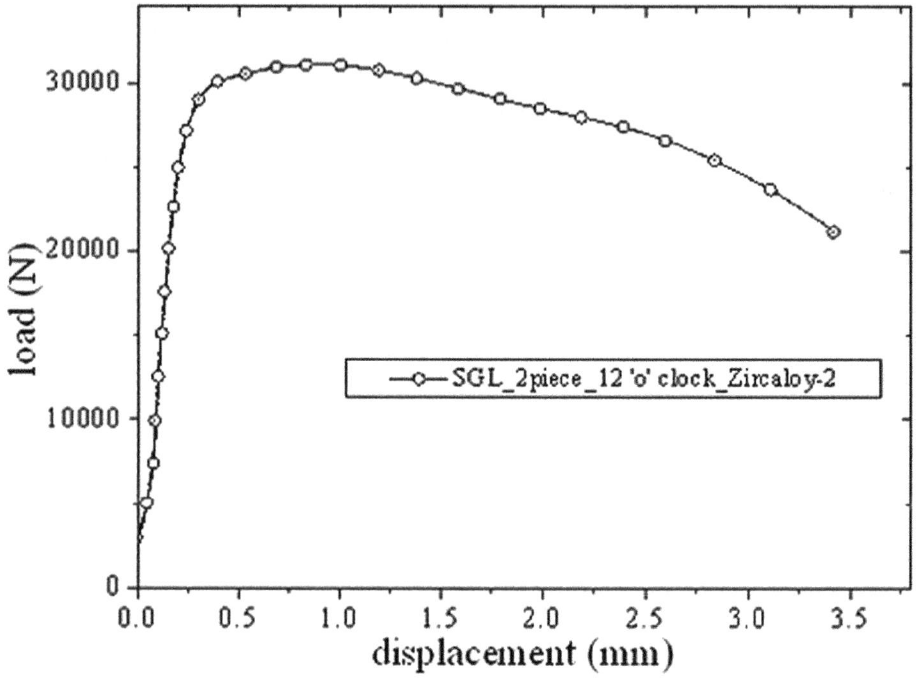

Figure 5. Load-displacement graph of Zr-2 from Ring tension test at room temperature with its GL positioned at 3'O clock using 2-piece mandrel

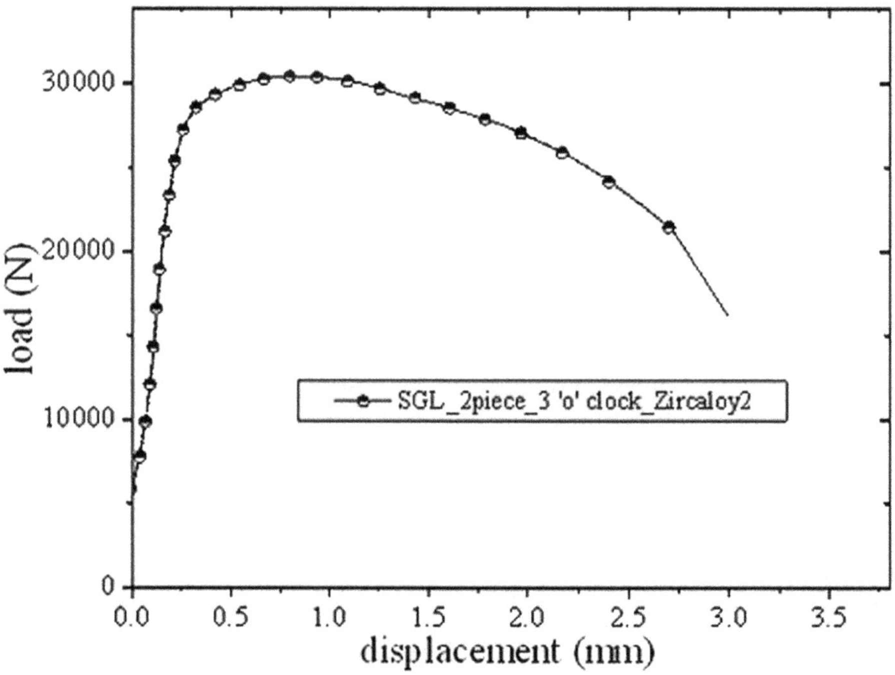

Figure 6. Load-displacement graph of Zr-2 from Ring tension test at room temperature with its GL positioned at 12' clock using 3-piece mandrel

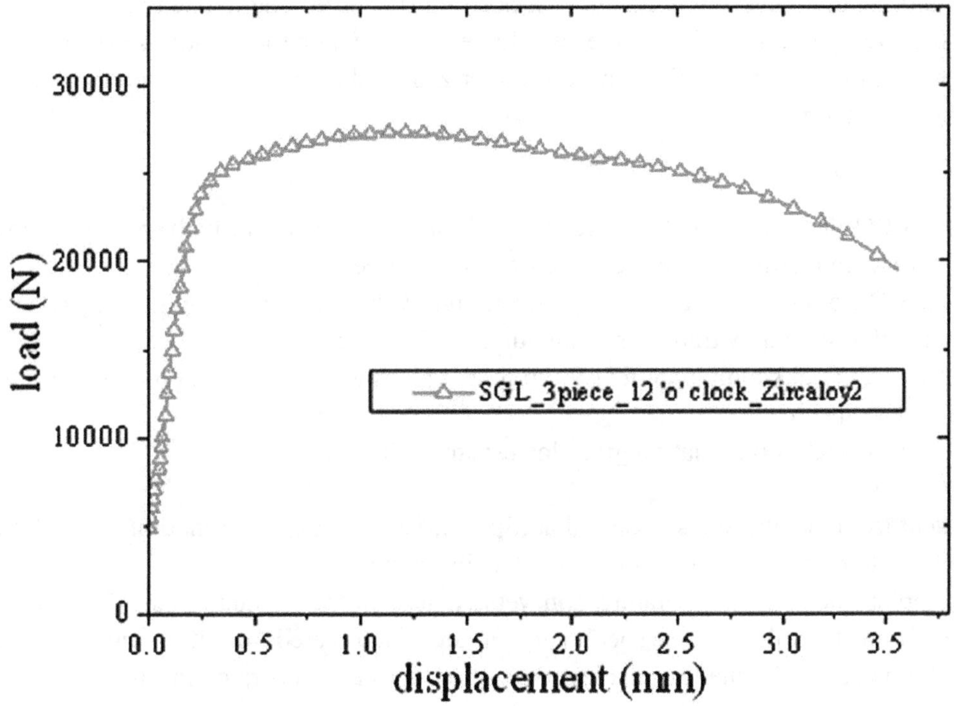

sample can be placed with respect to the mandrels. The GL can be placed at either 12'O clock position of the mandrels (i.e. at 90 degrees with respect to the loading direction) or at 3'O clock position of the mandrels ((i.e. at 0 degree with respect to the loading direction). All other locations with respect to the circle would be diagonally oriented positions of these only and hence are expected to provide identical output. Therefore, they are not being considered. In the case of samples fabricated from Zr-2 PT, the above mentioned two locations only were attempted.

Table 3 brings out the results obtained by conducting the ring tension tests on Zr-2 PT at ambient temperature using RTS type of samples extracted from it after orienting the gauge length of the test samples at various positions using 2 piece and 3 piece types of mandrels. It is clear from Table-3 that the grips used with 2- piece type of mandrels furnish relatively higher values of strength in comparison

Table 3. Results of ring tensile tests carried out on Zr-2 at ambient temperature

Type of Ring Specimen	Position of GL around the Mandrels	Type of Mandrels Used	0.2% YS (MPa)	UTS (MPa)	% Uni. Ductility	% Total Ductility
SGL	12'O clock	2 piece	551.14	632.07	4.21	24.27
SGL	3'O clock	2 piece	565.54	625.66	4.765	20.95
SGL	12'O clock	3 piece	531.53	581.6	7.2	27.61
SGL	3'O clock	3 piece	489.62	541.85	7.6	26.11

with 3-piece type of mandrels in the case of Zr-2 PT irrespective of the positions at which the GL of the sample was placed around the mandrels. It is also obvious from above that if the GL of the sample was placed around the mandrels at 12'O clock position (i.e. at 90 degrees with respect to the loading direction), the strength (YS & UTS) values would be highest when the mandrels are of 2-piece type.

The above results obtained at RT can be summarized as shown below for understanding the entire output in the simplest manner.

For Zr-2:

1. (2piece, 12'O clock) combination gives the highest strength values which are more than that shown by transverse miniature tensile specimens from the same tube.
2. (2piece, 3'O clock) combination also gives strength values closer to that given by (i) and hence are more than that shown by transverse miniature tensile specimens
3. (3piece, 12'O clock) combination gives strength values which are closer to that of transverse miniature tensile specimens
4. (3piece, 3'O clock) combination gives lowest strength values

It is evident from the above discussion that (3piece, 12'O clock) combination gives strength values which are closer to that of transverse miniature tensile specimens.

Table 4 depicts the room temperature transverse tensile property data of Zr-2 PT tube obtained by making use of miniature plate tensile specimens (Figure 7) fabricated from its circumferential direction. The stress-strain curves obtained from a few typical samples are illustrated in Figure 8.

RESULTS FROM RING TENSION AND MINIATURE SPECIMENS MACHINED FROM ZR-2.5NB PRESSURE TUBES

The results ring tensile specimens (tested at room temperature) machined from Zr-2.5Nb pressure tubes are summarized in Table 5. The load-displacement curve of the specimen with its gage length aligned in 12 O' clock position and loaded with 2-piece mandrel is shown in Figure 9. Table 6 depicts the room temperature transverse tensile property data of Zr-2.5Nb PT obtained by making use of miniature plate tensile specimens fabricated from its circumferential direction. The above results obtained on Zr-2.5%Nb

Table 4. Transverse plate tensile test results of Zr-2 at room temperature

Specimen No.	Type of Specimen	0.2% YS (MPa)	UTS (MPa)	% Uniform Ductility	% Total Ductility
1	Plate tensile	531.95	563.69	2.84	16.44
2	"	566.59	592.53	3.42	19.82
3	"	548.00	602.49	3.78	19.50
4	"	569.47	614.20	3.78	20.44
5	"	581.15	607.18	3.51	21.20
Average		559.43	596.02	3.47	19.48

Figure 7. Geometrical details of the transverse miniature tensile specimen used in the test

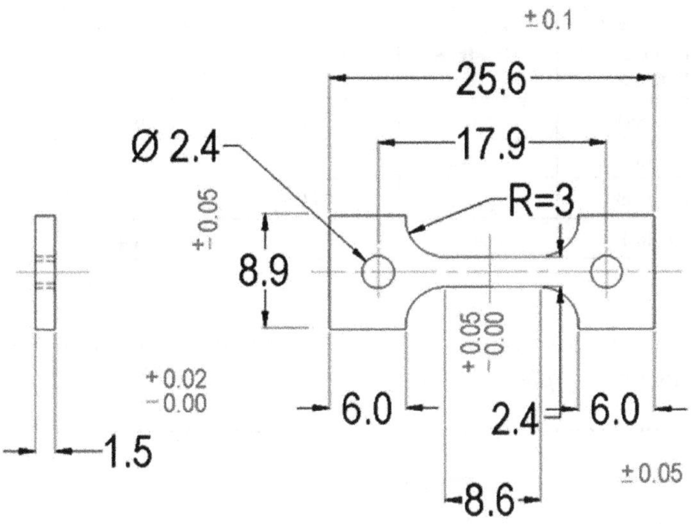

Figure 8. Engineering stress–strain graphs from Zr-2 transverse plate tensile specimens

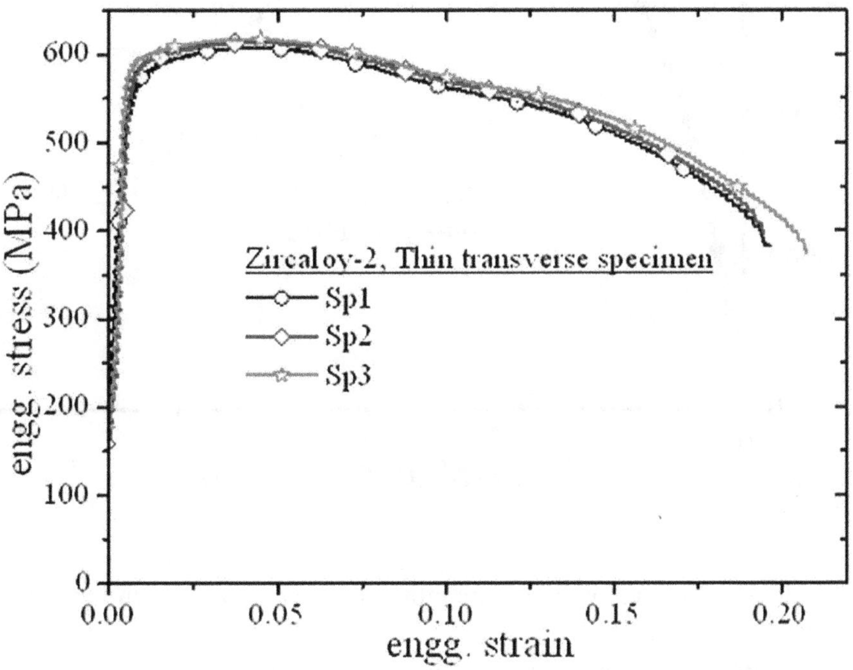

PT at room temperature can be summarized as shown below for understanding the entire output in the simplest manner.

For Zr-2.5Nb pressure tube:

1. (2piece, 12'O clock) combination gives the highest strength values, but is nearer to that shown by transverse miniature tensile specimen.

Table 5. Results of ring tensile test carried on Zr-2.5%Nb pressure tube material at ambient temperature

Type of Ring Specimen	Position of GL around the Mandrels	Type of Mandrels	0.2% YS (MPa)	UTS (MPa)	% Uniform Ductility	% Total Ductility
SGL	12'O clock	2 piece	759.23	916.82	2.70	21.21
SGL	3'O clock	2 piece	738.10	902.12	3.03	20.00
SGL	12'O clock	3 piece	711.66	799.54	3.94	25.45
SGL	3'O clock	3 piece	670.94	758.44	5.76	26.06

Figure 9. Load-displacement graph of Zr-2.5Nb from ring tension test at room temperature with its GL positioned at 12'O clock using 2 piece mandrel

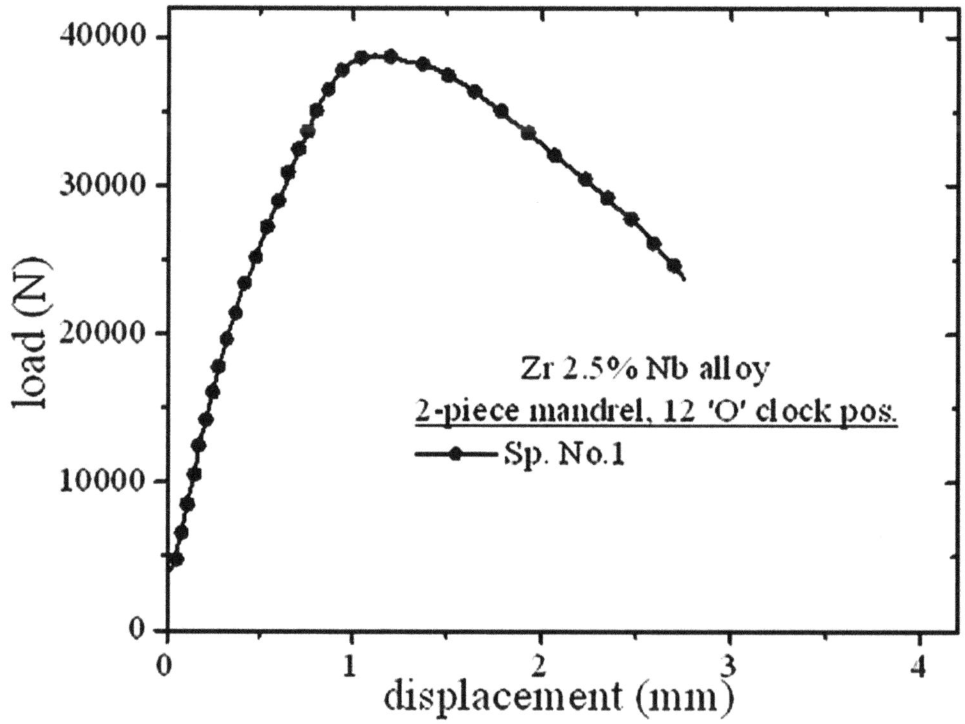

2. (2piece, 3'O clock) combination gives strength values which are much lower than that of trans. miniature tensile specimens.
3. (3piece, 12'O clock) combination gives strength values lower than those corresponding to that of case (i).
4. (3piece, 3'O clock) combination gives strength lowest of all.

It is clear from the above discussion that the grips used with 2-piece type of mandrels furnish relatively higher values of strength in comparison with 3-piece type of mandrels in the case of Zr-2.5Nb PT irrespective of the positions at which the GL of the sample was placed around the mandrels. It is evident

Table 6. Room temperature test results on Zr-2.5Nb transverse miniature tensile specimens

Specimen No.	Type of Specimen	0.2% YS (MPa)	UTS (MPa)	% Uniform Ductility	% Total Ductility
1	Plate tensile	768.00	913.51	1.7	14.94
2	"	777.96	939.38	1.5	14.25
3	"	768.84	915.37	1.4	12 .43
4	"	754.00	911.27	1.86	11.8
5	"	757.80	883.23	1.81	11.49
6	"	768.00	921.69	1.51	15.73
7	"	748.00	932.74	1.51	16.94
Average		763.23	916.74	1.61	13.94

that the 2-piece, 12'O clock GL position combination gives strength values which are closer to that of transverse miniature tensile specimens.

EVALUATION OF ENGINEERING STRESS-STRAIN DATA

From the experiment, the load-displacement curves are obtained directly after incorporating corrections for machine compliance in the displacement data. Uniform ductility is calculated from the displacement data as obtained from the experiment. The load-displacement data of the four types of experiments are presented in Figures 4-6 and 9 respectively. The uniform ductility has been calculated from the displacement corresponding to maximum load in the load-displacement diagram and the total or final ductility has been calculated from the final displacement corresponding to the point of fracture of the specimen. The expressions for calculation of these ductility values are given below.

$$Uniform\ ductility = \frac{2d_{\max load}}{l_{gauge}}$$

$$Total\ ductility = \frac{2d_{fracture\ load}}{l_{gauge}}$$

where $d_{\max load}$ is the displacement corresponding to maximum load, $d_{fracture\ load}$ is the displacement corresponding to fracture load, and l_{gauge} is the length of gauge length region of the specimen. The engineering stress σ_{engg} and engineering strain ε_{engg} values have been calculated from the experimental load P and displacement data δ, gauge length l_{gauge}, area of cross-section of the gauge length region A (6 mm x4.3 mm) using the following expressions.

$$\sigma_{engg} = \frac{P}{2A}$$
$$\varepsilon_{engg} = \frac{2\delta}{l_{gauge}}$$

FINITE ELEMENT ANALYSIS OF RING-TENSION SETUP

A 3-D finite element model of the ring-tension setup has been prepared. Keeping in mind of the symmetric boundary conditions, only one-half of the test-setup has been modeled as shown in Fig. 10(a) and 10(b) for the 2-piece and 3-piece type of loading mandrels respectively. The FE mesh uses 3-D 20-noded iso-parametric brick elements with tri-quadratic interpolation functions for the nodal variables (i.e., displacements in three Cartesian co-ordinate directions). The von Mises yield condition along with the Prandtl-Reuss flow rule have been used as the constitutive model for the elastic-plastic FE analysis.

The mandrels are in contact with the ring-specimen during loading and hence, surface-to-surface contact elements between the loading mandrels and the specimen have been used in the analysis. In the experiment, Teflon tapes are introduced between the mandrels and the specimen in order to reduce the co-efficient of friction between the contacting surfaces. A friction co-efficient value of 0.02 has been used in this FE analysis which is typical for the sliding condition existing between Teflon and the metallic materials. The mandrel is made up of SS403 stainless steel. The Young's modulus and Poisson's ratio for the mandrel material at room temperature has been taken as 210 GPa and 0.3 respectively. The Young's modulus and Poisson's ratio for the pressure tube materials at room temperature have been taken as 92 GPa and 0.35 respectively in this FE analysis.

Figure 10. FE mesh of the ring-tension setup: SGL specimen with (a) 2-piece loading mandrel, (b) 3-piece loading mandrel. 3-D elements for both specimen and mandrel with contact elements at the interface have been used in the analysis.

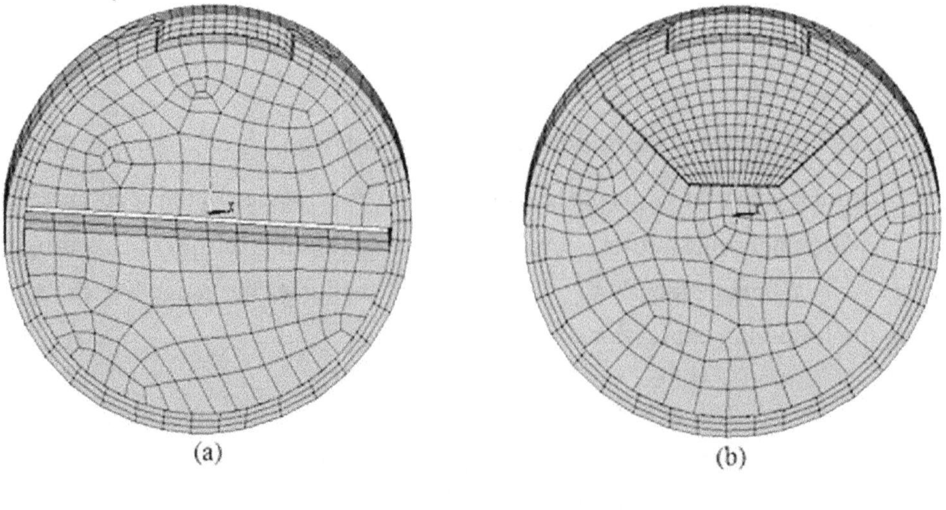

RESULTS AND DISCUSSION

In order to perform elastic-plastic finite element analysis, a material data in terms of true stress vs. true strain curve is required. However, the purpose of this work is to determine the transverse material stress-strain curve from the experimental load-displacement data and FE analysis. The 3-D model of FE analysis takes care of the specimen geometry and stiffness, mandrel geometry and stiffness, contact and sliding boundary conditions between specimen mandrel, contact and sliding boundary conditions between parts of 2-piece and 3-piece mandrel etc. on the load-displacement behavior of the test setup with only unknown being the material true stress vs. true plastic strain curve. For this purpose, an initial guess of the material stress strain curve is used as input and an optimization algorithm is employed for calculation of actual material data. This algorithm uses the modulus of difference between the experimental and FE calculated load for a given value of applied mandrel displacement as the optimization function and updates the true stress value for the given value of true plastic strain in the gauge length region of the specimen. Usually 8 to 10 iterations are needed in a particular time step for convergence. The tolerance in the ratio of error in load to experimental load has been taken as 1.0e-4.

It may be observed that the results with friction co-efficient of 0.02 closely compares with experimental data whereas results with friction co-efficient of 0.05 and 0.1 overestimate the maximum load as well as hardening behavior of the specimen in the test setup. This value of 0.02 is also similar to those reported for surfaces between Teflon and metallic materials in Samyn et al. (2006). Similarly, the 3-piece mandrel test setup was also analyzed with the FE model and the load-displacement data as obtained from FE analysis is optimized with that of experiment to determine the material stress-strain curve.

Figure 11. Engineering stress-strain curves of pressure tube material of Indian PHWR (Zircaloy-2 and Zr2.5%Nb) as obtained from inverse FE analysis

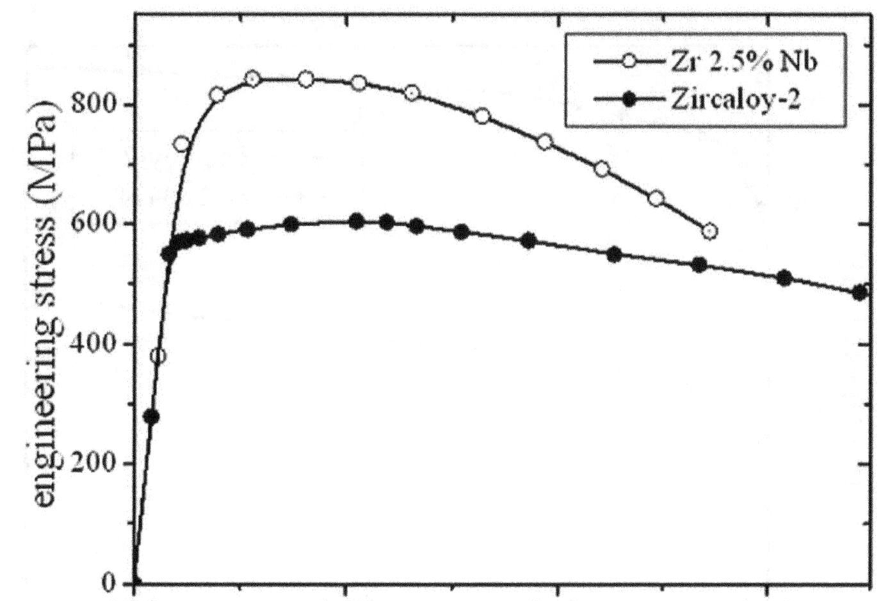

The engineering and true stress-strain curves as obtained from the above procedure are shown in Figures 11 and 12 respectively. The yield stress (YS) and ultimate tensile strength (UTS) values of Zircaloy-2 are given as 554 MPa and 605.8 MPa respectively. The values of the above parameters are 720 MPa and 842 MPa respectively for the material Zr-2.5%Nb (Figure 11). It may be noted from Figures 11 and 12 that Zr-2.5%Nb exhibits higher plastic strain hardening before the strength level reaches UTS compared to Zircaloy-2. However, plastic strain value at UTS for Zr-2.5%Nb is significantly lower (i.e., 0.028) compared to that of Zircaloy-2 (i.e., 0.047) which indicates that Zr-2.5%Nb has a lower uniform ductility compared to that of Zircaloy-2 (Figure 11). In the post-necking region of deformation, the plastic strain hardening behavior of Zr-2.5%Nb is lower compared to that of Zircaloy-2 as can be seen from the slopes of the material stress-strain curves (beyond the UTS region) in Figure 12. The material stress-strain curves as depicted in Figure 12 have been used in the FE analysis as an input.

The true-stress vs. true plastic strain curves as obtained from the optimization procedure followed for ring-type specimens have also been compared with the experimental true stress-strain curve as obtained from the flat tensile specimens in order to check the effect of loading conditions on the material data. The compared material stress-strain curves for Zircaloy-2 fuel-clad tubes (at room temperature) have been presented in Figure 13. The scatter-band is of the order of 2.5 to 3% for the ring specimens. The corresponding data from the flat tensile specimen falls within the scatter of results of the ring type specimens. Hence, it can be concluded that the procedure presented in this work can be used satisfactorily to evaluate the material stress-strain curve from ring tensile tests with the use of nonlinear finite element analysis and the results are same for flat and ring specimens within a small scatter band and hence, independent of the test methods. The repeatability of the data is also good with a very low scatter

Figure 12. True stress-strain curves of pressure tube material of Indian PHWR (Zircaloy-2 and Zr2.5%Nb) as obtained from inverse FE analysis

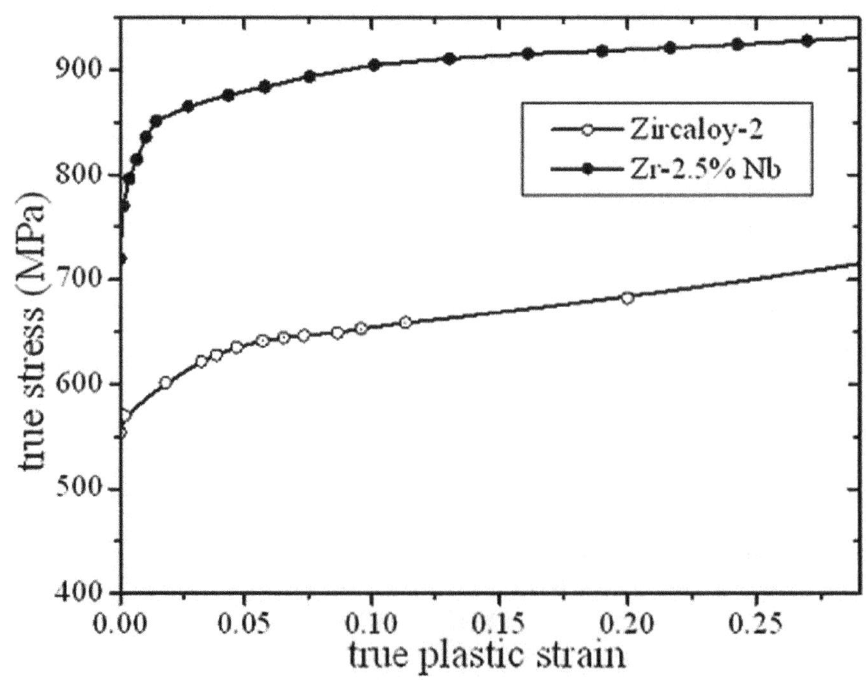

Figure 13. True stress vs. true plastic strain data (at room temperature) of the material Zircaloy-2 as obtained from the optimization procedure for different types of specimens and its comparison with flat tensile specimen

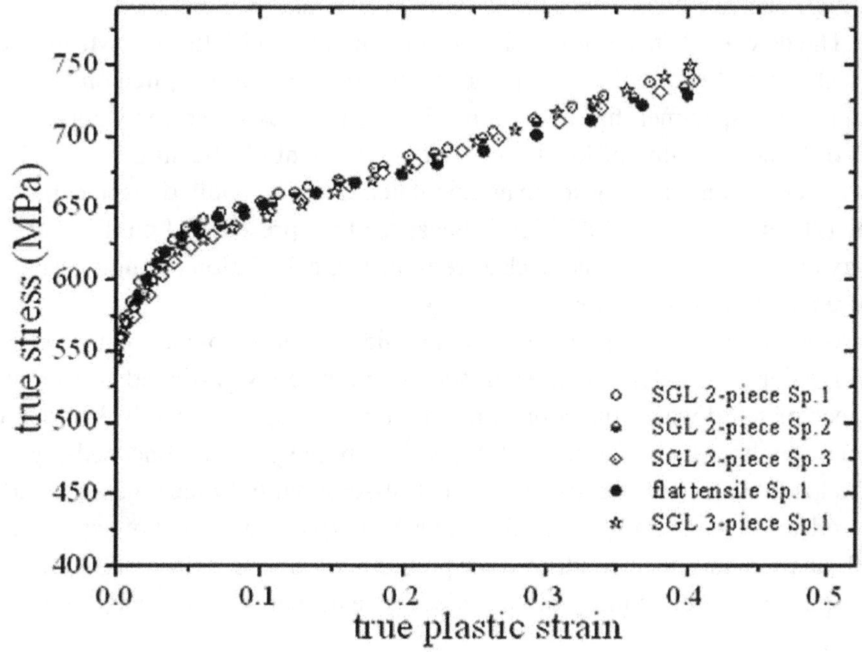

Figure 14. von Mises equivalent plastic strain contour in the Zr2.5%Nb ring-tensile specimen loaded with 3-piece mandrel for an applied displacement of 3 mm

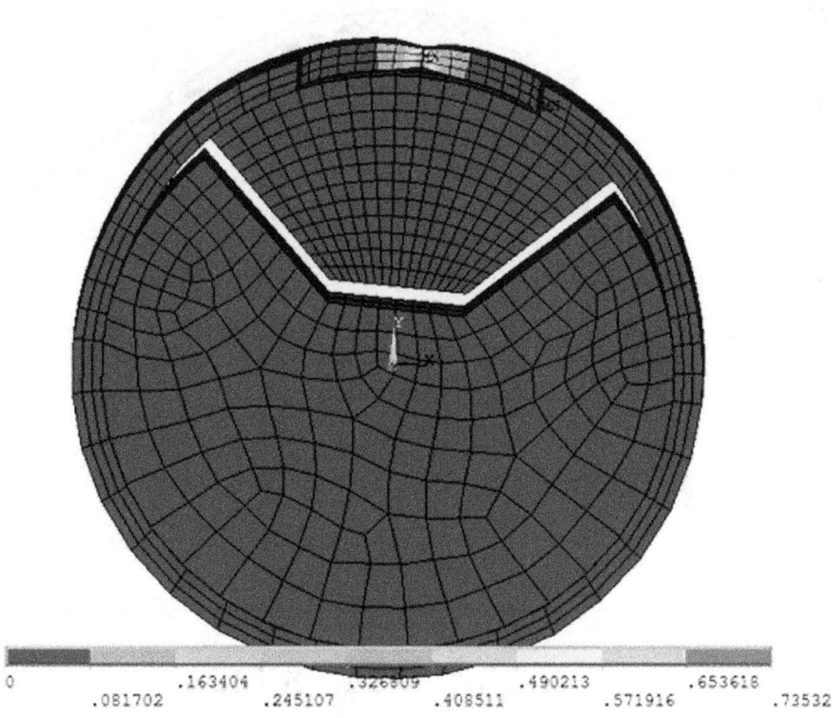

band. This data as presented in Fig. 13 can be used for integrity analysis and remaining life assessment of the nuclear reactor pressure tubes under different postulated loading conditions.

In order to validate the above procedure, the results of FE analysis is presented here for the Zr2.5%Nb pressure tube alloy. The results of ring loaded with 2-piece and 3-piece loading mandrels have been presented here. The deformed shape of the ring-specimen along with the von Mises equivalent plastic strain contour is shown in Figure 14 for a displacement value of 3 mm applied on the loading mandrel. It may be noted that the specimen has undergone significant necking at this value of applied mandrel displacement and the actual value of local von Mises equivalent plastic strain is 73.5% approximately (Figure 14). However, the engineering strain as calculated from the applied displacement (2x3 mm) and the gauge length (20 mm) is around 30%. This is because of the presence of multi-axial deformation and hence, a multi-axial state of stress in the necked-region. The calculation of engineering strain (as 30%) is valid for a pure uni-axial deformation case.

Figure 14 also shows that the contact between the middle loading mandrel and the specimen breaks on both sides of loading axis and hence, the contact conditions are significantly different in comparison to that of a 2-piece mandrel where the contact remain intact in approximately 180-degree angle of the loading mandrel throughout the deformation history (except the portion of necked-region after necking has started). The load-displacement response of the test-setup with 3-piece mandrel and the Zr2.5%Nb ring-specimen as determined from FE analysis has been compared with that of experiment in Figure 15.

Similarly, FE analysis has been conducted for predicting the load-deformation behavior of the SGL Zr2.5%Nb ring-specimen when loaded with a 2-piece type of mandrel, the gauge-length being oriented

Figure 15. Load-displacement response of the Zr2.5%Nb ring-tension specimen with 3-piece loading mandrel and gauge-length oriented perpendicular to loading axis (exp vs. FE analysis)

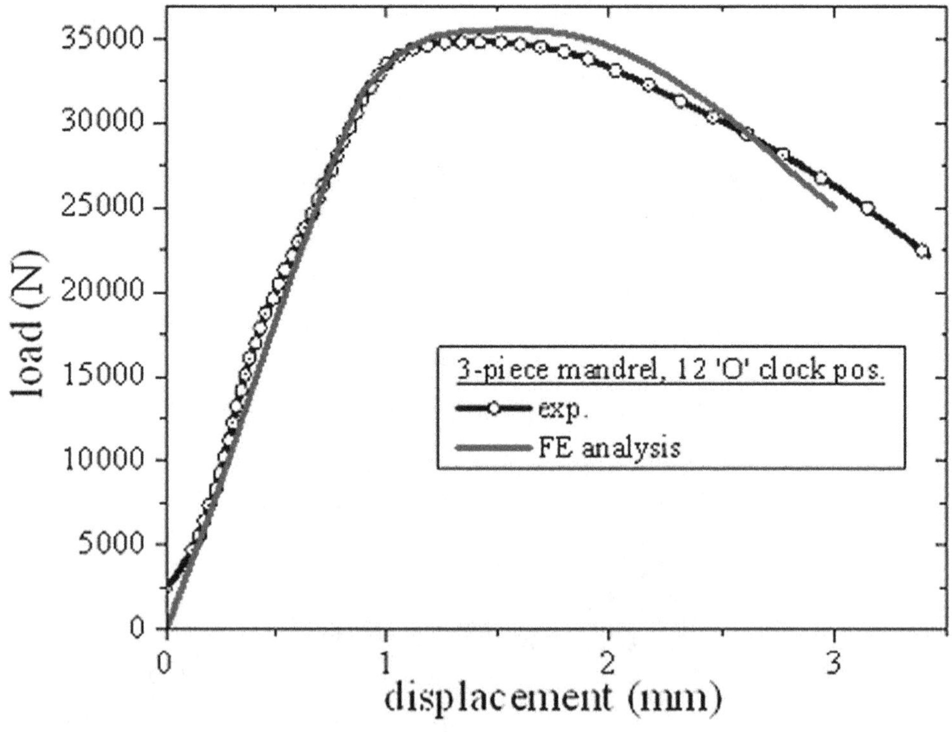

Figure 16. von Mises equivalent plastic strain contour in the Zr2.5%Nb ring-tensile specimen loaded with 2-piece mandrel for an applied displacement of 3 mm

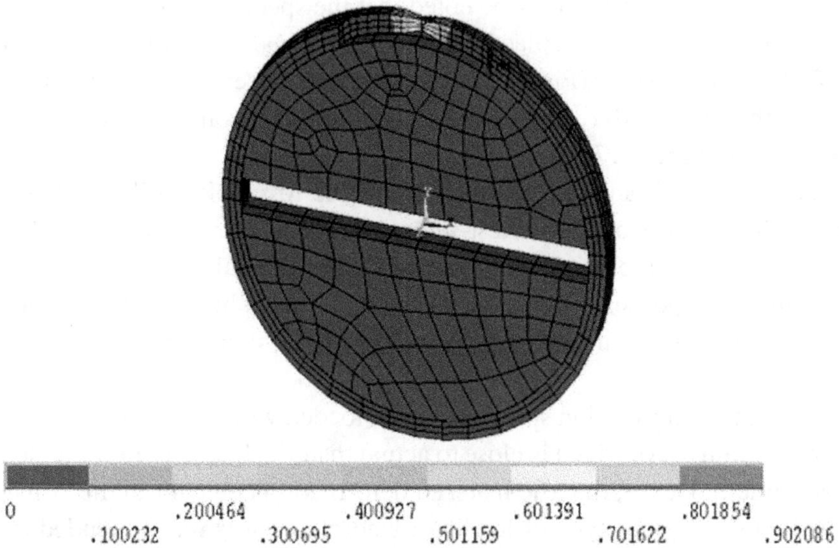

Figure 17. Load-displacement response of the Zr2.5%Nb ring-tension specimen with 2-piece loading mandrel and gauge-length oriented perpendicular to loading axis (exp vs. FE analysis)

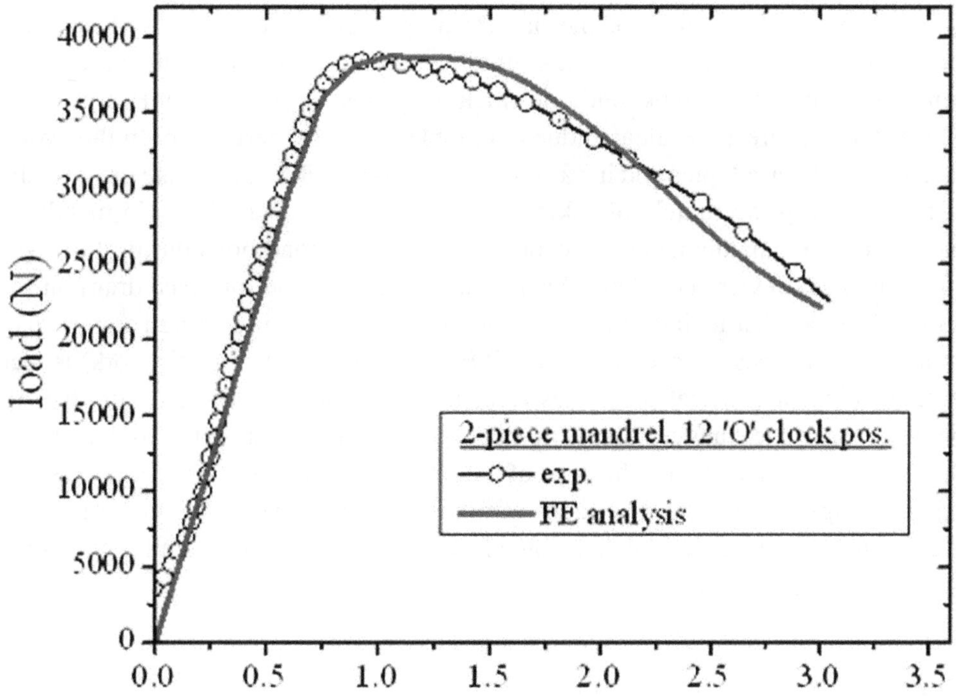

perpendicular to the loading axis. The deformed shape of the ring-specimen along with the von Mises equivalent plastic strain contour for the above case is shown in Figure 16 for a displacement value of 3 mm applied on the loading mandrel. It may be noted that the specimen has undergone significant necking at this value of applied mandrel displacement and the actual value of local von Mises equivalent plastic strain is 90.2% approximately (Figure 16). The load-displacement response of the test-setup with 2-piece mandrel and the Zr2.5%Nb ring-specimen as determined from FE analysis has been compared with that of experiment in Figure 17.

It may be noted that the FE analysis has also been able to satisfactorily predict the experimental load-displacement behavior of two types of test-setups. Hence, the procedure is validated and this gives confidence in the use of this combined FE analysis and experimental procedure for accurate estimation of transverse mechanical properties of the Zircaloy pressure tubes in the as-manufactured as well as in the service-exposed conditions, which is very crucial for the design as well as safety analysis of these components.

It was observed that the ring-tension specimen when loaded with a 3-piece type of mandrel provides an engineering stress-strain curve which is close to actual material behavior (as obtained from a flat pure uniaxial tensile specimen). Hence, the true material deformation behavior in the transverse direction is closely represented by the 3-piece type of loading mandrel and it is recommended to use this type of loading setup when the transverse material properties are obtained from ring-tension tests.

CONCLUSION

Use of ring-specimens for evaluation of mechanical properties in the as-received as well as service-exposed conditions of nuclear reactor pressure tubes is very attractive due to the ease of machining of the specimens and testing in different conditions including those of irradiation. However, the mechanical properties cannot be directly evaluated due to complexity of the test-setup. In this work, a 3-D FE analysis procedure has been adopted and it was shown that it is possible to estimate the load-displacement response of the test-setups accurately by taking into account of the variability in mandrel design and contact conditions. The dynamic nature of evolution of contact conditions during loading and hence, the effect of additional stresses (i.e., those due to lateral pressure and interface drag) on the yield and subsequent material strain-hardening behavior cannot be taken care of in standard empirical equations as used for uni-axial tensile tests. Hence, use of a 3-D FE analysis (as used in this work) is recommended for accurate estimation of material properties from these ring-tension tests. Again, the true material deformation behavior in the transverse direction is closely represented by the 3-piece type of loading mandrel and it is recommended to use this type of loading setup when the transverse material properties are obtained from ring-tension tests. The scope of future work includes assessment of the above procedure for other engineering materials such as low-alloy steels and for other tubular components used in different engineering applications.

REFERENCES

Arsene, S., & Bai, J. (1996). A New Approach to measuring transverse properties of structural tubing by a ring test. *Journal of Testing and Evaluation*, *24*(6), 386–391. doi:10.1520/JTE11461J

Bae, B.-K., Cho, S.-K., & Seok, C.-S. (2008). A study on ring tensile specimens. *Materials Science and Engineering A*, *483-484*, 248–250. doi:10.1016/j.msea.2006.10.203

Balakrishnan, K. S., Samal, M. K., Parashar, J., Tiwari, G. P., & Anantharaman, S. (2014). Suitability of miniature tensile specimens for estimating the mechanical property data of pressure tubes- An assessment. *Transactions of the Indian Institute of Metals*, *67*(1), 47–55. doi:10.1007/s12666-013-0316-0

Bertolino, G., Meyer, G., & Perez Ipiña, J. (2003). In-situ crack growth observation and fracture toughness measurement of hydrogen charged zircaloy-4. *Journal of Nuclear Materials*, *322*(1), 57–65. doi:10.1016/S0022-3115(03)00305-2

Bertsch, J., & Hoffelner, W. (2006). Crack resistance curve determination of tube cladding material. *Journal of Nuclear Materials*, *352*(1-3), 116–125. doi:10.1016/j.jnucmat.2006.02.045

Deshmukh, V., Singha, U., Tonpe, S., & Saibaba, N. (2010). Effect of process variables on texture and creep of pressurized heavy water reactor pressure tubes. *Transactions of the Indian Institute of Metals*, *63*(2-3), 397–402. doi:10.1007/s12666-010-0054-5

Edsinger, K., Davies, J. H., & Adamson, R. B. (2000). Degraded fuel cladding fractography and fracture behaviour. In *Ziconium in nuclear industry* (pp. 316-339). ASTM.

Fukuda, T., Itagaki, N., Kamimura, J., Mozumi, Y., & Furuya, T. (2003, March 16-19). Fracture toughness of hydrogen-absorbed zircaloy-2 cladding tube. Proceedings of Topfuel 2003, Wurzburg, Germany.

Grigoriev, V., & Jakobsson, R. (2000, November). Application of the Pin Loading Tension test to measurements of Delayed Hydride Cracking Velocity in Zircaloy cladding.

Grigoriev, V., & Jakobsson, R. (2005). Delayed Hydrogen Cracking Velocity and J-Integral Measurements on Irradiated BWR Cladding. *Journal of ASTM International*, *2*(8), 1–16. doi:10.1520/JAI12434

Grigoriev, V., Josefsson, B., Lind, A., & Rosborg, B. (1995). A pin-loading tension test for evaluation of thin-walled tubular materials. *Scripta Metallurgica et Materialia*, *33*(1), 109–114. doi:10.1016/0956-716X(95)00110-H

Grigoriev, V., Josefsson, B., & Rosborg, B. (1996). Fracture Toughness of Zircaloy Cladding Tubes. Zirconium In nuclear Industry. *Proc. of 11th Int. Symposium ASTM STP* (pp. 431-447). doi:10.1520/STP16184S

Grigoriev, V., Josefsson, B., Rosborg, B., & Bai, J. (1997). A novel fracture toughness testing method for irradiated tubing-experimental results and 3D numerical evaluation.*Transactions of the 14th Int. Conference on Structural Mechanics in Reactor Technology*, Lyon, France (pp. 57-64).

Hsu, H. H., Chien, K. F., Chu, H. C., & Liaw, P. K. (2001). In R. Chona (Ed.), *An X-specimen test for determination of thin walled tube fracture toughness. Fatigue and Fracture Mechanics: 32nd Volume, ASTM STP 1406* (pp. 214–226). West Conshohocken, PA: Americal Society for Testing and Materials.

Josefsson, B., & Grigoriev, V. (1996, May 22). Modified Ring Tensile Testing and a New Method for Fracture Toughness Testing of Irradiated Cladding. Studsvik Material, Nykoping, Sweden.

Leclercq, S., Parrot, A., & Leroy, M. (2008). Failure characteristics of cladding tubes under RIA conditions. *Nuclear Engineering and Design*, *238*(9), 2206–2218. doi:10.1016/j.nucengdes.2008.03.017

Lee, K. W., & Hong, S. I. (2002). Zirconium hydrides and their effect on the circumferential mechanical properties of Zr-Sn-Fe-Nb tubes. *Journal of Alloys and Compounds*, *346*(1-2), 302–307. doi:10.1016/S0925-8388(02)00527-3

Lee, K. W., Kim, S. K., Kim, K. T., & Hong, S. I. (2001). Ductility and strain rate sensitivity of Zircaloy-4 nuclear fuel claddings. *Journal of Nuclear Materials*, *295*(1), 21–26. doi:10.1016/S0022-3115(01)00509-8

Martín-Rengel, M. A., Gómez Sánchez, F. J., Ruiz-Hervías, J., Caballero, L., & Valiente, A. (2009, September 6-10). New Method to Calculate the Mechanical Properties of Unirradiated Hydrogen-Charged Fuel Cladding from Ring Tensile Tests. *Proceedings of the Water Reactor Fuel Performance Meeting (WRFPM Top Fuel 2009)*, Paris, France.

Martín-Rengel, M. A., Gómez Sánchez, F. J., Ruiz-Hervías, J., Caballero, L., & Valiente, A. (2012). Revisiting the method to obtain the mechanical properties of hydride fuel cladding in the hoop direction. *Journal of Nuclear Materials*, *429*(1-3), 276–283. doi:10.1016/j.jnucmat.2012.06.003

Murty, K. L. (1999). *Texture based physical, mechanical and corrosion characteristics of Zirconium alloys. Textures in Materials Research* (pp. 113–126). New Delhi: Oxford and IBH co. Ltd.

Murty, K. L. (2002, September 11-13). Deformation, texture and microstructure- property correlation and in-reactor performance of Zircaloy cladding in LWRs.*Proceedings of the international symposium on Advances in Zirconium*, Mumbai, India (pp. 28-45).

Nakatsuka, M., Kubo, T., & Hayashi, Y. (1991). Fatigue Behavior of Neutron Irradiated Zircaloy-2 Fuel Cladding Tubes. In *Ziconium in the Nuclear Industry* (pp. 230-245). American Society for Testing and Materials. doi:10.1520/STP25510S

Nikulina, A. V. (2003). Zirconium-Niobium Alloys for Core Elements of Pressurized Water Reactors. *Metal Science and Heat Treatment*, *45*(7-8), 287–292. doi:10.1023/A:1027388503837

Nikulina, A. V. (2004). Zirconium alloys in nuclear power engineering. *Metal Science and Heat Treatment*, *46*(11-12), 458–462. doi:10.1007/s11041-005-0002-x

O' Brien, D. M., & Ferguson, W. G. (1983). Modified manjoine WOL specimen for J_{IC} and δ_i tests. *Engineering Fracture Mechanics*, *18*(2), 453–465. doi:10.1016/0013-7944(83)90153-4

Rashid, Y. R., Lemaignan, C., & Strasser, A. (2000, April). Evaluation of fracture toughness and extension in fuel cladding. American Nuclear Society, Park City, UT.

Sainte Catherine, C., Le Boulch, D., Carassou, S., Ramasubramanian, N., & Lemaignan, C. (2006). An internal conical mandrel technique for fracture toughness measurements on nuclear fuel cladding. *ASTM Journal of Testing and Evaluation*, *34*(5), 373–382.

Samal, M. K., Balakrishnan, K. S., & Balakrishnan, S. (2015). A practical approach to evaluate stress-strain behavior of remotely handled pressure tubes of nuclear reactors using ring tension test. *Transactions of the Indian Institute of Metals*, *68*(2), 299–310. doi:10.1007/s12666-014-0461-0

Samal, M. K., Balakrishnan, K. S., Bhasin, V., Vaze, K. K., & Anantharaman, S. (2012, December 10-12). A combined experimental and FE analysis procedure to evaluate tensile behavior of Zircaloy pressure tubes. *Proceedings of theInternational Congress on Computational Mechanics and Simulation (ICCMS)*, IIT Hyderabad, India.

Samal, M. K., Balakrishnan, K. S., Parashar, J., & Tiwari, G. P. (2014). Investigation of deformation behaviour of ring-tensile specimens machined from pressure tubes of Indian PHWR. *Transactions of the Indian Institute of Metals*, *67*(2), 167–176. doi:10.1007/s12666-013-0314-2

Samal, M. K., Balakrishnan, K. S., Parashar, J., Tiwari, G. P., & Anantharaman, S. (2013). Estimation of tensile behaviour of Zirconium alloy pressure tubes using ring tensile test and finite element analysis. *Part C. Journal of Mechanical Engineering Science*, *227*(6), 1177–1186. doi:10.1177/0954406212460474

Samal, M. K., Sanyal, G., & Chakravartty, J. K. (2010a). Experimental and numerical study of fracture behaviour of tubular specimens in a Pin-Loading-Tension setup. *IMechE Part C Journal of Mechanical Engineering Science*, *224*(1), 1–12. doi:10.1243/09544062JMES1781

Samal, M. K., Sanyal, G., & Chakravartty, J. K. (2010b). Estimation of fracture behaviour of thin walled nuclear reactor fuel pins using pin-loading tension (PLT) test. *Nuclear Engineering and Design*, *240*(12), 4043–4050. doi:10.1016/j.nucengdes.2010.09.030

Samyn, P., De Baets, P., Schoukens, G., & Van Peteghem, A. P. (2006). Large-scale tests on friction and wear of engineering polymers for material selection in highly loaded sliding systems. *Materials & Design*, *27*(7), 535–555. doi:10.1016/j.matdes.2004.12.021

Seok, C. S., Bae, B. K., Koo, J. M., & Murty, K. L. (2006). The properties of the ring and burst creep of ZIRLO cladding. *Engineering Failure Analysis*, *13*(3), 389–397. doi:10.1016/j.engfailanal.2005.02.009

Shah, P. K., Samal, M. K., Dubey, J. S., Shriwastaw, R. S., Balakrishnan, K. S., Anantharaman, S., & Chakravartty, J. K. (2011, November6-11). Finite element study on mechanical behaviour of Zircaloy fuel clad tubes.*Proceedings of the 21st International conference on Structural Mechanics in Reactor Technology (SMiRT21)*, New Delhi, India.

Videm, K., & Lunde, L. (1979). Stress corrosion cracking and growth formation of pellet-clad interaction defects. In *Ziconium in nuclear industry* (pp. 230-237). Americal Society for Testing and Materials.

Wang, Y., & Murty, K. L. (1998, June 21-25). Effect of temperature on mechanical anisotropy of recrystallized Zircaloy-4 sheet.*Proc. 4th Asia-Pacific symposium on Advances in Engineering plasticity and its applications, AEPA 98*, Seoul, South Korea.

Yagnik, S. K., Ramasubramanian, N., Grigoriev, V., Sainte-Catherine, C., Bertsch, J., Adamson, R., & Oberländer, B. C. (2008). Round-Robin Testing of Fracture Toughness Characteristics of Thin-Walled Tubing. *Journal of ASTM International*, *5*(2), 1–21. doi:10.1520/JAI101140

Yoshitake, T., Abe, Y., Akasaka, N., Ohtsuka, S., Ukai, S., & Kimura, A. (2004). Ring-tensile properties of irradiated oxide dispersion strengthened ferritic/martensitic steel claddings. *Journal of Nuclear Materials*, *329-333*, 342–346. doi:10.1016/j.jnucmat.2004.04.084

Chapter 5
Dynamic Analysis of Offshore Wind Turbine Structures

Arundhuti Banerjee
Indian Institute of Technology (IIT), India

Vasant Matsagar
Indian Institute of Technology (IIT), India

Tanusree Chakraborty
Indian Institute of Technology (IIT), India

Tushar Kanti Datta
Indian Institute of Technology (IIT), India.

ABSTRACT

Wind turbines are slender flexible structures susceptible to strong wind fluctuations. The flexible wind turbine structure, when subjected to strong dynamic forces, it leads to an ideal condition for induced vibrations and resonance problems. Hence studying the dynamic response of these critical structures using the computational and experimental procedures becomes of utmost importance. This chapter reviews the theories used for the dynamic analysis of a modern day offshore wind turbine structure and applies these theories in analyzing realistic situations for offshore turbines under wave and wind action. The first half of the chapter gives a broad overview on the concepts of structural dynamics of wind turbine structures with illustrative examples that will enable the user to understand the methodology used to analyze these structures. The latter half of the chapter deals with the computational aspect of the analysis and focuses on the use of finite element software ANSYS 14 to model these critical structures.

INTRODUCTION

History of Onshore and Offshore Wind Turbines

The first windmills were constructed in Persia in the seventh century for irrigation purposes and for milling grain (Kaul et al., 2007). These old wind turbines were made of wood. The modern construction of windmills started at the time of industrial revolution. With the progress of industrial revolution, the industrialization sparked the development of larger windmills to generate electricity. The first electricity generating wind turbine was developed by Poul la Cour (2000). Later, larger and heavily engineered wind turbines were designed by Charles F. Brush from Cleveland, Ohio who designed a larger and heavily engineered wind turbine in 1888. The Brush wind turbine had a rotor of 17 m (56 feet) in diameter and it was mounted on an 18 m high (60 feet) tower.

DOI: 10.4018/978-1-5225-0588-4.ch005

Although gigantic by today's standards, the power rating of the turbine was only 12 kilowatt (kW); it rotated slowly as it had 144 blades. In the late 1930's, a megawatt-scale wind turbine generator was first time developed in the United States of America (USA) which made use of the latest technology. In 1941, a 1.25 megawatt (MW) Smith-Putnam wind turbine (Robert, 1996) was built in USA using the above mentioned turbine generator. Back in 1941 it was the largest wind turbine ever built and it kept its leading position for 40 years. In 1991, the first offshore wind farm was installed in Denmark. The wind farm consisted of eleven, 450 kW turbines.

At the dawn of 21st century, fossil fuel was still comparatively cheap. However, rising concerns over energy security, global warming, and eventual fossil fuel depletion led to development of interest in all available forms of renewable energy.

By 2014, over 240,000 commercial-sized wind turbines were operating in the world, producing 4% of the world's electricity. Total installed capacity exceeded 336 gigawatt (GW) in 2014 with China, the U.S., Germany, Spain and Italy leading in installations as reported by Anon, (1890), Global Wind Report (2014), American Bureau of Shipping (2014), American Petroleum Institute (2014), British Standard Europaische Norm. (2004). By 2015, the largest wind turbine of 8 MW capacity by Vestas named V164 was ready for offshore use (Global Wind Report, 2014).

Components of a Typical Wind Turbine

A wind turbine typically comprises of three blades connected by a rigid hub. The hub is attached to a nacelle containing the mechanical and electrical components of the turbine. The blade-hub-nacelle assembly is supported by a tower as shown in Figure 2. The nacelle houses some of the important electrical components used for wind energy conversion like rotor and shaft and is mounted on top of the tower. The electric current is then distributed by a transformer to the grid with the help of transmission lines.

Figure 1. (a) Charles Brush's windmill of 1888, used for generating (b) 1.250 MW Smith-Putnam wind turbine
Source: Anderson (2003).

Figure 2. Side-profile view of a typical wind turbine indicating main components

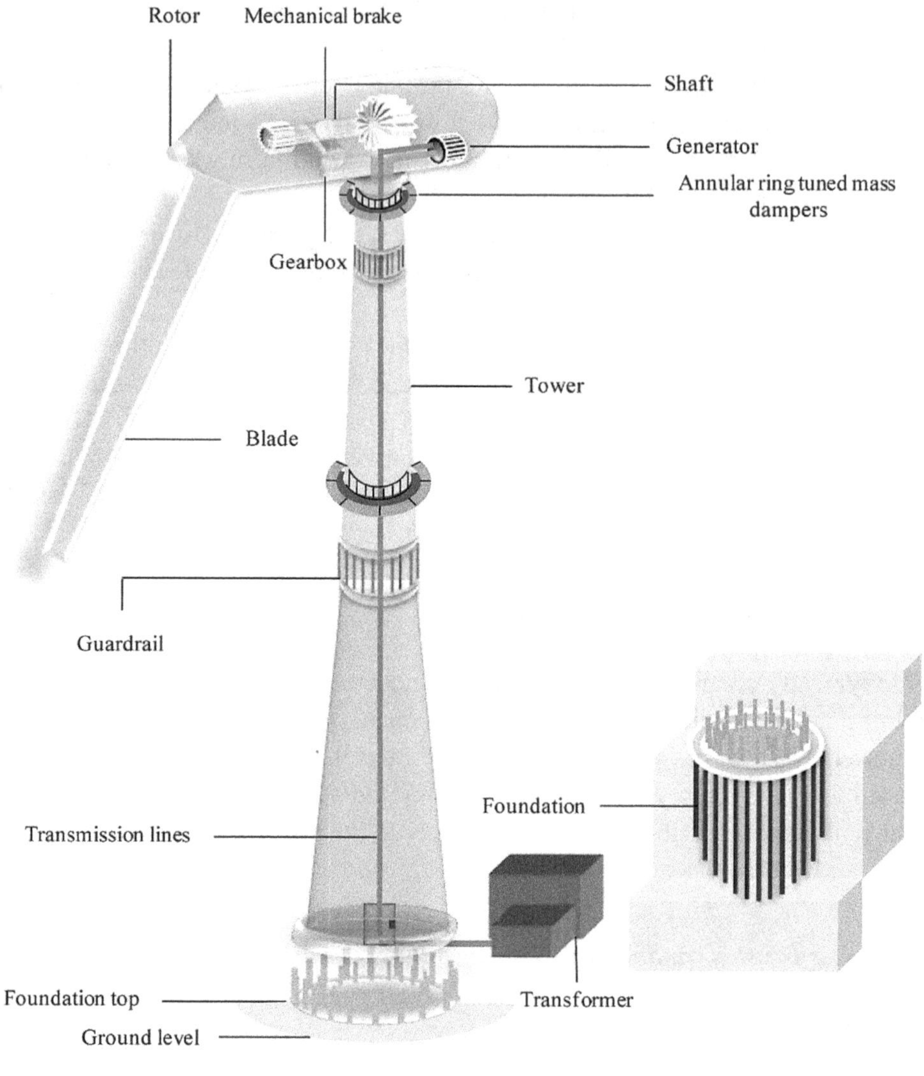

The hub connects the turbine blades to the main shaft. The nacelle is generally made of glass reinforced fiber or steel. Most modern wind turbines have conical towers made of steel. All modern wind turbines consist of a number of rotating blades looking like propeller blades. If the blades are connected to a vertical shaft, the turbine is called a vertical-axis wind turbine (VAWT), and if the shaft is horizontal, the turbine is called a horizontal-axis wind turbine (HAWT).

For commercial wind turbines the mainstream turbines mostly comprise HAWTs; the following study therefore focuses on this type of machine. Vibration control devices like annular ring tuned mass dampers as shown in Figure 2 are used to mitigate wind induced vibrations. The power in the wind is proportional to the cube of the wind speed. Considerable forces must therefore be controlled during high winds in order to attain safe operating conditions using brakes. Foundation arrangement for a typical wind turbine structure is presented in Figure 2.

Design Codes for Offshore and Onshore Wind Turbines

A number of design codes have been used over the past decade to model the wind turbine's dynamic behavior and to carry out design calculations. A short description and the features of the major design codes are presented as follows:

1. **ABS Guide (2014) :** Used for bottom founded offshore wind turbine installation.
2. **API (2014) :** Used for planning, designing and constructing fixed offshore platform.
3. **BS-EN50308 (2014)**: Used for wind turbines-protective measures-requirements for design operation and maintenance.
4. **DNV-OS-J101 (2004):** Used for design of offshore wind turbine structures.
5. **DNV-OS-J201 (2009):** Used for offshore substations for wind farms.
6. **GL Renewable Certifications (2005) :** Used for guideline for certifications of offshore wind turbines.
7. **IEC-61400-1 (2010):** Wind turbines design requirements.
8. **IEC-61400-3 (2010):** Wind turbines design requirements for offshore wind turbines.
9. **IEC-61400-22 (2010):** Wind turbines conformity testing and certification.
10. **IEC-61400-24 (2010):** Wind turbines lightning protection.
11. **ISO/CD 16079-1 (2014):** Condition monitoring and diagnostics of wind turbines.

LOADS ACTING ON AN OFFSHORE WIND TURBINE

Wind turbines are designed to be installed in an offshore environment several miles off the coast and in water depths even greater than 30 m. The loads on these systems, as demonstrated in Figure 3, are mainly dominated by aerodynamic and hydrodynamic effects. The general overview of aerodynamics, hydrodynamics and structural dynamics of an offshore wind turbine is provided here in this chapter. If readers wish to explore detailed information regarding the same, following textbooks given in Anderson (2003), Burton et al. (2001), Gasch (2007) and Rogers et al. (2002) are recommended. Out of all the loads acting on an offshore wind turbine system, the major loads which include wind and wave loadings have been discussed in detail in this chapter. For other loadings, the book by Faltinsen (1999) can be referred. The main loads acting on an offshore wind turbine are categorized under:

1. **Aerodynamic Loads:** The aerodynamic loads on the blades and the turbine can be subdivided into three categories given as:
   a. Steady aerodynamic forces generated by the mean wind speed of the wind,
   b. Periodic aerodynamic forces generated by wind shear, rotor rotation, off-axis winds and tower shadow
   c. Randomly fluctuating aerodynamic forces induced by wind turbulence and dynamic effects.
2. **Hydrodynamic Loads:** The significance of hydrodynamic load depends on the turbine component as well as on the wind and wave conditions. Aerodynamic and hydrodynamic effects are linked in terms of a long-term statistical correlation function in terms of wind speed, wave height, and wave period, which are expressed in the long-term joint probability density distribution (Jonkman,2007). The major loads acting on an offshore wind turbine as:

Figure 3. Different loads acting on an offshore wind turbine

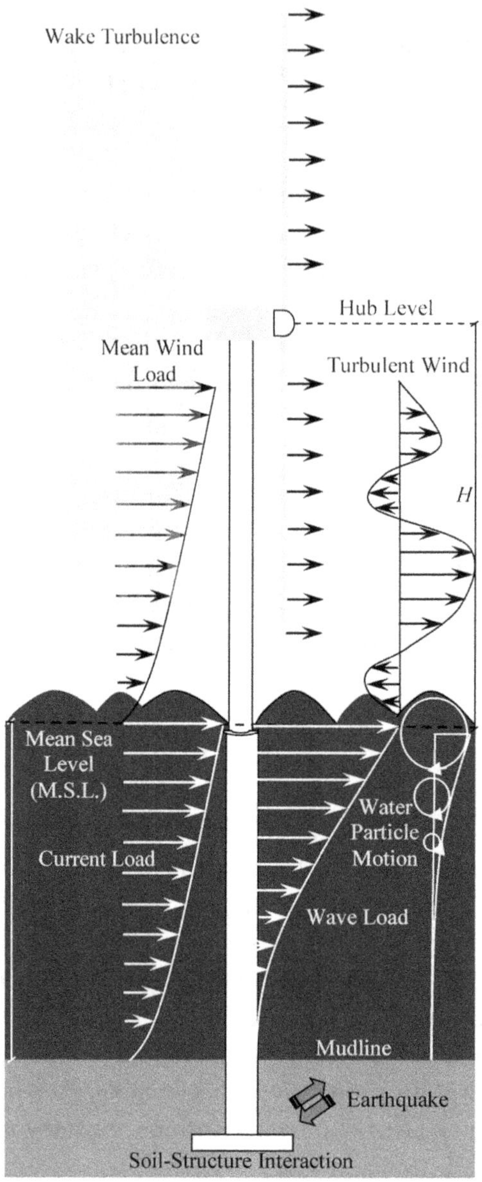

a. Steady hydrodynamic loads due to waves,
b. Randomly fluctuating hydrodynamic forces, and
c. Current loading.

Aerodynamics

Aerodynamics is necessary for modeling the loads and power output of a wind turbine. The governing equation for power extraction is given in Hansen (2008):

$$P = \vec{F} \cdot \vec{v} \tag{1}$$

where P is the power, F is the force vector and v is the velocity of the blades. The force F is generated by the wind transferring momentum to the blade. The primary focus of wind turbine aerodynamics is the magnitude and distribution of this force. Airflow over any surface creates two types of aerodynamic forces:

1. Drag forces which is in the direction of the airflow, and
2. Lift forces which is perpendicular to the airflow as shown in Figure 4. Either or both of these can be used to generate the forces needed to rotate the blades of a wind turbine.

When air is in motion, the kinetic energy in moving air can be evaluated as:

$$E_{\mathrm{k}} = \frac{1}{2} m\bar{v}^2 \tag{2}$$

where m is the mass of the air. The wind power can be obtained by differentiating the kinetic energy in wind with respect to time as given in Hansen (2008):

$$P = \frac{\partial E_{\mathrm{k}}}{\partial t} = \frac{1}{2} \dot{m}\bar{v}^2 \tag{3}$$

When wind passes through a wind turbine and drives the blades to rotate, the corresponding wind mass flow rate is given as Hansen (2008):

$$\dot{m} = \acute{A}a\bar{v} \tag{4}$$

where ρ is the air density and a is the swept area of blades. Substituting Equation 4 into Equation 3, the available power in wind P can be expressed as Hansen (2008):

Figure 4. Lift and drag force over an airfoil

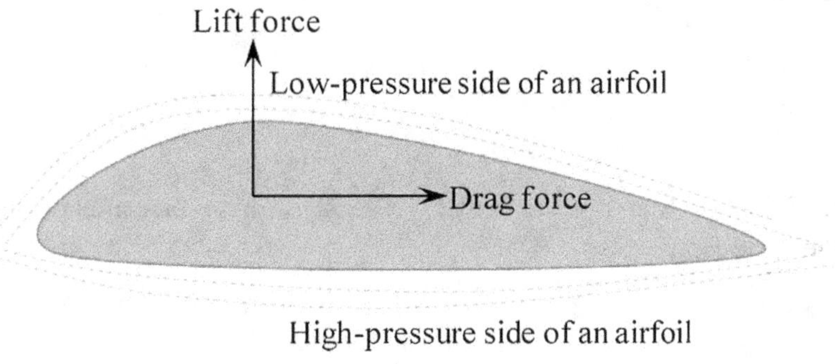

$$P = \frac{1}{2} m \bar{v}^3 \tag{5}$$

Aerodynamic Load

Steady Aerodynamic Load

- **Logarithmic Law:** Mean wind loads constitute the steady aerodynamic forces on a turbine. The wind speed is not constant over the height of the turbine, it increases with increasing height above ground and the rate of increase depends on the surface roughness parameter z_0. Due to the lower ground roughness parameter z_0 on seas (offshore), the wind speed $V(z)$ increases more rapidly with increasing height than it does onshore because of less friction encountered by the wind on seas. Mean wind speed is calculated using *logarithmic law* as given by Holmes (2010):

$$V_m = \frac{\upsilon_*}{K} \ln \frac{z}{z_0} \tag{6}$$

where V_m is the mean wind velocity at height z, υ_* is the friction velocity, K is Von Karman's constant (typically around 0.4), and z_0 is the roughness length provided in the codes.

- **Power Law:** Often the *power law* is used for wind-speed profile which is given in Holmes (2010):

$$V_m = V_{ref} \left(\frac{z}{z_{ref}} \right)^{\alpha} \tag{7}$$

where V_{ref} is the reference wind velocity at a particular height. It is usually obtained as the mean of the wind velocity on a 10 minutes' time period V_{10}. In the present research, according to the IEC 61400-3 offshore design standard [8,23], $\alpha = 0.14$ is used for all normal wind conditions and $\alpha = 0.11$ is used in extreme 1-year and 50-year return period wind conditions

The total drag force F along wind direction experienced by the wind turbine tower and the blade is given in Holmes (2010):

$$F = \frac{1}{2} C_D \rho A V_m^2 \tag{8}$$

where C_D is the coefficient of drag, ρ is the density of air, A is the surface area of the body and V_m is the mean velocity of the wind flow.

Periodic Aerodynamic Load

Wind shear generates loads on the rotor blades which vary periodically at the rotor frequency (rotation of the rotor due to blade movement) because each blade experiences periodically changing wind speeds during each full rotation (Burton et al., 2001). Constant off-axis winds cause a similar periodically varying load on the turbine and the blades. This oscillation excites the blade at a strong $1P$ (and multiples $2P$, $3P$, ...) frequency and the tower and nacelle at a $3P$ (and multiples $6P$, $9P$, ...) frequency. Here, the abbreviations $1P$, $2P$, $3P$, etc., refer to the rotor frequency and multiples thereof. Wind turbines generally are designed so that the major natural frequencies of the tower and platform modes are well above (usually 10%) or below these $1P$ and $3P$ frequencies to avoid resonances which affects the lifetime of these structures as shown in Figure 5.

Random Aerodynamic Load

The total wind force on any structural member is a summation of the mean and the fluctuating components. The total drag force along wind direction, $F(t)$, experienced by a slender structure is expressed by Davenport (1961):

$$F(t) = \frac{1}{2} C_D \rho A\, V(t)^2 = \frac{1}{2} C_D \rho A \left[V_m + v'(t) \right]^2 \qquad (10)$$

Figure 5. Wind turbine operation frequency range

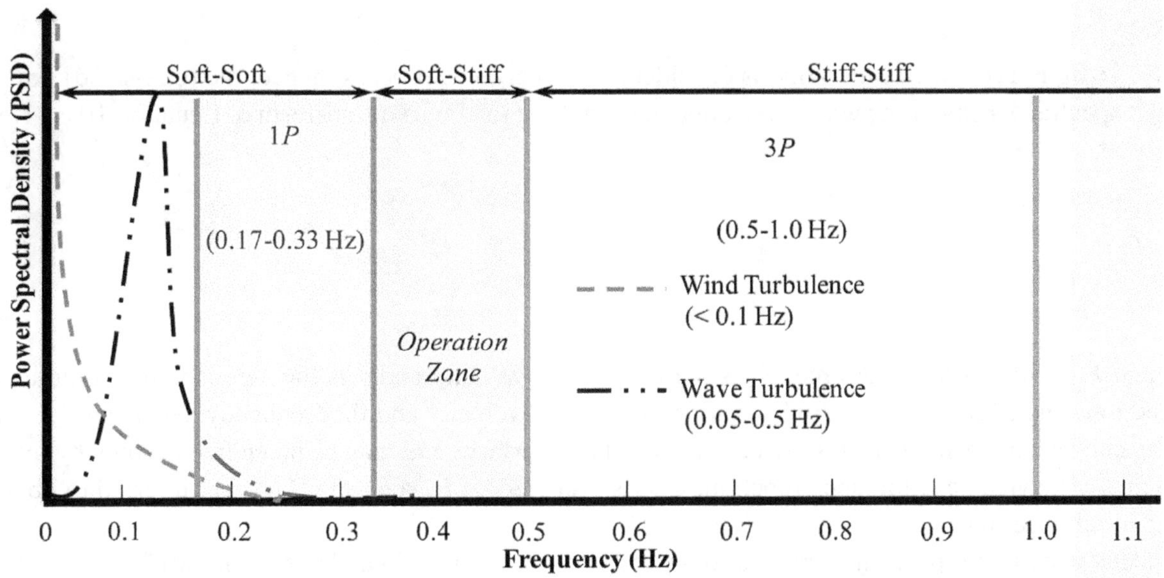

where C_D is the coefficient of drag, ρ is the density of air, A is the surface area of the body, and $V(t)$ is the total velocity of the wind flow. In realistic approach, the wind velocity $V(t)$, may be decomposed into a mean component V_m and a fluctuating component $v'(t)$. By neglecting higher powers of gust component in Equation 10, the total drag force in that case will be composed of a mean component F_m and a fluctuating component $F'(t)$ given in Davenport (1961):

$$F \cong \frac{1}{2} C_D \rho A (V_m)^2 + C_D \rho A V_m v'(t) = F_m + F'(t) \tag{11}$$

- **Kaimal Spectrum:** Different wind spectrums are used for the design of offshore structure to represent the random nature of the fluctuating wind velocity by a power spectral density function (PSDF) which is summarized in Holmes (2010). In the present chapter, Kaimal spectrum has been considered which is given as:

$$\frac{f S_{vv}(z,f)}{v_*^2} = \frac{200 n_m}{\left(1 + 50 n_m\right)^{5/3}} \tag{12}$$

where z is the elevation (m), $S_{vv}(z,f)$ is the power spectral density function of the fluctuating wind velocity as a function of elevation and frequency f in Hz, υ_* is the friction velocity, and n_m is known as the Monin coordinate. Monin coordinate is obtained from the expressions as:

$$n_m = \frac{f\,z}{V_m} \quad ; \quad V_m = \frac{\mathring{A}_*}{K} \ln \frac{z}{z_0} \tag{13}$$

- **Drag Force Spectrum:** Wind is considered as an ergodic process; hence, the (one-sided) power spectral density of the wind force obtained by taking the Fourier transform of Equation 10 is given as:

$$S_F(f) = 4 \left[\frac{F_m}{V_m} \right]^2 \chi S_{vv}(f) \tag{14}$$

where F_m is the mean wind force, V_m is the mean wind velocity and χ is the aerodynamic admittance function. The effects of the spatial variation in the wind velocity and the frequency dependence of the drag coefficient are important for structures with large surfaces and may be taken into account by introducing aerodynamic admittance function. Charts prepared by Davenport (1961) are referred to obtain the admittance function χ.

This wind force spectrum is related to the response as given in Chandrasekaran (2015):

$$S_F\left(f\right)=4\left[\frac{F_m}{V_m}\right]^2\left[\chi\left(\frac{f\sqrt{A}}{V_m}\right)\right]^2 S_U\left(f\right) \tag{15}$$

where A is the projected area of the structure subjected to wind, Vm is the mean wind speed and f is the frequency in Hz.

The aerodynamic admittance function χ varies as below:

$$\begin{aligned}&for\,\frac{f\sqrt{A}}{V_m}\Rightarrow 0 \quad \chi\frac{f\sqrt{A}}{V_m}\Rightarrow 1\\&for\,\frac{f\sqrt{A}}{V_m}\Rightarrow\propto \quad \chi\frac{f\sqrt{A}}{V_m}\Rightarrow 0\end{aligned} \tag{16}$$

- **Auto- and Cross-Spectrum:**Nigam and Narayanan (1994) presented an expression for the modal fluctuating drag force power spectrum, for a continuous line-like structure, in terms of the power spectral density function (PSDF). It can be used upon modifying for a discretized multi-degree of freedom system (MDOF) as:

$$S_{f,j}=\left(C_D A\rho\right)^2\sum_{k=1}^{n}\sum_{l=1}^{n}S_{vkvl}\left(f\right)v_m\left(k\right)v_m\left(l\right)\varphi_j\left(k\right)\varphi_j\left(l\right) \tag{17}$$

where A is the total area of the structure exposed to the wind, k and l are spatial nodes, S_{vkvl} is the velocity auto-PSDF when $k = l$ and the cross-PSDF when k is not equal to l, V_m (k) and V_m (l) are the mean wind velocities at nodes k and l, respectively, and ϕ_j (k) and ϕ_j (l) are the nodal k and l components of the jth mode shape. The auto- and cross-PSDF terms may be evaluated as:

$$S_{vkvl}\left(f\right)=e^{-\varphi}\sqrt{S_{vkvk}\left(f\right)S_{vlvl}\left(f\right)} \tag{18}$$

where $e^{-\varphi}$ is known as the narrow-band cross-correlation, is the square root of the coherence function $e^{-\varphi} = coh^2(f)$, and φ is given as:

$$\varphi=2f\frac{\sqrt{\left[C_x^2\left(x_j-x_k\right)^2+C_y^2\left(y_j-y_k\right)^2+C_z^2\left(z_j-z_k\right)^2\right]}}{V_m(z_j)+V_m(z_k)} \tag{19}$$

where the exponential decay coefficients $C_z = 10$ and $C_y = 16$. Different wind codes may however recommend other values for C_y and C_z than those given above.

The response of a MDOF (multi-degree of freedom system) is a function of both spectral and cross-spectral density functions. The structure is discretized in n nodal points, for which wind time histories are calculated, and each fluctuating component is represented as a vector of n points. The complete characterization of the fluctuating components of a wind field as stochastic processes is ensured by the knowledge of the power spectral density (PSD) function in the frequency domain as:

$$S_v(f) = \begin{matrix} S_{v1v1} & S_{v1v2} & \cdots & S_{v1vn} \\ S_{v2v1} & S_{v2v2} & \cdots & S_{v2vn} \\ \vdots & \vdots & \ddots & 0 \\ S_{vnv1} & S_{vnv2} & \cdots & S_{vnvn} \end{matrix} \tag{20}$$

where the diagonal terms are the auto-spectrum and the rest are the cross-spectrum terms as discussed previously. Hence, the tower can be modeled as a MDOF system, which includes a lumped mass at the top of the tower, to represent the mass of the nacelle and the blades.

Artificial Time History Conversion

Shinozuka (1972) has shown that it is possible to express the fluctuating velocity component $v'(t)$ of wind at any time t as:

$$v'(t) = \sqrt{2}\sum_{i=1}^{n}\left[S_v(f_i) \times \Delta f\right]\cos\left(2\pi f_i + \varphi_i\right) \tag{21}$$

where $S_v(f_i)$ is the value of the power spectral density function (PSDF) for the fluctuating component of wind at frequency f, $\Delta f = (f_i{+}1) - f_i$ and φ_i is the phase angle with a uniform probability distribution function that varies randomly between 0 and 2. Simulated fluctuating wind velocity spectra and its converted wind velocity time history using Shinozuka's method have been presented for the purpose of illustration in Figures 6(a) and 6(b) using MATLAB software (2010).

Figure 6. (a) PSDF of fluctuating wind velocity in frequency domain and (b) converted wind velocity in time domain

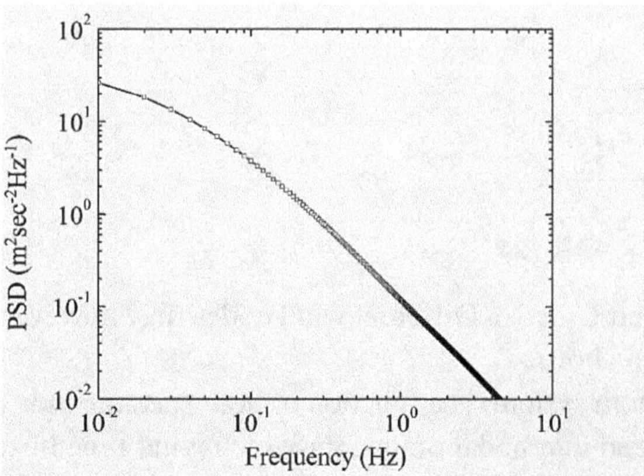

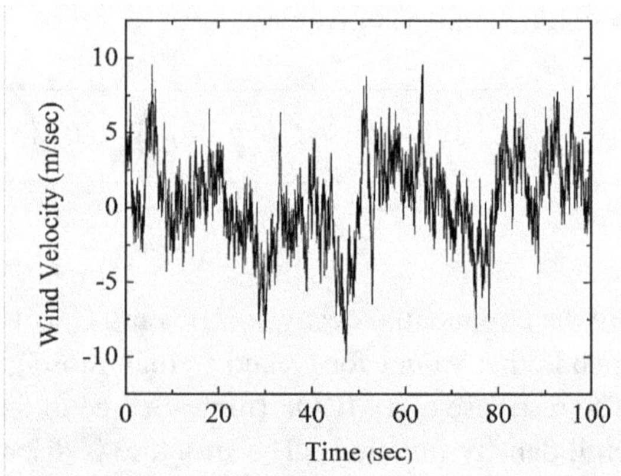

Hydrodynamics

Wave Theory

Several wave theories exist that allow the calculation of wave particle motion like velocity and acceleration beneath the sea level. Major wave theories that are used extensively in the analysis of offshore structures are given in Sarpkaya et al. (1981):

1. **Linear Wave Theory:**
    a. Airy's Theory.
2. **Non Linear Wave Theory:**
    a. Stokes's Wave Theory.
    b. Solitary Wave Theory.
    c. Cnoidal Wave Theory.

Linear wave theory is valid for waves with infinitely small amplitudes, whereas non-linear wave theories are required for finite amplitude waves. Non-linear waves have a different surface profile compared to linear waves, with sharper, higher crests and longer and shallower troughs. Detailed information about different wave theories could be obtained from the text book by Sarpkaya (1981).

Figure 7 gives a number of parameters that will be used in the characterization of a wave. The coordinate system to be used is the x, z coordinate system, with x-axis being placed along the mean sea surface and the origin of the z-axis (positive upward) being placed such that the crest of the wave is located at the origin. The sea bottom is located at a depth of d ($z = -d$) below the mean sea level (M.S.L.).

- **Airy's Theory:** Linear, regular wave theory, which also is termed Airy wave theory, is based on potential flow. Linear Airy wave theory considers the surface elevation to be described by a harmonic wave: as shown in Figure 7. The surface elevation of an Airy wave of height H, at any time t and horizontal position x in the direction of travel of the wave, is denoted by η (x,t) and is given in Sarpkaya (1981):

Figure 7. Wave characterization

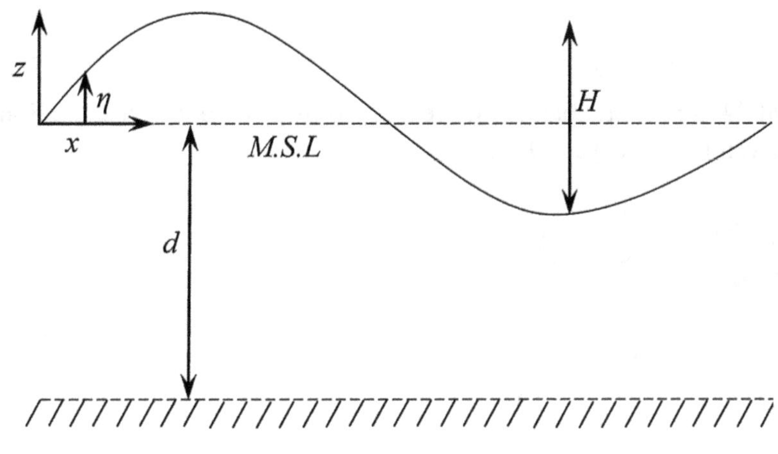

$$\eta\left(x,t\right) = \frac{H}{2}\cos\left(kx - \omega t\right) \tag{22}$$

where k is the wave number, H is the wave amplitude, and circular frequency $\omega = 2\pi/T$ in which T represents the period of the wave. The along wave particle velocity $u\ (x,t)$ and vertical water particle velocity $w\ (x,t)$ is given in Schwartz (1974):

$$u = \frac{\omega H}{2}\frac{\cosh k\left(z + d\right)}{\sinh kd}\cos\left(kx - \omega t\right) \tag{23}$$

$$w = \frac{\omega H}{2}\frac{\sinh k\left(z + d\right)}{\sinh kd}\sin\left(kx - \omega t\right) \tag{24}$$

- **Dispersion Relation:** As each wave is unique and has a unique wave number and frequency, this particular relation relates the wave number with the frequency. Here the wave number is obtained by iteration of the following equation which is given as:

$$\omega^2 = gk\tanh\left(kd\right) \tag{25}$$

where g is the acceleration due to gravity. Another interesting set of relationship is obtained when the relative depth ratio (d/L) is smaller than 0.05; this relative water depth is called the shallow water depth. For this depth, the term tanh $(kd) \cong kd$. Hence the above expression is given as:

$$\omega^2 = gk^2 d \tag{26}$$

After calculating the wave number k, the wavelength is given as:

$$L = \frac{gT^2}{2\pi} \tag{27}$$

Equation 23 and 24 are differentiated with respect to time to get the acceleration in horizontal and vertical direction given in Sarpkaya (1981):

$$a_x = \frac{\omega^2 H}{2}\frac{\cosh k\left(z + d\right)}{\sinh kd}\sin\left(kx - \omega t\right) \tag{28}$$

$$a_y = -\frac{\omega^2 H}{2}\frac{\sinh k\left(z+d\right)}{\sinh kd}\cos\left(kx-\omega t\right) \tag{29}$$

- **Stoke's Wave Theory:** It is an extension of the Airy theory. Stoke's assumed that all variation in the *x* direction can be represented by Fourier series and that the coefficients in these series can be written as perturbation expansions in terms of a parameter which increases with wave height, *H*.

The surface elevation η (*x*,*t*) and is given as:

$$\eta\left(x,t\right) = \frac{1}{k}\sum_{n=1}^{5} F_n \cos n\left(kx-\omega t\right) \tag{30}$$

where F_1, F_2 etc. are given by

$$F_1 = a$$

$$F_2 = a^2 F_{22} + a^4 F_{24}$$

$$F_3 = a^2 F_{32} + a^4 F_{34}$$

$$F_4 = a^4 F_{44}$$

$$F_5 = a^5 F_{55}$$

where $F_{22,}F_{24}$ denote wave profile parameters depending on *k*, *d* and *a* is obtained by iteration of the following equation:

$$kH = 2\left(a + a^3 F_{33} + a^5\left(F_{55} + F_{35}\right)\right) \tag{31}$$

The along wave horizontal water particle velocity *u* (*x*,*t*) and vertical water particle velocity *w* (*x*,*t*) is given as:

$$u = \frac{\omega}{k}\sum_{n=1}^{5} Gn\frac{\cosh nky}{\sinh nkd}\cos n(kx-\omega t) \tag{32}$$

$$v = \frac{\omega}{k}\sum_{n=1}^{5} Gn\frac{\sinh nky}{\sinh nkd}\sin n(kx-\omega t) \tag{33}$$

where G_1, G_2 …is given as:

$$G_1 = aG_{11} + a^3 G_{13} + a_5 G_{15}$$

$$G_2 = 2\left(a^2 G_{22} + a^4 G_{24}\right)$$

$$G_3 = 3\left(a^3 G_{33} + a^5 G_{35}\right) \tag{34}$$

$$G_4 = 4a^4 G_{44}$$

$$G_5 = 4a^5 G_{55}$$

where $G_{22,} G_{24}$ …denote wave profile parameters depending on *kd*. Explicit expressions and values for the parameters F_{22}, F_{24}, G_{11} etc. have been given by Skjelbreia and Hendrickson [33].

Hydrodynamic Load

Steady-State Hydrodynamic Load

- **Morison's Equation:** The wave loading acting on a cylinder submerged in water as shown in Figure 8 is given by Morrison's equation. The Morison equation is the sum of two force components, i) an inertia force in phase with the local flow acceleration, and ii) a drag force proportional to the square of the instantaneous flow velocity given by Morison et al. (1950):

$$F = \rho_w C_M V\dot{u} + \frac{1}{2}\rho_w C_D A\left|u\right| u \tag{35}$$

where *V* is volume of the body, *A* is a projected area, i.e. the cross-sectional area of the body perpendicular to the flow direction, ρ_w is the density of water, C_M is the inertia coefficient, C_D is the drag coefficient, *D* is the diameter of the body, *u* is the horizontal water particle velocity and $\dot{u}$ is the horizontal water particle acceleration. *F* (*t*) is the total force per unit height of cylinder length which is given by Morison et al. (1950):

$$F\left(t\right) = \rho_w C_M \frac{\pi}{4} D^2 \dot{u} + \frac{1}{2}\rho_w D C_D A\left|u\right| u = F_D + F_I \tag{36}$$

F_D and F_I are drag force and inertia force respectively given as:

Figure 8. Wave forces acting on the structure (Morrison's equation)

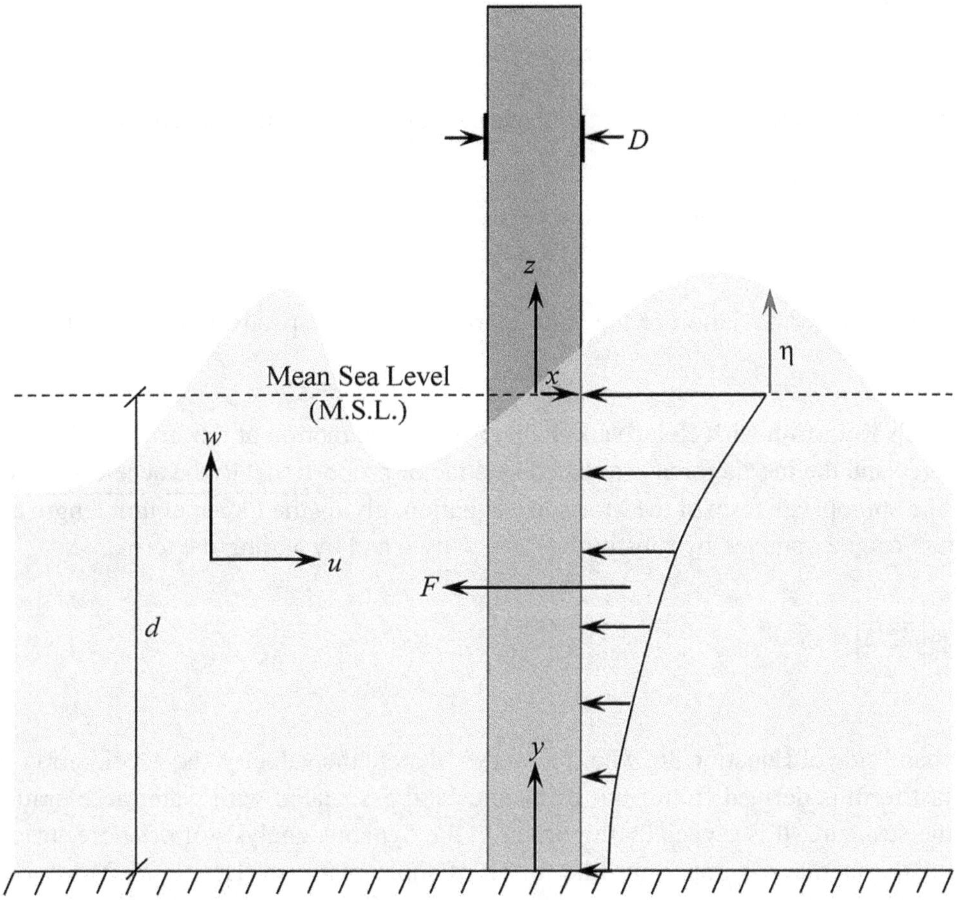

$$F_I = \rho_w C_M \frac{\pi}{4} D^2 \dot{u}, F_D = \frac{1}{2} \rho_w D C_D |u| u \tag{37}$$

- **Drag Linearization:** According to Morison equation, given in Equation 36, the drag force is nonlinear. This nonlinear formulation is for the determination of a transfer function (which will be discussed in later section) needed for frequency domain calculations. Hence the drag force has to be linearized in a suitable manner. Morrison's equation in form of the linearized drag term is given as:

$$F(t) = C_M \left[\rho_w \frac{\pi}{4} D^2 \right] \dot{u} + \frac{1}{2} \rho_w \bar{C}_D A |u| u = F_I + F_{D_{lineraized}} \tag{38}$$

The linearization of the drag component that is $\bar{C}_D |u| u$ obtained by equating the energy in a quarter of the oscillation period which is given as:

$$\int_0^{T/4} \bar{C}_D |u| u dt = \int_0^{T/4} (\bar{C}_D u) u dt \quad (39)$$

Hence, the above calculation leads to the linearized drag force which is given as:

$$F_{D_{lineraized}} = \frac{1}{2}\rho_w D\left(C_D\sqrt{8\pi/3}\right)\sigma_u u \quad (40)$$

where σ_u is the standard deviation of the water particle velocity spectrum which will be discussed in the next section.

- **Morison's Equation with Relative Velocity: When** the motion of the structure is considered, the drag force and the inertia force is reduced by a factor proportional to the acceleration of the structure. The appropriate form of the Morison's equation, giving the force per unit length acting on the structure can be obtained by substituting $u - \dot{x}$ by u and by adding the term

$$\rho_w (C_M - 1)\left[\frac{\pi D^2}{4}\right]\ddot{x} \quad (41)$$

to the right hand side of Equation 36, where $\dot{x}$ and $\ddot{x}$ denote the velocity and acceleration of the structure. This last term is derived from fluid mechanics and associated with water acceleration from the motion of the structure. It is especially important in the dynamic analysis of offshore structures where the inertia of the member is important like for an offshore platform. Hence the Morrison's equation becomes:

$$F(t) = C_M\left[\rho_w \frac{\pi}{4} D^2\right]\dot{u} + \frac{1}{2}\rho C_D\left[\frac{\pi D^2}{4}\right]|u - \dot{x}|(u - \dot{x}) - \rho_w (C_M - 1)\left[\frac{\pi D^2}{4}\right]\ddot{x} \quad (42)$$

Random Hydrodynamic Load

- **Pierson Moskowitz Spectrum:** To obtain random waves, multiple linear wave components are summed. This approach assumes that the sea can be described as a stationary random process, which generally is valid for the periods of no more than a few hours used in a typical wind turbine load analysis. Figure 9 illustrates the relationship between the time-domain solution of the waves and the frequency-domain representation of the waves by a wave spectrum $S(\omega)$. Basically, multiple regular waves of different frequencies and with random phase angles are summed with respect to their probability spectral density according to the wave spectrum used for the specific site. A typical wave spectrum used in the analysis of an offshore structure is the Pierson-Moskowitz spectrum which is represented by Pierson et al (1964) as:

$$S_{\eta\eta}(\omega) = 2\pi \frac{\alpha_{pm} \cdot g^2}{\omega^5} \cdot \exp\left[-\beta_{pm}\left(\frac{\omega_p}{\omega}\right)^4\right] \tag{43}$$

$$\alpha_{pm} = 0.0081, \beta_{pm} = \frac{5}{4}, \omega_p^4 = \frac{0.74}{\beta_{pm}}\left(\frac{g}{V_m}\right)^4 \tag{44}$$

- **JONSWAP Spectrum:** Hasselmann et al. (1973), after analyzing data collected during the Joint North Sea Wave Observation Project JONSWAP, modified the Pierson Moskowitz spectrum through an artificial factor which was added to the Pierson-Moskowitz spectrum in order to improve the fit to their measurements. The JONSWAP spectrum is thus Pierson-Moskowitz spectrum multiplied by an extra peak enhancement factor which is given as:

$$S_\eta(\omega) = \left[\frac{\alpha g^2}{(2\pi)^4 f^5}\right] \times e^{-1.25\left(\frac{f_p}{f^4}\right)} \times 3.3^a, a = e^{-\left(\frac{(f-f_p)^2}{2\sigma^2 f_p^2}\right)}$$

$$\sigma = \begin{cases} 0.07 & when\ f \leq f_p \\ 0.09 & when\ f \geq f_p \end{cases} \tag{45}$$

$$\alpha = 0.076\left(\frac{gF}{V_{10}^2}\right)^{-0.22}, f_p = \left(\frac{3.5g}{V_{10}}\right)\left(\frac{gF}{V_{10}^2}\right)^{-0.33}$$

- **Wave Force Spectrum:** The drag force spectrum acting per unit length on the submerged structure is obtained by taking Fourier transform of Equation 36 which is given as:

$$F(z,t) = \frac{\omega}{\sinh kd}\left[-C_M i\omega \cosh(kz) + C_D\sqrt{\frac{8}{\pi}} \cosh(kz)\sigma_u(z,t)\right]\eta(t) \tag{46}$$

After taking the Fourier transform of Equation 36, the force spectrum is obtained as a function of the wave energy spectral density as:

Figure 9. Wave elevation of irregular waves in the time domain as a combination of regular waves and a typical wave spectrum

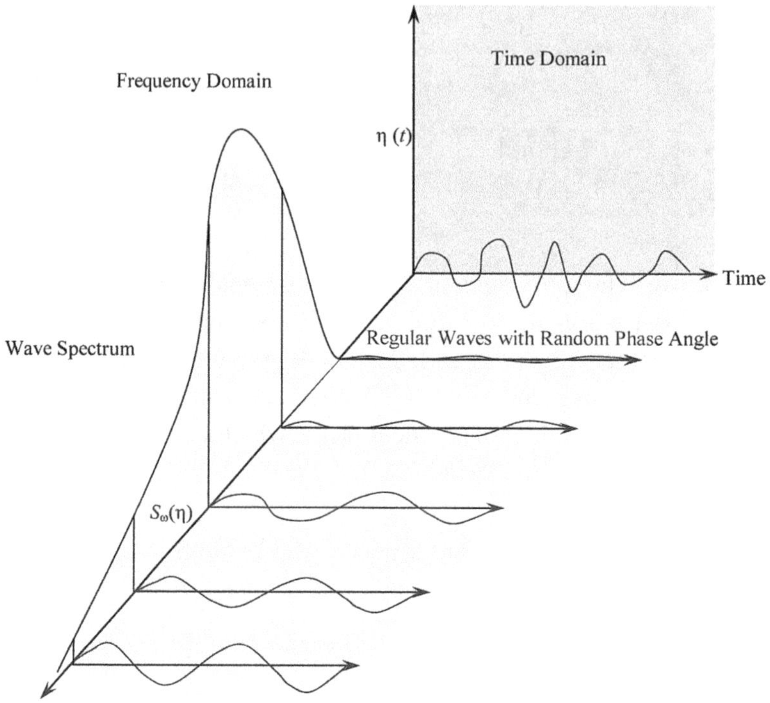

$$S_{FF}\left(\omega,z\right)=\left(\frac{\omega}{\sinh kd}\right)^{2}\left[\underbrace{\left[C_{M}\omega\cosh\left(kz\right)\right]^{2}+\left[C_{D}\sqrt{\frac{8}{\pi}}\cosh\left(kz\right)\sigma_{u}\left(z,t\right)\right]^{2}}_{Force\ Transfer\ Function}\right]S_{\eta\eta}\left(\omega\right) \tag{47}$$

$$\sigma_{u}^{2}\left(z\right)=\int_{0}^{\inf}S_{xx}\left(\omega,z\right)dw=\int_{0}^{\inf}\left(\omega\frac{\cosh kz}{\sinh kd}\right)^{2}S_{\eta\eta}\left(\omega\right)dw \tag{48}$$

where the expression between square braces in Equation 47 is the force transfer function, C_D is the drag coefficient, C_M is the inertia coefficient, S_{xx} is the PSDF of horizontal water particle velocity, σ_u is the standard deviation of the horizontal water-particle velocity component in *x* direction, *k* is the wave number and *z* is the elevation from the bottom most point.

The following relationship relates the spectral force $S_{FF}(\omega)$ with the response $S_{uu}(\omega)$ using a harmonic response function $H(\omega)$ given as:

$$\left|H\left(\omega\right)\right|^{2}S_{FF}\left(\omega\right)=S_{UU}\left(\omega\right) \tag{49}$$

where the harmonic response function $H(\omega)$ is given as:

$$\left|H(\omega)\right|^2 = \frac{1}{M^2\omega_n^4\left\{\left[1-\left(\frac{\omega}{\omega_n}\right)^2\right]^2 + \left(2\xi\frac{\omega}{\omega_n}\right)^2\right\}} \tag{50}$$

where ω_n is the natural frequency of the structure.

Current Load

Currents impose drag forces on fixed offshore structures. To include the effect of currents, current velocity should be added to the vertical water particle velocity before computing the drag force, because drag force depends on the square of the water particle velocity. Current decreases slowly with the increase in depth, but even a small magnitude of current velocity can cause significant drag force given as:

$$F_{current} = \frac{1}{2}\rho_w C_D D w\left|w\right| \tag{51}$$

where ρ is density of water, C_D is the drag coefficient, D is the diameter of the body, w is the vertical water particle velocity.

The presence of current is alternatively accounted by increasing the wave height to 10-15% and neglects the presence of current.

Numerical Example 1

Analyze a wind turbine tower for wind and wave loading with the following data: external diameter of tower, D = 324 mm; wall thickness, t = 9.5 mm; water depth, d = 8 m; tower mass, m = 73.8 kg/m; tip mass, M = 300 kg; second moment of area of tower, I = 0.116 × 10^{-3} m4; Young's modulus of pile, E = 205 × 109 N/m^2; the density of the tower is 7850 kg/m^3. In the absence of other data, the apparent pile fixity is taken as 6D below the sea bed (~2.0 m), and the effective pile length = 17 m; neglect the soil mass contributing to the motion. Assume the light tower to behave as a cantilever beam, bending its fundamental mode of vibration. Design wind speed (Wind speed at height 10 m above sea level) = 19.5 m/sec, z_0 = 0.001, wave height = 1.5 m, time period of wave = 1.3 sec.

1. Obtain the wind force spectrum for (i) SDOF and ii) MDOF system as shown in Figure 10.
2. Obtain the wave force spectrum for (i) SDOF and ii) MDOF system.

SDOF System: Natural Frequency

To calculate the mass of the SDOF system, the effective mass per unit length of the structure submerged in water has to calculated which is determined as:

Figure 10. Wind turbine tower discretized into two nodes

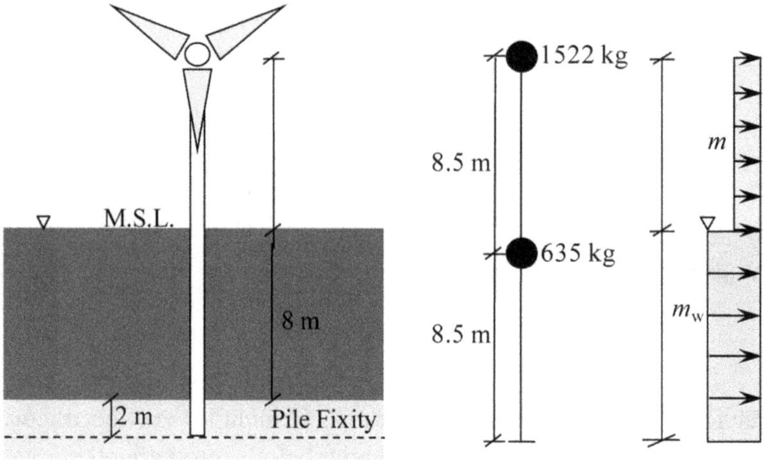

$$m_w = m + \rho_w \frac{\pi D^2}{4} + \rho_w \frac{\pi \left(D - 2t\right)^2}{4}$$
$$= 73.8 + \frac{1015 \times \pi \times 0.324}{4} + \frac{1015 \times \pi \times 0.305^2}{4} kg / m \tag{52}$$

$$m_w = 73.8 + 83.7 + 74.2 = 231.7 kg / m \tag{53}$$

Hence, the mass for the SDOF system by considering half of the length of the structure (18/2=8.5 m) is given as:

$$M = 73.8 \times 7 + 231.7 \times 1.5 + 300 = 1164.15 Kg \tag{54}$$

The bending stiffness of the structure which is in the form of a cantilever is given as:

$$K = \frac{3EI}{L^3} = 14520.6 N / m \tag{55}$$

Natural frequency of the structure is given as:

$$\omega = \sqrt{\frac{K}{M}} = \sqrt{\frac{14520.6}{1164.5}} = 0.6 Hz \tag{56}$$

$$v_* = \frac{V_m}{2.5\ln\frac{z}{z_0}} = \frac{19.5}{2.5\ln\frac{10}{0.001}} = 0.825m / \sec \tag{57}$$

Wind velocity at a height of 7 m above the sea level is given as:

$$V_m(7) = 2.5 \times 0.825 \times \ln\frac{7}{0.001} = 18.7m / \sec \tag{58}$$

Hence the wind force at the top node is given as:

$$F = \frac{1}{2}\rho C_D A V_m^2 = \frac{1}{2} \times 1.225 \times 1.2 \times 0.324 \times 7 \times 18.7^2 = 585.7N \tag{59}$$

Wind Force Spectrum

The spectral density function as explained in Equation 12 is given as:

$$S_{vv}(z,f) = \frac{200v_*^2 n_m}{\left(1+50n_m\right)^{5/3}} \tag{60}$$

For frequency i.e. 0.6 Hz,

$$n_m(7,0.6) = \frac{1\times 7}{18.66} = 0.2241 \tag{61}$$

$$S_{vv}(7,0.6) = \frac{200\times 0.825^2 \times 0.2241}{\left(1+50\times 0.2241\right)^{5/3}} = 0.8281m^2Hz \tag{62}$$

Hence, the force spectrum from Equation 14 is given as:

$$S_F\left(f\right) = \frac{4F^2 S\left(f\right)\chi}{V_m^2} \tag{63}$$

Aerodynamic admittance factor χ from chart of Davenport is 1. Hence the force spectrum for frequency of 0.6 Hz is given as:

$$S_F(f) = \frac{4 \times 582.9^2 \times 0.8281 \times 0.6}{18.7^2} = 3.234kNHz^{-1} \tag{64}$$

The variation of wind elevation spectrum and force spectrum with frequency is shown in Figure 11.

Figure 11. Variation of (a) wind velocity spectrum and (b) wind force spectrum with frequency

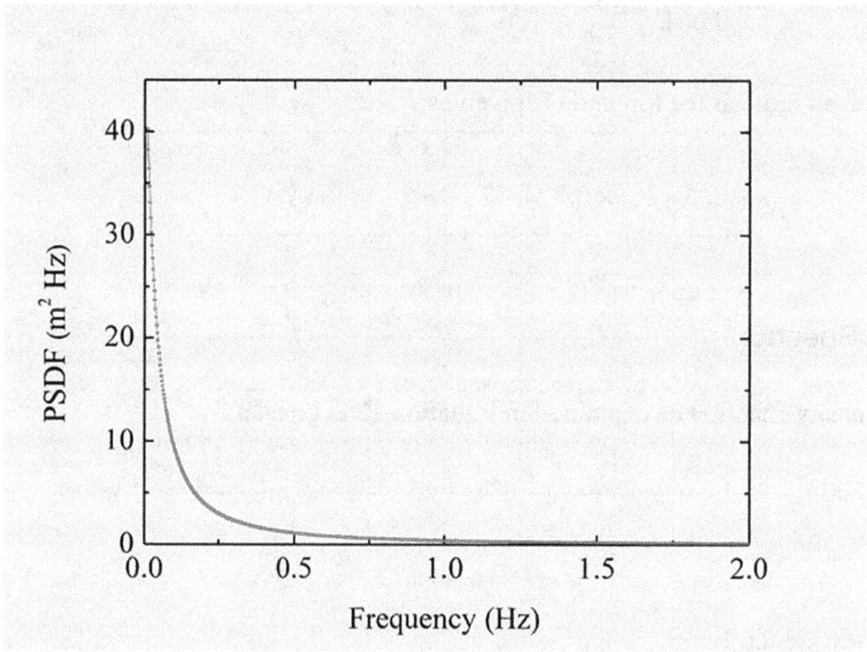

(a)

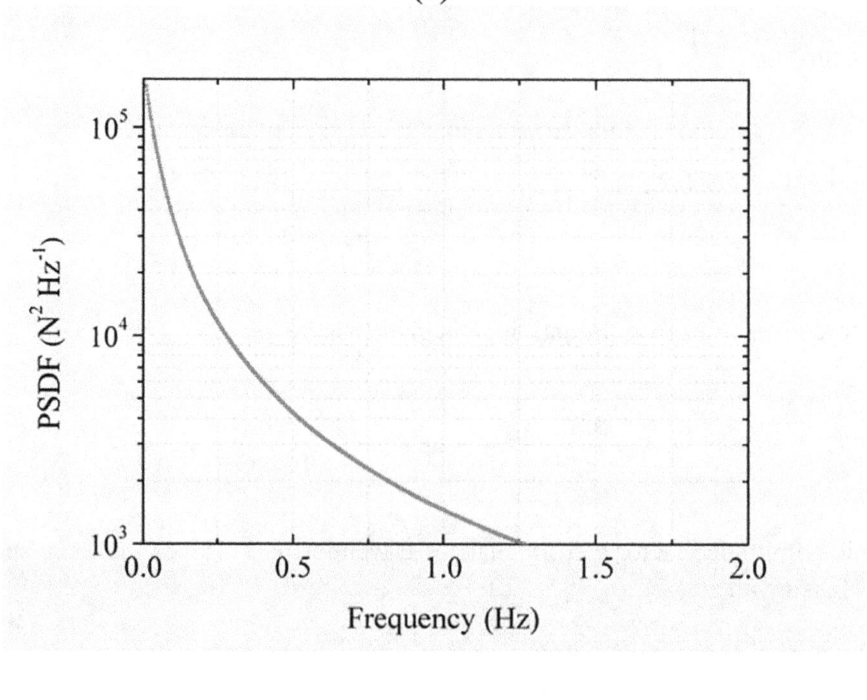

MDOF System

Natural Frequency

Discretizing the structure into lumped mass model as shown in Figure 10, the mass matrix is calculated from the effective mass and structural mass per unit length as:

$$[M]=\begin{bmatrix}1522 & 0\\ 0 & 635\end{bmatrix}kg \tag{65}$$

The stiffness matrix has been calculated as shown in Figure 12. The final stiffness matrix is given as:

$$[K]=\frac{6EI}{7L^3}\begin{bmatrix}16 & -5\\ -5 & 2\end{bmatrix}=10^3\times\begin{bmatrix}531 & -166\\ -166 & 66\end{bmatrix}N/m \tag{66}$$

Frequency is calculated using Eigen value characteristic equation and found as:

$$\begin{aligned}f_1 &= 0.67Hz\\ f_2 &= 3.31Hz\end{aligned} \tag{67}$$

Figure 12. Derivation of stiffness matrix

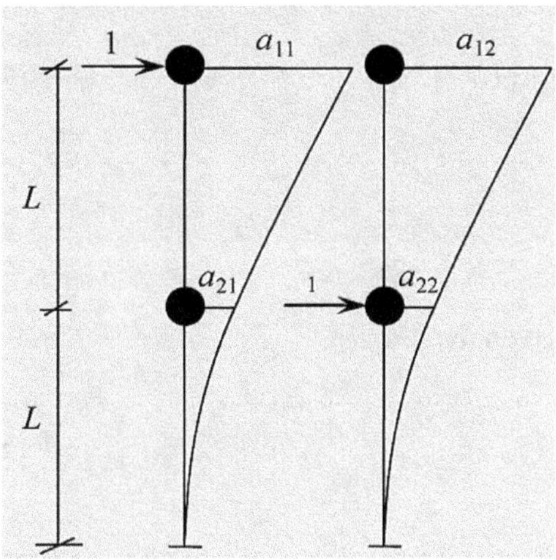

Flexibility Matrix: $[A]=\begin{bmatrix}a_{11} & a_{12}\\ a_{21} & a_{22}\end{bmatrix}$ (68)

$$a_{11}=\frac{L^3}{3EI}\quad a_{12}=a_{21}=\frac{L^3}{2EI}+\frac{L^3}{3EI}=\frac{5L^3}{6EI}\quad a_{22}=\frac{8L^3}{3EI}$$

(69)

$$[A]=\frac{L^3}{6EI}\begin{bmatrix}2 & 5\\ 5 & 16\end{bmatrix} \tag{70}$$

Stiffness Matrix: $[K]=[A]^{-1}$

$$[K]=\frac{6EI}{7L^3}\begin{bmatrix}16 & -5\\ -5 & 2\end{bmatrix} \tag{71}$$

Flexibility Matrix:

$$[A] = \begin{bmatrix} a_{11} & a_{12} \\ a_{21} & a_{22} \end{bmatrix} \tag{68}$$

$$a_{11} = \frac{L^3}{3EI}\ a_{12} = a_{21} = \frac{L^3}{2EI} + \frac{L^3}{3EI} = \frac{5L^3}{6EI}\ a_{22} = \frac{8L^3}{3EI} \tag{69}$$

$$[A] = \frac{L^3}{6EI}\begin{bmatrix} 2 & 5 \\ 5 & 16 \end{bmatrix} \tag{70}$$

Stiffness Matrix:

$$[K] = [A]^{-1}$$

$$[K] = \frac{6EI}{7L^3}\begin{bmatrix} 16 & -5 \\ -5 & 2 \end{bmatrix} \tag{71}$$

Wind Force Calculation

Hence the wind force at the top node is given as:

$$F = \frac{1}{2}\acute{A}C_D A\, V_m^2 = \frac{1}{2} \times 1.225 \times 1.2 \times 0.324 \times 4.25 \times 18.7^2 = 355.61\ \text{N} \tag{72}$$

Wind Force Spectrum

The spectral density function as explained in Equation 12 is given as:

$$S_{vv}(z,f) = \frac{200 v_*^2 n_m}{\left(1 + 50 n_m\right)^{5/3}} \tag{73}$$

For frequency i.e. 0.67 Hz,

$$n_m(7,0.67) = \frac{1 \times 7}{18.7} = 0.2502 \tag{74}$$

$$S_{vv}(7,0.67) = \frac{200 \times 0.825^2 \times 0.2241}{\left(1 + 50 \times 0.2241\right)^{5/3}} = 0.6990 m^2 Hz \tag{75}$$

Hence, the force spectrum from Equation 14 is given as:

$$S_F\left(f\right) = \frac{4 \times 355.61^2 \times 0.8281 \times 0.67}{18.7^2} = 1.006 kN / Hz \tag{76}$$

Similarly, through MATLAB simulation the wind velocity spectrum and force spectrum variation with frequency is given in Figure 13.

SDOF System

Natural Frequency

The equivalent mass M_s as shown in Figure 14 must be determined such that the natural frequency of the actual model is equal to that of the equivalent model. The equivalent mass for the SDOF system is given as:

$$\omega_S = \omega \rightarrow \sqrt{\frac{K_s}{M_s}} = \sqrt{\frac{K}{M}} \rightarrow M_s = M\frac{K_s}{K} = M\left(\frac{L}{L_s}\right)^3 = 1164.15\left(\frac{17}{10}\right)^3 = 5719.5 kg \tag{77}$$

Wave Elevation Spectrum

Spectral density function of the surface elevation $S_{\eta\eta}$ which is given, as a function of H_s and the zero-crossing period T_z, is:

$$S_\omega\left(\omega\right) = \frac{H_s^2}{4\pi T_z^4} e^{-\frac{16\pi^3}{\left(\omega T_z\right)^4}} \tag{78}$$

$$T_Z = \frac{1}{1.4F_m} \tag{79}$$

Figure 13. Variation of (a) wind velocity spectrum and (b) wind force spectrum with frequency

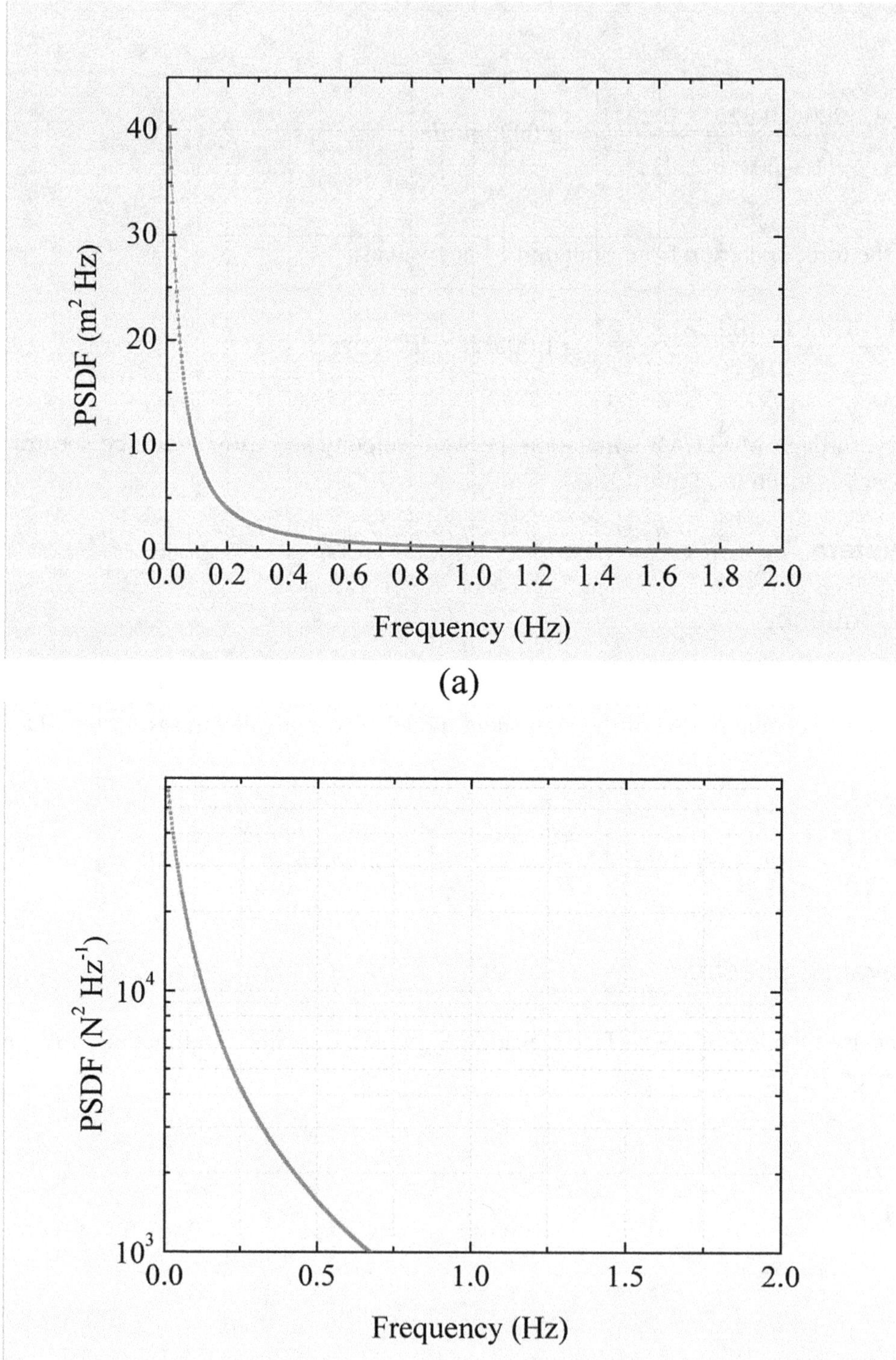

Figure 14. Equivalent mass calculation

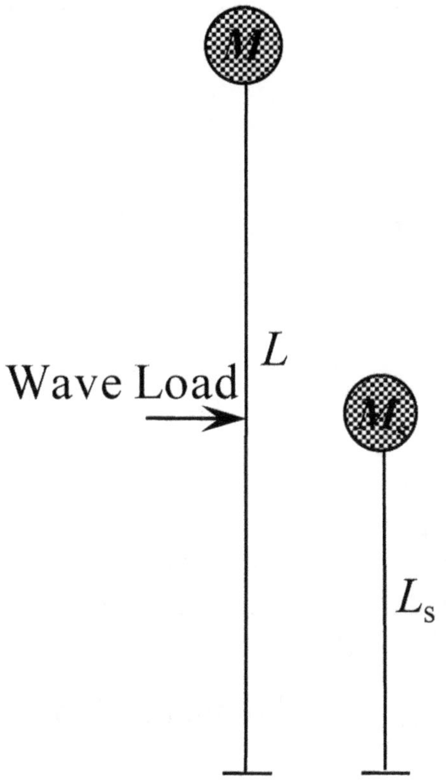

where F_m is the peak energy frequency given as 0.33 Hz. This wave elevation spectrum is plotted in the Figure 15 using MATLAB. The cut-off frequency, $f^*= 1.4$ Hz; beyond this value the waves have a short period and their energy content is negligible.

Wave Force Spectrum

The wave force spectrum is given in Equation 47 as:

$$S_{FF}\left(\omega,z\right)=\left(\frac{\omega}{\sinh kd}\right)^{2}\left[\left[C_{M}\omega\cosh\left(kz\right)\right]^{2}+\left[C_{D}\sqrt{\frac{8}{\pi}}\cosh\left(kz\right)\sigma_{u}\left(z,t\right)\right]^{2}\right]S_{\eta\eta}\left(\omega\right) \tag{80}$$

where C_I and C_D has been provided. Here, σ_u has been calculated as:

$$\sigma_{u}^{2}\left(z\right)=\int_{0}^{\inf}S_{u}\left(\omega,z\right)dw=\int_{0}^{\inf}\left(\omega\frac{\cosh kz}{\sinh kd}\right)^{2}S_{\eta\eta}\left(\omega\right)dw \tag{81}$$

Figure 15. Wave elevation spectrum

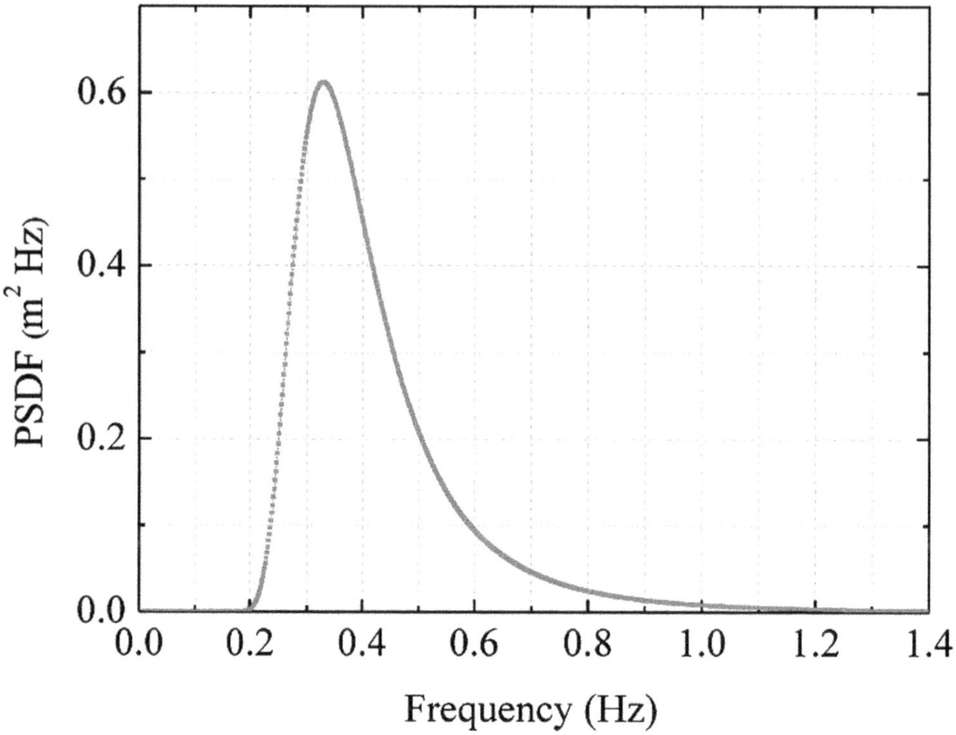

where σ_u is the standard deviation of water particle velocity spectrum and $S_u(\omega,z)$ is the water particle velocity spectrum.

Calculation of $S_u(\omega,z)$ involves determining the transfer function for the water particle velocity spectrum which is given as:

$$H_u(f) = 2\pi f \frac{\cosh kz}{\sinh kd} \tag{82}$$

This function depends upon the wave number k, which is related to the wave frequency by the dispersion equation of Airy's wave theory which relates the wave number with frequency as:

$$\omega^2 = gk \tanh kd \rightarrow f^2 = \frac{g}{2\pi L} \tanh\left(\frac{2\pi d}{L}\right)$$

$$\omega^2 \cong gk \rightarrow \frac{4\pi^2}{gf^2} \tag{83}$$

Hence, the expression of the mean square of the velocity wave force spectrum is given as:

$$\sigma_u^2(z) = \int_0^{f^*} \left|H_u(f)\right|^2 S_{\eta\eta}(f)\,df = \int_0^{f^*} \frac{\pi H_s^2}{T_Z^4}\frac{1}{f^3}\frac{\cosh^2\left(\dfrac{4\pi^2}{g}f^2 z\right)}{\sinh^2\left(\dfrac{4\pi^2}{g}f^2 d\right)} e^{\frac{-1}{\pi f^4 T_z^4}} \tag{84}$$

This integral can be calculated numerically using Trapz command in MATLAB which uses the trapezoidal rule of numeric integration for different values of z (in Figure 16). After determining the standard deviation of water particle velocity spectrum from Equation 59, the wave force spectrum per unit length is determined from Equation 51 and plotted in Figure 16. Since this method of determining the wave force spectrum involves tedious calculations, the author has chosen to only show the results using MATLAB simulations. Numerical example 2 explains step by step procedure to determine the wave force for this problem. Hence it is recommended for the readers to go through numerical example 2 for wave force calculations.

MDOF System

- **Natural Frequency:** The natural frequency of the structure is 0.6 Hz.
- **Wave Elevation Spectrum:** The wave elevation spectrum is independent of depth. Hence it remains the same as that for the SDOF system represented by Figure 15.

Figure 16. Wave force spectra for SDOF system

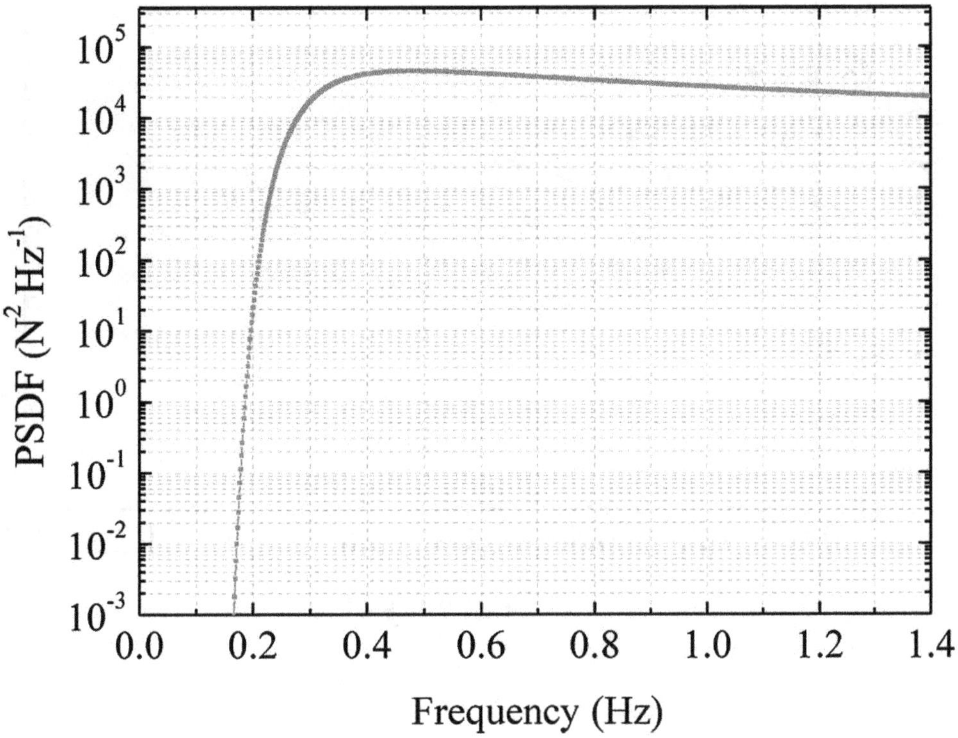

- **Wave Force Spectrum:** For the node under water, the wave force spectrum is calculated by substituting the depth z as -1.5 m, f = natural frequency of the structure and considering the lumped length and is given in Figure 17.

Analyze a wind turbine tower by Xiao et al. (2009) shown in Figure 24 for wind and wave loading with the following data: the base of the tower is 4.2 m in diameter, tapering to 2 m at the nacelle. the steel thickness at the bottom and the top of the structure is 50 mm; respectively; water depth, d = 20 m; tower mass, m = 3761.13 kg/m; tip mass due to blades and rotor = 19786 kg; second moment of area of tower, I = 0.1457 m^4; Young's modulus of pile, E = 205 × 10^9 N/m^2. the density of the tower is 7850 kg/m^3. Assume the light tower to behave as a cantilever beam, bending its fundamental mode of vibration. The foundation is assumed to be fully fixed into the ground, with no rotation allowed. Design wind speed (wind speed at height 10 m above sea level) = 18 m/sec, roughness length, z0 = 0.002, Cd = 1.2; wave height = 1.5 m; time period of wave = 6 sec.

1. Obtain the wind force spectrum for (i) SDOF and ii) MDOF system as shown in Figure 10.
2. Obtain the wave force spectrum for (i) SDOF and ii) MDOF system.

Figure 17. Wave force spectra for the node in water in MDOF system

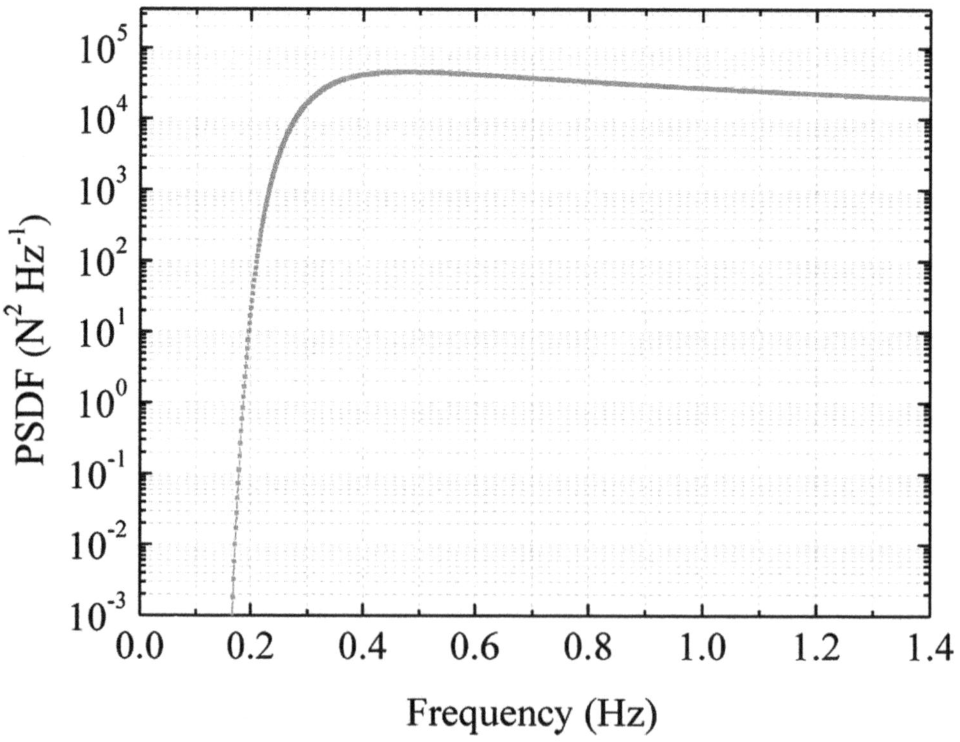

Figure 18. Structural idealization of wind turbine tower

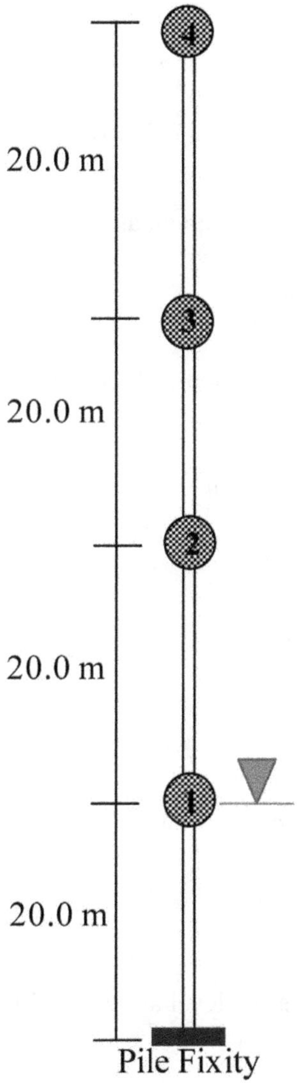

SOLUTION

SDOF System

Natural Frequency

Mass per unit length of the structure:

$$\left(m\right) = \frac{38.33 \times 7850}{80} = 3761.13kg \tag{85}$$

Effective mass per unit length in water given as:

$$m_w = m + \rho_w \frac{\pi D^2}{4} + \rho_w \frac{\pi \left(D - 2t\right)^2}{4} \tag{86}$$
$$= 3761.13 + 1015 \times 12.11 + 1015 \times 11.51 = 27735.43 kg / m$$

Hence the mass at tip for SDOF system is given as:

$$M = 3761 \times 40 + 19876 = 170.36 tonne \tag{87}$$

Taking the moment of inertia for the top node which is given as:

$$I = \frac{\pi \left(D_2^4 - D_1^4\right)}{64} = \frac{\pi \left(2^4 - 1.9^4\right)}{64} = 0.1457 m^4 \tag{88}$$

$$K = \frac{3 \times 205 \times 10^9 \times 0.1457}{80^3} = 175.01 kN / m \tag{89}$$

$$\omega = \sqrt{\frac{K}{M}} = \sqrt{\frac{175.01 \times 1000}{170.36 \times 1000}} = 1.01 rad / \sec = 0.16 Hz \tag{90}$$

Wind Force Calculation

Wind velocity at a height of 60 m above sea level above is 25 m/sec.
Shear velocity is:

$$v_* = \frac{V_m}{2.5 \ln \frac{z}{z_0}} = \frac{25}{2.5 \ln \frac{60}{0.002}} = 0.970 m / \sec \tag{91}$$

Drag force at the top node is given as:

$$F = \frac{1}{2} \rho C_D A V_m \left|V_m\right| = \frac{1}{2} \times 1.225 \times 1.2 \times 102 \times 25^2 = 4.6856 \times 10^4 N \tag{92}$$

Wind Force Spectrum

The spectral density function as given in Equation 12 is given as:

For first frequency i.e. 0.16 Hz,

$$n_m(60,0.16)=\frac{0.22\times 60}{25}=0.384 \quad (93)$$

$$S_{vv}(60,0.16)=\frac{200\times 0.970^2\times 0.384}{\left(1+50\times 0.384\right)^{5/3}}=1.8137m^2Hz \quad (94)$$

Force spectrum from Equation 14 is given as:

$$S_F\left(f\right)=\frac{4\times\left(4.68\times 10000\right)^2\times 1.184\times 0.75}{25^2}=1.91\times 10^7 N^2Hz^{-1} \quad (95)$$

Similarly, through MATLAB simulation the wind velocity spectrum and force spectrum variation with frequency is given in Figure 19.

MDOF System

- **Natural Frequency:** The natural frequency of the model as given in the paper is 0.2 Hz.
- **Wind Force Calculation:** For node 4:
    - Wind velocity at a height of 60 m above sea level above is 25 m/sec.
    - Drag force at the top node is given as:

$$F=\frac{1}{2}\acute{A}C_D A\,V_m\left|V_m\right|=\frac{1}{2}\times 1.225\times 1.2\times 21.375\times 25^2=9.82\times 10^3\ \mathrm{N} \quad (96)$$

Wind Force Spectrum

The spectral density function as given in Equation 15 is given as:

For first frequency i.e. 0.22 Hz,

$$n_m(60,0.2)=\frac{0.22\times 60}{25}=0.5280 \quad (97)$$

Figure 19. (a) Wind velocity spectrum and (b) wind force spectrum

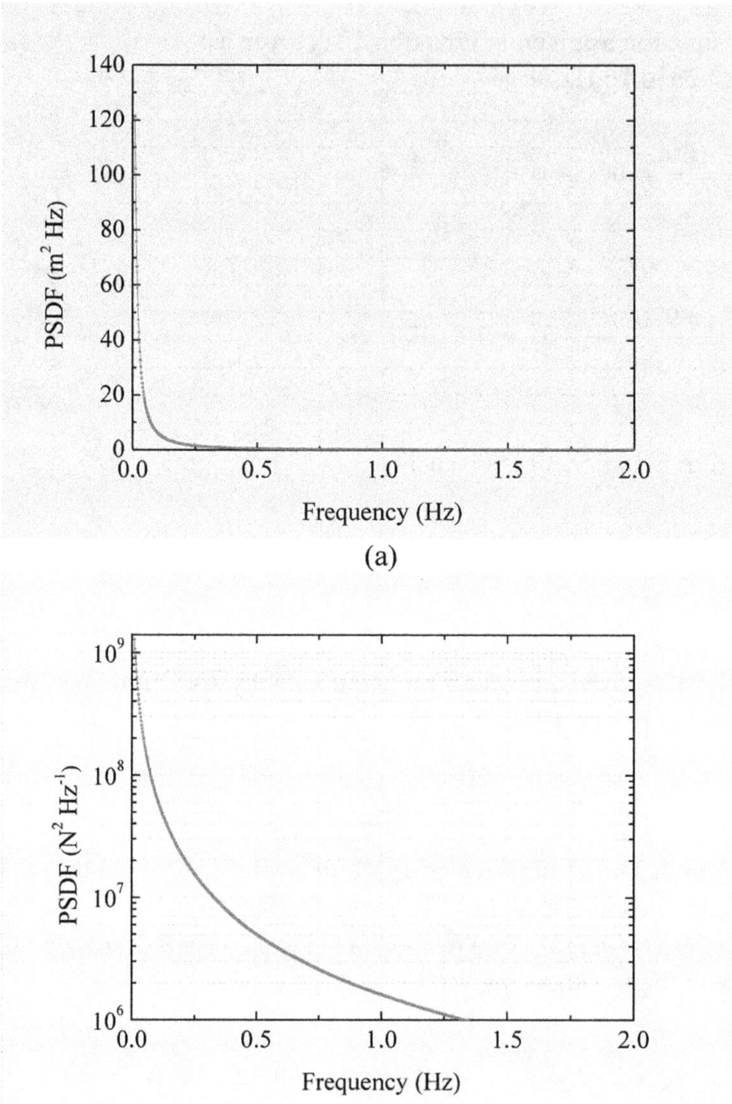

$$S_{vv}(60,0.22) = \frac{200 \times 0.970^2 \times 0.5280}{\left(1 + 50 \times 0.5280\right)^{5/3}} = 1.8137\ \mathrm{m^2/sec} \tag{98}$$

Force spectrum from Equation 14 is given as:

$$S_F\left(f\right) = \frac{4 \times \left(9.82 \times 1000\right)^2 \times 1.8137 \times 0.75}{25} = 8.39 \times 10^5\,\mathrm{N\,sec} \tag{99}$$

Similarly, through MATLAB simulation the wind velocity spectrum and force spectrum variation with frequency is given in Figure 26.

Wind Force Calculation for Node 3

Wind velocity at a height of 40 m above sea level above is

$$V_m = \frac{2.5}{0.970}\log\left(\frac{40}{0.002}\right) = 24.02m\,/\,\sec \tag{100}$$

Drag force at the top node is given as:

$$F = \frac{1}{2}\rho C_D A V_m \left|V_m\right| = \frac{1}{2}\times 1.225\times 1.2\times 51\times 24.02^2 = 2.16\times 10^4 N \tag{101}$$

Wind Force Spectrum

The spectral density function as given in Equation 12 is given as:

For first frequency i.e. 0.22 Hz,

$$n_m(40, 0.22) = \frac{0.22\times 40}{24.02} = 0.3664 \tag{102}$$

$$S_{vv}(40, 0.22) = \frac{200\times 0.970^2\times 0.3664}{\left(1+50\times 0.3664\right)^{5/3}} = 2.2531m^2Hz \tag{103}$$

Force spectrum from Equation 14 is given as:

$$S_F\left(f\right) = \frac{4\times\left(2.16\times 10000\right)^2\times 2.2531\times 0.75}{24.02^2} = 5.48\times 10^6 N^2Hz^{-1} \tag{104}$$

Similarly, through MATLAB simulation the wind velocity spectrum and force spectrum variation with frequency is given in Figure 26.

Wind Force Calculation for Node 2

Wind velocity at a height of 40 m above sea level above is

$$V_m = \frac{2.5}{0.970}\log\left(\frac{20}{0.002}\right) = 22.33m / \sec \quad (105)$$

Drag force at the top node is given as:

$$F = \frac{1}{2}\rho C_D A V_m \left|V_m\right| = \frac{1}{2} \times 1.225 \times 1.2 \times 62 \times 22.33^2 = 2.27 \times 10^4 N \quad (106)$$

Wind Force Spectrum

The spectral density function as given in Equation 12 is given as:
For first frequency i.e. 0.22 Hz,

$$n_m(20, 0.22) = \frac{0.22 \times 20}{22.33} = 0.1970 \quad (107)$$

$$S_{vv}(20, 0.22) = \frac{200 \times 0.970^2 \times 0.1970}{\left(1 + 50 \times 0.1970\right)^{5/3}} = 3.1691 m^2 / \sec \quad (108)$$

Force spectrum from Equation 14 is given as:

$$S_F\left(f\right) = \frac{4 \times \left(2.27 \times 10000\right)^2 \times 3.1691 \times 0.75}{22.33^2} = 9.85 \times 10^6 N \sec \quad (109)$$

Similarly, through MATLAB simulation the wind velocity spectrum and force spectrum variation with frequency is given in Figure 20.

Cross Spectrum

Cross force spectrum is calculated as given in Equation 30. Hence in the present problem it is given as:

$$S_{fD,Mj} = \left(C_D A\rho\right)^2 \sum_{k=1}^{n}\sum_{l=1}^{n} S_{vkvl}\left(\omega\right) v_m\left(k\right) v_m\left(l\right) \varphi_j\left(k\right) \varphi_j\left(l\right) \quad (110)$$

$$= \left(C_D A\rho\right)^2 \begin{pmatrix} S_{v5v5}\left(f\right) v_m\left(5\right) v_m\left(5\right) \varphi_j\left(5\right) \varphi_j\left(5\right) + S_{v6v6}\left(f\right) v_m\left(6\right) v_m\left(6\right) \varphi_j\left(6\right) \varphi_j\left(6\right) \\ + S_{v7v7}\left(f\right) v_m\left(7\right) v_m\left(7\right) \varphi_j\left(7\right) \varphi_j\left(7\right) + 2S_{v5v6}\left(f\right) v_m\left(5\right) v_m\left(6\right) \varphi_j\left(5\right) \varphi_j\left(6\right) \\ + 2S_{v6v7}\left(f\right) v_m\left(6\right) v_m\left(7\right) \varphi_j\left(6\right) \varphi_j\left(7\right) + 2S_{v5v6}\left(f\right) v_m\left(5\right) v_m\left(6\right) \varphi_j\left(5\right) \varphi_j\left(6\right) \end{pmatrix} \quad (111)$$

Figure 20. (a) Wind velocity spectrum and (b) wind force spectrum for the nodes subjected to wind load

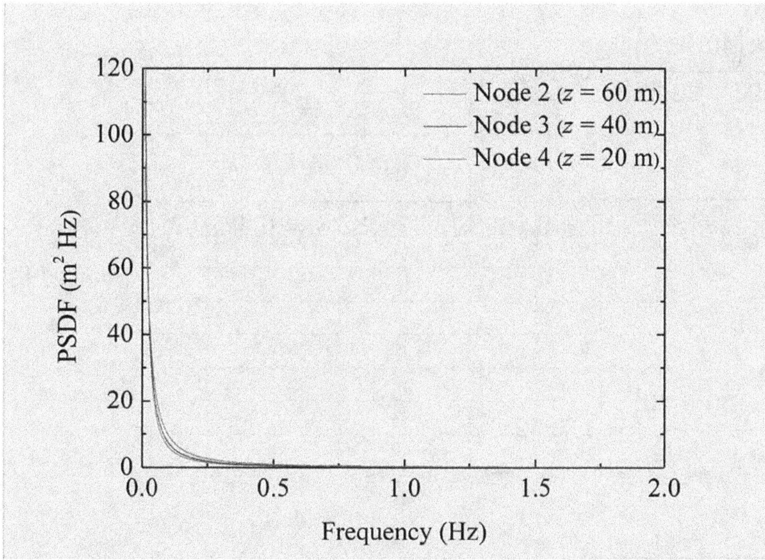

(a)

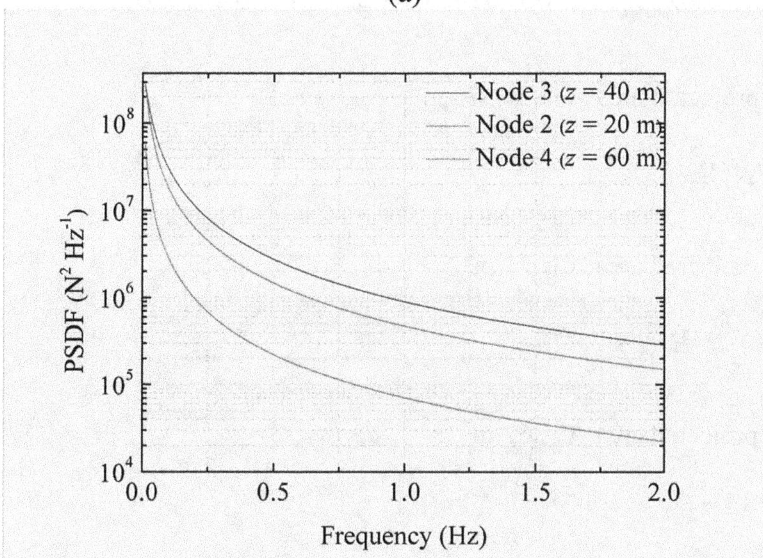

(b)

The modal coordinates at node 4 is 1, node 3 is 0.75 and node 1 is 0.50. For node 2 and 4:

$$\varphi = 2 \times 0.22 \times \frac{10 \times \left(60 - 20\right)}{25 + 22.33} = 3.72 \tag{112}$$

$$S_{v5v6}\left(f\right) = e^{-\varphi}\sqrt{S_{v5v5}\left(f\right)S_{v6v6}\left(f\right)} = 1.88 \times 10^{-10}\sqrt{2.1659 \times 3.7841} = 0.0695NHz^{-1} \tag{113}$$

For node 2 and 3:

$$\varphi = 2 \times 1.11 \times \frac{10 \times (40 - 20)}{24.02 + 22.33} = 1.8986 \tag{114}$$

$$S_{v5v6}(f) = e^{-\varphi}\sqrt{S_{v5v5}(f)S_{v6v6}(f)} = 0.1498\sqrt{2.69 \times 3.78} = 0.4779NHz^{-1} \tag{115}$$

For node 4 and 3:

$$\varphi = 2 \times 1.11 \times \frac{10 \times (60 - 40)}{25 + 24.02} = 1.7952 \tag{116}$$

$$S_{v5v6}(f) = e^{-\varphi}\sqrt{S_{v5v5}(f)S_{v6v6}(f)} = 0.1661\sqrt{2.69 \times 2.1659} = 0.4010NHz^{-1} \tag{117}$$

For node 4, the projected area A is 21.375 m^2

$$S_{FdM1} = 3.34 \times 10^6 N^2 Hz^{-1} \tag{118}$$

For node 6, the projected area A is 51 m^2

$$S_{FdM1} = 1.8994 \times 10^7 N^2 Hz^{-1} \tag{119}$$

For node 5, the projected area A is 62 m^2

$$S_{FdM1} = 2.81 \times 10^7 N^2 Hz^{-1} \tag{120}$$

The variation of modal force spectrum with frequency is plotted in Figure 21.

SDOF System

- **Natural Frequency:** The equivalent mass M_s for SDOF as shown in Figure 18 must be determined such that the natural frequency of the actual model is equal to that of the equivalent model.

$$\omega_s = \omega \rightarrow \sqrt{\frac{K_s}{M_s}} = \sqrt{\frac{K}{M}} \rightarrow M_s = M\frac{K_s}{K} = M\left(\frac{L}{L_s}\right)^3 = 170.36 \times \left(\frac{80}{20}\right)^3 = 10903tonne \tag{121}$$

Figure 21. Modal force spectrum considering correlation for MDOF system

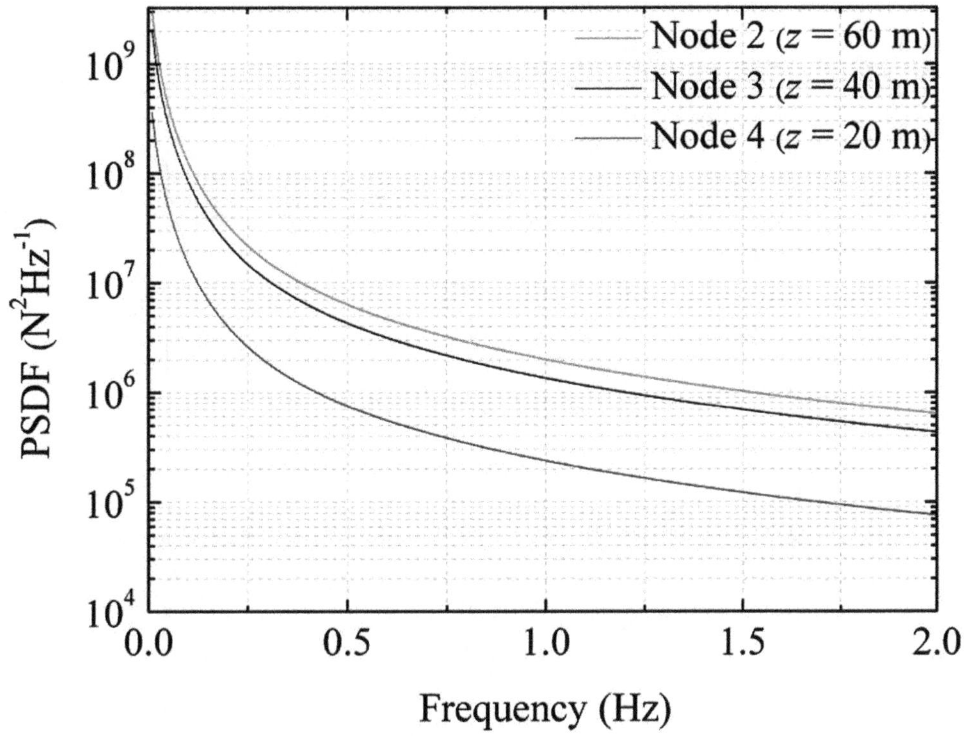

Hence, the natural frequency is same as that of the actual model.

- **Wave Elevation Spectrum:** The spectral density function of surface elevation $S_{\eta\eta}(f)$ at the point of application of wave load is presented in Figure 22.
- **Wave Force Spectrum:** For the node in water, the wave force spectrum is calculated by substituting the depth z as 20 m, natural frequency of the structure as 0.16 Hz and considering the lumped length and is given in Figure 22.

MDOF System

- **Natural Frequency:** The natural frequency of the model is 0.22 Hz
- **Wave Elevation Spectrum:** The wave elevation spectrum is the same as that of SDOF system.

Wave Force Spectrum

For node 1, the wave force spectrum remains the same as that of SDOF system.

Analyze a wind turbine tower by Basu et al. (2009) for wind and wave loading with the following data: external diameter of tower at base = 4.3 m, Young's modulus of pile, E = 205 × 10^9 N/m^2; Assume the light tower to behave as a cantilever beam, bending its fundamental mode of vibration. The foundation is assumed to be fully fixed into the ground, with no rotation allowed. The base of the tower is 4.3 m in diameter, tapering to 3.5 m at the nacelle. The steel thickness at the bottom and the top of the structure

Figure 22. (a) Wave elevation spectrum and (b) wave force spectrum for SDOF system

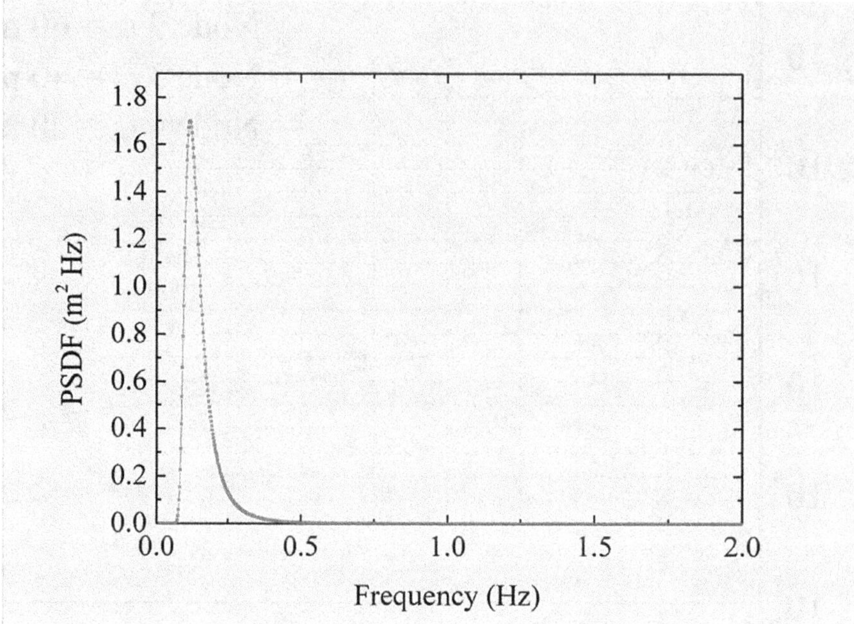

(a)

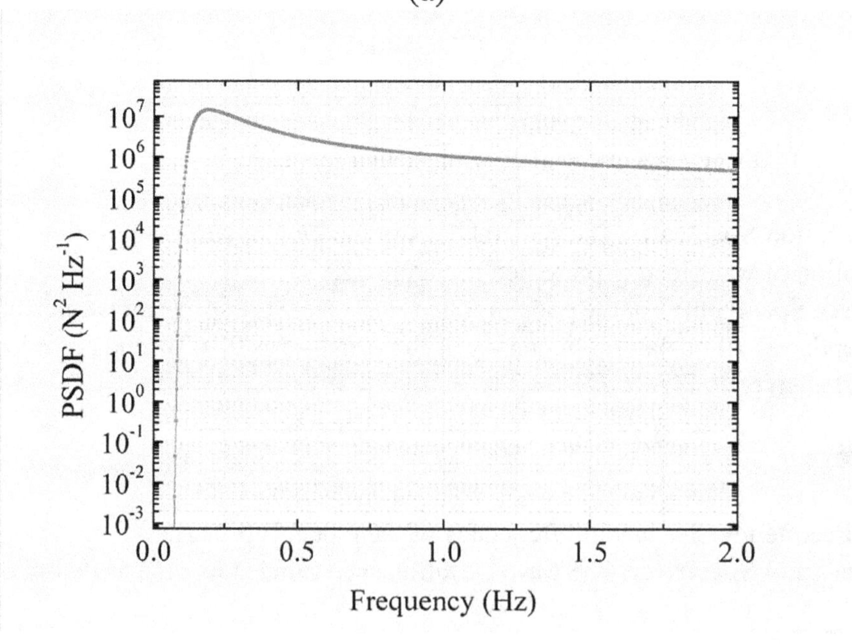

is 18 mm and 10 mm, respectively. Design wind speed (wind speed at height 10 m above sea level) = 18 m/sec, roughness length, z0 = 0.002, C_d = 1.2, wave height = 1.5 m, time period of wave = 6 sec.

1. Obtain the wind force spectrum for (i) SDOF and ii) MDOF system as shown in Figure 10.
2. Obtain the wave force spectrum for (i) SDOF and ii) MDOF system.

Figure 23. Structural idealization of an offshore wind turbine tower

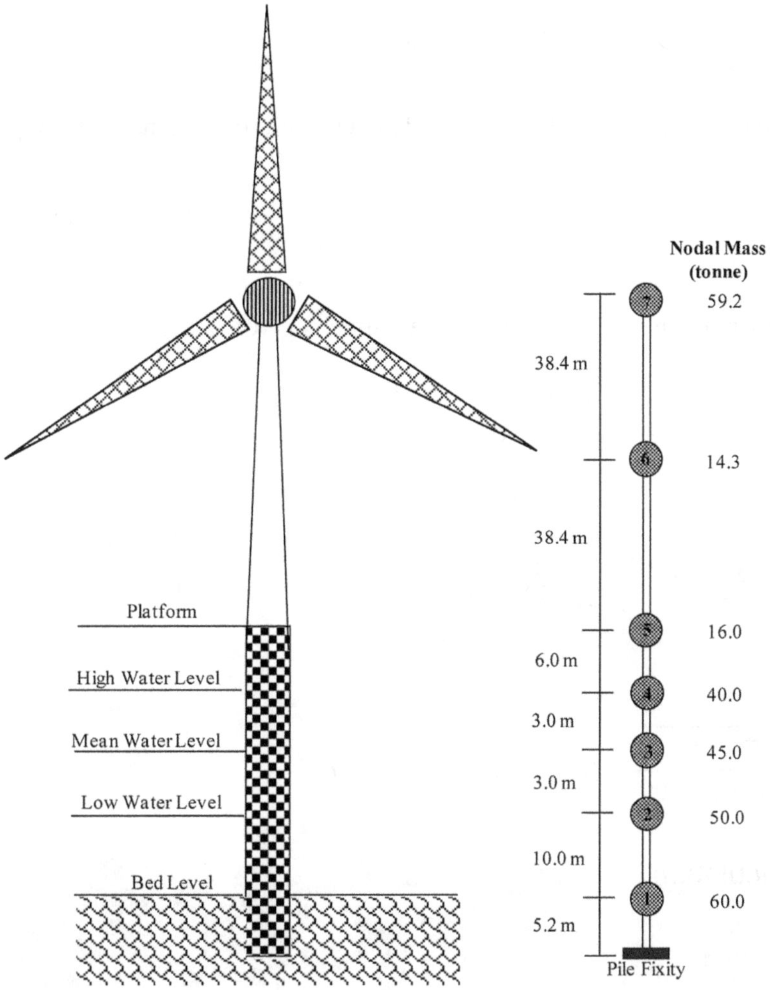

- **Solution:** SDOF System.
- **Natural Frequency:** Mass per length of the structure (considering volume of the tapered section + volume of the constant section) = 2155.73 kg/m.

Effective mass per unit length in water given as:

$$m_w = m + \rho_w \frac{\pi D^2}{4} + \rho_w \frac{\pi \left(D - 2t\right)^2}{4}$$
$$m_w = 2155.73 + \frac{1015 \times \pi \times (4.3 - 2 \times 0.018)^2}{4} + \frac{1015 \times \pi \times 4.3^2}{4} kg / m \tag{122}$$

$$m_w = 2155.73 + 14495.95 + 14741.75 = 31.3 tonne / m$$

Hence for the SDOF system, the mass at tip is given as:

$$M = 52 \times 2155.73 + 28.5 \times 1000 = 140.56 tonne \tag{123}$$

The mass of the blades and rotor has been added. The bending stiffness of the pile is

$$K = \frac{3EI}{L^3} \tag{124}$$

Taking the moment of inertia for the top node which is given as:

$$I = \frac{\pi\left(D_2^4 - D_1^4\right)}{64} = \frac{\pi\left(3.5^4 - 3.48^4\right)}{64} = 0.166 m^4 \tag{125}$$

$$K = \frac{3 \times 205 \times 10^9 \times 0.166}{104^3} = 90.76\ kN / m \tag{126}$$

$$\omega = \sqrt{\frac{K}{M}} = \sqrt{\frac{90.76 \times 1000}{140.56 \times 1000}} = 0.8\ rad / \sec = 0.13\ Hz \tag{127}$$

Wind Force Calculation

Shear velocity is:

$$v_* = \frac{V_m}{2.5 \ln \frac{z}{z_0}} = \frac{18}{2.5 \ln \frac{10}{0.002}} = 0.845\ m / \sec \tag{128}$$

Wind velocity at a height of 82.8 m (top node) above is given as:

$$V_m(82.8) = 2.5 \times 0.825 \times \ln \frac{82.8}{0.002} = 22.47\ m / \sec \tag{129}$$

Drag force at the top node is given as:

$$F = \frac{1}{2}\rho C_D A V_m \left|V_m\right| = \frac{1}{2} \times 1.225 \times 1.2 \times 196.04 \times 22.47^2 = 72.75 kN \tag{130}$$

Wind Force Spectrum

The spectral density function as given in Equation 12 is given as:

For first frequency i.e. 0.13 Hz,

$$n_m(82.8, 0.13) = \frac{0.265 \times 82.8}{22.47} = 0.4791 \tag{131}$$

$$S_{vv}(82.8, 0.13) = \frac{200 \times 0.845^2 \times 0.4791}{\left(1 + 50 \times 0.4791\right)^{5/3}} = 2471 m^2 Hz \tag{132}$$

Aerodynamic admittance factor from chart of Davenport is 1. Hence the force spectrum for frequency of 0.265 Hz is given as:

$$S_F\left(f\right) = \frac{4 \times \left(72.75 \times 1000\right)^2 \times 2.471 \times 0.75}{22.47^2} = 7.78 \times 10^7 N \sec \tag{133}$$

Similarly, through MATLAB simulation the wind velocity spectrum and force spectrum variation with frequency is given in Figure 24.

MDOF System

- **Natural Frequency:** The fundamental period is 0.18 Hz as given in the paper.

Wind Force Calculation for Node 7

Wind velocity at a height of 82.8 m (top node) above is given as:

$$V_m(82.8) = 22.47 m / \sec \tag{134}$$

Drag force at the top node is given as:

$$F = \frac{1}{2}\rho C_D A V_m \left|V_m\right| = \frac{1}{2} \times 1.225 \times 1.2 \times 142.08 \times 22.47^2 = 5.27 \times 10^4 N \tag{135}$$

Wind Force Spectrum

The spectral density function as given in Equation 12 is given as:

For first frequency i.e. 0.18 Hz,

Figure 24. (a) Wind velocity spectrum and (b) wind force spectrum for SDOF system

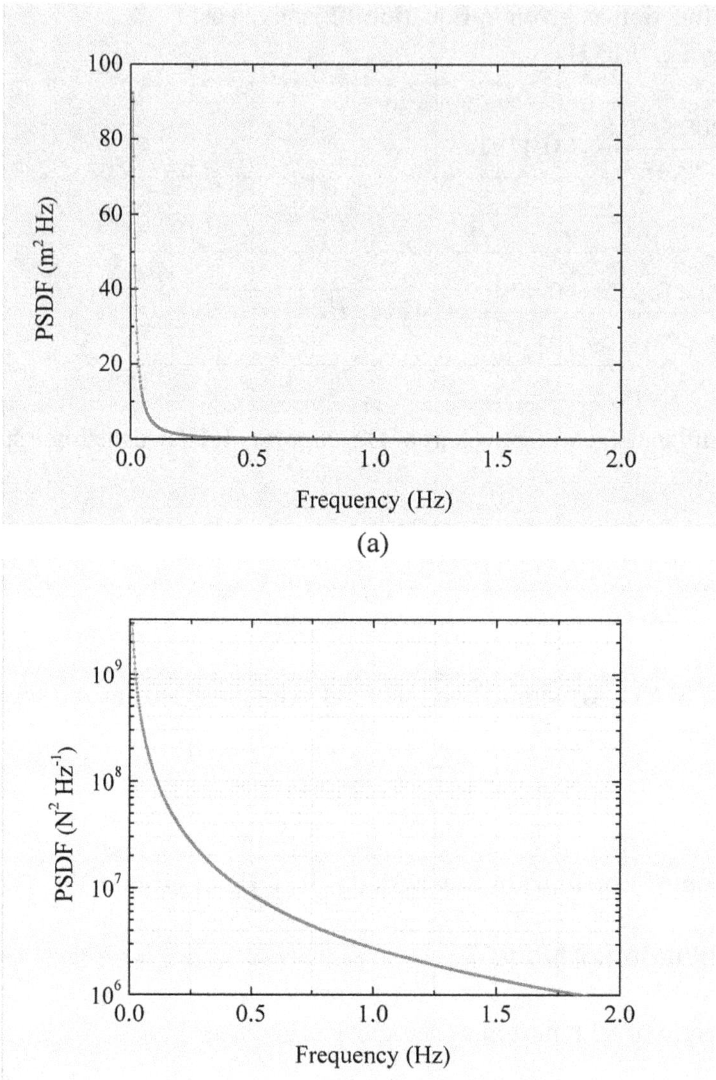

$$n_m(82.8, 0.18) = \frac{0.18 \times 82.8}{22.47} = 0.6634 \tag{136}$$

$$S_{vv}(82.8, 0.18) = \frac{200 \times 0.845^2 \times 0.6634}{\left(1 + 50 \times 0.6634\right)^{5/3}} = 1.464 m^2 / \sec \tag{137}$$

Aerodynamic admittance factor from chart of Davenport is 0.75. Hence the force spectrum for frequency of 0.18 Hz is given as:

$$S_F(f) = \frac{4 \times (5.27 \times 10000)^2 \times 1.464 \times 0.75}{22.47^2} = 2.417 \times 10^7 \; NHz \tag{138}$$

Similarly, through MATLAB simulation the wind velocity spectrum and force spectrum variation with frequency is given in Figure 25.

Figure 25. (a) Wind velocity spectrum and (b) wind force spectrum for node 7, node 6 and node 5 of the MDOF system subjected to wind load

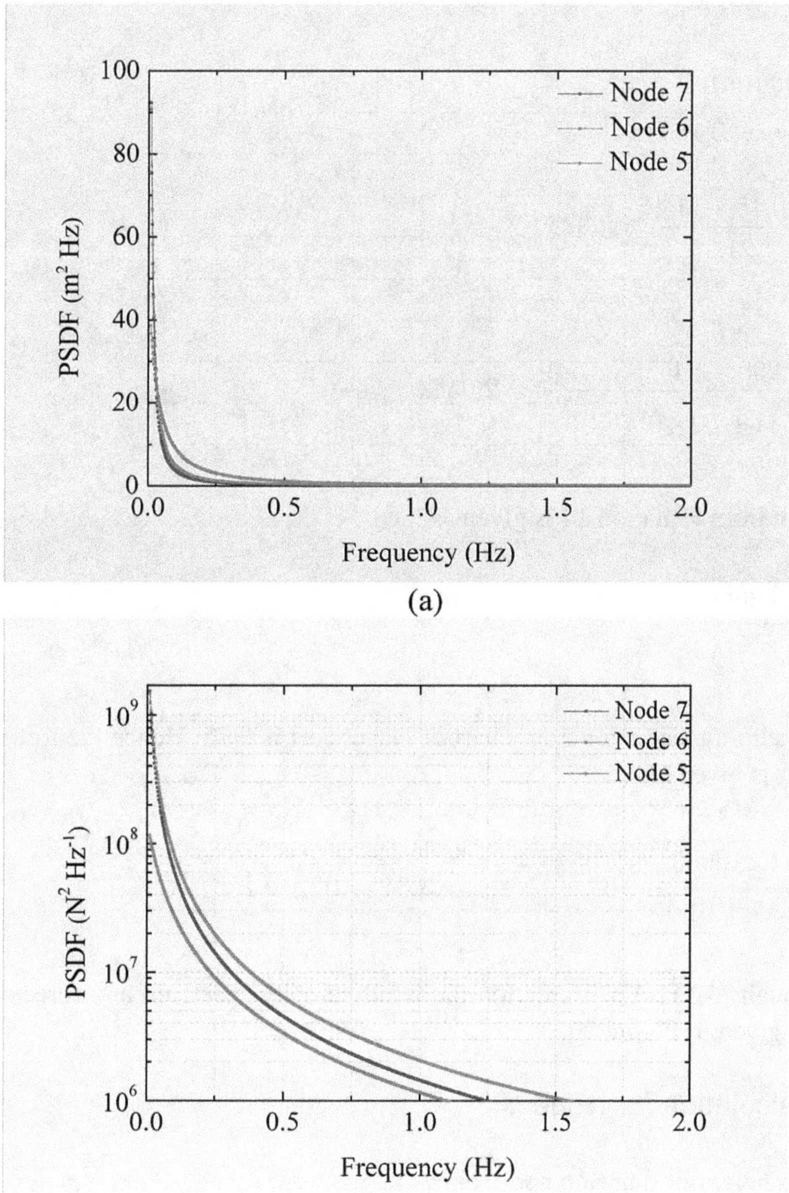

Wind Force Calculation for Node 6

Wind velocity at a height of 44.4 m (top node) above is given as:

$$V_m(82.8) = 2.5 \times 0.845 \times \ln \frac{44.4}{0.002} = 21.15\ m / \sec \tag{139}$$

Drag force at the top node is given as:

$$F = \frac{1}{2}\rho C_D A V_m \left|V_m\right| = \frac{1}{2} \times 1.225 \times 1.2 \times 149.76 \times 21.15^2 = 4.924 \times 10^4 N \tag{140}$$

Wind Force Spectrum

For first frequency i.e. 0.18 Hz,

$$n_m(44.4, 0.18) = \frac{0.18 \times 44.4}{21.15} = 0.3779 \tag{141}$$

$$S_{vv}(44.4, 0.18) = \frac{200 \times .845^2 \times 0.3779}{\left(1 + 50 \times 0.3779\right)^{5/3}} = 2.054 m^2 / \sec \tag{142}$$

Force spectrum from Equation 14 is given as:

$$S_F\left(f\right) = \frac{4F^2 S\left(f\right) A(f)}{V_m^2} \tag{143}$$

Aerodynamic admittance factor from chart of Davenport is 0.75. Hence the force spectrum for frequency of 1.11 Hz is given as:

$$S_F\left(f\right) = \frac{4 \times \left(4.94 \times 10000\right)^2 \times 2.054 \times 0.75}{21.15^2} = 3.34 \times 10^7\ N / Hz \tag{144}$$

Similarly, through MATLAB simulation the wind velocity spectrum and force spectrum variation with frequency is given in Figure 20.

Wind Force Calculation for Node 5

Wind velocity at a height of 6 m (top node) above is given as:

$$V_m(6) = 2.5 \times 0.845 \times \ln \frac{6}{0.002} = 16.92\ m / \sec \tag{145}$$

Drag force at the top node is given as:

$$F = \frac{1}{2}\rho C_D A V_m \left|V_m\right| = \frac{1}{2} \times 1.225 \times 1.2 \times 80.64 \times 16.92^2 = 1.69 \times 10^4\ N \tag{146}$$

Wind Force Spectrum

For first frequency i.e. 0.18 Hz,

$$n_m(6, 0.18) = \frac{0.18 \times 6}{16.92} = 0.0638 \tag{147}$$

$$S_{vv}(6, 0.265) = \frac{200 \times 0.845^2 \times 0.0638}{\left(1 + 50 \times 0.0638\right)^{5/3}} = 4.651\ m^2\ Hz \tag{148}$$

Aerodynamic admittance factor from chart of Davenport is 0.75. Hence the force spectrum for frequency of 0.18 Hz is given as:

$$S_F\left(f\right) = \frac{4 \times \left(1.69 \times 10000\right)^2 \times 4.6561 \times 0.75}{16.92^2} = 1.40 \times 10^7\ N / Hz \tag{149}$$

Similarly, through MATLAB simulation the wind velocity spectrum and force spectrum variation with frequency is given in Figure 25.

Cross Spectrum

Cross force spectrum is calculated as given in Equation 30. Hence in the present problem it is given as:

$$S_{fD,Mj} = \left(C_D A\rho\right)^2 \sum_{k=1}^{n}\sum_{l=1}^{n} S_{vkvl}\left(\omega\right) v_m\left(k\right) v_m\left(l\right) \varphi_j\left(k\right) \varphi_j\left(l\right) \tag{150}$$

$$\left(C_D A\rho\right)^2 \begin{pmatrix} S_{v5v5}\left(f\right) v_m\left(5\right) v_m\left(5\right) \varphi_j\left(5\right) \varphi_j\left(5\right) + S_{v6v6}\left(f\right) v_m\left(6\right) v_m\left(6\right) \varphi_j\left(6\right) \varphi_j\left(6\right) \\ +S_{v7v7}\left(f\right) v_m\left(7\right) v_m\left(7\right) \varphi_j\left(7\right) \varphi_j\left(7\right) + 2S_{v5v6}\left(f\right) v_m\left(5\right) v_m\left(6\right) \varphi_j\left(5\right) \varphi_j\left(6\right) \\ +2S_{v6v7}\left(f\right) v_m\left(6\right) v_m\left(7\right) \varphi_j\left(6\right) \varphi_j\left(7\right) + 2S_{v5v6}\left(f\right) v_m\left(5\right) v_m\left(6\right) \varphi_j\left(5\right) \varphi_j\left(6\right) \end{pmatrix}$$

Only the first mode shape is considered. The modal coordinates at node 7 is 1, node 6 is 0.45 and node 5 is 0.09.

For node 5 and 6:

$$\varphi = 2 \times 0.265 \times \frac{10 \times (44.4 - 6)}{16.92 + 21.15} = 5.3459 \tag{151}$$

$$S_{v5v6}(f) = e^{-\varphi}\sqrt{S_{v5v5}(f)\,S_{v6v6}(f)} = 0.0048\sqrt{2.2724 \times 0.9030} = 0.0068N / Hz \tag{152}$$

For node 6 and 7:

$$\varphi = 2 \times 1.11 \times \frac{10 \times (82.8 - 44.4)}{22.47 + 21.15} = 4.6657 \tag{153}$$

$$S_{v5v6}(f) = e^{-\varphi}\sqrt{S_{v5v5}(f)\,S_{v6v6}(f)} = 0.0094\sqrt{0.6364 \times 0.9030} = 0.0071\ N / Hz \tag{154}$$

For node 5 and 7:

$$\varphi = 2 \times 0.265 \times \frac{10 \times (82.8 - 6)}{22.47 + 16.92} = 10.336 \tag{155}$$

$$S_{v5v6}(f) = e^{-\varphi}\sqrt{S_{v5v5}(f)\,S_{v6v6}(f)} = 3.25 \times 10^{-5} \times \sqrt{2.2724 \times 0.6364} = 3.91 \times 10^{-5} NHz^{-1} \tag{156}$$

For node 7, the projected area A is 142.08 m^2

$$S_{FdM1} = 1.79 \times 10^{7} N^{2}Hz^{-1} \tag{157}$$

For node 6, the projected area A is 149.76 m^2

$$S_{FdM1} = 1.99 \times 10^{7} N^{2}Hz^{-1} \tag{158}$$

For node 5, the projected area A is 80.64 m^2

$$S_{FdM1} = 5.77 \times 10^{6} N^{2}Hz^{-1} \tag{159}$$

Similarly, through MATLAB simulation the cross wind force spectrum variation with frequency is given in Figure 26.

SDOF System

- **Natural Frequency:** The equivalent mass M_s as shown in Figure 18 must be determined such that the natural frequency of the actual model is equal to that of the equivalent model.

$$\omega_S = \omega \rightarrow \sqrt{\frac{K_s}{M_s}} = \sqrt{\frac{K}{M}} \rightarrow M_s = M\frac{K_s}{K} = M\left(\frac{L}{L_s}\right)^3 = 140.56 \times \left(\frac{104}{21.2}\right)^3 = 16594\ tonne \tag{160}$$

Hence, the natural frequency is same as that of the actual model.

- **Wave Elevation Spectrum:** The spectral density function of surface elevation $S_{\eta\eta}(f)$ at the point of application of wave load at high water level is given in Figure 33.
- **Wave Force Spectrum:** For the node in water, the wave force spectrum is calculated by substituting the depth z as 21.2 m, natural frequency of the structure as 0.13 Hz and considering the lumped length and is given in Figure 33.

Figure 26. Modal force spectrum considering correlation for MDOF system

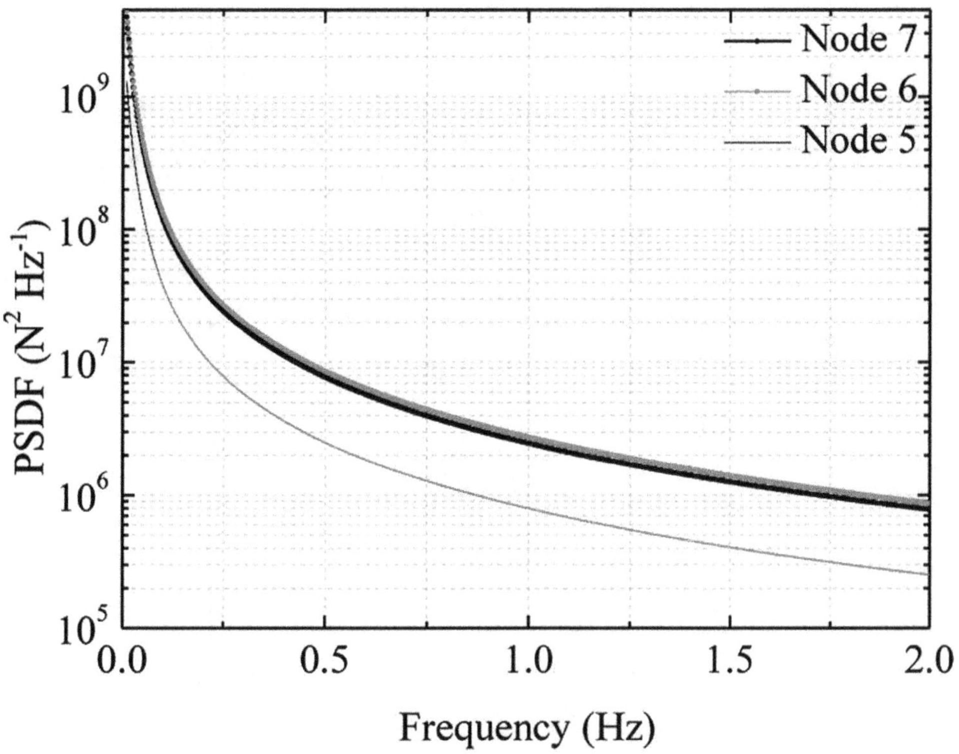

MDOF System

- **Natural Frequency:** The fundamental period is 0.18 Hz
- **Wave Elevation Spectrum:** The spectral density function of surface elevation $S_{\eta\eta}(f)$ remains the same as that in Figure 21.
- **Wave Force Spectrum:** For node 4, 3, 2, 1 the surface wave elevation spectrum remains the same as that of SDOF system as it is independent of depth z. Wave force spectra is determined in the same way as described in the previous section by altering the depth and the subsequent lumped length and is plotted in Figure 28.

Figure 27. (a) Wave elevation spectrum and (b) wave force spectrum for SDOF system

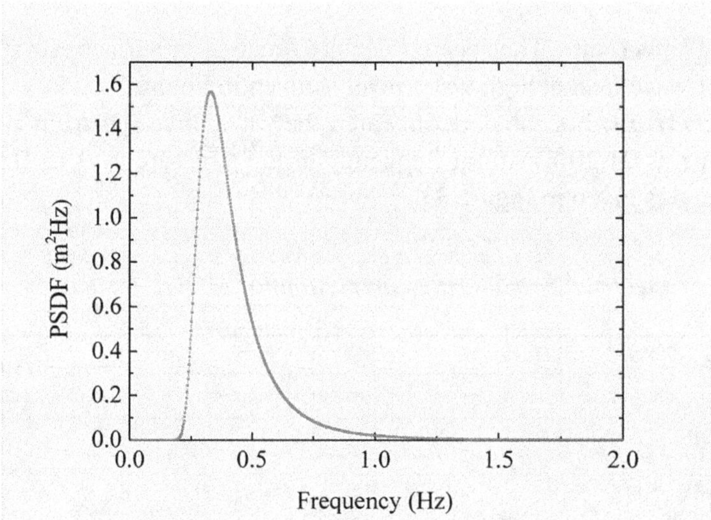

(a)

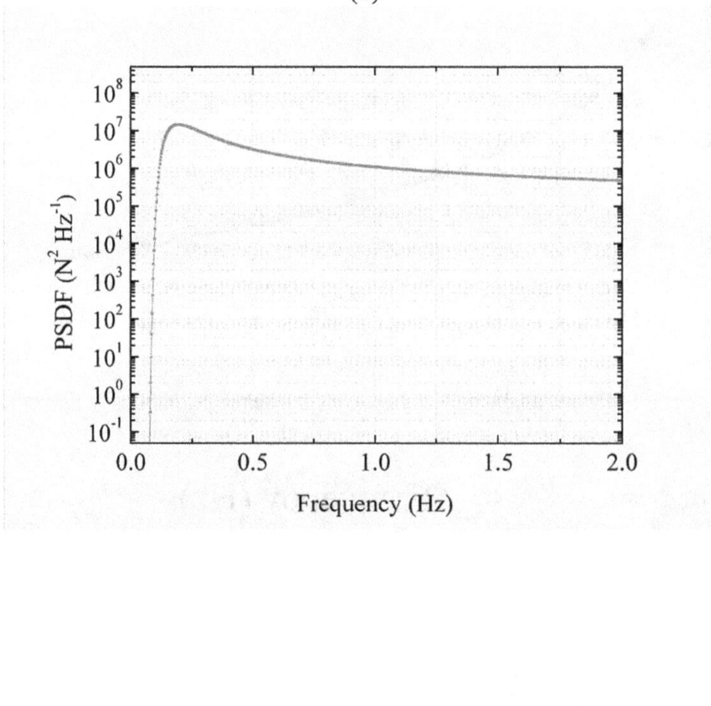

Figure 28. Wave force spectrum for submerged nodes for MDOF system

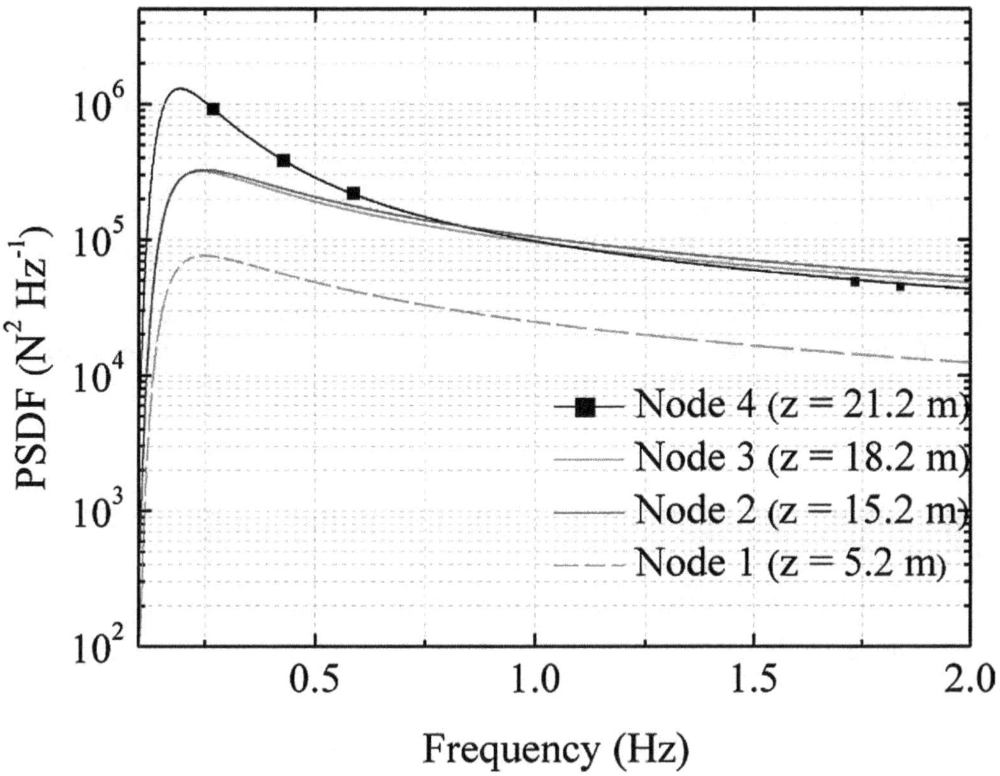

STRUCTURAL DYNAMICS OF WIND TURBINES

Finite Element Modeling (FEM)

Finite element (FE) modeling of slender towers are usually based on Euler-Bernoulli (EB) beam theory. Euler-Bernoulli beam theory (also known as classical beam theory) is a simplification of the linear theory of elasticity which provides a means of calculating the load-carrying and deflection characteristics of beams. Wind turbine towers and blades are slender in section compared to their lengths, and can thus be represented by Euler-Bernoulli beams whose structural response is characterized by flexural deformation which can be read in detail in (Xiao-bo, Jing, & Jian-Yo, 2009) The governing equation for displacement of a continuous Euler beam is obtained as:

$$m(x)\frac{\partial^2 u}{dt^2}+\frac{\partial^2}{dx^2}\left[EI(x)\frac{\partial^2 u}{dx^2}\right]=P(x,t) \quad (161)$$

where $m\ (x)$ is the mass density per unit length of the beam, $EI\ (x)$ is its flexural stiffness, $u\ (x,t)$ is its transverse displacement and $P\ (x,t)$ is the external load.

The displacement $u\ (x)$ can be represented as:

$$u(x,t) = \sum_{i=1}^{\infty} \varphi_i(x) q_i(t) \tag{162}$$

where φ(*x*) are the mode shapes of the structure and *q*(*t*) are known as the modal coordinates.

The mode shapes are determined by solving:

$$(EI(x)\varphi''(x))'' - \omega^2 m(x)\varphi(x) = 0 \tag{163}$$

where primes denote spatial derivatives. There are an infinite number of functions φ(*x*) and natural frequencies ω for the beam. The modal coordinates are determined by solving the equation given as:

$$M\ddot{q} + C\dot{q} + Kq = F \tag{164}$$

where overdots represent time derivatives and *M* and *K* are the $n \times n$ generalized mass and stiffness matrices, respectively.

$$M_{i,j} = \int_0^l \varphi_i(x) m(x) \varphi_j(x) dx \tag{165}$$

$$K_{i,j} = \int_0^l \varphi_i''(x) EI \varphi_j''(x) dx \tag{166}$$

$$F_{i,1} = \int_0^l \varphi(x) P(x,t) dx \tag{167}$$

where *l* is the length of the element and *n* is the number of modes.

Assuming appropriate cubic polynomial shape functions for the mode shape given as:

$$\varphi(x) = 1 - 3\xi + 2\xi^3 \tag{168}$$

A complete calculation gives the following local finite element matrices for the Euler beam:

$$[M] = \frac{\rho A l}{420}\begin{bmatrix} 156 & 22l & 54 & -13l \\ 22l & 4l^2 & 13l & -3l^2 \\ 54 & 13l & 156 & -22l \\ -13l & -3l^2 & -22l & 4l^2 \end{bmatrix} [K] = \frac{EI}{l^3}\begin{bmatrix} 12 & 6 & -12 & 6 \\ 6l & 4l^2 & -6l & 2l^2 \\ -12 & -6l & 12 & -6l \\ 6l & 2l^2 & -6l & 4l^2 \end{bmatrix} \tag{169}$$

The complete method of determining the local matrices are provided in (Robert, 1996). Readers are recommended to go through the above mentioned reference as a prerequisite to this section.

In general, a wind turbine structure can be classified and consequently modeled as a vertical cantilever structure i.e. the turbine tower supporting three cantilever sub-structures, e.g. the blades as shown in Figure 29 (a). If the masses of the nacelle and hub are assumed to be lumped at the top of the tower, the turbine can be modeled as shown in Figure 29 (c). The blade and tower elements have distributed stiffness and distributed mass. Since the blades and tower are long compared to their cross-sectional dimensions, they are assumed to be Euler-Bernoulli beams with dominant response contributed by the flexural modes. Figure 30 (b) demonstrates the dominant flexural response along the flap-wise direction of a wind turbine.

Finite Element Analysis

- **Free Vibration:** The structure may be divided into *multiple beam elements* as described in section 3.1, with local mass and stiffness matrices for each element derived using Equation 68. The local matrices are assembled into global mass and stiffness matrices, *M* and *K*, respectively after connecting each element. So equation of motion considering lateral motion *x* for the whole system becomes:

$$\left[M\right]\ddot{x}+\left[K\right]x=0 \tag{170}$$

Solving the Eigen value problem as:

Figure 29.

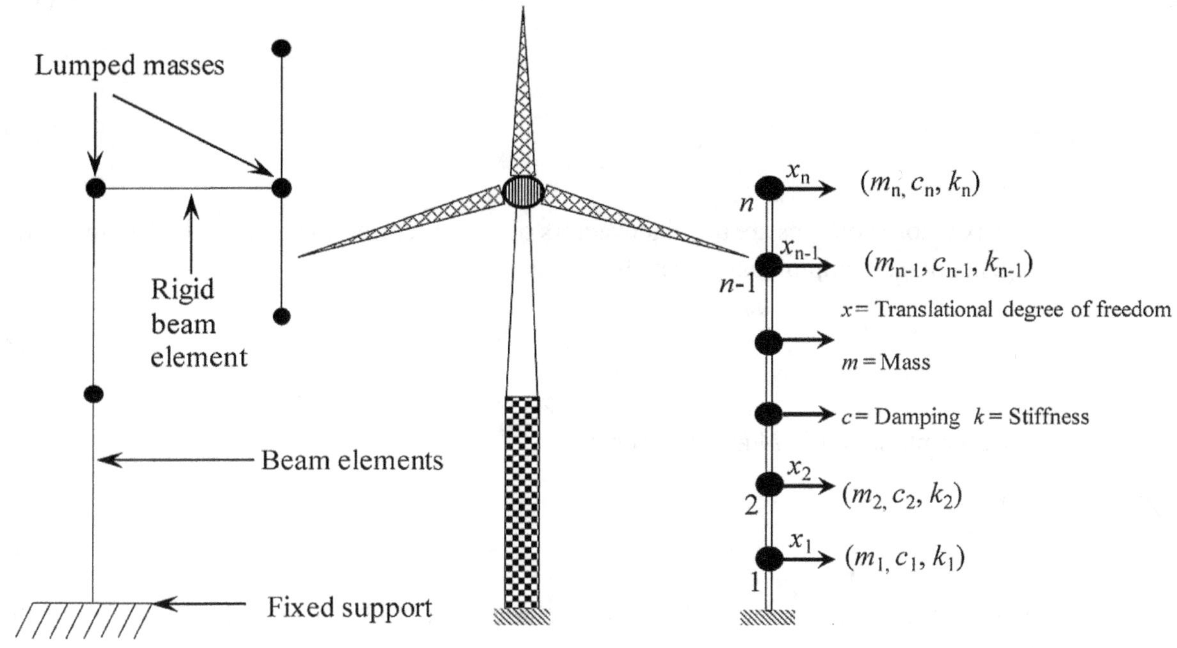

Figure 30. (a) Wind turbine model, and (b) its representation of flap-wise model of wind turbine

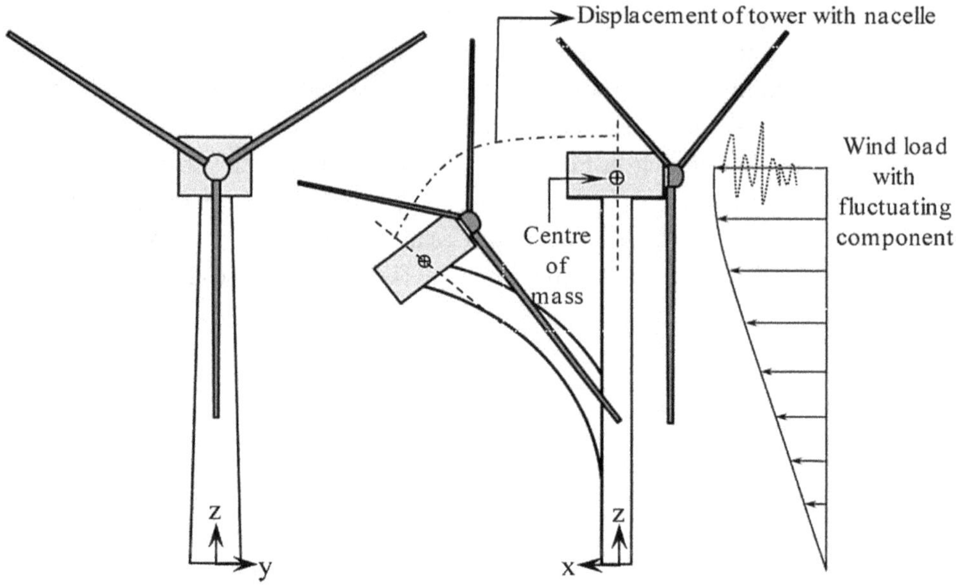

$$[K]-\omega_n^{\,2}[M]=0 \tag{171}$$

yields the natural frequencies and mode shapes.

- **Forced Vibration:** The generalized mass and stiffness matrices can be expressed as:

$$M_G=\varphi^{\mathrm{T}}M\varphi \tag{172}$$

$$K_G=\varphi^{\mathrm{T}}K\varphi \tag{173}$$

where, φ is a matrix whose columns are the eigenvectors of the structure, superscript 'T' denotes matrix transpose. The generalized damping matrix can be expressed as:

$$C_G=\varphi^{\mathrm{T}}C\varphi \tag{174}$$

where ***C*** is a linear combination of *M* and *K* given as:

$$C=\alpha[M]+\beta[K] \tag{175}$$

This is known as Rayleigh damping with the coefficients for *M* and *K* determined by assuming that the damping ratio ξ is known in two modes as:

$$\alpha = 2\xi_n \omega_n ; \beta = \frac{2\xi_n}{\omega_n} \tag{176}$$

The load vector can be expressed as:

$$F = \varphi^{\mathrm{T}} P \tag{177}$$

where *P* is a vector of nodal loads.

$$\varphi = \begin{bmatrix} \varphi_1 & \varphi_2 & \cdots & \varphi_n \end{bmatrix} = \begin{bmatrix} \varphi_{11} & \varphi_{12} & \cdots & \varphi_{1n} \\ \varphi_{21} & \varphi_{22} & \cdots & \varphi_{2n} \\ \vdots & \vdots & \ddots & \vdots \\ \varphi_{n1} & \varphi_{n2} & \cdots & \varphi_{nn} \end{bmatrix} \tag{178}$$

$$q = \begin{bmatrix} q_1 \\ q_2 \\ \vdots \\ q_n \end{bmatrix} \tag{179}$$

$$F = \begin{bmatrix} \varphi_1^{\mathrm{T}} P_1 \\ \varphi_2^{\mathrm{T}} P_2 \\ \vdots \\ \varphi_{\mathrm{n}}^{\mathrm{T}} P_{\mathrm{n}} \end{bmatrix} \tag{180}$$

In general, the equation of motion representing their wind as well as wave induced forced vibrational motion is as follows:

$$\begin{bmatrix} M_1 & 0 & \cdots & 0 \\ 0 & M_2 & \cdots & 0 \\ \vdots & \vdots & \ddots & 0 \\ 0 & 0 & 0 & M_n \end{bmatrix} \begin{Bmatrix} \ddot{x}_1 \\ \ddot{x}_1 \\ \vdots \\ \ddot{x}_n \end{Bmatrix} + \begin{bmatrix} C_1 & 0 & \cdots & 0 \\ 0 & C_2 & \cdots & 0 \\ \vdots & \vdots & \ddots & 0 \\ 0 & 0 & 0 & C_n \end{bmatrix} \begin{Bmatrix} \dot{x}_1 \\ \dot{x}_1 \\ \vdots \\ \dot{x}_n \end{Bmatrix} + \begin{bmatrix} K_1 & 0 & \cdots & 0 \\ 0 & K_2 & \cdots & 0 \\ \vdots & \vdots & \ddots & 0 \\ 0 & 0 & 0 & K_n \end{bmatrix} \begin{Bmatrix} x_1 \\ x_2 \\ \vdots \\ x_n \end{Bmatrix} = \begin{Bmatrix} P_1(t) \\ P_2(t) \\ \vdots \\ P_n(t) \end{Bmatrix} \tag{181}$$

where $\dot{x}$, $\ddot{x}$ are the time-dependent displacement, velocity and acceleration vectors respectively, and *P*(*t*) is the time varying load vector.

In this case, the mass matrix M includes the added mass of the structure up to the still water level for an offshore structure.

Numerical Example 2

Analyze the tower given in numerical problem 1.1 to calculate the following:

1. Calculate the forces at each node and write the equation of motion for the 2-DOF system.
2. Convert the system into a single degree of freedom (SDOF) system and calculate response spectrum under wave loading (CD = 1.3, CM = 2).

Solution: Wind Force Calculation

From Equation 6,

$$v_* = \frac{V_m}{2.5\ln\frac{z}{z_0}} = \frac{19}{2.5\ln\frac{10}{0.001}} = 0.825\ m/\sec \tag{182}$$

Wind velocity at a height of 8.5 m above is given as:

$$V_m(8.5) = 2.5\times 0.825\times \ln\frac{8.5}{0.001} = 18.66\ m/\sec \tag{183}$$

Drag force at the top node is given as:

$$F = \frac{1}{2}\rho C_D A u\left|u\right| = \frac{1}{2}\times 1.025\times 0.75\times 0.324\times 4.25\times 18.66^2 = 184.29\ N \tag{184}$$

Wave Force Calculation

Effective mass per unit length in water given as:

$$m_w = m + \rho_w\frac{\pi D^2}{4} + \rho_w\frac{\pi\left(D-2t\right)^2}{4} = 73.8 + \frac{1015\times\pi\times 0.324^2}{4} + \frac{1015\times\pi\times 0.305^2}{4} \tag{185}$$

$$m_w = 73.8 + 83.7 + 74.2 = 231.7\ kg/m \tag{186}$$

Discretizing the structure into lumped mass model as shown in Figure 10, the mass matrix is given as:

$$\left[M\right] = \begin{bmatrix} 1522 & 0 \\ 0 & 635 \end{bmatrix} m \tag{187}$$

The stiffness matrix has been calculated as shown in Figure 12. The final stiffness matrix is given as:

$$\left[K\right] = 10^3 \times \begin{bmatrix} 531 & -166 \\ -166 & 66 \end{bmatrix} N/m \tag{188}$$

Frequency is calculated using Eigen value characteristic equation as found as:

$$\begin{aligned} \omega_1 &= 4.27\ rad / \sec \\ \omega_2 &= 20.86\ \ rad / \sec \end{aligned} \tag{189}$$

The damping matrix is given by Raleigh damping and given as:

$$a_0 = 2\left[\frac{\omega_1\omega_2}{\omega_1 + \omega_2}\right] \times \xi = 2\left[\frac{4.27 \times 20.86}{4.27 + 20.86}\right] \times 0.008 = 0.056 \tag{190}$$

$$a_1 = \left[\frac{2\xi}{\omega_1 + \omega_2}\right] = \left[\frac{2 \times 0.008}{4.27 + 20.86}\right] = 6.37 \times 10^{-4} \tag{191}$$

$$\left[C\right] = a_0\left[M\right] + a_1\left[K\right] = 0.056 \times \begin{bmatrix} 1523 & 0 \\ 0 & 635 \end{bmatrix} + .637 \times \begin{bmatrix} 531 & -166 \\ -166 & 66 \end{bmatrix} \tag{192}$$

$$= \begin{bmatrix} 423.50 & -105.74 \\ -105.74 & 77.60 \end{bmatrix} Kg\ m / \sec \tag{193}$$

Taking CD = 1.3 and CM = 2. For the node under water, the wave force acting on it is given as:

$$F = \frac{1}{2}\rho C_D A u\left|u\right| + \rho C_M V\dot{u} \tag{194}$$

$$A = 0.324 \times 4.25 = 1.377 m^2 \tag{195}$$

$$V = \frac{\pi}{4} \times 0.324^2 \times 4.25 = 0.350 m^3 \tag{196}$$

$$y = z + d = -1.5 + 10 = 8.5 m \tag{197}$$

Water particle velocity is given as:

$$u = \frac{\pi H}{T} \times \frac{\cosh ky}{\sinh kd} \times \cos\left(kx - \omega t\right) \tag{198}$$

Water particle acceleration is given as:

$$\dot{u} = \frac{2\pi^2 H}{T^2} \times \frac{\cosh ky}{\sinh kd} \times \sin\left(kx - \omega t\right) \tag{199}$$

Wave length for shallow water condition is given as:

$$L = T\sqrt{gD} = 59.4 m \tag{200}$$

Wave number is given as:

$$k = \frac{2\pi}{L} = 0.106\ m^{-1} \tag{201}$$

Water Particle Velocity and Acceleration

Water particle velocity and acceleration for y = 8.5 m and x = 0 is given as:

$$u = \frac{\pi \times 1.5}{6} \times \frac{\cosh\left(0.106 \times 8.5\right)}{\sinh\left(0.106 \times 10\right)} \times \cos\left(kx - \omega t\right) = 0.88 \times \cos\left(\omega t\right) \tag{202}$$

$$\dot{u} = \frac{2\pi^2 \times 1.5}{6^2} \times \frac{\cosh\left(0.106 \times 8.5\right)}{\sinh\left(0.106 \times 10\right)} \times -\sin\left(\omega t\right) = -0.93 \times \sin\left(\omega t\right) \tag{203}$$

Water particle velocity and acceleration for y = 0 and x = 0 is given as:

$$u = \frac{\pi \times 1.5}{6} \times \frac{\cosh\left(0.106 \times 0\right)}{\sinh\left(0.106 \times 10\right)} \times \cos\left(kx - \omega t\right) = 0.61 \times \cos\left(\omega t\right) \tag{204}$$

$$\dot{u} = \frac{2\pi^2 \times 1.5}{6^2} \times \frac{\cosh\left(0.106 \times 0\right)}{\sinh\left(0.106 \times 10\right)} \times -\sin\left(\omega t\right) = -0.65 \times \sin\left(\omega t\right) \tag{205}$$

Assuming maximum response at ωt = 315°. Hence wave elevation at ωt=315° is given as:

$$\eta = \frac{H}{2} \cos\left(kx - \omega t\right) = \frac{1.5}{2} \cos\left(\omega t\right) = 0.53\,m \tag{206}$$

Hence, water particle velocity and acceleration for y = 10.53 m and x = 0 is given as:

$$u = \frac{\pi \times 1.5}{6} \times \frac{\cosh\left(0.106 \times 10.53\right)}{\sinh\left(0.106 \times 10\right)} \times \cos\left(\omega t\right) = 0.98 \times \cos\left(315\right) = 0.693 \; m / \sec \tag{207}$$

$$\dot{u} = \frac{2\text{À}^2 \times 1.5}{6^2} \times \frac{\cosh\left(0.106 \times 10.53\right)}{\sinh\left(0.106 \times 10\right)} \times -\sin\left(\text{É}t\right) = -1.095 \times \sin\left(\text{É}t\right) = 0.774 \text{ m/sec} \tag{208}$$

Now force per/length (f) at y = 0, 8.5 m and 10.53 m are given for y = 0 m:

$$f_D = \frac{1}{2}\rho_w C_D D \left|u\right| u = 1015 \times 0.5 \times 1.3 \times 0.324 \times 0.186 = 39.750 \; N / m \tag{209}$$

$$f_I = \rho_w C_M \frac{\pi D^2}{4} \dot{u} = 1015 \times 2 \times \frac{\pi \times 0.324^2}{4} \times 0.460 = 76.990 \; N / m \tag{210}$$

$$f = f_D + f_I = 116.740 \; N / m \tag{211}$$

For y = 8.5 m

$$f_D = \frac{1}{2}\rho_w C_D D \left|u\right| u = 1015 \times 0.5 \times 1.3 \times 0.324 \times 0.387 = 82.72 \; N / m \tag{212}$$

$$f_I = \rho_w C_M \frac{\pi D^2}{4} \dot{u} = 1015 \times 2 \times \frac{\pi \times 0.324^2}{4} \times 0.657 = 109.976 \ N / m \tag{213}$$

$$f = f_D + f_I = 192.696 \ N / m \tag{214}$$

For y = 10.53 m:

$$f_D = \frac{1}{2} \rho_w C_D D |u| u = 1015 \times 0.5 \times 1.3 \times 0.324 \times 0.480 = 102.604 \ N / m \tag{215}$$

$$f_I = \rho_w C_M \frac{\pi D^2}{4} \dot{u} = 1015 \times 2 \times \frac{\pi \times 0.324^2}{4} \times 0.774 = 129.560 \ N / m \tag{216}$$

$$f = f_D + f_I = 232.164 N / m \tag{217}$$

Assuming the (load)/(per unit length) acting on the elements 1-2 and 2-3 to be the average of the load distribution at the top and bottom of the elements as shown in Figure 31. Average load per unit length acting on the portion of top element = (232.164 + 192.696)/2 =212.430 N/m

- Average load per unit length acting on the bottom element = (192.696 + 116.740)/2 = 154.708 N/m.
- Force transferred at top node = Fwind + Fwave = 100 + 21.86=121.86 N
- Force transferred at second node = 409.370 + 657.509=1066 N

Figure 31. Equivalent load conversion at joints

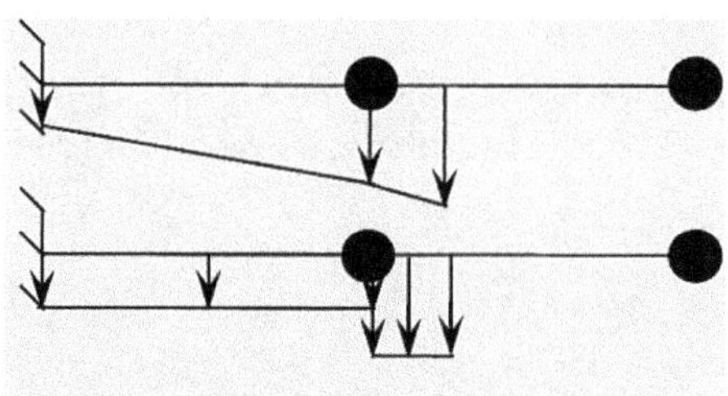

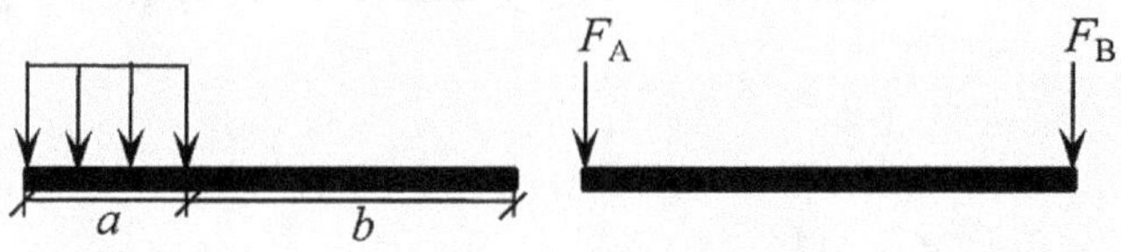

$l = a+b$

$$F_A = w \times a\left[1 - \left(\frac{a}{l}\right)^2 + \frac{1}{2}\left(\frac{a}{l}\right)^3\right]$$

$$F_B = w \times a\left[\left(\frac{a}{l}\right)^2 - \frac{1}{2}\left(\frac{a}{l}\right)^3\right]$$

The equation of motion is written as:

$$\begin{bmatrix} 1522 & 0 \\ 0 & 635 \end{bmatrix} \ddot{x} + \begin{bmatrix} 423.50 & -105.74 \\ -105.74 & 77.60 \end{bmatrix} \dot{x} + \begin{bmatrix} 531 & -166 \\ -166 & 66 \end{bmatrix} \times 10^3 x = \begin{bmatrix} 1066.02 \\ 121.86 \end{bmatrix} N \tag{218}$$

$$F_A = w \times a \left[1 - \left(\frac{a}{l} \right)^2 + \frac{1}{2} \left(\frac{a}{l} \right)^3 \right]$$

$$F_B = w \times a \left[\left(\frac{a}{l} \right)^2 - \frac{1}{2} \left(\frac{a}{l} \right)^3 \right]$$

SDOF System

- Mass at the tip is calculated as shown in Figure 14, M= 1006 as calculated in numerical example 1.1.

$$S_{xx}(f) = \left| H(if) \right|^2 S_{FF}(f) \tag{219}$$

With $S_{FF}(f)$ the wave force spectrum found above and the transfer function from Equation 50 in terms of frequency in Hz is given as:

$$\left| H(if) \right|^2 = \frac{1}{K^2 \left\{ \left[1 - \left(\frac{f}{f_n} \right)^2 \right]^2 + \left[2\xi \frac{f}{f_n} \right]^2 \right\}^2} \tag{220}$$

For the equivalent model, the fundamental period, $f = f_n$ = 0.6 Hz, damping ratio ξ is 0.008 and the stiffness of the equivalent model is given as:

$$K = \frac{3EI}{L^3} = \frac{3 \times 205 \times 10^9 \times 0.116 \times 10^{-3}}{10^3} = 71340\ N/m \tag{221}$$

Therefore the transfer function is calculated as:

$$\left| H(if) \right|^2 = \frac{1}{50893956000 \left\{ \left[1 - 2.7778 f^2 \right]^2 + 0.000711 f^2 \right\}} \tag{222}$$

The transfer function with varying frequency has been determined in MATLAB. The spectrum of the response is hence readily found in MATLAB and represented in Figure 32.

ANSYS MULTI-PHYSICS

The finite element method (FEM) is the most popular simulation method to predict the physical behaviour of systems and structures. Since analytical solutions are in general not available for most daily problems in engineering sciences numerical methods have been evolved to find a solution for the governing equations of the individual problem. ANSYS provides finite element solutions for several engineering disciplines like statics, dynamics, heat flow, fluid flow, electro-magnetics and also coupled field problems. ANSYS user is able to run simulations for linear and nonlinear problems in engineering where structural nonlinearities may occur due to nonlinear material behaviour, large deformations or contact boundary conditions. The procedure for modelling in ANSYS is summarised as:

1. Element Type,
2. Material Model,
3. Boundary Conditions and Loading,
4. ANSYS Solver and Convergence.

Figure 32. Response spectra for the SDOF system

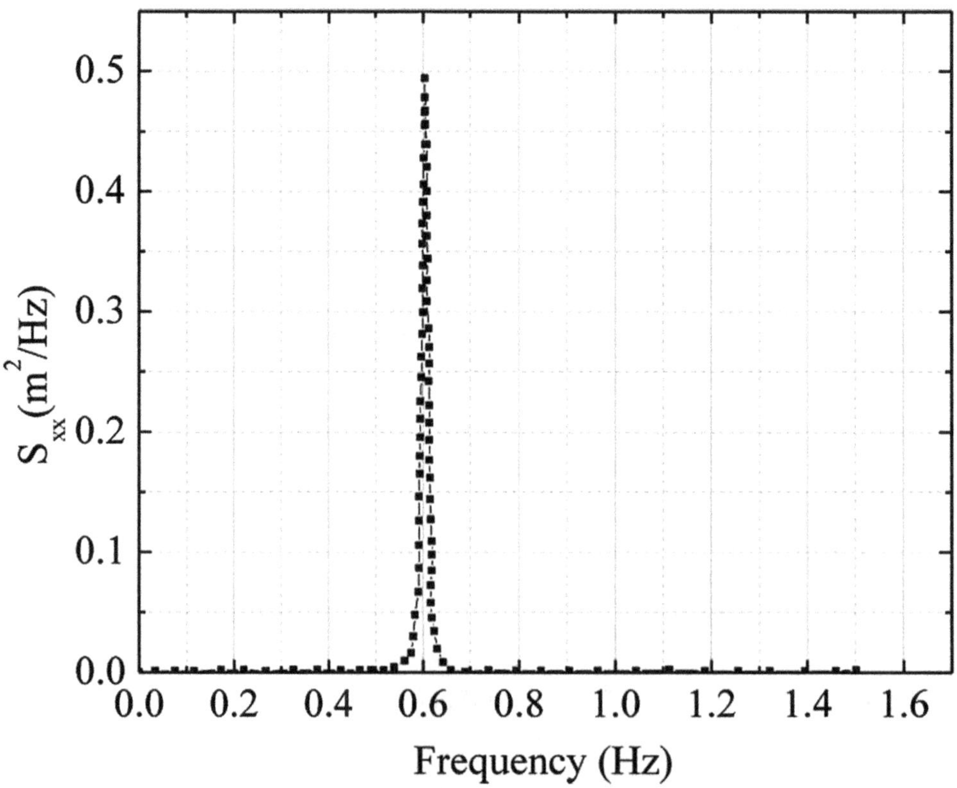

Element Type

In the current developed simulation models, the tower members for the offshore wind turbines were represented by BEAM188 elements. BEAM188 is a linear, two node beam element in 3D with six degrees of freedom at each node. The degrees of freedom at each node include translation in *x*, *y* and *z* directions, and rotation about *x*, *y* and *z* axes. The beam elements are well-suited for linear, large rotation, and large strain nonlinear applications. The lumped masses are represented using MASS21 elements. MASS21 is a point element having up to six-DOFs, e.g. translations in the nodal *x*, *y*, and *z* directions and rotations about the *x*, *y*, and *z* axes. The mass element is defined by a single node, concentrated mass components. Nodal displacements are included in the overall displacement solution. Figure 33 represents a finite element model of an offshore wind turbine tower using BEAM188 and MASS21 elements.

Material Model

For each element type, there are a minimum number of required material properties. The material properties may be:

Figure 33. (a) Lumped mass model and (b) 3D representation of the elements in ANSYS [40]

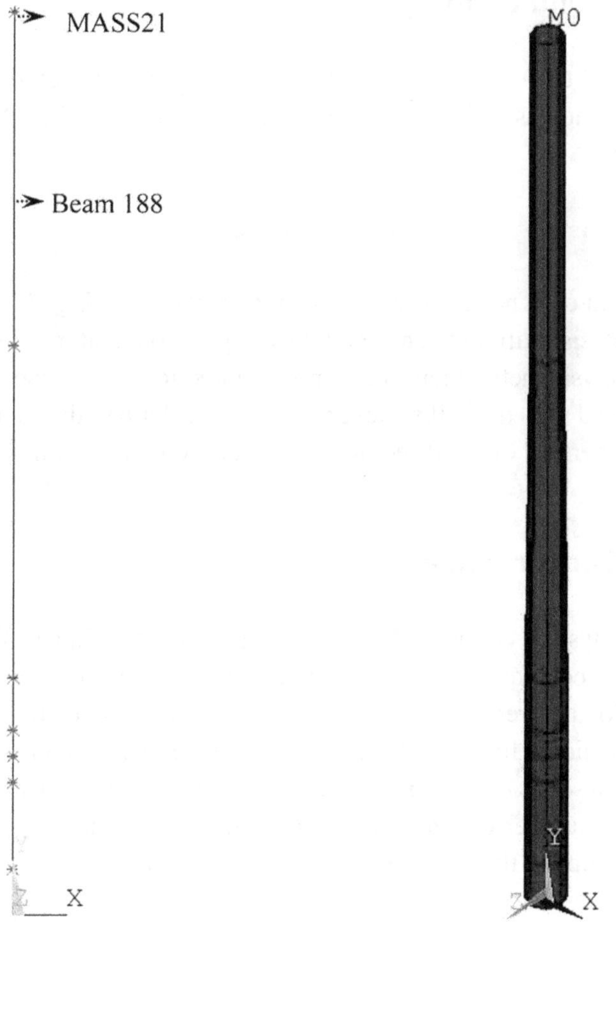

- Linear or non linear,
- Isotropic, orthotropic, or anisotropic,
- Temperature dependent or independent.

The material model is selected using the path:

```
Main Menu > Preprocessor> Material Props> Material Models
```

Yield strength of steel was assumed 345 MPa (ASTM 992 steel), and the material was considered *linear elastic*, which should apply to most offshore structures. In this way, the nonlinear effects (that are expected to grow with the load magnitude) were assessed up to a reasonable value of the external actions on the structure. Note that a structure may suffer from buckling before this level of loading is reached; therefore, our calculations should encompass the likely expected loading conditions for offshore turbines. Higher stresses would not be experienced by a reasonably designed structure and therefore, higher-than-assumed loads and their associated nonlinearities were ignored in this study. In the current analysis, linear isotropic elastic material model for steel tower has been used. Stress-strain relationship of the tower is linearly elastic. To define the elastic material model, the elasticity modulus and the density for the tower is defined in the material property module of ANSYS.

Boundary Condition and Loading

The foundation is assumed to be fully fixed into the ground, with no rotation allowed. Displacement constraints are applied at the base of the tower using the load module in ANSYS as shown in Figure. The path to the module is:

```
Main Menu > Preprocessor> Load> Displacement
```

The loads are applied at each node subjected to wind or wave loading. ANSYS offers three methods of transient dynamic analysis: Full transient, modal superposition, and reduced methods. In the present chapter, full transient analysis method has been implemented to analyze the structure subjected to random loads. The full method uses the full system matrices to calculate the transient response (no matrix reduction). It is the most general of the three methods because it allows all types of nonlinearities to be included.

Ansys Solver and Convergence

In ANSYS, nonlinear analysis is conducted using two different techniques. The first technique is load-control technique. In this technique, the total load is applied to the finite element model. The load is divided into a series of load increments during the analysis called load steps. The second technique is displacement-control technique. In this technique, the displacement is applied to the model and the displacement is divided into a series of increments called load steps. The load steps are defined by program user. With increasing the number of load steps, the accuracy of results is increased and the program needs a longer time to complete the solution. After completing each increment, the stiffness matrix of the model is adjusted to reflect nonlinear changes in structural stiffness. In the current analysis, load-control technique is used.

ANSYS uses Newton-Raphson method for updating the model stiffness. Before each solution, the Newton-Raphson method evaluates the out-of-balance load vector. The out-of-balance load vector is the difference between the restoring forces (the load corresponding to the element stresses) and the applied loads. The program then performs a linear solution, using the out-of-balance loads and checks for convergence. If convergence criteria are not satisfied, the out-of-balance load vector is revaluated, the stiffness matrix is updated, and a new solution is obtained. This iterative procedure continues until the solution converges. A number of convergence enhancement and recovery features, such as line search and automatic load stepping can be activated to help the problem to converge. If convergence cannot be achieved, then the program attempts to solve with a smaller load increment. Considering the time integration method for transient problems, ANSYS uses an implicit solution method by default.

FE SIMULATION OF ON ONSHORE WIND TURBINE

Wind Loading

Because a wind turbine responds non-linearly to different mean wind speeds as well as different heights and thickness of the tower, analysis considering these factors is necessary. The aim of the current section is to get the readers familiar with the simulation techniques using ANSYS software. For the parametric study is to examine the changes in maximal responses along the tower for variations in (a) frequency of load (b) the tower height, (c) the thickness of tower, and (d) the material property of the tower. For the purpose of this study, the wind turbine tower given in numerical problem 1.3 is considered herein for the wind load spectra obtained in the example problem and subjected to variation in frequency of loading, height of tower, thickness of tower and variation in material properties. The parametric study has been carried out in four phases as:

1. A sinusoidal load is applied at the top node of the structure with its time period is varied from 1 sec to 10 sec.
2. Thickness of tower is varied from 25 mm to 40 mm maintaining a constant taper ratio (*D/t*) of the tower for avoiding local buckling
3. Height is varied keeping a constant tapering ratio and increasing the base diameter in proportion to it
4. Elastic modulus is varied from 50 GPa to 250 GPa for varying height and thickness

Numerical Study

Wind Induced Response

To study the variation in time period of loading, an extreme wind load condition is considered i.e. 33186.4 N and applied at the top node (node 4) of the structure in the form of a sinusoidal load with its time period being varied from 1 to 10 seconds. The harmonic load is given as:

$$F = 33186.4 \sin\left(2\pi t \,/\, T\right) \tag{223}$$

where T is time period and F is force applied load at top node and 33186.4 N is the maximum value of random load applied at the top of the tower. Its variation of response in terms of displacement, acceleration, base shear and moment are presented in Figure 34 (a), 34 (b), 34 (c) and 34 (d). It shows a higher value of displacement at the natural time period (2.2 sec) of the structure due to resonance. As the time increases away from the fundamental period, the response decreases. The response at fundamental period of the structure is 99.8% higher than the response at 1 second which is 0.01683 m. The acceleration at the natural time period (1.14 m/sec^2) is 430% higher than the acceleration at 1 second which is 0.215 m/sec^2. Base shear and base moment at the fundamental period is 482% and 634% higher

Figure 34. Response in terms of (a) displacement (b) acceleration (c) base shear (d) base moment for harmonic loading with varying time periods

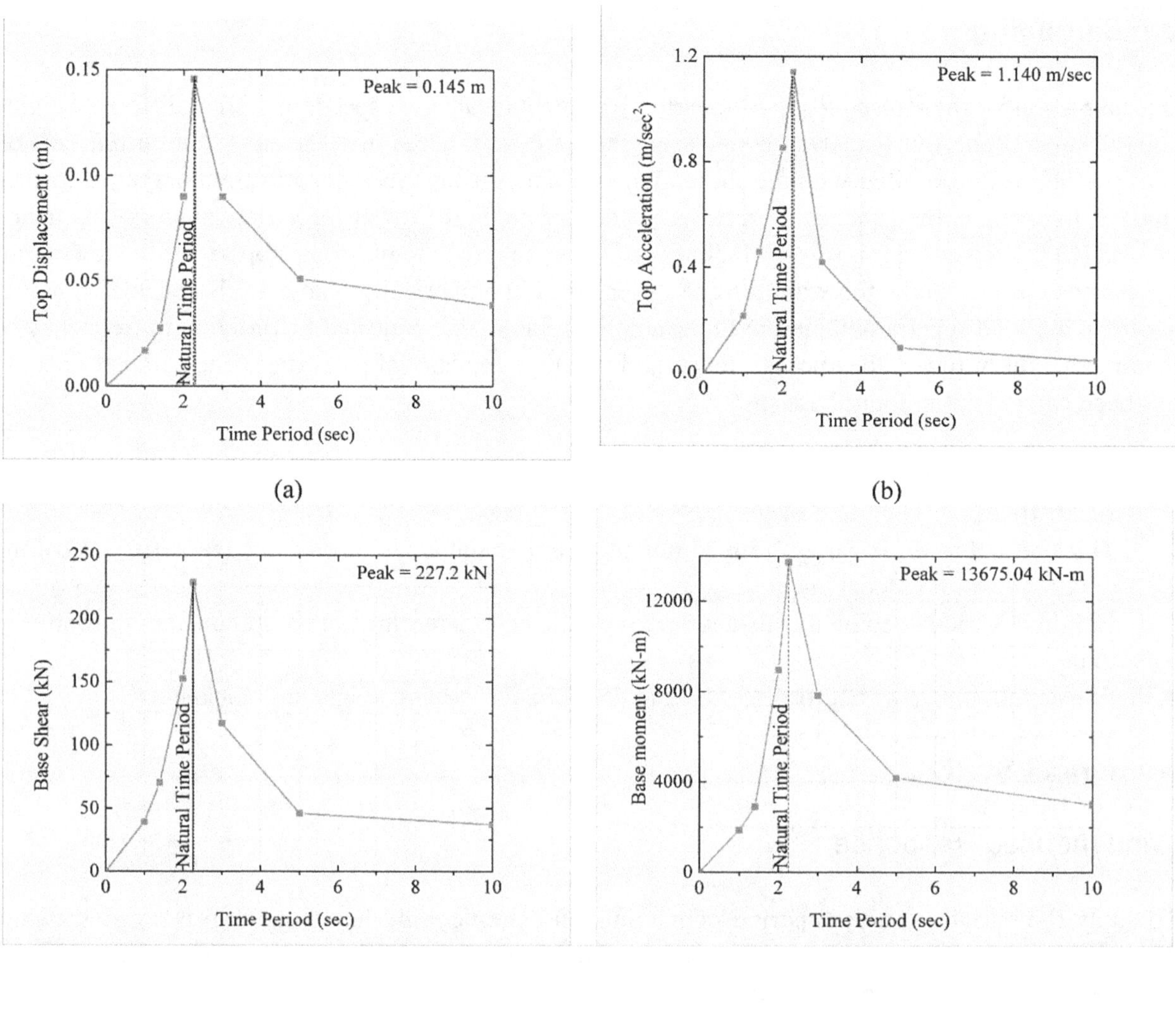

Variation of Thickness

Thickness is varied keeping the local buckling stress more than its yield stress at every cross section. For slender structure like windmill tower governing stress for failure is local buckling. So Brazier (Xiao-bo, Jing, & Jian-Yun, 2009) performed an experiment to study the buckling behavior of the cylindrical tower. The formula for local buckling stress is given by Brazier (1927) as:

$$\sigma_{cr} = 0.33E\frac{t}{R} \tag{224}$$

where σ_{cr} is critical local buckling stress, E is modulus of elasticity and R is mean radius and t is thickness of the tower. Obeying the critical buckling strength to be more than 250 MPa, the R/t ratio becomes less than 264. Hence keeping the diameter constant, thickness is varied in such a way that R/t ratio is always less than 264 to avoid local buckling. At the top, the minimum thickness comes out to be around 3.74 mm and at the bottom, the minimum thickness comes out to be 7.9 mm. Here, parametric studies are performed for uniform turbine thickness, hence thickness is varied from 8 mm to 40 mm. For more than 40 mm it is difficult to fabricate the cylindrical shells.

By applying the random loads at each node as obtained in the example problem 1.3 and converted to time domain as shown in Figure 35, the variation of displacement with thickness is studied. It can be seen from Figure 36, the maximum response decreases with increase in thickness as it makes the structure stiffer. The variation follows a parabolic law. The response at 10 mm thickness is 81% higher than the response at 50 mm thickness.

Figure 35. (a) Tower modeled using ANSYS 14 with elements represented in 3D with (b) random wind loading time history at node 1, node 2, node 3 and node 4

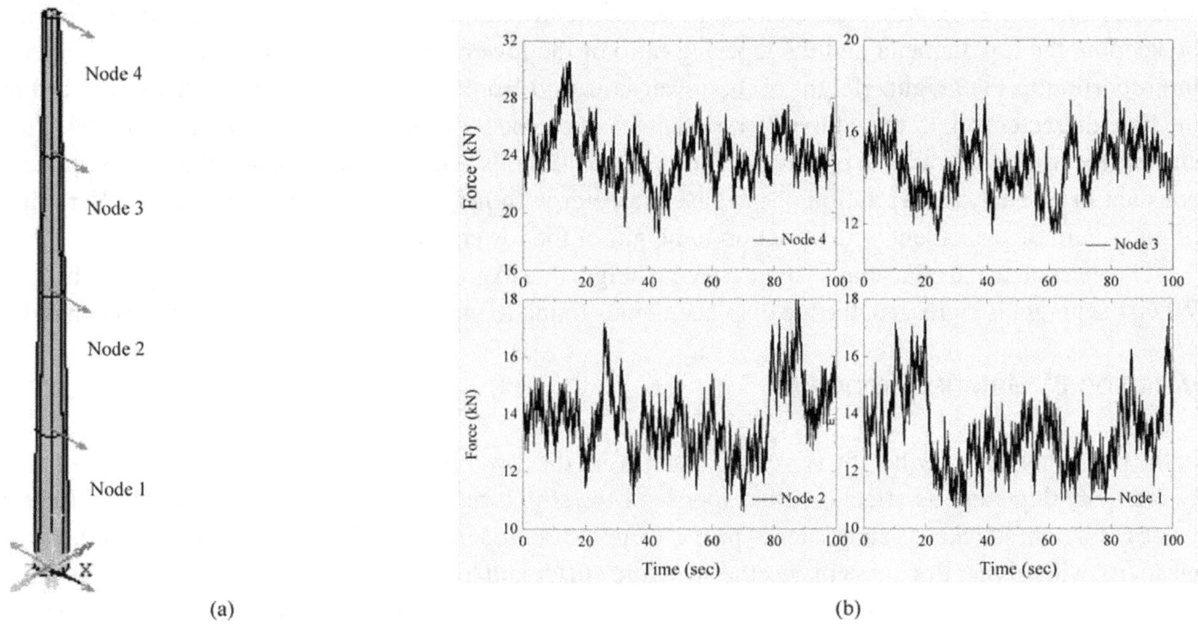

Figure 36. Variation of top displacement with thickness of tower

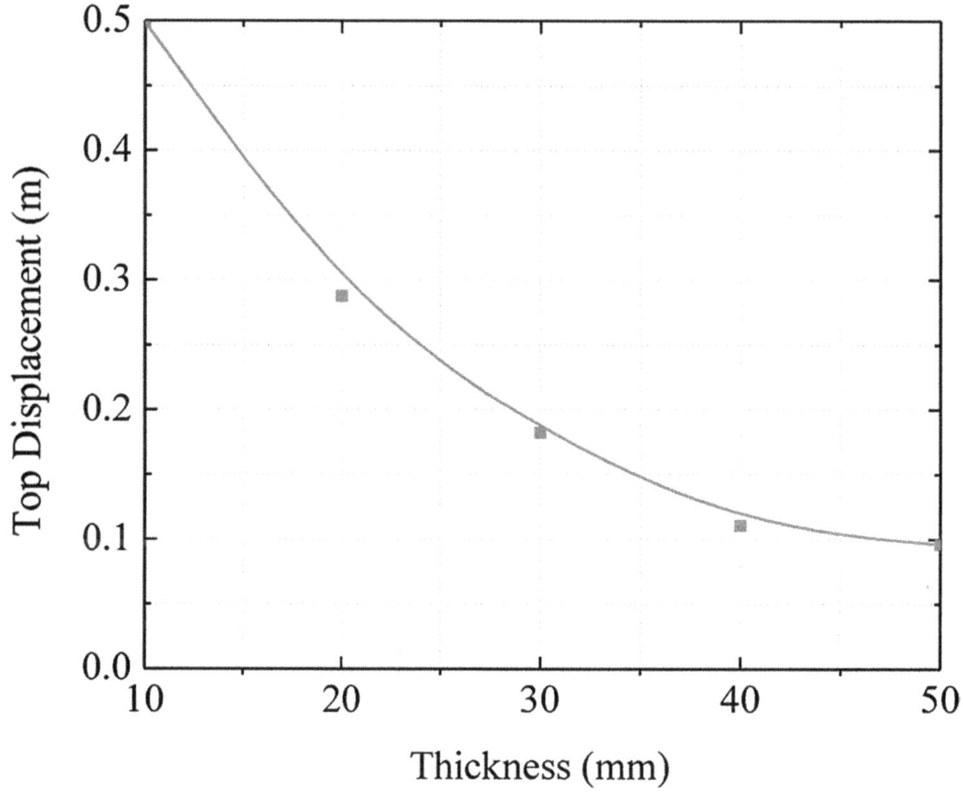

Variation of Height

The wind load increases with the increasing height of the structure. Here, parametric studies are performed by keeping the top diameter and the tapering ratio of the tower same and increasing the base diameter in proportion to the height. Height of the tower is varied from 60 m to 100 m. For a height of 100 m the base diameter is 4.75 m and the *R/t* ratio comes out to be 95 which lie within the range of 264. This ensures the avoidance of local bucking of the cylindrical shells. Mass of the rotor and the blades are kept constant in all analyses. Variation of the base diameter with height is also shown in Table 1. Variation of maximum displacement with variation in height of the tower is illustrated in Figure 37.

It is seen that the displacements increase with height due to greater values of wind loading with height. When the height is increased, the top displacement is found to increase almost linearly after 80 m height.

Variation of Material Property

In the following analysis, height as well as thickness of the tower is varied for different values of modulus of elasticity depicting its effect on the response of the structure. The response of the tower for different values of height, thickness and material property are presented in Figure 38. Higher value of modulus of elasticity with higher thickness makes the structure stiffer and hence its response reduces. The variation seems to follow a parabolic path. Different materials with their elasticity modulus have been presented

Table 1. Base diameter with increase in height

S. No.	Height (m)	Base Diameter (m)
1	60	3.650
2	70	3.925
3	80	4.200
4	90	4.475
5	100	4.750

Table 2. Different material grades with its elasticity modulus

Material	Young's Modulus (GPa)
Carbon Steel IS 2062	0.3
Steel SAE 950	20
Aluminium	69
Titanium Alloy	105 - 120
Carbon Fiber Reinforced Plastic	150
Steel, stainless AISI 302	180
Steel, Structural ASTM-A36	200
Low alloy steels	200 - 210

Figure 37. Top displacement variation with height

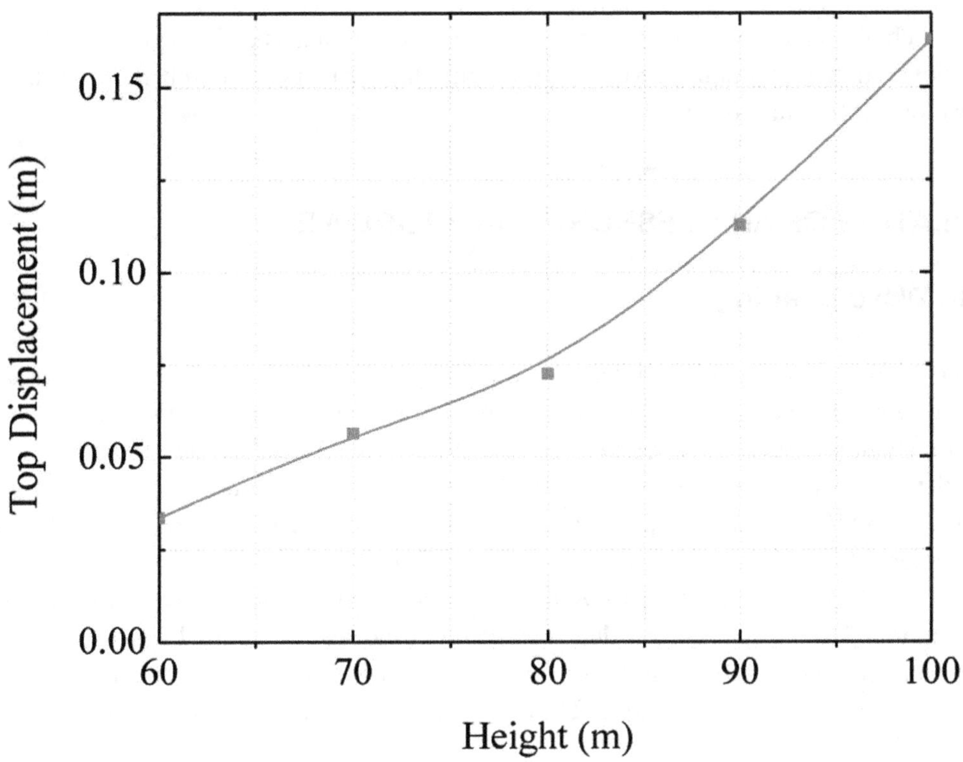

Figure 38. Maximum displacement for variation in modulus of elasticity, thickness and height of the tower

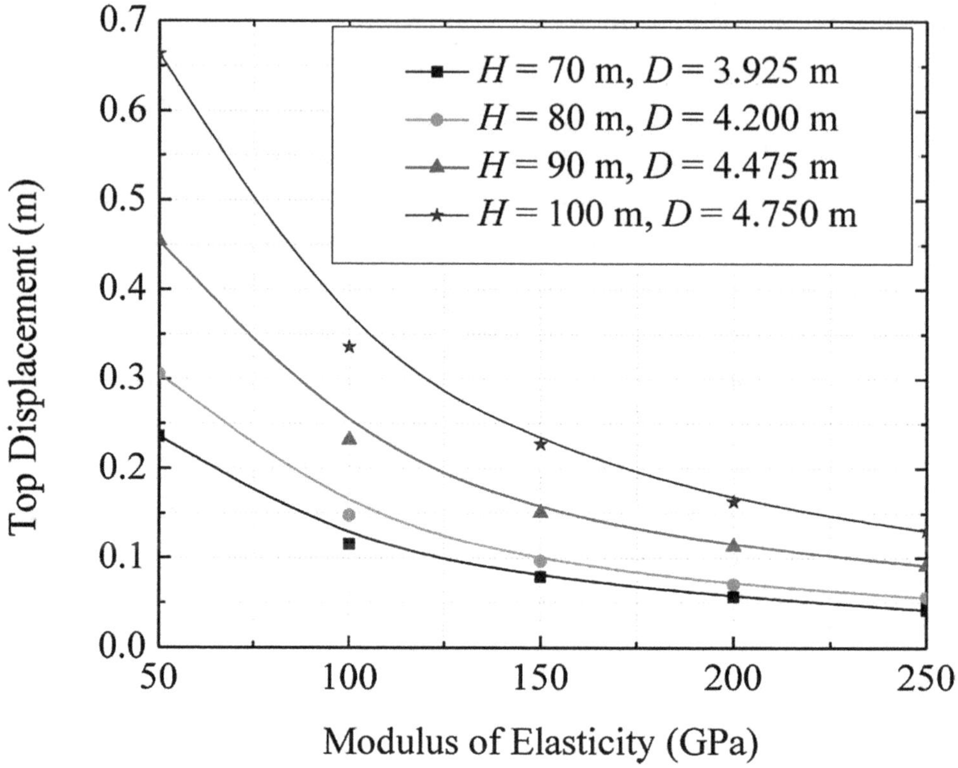

in Table 2 which gives a fairly good idea about the use of these materials in the manufacturing of these structures. For example, aluminum gives a significantly higher response compared to a structural steel (ASTM-36) for a 100 m tall tower.

FE SIMULATION OF AN OFFSHORE WIND TURBINE

Wind and Wave Loading

The offshore wind turbine tower illustrated in numerical problem 1.2 is considered for simulation under wind and wave loading in this present section. Figure 18 shows the structural model under consideration with the corresponding idealized lumped mass MDOF system. As the surface elevation increases and decreases, the total force from the sea is lumped into the appropriate nodes. The structure has been modeled in ANSYS, with each lumped mass represented using MASS21 element and connected by massless elements BEAM188 that has finite stiffness. The structural damping ratio is assumed to be 0.01. The time-history response of the MDOF system subjected to wind and wave loading is presented in this simulation. The power spectral densities of the wind and wave loadings were calculated and presented in numerical example 1.2 in Figure 39. The time histories for these loadings were calculated using Shinozuka's conventional technique. The converted time histories are shown in Figure 39. Finally, the MDOF model is subjected wind and wave loading.

Figure 39. (a) Wind drag force excitation at node 7 and (b) wave force excitation at node 4

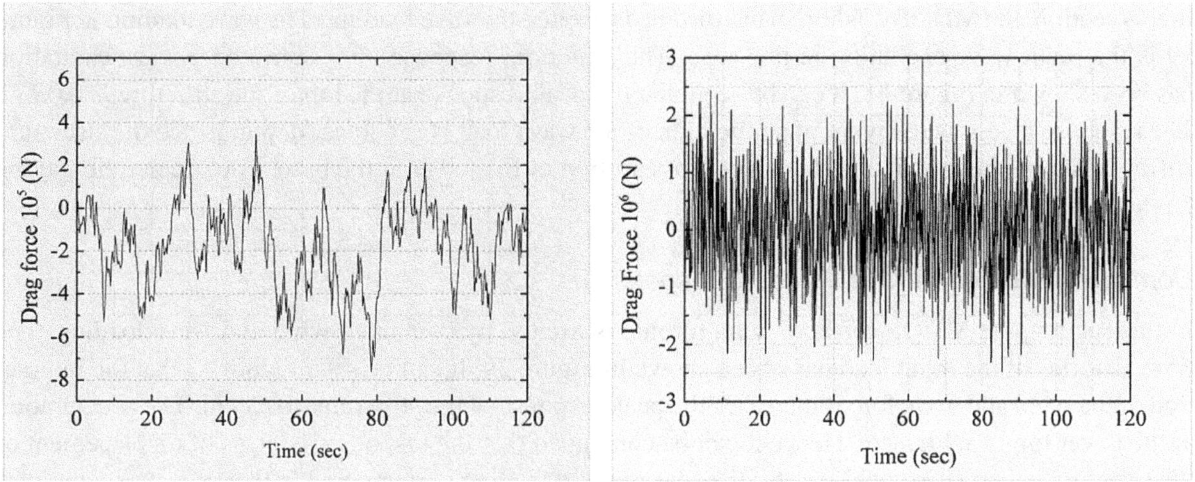

Numerical Study

Wind Induced Response

In this section, the MDOF offshore wind turbine is excited by wind loading only. The wind loading in Figure 39 is the input wind excitation in this case. Figure 40 shows the displacement and acceleration time history response for at node 8. It can be seen from the figure that the structure has a response of 0.467 m. The acceleration at the top node is 0.67 m/sec^2.

Figure 40. (a) Displacement and (b) acceleration response at top node due to wind load only

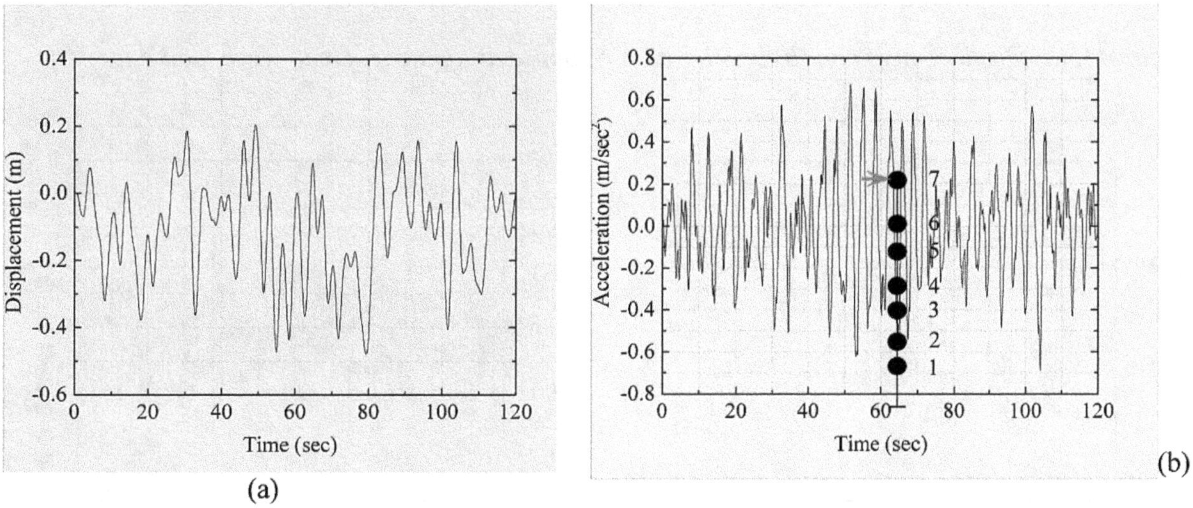

Wave Induced Response

In this section, the MDOF offshore wind turbine is excited by wave loading. The wave loading in Figure 39 is the input wave excitation in this case. The structural response of system under wave excitation has been shown in Figure 41. It can be seen that the peak response at the top of the structure is 0.09 m. The response is less as the point of application of the wave load is at a lower depth; hence the structural stiffness is quite high owing to the larger cross section of the tower at the base. The acceleration at the top node is 1.9 m/sec^2.

Combined Wind and Wave Induced Response

In this section, the MDOF offshore wind turbine is excited by combined wind and wave loading. The time histories of the input load have been shown in Figure 39. It can be seen Figure 42 that the interaction of the wind and wave forces increases the peak response of the structure to 0.59 m. The acceleration at the tower top is 3.3 m/sec^2. Hence it can be concluded that the response in terms of displacement of the structure is significantly higher for the case of wind loading as compared to that of wave loading for an offshore wind turbine structure.

CONCLUSION

Different aspects of computational analysis of wind turbine towers have been discussed in this chapter. To achieve the cost effective yet robust design criteria preferably by decrease in tower shaft diameter and increase in height of the tower makes the dynamic analysis of wind turbine towers extremely critical as it makes the structure more flexible. Flexibility makes the structure susceptible to random wind and wave forces, which has its frequency quite close to the natural frequency of the tower, which under resonance can trigger a catastrophic situation. Compared to other structures whose response is dominated by inertia, wind turbines are slender structures and fall on the "drag dominated" category of structures.

Figure 41. (a) Displacement and (b) acceleration response at top node due to wave load only

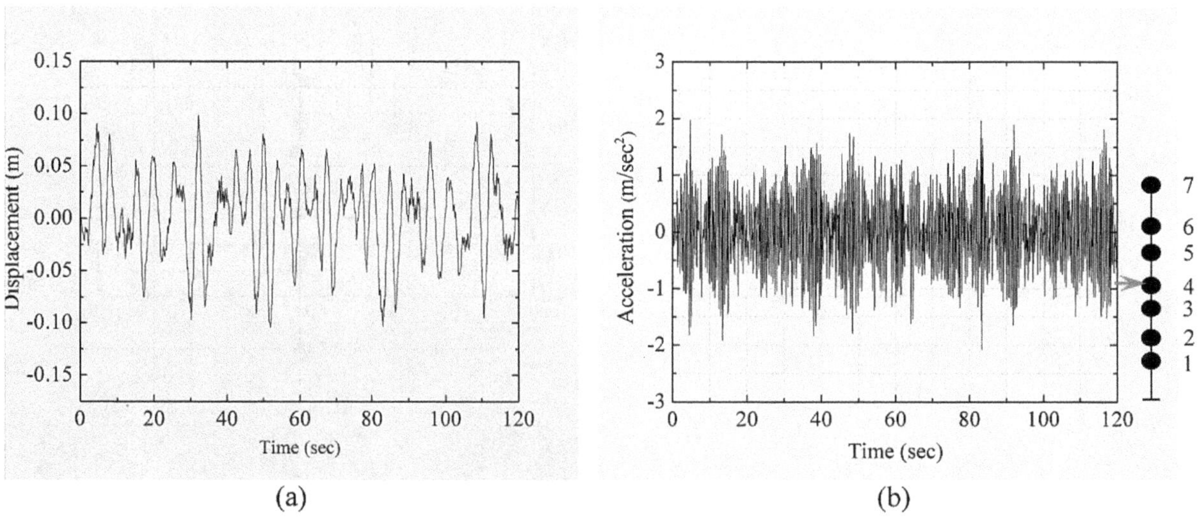

Figure 42. (a) Displacement and (b) acceleration response at top node due to combined wind and wave load

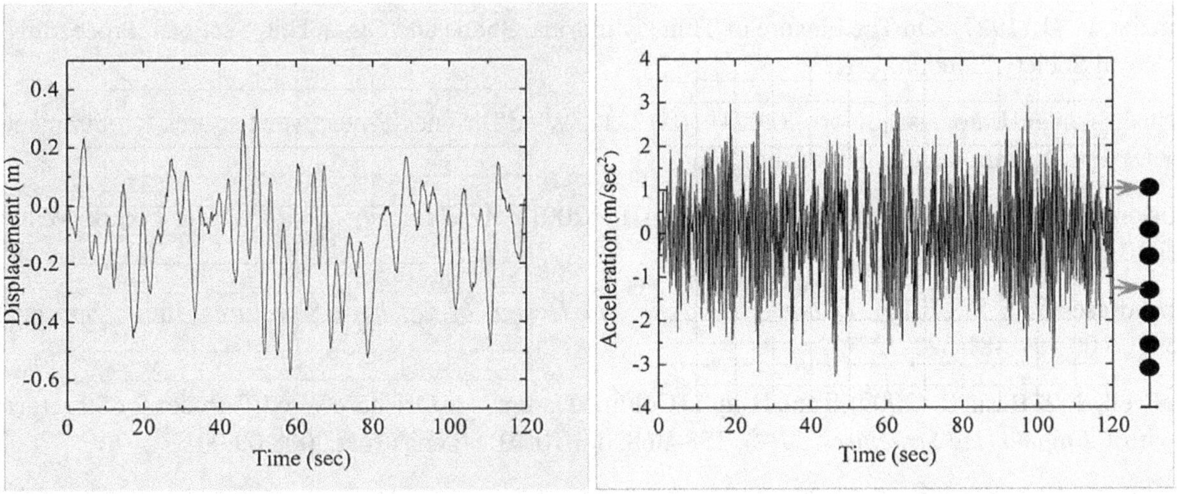

Hence it is essential to analyze these structures for drag induced wind and wave loading. A random load approach, which should be taken into account for wind as well as wave loading, has been discussed in detail in this chapter. Computational analysis capability of ANSYS software has been demonstrated through a parametric study of an onshore wind turbine tower which has been analyzed using a lumped mass mode using the software. The scope of the parametric study includes study of important parameters like thickness, height and material properties of the tower and studying its maximum response variation with respect to these parameters. It could be seen that the variation of thickness with displacement follows a parabolic path. The variation of displacement with height of the tower follows a linear path beyond 80 m height. In the similar way it was demonstrated from the variation of elasticity modulus with height and thickness also gave a parabolic pattern. To emphasize on the dynamic analysis of offshore wind turbine structures, an offshore wind turbine tower subjected to random wind and wave loading has been presented. It can be clearly interpreted from the results that compared to wave loading; structures tend to respond higher for wind loading. Response of the structure towards the loading depends upon the frequency content of the loading. The aim of the chapter is to give a broad overview to the readers about the computational analysis carried out for wind turbine towers using finite element software that takes into account the dynamics these slender structures.

REFERENCES

American Bureau of Shipping (ABS). (2014). *Bottom founded offshore wind turbine installation* (2014th ed.). New York.

American Petroleum Institute (API). (2014). *Recommended Practice for Planning*. Dallas: Design and Constructing Fixed Offshore Platform.

Anderson, J. D. (2003). *Fundamentals of Aerodynamics* (4th ed.). U.S.A.: McGraw-Hill.

ANSYS APDL 14, ANSYS Inc. (n. d.). Basic Analysis Procedures Guide Release 14. Retrieved from http://www.ansys.com

Brazier, L. G. (1927). On The Flexure of Thin Cylindrical Shells and Other Thin Sections. Proceedings of Royal Society, London, U.K.

British Standard Europaische Norm. (2004). (*BS*). U.K.: Wind Turbines-Protective Measures-Requirements for Design Operation and Maintenance.

Burton, T., Sharpe, D., Jenkins, N., & Bossanyi, E. (2001). *Wind Energy Handbook*. New York: Wiley. doi:10.1002/0470846062

Chandrasekaran, S. (2015). *Dynamic Analysis and Design of Offshore Structures*. India: Springer. doi:10.1007/978-81-322-2277-4

Colwell, S., & Basu, B. (2009). Tuned Liquid Column Dampers in Offshore Wind Turbines for Structural Control. *Engineering Structures*, *31*(2), 358–368. doi:10.1016/j.engstruct.2008.09.001

Craig, R. R., & Kurdila, A. J. (2006). *Fundamentals of Structural Dynamics* (2nd ed.). U.S.A.: John Wiley and Sons.

Danish Wind Turbine Manufacturers Association. (2000). *Wind Energy*. Pioneer Poul la Cour.

Davenport, A. G. (1961). The Spectrum of Horizontal Gustiness Near The Ground in High Winds. *Journal of Royal Metrological Society.*, *87*(372), 194–211. doi:10.1002/qj.49708737208

Det Norske Veritas. (2004). Design of Offshore Wind Turbine Structures. Offshore Standard DNV-OS-J101. Norway.

Det Norske Veritas. (2009). Offshore Substations for Wind Farms. Offshore Standard DNV-OS-J201. Norway.

Edinger, R., & Kaul, S. (2007). *Renewable Resources for Electric Power: Prospects and challenges*. Westport: Quorum.

Faltinsen, O. M. (1999). *Sea Loads on Ships and Offshore Structures*. U.S.A.: Cambridge University Press.

Gasch, R., & Twele, J. (2007). *Windkraftanlagen: Grundlagen, Entwurf, Planung und Betrieb*. Teubner Verlag.

Global Wind Report. (2014). Annual Market Update. Retrieved from http://www.gwec.net/publications/global-wind-report-2/global-wind-report-2014-annual-market-update/

Hansen, M. (2008). *Aerodynamics of Wind Turbines* (2nd ed.). U.K.: Earthscan.

Hasselman, K., Barnett, T. P., Bouws, E., et al. (1973) Measurement of Wind Wave Growth and Swell Decay During the Joint North Sea Wave Project (JONSWAP).

Holmes, J. D. (2010). *Wind Loading of Structures* (2nd ed.). U.S.A.: W.I.T. Press.

International Electro-technical Commission (IEC). (2005). Wind Turbine: Part-1. Switzerland, Edition 2005.

International Electro-technical Commission (IEC). (2009). Wind Turbine: Part-3. Switzerland, Edition 2009.

International Electro-technical Commission (IEC). (2010). Wind Turbine: Part-22. Switzerland, Edition 2010.

International Electro-technical Commission (IEC). (2010). Wind Turbine: Part-24. Switzerland, Edition 2010.

International Organization for Standardization (ISO). (2014). *Condition Monitoring and Diagnostics of Wind Turbines-Part 1: General Guidelines*. Switzerland: Edition.

Jonkman, J. (2007). *Dynamics Modeling and Loads Analysis of an Offshore Floating Wind Turbine. NREL/TP-500 - 41958*. Colden, Colorado: National Renewable Energy Laboratory. doi:10.2172/921803

Lloyd, G. (2005). Rules and Guidelines for the Certification of Offshore Wind Turbines (2005th ed.). Germany.

Manwell, J., McGowan, J., & Rogers, A. (2002). *Wind Energy Explained*. New York: Wiley. doi:10.1002/0470846127

MATLAB. (2010). Mathworks Inc., Basic Guidelines. Retrieved from http://www.matlab.com

Morison, J. R., O'Brien, M. P., Johnson, J. W., & Schaaf, S. A. (1950). *The Force Exerted by Surface Waves on Piles. Petroleum Transactions*. American Institute of Mining Engineers.

Mr. Brush's Windmill Dynamo. (1890). *Scientific American*, 63(25), 54.

Nigam, N. C., & Narayanan, S. (1994). *Applications of Random Vibrations*. India: Springer.

Pierson, W. J. Jr, & Moskowitz, L. (1964). A Proposed Spectral Filter For Fully-Developed Wind Seas Based on The Similarity Theory of S.A. Kitaigorodski. *Journal of Geophysics*, *69*(24), 181–203. doi:10.1029/JZ069i024p05181

Robert, W. (1996). *Wind Energy in America: A History*. U.S.A.: University of Oklahoma Press.

Sarpkaya, T., & Isaacson, M. (1981). *Mechanics of Wave Forces on Offshore Structures*. New York: Van Nostrand Reinhold.

Schwartz, L. W. (1974). Computer Extension and Analytic Continuation of Stokes' Expansion for Gravity Waves. *Journal of Fluid Mechanics*, *62*(3), 553–578. doi:10.1017/S0022112074000802

Shinozuka, M. (1972). Monte Carlo Solution of Structural Dynamics. *International Journal of Computers and Structures.*, *2*(5-6), 855–874. doi:10.1016/0045-7949(72)90043-0

Xiao-bo, C., Jing, L., & Jian-yun, C. (2009). Wind-Induced Response Analysis of a Wind Turbine tower including the blade-tower coupling effect. *Journal of Zhejiang University Applied Physics & Engineering*, *10*(11), 1573–1580.

Chapter 6
Seismic Reliability Analysis in the Framework of Metamodelling Based Monte Carlo Simulation

Shyamal Ghosh
Indian Institute of Engineering Science and Technology (IIEST), Shibpur, India

Swarup Ghosh
Indian Institute of Engineering Science and Technology (IIEST), Shibpur, India

Soham Mitra
Indian Institute of Engineering Science and Technology (IIEST), Shibpur, India

Subrata Chakraborty
Indian Institute of Engineering Science and Technology (IIEST), Shibpur, India

ABSTRACT

A comparative study of various metamodelling approaches namely the least squares method (LSM), moving least squares method (MLSM) and artificial neural network (ANN) based response surface method (RSM) are presented to demonstrate the effectiveness to approximate the nonlinear dynamic response of structure required for efficient seismic reliability analysis (SRA) of structures. The seismic response approximation by the LSM, MLSM and ANN based RSMs are explained with a brief note on the important issue of ground motion bin generation. The procedure adopted herein for SRA is based on the dual response surface approach. In doing so, the repetition of seismic intensity for SRA at different intensity levels is avoided by including this as one of the predictors in the seismic response prediction model. A nonlinear SDOF system has been taken up to elucidate the effectiveness of various metamodels in SRA.

INTRODUCTION

Seismic Reliability Analysis (SRA) of structures is basically a structural reliability analysis problem in which the limit state of interest is the difference between seismic demand (D) and capacity (C) considering uncertainty due to earthquake motions, structural properties, physical damage, economic and human losses etc. The problem can be envisaged as a time dependent structural reliability analysis problem in which the limit state of interest can be expressed as,

DOI: 10.4018/978-1-5225-0588-4.ch006

$$Z(\boldsymbol{X}_C, \boldsymbol{X}_D, t) = C(\boldsymbol{X}_C, t) - D(\boldsymbol{X}_D, t) \tag{1}$$

where, $\boldsymbol{X}_C$ and $\boldsymbol{X}_D$ are the variables governing the capacity and demand and t is the time parameter. The computation of probability that the limit state function is negative means to evaluate the seismic risk of the structure i.e.

$$Z < 0 \rightarrow Failed, Z = 0 \rightarrow Limiting\ and\ Z > 0 \rightarrow Safe \tag{2}$$

Thus, the seismic risk analysis is the evolution of the following multi-dimensional integral to obtain the probability of failure (P_f),

$$P_f = \int_{Z<0} f_Z(\boldsymbol{X}) dZ \tag{3}$$

where $\boldsymbol{X}$ is an 'n' dimensional vector having variables involving $\boldsymbol{X}_C$ and $\boldsymbol{X}_D$, $f_Z(\mathbf{X})$ is the joint probability density function (pdf) of the involved random variables. The exact computation of the above is often computationally demanding. In fact, the joint pdf of $f_Z(\boldsymbol{X})$ is hardly available in closed form. Various approximations are usually adopted to obtain the probability of exceeding a response parameter for different limit states of damage.

The more accurate and conceptually straightforward but computationally demanding methodology for SRA is based on Monte Carlo Simulation (MCS) technique. The robustness and better accuracy of MCS based SRA is well known (e.g. Dymiotis et al. 1999; Kwon & Elnashai, 2006; Kazantzi, Righiniotis & Chryssanthopoulos, 2008). The SRA by non-linear performance based analysis using random field theory in the framework of statistical simulation is notable (Au & Beck, 2003; Kwon & Elnashai, 2006). The approach is always preferable as it does not require an assumption about the shape of the failure surface. However, such full simulation approach needs a large number of replications to obtain acceptable confidence in probabilities of failures of structures, which are typically very small in magnitude. For each replication in the simulation process, the computation of maximum response requires to perform complete nonlinear time history analysis (NLTHA) which is computationally demanding for large complex structural models. Various metamodelling techniques are found to be useful in this regard to replace the complex model. In fact, response surface method (RSM) based MCS simulation using various sampling technique for reliability analysis of structures is studied extensively (Faraveli, 1989; Bucher, & Bourgund, 1990; Rajashekhar& Ellingwood; 1993; Bucher and Most, 2008). This approach is found to be useful to replace a complex mechanical model of the actual structure for seismic response analysis involving large computational need (Iervolino, Fabbrocino & Manferedi, 2004; Towashiraporn, 2004; Park & Towashiraporn, 2014). A comparative study of various RSMs for SRA of structures involving NLTHA is performed by Möller et al. (2009). Simulation based SRA using Latin hypercube sampling technique for generation of RSM is also notable (Unnikrishnan et al., 2013). Generally, the RSM is based on the global approximation of scatter position data, usually through a second order polynomial function obtained by using the least squares method (LSM) of regression. The efficiency of the approach largely depends on the selection of the basic functions and should be chosen to resemble the basis function as closely as possible to the actual variation of the response within the domain. Such a selection is not straightforward (Gavin & Yau, 2007) e.g. typically in LSM based RSM, the basic functions are fixed second order polynomial. The moving least square method (MLSM) circumvents such problem by establishing a local approximation for polynomial coefficients around each point in the interpolation domain through a varying weight functions with respect to the position of the approximation. This has been found to be more efficient and successfully applied in approximation of complex responses in reliability analysis (Kim & Wang, 2005; Kang et al. 2010; Bhattacharjya & Chakraborty, 2011, Taflanidis & Cheung, 2012; Chakraborty & Sen, 2014). These response approximation approaches are usually termed as polynomial based RSM. Application artificial neural network (ANN) is also found to be effective for approximation of implicit limit state function for reliability analysis of structures (Chojaczyk et al., 2015, Elhewy and Mesbahi, 2006). In fact, the approach has been applied for nonlinear seismic response approximation (Zhang & Foschi, 2004; Moller et al., 2009). It may be noted that the ANN based approach also does not require to use of any specific form of basic functions like polynomial-based RSMs.

A comparative study of the three metamodelling methods as mentioned above is performed in the present study to demonstrate the effectiveness to approximate the nonlinear dynamic response of structure for efficient SRA of structures. For ease in presentation, the three metamodel approaches of response approximations are termed as LSM, MLSM and ANN based RSM in the text. The core numerical simulation in the frame work of metamodel is proposed to adopt in an innovative form to approximate the nonlinear dynamic response of the structure. In doing so, the repetition of seismic intensity for estimating seismic reliabilities at different intensities is avoided judiciously by including this as one of the predictors in the seismic response prediction model. For NLTHA, a representative ground motion bin corresponding to the specified hazard level for the location of the structures is necessary. The structure is considered to be located in the Guwahati City of the Northeastern (NE) region of India. As the recorded accelerograms in the region are very scarce, the accelerograms are generated artificially compatible to the acceleration response spectra of the location and also synthetically by identifying the most vulnerable magnitude distance combination for the specific hazard level of the location. The accuracy possible to achieve to approximate nonlinear seismic response and subsequently to estimate the seismic reliability is demonstrated by considering a single degree of freedom (SDOF) non-linear model.

SRA BY METAMODEL BASED MCS

The SRA in the framework of metamodel technique starts with defining the input system parameters and desired output response variables at a set of carefully selected points referred as design of experiment (DOE), normally obtained from experimental investigation or numerical simulation. The DOE issue in connection with the present study is further discussed in section 3.4. Uncertainty due to earthquake is implicitly incorporated in the analysis by using a suite of earthquake accelerograms; details of which are discussed in section 4. The procedure adopted herein for SRA is based on the dual response surface approach (Lin & Tu, 1995). The responses are evaluated at each DOE point for all the input ground motions in the suit. The vector of the mean, y_μ and standard deviation (SD), y_σ of any desired response, y is then computed at each peak ground acceleration (PGA) level, the considered intensity measure in the present study. Then the response surface for mean and SD can be obtained for the considered response as,

$$y_\mu = g(\boldsymbol{X}) \text{ and } y_\sigma = h(\boldsymbol{X}) \tag{4}$$

A lognormal distribution for statistical description of seismic response is a reasonable assumption (Cornell et al., 2002). The lognormal distribution parameters of the response y i.e. $\mu_{\ln y}$ and $\sigma_{\ln y}$ can be readily estimated using the mean and SD of response as approximated by the response surface models through Equation 4. The overall response can be obtained by using these parameters. It may be noted that the response surface model obtained by such procedure is for a specific level of earthquake intensity and generation of response surface needs to be repeated for all intensity levels. This will require computationally involved NLTHA at each PGA level. Thus, the computational time and effort required for SRA for different earthquake intensity will be enormous and may not be feasible for real structures. To circumvent this, the intensity measure i.e. PGA can be included within the response surface model as one of the predictors (Towashiraporn, 2004). Then, Equation 4 can be re-written as following:

$$y_\mu = g(\boldsymbol{X}, PGA) \ and \ y_\sigma = h(\boldsymbol{X}, PGA) \tag{5}$$

The evaluation of above equation requires the response (y) at all DOE points. It is to be noted that the input variables in the response surface models are now composed of random variables and a control variable which defines earthquake intensity level. The control variable (PGA, herein) is deterministic with its fixed values characterizing different response prediction models. The evaluations of response surface at different values of PGA yield models for predicting the responses due to specific levels of PGA. The overall response is now obtained as earlier based on lognormal assumption. However, the response approximation as described above is no longer conditional to a specific intensity of earthquake.

RESPONSE APPROXIMATIONS

As already mentioned, the present work is intended to make a comparative study with regard to the effectiveness of three metamodelling approaches namely the LSM, MLSM and ANN based RSMs to approximate the nonlinear dynamic response of structure required for SRA of structures. These are briefly presented in the following subsections.

LSM RSM based Response Approximation

If there are n response values y_i corresponding to n numbers of observed data, x_{ij} (denotes the i-th observation of the j-th input variable x_j in a DOE), the relationship between the response and the input variables can be expressed as:

$$\boldsymbol{y} = \boldsymbol{X}\beta + \varepsilon_y \tag{6}$$

In the above multiple non-linear regression model, $\boldsymbol{X}, \boldsymbol{y}, \boldsymbol{\beta}$ and $\boldsymbol{\varepsilon}_y$ are the design matrix containing the input data from the DOE, the response vector, the unknown co-efficient vector and the error vector, respectively. Typically, the quadratic polynomial form used in the RSM is as following:

$$y = \beta_0 + \sum_{i=1}^{k} \beta_i x_i + \sum_{i=1}^{k}\sum_{j=1}^{k} \beta_{ij} x_i x_j \tag{7}$$

In the LSM of estimation technique, the unknown polynomial coefficients are obtained by minimizing the error norm defined as:

$$L = \sum_{i=1}^{n} \left(y_i - \beta_0 - \sum_{i=1}^{k} \beta_i x_i - \sum_{i=1}^{k}\sum_{j=1}^{k} \beta_{ij} x_i x_j) \right)^2 = \left(\boldsymbol{y} - \boldsymbol{X\beta} \right)^T \left(\boldsymbol{y} - \boldsymbol{X\beta} \right) \tag{8}$$

And the least squares estimate of $\boldsymbol{\beta}$ is obtained as,

$$\beta = \left[\boldsymbol{X}^T \boldsymbol{X} \right]^{-1} \left\{ \boldsymbol{X}^T \boldsymbol{y} \right\} \tag{9}$$

Once the polynomial coefficients β are obtained from the above equation, the response y can be readily evaluated for any set of input parameters.

MLSM-RSM Based Response Approximation

The MLSM based RSM is a weighted LSM that has varying weight functions with respect to the position of approximation. The weight associated with a particular sampling point x_i decays as the prediction point x moves away from x_i. The weight function is defined around the prediction point x and its magnitude changes with x. The least-squares function $L_y(x)$ can be defined as the sum of the weighted errors as following,

$$L_y(x) = \sum_{i=1}^{n} w_i \varepsilon_i^2 = \boldsymbol{\varepsilon}^T \boldsymbol{W}(x) \varepsilon = (\boldsymbol{y} - \boldsymbol{X\beta})^T \boldsymbol{W}(x) (\boldsymbol{y} - \boldsymbol{X\beta}) \tag{10}$$

where, $\boldsymbol{W}(x)$ is the diagonal matrix of the weight function and can be obtained by utilizing the weighting function such as constant, linear, quadratic, higher order polynomials, exponential functions. In the present study, the following exponential function has been adopted (Taflanidis & Cheung, 2012):

$$w(d) = \frac{e^{-(d/cD)^{2k_0}} - e^{-(1/c)^{2k_0}}}{1 - e^{-(1/c)^{2k_0}}} \text{ if d < D, else '0'} \tag{11}$$

In the above, d is the distance of the point where approximate response is required to the origin of the approximating domain and D is the radius of the sphere of influence. It should be chosen so that it contains sufficient numbers of DOE points to avoid singularity in the solution. The parameters, c and k_0, are used for better efficiency. Eventually, a weight matrix $\mathbf{W}(x)$ can be constructed by using the weighting function in the diagonal terms as follows:

$$\boldsymbol{W}(x) = \begin{bmatrix} w(x - x_1) & 0 & 0 & 0 \\ 0 & w(x - x_2) & .. & 0 \\ .. & .. & .. & .. \\ .. & .. & .. & .. \\ 0 & 0 & .. & w(x - x_n) \end{bmatrix} \tag{12}$$

The weighting function has its maximum value of unity at a normalized distance of zero and zero value (minimum) outside of unit normalized distance, i.e. w (0.0) = 1.0 and w (d/D >1.0) = 0.0. The function decreases smoothly from 1.0 to 0.0. By minimizing the least-squares estimator $L_y(x)$, the coefficients $\boldsymbol{\beta}(\boldsymbol{x})$ can be obtained as,

$$\boldsymbol{\beta}(x) = \left[\boldsymbol{x}^T \boldsymbol{W}(x)\boldsymbol{x}\right]^{-1} \boldsymbol{x}^T \boldsymbol{W}(x)\boldsymbol{y} \tag{13}$$

It is important to note here that the coefficients $\boldsymbol{\beta}(\boldsymbol{x})$ are the function of location *x*, where the approximation is sought. Thus, the procedure to calculate $\boldsymbol{\beta}(\boldsymbol{x})$ is a local approximation and 'moving' processes performs a global approximation throughout the whole design domain. In this regard, it may be noted that the response evaluation at each point needs equation (13) to be solved each time. Thus, computational time requirement increase marginally with respect to LSM based evaluation. However, this is negligible compare to time requirement of each NLTHA.

ANN Based Response Approximation

ANNs are relatively crude electronic models based on the neural structure of brain which can establish a functional relationship between two data spaces during learning process and reproduce these data during a recall process. On successful training, ANNs can provide the correlating mathematical relationship between set of input and associated output data as:

$$f : \boldsymbol{X} \rightarrow \boldsymbol{Y} \tag{14}$$

where, f is the relationship function, $X = (x_1, x_2, .., x_n)$ and $Y = (y_1, y_2, .., y_n)$ are the real input and output vectors. In general, an ANN structure consists of several layers and each layer consists of several neurons usually termed as processing elements (PE). A multi-layer feed-forward neural network structure, trained by back-propagation, is used in this study. Figure 1 shows a typical structure of a feed-forward ANN model. The left column is the input layer, the right most column is the output layer. In between input and output layer there is hidden layer. Generally, there could be more than one hidden layer. More details may be seen elsewhere (Elhewy, Mesbahi & Pu, 2006).

The procedure to establish an ANN model involves three stages:

1. Determining the topology (architecture) of the model,
2. Training the model with a sub-set of available data and finally

Figure 1. General structure of an ANN

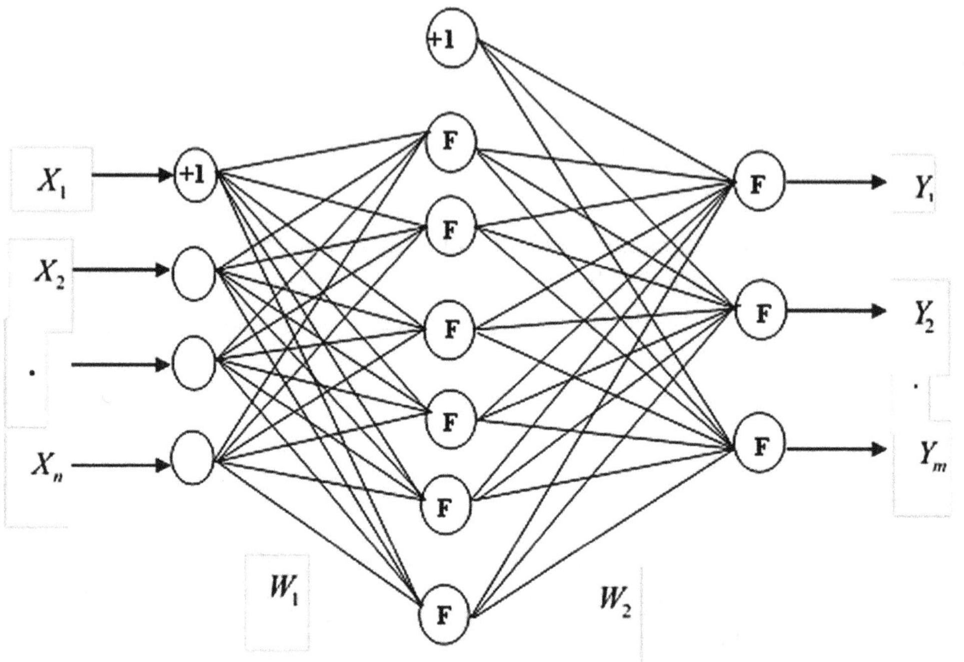

3. Cross-validation and testing the model with all the remaining data.

Determination of topology consists of deciding the number of hidden layers and determining the number of processing elements in each hidden layers. The number of PEs in the hidden layer is a very important parameter, which determines the accuracy of ANN models. Unfortunately, there is no general rule or algorithm to determine this number and need to be determined by a trial-and-error process. The training of a network is an iterative process to obtain the optimum unknown weights and biases required to establish a functional relationship between two data spaces which are implicitly related. The initial weights and biases are set to random values (typically between 0 and 0.1) and subsequently updated. The most widely used back-propagation training algorithm is adopted here. During the training of a network the same set of data is processed many times as the connection weights are adjusted by back-propagation. The iterative algorithm basically performs an error minimization procedure that is repeated until the network output converges to the target desired values. The training stops when the model reaches some statistically desired level of accuracy. If a training procedure is not successful, one need to repeat the procedure while varying the randomized weights, number of neurons in the hidden layer and applying activation function. For cross-validation and testing of ANN, the data is divided into three different sub-sets: (a) training, (b) cross validation and (c) test. Corresponding error when the network was presented with the cross-validation sub-set during the training procedure is used as a measure of model accuracy and as a signal to stop the training. Consequently, the ANN model accuracy is verified by introducing the test sub-set. If the error with respect to this sub-set is not acceptable, the training may be repeated. Indeed, this testing is critical to ensure that the network has successfully learned the correct functional relationship within the whole set of data.

As already mentioned, the dual response surface approach is adopted in the present study for approximation of nonlinear seismic responses. Thus, two separate ANN models are developed to approximate the mean (y_μ) and SD (y_σ) of seismic response. In this regard, it may be noted that the ANNs do not require assumptions of any basis functions. This may be considered as an advantage of using ANN when compared with the polynomial-based RSM. Once the ANN model is created, linking it to MCS based reliability analysis is straightforward. For this, a set of values for all random variables are generated. Now to obtain the value of the limit state function at the randomly generated values of the variables, the trained ANN is called to provide the desired responses. The selected response is now obtained as earlier based on lognormal assumption.

Design of Experiment for Response Surface Model

For constructing the metamodel, various DOE are used e.g. Saturated Design, Factorial Design, Central Composite Design etc. But, such classical designs are more appropriate for physical experiments where replication errors exist. It is to be noted here that the experiment in connection with SRA will be a computer analysis of nonlinear seismic response in which the random or replicating error term is absent. Thus, the absence of random error term leaves the least-square fit of a model without obvious statistical meaning (Simpson, Peplinski, Koch & Allen, 2001). Park and Towashiraporn (2014) mentioned that for constructing the response surface where experiments are performed artificially through computer simulation, DOE should have its design points filling the design space and treat all regions of the design space equally. Thus, in order to construct an efficient metamodel, a uniform design (UD), basically a space filling design appropriate for deterministic computer experiment (Fang, 1980) is adopted in the present study. The UD table used for numerical study is readily available at http://uic.edu.hk/isci/UniformDesign/UD%20Tables.html for different levels of sampling points for a given numbers of factors.

SELECTION OF GROUND MOTION BIN

The ground motion record is the most important parameter governing the seismic response outcome obtained through NLTHA. The most acceptable form to obtain seismic reliability corresponds to the use of recorded accelerograms. However, due to the limited resource of recorded accelerograms specific to a hazard level for the focused region (Guwahati City of the NE region of India) of the present study, the accelerograms are generated artificially and also synthetically by identifying the most vulnerable M_j and R_i combination for the specific hazard level of the location under consideration to ensure the variability in the input ground motion.

The artificial accelerograms are generated which are compatible to the acceleration response spectra for rock and hard soil for 5% damping (IS 1893: 2002). The power spectral density function is obtained following Kaul (1978) and the accelerograms are generated accordingly following the methodology proposed by Gasparini and Vanmarcke (1976). Defining a vector of amplitudes and simulating different arrays of phase angles, stationary stochastic process is obtained with the same general aspect but with different characteristics. These amplitudes are obtained using the power spectral density and the random phase angles are generated from uniform distribution in the interval of 0 to 2π. Further, to simulate the transient nature of the earthquakes, the steady state motions are multiplied by a deterministic envelope function (Saragoni & Hart, 1974).

The synthetic ground motion model as proposed by Boore (2003) is used for generation of synthetic acceleration time histories. The Fourier amplitude spectrum of ground motion at the location is obtained in terms of source and wave propagation functions. For generating synthetic accelerogram, white Gaussian noise is generated with zero mean and unit SD. This is filtered to retain the frequencies between 0 to 50 Hz in order to avoid any aliasing distortion as the sampling frequency is 50 Hz. The filtered Gaussian noise is further windowed by envelope function (Saragoni & Hart, 1974). This windowed noise is now Fourier transformed and normalized with its root mean square value. The normalized ordinates of the Fourier amplitudes are then multiplied with the Fourier amplitude as obtained from the stochastic point source model to get the Fourier spectrum of the ground motion. Finally, this Fourier spectrum is inverse Fourier transformed to obtain the synthetic accelerogram. Following this procedure, twenty accelerograms are generated for different magnitudes between 6.0 to 8.0 and epicentral range within 300km around the study area.

EXAMPLE APPLICATION

The example problem taken up for demonstration is an SDOF system characterized by a nonlinear spring connecting a lumped mass (m) to the ground as shown in Figure 2a. The nonlinear spring behaviour is described in Figure 2b. The damping is assumed to be Rayleigh-type i.e. proportional to mass and stiffness of the structure. The system is subjected to seismic acceleration at the base. Because of its simplicity, it is feasible to obtain large number of non-linear dynamic response of the SDOF system necessary for reliability computation by the direct MCS with reasonable time. Thus, this problem has been taken up for comparative study of the three metamodelling approaches presented in the previous sections.

The stochastic nature of the earthquake motion i.e. record to record variation is taken care by considering a suite of ground motion records. As detailed in section 4, the ground motion bin is formed with twenty artificially generated accelerograms and another twenty synthetically generated for the Guwahati city. The mean and SD of the nonlinear response of the system, is thus obtained based on these forty ground motions. Randomness in the structural characteristics is described by the pdf of each parameter, which in turn is used as input variables of the three metamodels. The random variables considered are the frequency (ω rad/s), damping (ξ in percentage), yield force (F_y in N) and ratio of the post-yield to

Figure 2. (a) The spring mass system; (b) the force deformation behaviour of the nonlinear spring

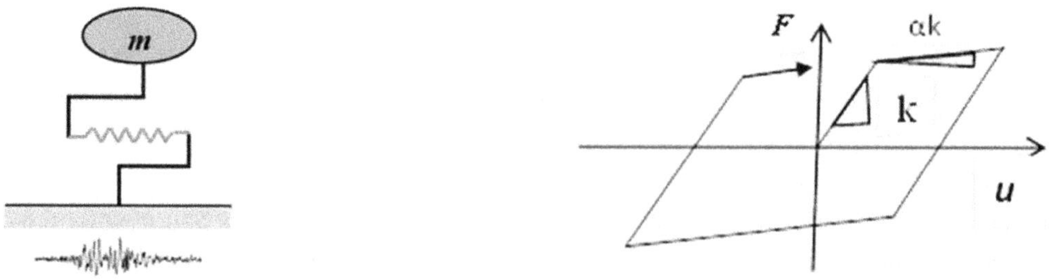

Figure 2(a) The spring mass system

Figure 2(b) The force deformation behaviour of the nonlinear spring elastic stiffness (α) which are assumed to be statistically independent normal. The ranges of the variables are depicted in Table 1. The PGA value is considered as control variables whose range varies from 0.1g to 1.0g. The maximum displacement of the mass is taken as the output response variable.

It is to be noted that the input variables in the metamodels are composed of two components: random variables and the control variable. The random variables are those representing uncertainties in structural properties i.e. X1 to X4 in Table 1 and the control variable is the PGA that represents earthquake intensity level. For developing the metamodels, the control variable is treated in a similar way as other random variables. The evaluation of response surface models at specific PGA values yield models for predicting the mean and SD of the response to obtain reliability of the system for that specific PGA.

As already discussed in section 3.4, the UD is adopted in the present study for DOE. For this, total thirty levels of sampling points for five factors are prepared following the UD table and transformed into real values of factors to implement the experiment. For each random variable its design points fill the design space and treat all regions of the design space equally. In order to capture the expectation as well as the record-to-record variability due to earthquakes, the responses sought are the mean and SD of the response. For this, the NLTHA is carried out for forty scaled ground acceleration records (twenty artificial and twenty synthetic) at each of these thirty sample points and the mean and SD of the responses at these points are the input responses at the design points for constructing response surface models.

Approximation by RSM Based Metamodels

To construct the LSM and MLSM based response surface models, the quadratic polynomial with cross terms has been adopted which contains twenty-one unknown coefficients. The design matrices and the response vector for the LSM and MLSM based RSM remain same. However, the coefficient vector of the MLSM based approach changes at each point of interest due to the change in the weight matrix. The influence radius D for computing the weight matrix is taken such that sufficient numbers of neighbouring supporting points are included and the values of c and k_0 in Equation 11 are taken as 0.15 and 1.0, respectively. The mean and SD at any sample point of interest can be evaluated with the help of these response surfaces and the overall response is evaluated following the lognormal assumption of seismic responses.

Approximation by ANN Based Metamodel

As already mentioned, two separate ANN models are developed in the present study to compute the mean and SD of nonlinear seismic response and the overall response is evaluated from these. The architecture of the network for mean response approximation is shown in Figure 3. It has two hidden layers

Table 1. The details of the random input variables

Parameters	ω (X_1)	$\xi(X_2)$	$F_y(X_3)$	$\alpha(X_4)$
Upper limit	9.27	0.03	2.913	0.075
Lower limit	3.29	0.01	1.035	0.025
Mean	6.28	0.02	1.974	0.05
CoV	0.2	0.25	0.2	0.25

with twelve neurons and five neurons respectively, five inputs and one output. For SD approximation, the similar architecture is used but instead of two hidden layers, one hidden layer is taken. The available Levenberg-Marquadt algorithm in MatLab is used for training the network. Both of the hidden and output layer use hyperbolic tangent activation functions as it is noted to produce better results than sigmoid function in this application. A set of forty levels of sampling points for five factors are prepared following the UD table for developing ANN models. A sub-set of data (70%) is used for training. For cross-validation 15% and for testing the model 15% of the total data are used. The data division is done in a random manner. The performance is studied using mean squared error, regression coefficients and histogram of error values. The mean square error is studied with respect to iterations (epochs) for training, validation and testing. It is found that the best performance for all the data sets is obtained for nine epochs. The regression models are prepared and the data points are fitted into the model to check the accuracy of the regression model for training, validation, test and overall performance. The regression coefficient, R for training, validation, test and overall are very close to one indicating the accuracy of the ANN architecture. The histograms of error values i.e. the difference of targets and the actual outputs obtained from the network for training, validation and testing are also studied (not shown here) and the performance of the network is noted to be 0.0262. These clearly indicate the accuracy of the ANN model.

Comparative Study

To study the effectiveness of the metamodels in approximating seismic responses, the mean and SD of the displacement of the SDOF system are computed for different PGA values with mean values of the random structural parameters by all the three metamodels. The results are shown in Figures 4 and 5. For comparative study, the mean and SD of the responses are also obtained at these points by direct MCS method. For this, the NLTHA is performed at these checking points considering all forty ground motions of the bin and the mean and SD of the responses are computed accordingly. The results are denoted as DMS and shown in the same plot. The improvement achieved by the MLSM and ANN based approaches compared to the LSM based RSM to estimate the mean and SD of nonlinear dynamic responses can be readily observed in the plots by comparing those with the direct MCS based responses.

Now, the SRA is performed by all the three metamodels and also by the direct MCS method. For SRA by the metamodels, the simulation is performed on the metamodels of random variables for any desired level of PGA (X5). The random structural parameters (i.e. X1 to X4) are simulated corresponding to their respective pdf and are combined at random to generate a large number (thirty thousand herein) of SDOF system. The mean and SD values are computed from the respective metamodels at each simulated random variables combination. The overall response is obtained accordingly for each such SDOF system. Thereby, an ensemble of random responses for the considered seismic intensity level (X5) is obtained.

Figure 3. The architecture of the Artificial Neural Network

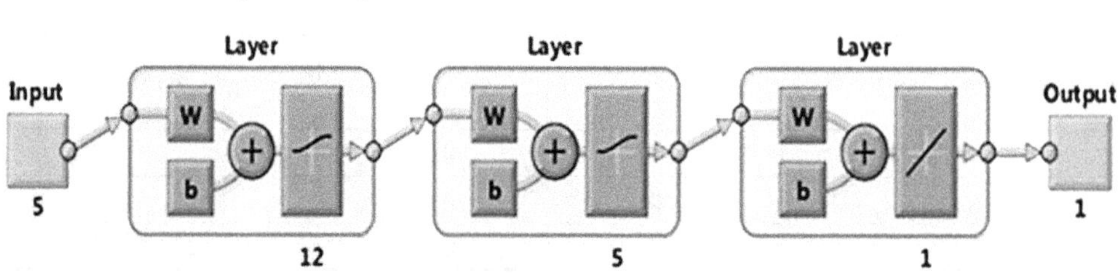

Figure 4. The comparison of mean displacement by different metamodels

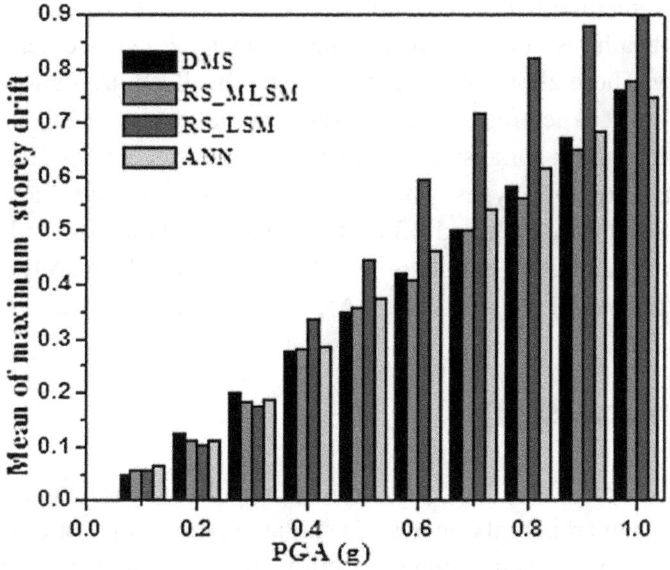

Figure 5. The comparison of SD of displacement by different metamodels

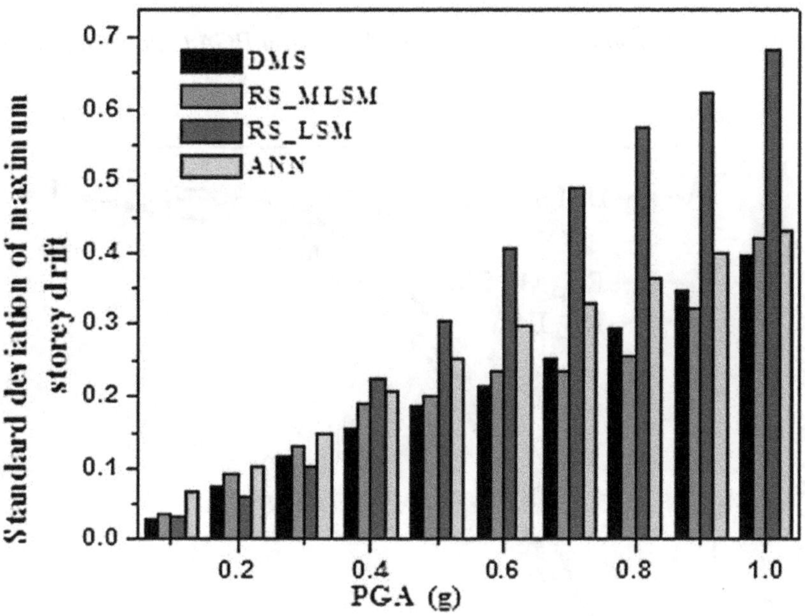

The probability of exceeding a given threshold displacement is obtained accordingly from the ensemble yielding the probability of failure of the system for the considered level of seismic intensity. The process is repeated for different PGA level. To obtain the seismic reliability by direct MCS, following the assumption that earthquake in a suit (forty records in the present study) are equally likely to occur, the ground motions are selected randomly from the suit to associate with the thirty thousand SDOF systems.

The NLTHA is performed on each earthquake structure combination to obtain the displacement and the probability of failure is computed with respect to the given threshold displacement. Figures 6, 7 and 8 show the probabilities of failures obtained for different allowable displacements i.e. 0.203m, 0.406m and 0.812m, respectively. These allowable values are taken for illustration purpose only. However, for reliability analysis of actual structures, these allowable values for different performance levels e.g. immediate occupancy, life safety; collapse prevention, etc. codal recommended values should be used depending on the specific structural types. The improvement possible to achieve in seismic reliability computation by the MLSM and ANN based RSMs with that of obtained by the LSM based RSM when compared with the direct MCS based seismic reliability estimate can be readily noted from these plots. And the observation is true for all the levels of PGA considered in the present study.

SUMMARY AND CONCLUSION

The effectiveness of LSM, MLSM and ANN based metamodels to approximate nonlinear dynamic response of structure for seismic reliability analysis is studied in the framework of MCS. In doing so, the repetition of intensity for complete generation of fragility curve is avoided judiciously by including this as one of the predictors in the seismic response prediction model. The improvement possible to achieve by the MLSM and ANN based approach in nonlinear dynamic response approximation is quite apparent

Figure 6. The comparison of probability of failure with varying PGA (allowable displacement 0.203m)

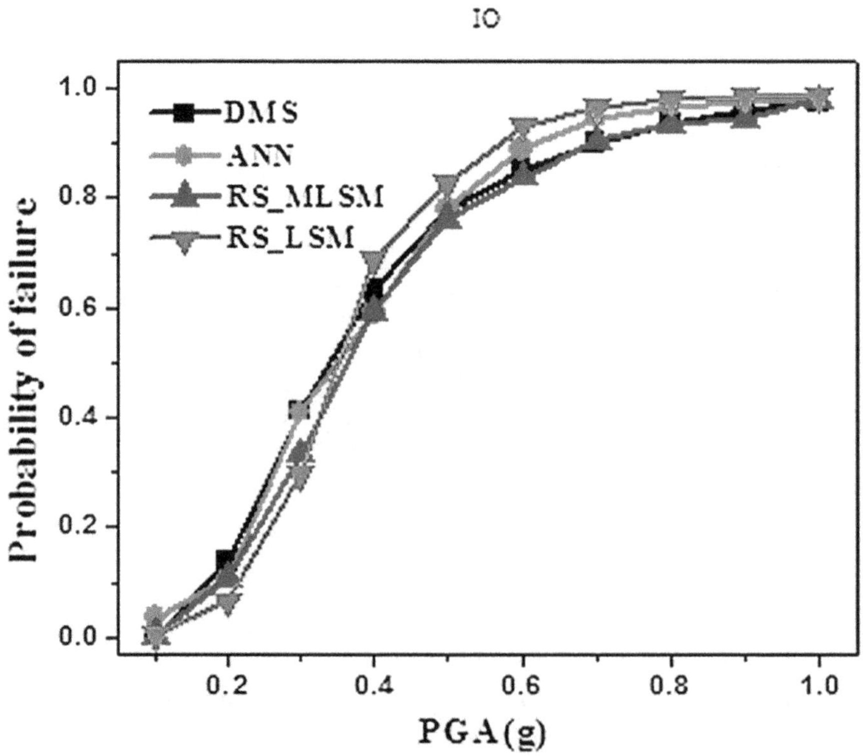

Figure 7. The comparison of probability of failure with varying PGA (allowable displacement 0.406m)

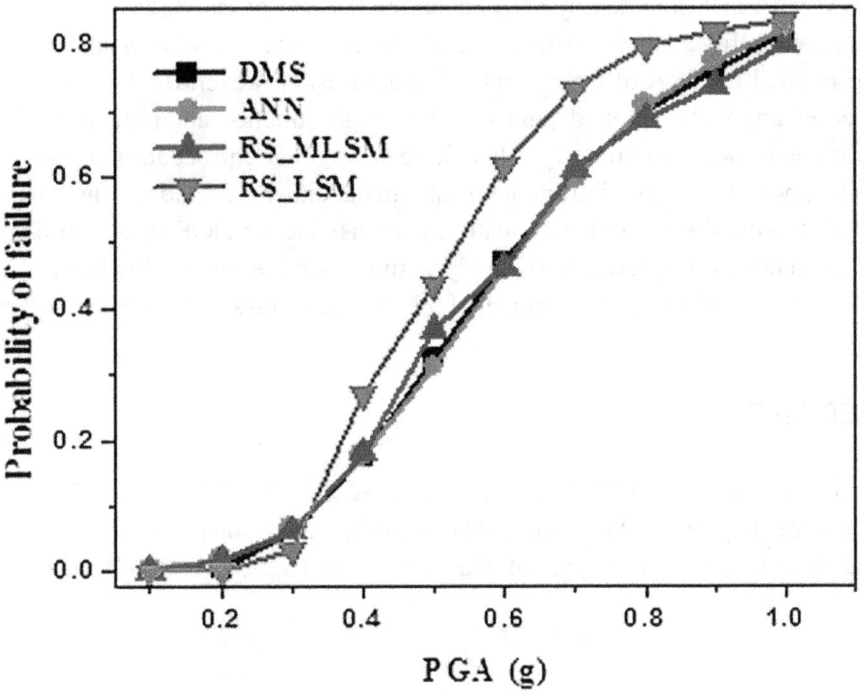

Figure 8. the comparison of probability of failure with varying PGA (allowable displacement 0.812m)

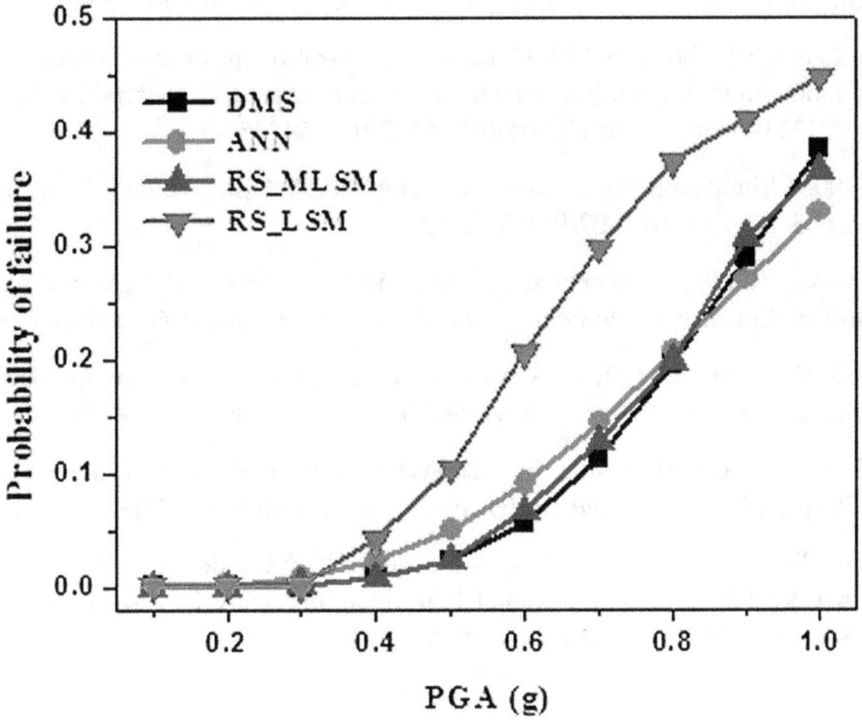

from the results of numerical study. The accuracy of computed seismic reliability by the MLSM and ANN based RSM with that obtained by the conventional LSM based RSM when compared with the direct MCS based reliability analysis results is clearly noted. Thus, the MLSM and ANN based adaptive RSMs could be applied for SRA of large complex structures. However, the level of accuracy achieved by the MLSM based approach largely depends on the weight function and radius of sphere of influence which are required to be selected properly. Also, there is no generic procedure to select a specific ANN architecture for response approximation and it is problem dependent. To address these aspects more study is felt essential. Although the example application is focused to a SDOF system and specific location, the basic steps are generic enough to readily apply to more complex structures located in other regions. One only needs to change the mechanical model of the structure and seismic hazard data of the locations.

ACKNOWLEDGMENT

The financial support received in TSD Scheme No. DST/TSG/STS/2012/45, 21.1.13 from the DST, Govt. of India (for the project entitled "Seismic vulnerability assessment of existing building to supplement rehabilitation practices with special emphasis to North Eastern Region") in connection with this work is gratefully acknowledged.

REFERENCES

Au, S. K., & Beck, J. L. (2003). Subset Simulation & its application to seismic risk based on dynamic analysis. *Journal of Engineering Mechanics*, *129*(8), 901–917. doi:10.1061/(ASCE)0733-9399(2003)129:8(901)

Bhattacharjya, S., & Chakraborty, S. (2011). Robust optimization of Structures subjected to stochastic earthquake excitation under limited information on system parameters uncertainty. *Engineering Optimization*, *43*(12), 1311–1330. doi:10.1080/0305215X.2011.554545

Boore, D. M. (2003). Simulation of ground motion using the stochastic method. *Pure and Applied Geophysics*, *160*(3), 635–676. doi:10.1007/PL00012553

Bucher, C., & Most, T. (2008). A comparison of approximate response functions in structural reliability analysis. *Probabilistic Engineering Mechanics*, *23*(2-3), 154–163. doi:10.1016/j.probengmech.2007.12.022

Bucher, G. C., & Bourgund, U. (1990). A fast and efficient response surface approach for structural reliability problems. *Structural Safety*, *7*(1), 57–66. doi:10.1016/0167-4730(90)90012-E

Chakraborty, S., & Sen, A. (2014). Adaptive response surface based efficient finite element model updating. *Finite Elements in Analysis and Design*, *80*, 33–40. doi:10.1016/j.finel.2013.11.002

Chojaczyk, A. A., Teixeira, A. P., Neves, L. C., Cardoso, J. B., & Guedes Soares, C. (2015). Review and application of Artificial Neural Networks models in reliability analysis of steel structures. *Structural Safety*, *52*, 78–89. doi:10.1016/j.strusafe.2014.09.002

Cornell, C. A., Jalayer, F., Hamburger, R. O., & Foutch, D. A. (2002). The probabilistic basis for the 2000 SAC Federal Emergency Management Agency steel moment frame guidelines. *Journal of Structural Engineering*, *128*(4), 526–533. doi:10.1061/(ASCE)0733-9445(2002)128:4(526)

Dymiotis, C., Kappos, A. J., & Chryssanthopoulos, M. K. (1999). Seismic reliability assessment of RC frames with uncertain drift and member capacity. *Journal of Structural Engineering*, *125*(9), 1038–1047. doi:10.1061/(ASCE)0733-9445(1999)125:9(1038)

Elhewy, A. H., Mesbahi, E., & Pu, Y. (2006). Reliability analysis of structures using neural network method. *Probabilistic Engineering Mechanics*, *21*(1), 44–53. doi:10.1016/j.probengmech.2005.07.002

Fang, K. T. (1980). Experimental design by uniform distribution. *Acta Mathematice Applicatae Sinica*, *3*, 363–372.

Faravelli, L. (1989). Response surface approach for reliability analyses. *Journal of Engg. Mech. ASCE*, *115*(2), 2763–2781. doi:10.1061/(ASCE)0733-9399(1989)115:12(2763)

Gasparini, D. A., & Vanmarcke, E. H. (1976). *SIMQKE, a program for artificial motion generation, user's manual and documentation. Publication R76-4*. Cambridge, Massachusetts: MIT Press.

Gavin, H. P., & Yau, S. C. (2007). High-order limit state functions in the response surface method for structural reliability analysis. *Structural Safety*, *30*(2), 162–179. doi:10.1016/j.strusafe.2006.10.003

IS1893. (2002). *Criteria for Earthquake Resistant Design of Structures, Part 1: General Provisions and Buildings*. Fifth Revision.

Iervolino, I., Fabbrocino, G., Manferedi, G. (2004). Seismic vulnerability of standardized industrial components: application to oil storage tanks. *Proceedings of the 13th WCEE*, Vancouver.

Kang, S.-C., Koh, H.-M., & Choo, J. F. (2010). An efficient response surface method using moving least squares approximation for structural reliability analysis. *Probabilistic Engineering Mechanics*, *25*(4), 365–371. doi:10.1016/j.probengmech.2010.04.002

Kaul, M. K. (1978). Stochastic characterization of earthquakes through their response spectrum. *Earthquake Engineering & Structural Dynamics*, *6*(5), 497–509. doi:10.1002/eqe.4290060506

Kazantzi, A. K., Righiniotis, T. D., & Chryssanthopoulos, M. K. (2008). Fragility and hazard analysis of a welded steel moment resisting frame. *Journal of Earthquake Engineering*, *12*(4), 596–615. doi:10.1080/13632460701512993

Kim, C., Wang, S., & Choi, K. K. (2005). Efficient response surface modeling by using moving least-squares method and sensitivity. *AIAA Journal*, *43*(1), 2404–2411. doi:10.2514/1.12366

Kwon, O. S., & Elnashai, A. S. (2006). The effect of material and ground motion uncertainty on the seismic vulnerability curves of RC structure. *Engineering Structures*, *28*(2), 289–303. doi:10.1016/j.engstruct.2005.07.010

Lin, D. K. J., & Tu, W. (1995). Dual Response Surface Optimization. *Journal of Quality Technology*, *21*(1), 34–39.

Möller, O., Foschi, R. O., Rubinstein, M., & Quiroz, L. (2009). Seismic structural reliability using different nonlinear dynamic response surface approximations. *Structural Safety*, *31*(5), 432–442. doi:10.1016/j.strusafe.2008.12.001

Park, J., & Towashiraporn, P. (2014). Rapid seismic damage assessment of railway bridges using the response-surface statistical model. *Structural Safety*, *47*, 1–12. doi:10.1016/j.strusafe.2013.10.001

Rajashekhar, M. R., & Ellingwood, B. R. (1993). A new look at the response surface approach for reliability analysis. *Structural Safety*, *12*(3), 205–220. doi:10.1016/0167-4730(93)90003-J

Saragoni, G. R., & Hart, G. C. (1973). Simulation of artificial earthquakes. *Earthquake Engineering & Structural Dynamics*, *2*(3), 249–267. doi:10.1002/eqe.4290020305

Simpson, T. W., Poplinski, J. D., Koch, P. N., & Allen, J. K. (2001). Metamodels for computer-based engineering design: Survey and recommendations. *Engineering with Computers*, *17*(2), 129–150. doi:10.1007/PL00007198

Taflanidis, A. A., & Cheung, S. H. (2012). Stochastic sampling using moving least squares response surface approximations. *Probabilistic Engineering Mechanics*, *28*, 216–224. doi:10.1016/j.probengmech.2011.07.003

Towashiraporn, P. (2004). Building seismic fragility using response surface metamodel [PhD Thesis]. Georgia Inst. of Tech.

Unnikrishnan, U., Prasad, A. M., & Rao, B. N. (2013). Development of fragility curves using high-dimensional model representation. *Earthquake Engineering & Structural Dynamics*, *42*(3), 419–430. doi:10.1002/eqe.2214

Zhang, J., & Foschi, R. O. (2004). Performance-based design and seismic reliability analysis using designed experiments and neural networks. *Probabilistic Engineering Mechanics*, *19*(3), 259–267. doi:10.1016/j.probengmech.2004.02.009

Chapter 7
Finite Element Analysis of Pipe Bends under External Loads

Sumesh S.
National Institute of Technology, Tiruchirappalli, India

A. R. Veerappan
National Institute of Technology, Tiruchirappalli, India

S. Shanmugam
National Institute of Technology, Tiruchirappalli, India

ABSTRACT

Pipelines are being used to convey different sorts of fluids from hazardous and toxic substances to high pressure steam. Piping systems are subjected to various external loads leading to major failures with gross plastic deformation. Pipe bends are incorporated into piping systems not only to change the direction of flow but also to provide flexibility, hence they are considered to be critical components and its safe design under various loads becomes important. Earlier studies of pipe bends utilized analytical methods to determine the plastic loads. The evolution of FEM and the advancements in computational capabilities have enabled analysts to generate large number of data which is expensive and time consuming with experimental investigations. In this chapter, the major studies on pipe bends by various researchers are explored. Different studies on pipe bends namely stress analysis and the influence of geometric shape imperfections are also presented.

INTRODUCTION

Piping Systems

Within an industry, the term piping relates to a system of pipes used to convey fluids (liquids and gases) from one location to another. The engineering discipline of piping design studies the efficient transport of various fluids under varied loading conditions. In urban areas like in cities, piping systems are used to transport water from the sources of water supply to the points of distribution; convey waste from

DOI: 10.4018/978-1-5225-0588-4.ch007

residential and commercial buildings and other civic facilities to the treatment facility or to the point of discharge. The fire protection piping networks in residential, commercial, industrial, and other buildings carry fire suppression fluids, such as water, gases, and chemicals to provide protection to life and property. Similarly, pipelines are also used to carry crude oil from oil wells to tank farms for storage or to refineries for processing. The natural gas transportation and distribution lines convey natural gas from the source and storage tank to points of utilization. In chemical plants, paper mills, food processing plants, and other similar industrial establishments, the piping systems are utilized to carry liquids, chemicals, mixtures, gases, vapors, and solids from one location to another. The piping systems in thermal power plants convey high-pressure and high-temperature steam to generate electricity. Other piping systems in a power plant, transport high and low pressure water, chemicals, low pressure steam and condensate. Sophisticated piping systems are used to process and carry hazardous and toxic substances.

In health facilities, piping systems are used to transport gases and fluids for medical purposes. The piping systems in laboratories carry gases, chemicals, vapors, and other fluids that are critical for conducting research and development. In short, the piping systems are an essential and an integral part of modern civilization. The design, construction, operation, and maintenance of various piping systems involve understanding of piping fundamentals, materials, generic and specific design considerations, fabrication and installation, examinations, and testing and inspection requirements, in addition to the local, state and federal regulations.

Piping includes pipe, flanges, fittings, bolting, gaskets, valves, and the pressure containing portions of other piping components. It also includes pipe hangers and supports and other items necessary to prevent over pressurization and overstressing of the pressure-containing components. Pipe sections when joined with fittings, valves, and other mechanical equipment and properly supported by hangers and supports, are called piping systems.

Piping for most process units represents the major item of investment. Typically, the total erected piping cost ranges from 25 to 50 percent of the total cost of a unit. Consequently, the piping engineer often faces the necessity of making careful and realistic compromises between design features and cost without sacrificing minimum safety standards (Figure 1).

Materials

The material with which a pipe is manufactured often forms the basis for choosing any pipe. Materials that are used commonly for manufacturing pipes include:

- Carbon steel.
- ASTM A252 Spec Grade 1, Grade 2, Grade 3 Steel Pile Pipe.
- Low temperature service carbon steel.
- Stainless steel.
- Nonferrous metals, e.g. cupro-nickel.
- Nonmetallic, e.g. tempered glass.

Pipe Bends

Pipe bends are critical components in piping systems. Pipe bends are used to change the direction of the fluid flow. (Veerappan, 2008). The bend section may be a potential source of damage for the duration

Figure 1. Industrial piping system

of service when considerable ovality and wall thickness variation (thinning/thickening) exist, which are introduced during the manufacturing process. Hence the acceptability of pipe bends depends on the magnitude of these shape irregularities induced during the manufacturing process. In the manufacturing process of pipe bends it is difficult to avoid thickening on the intrados and thinning on the extrados. The satisfactoriness of pipe bends is based on the induced level of these shape imperfections.

Ovality is a deviation of a pipe from an ideal round shape caused by imperfection during milling or flattening from storage. Ovality and thinning are the shape imperfections have been considered for structural analysis using FEM. It has been reported that thinning and ovality are to be taken into description together to decide the suitability of these bends. The probable flexibility is optional in the selection of ovality and thinning limits of pipe bends to reduce their elimination. Hence the suitability of pipe bends depends on the scale of these shape irregularities induced during the developed process (Figure 2).

Definitions

When ovality is included, the cross sections assumed, in the analysis of pipe bend, are elliptic (Dan, 2009) and semi oval (Figure 3) (Boyle, 1981).

Percent ovality C_0, thinning C_t, and thickening C_{th}, are defined as follows (Veerappan, 2008). For elliptical cross section:

$$C_o^{EL} = \frac{\left(D_{\max} - D_{\min}\right)}{D} \times 100 \tag{1}$$

where

Figure 2. Pipe bend cross section before and after bending

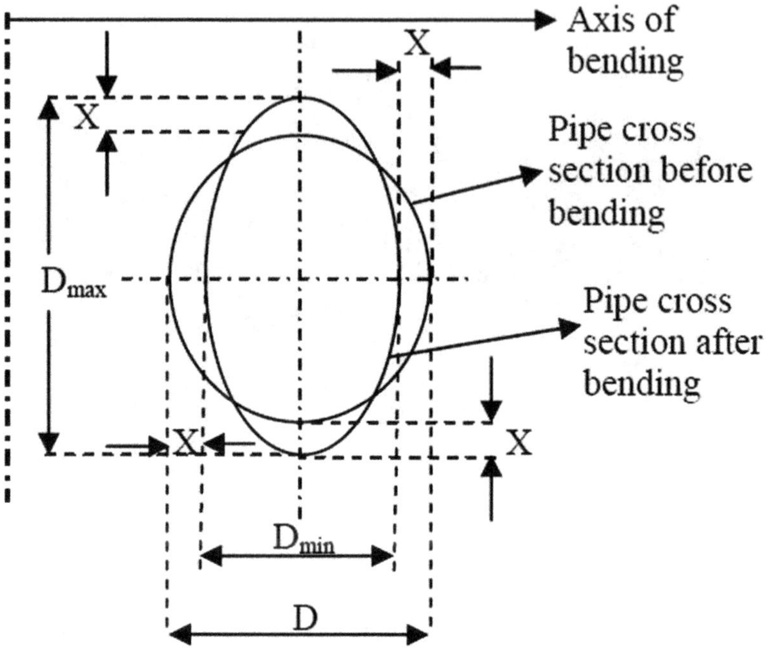

$$D = \frac{D_{\max} + D_{\min}}{2}$$

For semi oval cross section:

$$C_o^{SO} = \frac{\left(r_o - r_{\min}\right)}{r_o} \times 100 \tag{2}$$

$$C_t = \frac{\left(t - t_{\min}\right)}{t} \times 100 \tag{3}$$

$$C_{th} = \frac{\left(t_{\max} - t\right)}{t} \times 100 \tag{4}$$

For modeling of pipe bend cross section, the value of X is required which can be determined as follows

$$\text{Dmax} = \text{D} + 2[\text{X}] \tag{5}$$

Figure 3. Pipe bend with attached straight pipe showing elliptic and semi oval cross sections
Christo et.al 2011.

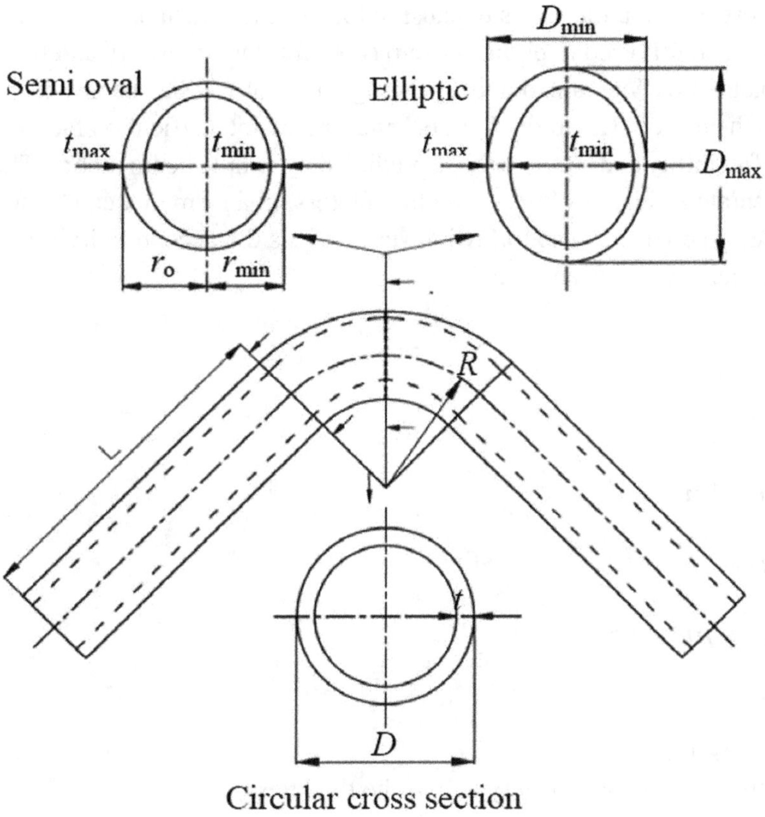

Dmin = D – 2[X] (6)

Substituting Equation 5 and Equation 6 in Equation 1

Co = 400[X]/D (7)

The divergence of the pipe bends cross division for circularity can be calculated for any ovality using Equation 7 and hence the maximum and minimum pipe bends diameters can be obtained. Thinning at the extrados is equal to thickening at the intrados, and accordingly the proportions of the geometry are calculated and created in commercial finite element analysis software.

Theoretical Analysis of Pipe Bends

Systematic analysis of pipe bends began in 1910 (Wiley, 1956), with reports on the occurrence of ovalization, and on the information that it leads to greater suppleness than could be accounted for by bar theories. A year later, the first theoretical handling of the subject was published by Von Kármán, who investigated the stress allocation in curved tube subjected to in-plane bending. At about the same time Lorenz and Marbec furnished a solution of the problem, using castigliano's theorem in their work instead of the principal of minimum potential energy as used by Kármán. Hovgard continued Kármán work and arrived at a solution through a different approach while Kari refined the solution by making an allowance for more terms in the series expansion for the basic variables.

The general approach followed in piping system flexibility in the ovality and thinning limits calculations has been explained by Veerappan and Shanmugam (2008). It has been widely adopted in design codes and relies on the use of "flexibility factors" and "stress-intensification factors", which are simply the ratios of actual flexibility and stress to those predicted by simple beam theory. The approaches aim at providing more accurate estimates of the correction factors, in a form that can be simplified, to become readily usable in design contexts. The following summarizes the acceptability criteria used to evaluate the results of finite element stress analysis

$$Pm < kSm$$

$$Pm + PL < 1.5kSm$$

$$Pm + PL + Pb < 1.5kSm$$

$$Pm + PL + Pb + Q < 3Sm$$

$$Pm + PL + Pb + Q + PF < 2Sa$$

P – Pipe Internal Pressure;
P_m – Average memberance stress intensity across the thickness;
P_L – Local memberance stress intensity across the thickness;
P_b – Average bending stress intensity across the thickness K – Load factor;
S_m – Allowable stress intensity; Q – Secondary stress intensity.

A. Ostsemin and V. L. Dil'man (2002) expressed that manipulative wall thickness for major oil product pipelines. The basic purpose of the design estimate is to make the pipeline viable for the calculated working period. B Lynn Ferguson and Metallurgical; Zhichao presented wall thinning during tube bending. Pipe bending machines are equipped with Pressure Die Assist (PDA) that applies the lateral pressure and axial push to meet the design requirement. The study showed that pipe bending could be simulated successfully, given accurate data is available for the material, interfacial friction, and bending conditions. Machines need to be capable of varying booster pressure during the bend cycle to minimize wall thinning while avoiding tube buckling. Carriage boosters and normal PDAs do not provide sufficient boost to overcome the material's natural yield strength, especially as the inside wall thickness increases and the force required to deform the material becomes higher. Orynyak and Radchenko (2004) analyzed a problem for determination of stress and flexibility of a pipe bend under a combined action of internal pressure and bending moments applied both in and out of plane of the bend curvature. The problem of allowing for internal pressure in three- dimensional bending falls into the category of geometrically nonlinear problems. To solve it, a very simple technique is put forward, which implies that the pressure-induced circumferential force is much higher than that caused by ovalization and can be considered as an additional external loading constant in the x coordinate. This supposition makes it possible to allow for internal pressure by entering a proportional term proportional and multiplying by the curvature associated with the contour deformation into the first equilibrium equation.

Pipe Stress Analysis

Process piping systems and power piping systems are typically checked by pipe stress engineers to verify that the routing, nozzle loads, hangers, and supports are properly placed and selected such that allowable pipe stress is not exceeded under different loads such as sustained loads, operating loads, pressure testing loads, etc., as stipulated by the ASME B31, EN 13480 or any other applicable codes and standards. It is necessary to evaluate the mechanical behavior of the piping under regular loads (internal pressure and thermal stresses) as well under occasional and intermittent loading cases such as earthquake, high wind or special vibration, and water hammer. This evaluation is usually performed with the assistance of a specialized (finite element) pipe stress analysis computer programs such as CAESAR.

In cryogenic pipe supports, most steel become more brittle as the temperature decreases from normal operating conditions, so it is necessary to know the temperature distribution for cryogenic conditions. Steel structures will have areas of high stress that may be caused by sharp corners in the design, or inclusions in the material. The impact of high piping stress on operating piping systems can be dramatic and costly. Yet for many piping designers, piping stress analysis is the least understood area of pipe knowledge. Neglecting the impact of the weight of the pipe and thermal expansion of hot pipe can cause significant maintenance problem due to high piping stress. Some of the associated issues are pump bearings worn-out, hairline cracks developing in vessel and nozzle junctures, flanges leaking flammable liquids and pipes permanently deforming.

In many cases, designers will prefer to utilize smooth pipe bends due to the lack of discontinuity, the relatively smaller stress concentration and the more favorable flow characteristics. However, in large-diameter applications or applications with restricted space, mitered bends still find application (Wood, 2008). Both types of pipe bends develop an increase in flexibility and local stress levels due to cross-section ovalisation. The mitered bends obviously has the added complexity of discontinuity stresses at the intersection. In the early theoretical work, an edge solution was developed for this case, which was then superimposed on the flattening analysis.

Experiments on the collective effects of pressure and both in plane bending (IPB) and put of plane bending (OPB) for two single un-reinforced mitres have been documented (So, 1971). The effect of pressure was more significant with larger a/t ratio. In addition, there was a progressive association of the maximum stress location to the 'crotch' position, with internal pressure. This behavior is more akin to simple bending theory.

Markl (1947) investigated the consequence of internal pressure on the cyclic IPB and OPB fatigue behavior of multi-mitred bends. Markl reported that the resulting mean stress level had little effect on the magnitude of the fatigue stress concentrations. Bond (1971) and Kitching (1972) reported the results of tests on a multi-mitred pipe bend when subjected to an IPB load at various internal pressures. The results of these tests were compared with those derived from the theory produced by the same authors. The theoretical approach was similar to that used by Rodabaugh and George (1957), for the analysis of smooth bends. The authors concluded that the flexibility factor was not linearly related to pressure and that stress concentration factor due to bending were related to pressure in much the same way as flexibility factors. Furthermore, it was found that the maximum stresses due to bending did not necessarily occur at the same location or in the same directions, for all pressures. Results obtained using the above theory, also compared favorably with the experimental results of Lane and Rose (1960).

Bond and Kitching (1972), while investigating out-of-circularity effects, also considered the combined loading case of internal pressure and IPB. The authors reported good agreement between theoretical and experimental results for stresses due to the bending moment and also for flexibility factors, when the specimen was subjected to IPB in the presence of a constant internal pressure. Strains and deflections, in this case, were linear with bending moment for any given pressure. It was also reported that, when the specimen was subjected to internal pressure, or to internal pressure in the presence of a constant bending moment, the relationships between strains and pressure were non-linear. This non-linearity was thought to be caused by initial out-of-circularity of the pipe cross-section. The theoretical distributions bore only partial resemblance to the experimental distributions in this case. Finally, it was concluded that the total strains, due to combinations of internal pressure and IPB, were found to be independent of the sequence of loading for the range of loads used. While this conclusion is true for the bends examined, it is generally dependent on the bend geometry and loading magnitudes (Babaii, 1994).

Flexibility Analysis

Pipe bends are critical components in piping systems and acknowledged to be the most economical means of changing flow on the condition that flexibility and end reactions are within the acceptable restrictions. Various researchers have studied the effect of limit/collapse load of pipe elbows or pipe bends with circular cross sections under in-plane bending moment with or without internal pressure (Shalaby, 1998; Chattopadhyay, 2000; Kim, 2006).

Gresnigt (2002) considered the elastic and plastic performance of single mitres with two dissimilar bend angles and radius to thickness ratios. The specimens were subjected to IPB moments with opening and closing, at different bend angles under with and without pressure. Bond and Kitching (1972) examined a 90^0, two-segment mitre subjected to a grouping of bending moments and internal pressure. When the bending moments were applied, the stress distributions and flexibility factors compared well with hypothetical predictions. When internal pressure was applied and then in combination with bending loads non-linear elastic performance was observed. This performance was credited to the out-of-circular shape of the pipe segment.

Finite Element Analysis

Veerapan et al (2006) investigated stresses introduced in pipe bends with different ovality and thinning for a particular internal pressure using the finite element method. As ovality is increased to 5 per cent from 0 per cent, the membrane stress decreases as the bend radius is increased for 20 per cent thinning (Figure 4). At the neutral section the membrane stress decreases as ovality is increased for all bend radii [Figure 4(b)]. As the ovality is increased from 5 to 20 per cent, assuming there is no thinning, the membrane stress increases at intrados and extrados. The outer surface is the most highly stressed.

The constant allowable pressure ratio for different ovality and thinning at different bend radii was presented by Veerappan et al. (2006). The allowable pressure ratio increases, attains a maximum, and then decreases as the values of ovality and thinning are increased. Figure 5 shows the variation in allowable

Figure 4. Variation in average membrane stress for 20 per cent thinning

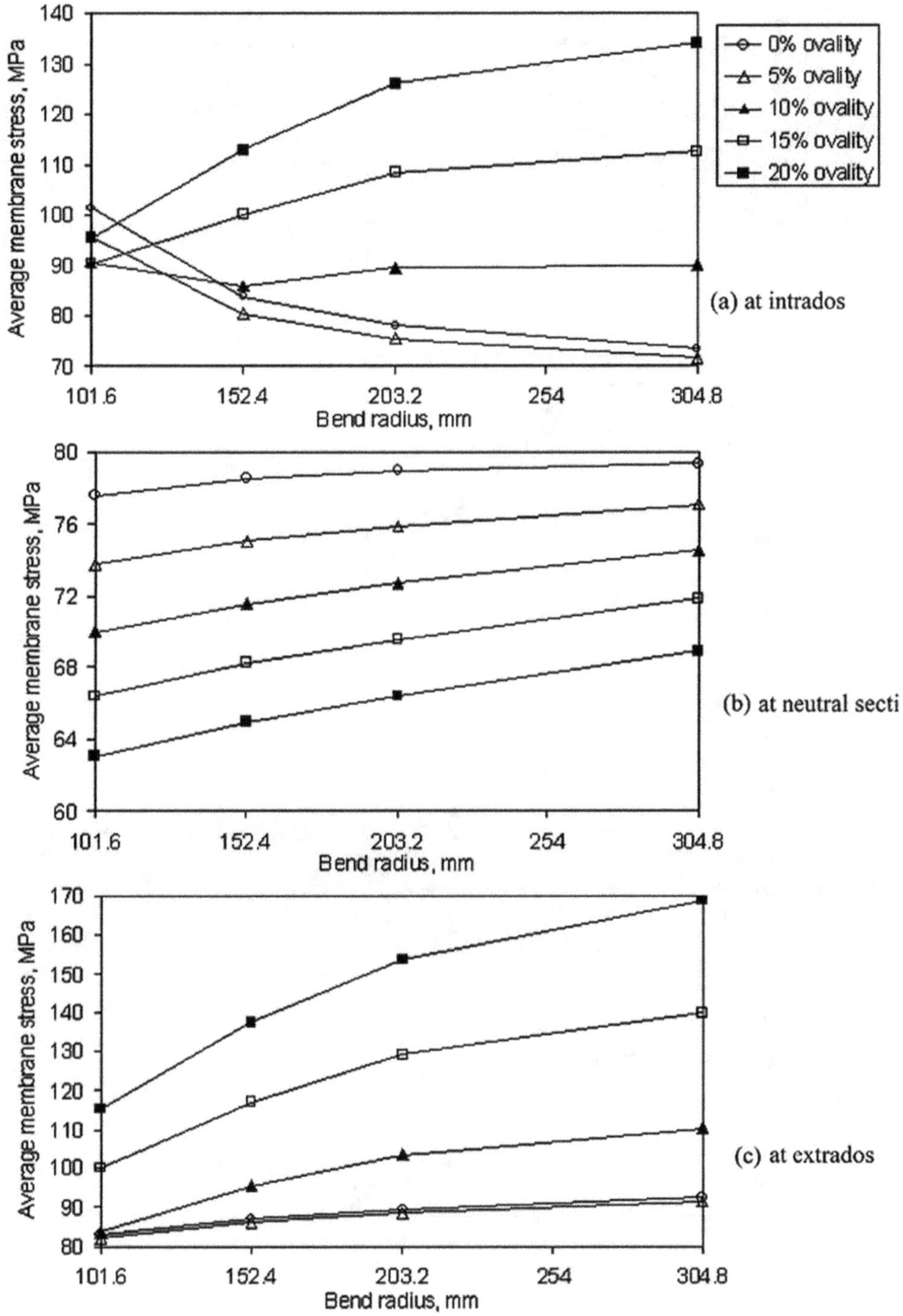

Figure 5. Interdependence of allowable pressure ratio with shape irregularity

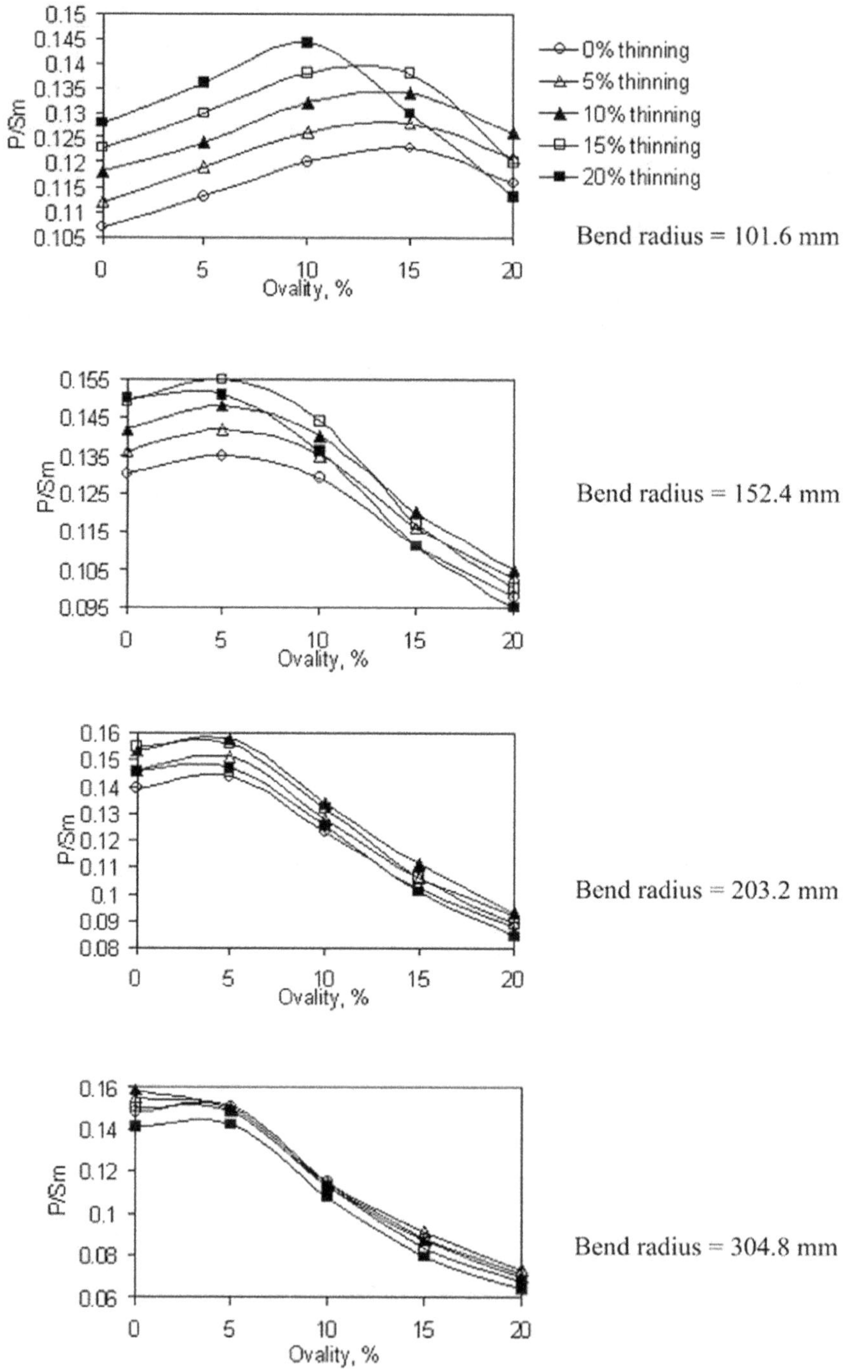

pressure ratio with shape irregularity for different bend radii. It can be seen that the allowable pressure ratio increases, attains a maximum value, and then decreases as the ovality and thinning are increased. As the ovality increases, the net vertical force on the pipe bend decreases because the minor axis of the elliptical shape decreases. Hence, the induced stress is found to decrease up to 6 per cent ovality. Other effects such as the bending stress become predominant after 6 per cent ovality. Hence, this behaviour is not observed in higher ovality. It can also be observed from Figure 3 that the maximum value of allowable pressure ratio shifts towards lower per cent ovality as the bend radius is increased. As the thinning increases, the load on the inner side of the bend is resisted by higher thickness, hence the stress decreases. Similarly, as the thinning increases, the load on the outer side of the bend is resisted by lower thickness.

As shown in Figure 6, the area enveloped by constant pressure ratio becomes larger and larger as the pressure ratio is smaller and less important if the pressure ratio obtained as per codes for influential allowable ovality and thinning is 0.27, the possible amalgamation of ovality and thinning of a pipe bend is recognized at any point in the shaded area. The major conclusions that could be drawn from the analysis is the dependence of allowable pressure ratio of the pipe bend on bend ratio, tube ratio, ovality and thinning. There is more than one value of ovality and thinning for a pipe bend manufactured, which would reduce rejection of pipe bends after they have been made. Given the values of ovality and thinning of pipe bend, it can be checked whether the bend can withstand the designed pressure. Yun-Jae Kim (2007) explained the effects of attached straight pipes on finite element limit analysis for pipe bends. The effect of the length of straight pipe on the plastic boundary load of a 90^0 pipe bend under collective

Figure 6. Variation of constant pressure ratio with shape imperfections for pipe material SA 106 Gr B, D=50.8mm, t=6.35mm R=75mm and standalone 90° pipe bend
Veerappan, 2008.

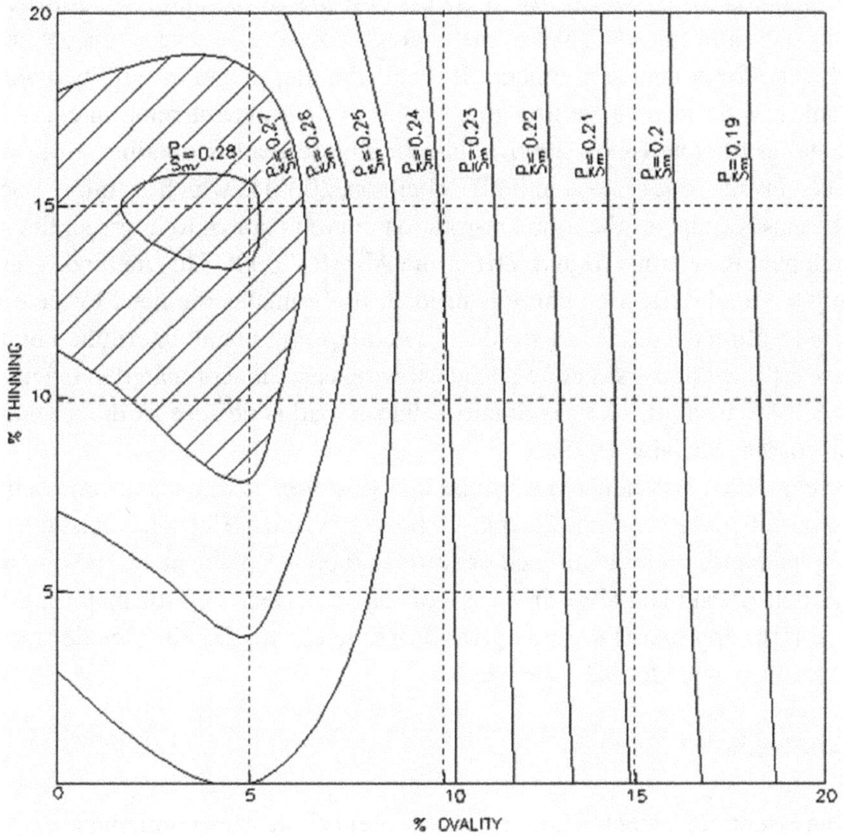

pressure and bending was quantified, based on finite element (FE) limit analyses with the small geometry transform selection. Systematic FE limit analyses of pipe bends with various lengths of the attached pipe were performed. It was inferred that the consequence of the length of the attached pipe on plastic limit loads can be significant, and the limit loads tend to decrease with decrease of the length of straight pipe.

Yun-Jae Kim (2007) presented effects of bend angle on plastic loads of pipe bends under interior pressure and in-plane bending, by means of FE method. The consequence of a bend angle of a pipe bend on plastic loads was done via diminutive strain and large strain Finite Element (FE) limit analysis considering elastic–perfectly plastic materials. At the same time of making an allowance for the consequence of the attached straight pipe, two limiting cases were well thought out. One case corresponds to the pipe bend lacking the attached straight pipe, and the other with a satisfactorily long attached straight pipe. For the former case, the FE results put forward that the limit load is not exaggerated by the bend angle for both in-plane bending and internal pressure. For the concluding case, the bend angle affected the plastic loads. It was found that the plastic load effortlessly changes from the limit load of the straight pipe when the bend angle approaches zero to the plastic load of the 90^0 pipe bend when the bend angle approaches 90^0. Based on such explanation, closed-form plastic load solutions were proposed for the pipe bend under in-plane bending and internal pressure.

Definitions of Plastic Loads

Plastic load is the common term used to indicate limit, collapse and instability loads or moments. The plastic loads are obtained from moment-rotation curve drawn for each pipe bend model. The curves of a typical bend are shown in Figure 7. When the analysis considers linear geometry, the plastic load is called limit load while for nonlinear geometry, it is called collapse and instability loads.

The limit moment is defined as the maximum point in the moment-rotation curve as the curve approaches horizontal asymptote (Kim, 2006). Plastic collapse load is obtained from moment-rotation curves using twice-elastic-slope (TES) method (Robertson, 2005) in which a straight line from the origin equal to twice the elastic slope of the moment-rotation curve is drawn to intersect the same curve. The TES method is adopted in sections III and VIII of the ASME (2000). This method is also convenient in its use, since only the load-deflection curve is needed. It eliminates the need for determining the first yield point, or the location of maximum strains. The collapse moment determined by this method is a more conservative estimate of the structure's load carrying capacity compared to the instability moment (Mourad, 1999). Plastic instability or maximum collapse load is defined as the maximum point in the moment-rotation curves (Shalaby, 1998).

Kuk-Hee Lee et al. (2012) conducted a similar analysis of plastic loads of pipe bends under collective pressure and out-of-plane bending. The plastic boundary and TES (twice-elastic-slope) plastic load solutions for 90^0 pipe bends under combined pressure and out-of-plane bending were carried out using three dimensional non-linear FE analysis by means of elastic-perfectly plastic materials. Without internal pressure, a closed-form approximation was given In-plane closing bending mode was found to be the most significant loading mode for 90^0 pipe bends.

Limit Load Analysis

The limit load is an essential tool for the integrity assessment of pressure components. Analytical methods have been developed for simple components. Elastic-plastic analysis is widely used in complex problems

Figure 7. Typical moment–end rotation curve

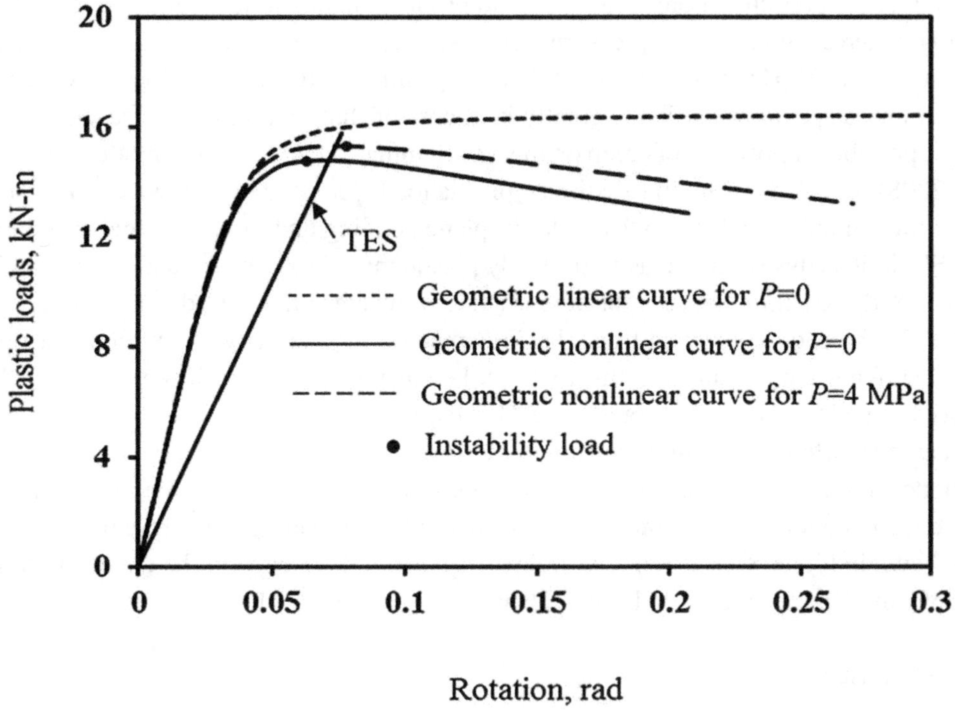

for evaluation of limit load. In piping systems, allowing some measure of controlled plastic deformation without endangering the safety of components are sometimes considered as a good design practice. Limit analysis is one method of avoiding undesirable structural response in the components of such systems.

Yun-Jae Kim (2007) presented limit loads for pipe bends under combined pressure and in-plane bending based on finite element limit analysis. The detailed three-dimensional (3–D) FE limit analysis for pipe bends under combined pressure and bending were performed based on elastic–perfectly plastic materials with the small geometry change option. A wide range of parameters related to the bend geometry was considered. The FE results shows that existing limit load solutions for pipe bends are indeed lower bounds. However, for bending and for combined bending and pressure, existing solutions are found to differ significantly from the present FE results. Accordingly, based on the FE results, closed-form approximations of plastic limit loads for pipe bends under internal pressure, under bending and under combined bending and pressure were proposed.

Yun-Jae Kim's (2007) work quantifies the effect of a bend angle of a pipe bend on plastic loads, via small strain and large strain finite element (FE) limit analyses using elastic–perfectly plastic materials. While considering the effect of the attached straight pipe, two limiting cases were considered. One case corresponds to the pipe bend without the attached straight pipe, and the other with a sufficiently long attached straight pipe. For the former case, the FE results suggest that the limit load is not affected by the bend angle for both in-plane bending and internal pressure.

Donald Mackenzie (2004) conducted analysis for plastic collapse of pipe bends under combined internal pressure and in- plane bending. The results of the investigation showed that geometric nonlinearity was a significant consideration when calculating plastic failure loads of pipe bends subject to combined loading. Significant geometric weakening is observed when the closing bending moment dominates. At higher pressure, both pressure, moment loading and proportional loading cause considerable geometric strengthening, whereas the moment–pressure load case exhibits significant geometric weakening.

Several investigators (Bond, 1972; So, 1971) during 'limit load' investigations, have reported 'cold creep' of the mitred specimens at loads greatly in excess of that required to cause yield. Hose (1988) discussed the possible importance of creep on the inter laminar shear stress and modulus for GRP bends. Kim et al. (2008) provided a method to estimate plastic loads [defined by the twice-elasticslope (TES)] for elbows with non-uniform thicknesses under in-plane bending and under internal pressure, based on systematic FE limit analyses using elastic-perfectly plastic materials. The intrados thickness is assumed to be up to 30% thicker than the straight pipe thickness, whereas the extrados thickness is up to 30% thinner. The plastic load determined from the TES method for the above maximum case is the same as that for the reference case (uniform thickness), whereas for the case of 30% increase in thickness at intrados and nominal thickness at extrados, the TES load is larger.

The percentage difference of limit moment between the reference (without ovality and thinning) and irregular models (with varying degrees of ovality and thinning) for various thicknesses and bend radii under moment and combined moment and an internal pressure loading was presented by Christo et al. (2012) (as shown in Figure8). The negative values of percent difference in the figures indicate that the magnitude of limit load for the irregular models is more than that of the reference model.

In-Plane Bending

As shown in Figure 8 the percentage difference increases with ovality for all the models considered. The cross section deviation from circularity may be the reason for this increase of percent difference. The percentage difference of limit load for all the models considered ranges from 1.7% to 2.5% when the ovality is 5% and the values vary from 3.4% to 5% as the ovality is increased to 10%. The variation is from 5.3% to 7.6% for the models with 15% ovality and the range is from 7.2% to 10.2% when the ovality is further increased to 20%.

Combined Moment and Internal Pressure

The dotted lines in Figure 8 exemplify the effect of ovality under combined bending and internal pressure of 5 MPa. The negative and positive percent difference values are directly attributed to the internal pressure load and the variation in the geometry of the pipe bends as the limit loads are obtained using geometrically linear analysis.

Collapse Load in Pipe Bends

The collapse load (TES) is obtained using the twice-elastic-slope method in which a straight line from the source with twice the gradient of the initial elastic reaction (with respect to the load axis) of the instant– regular change curve is drawn to interconnect the equivalent curve (Robertson, 2005). The greatest crumple load or collapse load is defined as the highest load in the load– deflection curve. Christo et. al. (2012) explains the effects of ovality and thinning/thickening on crumple loads of pipe bends when subjected to in-plane concluding bending moment using nonlinear finite element analysis. Analyses in

Figure 8. Variation of limit load with respect to ovality

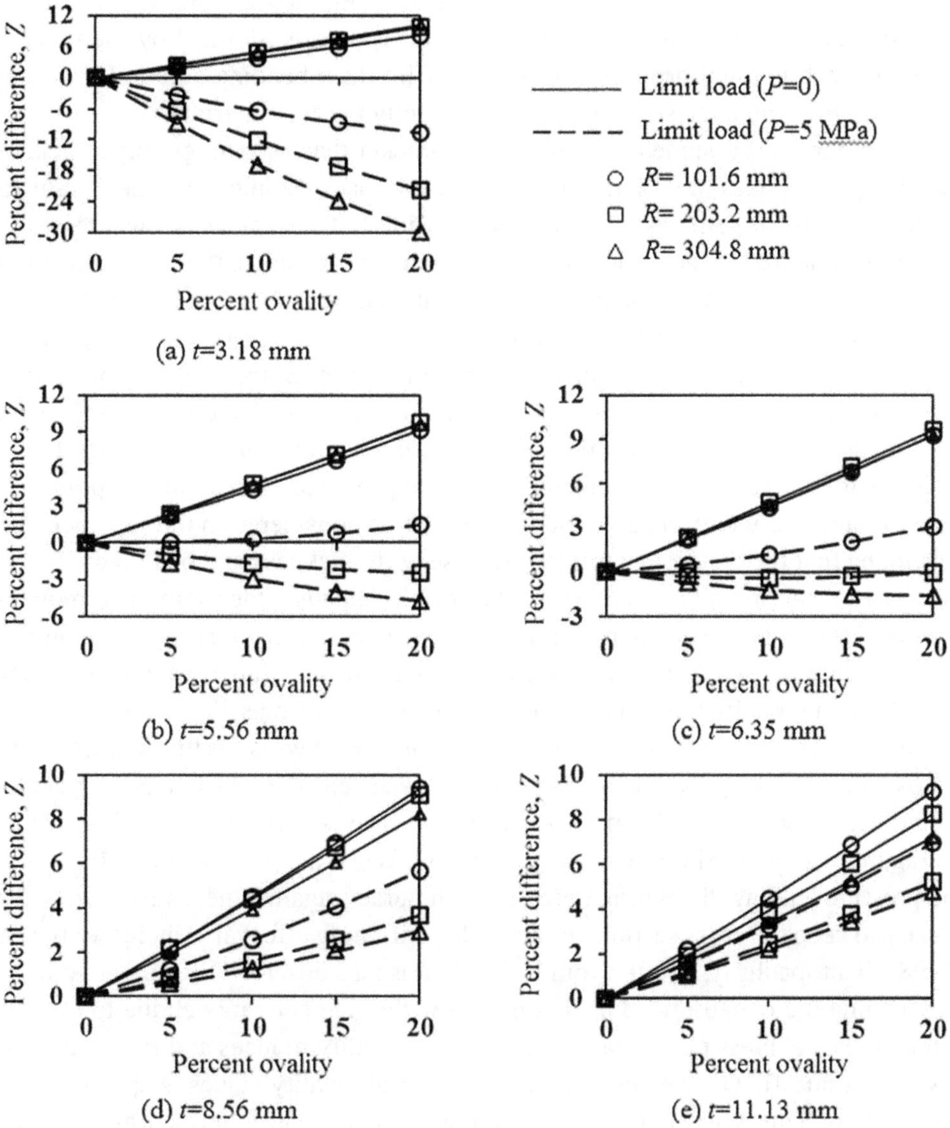

pipe bends rely on the assumptions of invariable wall thickness along the contour of the pipe's irritated segment and no preliminary ovality (Bolt, 1971).

The cross segment of pipe bends is not absolutely round but exhibits ovality, wall thickness dissimilarity (thinning/thickening) and other shape imperfections such as wrinkling and destruction which are introduced during manufacturing procedure (Griffiths, 1978). Ovality is a most important imperfection in all pipe twisting techniques (Moore, 1978). The satisfactoriness of pipe bend depends on the importance of these shape imperfections, predominantly ovality and thinning/thickening (Hilsenkopf, 1998). Thinning and ovality are the frequently experimented irregularities in pipe bends, which encourage higher stress than the absolutely circular cross sections (Hassan, 2000). Dan (2009) carried out linear and nonlinear cycling finite element analyses to get evaluate hoop stress under internal pressure, bending moment and temperature loads, as well as the progressive strain during cycling loading environment for elbows with ovality of 8% and compared the results for the elbows with perfectly circular cross section. Kim et al. (2006) provided a method to calculate approximately plastic loads [defined by the twice-elasticslope (TES)] for elbows with non-uniform thicknesses under in-plane bending and under internal pressure, based on organized FE limit analyses using elastic-perfectly plastic materials.

The intrados thickness is assumed to be up to 30% thicker than the straight pipe thickness, whereas the extrados thickness awakes to 30% thinner. The plastic load resolute from the TES method for the above maximum case is the same as that for the case with uniform thickness, whereas for the case of 30% increase in thickness at intrados and nominal thickness at extrados, the TES load is larger.

Veerappan (2013) demonstrated using finite element analyses the effect of ovality and thinning on collapse moments of pipe bends subjected to in-plane bending closing moment. The presence of ovality produces significant effect on collapse moments of the pipe bend. Since the value of collapse moments decrease, as the ovality is increased, the effect of ovality must be considered for the design of pipe bends. A mathematical equation is proposed for the collapse moment which includes the effect of ovality. This equation covers wide range of bend characteristics. The presence of thinning in pipe bend produces negligible effect on the results. Hence, thinning need not be considered in the pipe bend geometry for analysis. It is found that the collapse moments increase with thickness and bend radii.

The effect of internal pressure on collapse loads of pipe bends under combined moment and pressure loading considering geometric nonlinearity, has been studied by various researchers (Shalaby & Younan, 1998; Chattopadhyay, 2000) found that the collapse load, for any model, increases first, reaches a maximum value and then decreases as the internal pressure is increased.

The pipe bend under in-plane bending closing moment ($P = 0$) weakens the geometry due to ovalization of the cross section. The presence of ovality further weakens the pipe bends as the collapse load is below the reference models for all the models considered. The positive percent difference is maximum, as shown in Figure 9, for all ovalities under moment load. When the pipe bend is subjected to combined moment and pressure loading, the internal pressure counteracts against the ovalization of the pipe cross section and tries to keep the cross section circular. This stiffening effect also increases the collapse load (Shalaby, 1998; Chattopadhyay, 2000). From Figure 9, it is also observed that under combined moment and pressure loading the collapse load decreases below the reference model due to the effect of ovality for low pressures. As pressure increases, the effect of ovality reduces and for a particular pressure, this effect is almost nullified as the percent difference for all ovality values is approximately the same and is closer to zero. This particular pressure, in the present work, is termed as nullifying pressure. The negative percent difference increases as the internal pressure increases further, which reveal that the presence of ovality increases the stiffness of the pipe bends for higher pressures. It is also observed from the Figure 9. that for moment loading, increase in ovality increases the percent difference for all the models considered. When the pressure load is also applied, this trend continues up to the nullifying pressure beyond which the increase in ovality produces a positive effect of increasing the collapse load.

Plastic Load Solutions

Analytical Solutions

Analytical developments for pipe bends are based on idealized concepts of limit analysis. According to these concepts, the theoretical limit loads for pipe bends are based on the theorems of limit analyses.

Figure 9. Effect of internal pressure and ovality on collapse load for t=3.18 mm

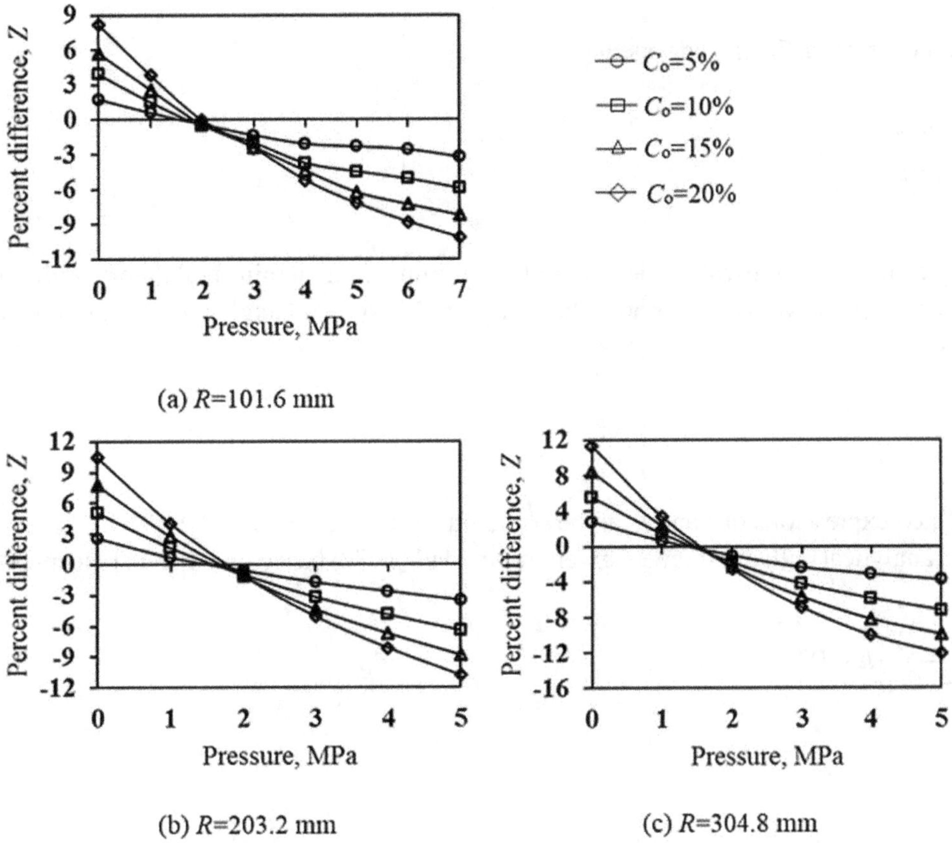

(a) R=101.6 mm

(b) R=203.2 mm

(c) R=304.8 mm

Lower-bound and upper-bound solutions were developed to bracket the theoretical limit load because bounding solutions are easier to obtain than exact solutions (Shalaby, 1998). Some of the theoretical models for pipe bends under in-plane bending moment without and with internal fluid pressure are given below:

Based on small displacement analysis, Spence, (1973) and Callidine (1974) expressed the lower-bound in-plane limit moments as given in Equations 8 and 9, respectively.

$$\frac{M_L}{M_0^s} = 0.8h^{0.6}\,(h < 1.45) \tag{8}$$

$$\frac{M_L}{M_0^s} = 0.935h^{2/3}\,(h < 0.5) \tag{9}$$

where M_L is the limit moment of pipe bend under in-plane bending moment based on small displacement analysis and M_0^s is the limit moment of straight pipe (Kim, 2006) which is given by

$$M_0^s = 4r^2 t\sigma_0 \tag{10}$$

The bend characteristic, *h*, is defined as

$$h = \frac{Rt}{r^2} = \frac{R/t}{r/t} \tag{11}$$

Kitching et al. (1979) presented a lower bound solution based on thin shell theory with no restriction on the geometric proportions but without the consideration of bend angle and straight pipe attachment

$$\frac{M_L}{M_0^s} = \frac{I}{f} \tag{12}$$

The defined expressions of *I* and *f* can be obtained from Kitching et al (1979).

The first analytical collapse moment given by Goodall (1978) based on large displacement analysis is

$$\frac{M_0}{M_0^s} = \frac{1.04h^{2/3}}{1+\beta}\,(h < 0.5) \tag{13}$$

where

$$\beta = \left(2 + \frac{(3h)^{2/3}}{3}\right)\left(\frac{4\sqrt{3\left(1-v^2\right)}}{\pi}\frac{\sigma_0}{E}\frac{r}{t}\right)$$

M_0 is the collapse in-plane moment based on large displacement analysis, *E* is the Young's modulus, and v is the Poisson's ratio.

The plastic collapse moment equations given by ASME (2000) and Zahoor (1991) are given in Equations 14 and 15.

$$\frac{M_L}{M_0^s} = 0.906h^{2/3} \tag{14}$$

$$M_L = 0.935h^{2/3}\left(2r\right)^2 t\sigma_f \tag{15}$$

where σ_f denotes the flow strength, usually defined as the average of yield and ultimate strengths.

Based on the experimental study, Touboul (1988) proposed the following collapse moment equation for elbows.

$$\frac{M_0}{M_0^s} = 0.715h^{2/3} \tag{16}$$

The first analytical solution for elbows under combined in-plane bending and internal pressure loading based on small displacement analysis was proposed by Goodall (1978) as

$$M_L^P = 1.04h^{2/3}\left(1 - \frac{P}{2\sigma_0}\frac{r}{t}\right)^{1/3}\left(D^2 t\sigma_0\right)(h < 0.5) \tag{17}$$

M_L^P is the limit moment under combined moment and pressure loading.

Numerical Analysis Based Solutions

Based on FE results through small displacement analysis, Kim and Oh (2006) modified the limit moments given in Equations 8 and 9 with an additional dependence of *r*/*t*.

$$\frac{M_L}{M_0^s} = A\left(h + k\right)^n \; (0.1 \le h \le 0.5) \tag{18}$$

with $A = 0.6453\left(\frac{r}{t}\right)^{0.0772}$, $k = 1.5398\left(\frac{r}{t}\right)^{-0.6755}$ and $n = 0.5157\left(\frac{r}{t}\right)^{0.0601}$.

Equations 19 and 20 were proposed by Chattopadhyay et al. (2000) and Kim and Oh (2006) respectively, based on finite element limit analyses considering geometric nonlinearity as

$$\frac{M_0}{M_0^s} = 1.075h^{2/3} \; (0.24 \le h \le 0.6) \tag{19}$$

$$\frac{M_0}{M_0^s} = A_C\left(h + k_c\right)^{n_c} \; (0.1 \le h \le 0.5) \tag{20}$$

with $A_C = 0.800\left(\frac{r}{t}\right)^{-0.017}$; $k_C = 1.460\left(\frac{r}{t}\right)^{-0.911}$; $n_C = 0.423\left(\frac{r}{t}\right)^{0.127}$

Kim and Oh (2006) modified Equation 17 based on the FE results and proposed the following solution

$$\frac{M_L^P}{M_L} = 1 - \left[\frac{1 - r/R}{1 - r/(2R)} \frac{P}{P_0}\right]^3 = 1 - \left(\frac{P}{P_0^s}\right)^3 \ (0.1 \le h \le 0.5) \tag{21}$$

M_L is taken from Equation 18. The values of P/P_0 were kept between 0 and 1, where P_0 is the theoretical limit pressure of the pipe bend (based on the von Mises yield condition) given by Kim and Oh (2006) as

$$P_0 = \left(\frac{2}{\sqrt{3}} \sigma_0 \frac{t}{r}\right)\left[\frac{1 - r/R}{1 - r/(2R)}\right] = P_0^s \left[\frac{1 - r/R}{1 - r/(2R)}\right] \tag{22}$$

The closed-form solution based on large displacement analysis using finite element method was proposed by Chattopadhyay et al. (2000) as

$$\frac{M}{M_0^s} = \left[1.075 h^{2/3} + \frac{2.071 p^{1.418}}{h^{0.223}} + 8.41 p^{12.129}\right]\left[1 - p\right] (0.1 \le h \le 0.6) \tag{23}$$

Kim and Oh (2006) developed the following equation based on large displacement analysis

$$\frac{M}{M_0} = 1 + \alpha p + \beta p^2 + \gamma p^3 \tag{24}$$

where M_0 is taken from Equation 20,

$$p = \frac{P}{\sigma_0} \frac{r}{t}$$

$$\left.\begin{aligned} \alpha &= -0.18 + 0.07\left(\frac{r}{t}\right), \\ \beta &= 0.0005\left(\frac{r}{t}\right)^3 - 0.0075\left(\frac{r}{t}\right)^2 + 0.0375\left(\frac{r}{t}\right) - 0.05 \\ \gamma &= -0.011\left(\frac{r}{t}\right)^2 + 0.110\left(\frac{r}{t}\right) - 0.880 \end{aligned}\right\} for\ r/t = 5\ and\ 10$$

$$\left.\begin{array}{l} \alpha = \begin{cases} 135h^3 - 202.5h^2 + 101.25h - 14.575 & for\ 0.3 \le h \le 0.5 \\ 1.22 & for\ 0.3 \le h \le 0.5 \end{cases} \\ \beta = \begin{cases} 105h^2 + 105h - 23.75 & for\ 0.3 \le h \le 0.5 \\ 1.70 & for\ 0.1 \le h \le 0.3 \end{cases} \\ \gamma = \begin{cases} -75h^2 + 75h - 18.83 & for\ 0.3 \le h \le 0.5 \\ -3.08 & for\ 0.1 \le h \le 0.3 \end{cases} \end{array}\right\} for\ r/t = 20$$

Based on the finite element limit analysis, a closed form solution as given in Equation 25 was proposed by Christo et al (2012) to include the effect of ovality to determine collapse moment of the pipe bend under in-plane bending closing moment.

$$\frac{M_0}{M_0^s} = \frac{72\left(\dfrac{R}{r\left(100 + 2C_o\right)}\right)^{1/3}}{\left(\dfrac{r}{t} + 10\right)^{1.2675}} \tag{25}$$

The effect of ovality on pipe bends under internal pressure and moment loading has been studied by Christo et al (2013). The range of internal pressure applied for each model based on which the equations are developed is from zero to 32 MPa.

Based on the collapse load results, closed-form solutions are obtained as follows:

For $0.105 \le h \le 0.573$

$$\frac{M}{M_0} = 1 + 0.001\left\{\frac{R}{r\left(1 + 0.14C_o\right)}\right\}^{-0.6}\left(\frac{r}{t} + 4\right)^{2.97}\left(p\right)^{0.712} h^{0.9} \tag{26}$$

and for $0.573 < h \le 1.275$

$$\frac{M}{M_0} = 1 + 0.07\left\{\frac{R}{r\left(1 + 0.3C_o\right)}\right\}^{-0.7}\left(\frac{r}{t} - 3\right)^{1.2}\left(p\right)^{0.6} h^{-0.4} \tag{27}$$

M_0 is the collapse moment of the pipe bends under in-plane bending closing moment

Strain Analysis

Murray (1979) conducted experiments for four identical bends subjected to internal pressure. Two of the bends were subjected to fatigue cyclic loading. The importance of some discontinuous pressure components, superimposed on top of a mean pressure level, would have been sufficient to cause yielding at the bend intersection. These two bends were then subjected to a static burst pressure. The repeated pressure test characteristically demonstrated a non-linear initial loading curve and a linear unloading curve, with a resulting residual strain. For the second and all subsequent loadings, the curves followed the first unloading curves, indicating that the residual stresses had shaken down. No discussion is presented on the presence of fatigue damage. The author's conclusion that the fatigue life of a welded mitre bend appears to be comparable to that of an equivalent straight pipe.

Bending Processes

Luo (1992) analyzed the effects of outside diameter, wall thickness and bend radius on ovality of nylon tube during bending operation and inferred that wall thickness was the most significant factor for the ovality of the tubing. Yuan et al. (1998) proposed a hydro bulging technology to overcome the difficulties in manufacturing large diameter elbows using traditional technology. They also demonstrated experimentally the feasibility of the process. Hu and Li (1999) developed a computer simulation system based on Ansys and elasto-plastic theory from which the results, such as the thinning and thickening ratio of the pipe wall thickness, the pushing force with or without a reverse moment acting on the bending arm, the proper reverse moment that is to be applied on the bending arm to control the wall thinning ratio of the bending outside to within a particular value, the ovality of the cross-section of the pipe and the spring-back angle after unloading could be obtained. Simulation was done on pipe-bending process with a relative bending radius $R/D=1.5$ using local induction heating and with or without a reverse moment acting on the bending arm. Al-Qureshi (1999) derived theoretical expressions in an attempt to provide a quantitative method with practical utility for predicting the spring back behaviour and residual stress distributions in a bent tube section as a function of die radius, wall thickness, and stress-strain characteristics of the materials.

Li and Yao (2001) examined mechanisms of the laser bending process to better understand the deformation characteristics such as wall thickness variation, cross-section ovalization, bending radius, and asymmetry. Factors important to these characteristics were experimentally and numerically investigated. Appreciable wall thinning was absent in laser bent tubes. Cross section ovalization of the laser-bent tubes was also much smaller than that observed in comparable mechanical bent tubes due to lack of bending die and appreciable tensile stress/strain in the extrados. The laser beam diameter governed the bending radius, not the tensile failure at the extrados. The curvature radii of the bent tube increased with the beam diameter and the ratio of the tube outer diameter to the wall thickness. The bending efficiency increased with the maximum scanning angle up to a critical value. Jin et al. (2001) developed a knowledge-based system (KBS), which are beneficial to increase efficiency, reduce production costs, and improve quality, to aid the design of tube bending processes, including bending methods selection, tool/die design, and process parameters setup. The KBS would also provide a guide to select a correct method of tube bending production and to reduce the percentage of defective parts, so as to increase the production efficiency, to reduce the labor time and material waste, and save the overall costs significantly. The object-oriented programming techniques and the goal-driven search mechanism, featured by an interactive graphic user interface, have been applied in development of the proposed KBS. The system developed has shown effective in tube parts production with significantly reduced number of potential defects and failures, such as inaccuracy of bend angles or linear dimensions, wall thinning, flattening, wrinkling, twisting, dents or cracks, etc.

Heng (2006) developed a complete 3 dimensional elastic-plastic FEM model of the thin-walled tube numerical control (NC) bending process using ABAQUS Explicit code, including bending process, balls retracting and unloading process, and thus the plastic deformation characteristics with small bending radius were investigated. The main results showed that the utmost deformation feature of the NC bending process is its continuous progressive deformation and the wrinkling is traditional on the double compressive stresses state and the tension instability is on the double tension stresses state. For one dimensional small bending process, a new method - *stepped mandrel retraction* is proposed to improve the bending quality in experiment according to the FE simulation. The simulation results were verified by experiment. Kale and Thorat (2009) presented a factual method to reduce thinning of the wall thickness of pipe during bending. They developed a new mechanism for bending of pipes. This mechanism has a provision of precompression (radial squeeze) of the pipe along the directrix of maximum deformation during bending. This is achieved by clamping the pipe using two parallel plates from top and bottom. In fact, the pipe is wrapped using two rollers—one from inside and one from outside in the horizontal plane—and two plates parallel to the horizontal plane—one from the top and one from the bottom. They also carried out experimentation on this mechanism, and thicknesses are measured at the grid points along the length of the pipe. From the experimental values of thicknesses on the tension and compression sides, dimensionless variations in wall thickness of various groups of pipes are computed for different precompression values. In order to represent the thickness at any point, a mathematical equation is derived. Analytical values of thickness variations on tension and compression sides are computed using this equation. They compared experimental and analytical results, and presented its methodical approach. Results show that precompression reduces thickness variation of the pipe after bending.

Comparison Study in Pipe Bends

The detailed three-dimensional FE limit analyses for pipe bends under combined pressure and bending are performed based on different percentage of ovality for Incoloy 800 ASTM B407 / Stainless Steel materials with the small geometry change option. A wide range of parameters related to the bend geometry was measured. The presented FE results show that there is more than one value of ovality and thinning for a pipe bend manufactured, which would reduce rejection of pipe bends after they have been made. Given the values of ovality and thinning of the pipe bend, it can be checked whether the bend can withstand the designed pressure.

FUTURE RESEARCH DIRECTIONS

Burst test analysis, is the field of analysis of pipe bends which is relatively new and only few researchers have analyzed on this. The burst test analysis under different weather conditions need to be explored because it can present the crack and failures of pipe bends thus resulting in saving time, money and material. Moreover, fracture mechanics is another field of study involving the use of XFEM (extended finite element method) for consideration of crack growth a propagation in pipe bends. The pipe bends with postulated cracks based on configuration (axial, circumferential) types (through wall, surface) and location (intrados, extrados, crown) have been analysed for estimation of plastic loads for leak before break analysis. A wide-ranging review of knowledge obtainable on all aspects of pipe bend performance shows that there are limitations in the solutions available, which can be used in industry. The assessment shows that all theories are limited from practical usage with limited implications on loading modes and bend types. At the moment, the finite element method is conceivably the only technique that is obviously appropriate to the complete continuum of analysis pipe bends with varying materials and loading environment.

The existence of wide-ranging design curves for the elastic and plastic behavior of all bend types is impossible, over the range of geometry and loading parameters. Correspondingly, there is still possibility for further employment of FEA on the effect of combined loading, end effects and non-circularity. Creep and optimization represent the territory that is less explored as far as pipe bends are concerned, and when taken into account can be advantageous for high-cycle fatigue failure in pipe work proving wider scope for vibration-induced studies.

REFERENCES

Al-Qureshi, H. A. (1999). Elastic-plastic analysis of tube bending. *International Journal of Machine Tools & Manufacture, 39*(1), 87–104. doi:10.1016/S0890-6955(98)00012-1

American Society of Mechanical Engineers Boiler and Pressure Vessel Code. (2000). *Sec. III, Nuclear Power Plant Components.*

Bolt, S.E., & Greenstreet, W.L. (1971). Experimental determination of plastic collapse loads for pipe elbows. ASME Paper 71-PVP-37.

Bond, M. P. (1971). *A theoretical and experimental investigation of multi mitred pipe bends subjected to various types of loading* [PhD thesis]. Institute of Science & Technology, Victoria University of Manchester.

Bond, M. P., & Kitching, R. (1972). Stress and flexibility factors for multi-mitred pipe bends subjected to internal pressure combined with external loadings. *J Strain Anal, 7*(2), 97–108. doi:10.1243/03093247V072097

Boyle, J. T., & Spence, J. (1981). A simple stress analysis for out-of-round pressurised pipe bends. *International Journal of Pressure Vessels and Piping, 9*(4), 251–261. doi:10.1016/0308-0161(81)90027-2

Brabin, T. A., Christopher, T., & Rao, B. N. (2011). Bursting pressure of mild steel cylindrical vessels. *International Journal of Pressure Vessels and Piping, Vol, 88*, 11–22.

Calladine, C. R. (1974). Limit analysis of curved tubes. *Journal of Mechanical Engineering Science, 16*(2), 85–87. doi:10.1243/JMES_JOUR_1974_016_016_02

Chattopadhyay, J., Natahani, D. K., Dutta, B. K., & Kushwaha, H. S. (2000). Closed-form collapse moment quations of elbows under internal pressure and in-plane bending moment. *J. Pressure Vessel Technol., 6*(4), 431–436. doi:10.1115/1.1285988

Christo Michael, T., Veerappan, A. R., & Shanmugam, S. (2011). Effect of cross section on collapse load in pipe bends subjected to in-plane closing moment. *International Journal of Engineering Science and Technology, 3*(6), 247–256.

Christo Michael, T., Veerappan, A. R., & Shanmugam, S. (2012). Comparison of plastic limit and collapse loads in pipe bends with shape imperfections under in-plane bending and an internal pressure. *International Journal of Pressure Vessels and Piping*, *99-100*, 23–33. doi:10.1016/j.ijpvp.2012.07.013

Dan, V. (2009). The Influence Of The Initial Ovality Tolerance On The Nonlinear Cycling analysis Of Piping Bends. *Journal of Pressure Vessel Technology*, *131*(4).

Goodall, I. W. (1978). Large deformations in plastically deforming curved tubes subjected to in-plane bending. *Research Division Report RD/B/N4312, Central Electricity Generating Board*, England.

Goodall, I. W. (1978). *Lower bound limit analysis of curved tubes loaded by combined internal pressure and in-plane bending moment, Research Division report RD/B/N4360*. England: Central Electricity Generating Board.

Gresnigt, A. M. (2002). Elastic and plastic design of mitred bends.*Proceedings of the 12th international offshore and polar engineering conference*, Kitakyushu, Japan (pp. 26–31).

Griffiths, J. E. (1978). The effect of cracks on the limit load of pipe bends under in-plane bending experimental study. *International Journal of Mechanical Sciences*, *21*(2), 119–130. doi:10.1016/0020-7403(79)90038-9

Han, J. J., Lee, K. H., Kim, N. H., Kim, Y. J., Jerng, D. W., & Budden, P. J. (2012). Comparison of existing plastic collapse load solutions with experimental data for 90° elbows. *International Journal of Pressure Vessels and Piping*, *89*, 19–27. doi:10.1016/j.ijpvp.2011.09.001

Hassan, T., Modlin, M., & Matzen, V. C. (2000). *Monotonic in-plane tests on 2-inch, schedule 10* elbows. In *Data developed at the center for nuclear power plant structures, equipment and piping*. NC: North Carolina State University.

Heng, L., He, Y., Mei, Z., & Jie, G. R. (2006). Forming characteristics of thin-walled tube bending process with small bending radius. *Transactions of Nonferrous Metals Society of China*, *16*, 613–623. doi:10.1016/S1003-6326(06)60266-5

Hilsenkopf, P., Boneh, B., & Sollogoub, P. (1998). Experimental study of behavior and functional capability of ferric steel elbows and austenitic stainless steel thin walled elbows. *International Journal of Pressure Vessels and Piping*, *33*(2), 111–128. doi:10.1016/0308-0161(88)90065-8

Hose, D.R. (1988). *Glass reinforced plastic pipework of mixed wall construction* [PhD thesis]. University of Manchester.

Hu, Z., & Li, J. Q. (1999). Computer simulation of pipe-bending processes with small bending radius using local induction heating. *Journal of Materials Processing Technology*, *91*(1-3), 75–79. doi:10.1016/S0924-0136(98)00425-7

Jin, Z., Luo, S., & Fang, X. D. (2001). KBS-aided design of tube bending processes. *Engineering Applications of Artificial Intelligence*, *14*(5), 599–606. doi:10.1016/S0952-1976(01)00016-1

Kale, A. V., & Thorat, H. T. (2009). Effect of precompression on thickness of pipe during bending. *Journal of Pressure Vessel Technology*, *131*(031201), 1–8.

Kim, Y. J., Kim, N. H., Oh, C. Y., & Oh, C. S. (2008). A method to estimate plastic loads for elbows with non-uniform thicknesses. *Fatigue Fract. Eng. Mater. Struct.*, *31*, 822–837.

Kim, Y. J., Lee, K.-H., Oh, C.-S., Yoo, B., & Park, C.-Y. (2007). Effect of bend angle on plastic loads of pipe bends under internal pressure and in-plane bending. *International Journal of Mechanical Sciences*, *49*(12), 1413–1424. doi:10.1016/j.ijmecsci.2007.03.006

Kim, Y. J., & Oh, C. S. (2006). Closed-form plastic collapse loads of pipe bends under combined pressure and in-plane bending. *Engineering Fracture Mechanics*, *73*(11), 1437–1754. doi:10.1016/j.engfracmech.2006.02.001

Kim, Y. J., & Oh, C. S. (2006). Limit loads for pipe bends under combined pressure and in-plane bending based on finite element limit analysis. *International Journal of Pressure Vessels and Piping*, *83*(2), 148–153. doi:10.1016/j.ijpvp.2005.11.001

Kim, Y. J., & Oh, C.-S. (2007). Effect of attached straight pipes on finite element limit analysis for pipe bends. *International Journal of Pressure Vessels and Piping*, *84*(3), 177–184. doi:10.1016/j.ijpvp.2006.09.017

Kima, J. W., Leeb, S. H., & Parkb, C. Y. (2009). *Experimental Evaluation Of The Effect of Local Wall Thinning On The Failure Pressure Of Elbows. Nuclear Engineering and Design, 239*, 2737–2746.

Kitching, R., & Bond, M. P. (1972). *Out-of-circularity effects in a pipe bend subjected to pressure and bending load*. J Strain Anal.

Kitching, R., Zarrabi, K., & Moore, M. A. (1979). Limit load for a smooth pipe bend under in-plane bending. *International Journal of Mechanical Sciences*, *21*(12), 731–738. doi:10.1016/0020-7403(79)90053-5

Kochekseraii, S. B. (1994). *Mitred pipe bends subject to internal pressure and in-plane bending moment* [PhD thesis]. University of Manchester.

Lane, P.H.R. & Rose, T.T. RT. (1960). Experiments on fabricated pipe bends. B.W.R.A, report no. D5. 12-60.

Law, M & Bowie, G. (2006). Failure strain in high yield-to-tensile ratio linepipes. *Journal of Pipeline Integrity,*5, 25-36.

Law, M., & Bowie, G. (2007). Prediction of failure strain and burst pressure in high yield to tensile strength ratio line pipe. *International Journal of Pressure Vessels Piping, 84*(8), 487–492. doi:10.1016/j.ijpvp.2007.04.002

Li, Y. W., & Yao, L. (2001). Laser bending of tubes: Mechanism, analysis, and prediction. *Journal of Manufacturing Science and Engineering*, *123*(4), 674–681. doi:10.1115/1.1392992

Luo, D. Z. (1992). Effect of outside diameter, wall thickness and bend radius on ovality of nylon tube during bending operation. *International Journal of Pressure Vessels and Piping*, *52*(1), 145–148. doi:10.1016/0308-0161(92)90065-N

Mackenzie, D. (2004). Plastic collapse of pipe bends under combined internal pressure and in-plane bending. *International Journal of Pressure Vessels and Piping*, *82*, 407–416.

Markl, A. R. C. (1947). *Fatigue tests of welding elbows and comparable double-mitre bends* (Vol. 69). Trans ASME.

Michael, T. C., Veerappan, A. R., & Shanmugam, S. (2012). Effect of ovality and variable wall thickness on collapse loads in pipe bends subjected to in-plane bending closing moment. *Engineering Fracture Mechanics*, *79*, 138–148. doi:10.1016/j.engfracmech.2011.10.009

Michael, T. C., Veerappan, A. R., & Shanmugam, S. (2013). Effect Of Internal Pressure And Shape Imperfections On Plastic Loads Of Pipe Bends Under In-Plane Closing Moment. *Engineering Fracture Mechanics*, *105*, 1–15. doi:10.1016/j.engfracmech.2013.03.031

Mourad, H. M. (1999). *Elastic-plastic behaviour and limit load analysis of pipe bends under out-of-plane moment loading and internal pressure* [Thesis, Master of Science in Engineering]. The American University in Cairo, School of Sciences and Engineering.

Murray, , N.W. (1979). Bursting Tests On Welded Mitred Bends For Gas Pipelines. *Aust. Weld Res.*, *7*, 1–26.

Neilson, R., Wood, J., Hamilton, R., & Li, H. (2010). A Comparison Of Plastic Collapse And Limit Loads For Single Mitred Pipe Bends Under In-Plane Bending. *International Journal of Pressure Vessels and Piping, 87*(10), 550–558. doi:10.1016/j.ijpvp.2010.08.015

Orynyak, V., & Radchenko, S. A. (2004). A Mixed-approach analysis of deformations in pipe bends. *Strength of Materials*, *36*(4), 365–376. doi:10.1023/B:STOM.0000041537.43590.f9

Ostemin, A. A., & Dil'man, V. L. (2002). Calculating wall thickness for major oil product pipelines: Analysis of regulation documents. *Chemical and Petroleum Engineering*, *38*(1/2), 58–63. doi:10.1023/A:1015278306138

Qingquan, D., Hong, Z., Feng, Y., & Changyi, D. (2009). Hydrostatic Burst Test of X80 Grade steel pipe. *Journal of Loss Prevention in the Process Industries, 22*(6), 897–900. doi:10.1016/j.jlp.2009.09.003

Robertson, A., Li, H., & Mackenzie, D. (2005). Plastic collapse of pipe bends under combined internal pressure and in-plane bending. *International Journal of Pressure Vessels and Piping*, *82*(5), 407–416. doi:10.1016/j.ijpvp.2004.09.005

Rodabaugh, E. C. (1975). *Review of data on mitre joints in piping to establish maximum angularity for fabrication of girth butt welds* (Vol. 208). WRC Bulletin.

Rodabaugh, E. C., & George, H. H. (1957). *Effect of internal pressure on flexibility and stress intensification factors of curved pipe or welding elbows* (Vol. 79). Trans ASME.

Rodabaugh, E.C. & Moore, S.E. (1978). Evaluation of the plastic characteristics of piping products in relations to ASME Code *Criteria. NUREG Report* CR-0261.

Shalaby, M. S., & Younan, M. Y. A. (1998). Limit loads for pipe elbows with internal pressure under in- plane closing moments. *J Press Vessel Technol.*, *120*(1), 35–42. doi:10.1115/1.2841882

So, H. S. (1971). *Single mitred pipe bends under bending and internal pressure loads* [MSc thesis]. Victoria University of Manchester.

So, H. S. (1971). *Single mitred pipe bends under bending and internal pressure loads* [MSc thesis]. Victoria University of Manchester.

Spence, J., & Findlay, G. E. (1973). Limit load for elbows under in-plane bending. *Proceedings of second international conference on pressure vessel technology*, San Antonio (Vol. 28, pp. 393–399).

Suneel, K. G., Vivek, B., Vaze, K. K., Ghosh, A. K., & Kushwaha, H. S. (2007). General limit load and B2 stress index equation for pipe bends under in-plane bending. Trans, SMiRT (Vol. 19).

Touboul, F., Ben, D. M., & Acker, D. (1988) Design criteria for piping components against plastic collapse: application to pipe bend experiments, *Pressure vessel technologyProceedings of the 6th international conference*, Beijing.

Tsuji, M., & Meshii, T. (2013). Influence Of Circumferential Flaw Length On Internal Burst Pressure Of A Wall-Thinned Pipe. *Nuclear Engineering and Design, 255*, 59–67. doi:10.1016/j.nucengdes.2012.10.002

Veerappan, A. R., & Shanmugam, S. (2006). Stress analysis of pipe bends subjected to internal fluid pressure using the finite element technique. *Journal of Strain Analysis*, *41*(8), 561–573. doi:10.1243/03093247JSA213

Veerappan, A. R., & Shanmugam, S. (2008). Analysis for flexibility in the ovality and thinning limits of pipe bends. *ARPN J Eng Appl Sci.*, *1*(1), 31–41.

Wang, L., & Zhang, Y. (2011). Plastic collapse analysis of thin-walled pipes based on uniform yield criterion. *International Journal Mechanical Sciences, 53*(5), 348–354. doi:10.1016/j.ijmecsci.2011.02.004

Wiley, A. (1956). Design of Piping System (2nd ed.). Interscience Publication.

Wilkins, K., Tan, Y., & Matzen, V.C. (2001). Monotonic Out-Of Plane Tests On 2-Inch, Schedule 10, Stainless Steel, 90° Elbows. *Proceedings of the 2003 ASME pressure vessels and piping conference.*

Wood, J. (2008). A review of literature for the structural assessment of mitred bends. *International Journal of Pressure Vessels and Piping*, *85*(5), 275–294. doi:10.1016/j.ijpvp.2007.11.003

Xian-Kui, Z., & Leis, B. N. (2012). Evaluation Of Burst Pressure Prediction Models For Line pipes. *International Journal of Pressure Vessels and Piping, 89*, 85–97. doi:10.1016/j.ijpvp.2011.09.007

Yuan, S. J., Xu, Z., Wang, Z. R., & Hai, W. (1998). The integral hydro-forming process of pipe elbows. *International Journal of Pressure Vessels and Piping*, *75*(1), 7–9. doi:10.1016/S0308-0161(97)00107-5

Zahoor, A. (1991). *Ductile fracture handbook* (Vol. 3). USA: Novotech Corp.

Zhu, X. K., & Leis, B. N. (2004). Accurate prediction of burst pressure for linepipes. *Journal of Pipeline Integrity, 4*, 195–206.

Zhu, X. K., & Leis, B. N. (2007). Deterministic and probabilistic of pipeline burst failure.*Proceedings of ASME Pressure Vessels and Piping Conference*, San Antonio, Texas, USA (pp. 22-26). doi:10.1115/PVP2007-26272

APPENDIX

Abbreviations

ASME: American Society of Mechanical Engineers
BPVC: Boiler and Pressure Vessel Code
FE: Finite Element
FEA: Finite Element Analysis
FEM: Finite Element Method
TES: Twice Elastic Slope

List of Symbols

***Co*:** Ovality of the bend at the midsection for non-circular cross section (%)
C_t: Thinning at the extrados of the bend section (%)
***D*:** Nominal outside diameter of the pipe (mm)
D_{max}: Maximum diameter at the bend section of the pipe (mm)
D_{min}: Minimum diameter at the bend section of the pipe (mm)
***E*:** Young's modulus of the material (MPa)
***h*:** Bend characteristic or bend factor (or λ)
***L*:** Length of straight pipe attached with pipe bend (mm)
***M*:** Collapse moment of the pipe bend under combined in-plane closing bending moment and internal fluid pressure (kN-m)
M_0: Collapse in-plane moment of the pipe bend for the closing bending mode (kN-m)
M_L: Limit in-plane moment of pipe bend (kN-m)
M_L^P : Limit moment of the pipe bend under combined in-plane closing bending moment and internal fluid pressure (kN-m)
M_0^s : Limit moment of the straight pipe (kN-m)
***P*:** Internal fluid pressure (MPa)
P_0: Limit pressure of the pipe bend (MPa)
P_0^s : Limit pressure of the straight pipe (MPa)
***R*:** Bend radius of the pipe bend (mm)
***r*:** Mean radius of the pipe (mm)
r_o: Nominal outside radius of the pipe (mm)
r_{min}: Minimum outside radius at bend section of the semi oval cross section (mm)
***t*:** Nominal thickness of the pipe (mm)
t_c: Thickness at the crown of the bend section (=*t*) (mm)
t_i: Thickness at the intrados of the pipe bend section (mm)
t_e: Thickness at the extrados of the pipe bend section (mm)

t_{max}: Maximum thickness at the intrados of the bend section ($=t_i$) (mm)
t_{min}: Minimum thickness at the extrados of the bend section ($=t_e$) (mm)
φ: Bend angle of the pipe bend (°)
σ_0: Yield stress of the material (MPa)
σ_f: Average of yield and ultimate strengths (MPa)
υ: Poisson's ratio

Chapter 8
Fuzzy Structural Analysis Using Surrogate Models

A. S. Balu
National Institute of Technology Karnataka, India

B. N. Rao
Indian Institute of Technology Madras, India

ABSTRACT

The exponential growth of computational power during the last few decades has enabled the finite element analysis of many real-life engineering systems which are too complex to be analytically solved in a closed form. In the traditional deterministic finite element analysis, system parameters such as mass, geometry and material properties are assumed to be known precisely and defined exactly. However, in practice most of the data used in the solution process of many practical engineering systems are either collected from experiments or acquired as empirical data from the past, which are usually ill defined, imprecise and uncertain in nature. This work presents a practical approach based on High Dimensional Model Representation (HDMR) for analyzing the response of structures with fuzzy parameters. The proposed methodology involves integrated finite element modelling, HDMR based surrogate model, and explicit fuzzy analysis procedures.

INTRODUCTION

The uncertainties present in the system can be categorized into randomness and fuzziness. Probabilistic methods are especially suitable in case of uncertainty due to randomness for which the information on both the range and the probability density function are available. The probabilistic concept is already well established for the extension of the deterministic finite element method towards uncertainty assessment. This has led to a number of probabilistic finite element procedures (Haldar & Mahadevan, 2000; Lernout et. al., 2006; Pellissetti & Schueller, 2007; Schueller et. al., 2004; Soize, 2009) that are often used in conjunction with Monte Carlo simulations. However, the use of a probabilistic approach can lead to subjective results if the amount of statistical data on the uncertainties is limited, and fuzzy information that is vague, imprecise, qualitative, linguistic or incomplete need to be included (Elisha-

DOI: 10.4018/978-1-5225-0588-4.ch008

koff, 1995). The need to incorporate subjective information into the design and analysis of engineering systems has led to the development of fuzzy finite element methodology that began in the early 1990s.

Fuzzy set theory initiated by Zadeh (1965) provides a concept for the description of linguistic or subjective knowledge and incomplete data in a non-probabilistic manner. The use of Zadeh's concept in the finite element context has led to the development of the fuzzy finite element method for the solution of engineering problems (Dhingra et. al., 1992; Valliappan & Pham, 1995; Chen & Rao, 1997; Akpan et.al., 2001a; Akpan et.al., 2001b). Its aim is to obtain the membership function of an output quantity, based on the fuzzy description of the uncertain input parameters. By using the α -level technique, the interval finite element method forms the core of the fuzzy procedure (Gersem et. al., 2007). Over the last decade, several interval FE procedures have been reported, such as the vertex method (Dong & Shah, 1987; Hanss, 2002; Donders et. al., 2005; Gauger et al., 2008), the global optimization approaches (Valliappan & Pham, 1995; Möller et. al., 2000; Farkas et. al., 2008) and the interval arithmetic approaches (Rao & Berke, 1997; Qiu & Elishakoff, 1998; Dessombz et. al., 2001; Muhanna & Mullen, 2001). However, each of these techniques has its specific limitations when applied to realistic engineering problems, and none of them has been established as standard interval procedure, and the choice of technique depends on the type of the conducted FE analysis.

This work presents a practical approach based on High Dimensional Model Representation (HDMR) (Li et. al., 2001a; Li et. al., 2001b; Sobol, 2003; Chowdhury et. al., 2008; Balu & Rao, 2012a; Balu & Rao, 2012b; Balu & Rao, 2012c) for analyzing the response of structures with fuzzy parameters. The proposed methodology involves integrated finite element modelling, HDMR based response surface generation, and explicit fuzzy analysis procedure. The uncertainties in the material, geometric, loading and structural parameters are represented using fuzzy sets. To facilitate efficient computation, a HDMR based response surface generation is employed for the approximation of the fuzzy finite element response quantity, and the response is expressed in a separable closed-form expression as a linear combination of the fuzzy variables through the definition of intervening variables, then using the transformation technique the bounds on the response at each α-level are obtained.

FUZZY FINITE ELEMENT APPROACH

Fuzzy Sets

The concept of fuzzy sets, introduced by Zadeh (1965), has gained an increasing popularity during the last two decades. Its most important property is that it is capable of describing linguistic and therefore incomplete information in a non-probabilistic manner. While a classical set clearly distinguishes between members and non-members, the fuzzy set introduces a degree of membership, represented by the membership function. For a fuzzy set $\tilde{x}$, the membership function $\mu_{\tilde{x}}\left(x\right)$ describes the grade of membership to the fuzzy set for each element x in the domain X:

$$\tilde{x} = \left\{\left(x, \mu_{\tilde{x}}\left(x\right)\right) \middle| \left(x \in X\right), \left(\mu_{\tilde{x}}\left(x\right) \in \left[0,\ 1\right]\right)\right\} \tag{1}$$

If $\mu_{\tilde{x}}(x) = 1$, x is definitely a member of the set $\tilde{x}$, whereas if $\mu_{\tilde{x}}(x) = 0$, x is definitely not a member of the set $\tilde{x}$. For all x with $0 < \mu_{\tilde{x}}(x) < 1$, the membership is not certain. The most frequently applied membership function shapes are the triangular and Gaussian (Gersem et. al., 2007).

Fuzzy Numerical Analysis

The fuzzy finite element method (FFEM) aims to obtain a fuzzy description of an FE analysis result, starting from the fuzzy descriptions of all non-deterministic FE model parameters. In practice, the description of uncertain parameters and quantities using fuzzy sets can be implemented using the α-level strategy. This approach subdivides the membership function range into a number of α-levels. The intersection with the membership function of the input uncertainties at each α-level results in an interval $x_{i,\alpha}^{I} = \left[\underline{x}_i,\ \overline{x}_i\right]_{\alpha}$. With these input intervals of the α-sublevel, an interval finite element (IFE) analysis is performed, resulting in an interval for the analysis result at the considered α-level. Finally, the fuzzy solution is assembled from the resulting intervals at each sublevel.

The fuzzy finite element technique consists of the application of the α-level strategy on the numerical procedure of the deterministic FE analysis. Through this procedure, the fuzzy FE analysis can be interpreted as a large-scale sensitivity analysis, which can be used to study the combined effect of the interval bounds of design variables on critical design properties. More comprehensive overview of fuzzy analysis for non-deterministic analysis in engineering design is available in reference (Moens & Vandepitte, 2004). Through the α-level strategy, the interval problem corresponding to the considered FE analysis forms the core of the fuzzy procedure. A number of general interval solution strategies have been proposed as the basis for fuzzy FE implementations. The most popular one is the interval arithmetic approach, in which all basic deterministic algebraic operations are replaced by their interval arithmetic counterparts. Although this approach is computationally very inexpensive, it is of little practical use as it tends to largely overestimate the interval outcome of each operation, accumulating in a huge amount of conservatism in the interval analysis result of realistic problems. Therefore, other approaches such as the vertex method (Dong & Shah, 1987; Dong & Wong, 1987; Hanss, 2002; Donders et. al., 2005), the global optimization method (Valliappan & Pham, 1995; Möller et. al., 2000; Farkas et. al. 2008), and methods based on interval arithmetic (Rao & Berke, 1997; Qiu & Elishakoff, 1998; Dessombz et. al., 2001; Muhanna & Mullen, 2001) have been proposed for the solution of the underlying interval FE problem. Each of these techniques has its specific disadvantages when applied to realistic engineering problems, such that none of them has been established as standard interval procedure, and the choice of technique depends on the type of the conducted FE analysis.

The vertex method evaluates the function value at each of the vertices of the design space, represented by the bounds on the variables, to obtain the minimum and maximum values of the response. This approach requires 2^N function evaluations, where N is the number of uncertain parameters. The easy implementation and the limited computational cost if the number of uncertainties is not too high form the major advantages of this method. The important limitation of the vertex method is that the exact interval results can be obtained only if the variation of the outputs is monotonic with respect to a variation of the uncertain parameters. This is difficult to predict in case of a general structural/mechanical problem. In case of nonlinear non-monotonic responses, the vertex method does not produce conservative results. This means that the real interval results are underestimated. Especially in design validation applications, this forms a severe drawback. The global optimization techniques require the function value and/or gradient information at several points in the iterative process at each α-cut, to obtain the minimum and maximum values of a nonlinear response. Taking into account the number of optimizations, the number of function evaluations required for one optimization and the computational cost of one function evaluation, this procedure is computationally expensive for most practical applications with implicit response functions. However, the computational efficiency of this approach can be improved by adopting the reduced optimization (Farkas et. al., 2008).

As an alternative to the optimization procedures, response surface based optimization methods (Akpan et. al., 2001a; Munck et. al., 2008) use a different approach: instead of evaluating the objective function every time during the optimization procedure, they first create an approximation of the objective function based on objective function evaluations in a number of so-called response points. This approximation, whose evaluation should be computationally less expensive than the evaluation of the objective function itself, is then used to find the optimum. Although more advanced approximation techniques such as Kriging methods (Jones et. al., 1998), neural networks (Chen, 1996) and support vector regression (Drucker, 1997) can be used for this application, response surfaces are chosen because they adequately describe the objective function behaviour. Response surface behaviour is easy to understand and to visualize, making it much easier to take corrective actions if the accuracy of the approximation is insufficient.

Response surface based optimization techniques prove to be extremely useful in the context of fuzzy analysis. A fuzzy analysis requires the same objective functions to be minimised and maximised on different α-levels or, in optimization terms, with different bound constraints. Since the uncertain parameter intervals at higher α-levels are subsets of the uncertain parameter interval at the lowest α-level, a response surface suitable for the lowest α-level can be used at all α-levels. Since the construction of the response surfaces is the computationally most expensive part of these techniques, the computational cost of a fuzzy analysis is only slightly higher than the computational cost of an interval analysis based on fuzzy arithmetic operations. However, response surface based optimization techniques avoid other practical problems associated with interval analysis, which include the possibility of obtaining unbounded, unrealistic and non-unique solutions if the dependent relations between the fuzzy variables and the constrained fuzzy operation procedures are not considered in the formulation.

HIGH DIMENSIONAL MODEL REPRESENTATION

In recent years there have been efforts to develop efficient methods to approximate multivariate functions in such a way that the component functions of the approximation are ordered starting from a constant and gradually approaching to multi-variance as we proceed along the terms like first-order, second-order and so on. One such method is High Dimensional Model Representation (HDMR) (Li et. al., 2001a; Li et. al., 2001b; Sobol, 2003; Balu & Rao, 2013; Balu & Rao, 2014). HDMR is a general set of quantitative model assessment and analysis tools for capturing the high-dimensional relationships between sets of input and output model variables. It is a very efficient formulation of the system response, if higher-order variable correlations are weak, allowing the physical model to be captured by the first few lower-order terms. Practically for most well-defined physical systems, only relatively low order correlations of the input variables are expected to have a significant effect on the overall response. HDMR expansion utilizes this property to present an accurate hierarchical representation of the physical system. The notion of "high" dimensionality is system-dependent, with some situations being considered high for practical reasons at $N \sim 3$–5, while others will only reach that level of complexity for $N > 10$ or more. Recently potential of HDMR concepts is explored to find an equivalent continuous function to replace a univariate or multivariate piece wise continuous function, rather than seeking an exact continuous function (Chowdhury et. al., 2008), and the HDMR concepts are extended to predict the failure probability of structural or mechanical systems subjected to random loads and material properties.

Let the N–dimensional vector $\boldsymbol{x} = \left\{x_1, x_2, \cdots, x_N\right\}$, represent the input variables of the model under consideration, and the response function as $g\left(\boldsymbol{x}\right)$. Since the influence of the input variables on the response function can be independent and/or cooperative, HDMR expresses the response $g\left(\boldsymbol{x}\right)$ as a hierarchical correlated function expansion in terms of the input variables as

$$\begin{aligned} g\left(\boldsymbol{x}\right) &= g_0 + \sum_{i=1}^{N} g_i\left(x_i\right) + \sum_{1 \le i_1 < i_2 \le N} g_{i_1 i_2}\left(x_{i_1}, x_{i_2}\right) + \ldots \\ &+ \sum_{1 \le i_1 < \ldots < i_l \le N} g_{i_1 i_2 \ldots i_l}\left(x_{i_1}, x_{i_2}, \ldots, x_{i_l}\right) + \ldots + g_{12\ldots N}\left(x_1, x_2, \ldots, x_N\right) \end{aligned} \quad (2)$$

where g_0 is a constant term representing the zeroth-order component function or the mean response of $g\left(\boldsymbol{x}\right)$. The function $g_i\left(x_i\right)$ is a first-order term expressing the effect of variable x_i acting alone, although generally nonlinearly, upon the output $g\left(\boldsymbol{x}\right)$. The function $g_{i_1 i_2}\left(x_{i_1}, x_{i_2}\right)$ is a second-order term which describes the cooperative effects of the variables x_{i_1} and x_{i_2} upon the output $g\left(\boldsymbol{x}\right)$. The higher order terms give the cooperative effects of increasing numbers of input variables acting together to influence the output $g\left(\boldsymbol{x}\right)$. The last term $g_{12,\cdots,N}\left(x_1, x_2, \cdots, x_N\right)$ contains any residual dependence of all the input variables locked together in a cooperative way to influence the output $g\left(\boldsymbol{x}\right)$. Once all the relevant component functions in Equation 2 are determined and suitably represented, then the component functions constitute HDMR, thereby replacing the original computationally expensive method of calculating $g\left(\boldsymbol{x}\right)$ by the computationally efficient model. Usually the higher order terms in Equation 2 are negligible such that HDMR with only few low order correlations, amongst the input variables are typically adequate in describing the output behavior resulting in rapid convergence of HDMR expansion (Li et. al., 2001a). However, importance of higher order terms in HDMR is problem dependent and an independent analysis such as for example, the Sobol' sensitivity indices method (Sobol, 1993) should be made beforehand to establish which order terms in the HDMR expansion are not important.

Depending on the method adopted to determine the component functions in Equation 2 there are two particular HDMR expansions: ANOVA-HDMR and cut-HDMR. ANOVA-HDMR is useful for measuring the contributions of the variance of individual component functions to the overall variance of the output. On the other hand, cut-HDMR expansion is an exact representation of the output $g\left(\boldsymbol{x}\right)$ in the hyperplane passing through a reference point in the variable space. In this work, cut-HDMR procedure is used to approximate the original response function with an explicit performance function to estimate the response of structural/mechanical systems subjected to fuzzy variables. With cut-HDMR method, first a reference point $\boldsymbol{c} = \left\{c_1, c_2, \cdots, c_N\right\}$ is defined in the variable space. In the convergence limit that all correlated functions in Equation 2 are considered, cut-HDMR is invariant to the choice of reference point $\boldsymbol{c}$. However, in practice the choice of reference point $\boldsymbol{c}$ is very important for cut-HDMR especially if only terms up to first- and second-order in Equation 2 are considered, and it is shown that it is optimal to choose reference point $\boldsymbol{c}$ as mean values of the input variables (Li et. al., 2001b). The expansion functions are determined by evaluating the input-output responses of the system relative to the defined reference point $\boldsymbol{c}$ along associated lines, surfaces, subvolumes, etc. (i.e. cuts) in the input variable space. This process reduces to the following relationship for the component functions in Equation 2

$$g_0 = g(\boldsymbol{c}) \tag{3}$$

$$g_i(x_i) = g(x_i, \boldsymbol{c}^i) - g_0 \tag{4}$$

$$g_{i_1 i_2}(x_{i_1}, x_{i_2}) = g(x_{i_1}, x_{i_2}, \boldsymbol{c}^{i_1 i_2}) - g_{i_1}(x_{i_1}) - g_{i_2}(x_{i_2}) - g_0 \tag{5}$$

where the notation $g(x_i, \boldsymbol{c}^i) = g(c_1, c_2, \ldots, c_{i-1}, x_i, c_{i+1}, \ldots, c_N)$ denotes that all the input variables are at their reference point values except x_i. The g_0 term is the output response of the system evaluated at the reference point $\boldsymbol{c}$. The higher order terms are evaluated as cuts in the input variable space through the reference point. Therefore, each first-order term $g_i(x_i)$ is evaluated along its variable axis through the reference point. Each second-order term $g_{i_1 i_2}(x_{i_1}, x_{i_2})$ is evaluated in a plane defined by the binary set of input variables x_{i_1}, x_{i_2} through the reference point, etc. The process of subtracting off the lower order expansion functions removes their dependence to assure a unique contribution from the new expansion function.

Considering the terms up to first- and second-order in Equation 2 yields, respectively

$$g(\boldsymbol{x}) = g_0 + \sum_{i=1}^{N} g_i(x_i) + \mathcal{R}_2 \tag{6}$$

and

$$g(\boldsymbol{x}) = g_0 + \sum_{i=1}^{N} g_i(x_i) + \sum_{1 \le i_1 < i_2 \le N} g_{i_1 i_2}(x_{i_1}, x_{i_2}) + \mathcal{R}_3 \tag{7}$$

Substituting Equations 3–5 into Equations 6 and 7 leads to

$$g(\boldsymbol{x}) = \sum_{i=1}^{N} g(c_1, \ldots, c_{i-1}, x_i, c_{i+1}, \ldots, c_N) - (N-1) g(\mathbf{c}) + \mathcal{R}_2 \tag{8}$$

and

$$\begin{aligned} g(\boldsymbol{x}) = & \sum_{\substack{i_1=1, i_2=1 \\ i_1<i_2}}^{N} g(c_1, \ldots, c_{i_1-1}, x_{i_1}, c_{i_1+1}, \ldots, c_{i_2-1}, x_{i_2}, c_{i_2+1}, \ldots, c_N) \\ & -(N-2) \sum_{i=1}^{N} g(c_1, \ldots, c_{i-1}, x_i, c_{i+1}, \ldots, c_N) + \frac{(N-1)(N-2)}{2} g(\mathbf{c}) + \mathcal{R}_3 \end{aligned} \tag{9}$$

respectively. Now consider first- and second-order approximation of $g(\boldsymbol{x})$, denoted respectively by

$$\tilde{g}(\boldsymbol{x}) = \sum_{i=1}^{N} g(c_1, \ldots, c_{i-1}, x_i, c_{i+1}, \ldots, c_N) - (N-1) g(\mathbf{c}) \tag{10}$$

and

$$\begin{aligned} \tilde{g}(\boldsymbol{x}) = & \sum_{\substack{i_1=1, i_2=1 \\ i_1<i_2}}^{N} g(c_1, \ldots, c_{i_1-1}, x_{i_1}, c_{i_1+1}, \ldots, c_{i_2-1}, x_{i_2}, c_{i_2+1}, \ldots, c_N) \\ & -(N-2) \sum_{i=1}^{N} g(c_1, \ldots, c_{i-1}, x_i, c_{i+1}, \ldots, c_N) + \frac{(N-1)(N-2)}{2} g(\mathrm{c}) \end{aligned} \tag{11}$$

Comparison of Equations 8 and 10 indicates that the first-order approximation leads to the residual error $g(\boldsymbol{x}) - \tilde{g}(\boldsymbol{x}) = \mathcal{R}_2$, which includes contributions from terms of second and higher order component functions. Similarly, the second-order approximation leads to the residual error $g(\boldsymbol{x}) - \tilde{g}(\boldsymbol{x}) = \mathcal{R}_3$, which includes contributions from terms of third and higher order component functions.

The notion of 0th, 1st, 2nd order, etc. in the HDMR expansion should not be confused with the terminology used either in the Taylor series or in the conventional polynomial based approximation formulas. It can be shown that, the first-order component function $g_i(x_i)$ is the sum of all the Taylor series terms which contain and only contain variable x_i. Similarly, the second-order component function $g_{i_1 i_2}(x_{i_1}, x_{i_2})$ is the sum of all the Taylor series terms which contain and only contain variables x_{i_1} and x_{i_2}. Hence first- and second-order HDMR approximations should not be viewed as first- or second-order Taylor series expansions nor do they limit the nonlinearity of $g(\boldsymbol{x})$. Furthermore, the approximations contain contributions from all input variables. Thus, the infinite number of terms in the Taylor series is partitioned into finite different groups and each group corresponds to one cut-HDMR component function. Therefore, any truncated cut-HDMR expansion provides a better approximation and convergent solution of $g(\boldsymbol{x})$ than any truncated Taylor series because the latter contains only a finite number of terms of

Taylor series. Furthermore, the coefficients associated with higher dimensional terms are usually much smaller than that with one-dimensional terms. As such, the impact of higher dimensional terms on the function is less, and therefore, can be neglected. Compared with the first- and second-order Taylor series expansions, which retain only linear and quadratic terms, respectively, the first-order HDMR provides more accurate approximation $\tilde{g}\left(\boldsymbol{x}\right)$ of the original response function $g\left(\boldsymbol{x}\right)$. If the first-order HDMR approximation is not sufficient, the second-order HDMR approximation may be adopted at the expense of additional computational cost.

SURROGATE MODEL GENERATION

The HDMR in Equation 2 is exact along any of the cuts, and the output response $g\left(\boldsymbol{x}\right)$ at a point $\boldsymbol{x}$ off of the cuts can be obtained as follows.

Step 1: Interpolate each of the low dimensional HDMR expansion terms with respect to the input values of the point $\boldsymbol{x}$. For example, consider the first-order component function

$$g\left(x_i,\mathbf{c}^i\right) = g\left(c_1,c_2,\ldots,c_{i-1},x_i,c_{i+1},\ldots,c_N\right).$$

If for $x_i = x_i^j$, n function values

$$g\left(x_i^j,\mathbf{c}^i\right) = g\left(c_1,\ldots,c_{i-1},x_i^j,c_{i+1},\ldots,c_N\right); j = 1,2,\ldots,n \tag{12}$$

are given at $n\left(= 3,5,7\ or\ 9\right)$ sample points x_{iL},

$x_{iM} - \left(n-3\right)\left(x_{iM} - x_{iL}\right)/\left(n-1\right)$,

$x_{iM} - \left(n-5\right)\left(x_{iM} - x_{iL}\right)/\left(n-1\right)$,

$\ldots,\ x_{iM},\ \ldots,$

$x_{iM} + \left(n-5\right)\left(x_{iU} - x_{iM}\right)/\left(n-1\right)$,

$x_{iM} + \left(n-3\right)\left(x_{iU} - x_{iM}\right)/\left(n-1\right)$,

x_{iU}, along the fuzzy variable x_i having triangular membership function with the triplet number $\left[x_{iL}, x_{iM}, x_{iU}\right]$. The function value for arbitrary x_i can be obtained by the Lagrange interpolation as

$$g\left(x_i, \mathbf{c}^i\right) = \sum_{j=1}^{n} \phi_j\left(x_i\right)\ g\left(c_1, \ldots, c_{i-1}, x_i^j, c_{i+1}, \ldots, c_N\right); j = 1, 2, \ldots, n \tag{13}$$

Similarly, consider the second-order component function

$$g\left(x_{i_1}, x_{i_2}, \mathbf{c}^{i_1 i_2}\right) = g\left(c_1, \ldots, c_{i_1-1}, x_{i_1}, c_{i_1+1}, \ldots, c_{i_2-1}, x_{i_2}, c_{i_2+1}, \ldots, c_N\right).$$

If for $x_{i_1} = x_{i_1}^{j_1}$, and $x_{i_2} = x_{i_2}^{j_2}, n^2$ function values

$$\begin{aligned} & g\left(x_{i_1}^{j_1}, x_{i_2}^{j_2}, \mathbf{c}^{i_1 i_2}\right) = g\left(c_1, \ldots, c_{i_1-1}, x_{i_1}^{j_1}, c_{i_1+1}, \ldots, c_{i_2-1}, x_{i_2}^{j_2}, c_{i_2+1}, \ldots, c_N\right); \\ & j_1 = 1, 2, \ldots, n, \\ & j_2 = 1, 2, \ldots, n \end{aligned} \tag{14}$$

are given on a grid formed by taking $n\left(= 3, 5, 7\ or\ 9\right)$ sample points along x_{i_1} and x_{i_2} axes as explained above, then the function value for arbitrary $\left(x_{i_1}, x_{i_2}\right)$ can be obtained by the Lagrange interpolation as

$$g\left(x_{i_1}, x_{i_2}, \mathbf{c}^{i_1 i_2}\right) = \sum_{j_1=1}^{n} \sum_{j_2=1}^{n} \phi_{j_1 j_2}\left(x_{i_1}, x_{i_2}\right) g\left(c_1, \ldots, c_{i_1-1}, x_{i_1}^{j_1}, c_{i_1+1}, \ldots, c_{i_2-1}, x_{i_2}^{j_2}, c_{i_2+1}, \ldots, c_N\right) \tag{15}$$

where the interpolation functions $\phi_j\left(x_i\right)$ and $\phi_{j_1 j_2}\left(x_{i_1}, x_{i_2}\right)$ can be obtained using the Lagrange interpolation.

By using Equation 13, $g_i\left(x_i\right)$ can be generated if *n* function values are given at corresponding sample points. Similarly, by using Equation 15, $g_{i_1 i_2}\left(x_{i_1}, x_{i_2}\right)$ can be generated if n^2 function values at corresponding sample points are given. The same procedure shall be repeated for all the first-order component functions, i.e., $g_i\left(x_i\right); i = 1, 2, \ldots, N$ and the second-order component functions, i.e., $g_{i_1 i_2}\left(x_{i_1}, x_{i_2}\right);\ i_1, i_2 = 1, 2, \ldots, N$.

Step 2: Sum the interpolated values of HDMR expansion terms from zeroth-order to the highest order retained in keeping with the desired accuracy. This leads to first-order approximation of the function $g\left(\boldsymbol{x}\right)$ as

$$\tilde{g}(\boldsymbol{x}) = \sum_{i=1}^{N}\sum_{j=1}^{n}\phi_j(x_i)\,g(c_1,\ldots,c_{i-1},x_i^j,c_{i+1},\ldots,c_N) - (N-1)\,g_0 \tag{16}$$

and second-order approximation of the function $g(\boldsymbol{x})$ as

$$\begin{aligned}\tilde{g}(\boldsymbol{x}) = &\sum_{\substack{i_1=1,i_2=1\\ i_1<i_2}}^{N}\sum_{j_1=1}^{n}\sum_{j_2=1}^{n}\phi_{j_1j_2}(x_{i_1},x_{i_2})\,g(c_1,\ldots,c_{i_1-1},x_{i_1}^{j_1},c_{i_1+1},\ldots,c_{i_2-1},x_{i_2}^{j_2},c_{i_2+1},\ldots,c_N)\\ &-(N-2)\sum_{i=1}^{N}\sum_{j=1}^{n}\phi_j(x_i)\,g(c_1,\ldots,c_{i-1},x_i^j,c_{i+1},\ldots,c_N) + \frac{(N-1)(N-2)}{2}g_0\end{aligned} \tag{17}$$

The interpolation functions $\phi_j(x_i)$ and $\phi_{j_1j_2}(x_{i_1},x_{i_2})$ by using the Lagrange interpolation are defined as

$$\phi_j(x_i) = \frac{\left(x_i - x_i^{(1)}\right)\cdots\left(x_i - x_i^{(j-1)}\right)\left(x_i - x_i^{(j+1)}\right)\cdots\left(x_i - x_i^{(n)}\right)}{\left(x_i^{(j)} - x_i^{(1)}\right)\cdots\left(x_i^{(j)} - x_i^{(j-1)}\right)\left(x_i^{(j)} - x_i^{(j+1)}\right)\cdots\left(x_i^{(j)} - x_i^{(n)}\right)} \tag{18}$$

$$\begin{aligned}\phi_{j_1j_2}(x_{i_1},x_{i_2}) = &\left[\frac{\left(x_{i_1} - x_{i_1}^{(1)}\right)\cdots\left(x_{i_1} - x_{i_1}^{(j_1-1)}\right)\left(x_{i_1} - x_{i_1}^{(j_1+1)}\right)\cdots\left(x_{i_1} - x_{i_1}^{(n)}\right)}{\left(x_{i_1}^{(j_1)} - x_{i_1}^{(1)}\right)\cdots\left(x_{i_1}^{(j_1)} - x_{i_1}^{(j_1-1)}\right)\left(x_{i_1}^{(j_1)} - x_{i_1}^{(j_1+1)}\right)\cdots\left(x_{i_1}^{(j_1)} - x_{i_1}^{(n)}\right)}\right]\\ \times&\left[\frac{\left(x_{i_2} - x_{i_2}^{(1)}\right)\cdots\left(x_{i_2} - x_{i_2}^{(j_2-1)}\right)\left(x_{i_2} - x_{i_2}^{(j_2+1)}\right)\cdots\left(x_{i_2} - x_{i_2}^{(n)}\right)}{\left(x_{i_2}^{(j_2)} - x_{i_2}^{(1)}\right)\cdots\left(x_{i_2}^{(j_2)} - x_{i_2}^{(j_2-1)}\right)\left(x_{i_2}^{(j_2)} - x_{i_2}^{(j_2+1)}\right)\cdots\left(x_{i_2}^{(j_2)} - x_{i_2}^{(n)}\right)}\right]\end{aligned} \tag{19}$$

If n is the number of sample points taken along each of the variable axis and s is the order of the component function considered, starting from zeroth-order to l-th order, then total number of function evaluation for interpolation purpose is given by, $\sum_{s=0}^{l}\left[N!(n-1)^s\right]/\left((N-s)!s!\right)$ which grows polynomially with N and n, and exponentially with s. As a few low order component functions of HDMR are used, the sample savings due to HDMR are significant compared to traditional sampling. Hence uncertainty analysis using HDMR relies on an accurate reduced model being generated with a small number of full model simulations. The tremendous computational savings result from just having to perform interpolation instead of full model simulations for output determination.

EXPLICIT FUZZY ANALYSIS

The original response function $g(\boldsymbol{x})$ is approximated using HDMR. Once the response function is expressed as a linear combination of the fuzzy variables, the bounds of the response at each α-cut can be obtained by using interval arithmetic algorithm. In order to apply interval arithmetic algorithm, the first-order HDMR approximation $\tilde{g}(\boldsymbol{x})$ of the original nonlinear response function $g(\boldsymbol{x})$ is transformed into a linear combination of intervening variables as follows.

$$\tilde{g}(\boldsymbol{z}) = z_1 + z_2 + \ldots + z_N + \beta_0 \tag{20}$$

where $z_i = (\beta_i x_i + \gamma_i)^\kappa$ is the relation between the intervening and the original variables with κ being order of the first-order HDMR component functions approximation taking values $\kappa = 1$ for linear approximation, $\kappa = 2$ for quadratic approximation, $\kappa = 3$ for cubic approximation, and so on. If the membership functions of the intervening variables are available, then at each α-cut, interval arithmetic techniques (Moore, 1979) can be applied to estimate the response bounds at that level.

Let the intervening variable, *z*, be a function of *x* given by, $z = G(x)$. The membership function of *z* is obtained by using the membership function of *x*. The membership function gives the possibility of occurrence. For this reason, the possibility of occurrence of a point z_0 is equal to the possibility of occurrence of x_0, where $x_0 = G^{-1}(z_0)$. Therefore, the possibility function of *z* can be obtained using the following equation and appropriate transformations is shown in Figure 1:

$$\mu_z(z) = \mu_x(g^{-1}(z)) \tag{21}$$

The membership function of *z* can be obtained numerically by the use of transformations.

Figure 1. Transformation to obtain membership function of intervening variable

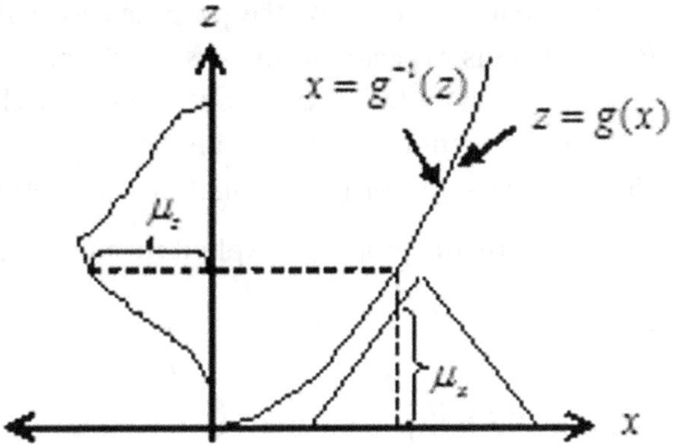

Once the first-order HDMR approximation $\tilde{g}(x)$ of the original nonlinear response function $g(x)$ is expressed as a linear combination in terms of intervening variables (Equation 20), interval arithmetic techniques (Moore, 1979; Alefeld and Herzberger, 1983) can be used to estimate the response bounds. The drawback of using interval arithmetic (Hansen 1992) for a closed-form expression is the problem with dependency. In general, if any variable appears more than once in an expression, it is treated as a different variable for each occurrence, resulting in wider bounds. By using intervening variables, each variable appears only once in the expression, resulting in accurate bounds on the response.

Similar transformation techniques are adopted by Adduri and Penmetsa (2008) to obtain minimum and maximum values of a nonlinear response at each α-cut, except that a second-order response surface model is used for the original response approximation, however the proposed first-order HDMR provides better approximation $\tilde{g}(x)$ of the original response function $g(x)$. The use of transformation techniques facilitates the determination of the minimum and maximum values of the fuzzy variables that correspond to the extreme values of the response at a particular level without the use of optimization techniques. Moreover, this is an analytical procedure in which there is little room for errors. In all the numerical examples presented in this work, first-order HDMR approximation is adopted. A flow diagram for explicit fuzzy analysis using HDMR approximation is shown in Figure 2.

NUMERICAL EXAMPLES

Three numerical examples are presented to demonstrate the accuracy and efficiency of the proposed methodology. The proposed explicit formulations can be applied to fuzzy variables with any type of membership functions to result in an accurate estimation of the bounds of the response. If *N, n,* and *m* respectively denote the number of fuzzy variables, the number of sample points taken along each of the variable axis, and the number of α-cuts, then using first-order HDMR approximation the total cost of original function evaluation entails a maximum of $N \times (n-1)+1$ by the proposed explicit formulation. It can be easily verified that the total computational cost of the proposed formulations is far less than $2^N m+1$ number of function evaluations required for the crude vertex method fuzzy analysis in conjunction with combinatorial optimization (in which all possible combinations of the fuzzy variables are considered), especially for large number of fuzzy input parameters and large finite element models. This clearly demonstrates the computational efficiency of the proposed formulations. The efficiency and robustness of the proposed formulation is expected to increase with increase in the complexity of the structure, number of fuzzy variables and number of α-cuts. If first-order HDMR approximation is not sufficient second-order HDMR approximation may be adopted at the expense of additional computational cost. In that case, the total cost of original function evaluation entails a maximum of $(n-1)^2 (N-1)N/2+(n-1)N+1$ by the proposed explicit formulation.

Closed-Form Example

To illustrate the accuracy and the applicability of the transformation technique in the proposed explicit formulation to obtain the bounds on the response, the following nonlinear response function is considered.

Figure 2. Flowchart of explicit fuzzy analysis using HDMR approximation

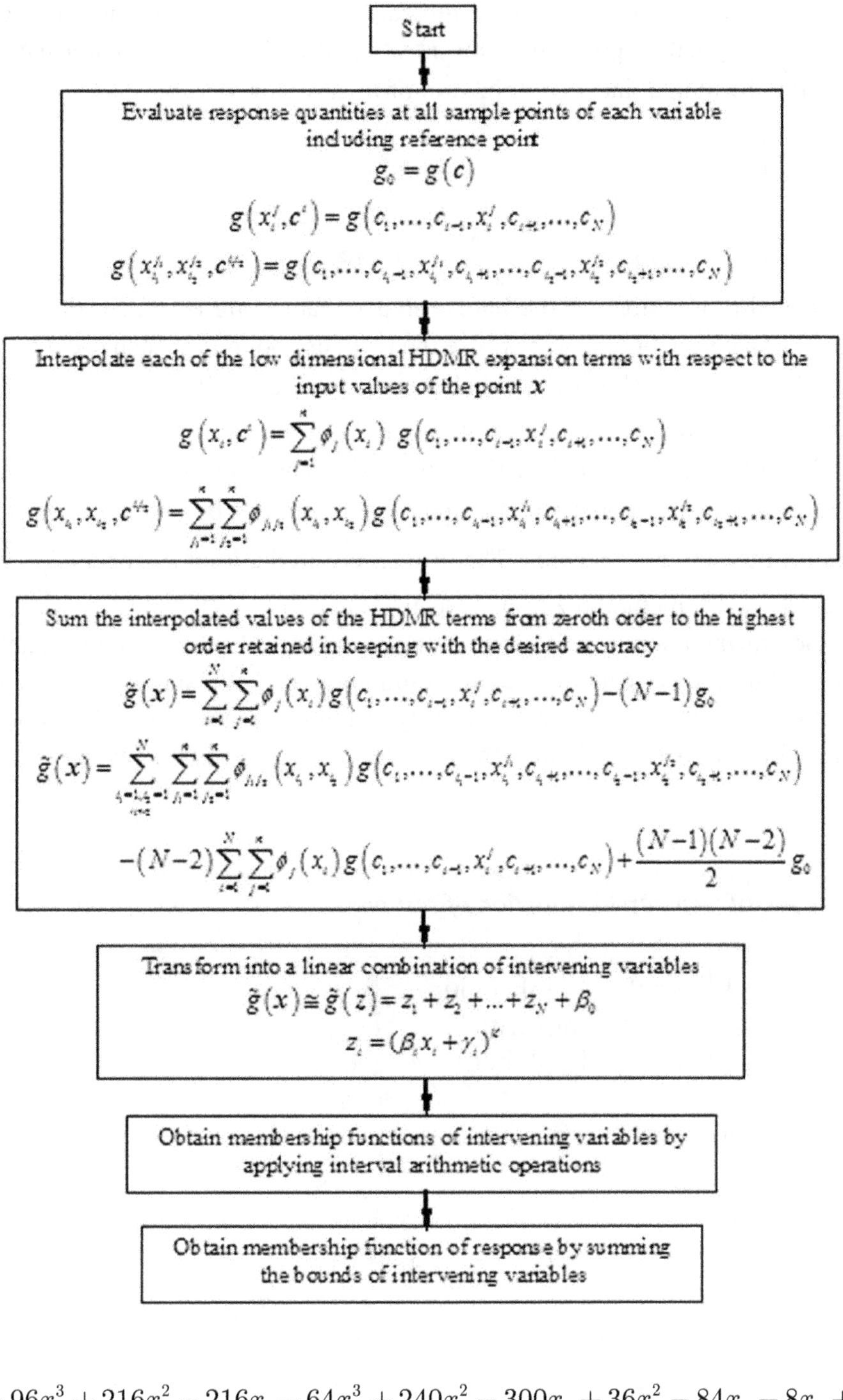

$$f(\boldsymbol{x}) = 16x_1^4 - 96x_1^3 + 216x_1^2 - 216x_1 - 64x_2^3 + 240x_2^2 - 300x_2 + 36x_3^2 - 84x_3 - 8x_4 + 264 \tag{22}$$

where $x_1 = [1,\ 2,\ 3]$, $x_2 = [-1,\ 0,\ 1]$, $x_3 = [-5,\ 0,\ 5]$, and $x_4 = [10,\ 15,\ 20]$ are assumed to have triangular membership functions and the triplet number for each fuzzy variable $\left[x_L,\ \ x_M,\ \ x_U\right]$ defines a triangular membership function, where x_L and x_U are the lower and upper bounds at $\alpha = 0$ and x_M is the nominal value at $\alpha = 1$. As Equation 22 represents closed form nonlinear function, the HDMR approximation of the original response function is skipped and suitable transformations of variables $x_1 - x_4$ are applied to form in a linear combination of intervening variables, as follows

$$f(\mathbf{z}) = z_1 + z_2 + z_3 + z_4 \tag{23}$$

where $z_1 = (2x_1 - 3)^4$, $z_2 = -(4x_2 - 5)^3$, $z_3 = (6x_3 - 7)^2$ and $z_4 = -8x_4 + 9$.

Once the membership functions of the intervening variables are obtained, the interval arithmetic techniques are applied on Equation 22 to estimate the response bounds at each α-cut to get the membership function of $f(\mathbf{x})$. Figure 3 shows the membership function of $f(\mathbf{x})$ obtained by the crude vertex method fuzzy analysis in conjunction with combinatorial optimization, by adopting interval arithmetic techniques directly on Equation 21, and by the proposed explicit formulation. As expected due to the problem of dependency the membership function of $f(\mathbf{x})$ obtained by directly adopting interval arithmetic techniques resulted in wider bounds. The differences in the joint membership function of $f(\mathbf{x})$ obtained by the crude vertex method fuzzy analysis in conjunction with combinatorial optimization and the proposed explicit formulation is mainly due to the reason that the membership function obtained by using the vertex method fails to capture the minimum value of the response because it calculates the minimum and maximum values of the response only at the lower and upper limits of the variables at each α-level. Therefore, it does not identify the minimum of the response, which is located inside the design space.

Figure 3. Comparison of membership function of response

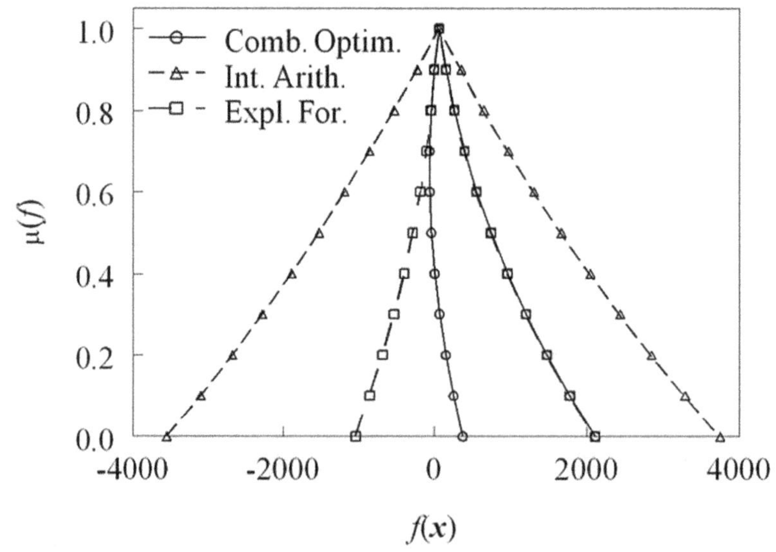

Three Steeped Bar

In this example, the static and dynamic responses of a three stepped bar (Akpan et. al., 2001a) shown in Figure 4, are considered.

Static Analysis

The following three cases of fuzzy input variables are considered for the static analysis:

1. Case A in which only the load P is fuzzified;
2. Case B in which both the load and the elastic modulus are fuzzified; and
3. Case C in which the load, elastic modulus, cross sectional areas and the lengths of the bar are taken as fuzzy variables.

The details of fuzzy input variables for all three cases are presented in Table 1. The possibility distributions of the fuzzy axial displacement at the free end and the strain energy of the system are computed using the proposed first-order HDMR based explicit formulation. The effect of number of sample points used for first-order HDMR approximation of the original response function is examined by varying n from 3 to 9. It is observed that $n = 7$ results in the convergent estimation of the bounds of the response using the proposed explicit formulation. In addition, the effect of order of approximation κ is examined by taking $\kappa = 2$, 3 and 4. The possibility distributions of the fuzzy axial displacement at the free end for Case A obtained using the proposed first-order HDMR based explicit (with $n = 7$) formulations and the crude vertex method fuzzy analysis in conjunction with combinatorial optimization are presented in Figure 5. Similarly Figure 5 compares the possibility distributions of the strain energy of the system for Case A.

Figure 6 respectively presents the possibility distributions of the fuzzy axial displacement at the free end and the strain energy of the system for Case B. Similarly Figure 7 respectively presents the possibility distributions of the fuzzy axial displacement at the free end and the strain energy of the system for Case C. Similar to the observations reported by Akpan et al. (2001a), the present study shows that as the number of fuzzy variables increases the bounds of the possibility distributions of both the displacements and the strain energy are also increased. Also it shows that for Case B and Case C the possibility distributions of the response are skewed to the right because the displacement and the strain energy are

Figure 4. Three-stepped bar configuration

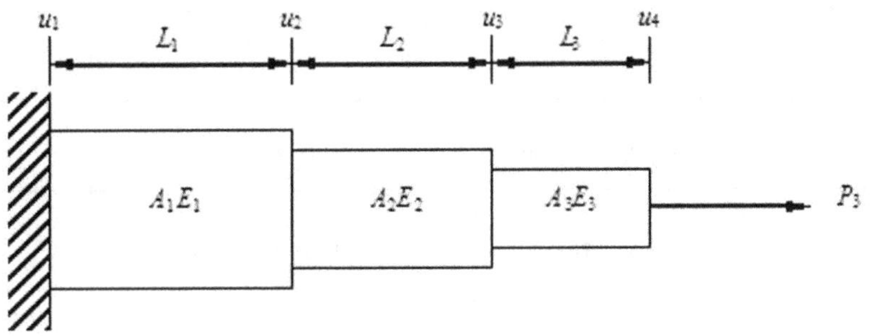

Table 1. Geometrical and material properties of three stepped bar

Parameter	Case A	Case B	Case C
Data for Fuzzy Static Analysis of Three Stepped Bar			
A_1 (in.^2)	[3.00, 3.00, 3.00]	[3.00, 3.00, 3.00]	[2.99, 3.00, 3.01]
A_2 (in.^2)	[2.00, 2.00, 2.00]	[3.00, 3.00, 3.00]	[1.99, 2.00, 2.01]
A_3 (in.^2)	[1.00, 1.00, 1.00]	[1.00, 1.00, 1.00]	[0.99, 1.00, 1.01]
L_1 (in.)	[12.00, 12.00, 12.00]	[12.00, 12.00, 12.00]	[11.95, 12.00, 12.05]
L_2 (in.)	[10.00, 10.00, 10.00]	[10.00, 10.00, 10.00]	[9.95, 10.00, 10.05]
L_3 (in.)	[6.00, 6.00, 6.00]	[6.00, 6.00, 6.00]	[5.95, 6.00, 6.05]
$E_1, E_2, E_3\,(\text{psi})$	[3.0e7, 3.0e7, 3.0e7]	[2.8e7, 3.0e7, 3.1e7]	[2.8e7, 3.0e7, 3.1e7]
$P_1, P_2\,(\text{lb})$	[0.0, 0.0, 0.0]	[0.0, 0.0, 0.0]	[0.0, 0.0, 0.0]
P_3 (lb)	[7500, 10000, 12500]	[7500, 10000, 12500]	[7500, 10000, 12500]
Data for Fuzzy Dynamic Analysis of Three Stepped Bar			
A_1 (in.^2)	[3.00, 3.00, 3.00]	[3.00, 3.00, 3.00]	[2.99, 3.00, 3.01]
A_2 (in.^2)	[2.00, 2.00, 2.00]	[3.00, 3.00, 3.00]	[1.99, 2.00, 2.01]
A_3 (in.^2)	[1.00, 1.00, 1.00]	[1.00, 1.00, 1.00]	[0.99, 1.00, 1.01]
L_1 (in.)	[12.00, 12.00, 12.00]	[12.00, 12.00, 12.00]	[11.95, 12.00, 12.05]
L_2 (in.)	[10.00, 10.00, 10.00]	[10.00, 10.00, 10.00]	[9.95, 10.00, 10.05]
L_3 (in.)	[6.00, 6.00, 6.00]	[6.00, 6.00, 6.00]	[5.95, 6.00, 6.05]
$E_1, E_2, E_3\,(\text{psi})$	[3.0e7, 3.0e7, 3.0e7]	[2.8e7, 3.0e7, 3.1e7]	[2.8e7, 3.0e7, 3.1e7]
$\rho_1, \rho_2, \rho_3\,(\text{lb}/\text{in.}^3)$	[0.27, 0.28 0.29]	[0.27, 0.28 0.29]	[0.27, 0.28 0.29]

Figure 5. Possibility distributions for Case A; fuzzy axial displacement at free end; and strain energy of system

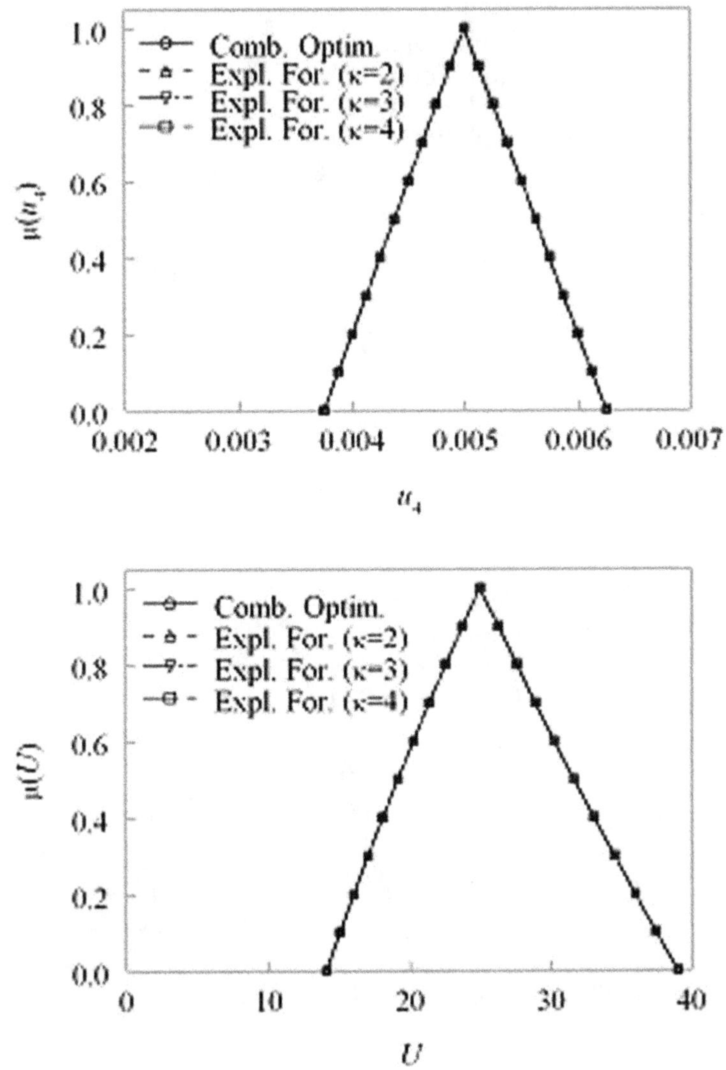

inversely proportional to the modulus of elasticity, which is skewed to the left. Since in Case A only one fuzzy variable is present the dependency problem is not encountered and the proposed first-order HDMR based explicit formulation, and the combinatorial approach yield exactly the same results for all the possibility distributions of the responses. For Case B and Case C the proposed first-order HDMR based explicit formulation estimates a narrow bound for the response and with increase in the order of approximation κ, the bounds for the response are observed to become still narrower.

Natural Frequency Analysis

The effect of input parameter fuzzification on the natural frequencies is studied by considering the following three cases:

Figure 6. Possibility distributions for Case B; fuzzy axial displacement at free end; and strain energy of system

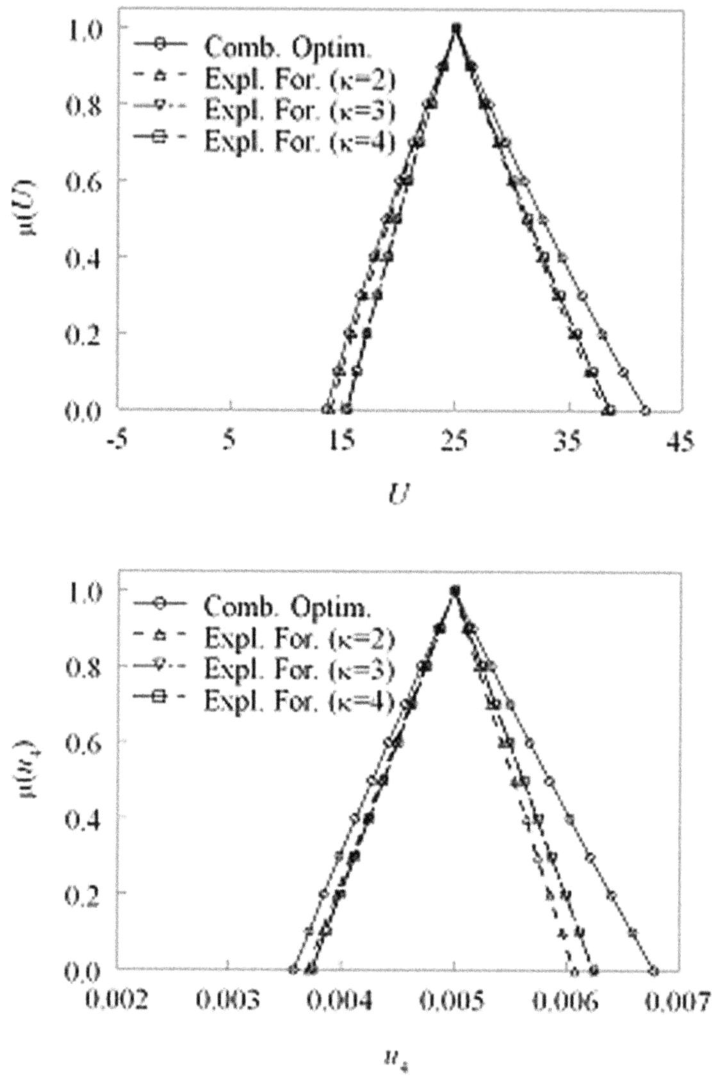

1. Case A in which only the density of the bars as fuzzy;
2. Case B in which both the density and the modulus of elasticity are taken as fuzzy variables; and
3. Case C in which the density, modulus of elasticity, cross-sectional area and length of each section are considered as fuzzy.

The triangular fuzzy numbers for each parameter considered are shown in Table 1. The fuzzy natural frequencies of modes 1–3 for Case A, Case B, and Case C are respectively shown in Figures 8–10. Again, for Case A since only one fuzzy variable is present, the dependency problem is not encountered and the proposed first-order HDMR based explicit formulation, and the combinatorial approach yield exactly the same results for all the possibility distributions of the responses. For Case B and Case C trends for

Figure 7. Possibility distributions for Case C; fuzzy axial displacement at free end; and strain energy of system

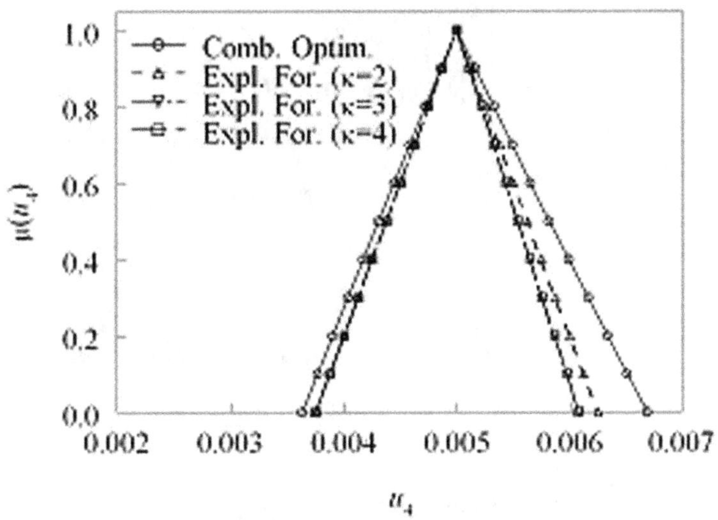

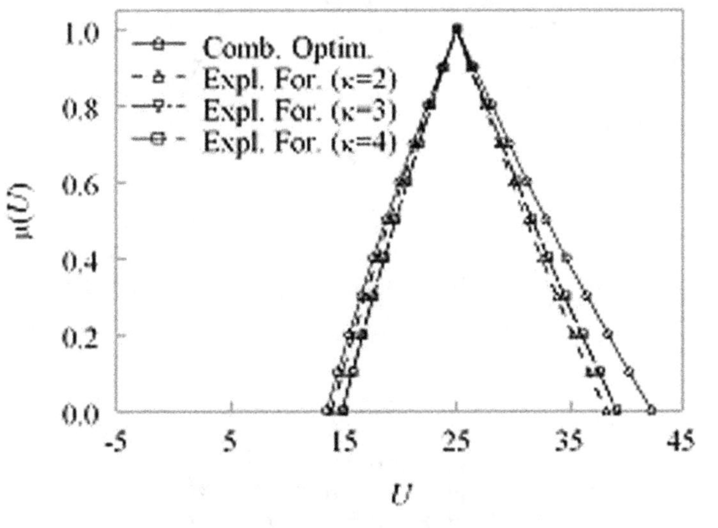

the possibility distributions of the fuzzy natural frequencies are similar to the possibility distributions of the static responses. Similar to the possibility distributions of the static responses as the number of input fuzzy variables increases from Case A to Case C the spread of fuzzy frequencies also increases. The fuzzy frequencies are skewed to the left as is the fuzzy number for the modulus of elasticity, since the frequency solution is directly related to the modulus of elasticity.

Soil Settlement

The settlement of a point A in Figure 11 due to the construction of a structure is primarily caused by the consolidation of the clay layer. Suppose the contribution of settlement due to secondary consolidation is negligible. For normally loaded clay, the settlement S is given by

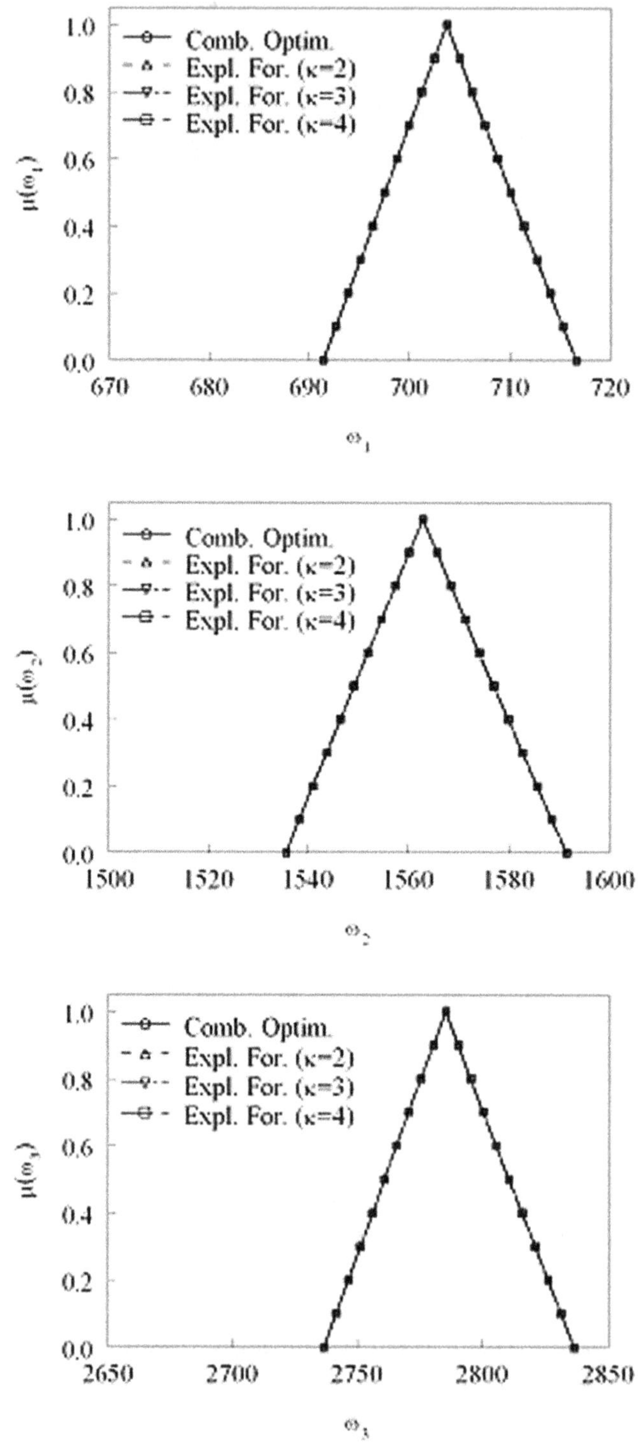

Figure 8. Possibility distributions of natural frequencies for Case A; modes 1-3

Figure 9. Possibility distributions of natural frequencies for Case B; modes 1-3

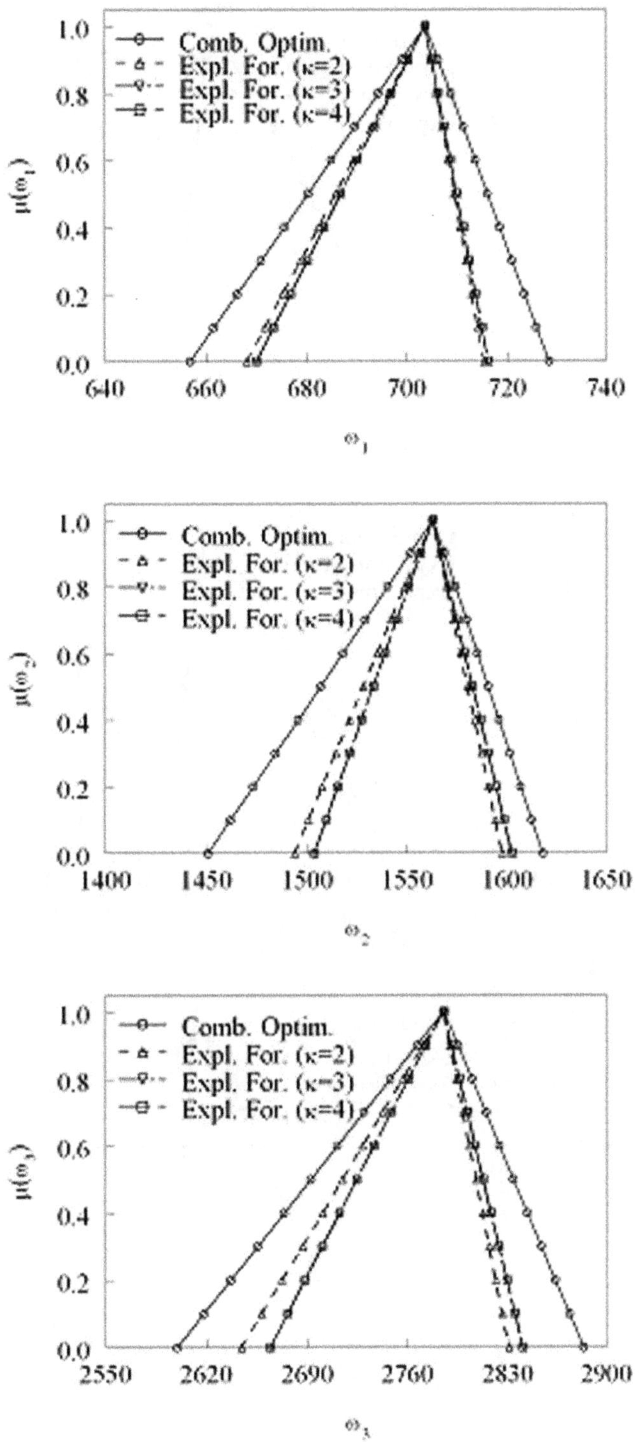

Figure 10. Possibility distributions of natural frequencies for Case C; modes 1-3

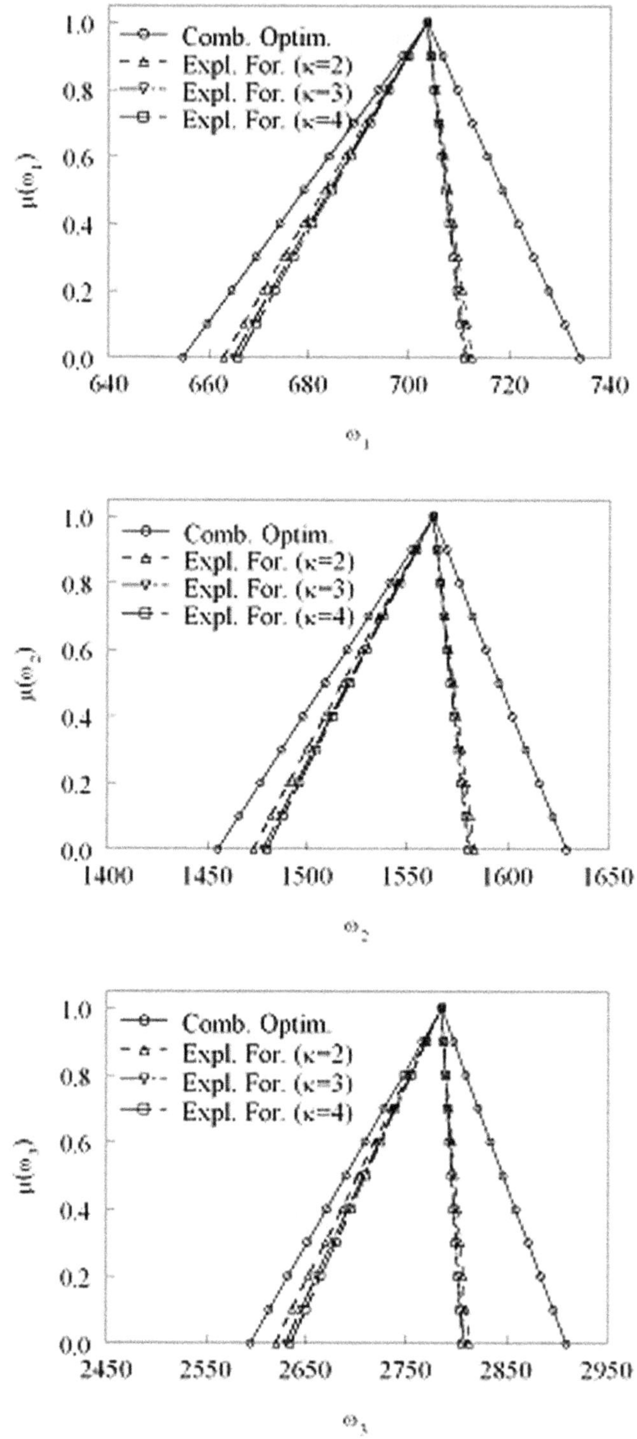

$$S = \frac{C_c}{1+e_0} H \log \frac{p_0 + \Delta p}{p_0} \tag{24}$$

where C_c is the compression index of the clay; e_0 is the void ratio of the clay layer before loading; H is the thickness of the clay layer; p_0 is the original effective pressure at point B before loading; and Δp is the increase in pressure at point B caused by the construction of the structure; "log" denotes logarithm to the base 10. Because of the non-uniform thickness and lack of homogeneity of the clay layer, the settlement predicted by the empirical formula could be subject to uncertainty in predicted settlement. This is a practical engineering example having the triangular fuzzy variables as follows. C_c =[0.099, 0.396, 0.693], e_0=[0.6545, 1.19, 1.7255], H=[142.8, 168.0, 193.2], p_0=[3.162, 3.72, 4.278], and Δp=[0.2, 0.5, 0.8]. The possibility distributions of the fuzzy soil settlement obtained using the proposed first-order HDMR based explicit formulation and the crude vertex method fuzzy analysis in conjunction with combinatorial optimization are presented in Figure 12. Similar to the previous example, the effects of number of sample points and the order of approximation κ are examined. Again the proposed first-order HDMR based explicit formulation estimates a narrow bound for the response and with increase in the order of approximation κ the bounds for the response are observed to become still narrower.

CONCLUSION

In this work, a practical approach based on HDMR for analyzing the response of structures with fuzzy parameters is presented. The proposed methodology involves integrated finite element modelling, HDMR based surrogate model generation, and explicit fuzzy analysis procedure. The uncertainties in the material, geometric, loading and structural parameters are represented using fuzzy sets. To facilitate efficient computation, a HDMR based response surface generation is employed for the approximation of the fuzzy finite element response quantity. In the proposed method, using HDMR the response is expressed in a separable closed-form expression as a linear combination of the fuzzy variables through the definition of intervening variables, then using the transformation technique the bounds on the response at each

Figure 11. Soil profile

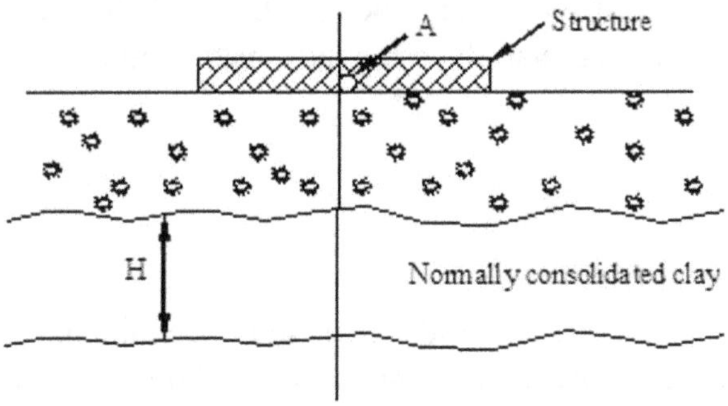

Figure 12. Possibility distributions of soil settlement

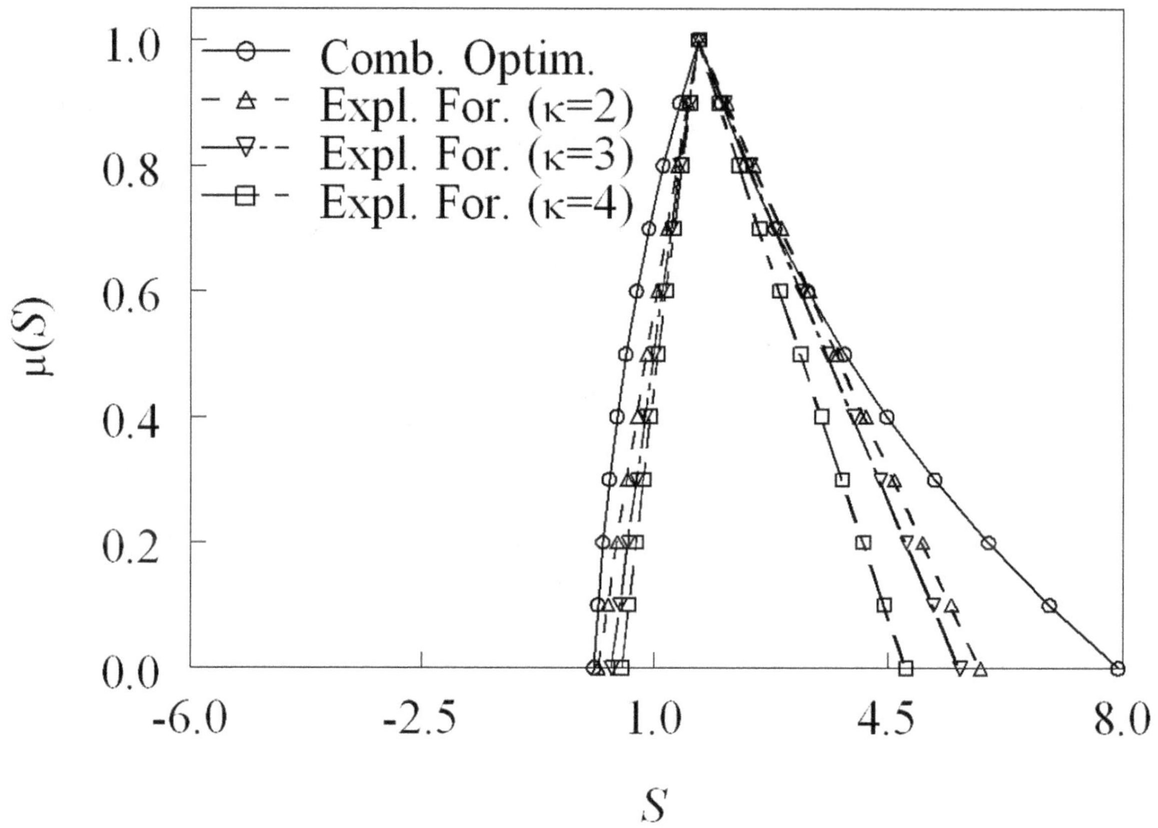

α-level are obtained. The advantage of this approach is that the bounds of the fuzzy response at an α-level can be obtained by solving the resulting set of interval equations once. This approach is mathematically elegant and not computationally expensive. The use of transformations facilitates the determination of the minimum and maximum values of the fuzzy variables that correspond to the extreme values of the response at a particular α-level without the use of optimization techniques. Moreover, this is semi-analytical procedure making use of deterministic finite element software and hence there is little room for errors. The proposed explicit formulation is free from the problem associated with dependency of the fuzzy variables, unlike most of the reported explicit formulations which are plagued with practical problems like the possibility of obtaining unbounded, unrealistic and non-unique solutions if the dependent relations between the fuzzy variables and the constrained fuzzy operation procedures are not considered in the formulation.

REFERENCES

Adduri, P. R., & Penmetsa, R. C. (2008). Confidence bounds on component reliability in the presence of mixed uncertain variables. *International Journal of Mechanical Sciences*, *50*(3), 481–489. doi:10.1016/j.ijmecsci.2007.09.015

Akpan, U. O., Koko, T. S., Orisamolu, I. R., & Gallant, B. K. (2001a). Practical fuzzy finite element analysis of structures. *Finite Elements in Analysis and Design*, *38*(2), 93–111. doi:10.1016/S0168-874X(01)00052-X

Akpan, U. O., Koko, T. S., Orisamolu, I. R., & Gallant, B. K. (2001b). Fuzzy finite element analysis of smart structures. *Smart Materials and Structures*, *10*(2), 273–284. doi:10.1088/0964-1726/10/2/312

Alefeld, G., & Herzberger, J. (1983). *Introduction to Interval Computations*. New York, USA: Academic Press.

Balu, A. S., & Rao, B. N. (2012a). High dimensional model representation based formulations for fuzzy finite element analysis of structures. *Finite Elements in Analysis and Design*, *50*, 217–230. doi:10.1016/j.finel.2011.09.012

Balu, A. S., & Rao, B. N. (2012b). Multicut-HDMR for structural reliability bounds estimation under mixed uncertainties. *Computer-Aided Civil and Infrastructure Engineering*, *27*(6), 419–438. doi:10.1111/j.1467-8667.2012.00763.x

Balu, A. S., & Rao, B. N. (2012c). Inverse structural reliability analysis under mixed uncertainties using high dimensional model representation and fast Fourier transform. *Engineering Structures*, *37*, 224–234. doi:10.1016/j.engstruct.2011.12.043

Balu, A. S., & Rao, B. N. (2013). Confidence bounds on design variables using HDMR based inverse reliability analysis. *Journal of Structural Engineering*, *139*(6), 985–996. doi:10.1061/(ASCE)ST.1943-541X.0000709

Balu, A. S., & Rao, B. N. (2014). Efficient assessment of structural reliability in presence of random and fuzzy uncertainties. *Journal of Mechanical Design*, *136*(5), 051008. doi:10.1115/1.4026650

Chen, C. (1996). *Fuzzy Logic and Neural Network Handbook*. New York: McGraw-Hill.

Chen, L., & Rao, S. S. (1997). Fuzzy finite element approach for the vibration analysis of imprecisely defined systems. *Finite Elements in Analysis and Design*, *27*(1), 69–83. doi:10.1016/S0168-874X(97)00005-X

Chowdhury, R., Rao, B. N., & Prasad, A. M. (2008). High dimensional model representation for piece wise continuous function approximation. *Communications in Numerical Methods in Engineering*, *24*(12), 1587–1609. doi:10.1002/cnm.1053

Dessombz, O., Thouverez, F., Laîné, J. P., & Jézéquel, L. (2001). Analysis of mechanical systems using interval computations applied to finite element methods. *Journal of Sound and Vibration*, *239*(5), 949–968. doi:10.1006/jsvi.2000.3191

Dhingra, A. K., Rao, S. S., & Kumar, V. (1992). Nonlinear membership function in multi-objective fuzzy optimization of mechanical and structural systems. *AIAA Journal*, *30*(1), 251–260. doi:10.2514/3.10906

Donders, S., Vandepitte, D., Van de Peer, J., & Desmet, W. (2005). Assessment of uncertainty on structural dynamic responses with the short transformation method. *Journal of Sound and Vibration*, *288*(3), 523–549. doi:10.1016/j.jsv.2005.07.003

Dong, W. M., & Shah, H. C. (1987). Vertex method for computing functions of fuzzy variables. *Fuzzy Sets and Systems*, *24*(1), 65–78. doi:10.1016/0165-0114(87)90114-X

Dong, W. M., & Wong, F. S. (1987). Fuzzy weighted averages and implementation of the extension principle. *Fuzzy Sets and Systems*, *21*(2), 183–199. doi:10.1016/0165-0114(87)90163-1

Drucker, H., Burges, C. J. C., Kaufman, L., Smola, A., & Vapnik, V. (1997). Support vector regression machines. *Advances in Neural Information Processing Systems*, *9*, 155.

Elishakoff, I. (1995). Essay on uncertainties in elastic and viscoelastic structures: From A.M. Freudenthal's criticism to modern convex modelling. *Computers & Structures*, *56*(2), 871–895. doi:10.1016/0045-7949(94)00499-S

Farkas, L., Moens, D., Vandepitte, D., & Desmet, W. (2008). Application of fuzzy numerical techniques for product performance analysis in the conceptual and preliminary design stage. *Computers & Structures*, *86*(10), 1061–1079. doi:10.1016/j.compstruc.2007.07.012

Gauger, U., Turrin, S., Hanss, M., & Gaul, L. (2008). A new uncertainty analysis for the transformation method. *Fuzzy Sets and Systems*, *159*(11), 1273–1291. doi:10.1016/j.fss.2007.12.027

DeGersem, H., Moens, D., Desmet, W., & Vandepitte, D. (2007). Interval and fuzzy dynamic analysis of finite element models with superelements. *Computers & Structures*, *85*(5–6), 304–319. doi:10.1016/j.compstruc.2006.10.011

Haldar, A., & Mahadevan, S. (2000). *Reliability Assessment Using Stochastic Finite Element Analysis*. New York: John Wiley & Sons, Inc.

Hansen, E. (1992). *Global Optimization Using Interval Analysis*. New York, USA: Marcel Dekker, Inc.

Hanss, M. (2002). The transformation method for the simulation and analysis of systems with uncertain parameters. *Fuzzy Sets and Systems*, *130*(3), 277–289. doi:10.1016/S0165-0114(02)00045-3

Harvey, R. L. (1994). *Neural network principles*. NJ, USA: Prentice-Hall Inc.

Jones, D. R., Schonlau, M., & Welch, W. J. (1998). Efficient global optimization of expensive black-box functions. *Journal of Global Optimization*, *13*(4), 455–492. doi:10.1023/A:1008306431147

Lernout, E.C., Pellissetti, M., Pradlwarter, H., Schueller, G.I., & Soize, C. (2006). Data and model uncertainties in complex aerospace engineering systems. *Journal of Sound and Vibration,* 295(3–5), 923–938.

Li, G., Rosenthal, C., & Rabitz, H. (2001a). High dimensional model representations. *The Journal of Physical Chemistry A*, *105*(33), 7765–7777. doi:10.1021/jp010450t

Li, G., Wang, S. W., & Rabitz, H. (2001b). High dimensional model representations generated from low dimensional data samples–I. mp–Cut–HDMR. *Journal of Mathematical Chemistry*, *30*(1), 1–30. doi:10.1023/A:1013172329778

Moens, D., & Vandepitte, D. (2004). Non-probabilistic approaches for non-deterministic dynamic FE analysis of imprecisely defined structures.*Proceedings of the International Conference on Noise and Vibration Engineering*, Leuven (pp. 3095–3119).

Möller, B., Graf, W., & Beer, M. (2000). Fuzzy structural analysis using α-level optimization. *Computational Mechanics*, *26*(6), 547–565. doi:10.1007/s004660000204

Moore, R. E. (1979). *Methods and Applications of Interval Analysis*. Philadelphia, USA: SAIM Publication. doi:10.1137/1.9781611970906

Muhanna, R. L., & Mullen, R. L. (2001). Uncertainty in mechanics problems interval-based approach. *Journal of Engineering Mechanics*, *127*(6), 557–566. doi:10.1061/(ASCE)0733-9399(2001)127:6(557)

Munck, M. D., Moens, D., Desmet, W., & Vandepitte, D. (2008). A response surface based optimisation algorithm for the calculation of fuzzy envelope FRFs of models with uncertain properties. *Computers & Structures*, *86*(10), 1080–1092. doi:10.1016/j.compstruc.2007.07.006

Pellissetti, M. F., & Schueller, G. I. (2007). The effects of uncertainties in structural analysis. *Journal of Structural Engineering and Mechanics*, *25*(3), 311–330. doi:10.12989/sem.2007.25.3.311

Qiu, Z., & Elishakoff, I. (1998). Antioptimization of structures with large uncertain-but-non-random parameters via interval analysis. *Computer Methods in Applied Mechanics and Engineering*, *152*(3–4), 361–372. doi:10.1016/S0045-7825(96)01211-X

Rao, S. S., & Berke, L. (1997). Analysis of uncertain structural systems using interval analysis. *AIAA Journal*, *34*(4), 727–735. doi:10.2514/2.164

Schueller, G. I., Pradlwarter, H. J., & Koutsourelakis, P. (2004). A critical appraisal of reliability estimation procedures for high dimensions. *Probabilistic Engineering Mechanics*, *19*(4), 463–474. doi:10.1016/j.probengmech.2004.05.004

Sobol', I. M. (1993). Sensitivity estimates for nonlinear mathematical models. *Mathematical Modeling and Computational Experiment*, *1*, 407–414.

Sobol', I. M. (2003). Theorems and examples on high dimensional model representations. *Chemical Reliability Engineering and System Safety*, *79*(2), 187–193. doi:10.1016/S0951-8320(02)00229-6

Soize, C. (2009). Nonparametric probabilistic approach of uncertainties for elliptic boundary value problem. *International Journal for Numerical Methods in Engineering*, *80*(6&7), 673–688. doi:10.1002/nme.2563

Valliappan, S., & Pham, T. D. (1995). Fuzzy logic applied to numerical modeling of engineering problems. *Computational Mechanics Advances*, *2*, 213–281.

Yau, J. F., Wang, S. S., & Corten, H. T. (1980). A mixed-mode crack analysis of isotropic solids using conservation laws of elasticity. *Journal of Applied Mechanics*, *47*(2), 335–341. doi:10.1115/1.3153665

Zadeh, L. (1965). Fuzzy sets. *Information and Control*, *8*(3), 338–353. doi:10.1016/S0019-9958(65)90241-X

Chapter 9
FE Analysis and Experimental Investigation of Cracked and Un-Cracked Thin-Walled Tubular Components to Evaluate Mechanical and Fracture Properties

M.K. Samal
Bhabha Atomic Research Centre, India

ABSTRACT

For investigation of fracture behavior and assessment of remaining life of critical thin-walled tubular components in industry, the transverse mechanical property and the axial fracture properties are essential. However, evaluation of these properties by machining suitable standard specimens from these components directly and subjecting them to standard tests is usually not feasible. In this chapter, the nonlocal version of the Rousselier's damage model has been used to predict the fracture resistance behavior of double-edged-notched-tensile specimens made from Zircaloy-4 material. Initially, the micro-mechanical parameters have been determined from the testing of ring-type specimens. Subsequently, these parameters were used in finite element analysis of the double-edged-notched-tensile specimen in order to predict the crack growth behavior and the crack path under applied displacement-controlled loading conditions. The fracture resistance behavior obtained in terms of J-R curve was also compared with the corresponding J-R curves of axially-cracked pin-loading-tension specimens.

INTRODUCTION

Zirconium alloys are used for manufacturing of various types of components in nuclear reactors, such as fuel-clad tubes, pressure tubes, spacer grids etc., because of the excellent mechanical properties, corrosion resistance behavior and low neutron absorption cross-section of these materials. In order to design for maximum burn-up and higher resident times of fuel-bundles in the nuclear reactor, one needs to know the

DOI: 10.4018/978-1-5225-0588-4.ch009

fracture properties along with the mechanical properties (Leclercq et al., 2008; Link et al., 1998; Daum et al., 2002). Unlike mechanical properties, the fracture initiation toughness and the fracture-resistance curves depend upon the geometry of the components (Needleman & Tvergaard, 1984; Kussmaul et al., 1995; Kim et al., 2016). The fracture behavior of a thin-walled tube is significantly different from that of a thick-walled tube and a thick-walled standard fracture mechanics specimen. Experiments are usually conducted on standard fracture mechanics specimens (thickness being of the order of 25 mm) in the laboratory to determine the J-R curves. However, these thick-walled specimens cannot be machined from the thin-walled tubes, such as nuclear reactor fuel-clad tubes, where the thickness is of the order of 0.4 to 0.9 mm.

It may also be noted that these tubes undergo various mechanical working and heat-treatment processes during their manufacturing process and hence, the fracture properties of as-manufactured fuel-clad tubes are important from the point of view of designers and safety analysts. The main priority in the nuclear industry is the safe, reliable, and economic operation of different types of nuclear reactors. Recently, the development of accident-tolerant fuel has become very important after the 2011 Fukushima accident (Kim et al., 2006, 2007, 2008, 2013, 2016; Youinou & Sen, 2014; Zinkle et al., 2014; Ott et al., 2014; Terrani et al., 2014). The emphasis is to increase fuel safety and reliability during normal operations, operational transients, and also accident events. The microcell UO_2 and high-density composite pellet concepts are being considered in order to achieve enhanced retention capabilities of highly radioactive and corrosive fission products within the fuel-pellets (Kim et al., 2016; Youinou & Sen, 2014; Zinkle et al., 2014). High-density pellets are being planned to be used along with new types of cladding, e.g., surface-modified Zr-based alloy (Kim et al., 2016), SiC composite material (Kim et al., 2013; Stone et al., 2015; Stempien et al., 2013), and oxidation-resistance iron-based alloys (Terrani et al., 2013; Pint et al., 2013) are the important concepts which are being pursued by researchers which can effectively suppress hydrogen explosions and the release of radio-nuclides into the environment.

For safety-critical components made up of ductile materials (e.g., Zircaloy, ferritic pressure vessel steel, austenitic stainless steel, aluminum alloys etc.), finite element (FE) approaches which use material damage constitutive models (Needleman & Tvergaard, 1984; Kussmaul et al., 1995; Pitard-Bouet et al., 1999; Rousselier, 1987; Gurson, 1977; Tvergaard & Needleman, 1984; Rudnicki & Rice, 1975) are extremely useful in predicting the failure behavior and the corresponding process with high accuracy. For structural integrity assessment accounting for design-basis accident conditions (such as loss-of-coolant-accident and reactivity-insertion-accidents), the fracture resistance data in terms of J-R curves are also required to be evaluated at different temperatures.

Continuum damage mechanics approaches (Rousselier, 1987; Gurson, 1977) offer an impressive alternative to predict the J-R curves of the components with various postulated cracks and loading conditions. In the last two decades, damage mechanics based models like Rousselier's model (Rousselier, 1987), Gurson's model (and its modified versions) (Gurson, 1977; Tvergaard & Needleman, 1984) have become powerful tools in the safety analysis of nuclear reactor components, e.g., pressure vessels, shell-nozzle junctions, pipings, headers and elbows etc. The above reactor components are usually made up of low-alloy ferritic grade steels and austenitic stainless steels. However, the fuel-clad tubes and pressure tubes are made up of Zirconium-based alloys and the study of their fracture behavior in the context of damage mechanics has not received much attention in literature.

These damage models predict the material behavior on the basis of the micromechanical processes (i.e., void nucleation, growth and coalescence) leading to ductile fracture. In ductile materials, voids tend to nucleate in the regions of plastic strain localization in the presence of dominant hydrostatic stresses.

These micro-voids grow under the influence of increased loading in terms of stress and strain field and subsequently, the micro-voids coalesce by plastic deformation and necking of the remaining ligament. In this way, micro-cracks can form in the regions of high stress and strain concentration and these tend to grow leading to stable ductile crack growth. All the above processes are usually captured through suitable mathematical forms in the continuum damage mechanics models and hence, these models are able to predict the extent of crack growth and crack path under a given loading condition in a component.

However, the conventional local damage mechanics approaches suffer from the problem of mesh-dependency of the results (Rudnicki & Rice, 1975; Benallal et al., 1993; Bazant & Pijaudier-Cabot, 1988; Geers et al., 1998; Peerlings et al., 2002; Reusch et al., 2003a,b; Samal et al., 2008, 2010a,b, 2011) and it does not allow the use of small size elements near the crack-tip (for a converged solution) to simulate the presence of large stress gradients. Nonlocal regularization of the material state variables can alleviate this problem and this has been investigated by various researchers over the years (Peerlings et al., 2002; Reusch et al., 2003a,b; Samal et al., 2008, 2010a,b, 2011). Recently, the author has developed a nonlocal version of the Rousselier's damage model and showed that the results of this model are mesh-independent (Samal et al., 2008). For evaluation of mechanical and fracture properties of Zirconium alloys under different environment and loading conditions conditions (including hydriding atmosphere, irradiation, corrosive water environments etc.), ring type specimens and axially-cracked tubular specimens have been very popular and widely used in literature (Jiang et al., 2008; Ishida, 1984; Yamada et al., 1977; Reddy & Reid, 1979; Laiarinandrasana et al., 2011; Arsène & Bai, 1996,1998; Janiszewski, 2012; Bertolino et al., 2003; Grigoriev & Jakobsson, 2005; Shah et al., 2011; Sanyal et al., 2012; Nemat-Alla, 2003; Desquines et al., 2011; Kaynak et al., 2005; Wang et al., 2002; Price, 1972). Various types of loading configurations, such as, ring-tension, ring-compression etc. have been used by researchers in order to evaluate the mechanical properties of thin-walled tubular components. However, research efforts regarding fracture resistance evaluation of the Zircaloy components using continuum damage mechanics models are virtually absent in literature.

In this work, the author has used the nonlocal version of Rousselier's damage model in the finite element analysis in order to simulate the evolution of damage in ring-type specimens machined from Zircaloy-4 fuel-clad tubes of Indian pressurized heavy water reactors (PHWRs). The most important variables in the micro-mechanical model are the micro-mechanical parameters of the model and these have been determined for the material from experiments carried out on ring-type tensile specimens. These parameters have been used in the nonlocal Rousselier's damage model in order to predict the fracture resistance behavior of a standard double-edged-notched-tensile (DENT) specimen. In order to validate the predicted results, experiments have been conducted on axially-cracked Pin-Loading-Tension (PLT) type of specimens machined from the same Zircaloy-4 fuel-clad tubes. The fracture resistance curve for these PLT specimens have been evaluated using a load-normalization technique. Finally, the predicted result of J-R curve of the DENT specimen, as obtained from the nonlocal damage model, has been compared with that of the PLT specimen. The fracture resistance behaviors of the above specimens were also compared with data of similar material from literature.

The chapter is divided into six sections. The information regarding the material used in this work, experimental details on testing of ring-type specimens etc. are presented in Section-2 and these are followed by a brief description of the nonlocal damage model in Section-3. The details of FE analysis, results on load-deformation and fracture resistance behavior of DENT fracture specimens are presented in Section-4. The results of FE analysis of axially-cracked tubular specimens in Pin-Loading-Tension test setup is presented in Section-5. The discussion regarding comparison of fracture resistance behavior of various types of specimens is presented in Section-5. The concluding remarks and the scope of future research are presented in Section-6.

MATERIAL, SPECIMEN, AND EXPERIMENTAL DETAILS FOR RING SPECIMENS

Ring tensile test setup has been modeled in this work to determine the micromechanical parameters. The ring-type specimens have been directly machined from the thin-walled fuel-clad tubes of 220 MWe Indian pressurized heavy water reactors (PHWRs) and tested in a tensile mode. The experimental data are taken from (Shah et al., 2011) and details of the tests are described in (Shah et al., 2011). These test-data have been used subsequently by the nonlocal damage model in order to estimate the micro-mechanical parameters which can be used in prediction of crack growth behavior of a standard DENT specimen. The 220 MWe Indian PHWRs use 19-element fuel-bundle as shown in Figure 1. The internal diameter of each fuel-clad tube is of the 14.4 mm and the thickness is of the order to 0.4 mm, whereas the length is of the order of 495 mm. Fuel-pellets are introduced into these tubes and these are arranged in the shape of a fuel-bundle, which are ready to be introduced into the pressure tubes of the PHWRs.

The material of these fuel-clad tubes is stress-relief-annealed Zircaloy-4. The chemical composition of the material is shown in Table 1. The strength and ductility values of Zircaloy-4 at room temperature are provided in Table 2.

Figure 1. Photograph of 19-pin fuel-bundle used in Indian PHWRs

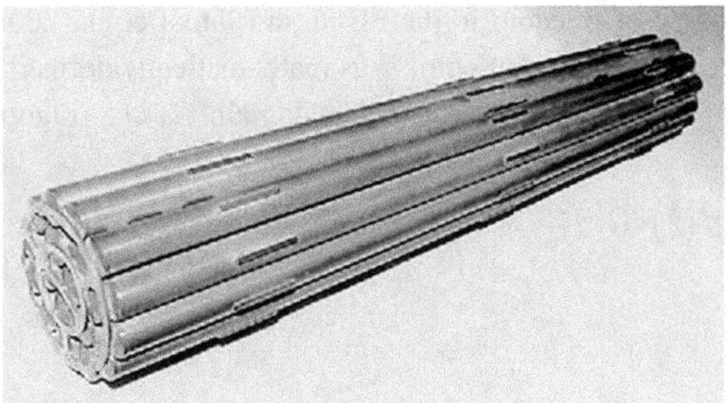

Table 1. Chemical compositions of Zircaloy-4

Element	Sn (%)	Fe (%)	Cr (%)	Ni (%)	O (%)	Zr (%)
Wt. %	1.50	0.21	0.10	0.007	0.12	Rest

Table 2. Strength and ductility values of Zircaloy-4 at room temperature

Yield Stress (MPa)	UTS (MPa)	Uniform Ductility	% Elongation
530	680.7	0.065	28

Experiments have been carried on ring-type specimens. These have been machined from Zircaloy-4 fuel-clad tubes and a typical specimen is shown in Figure 2(a). The specimen width is of the order of 3 mm. The loading mandrel consists of two semi-cylindrical halves attached to two prismatic loading rods as shown in Figure 2(b). The ring-type specimen has been loaded into the loading grip and the arrangement is shown in Figure 2(c). The loading rods are in turn attached to the loading frames of the machine and the tests are conducted in a tensile mode. The load-displacement curves as obtained from the test (after necessary correction for the machine compliance) have been used in subsequent FE analysis in order to evaluate the micro-mechanical parameters for Zircaloy-4.

FINITE ELEMENT ANALYSIS OF DUCTILE FRACTURE USING NONLOCAL DAMAGE MODEL

The conventional damage mechanics models are also known as local damage models as the material damage parameter is dependent upon the stress and strain fields at a given material point of interest. However, it is well know that diffusion of damage (in terms of vacancies and motion of dislocations) is a common phenomenon in ductile materials and hence, there occurs considerable redistribution of plastic strain and damage around a stress-concentration region. Hence, the development of material damage is a nonlocal phenomenon and the damage parameter is associated with a characteristic length scale in the material. The local damage models suffer from the pathological mesh-dependency and hence, these are being replaced with sophisticated nonlocal versions of the damage mechanics based models. In a previous work, the author has developed a nonlocal formulation of Rousselier's model using nonlocal damage d as a nodal degree of freedom in the FE mesh (Samal et al., 2008). The increment of the nonlocal damage variable $\dot{d}$ in a material point $\vec{x}$ is mathematically defined as a weighted average of the increment of the local void volume fraction $\dot{f}$ in a domain Ω, i.e., [Figure 3]

$$\dot{d}\left(\vec{x}\right)=\frac{1}{\Psi\left(\vec{x}\right)}\int_{\Omega}\Psi\left(\vec{y};\vec{x}\right)\dot{f}\left(\vec{y}\right)d\Omega\left(\vec{y}\right) \tag{1}$$

Figure 2. Sketch of the ring specimen, mandrel and the test setup

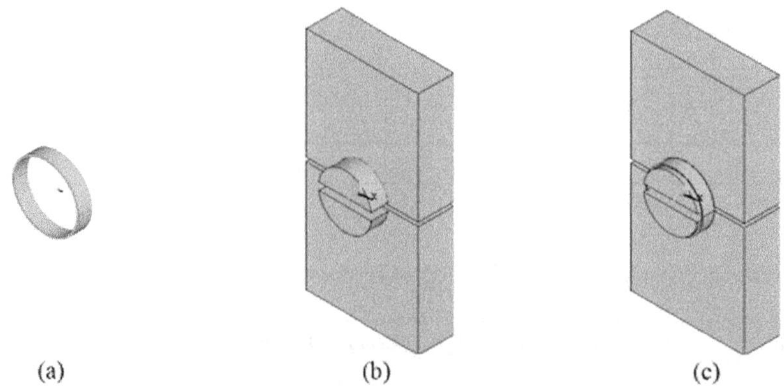

Figure 3. Nonlocalization of the rate of damage variable as a function of rate of local void volume fraction $\dot{f}$ in a domain Ω

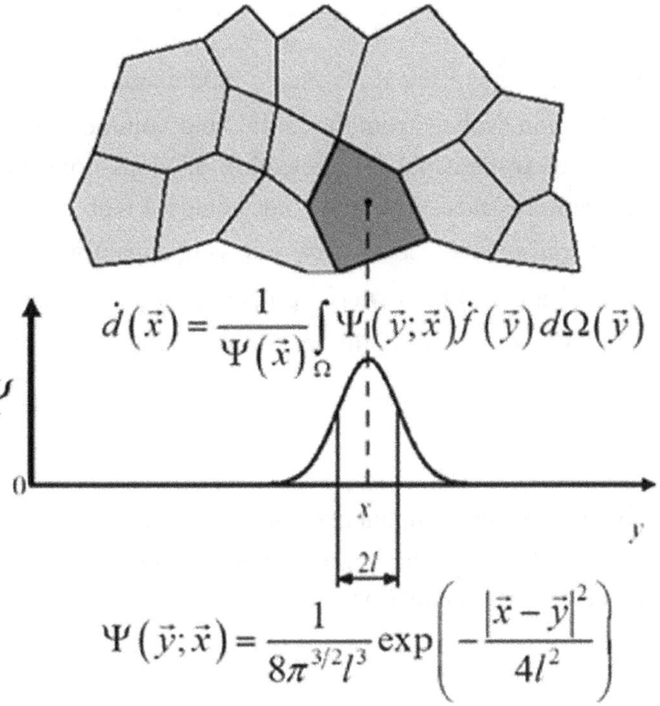

where $\vec{y}$ is the position vector of the infinitesimally small volume $d\Omega$ and $\Psi\left(\vec{y};\vec{x}\right)$ is the Gaussian weight function given by

$$\Psi\left(\vec{y};\vec{x}\right) = \frac{1}{8\pi^{3/2}l^3}\exp\left(-\frac{\left|\vec{x}-\vec{y}\right|^2}{4l^2}\right) \tag{2}$$

The length parameter l in Equation 2 determines the size of the volume, which effectively contributes to the nonlocal quantity and is related to the scale of the microstructure. The above integral nonlocal kernel holds the property that the local continuum is retrieved if $l \to 0$.

By expanding $f\left(\vec{y}\right)$ in Taylor's series and substituting in Equation 1 and doing some algebra, one can obtain the damage diffusion equation as

$$\dot{d} - \dot{f} - C_{\text{length}}\nabla^2\dot{d} = 0 \tag{3}$$

where C_{length} is the characteristic length parameter of the material (Samal et al., 2008). The yield function of the Rousselier's model (Rousselier, 1987) is modified by substituting the nonlocal damage d in place of the local ductile void volume fraction f as (Samal et al., 2008)

$$\phi = \frac{q}{1-d} + D\sigma_k d \exp\left(\frac{-p}{(1-d)\sigma_k}\right) - R(\varepsilon_{eq}) = 0 \tag{4}$$

where D and σ_k are the parameters of the Rousselier's model and are constants for a material. With loading, the void volume fraction evolves from the initial void volume fraction f_0 (volume fraction of eligible second phase particles responsible for nucleation of voids upon plastic deformation) in the material. At a critical void volume fraction f_c, the voids coalesce with each other and at the final void volume fraction f_f, the material points lose its stress carrying capability. Hence, the above three void parameters (i.e., f_0, f_c and f_f) are also the material properties of the damage model. The void growth rate is obtained using the plastic incompressibility condition of the matrix material as

$$\dot{f}_{growth} = (1-f)\dot{\varepsilon}^p_{kk} \tag{5}$$

where $\dot{\varepsilon}^p_{kk}$ is the change in hydrostatic component of the plastic strain tensor. For solving the boundary value problem of the nonlocal damage continuum, one needs to solve the partial differential equation along with the mechanical equilibrium equation

$$\nabla.\sigma + F_b = 0 \tag{6}$$

and the associated boundary conditions, where σ is the Cauchy stress tensor and F_b is the body force per unit volume. For the nonlocal damage degree of freedom, the additional (i.e., Neumann) boundary condition is used and is expressed as

$$\nabla \dot{d}.n\big|_{\Gamma_d} = 0 \tag{7}$$

where $n\big|_{\Gamma_f}$ is the normal to the boundary Γ_f. By discretizing the weak forms of the governing differential equations and, we obtain the FE equations in matrix form as (Samal et al., 2008)

$$\begin{bmatrix} K_{uu} + K_{NL} & K_{ud} \\ K_{du} & K_{dd} \end{bmatrix} \begin{Bmatrix} \Delta\hat{u} \\ \Delta\hat{d} \end{Bmatrix} = \begin{Bmatrix} f_m^{ext} - f_m^{\text{int}} \\ -f_d^{\text{int}} \end{Bmatrix} \tag{8}$$

It may be noted that the stiffness terms K_{ud}, K_{du} and K_{dd} in the element stiffness matrix are contributions of the nonlocal damage mechanics formulation. The matrix K_{uu} represents the conventional stiffness of the finite element (with displacements u, v and w as the degrees of freedom of the element) and it represents the mechanical stiffness, which corresponds to the relationship of the nodal mechanical forces and nodal displacement vectors.

Similarly, K_{ud} represents the matrix, which produces nodal damage vector corresponding to applied mechanical force vector. The matrix K_{du} produces nodal displacement vector corresponding to an applied damage force vector and K_{dd} produces nodal damage vector corresponding to applied damage force vector respectively. f_m^{int} and f_m^{ext} are the internal and external mechanical force vectors respectively, whereas f_d^{int} is the internal damage force vector. The above-described nonlocal formulation of the Rousselier's damage model was implemented in an in-house FE-based code and used for numerical simulation of fracture behaviour of various types of specimens as discussed in the later sections.

RESULTS OF FE ANALYSIS FOR RING TENSILE AND FRACTURE SPECIMENS OF ZIRCALOY-4

For finite element analysis of Zircaloy-4 ring-type specimens as shown in Figure 3(a), a 3D finite element model of the test setup was prepared. By considering the symmetry of the ring-specimen from the point of view of material, loading and boundary conditions, only 1/8th of the specimen and the loading grips were modeled. The material properties required for FE analysis are in terms of true stress vs. true plastic strain curve and the same for the Zircaloy-4 at room temperature is shown in Figure 4. The yield stress, ultimate tensile stress, uniform ductility and % elongation for the material at room temperature are shown in Table 2. The Young's modulus and Poisson's ratio of the material are provided in Table 3.

Figure 4. Material stress-strain curve of Zircaloy-4 used in FE analysis

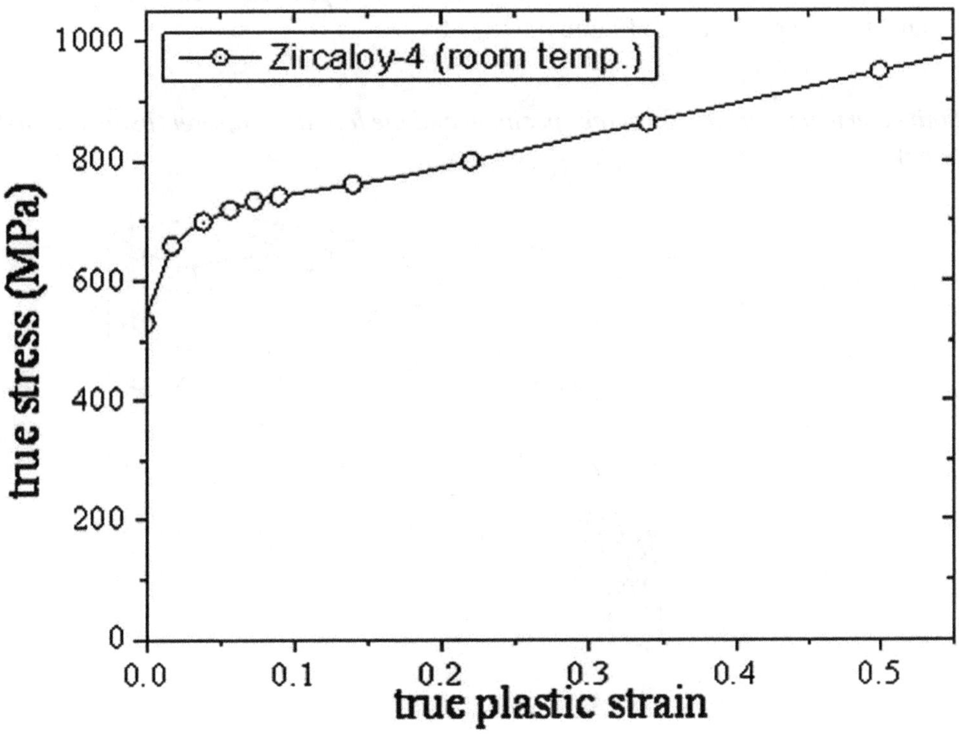

Table 3. Material properties of Zircaoy-4

Rousselier's Constants		Void Parameters			Char. Length Parameter	E (Young's Mod. in GPa)	ν (Poisson's Ratio)
D	σ_k (MPa)	Initial Void Volume Fraction f_0	Void Volume fraction at Coalescence f_c	Final Void Volume Fraction f_f	C_{length} (mm²)		
2	600	0.001	0.05	0.3	0.05	90	0.35

The ring-type specimens were loaded in tensile mode till they fractured after severe necking. The fracture surface was examined under scanning electron microscope (SEM) in order to understand the microscopic mode of fracture. It may be noted that the fracture surface is filled with numerous microscopic holes or voids and the ligament between the voids have undergone severe plastic deformation and inter-void necking. This explains that the material is highly ductile at room temperature and the ductile fracture process involves the stages of void nucleation, growth and coalescence. The voids usually nucleate at the interface of matrix and intermetallic particle phases in this material and hence, the volume fraction and the distribution of these particles control the ductility, extent and stability crack growth and fracture resistance behavior of these alloys.

In order to use the continuum damage mechanics models for simulation of crack initiation and propagation in these materials, the results from the experiments on ring-type specimens were used. The 3D FE model of the test setup (considering symmetry) is shown in Figure 5. Analysis was carried out using elastic-plastic material constitutive models (without considering evolution of material damage) and continuum damage mechanics based models. The load-displacement responses as obtained from both type of analysis are presented in Figure 6.

Figure 5. Finite element mesh of 1/8th of the specimen and the loading mandrel (using symmetric boundary conditions)

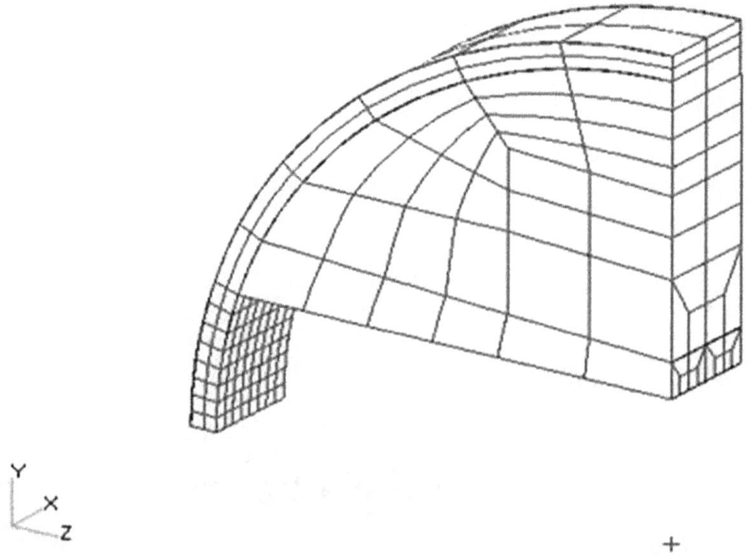

Figure 6. Load-displacement curve of ring-specimen machined from fuel-clad tube (elastic-plastic vs. damage mechanics analysis)

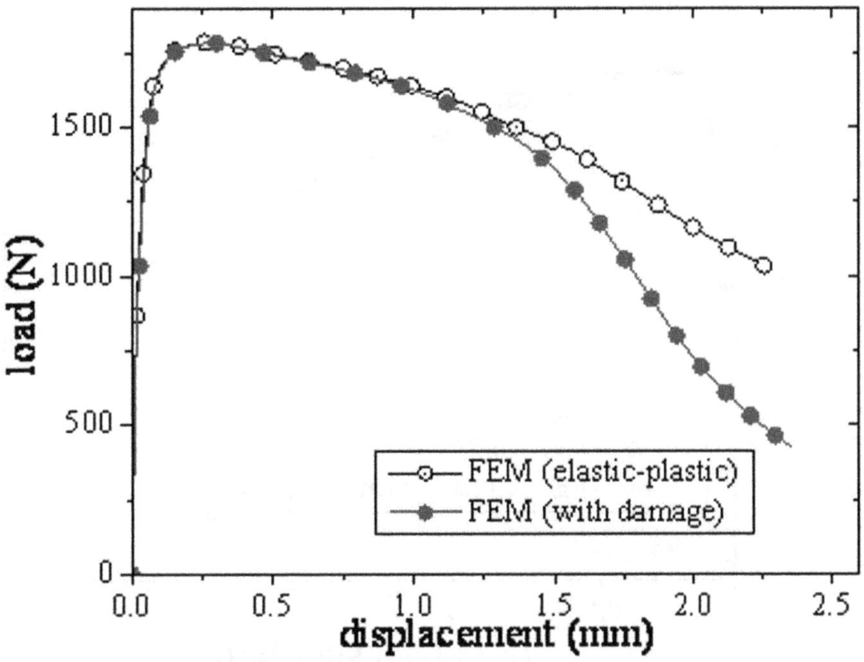

It may be noted that the load-displacement response remains almost same till material damage develops in the centre of necked-region of the specimens. Once damage starts growing due to micro-void growth and coalescence, the material stress carrying capability reduces drastically. This in turn results in drastic load-drop in case of the analysis with continuum damage models. From Figure 6, the difference between the load-deformation responses as obtained from both types of analysis can be observed. The load-deformation response as obtained for the FE analysis with nonlocal damage model is realistic and it is able to model the drastic load-drop in the final phases of the loading as observed in experiments.

Hence, one can compare the experimental load-deformation response with the response obtained from FE analysis and optimize the micro-mechanical parameters. Such a procedure was followed in this work and the micro-mechanical parameters of the Rousselier's model were obtained. These parameters are presented in Table 3. It may be noted that some of the parameters, e.g., D, f_c and f_f are usually constant for a class of ductile materials and these values have been taken from previous experience of FE analysis (Samal et al., 2008). Figure 7 shows the comparison of load-displacement response as obtained from FE analysis and those from experiments.

In order to validate the FE analysis, another set of experiments were carried out on ring tensile specimens and the same were analyzed by the method described earlier. The load-displacement results of analysis along with those of experiment are presented in Figures 8 and 9 for the two different ring tensile specimens. The comparison between experiments and FE analysis is very satisfactory and hence, the micro-mechanical parameters can now be used to simulate the fracture resistance behavior of other types of fracture mechanics specimens made from the same material. This type of approach has the inherent advantages that the parameters are truly material properties and hence, these can be transferred to FE analysis of various other types of components once these are estimated from simple experiments.

Figure 7. Comparison of load-displacement curve as predicted by damage mechanics analysis with experimental data sets

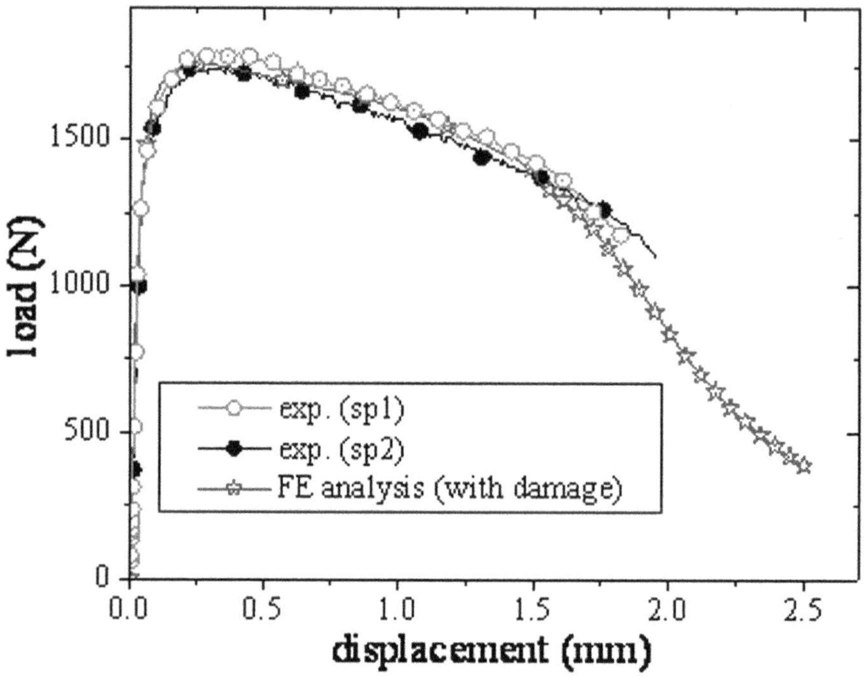

Figure 8. Comparison of load-displacement response as obtained from FE analysis with that of experiment for a ring tension specimen (Sp. No. 2_01)

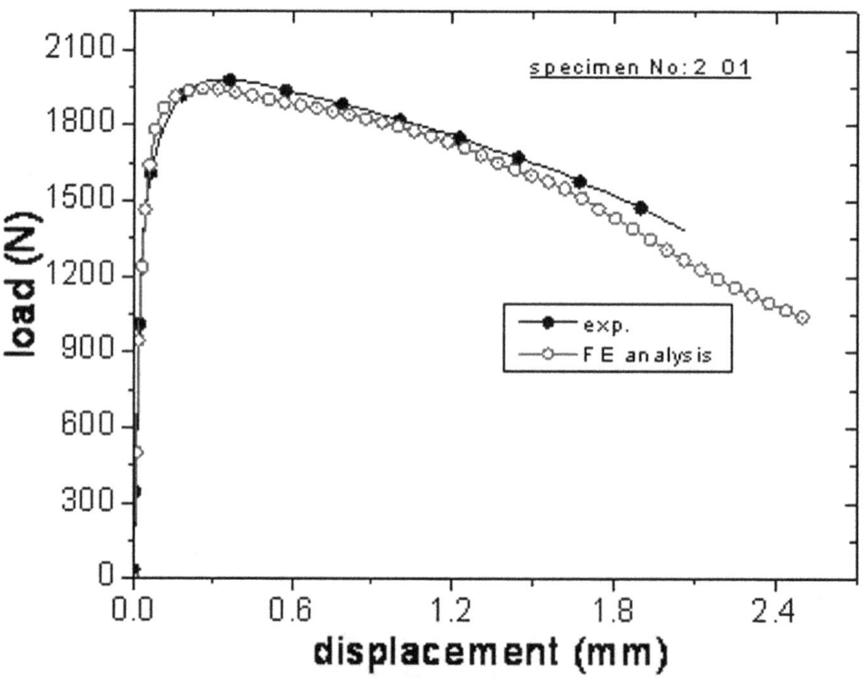

Figure 9. Comparison of load-displacement response as obtained from FE analysis with that of experiment for a ring tension specimen (Sp. No. 2_05)

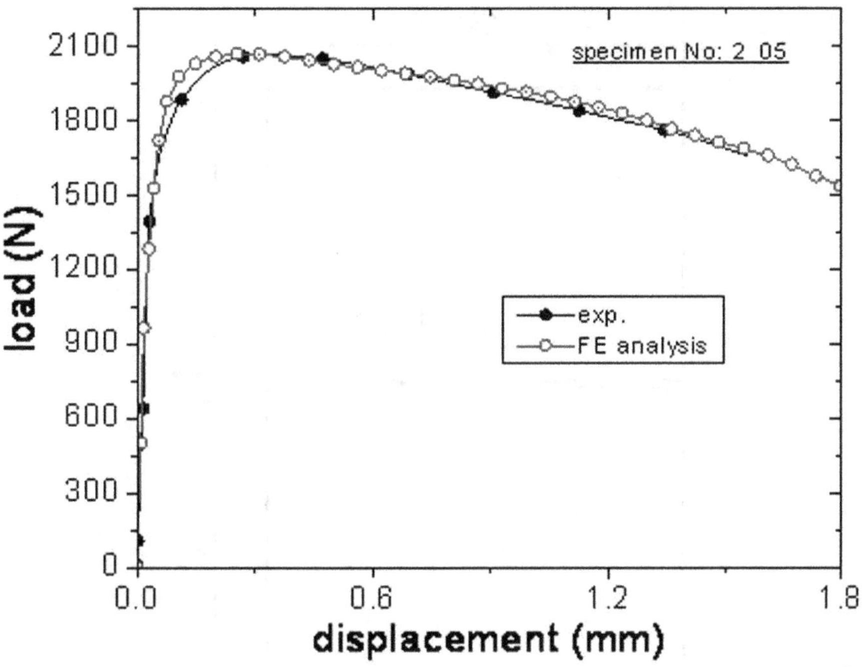

The nonlocal Rousselier's damage model was now used to predict the crack growth, crack path and fracture resistance behavior of a DENT specimen. The dimensions of the DENT specimen are shown in Figure 10. The initial a/W ratio of the specimen is 0.52. Once the specimen is loaded in tensile mode, the cracks on both sides of the specimen start propagating. This is after the process of initiation of cracks from the pre-existing crack-tip following a crack-blunting phase. The specimen thickness is 6.25 mm. For FE analysis, symmetry in geometry, material, loading and boundary conditions have been employed. Hence, $1/4^{th}$ of the specimen have been modeled assuming the state of stress to be of plane strain. This is valid as the thickness of the specimen comparable to other dimensions of the specimen.

Two-dimensional (2-D) plane-strain 8-noded iso-parametric quadrilateral elements have been used in the FE mesh. The FE mesh is shown in Figure 11(a) along with a zoomed portion of the mesh near the crack-tip in Figure 11(b). The mesh is very fine near the crack-tip and it is of the order of 0.05 mm. 3x3 integration points are used in each element and hence, the crack growth modeled is in increments of 0.0166 mm.

It may be noted that the crack initiates from the crack-tip after an applied displacement of 0.14 mm [Figure 12(a)]. Figure 12 (a-f) presents the snapshots of crack growth in the DENT specimen at various values of applied displacements, i.e., (a) 0.14 mm; (b) 0.2 mm; (c) 0.3 mm; (d) 0.4 mm; (e) 0.5 mm; (f) 0.6 mm respectively. The regions of crack growth near the crack-tip are represented by contour of damage with values exceeding 0.4. The corresponding pairs of crack growth and applied displacement values are shown in Figure 13.

From Figure 13, it may be noted that the crack-tip blunting occurs up to applied displacements of 0.14 mm and after that stable crack extension takes place. The corresponding load-displacement response of the DENT specimen is shown in Figure 14. Figure 14 presents the results of FE analysis

Figure 10. Schematic diagram of the DENT specimen used in FE analysis (a=3.25 mm, 2W=12.5 mm. H=50 mm)

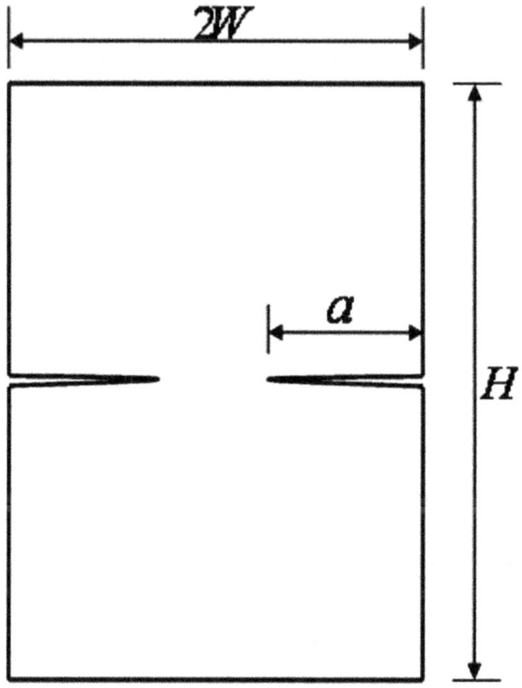

Figure 11. FE-mesh of the DENT specimen and the mesh at the crack-tip

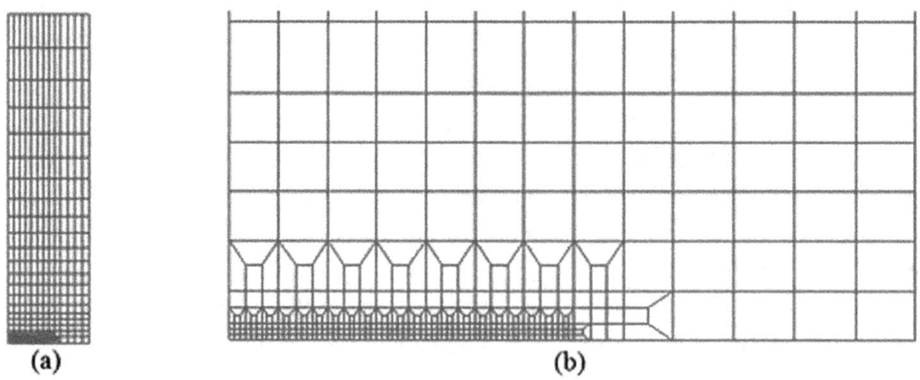

for both elastic-plastic material constitutive models (without accounting for damage) and the nonlocal Rousselier's damage model.

It may be noted that there is a drastic reduction in load-carrying capability of the specimens once damage develops in the crack-tip and hence, the load goes on reducing with more and more crack growth. However, the remaining ligament continues to get damaged in a stable manner and hence, the load-drop is not sudden as experienced by the ring-type specimen. However, the effect of damage on reduction of

Figure 12. Snapshots of crack growth in DENT specimen at various values of applied displacements, i.e., (a) 0.14 mm; (b) 0.2 mm; (c) 0.3 mm; (d) 0.4 mm; (e) 0.5 mm; (f) 0.6 mm

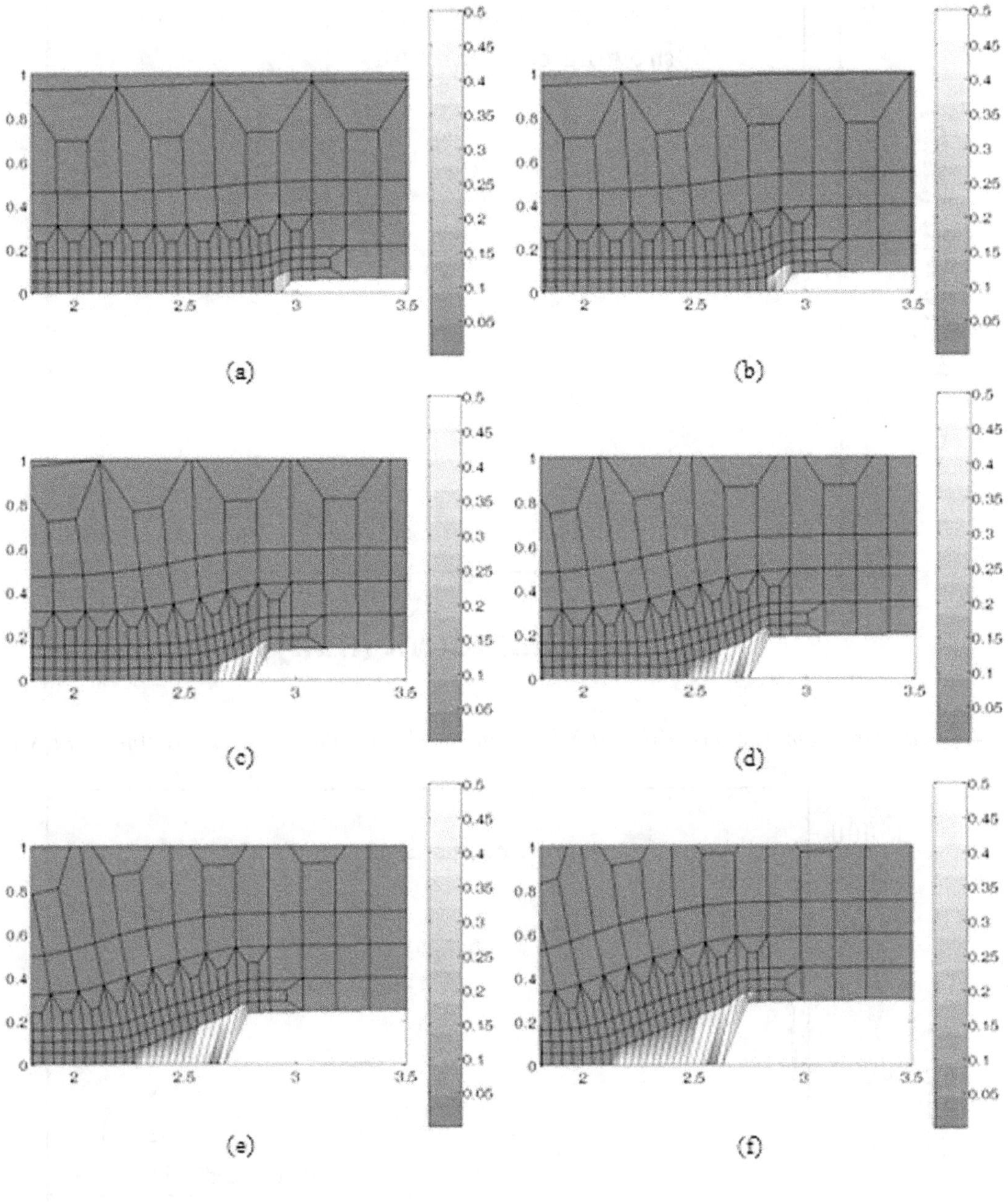

load-carrying capacity is evident in both the cases. The information from Figure 13 has been used along with the area under the load-displacement curve (as presented in Figure 14) to calculate the fracture resistance behavior in terms of J-R curve. The standard η and γ functions of the fracture mechanics procedure have been used to calculate the J-R curve of the DENT specimen and it is presented in Figure 15. It may be noted that the values of J-integral increases rapidly initially due to crack-tip blunting and hence, the initiation value of J-integral can be calculated to be around 100 N/mm (Figure 15). This

Figure 13. Crack growth vs. applied displacement for the DENT specimen

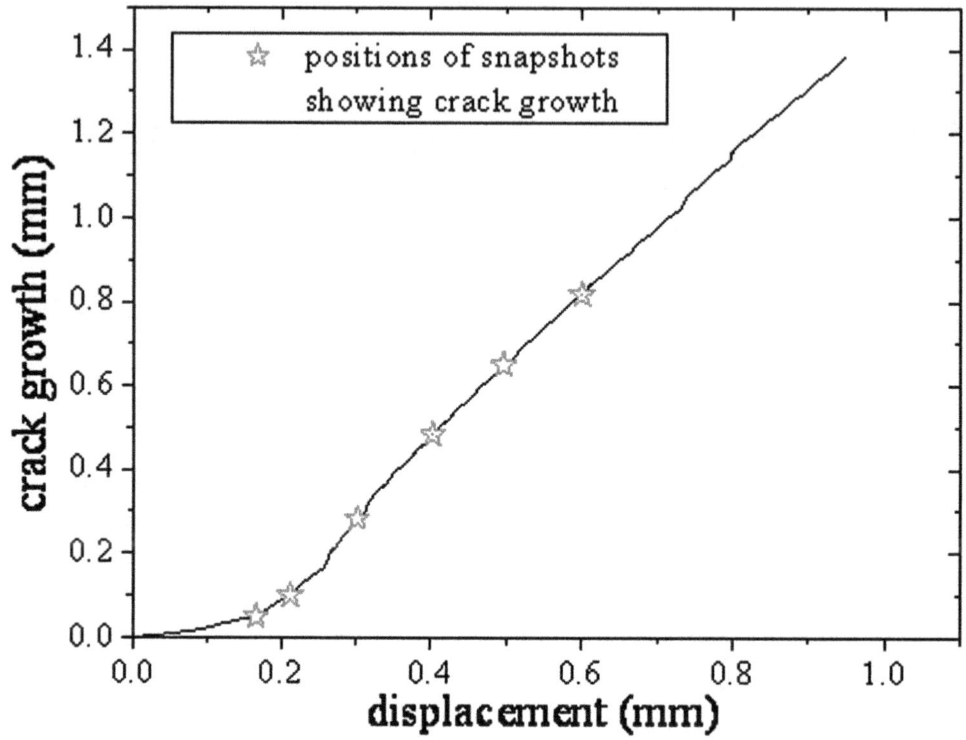

Figure 14. Load-displacement curve of the DENT specimen (elastic-plastic vs. damage mechanics analysis)

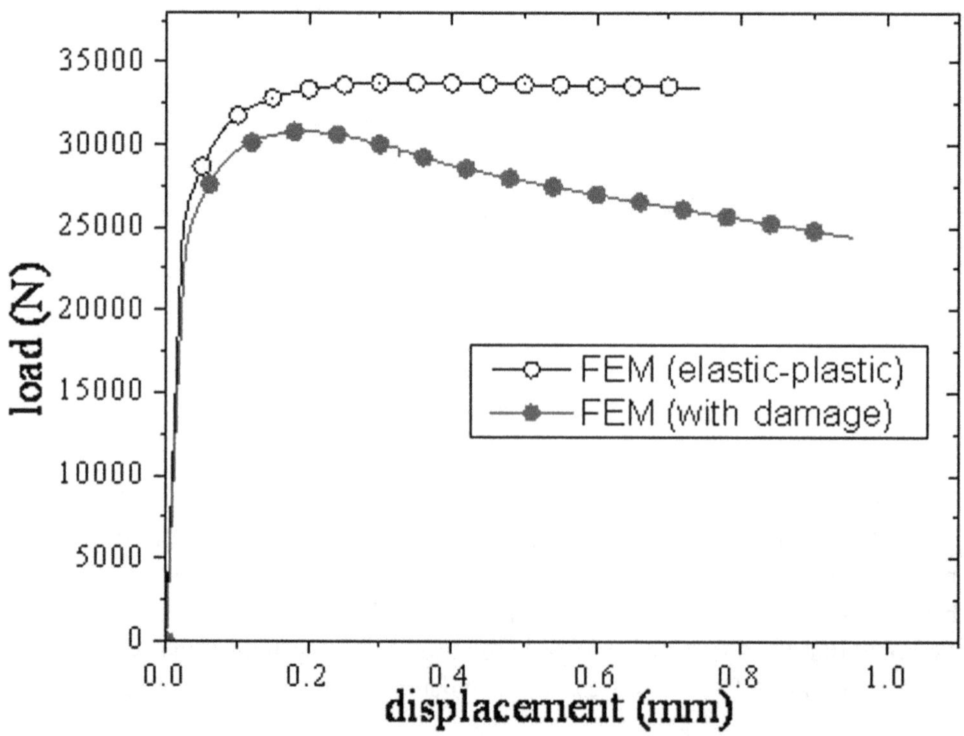

Figure 15. J-R curve of the DENT specimen at room temperature

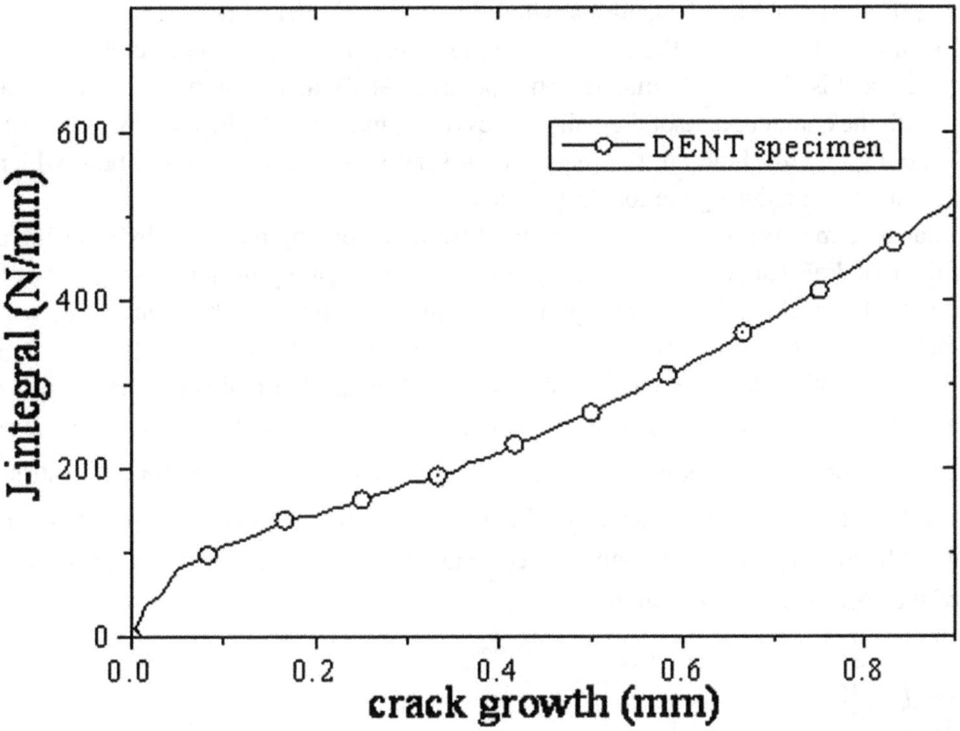

value is consistent with J-initiation values of similar Zirconium alloys from literature (Bertolino et al., 2003). The subsequent J-R curve in the stable crack growth region has a slope (dJ/da) of around 400 N/mm^2.

RESULTS OF FE ANALYSIS OF AXIALLY-CRACKED TUBULAR SPECIMENS IN PIN-LOADING-TENSION TEST SETUP

The fracture resistance behavior of DENT specimens has been compared with those of axially-cracked tubular specimens tested in a Pin-Loading-Tension (PLT) test setup. The details of the test setup are described in (Samal et al., 2010a,b, 2011). The test method and the FE analysis procedure is described here in brief before the results of DENT specimens are compared with those of PLT test specimens. The PLT test specimens are axially-cracked tubular specimens of 13 mm length. The diameter and thickness values are same as that of the ring specimens discussed in the previous section. A set of seven specimens with notch lengths (a_0) varying from 1 mm to 7 mm with an interval of 1 mm has been fabricated for compliance measurements for deriving geometric shape function required for stress intensity function (SIF) calculation. The initial notch length to width $\left(a_0/W\right)$ ratios of these specimens are given as: 0.474, 0.526, 0.579, 0.632, 0.684, 0.737 and 0.789 respectively. Another set of six specimens (i.e. the second set) with constant notch length of 1.5 mm so as to get specimens with almost same a_0/W ratios have been fabricated for the fracture toughness testing. The experimental details are presented in (Sanyal et al., 2012).

The finite element model of the PLT test setup is presented in Figure 16. The loading fixture consists of a pair of split semi-cylindrical mandrels which form the cylindrical holder when put together. The fixture halves are loaded through the pins after being inserted into the tubular cracked specimen. The loading arrangement is shown in Figure 16. The specimen is aligned such that the plane of axial cracks is co-planar with the contact plane of the split-halves of the mandrel. A pin at the back end of the fixture is inserted into a grove machined in the mandrels. It serves as an anchor point about which the semi-cylindrical fixtures rotate during the loading process.

The tubular specimen is prevented from slipping from the loading mandrels by inserting a clip at the rear end of the mandrels (opposite to loading grooves). The loading pins at the right end of the fixtures are connected to the machine frames through the loading attachments such as clevis grips. The fracture resistance behavior for fuel pin specimens made of ductile materials like Zircaloy-4 is represented in terms of J-R (J-integral vs. crack extension) curves. The J-integral contains two terms, i.e., elastic part J_{el} and plastic part J_{pl}. The elastic part is calculated from the expression of linear elastic stress intensity factor K_I (the crack tip is subjected to mode-I loading) as $J_{el} = K_I^2/E$ for plane stress condition (E being the Young's modulus of elasticity which is 90 GPa for zirconium at room temperature). The stress intensity factor, K_I, depends upon applied load P, thickness t, crack length a, width W and geometry of the specimen and is written as

$$K_I = \frac{P}{2t\sqrt{W}} f\left(a/W\right) \tag{9}$$

The effect of finite specimen size is accommodated through the term, $f\left(a/W\right)$, known as 'geometric shape function'.

Figure 16. Finite element model of the pin-loading tension test setup

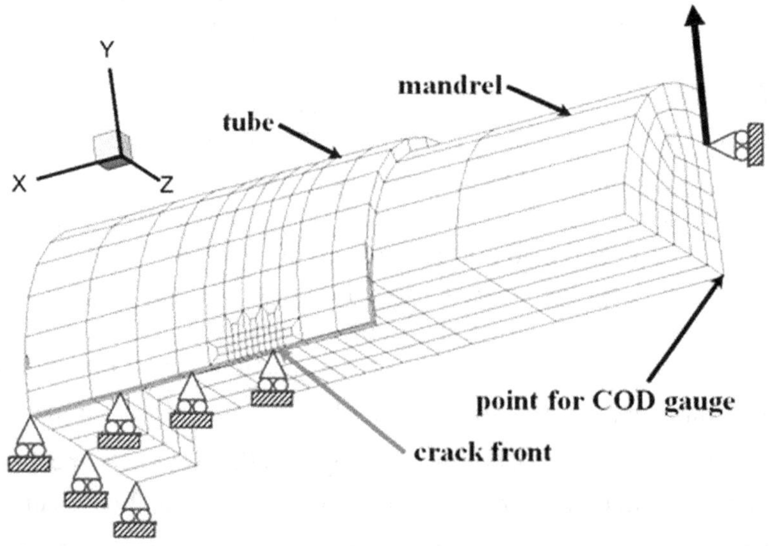

The stress intensity factor K_I for axially cracked fuel pin specimens can also be evaluated from results of finite element (FE) analysis (Samal et al., 2010a). Several methods exist, such as displacement extrapolation, stress extrapolation, J-integral etc. Once the K_I values are numerically evaluated, the function $f\left(a/W\right)$ can be estimated from Equation 9. For this purpose, a 3D FE analysis of the fuel pin specimen along with the loading mandrel has been carried out.

Taking into account of the symmetry in the specimen geometry, crack and loading configuration, one quarter of the test setup has been modeled as shown in Figure 16. A prescribed displacement is applied at the end of the mandrel in y-direction and the movement of the loading point is restricted in the x-direction (i.e., direction of length of the tube) as this point is attached to the loading axis of the machine. The un-cracked ligament of the tubular specimen is fixed in y-direction (however, this is free to move in x and z-directions) in order to take into account of the symmetry condition. The line of the loading pin (which acts like a fulcrum) is also fixed in y-direction only. The symmetric plane of the tubular specimen as well as of the mandrel is fixed in z-direction only in order to consider the symmetric boundary condition. The FE discretization consists of 20-noded iso-parametric brick elements with 3×3×3 Gauss point integration (full integration scheme). Elastic-plastic analysis with von-Mises yield surface and isotropic plastic hardening constitutive equations have been carried out in this work for evaluation of limit load of the specimens which will be discussed later. The analysis has been carried out using an in-house FE code.

For evaluation of stress intensity factor K_I, the displacement extrapolation method has been used. Neglecting higher order terms, the expression for K_I with respect to v (opening displacement ahead of crack tip) can be written for plane stress condition as

$$K_I = \frac{E}{4}\sqrt{\frac{2\pi}{r}}\mathrm{v} \tag{10}$$

In Equation 10, r is the distance of the material point from the crack tip where the displacement v is evaluated from the FE analysis results. For an applied displacement of 0.01 mm, the stress intensity factor K_I has been evaluated as a function of a/W at different points along the crack front (in the thickness direction of the fuel pin specimen). The results are presented in Figure 17. It can be observed that for lower values of a/W, there is significant difference among the values of K_I at different points of crack front (i.e., inner, middle and outer points).

At lower values of a/W, the inner side of the crack front is subjected to higher values of stress intensity factor for the given applied displacement compared to the outer point of the crack front. However, this situation is reversed for a/W values greater than 0.6 approximately. This is because of change in crack-tip loading conditions as the crack length changes. The function $f\left(a/W\right)$ has been indirectly evaluated from the results of Figure 17 using Equation 9. It can be observed that these results compare very well with those obtained by the analytical schemes (Sanyal et al., 2012) used for finding geometric shape function from experiment. Once the solutions for $f\left(a/W\right)$ are known, the stress intensity factor K_I can be evaluated for a given load P and hence the elastic part of J-integral J_{el}.

Figure 17. Variation of linear elastic stress intensity factor with a/W at different points of the crack front across the tube thickness direction for an applied displacement of 0.01 mm, as derived through FEM

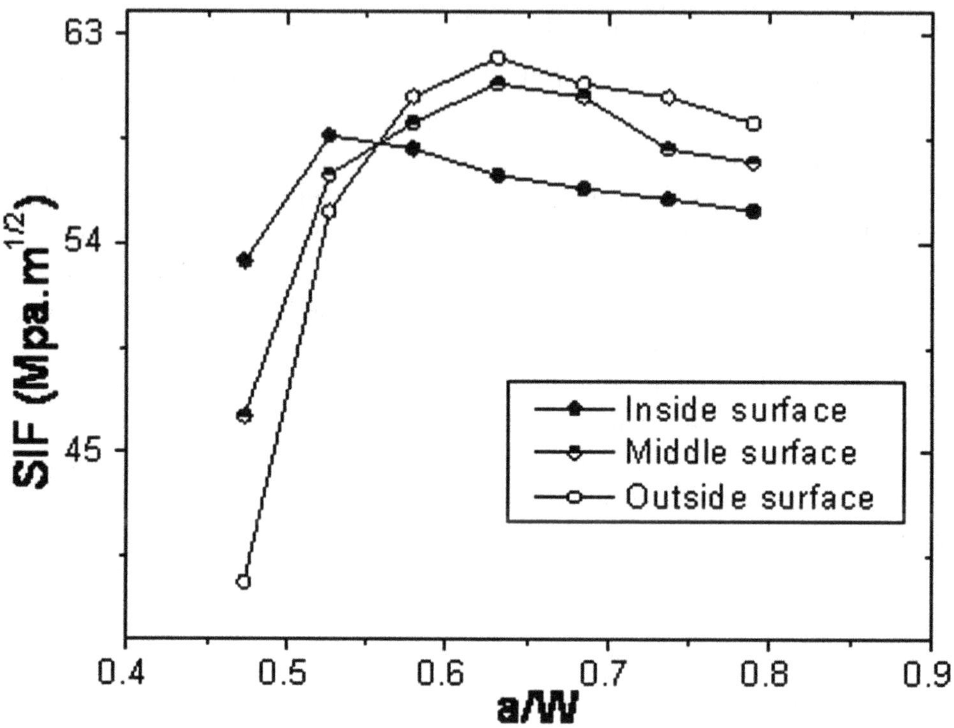

The plastic part of J integral, if required to be evaluated from experimental load-load line displacement data can be derived from the equation (ASTM standard E 1820-01 2000)

$$J_{pl} = \frac{\eta A_{pl}}{b_0 B} \tag{11}$$

where η is a function in a / W, A_{pl} is the plastic area under load- load line displacement curve, b_0 is length of initial unbroken ligament and B is specimen thickness.

From the load-displacement data and crack size estimates at each data point, the crack growth corrected plastic part of *J*-integral J_{pl} at each data point i is evaluated using the following expression (ASTM standard E 1820-01 2000)

$$J_{pl(i)} = \left[J_{pl(i-1)} + \left(\frac{\eta_{(i-1)}}{b_{(i-1)}} \right) \frac{A_{pl(i)} - A_{pl(i-1)}}{2t} \right] \left[1 - \gamma_{(i-1)} \frac{a'_{(i)} - a'_{(i-1)}}{b_{(i-1)}} \right] \tag{12}$$

with γ as the correction factor for a growing crack situation.

The η factor may also be defined as a non-dimensional parameter that takes into account of the effect of geometry and loading condition at the crack-tip of a cracked specimen on the differential energy that is required for crack growth. It can be derived as a function of limit load function $F_L\left(a/W\right)$ as (Samal et al., 2010b)

$$\eta = -\left(1 - \frac{a}{W}\right)\frac{1}{F_L}\frac{\partial F_L}{\partial\left(a/W\right)} \tag{13}$$

γ factor can also be defined as a non-dimensional parameter that accounts for correction in the J-integral calculation due to crack propagation, which is otherwise defined for a stationary crack. The expression for γ factor as a function of $\eta\left(a/W\right)$ can be written as (Samal et al., 2010b)

$$\gamma = \eta - 1 - \left(1 - \frac{a}{W}\right)\frac{1}{\eta}\frac{\partial \eta}{\partial\left(a/W\right)} \tag{14}$$

For evaluation of the above η and γ functions, the limit load solutions are required for the different geometries which are obtained using an elastic-plastic FE analysis. The following material properties (of SRA Zircaloy-4 at room temperature) have been used for elastic-plastic FE analysis, i.e., yield stress of 535 MPa; ultimate tensile stress of 733 MPa and plastic strain hardening exponent of 0.1135.

The limit load F_L has been obtained as the starting load (corresponding to the applied displacement) at which the remaining ligament of the specimen goes into plastic deformation. The limit load decreases with increase in a/W ratio. For the seven specimen tested with different a/W ratios, i.e. first set of specimens, the limit load values obtained by FE analysis are: 1626 N, 1354 N, 1157 N, 946 N, 735 N, 552 N and 280 N for the a/W values of 0.474, 0.526, 0.579, 0.632, 0.684, 0.737 and 0.789 respectively. The variation of limit load F_L with a/W ratio is almost linear with a slope of -4114.9 N and a y axis intercept value (i.e., corresponding to a/W equals to zero) of 3548.9 N respectively. Using these data, the expressions for $\eta\left(a/W\right)$ and $\gamma\left(a/W\right)$ have been derived using Eqs. and respectively.

The experimental load-displacement curves (Sanyal et al., 2012) of the PLT specimens have been normalized in order to obtain a normalized load-displacement curve which can be later used to calculate the crack growth during loading of these specimens using load-normalization technique (Sharobeam & Landes, 1991, 1993). The load-normalization procedure followed in this work is described here briefly. The load P_i at any point *i*, up to the maximum load is converted to a normalized load P_{Ni} as (Sharobeam & Landes, 1991, 1993)

$$P_{Ni} = \frac{P_i}{2tW\left[\left(W - a'_{bi}\right)/W\right]^{\eta}} \tag{15}$$

The term a'_{bi} is the blunting-corrected crack size at the ith data point and is written as

$$a'_{bi} = a'_0 + J_i / 2\sigma_Y \tag{16}$$

where a'_0 is the crack length at the start of the test and σ_Y is the material flow stress (i.e., average of the yield and ultimate stress. Calculation of J_i is based on a'_0. The normalized plastic displacement v_{pli} at point i is evaluated as

$$v'_{pli} = \left(v_i - P_i C_i\right) / W \tag{17}$$

where C_i is the elastic compliance of the specimen based on the blunting corrected crack size a'_{bi}. The final load-displacement pair corresponding to the end of loading is also normalized using the same equations as described above except that the final measured crack size a'_f (after the specimen is break-open) is used instead of the crack size evaluated using the normalization curve.

These normalized load-displacement data points are fitted with the following standard normalization function, i.e.,

$$P_N = \frac{a + b\mathrm{v}'_{pl} + c\mathrm{v}'^2_{pl}}{d + \mathrm{v}'_{pl}} \tag{18}$$

where *a*, *b*, *c* and *d* are fitting coefficients. Using these values of the coefficients, the crack growth information during the course of loading are evaluated. The process involves an iterative procedure.

The J-R curves for axially-cracked tubular specimen in PLT test setup were evaluated using the standard approach of area under load-displacement diagram and the data are presented in Figure 18. The J-R curve as obtained from continuum damage mechanics analysis of DENT specimen is also presented in Figure 18 along with the data from PLT specimens. It may be noted that the initiation values of fracture toughness are very similar for both types of specimens and the curves are close to each other up to a crack growth value of 0.2 mm. This value is typically considered to be region of crack initiation after blunting of the crack-tip and the corresponding J-value can be considered to be the initiation fracture toughness.

This value of initiation value of fracture toughness seems to be independent of specimen type and hence, it may be considered as representative of the Zircaloy-4 at room temperature. The slope of J-R curve for the DENT specimen is however small compared to that of the PLT specimens. It may be noted that the state of stress at the crack-tip of the DENT specimen corresponds to that of plane-strain, whereas the state of stress at the crack-tip of the PLT specimen is more of plane-stress nature due to small thickness of the tube (~0.4 mm). As expected, the fracture resistance slopes are higher for a specimen with crack-tip loaded in plane stress compared to that of a specimen with crack-tip loaded in plane-strain.

Nevertheless, the J-R curve of the DENT specimen as predicted by the nonlocal damage model can be considered as a conservative value for structural integrity analysis of these fuel-clad tubes. The J-R

Figure 18. Comparison of J-R curves of different types of fracture specimens machined from Zircaloy-4 fuel clad tubes

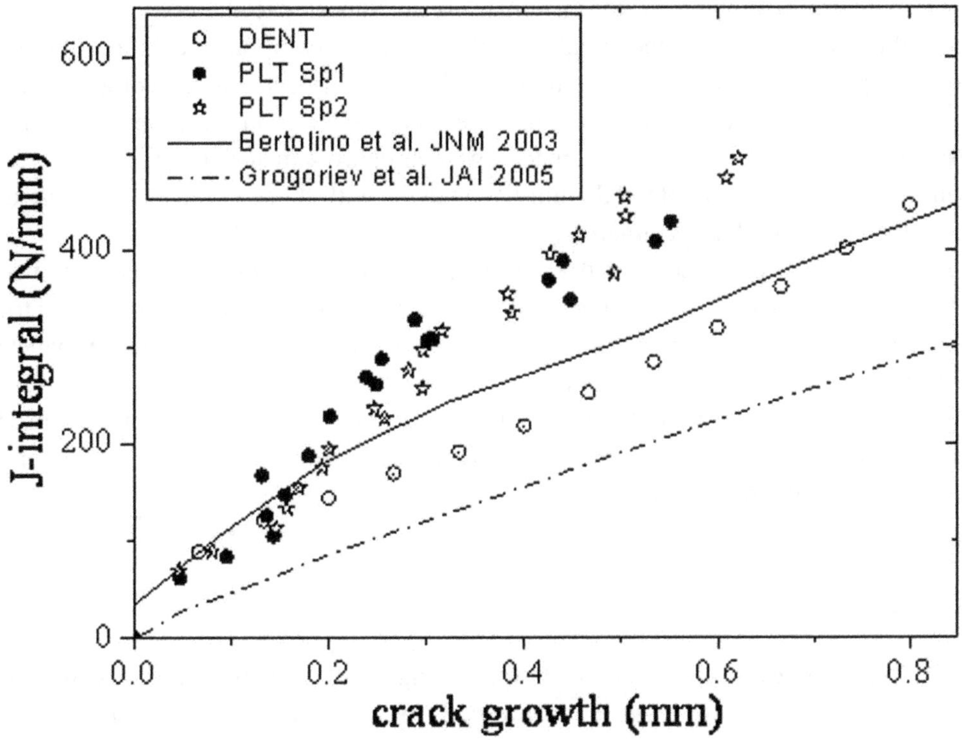

curves of PLT and DENT specimens were also compared with similar data from literature (Bertolino et al., 2003; Grigoriev & Jakobsson, 2005). It may be noted that J-R curve as obtained from the current FE analysis of DENT specimen is close to that of reported values in literature.

The close agreement between the results of finite element analysis and experiment gives confidence in the use of nonlocal damage mechanics based models in prediction of crack growth and crack paths in various types of Zirconium alloy based nuclear reactor components. Nevertheless, the method discussed in this chapter is applicable for other materials commonly used in engineering applications such as aluminum alloy, low alloy steel, ferritic steel, stainless steel etc. to name a few. The method to evaluate the fracture properties for thin-walled tubular components is also applicable for similar components in power plants, refineries, chemical, aerospace and automobile industries etc.

For other load cases such as bending, compression and torsion, the design and test setup will be different. The modifications have to be carried out in the test setup such that the crack-tip is loaded predominantly in the stress-state relevant to the application. For mechanical properties, the specimens in compressions and torsion are usually solid cylindrical specimens with relevant diameter to length values. Mechanical properties in bending can be known once we know the mechanical properties in tension and compression and hence, no special design of specimens is required for bending. However, for evaluation of fracture properties, specimens such as three-point and four-point cracked bend specimens can be machined from the components and tested in test setup which are discussed in relevant ASTM standards.

CONCLUSION

Continuum damage mechanics based nonlocal models are superior to the conventional local damage approaches as the results are mesh-independent and it is well known that prediction of extent of crack growth and crack path in components using these models are very accurate and close to that of experimental observation. In this work, it was demonstrated that such a model can be used in prediction of fracture resistance behavior of components made up of Zircaloy-4 material. The results were validated by comparing the J-R curves as obtained from FE analysis with those obtained from the use of load-normalization technique in PLT specimens.

The micro-mechanical parameters of the Rousselier's damage model were estimated from simple experiments on ring-type specimens. As the parameters are transferable, these can be used to predict the fracture resistance behavior of components with postulated cracks made from these materials. It may be mentioned that these models are of significant interest to the engineers associated with design and safety analysis of high burn-up fuel-clad tubes of nuclear reactors. These may also be used in design of thin-walled components in various other engineering applications such as steam generator and heat-exchanger tubes in thermal power plants.

REFERENCES

Arsène, S., & Bai, J. (1996). A new approach to measuring transverse properties of structural tubing by a ring test. *ASTM Journal of Testing and Evaluation*, *24*(6), 386–391. doi:10.1520/JTE11461J

Arsène, S., & Bai, J. (1998). A new approach to measuring transverse properties of structural tubing by a ring test–experimental investigation. *ASTM Journal of Testing and Evaluation*, *26*(1), 26–30. doi:10.1520/JTE11966J

Bazant, Z. P., & Pijaudier-Cabot, G. (1988). Nonlocal continuum damage, localization instability and convergence. *Journal of Applied Mechanics*, *55*(2), 287–293. doi:10.1115/1.3173674

Benallal, A., Billardon, R., & Geymonat, G. (1993). Bifurcation and localization in rate-independent materials. Some general considerations. In Q. S. Nguyen (Ed.), *Bifurcation and Stability of dissipative Systems 1993* (pp. 1–44). Berlin: Springer Verlag. doi:10.1007/978-3-7091-2712-4_1

Bertolino, G., Meyer, G., & Perez, I. J. (2003). Effects of hydrogen content and temperature on fracture toughness of Zircaloy-4. *Journal of Nuclear Materials*, *320*(3), 272–279. doi:10.1016/S0022-3115(03)00193-4

Daum, R. S., Majumdar, S., Tsai, H., Bray, T. S., Koss, D. A., Motta, A. T., & Billone, M. C. (2002). Mechanical property testing of irradiated Zircaloy cladding under reactor transient conditions. In M. Sokolov, J. D. Landes, & G. E. Lucas (Eds.), *ASTM STP 1418. Small Specimen Test Techniques: fourth volume*. West Conshohocken, PA, USA: ASTM International. doi:10.1520/STP10821S

Desquines, J., Koss, D. A., Motta, A. T., Cazalis, B., & Petit, M. (2011). The issue of stress state during mechanical tests to assess cladding performance during a reactivity-initiated accident (RIA). *Journal of Nuclear Materials*, *412*(2), 250–267. doi:10.1016/j.jnucmat.2011.03.015

Geers, M. G. D., De Borst, R., Brekelmans, W. A. M., & Peerlings, R. H. J. (1998). Strain-based transient-gradient damage model for failure analyses. *Computer Methods in Applied Mechanics and Engineering*, *60*(1-2), 133–153. doi:10.1016/S0045-7825(98)80011-X

Grigoriev, V., & Jakobsson, R. (2005). Delayed Hydrogen Cracking Velocity and J-Integral Measurements on Irradiated BWR Cladding. *Journal of ASTM International*, *2*(8), 1–16. doi:10.1520/JAI12434

Gurson, A. L. (1977). Continuum theory of ductile rupture by void nucleation and growth: Part-I: Yield criteria and flow rules for porous ductile media.*Transactions of ASME Journal Engineering Materials and Technology* (*Vol. 99*, pp. 2-15). doi:10.1115/1.3443401

Ishida, T. (1984). Ring test of 304 stainless steel tube oxidized in high temperature steam. *Journal of Nuclear Materials*, *125*(1), 33–51. doi:10.1016/0022-3115(84)90513-0

Janiszewski, J. (2012). Ductility of selected metals under electromagnetic ring test loading conditions. *International Journal of Solids and Structures*, *49*(7-8), 1001–1008. doi:10.1016/j.ijsolstr.2012.01.005

Jiang, L., Jonas, J. J., Boyle, K., & Martin, P. (2008). Deformation behavior of two Mg alloys during ring hoop tension testing. *Materials Science and Engineering A*, *492*(1–2), 68–73. doi:10.1016/j.msea.2008.04.028

Kaynak, C., Erdiller, E. S., Parnas, L., & Senel, F. (2005). Use of split-disk tests for the process parameters of filament wound epoxy composite tubes. *Polymer Testing*, *24*(5), 648–655. doi:10.1016/j.polymertesting.2005.03.012

Kim, H.-G., Yang, J.-H., Kim, W.-J., & Koo, Y.-H. (2016). Development status of accident-tolerant fuel for light water reactors in Korea. *Nuclear Engineering and Technology*, *48*(1), 1–15. doi:10.1016/j.net.2015.11.011

Kim, J. H., Lee, M. H., Choi, B. K., & Jeong, Y. H. (2006). Effects of oxide and hydrogen on the circumferential mechanical properties of Zircaloy-4 cladding. *Nuclear Engineering and Design*, *236*(18), 1867–1873. doi:10.1016/j.nucengdes.2006.02.010

Kim, J. H., Lee, M. H., Choi, B. K., & Jeong, Y. H. (2007). Effect of the hydrogen contents on the circumferential mechanical properties of zirconium alloy claddings. *Journal of Alloys and Compounds*, *431*(1-2), 155–161. doi:10.1016/j.jallcom.2006.05.074

Kim, J. H., Lee, M. H., Jeong, Y. H., & Lim, J. G. (2008). Behavior of zirconium fuel cladding under fast pressurization rates. *Nuclear Engineering and Design*, *238*(6), 1441–1447. doi:10.1016/j.nucengdes.2007.10.020

Kim, W.-J., Kim, D.-J., & Park, Y. (2013). Fabrication and material issues for the application of SiC composites to LWR fuel cladding. *Nuclear Engineering and Technology*, *45*(4), 565–572. doi:10.5516/NET.07.2012.084

Kussmaul, K., Eisele, U., & Seidenfuss, M. (1995). On the applicability of local approach models for the determination of the failure behaviour of steels with different toughness. In Mehta et al (Eds.), Fatigue and fracture mechanics in pressure vessel and piping (Vol. 304, pp. 17-25). New York: ASME.

Laiarinandrasana, L., Devilliers, C., Oberti, S., Gaudichet, E., Fayolle, B., & Lucatelli, J. M. (2011). Ring tests on high density polyethylene: Full investigation assisted by finite element modeling. *International Journal of Pressure Vessels and Piping*, *88*(1), 1–10. doi:10.1016/j.ijpvp.2010.12.002

Leclercq, S., Parrot, A., & Leroy, M. (2008). Failure characteristics of cladding tubes under RIA conditions. *Nuclear Engineering and Design*, *238*(9), 2206–2218. doi:10.1016/j.nucengdes.2008.03.017

Link, T. M., Koss, D. A., & Motta, A. T. (1998). Failure of Zircaloy cladding under transverse plane-strain deformation. *Nuclear Engineering and Design*, *186*(3), 379–394. doi:10.1016/S0029-5493(98)00284-2

Needleman, A., & Tvergaard, V. (1984). An analysis of ductile rupture in notched bars. *Journal of the Mechanics and Physics of Solids*, *32*(6), 461–490. doi:10.1016/0022-5096(84)90031-0

Nemat-Alla, M. (2003). Reproducing hoop stress–.strain behavior for tubular material using lateral compression test. *International Journal of Mechanical Sciences*, *45*(4), 605–621. doi:10.1016/S0020-7403(03)00115-2

Ott, L. J., Robb, K. R., & Wang, D. (2014). Preliminary assessment of accident-tolerant fuels on LWR performance during normal operation and under DB and BDB accident conditions. *Journal of Nuclear Materials*, *448*(1-3), 520–533. doi:10.1016/j.jnucmat.2013.09.052

Peerlings, R. H. J., De Borst, R., Brekelmans, W. A. M., & Geers, M. G. D. (2002). Localisation issues in local and nonlocal continuum approaches to fracture. *European Journal of Mechanics. A, Solids*, *21*(2), 175–189. doi:10.1016/S0997-7538(02)01211-1

Pint, B. A., Terrani, K. A., Brady, M. P., Cheng, T., & Keiser, J. R. (2013). High temperature oxidation of fuel cladding candidate materials in steam-hydrogen environments. *Journal of Nuclear Materials*, *440*(1-3), 420–427. doi:10.1016/j.jnucmat.2013.05.047

Pitard-Bouet, J.-M., Seidenfuss, M., Bethmont, M., & Kussmaul, K. (1999). Experimental investigations on the shallow crack effect on the 10MnMoNi55 steel and computational analysis in the upper shelf by means of the global and local approaches. *Nuclear Engineering and Design*, *190*(1-2), 171–190. doi:10.1016/S0029-5493(98)00322-7

Price, E. G. (1972). Hydride orientation and tensile properties of Zr-2.5 wt % Nb pressure tubing hydrided while internally pressurized. *Canadian Metallurgical Quarterly*, *11*(1), 129–136. doi:10.1179/cmq.1972.11.1.129

Reddy, T. Y., & Reid, S. R. (1979). On obtaining material properties from the ring compression test. *Nuclear Engineering and Design*, *52*(2), 257–263. doi:10.1016/0029-5493(79)90055-4

Reusch, F., Svendsen, B., & Klingbeil, D. (2003a). A non-local extension of Gurson-based ductile damage modelling. *Computational Materials Science*, *26*, 219–229. doi:10.1016/S0927-0256(02)00402-0

Reusch, F., Svendsen, B., & Klingbeil, D. (2003b). Local and non-local Gurson-based ductile damage and failure modelling at large deformation. *European Journal of Mechanics. A, Solids*, *22*(6), 779–792. doi:10.1016/S0997-7538(03)00070-6

Rousselier, G. (1987). Ductile fracture models and their potential in local approach of fracture. *Nuclear Engineering and Design*, *105*(1), 97–111. doi:10.1016/0029-5493(87)90234-2

Rudnicki, J. W., & Rice, J. R. (1975). Conditions for the localization of deformation in pressure-sensitive dilatant materials. *Journal of the Mechanics and Physics of Solids*, *23*(6), 371–394. doi:10.1016/0022-5096(75)90001-0

Samal, M. K., Sanyal, G., & Chakravartty, J. K. (2010a). Experimental and numerical study of fracture behaviour of tubular specimens in a Pin-Loading-Tension setup. *IMechE Part C Journal of Mechanical Engineering Science*, *224*(1), 1–12. doi:10.1243/09544062JMES1781

Samal, M. K., Sanyal, G., & Chakravartty, J. K. (2010b). Estimation of fracture behaviour of thin walled nuclear reactor fuel pins using pin-loading tension (PLT) test. *Nuclear Engineering and Design*, *240*(12), 4043–4050. doi:10.1016/j.nucengdes.2010.09.030

Samal, M. K., Sanyal, G., & Chakravartty, J. K. (2011). Investigation of failure behavior of two different types of Zircaloy clad tubes used as nuclear reactor fuel pins. *Engineering Failure Analysis*, *18*(8), 2042–2053. doi:10.1016/j.engfailanal.2011.06.009

Samal, M. K., Seidenfuss, M., Roos, E., Dutta, B. K., & Kushwaha, H. S. (2008). Finite element formulation of a new nonlocal damage model. *Finite Elements in Analysis and Design*, *44*(6-7), 358–371. doi:10.1016/j.finel.2007.12.002

Sanyal, G., & Samal, M. K. (2012). Fracture behavior of thin-walled Zircaloy fuel clad tubes of Indian pressurized heavy water reactor. *International Journal of Fracture*, *173*(2), 175–188. doi:10.1007/s10704-012-9678-2

Shah, P. K., Samal, M. K., Dubey, J. S., Shriwastaw, R. S., Balakrishnan, K. S., Anantharaman, S., & Chakravartty, J. K. (2011, November 6-11). Finite element study on mechanical behaviour of Zircaloy fuel clad tubes.*Proceedings of the 21st International conference on Structural Mechanics in Reactor Technology (SMiRT21)*, New Delhi, India.

Sharobeam, M. H., & Landes, J. D. (1991). The load separation criterion and methodology in ductile fracture mechanics. *International Journal of Fracture*, *47*(2), 81–104. doi:10.1007/BF00032571

Sharobeam, M. H., & Landes, J. D. (1993). The load separation and η_{pl} development in precracked specimen test records. *International Journal of Fracture*, *59*(3), 213–226. doi:10.1007/BF02555184

Stempien, J. D., Carpenter, D. M., Kohse, G., & Kazimi, M. S. (2013). Characteristics of composite silicon carbide fuel cladding after irradiation under simulated PWR conditions. *Nuclear Technology*, *183*, 13–29.

Stone, J. G., Schleicher, R., Deck, C. P., Jacobsen, G. M., Khalifa, H. E., & Back, C. A. (2015). Stress analysis and probabilistic assessment of multi-layer SiC-based accident tolerant nuclear fuel cladding. *Journal of Nuclear Materials*, *466*, 682–697. doi:10.1016/j.jnucmat.2015.08.001

Terrani, K. A., Wang, D., Ott, L. J., & Montgomery, R. O. (2014). The effect of fuel thermal conductivity on the behavior of LWR cores during loss-of-coolant accidents. *Journal of Nuclear Materials*, *448*(1-3), 512–519. doi:10.1016/j.jnucmat.2013.09.051

Terrani, K. A., Zinkel, S. J., & Snead, L. L. (2013). Advanced oxidation-resistance iron-based alloys for LWR fuel cladding. *Journal of Nuclear Materials*, *448*(1-3), 420–435. doi:10.1016/j.jnucmat.2013.06.041

Tvergaard, V., & Needleman, A. (1984). Analysis of cup-cone fracture in a round tensile bar. *Acta Metallurgica*, *32*(1), 157–169. doi:10.1016/0001-6160(84)90213-X

Wang, H., Bouchard, R., Eagleson, R., Martin, P., & Tyson, W. R. (2002). Ring hoop tension test (RHTT): A test for transverse tensile properties of tubular materials. *Journal of Testing and Evaluation*, *30*(5), 381–391.

Yamada, R., Suzuki, M., & Harayama, Y. (1977). Application of finite element method to ring compression test. *Nuclear Engineering and Design*, *44*(1), 75–85. doi:10.1016/0029-5493(77)90124-8

Youinou, G. J., & Sen, R. S. (2014). Impact of accident-tolerant fuels and claddings on the overall fuel cycle: A preliminary systems analysis. *Nuclear Technology*, *188*, 123–128.

Zinkle, S. J., Terrani, K. A., Gehin, J. C., Ott, L. J., & Snead, L. L. (2014). Accident tolerant fuels for LWRs: A perspective. *Journal of Nuclear Materials*, *448*(1-3), 374–379. doi:10.1016/j.jnucmat.2013.12.005

APPENDIX: NOMENCLATURE

D, σ_k : Rousselier's material constants
f : Void volume fraction
$\dot{f}_{\text{growth}}$: Increment in void volume fraction due to growth
f_0 : Initial void volume fraction
f_c : Void volume fraction at coalescence
f_f : Final void volume fraction at fracture
F_b : Body force per unit volume
q : von Mises equivalent stress
p : Mean or hydrostatic component of stress tensor
ε_{eq} : von Mises equivalent strain
$\dot{\varepsilon}^{p}_{kk}$: Increment in mean or hydrostatic component of plastic strain tensor
$R\left(\varepsilon_{eq}\right)$: Material resistance in terms of true stress (as a function of equivalent strain)
ϕ : Yield potential function of Rousselier's model
∇ : Gradient operator
$\nabla.\sigma$: Divergence of Cauchy stress σ

Chapter 10
MARS and Neural Network Models for Shear Strength Prediction of Squat Reinforced Concrete Walls:
Shear Strength of Squat Reinforced Concrete Walls

Anthony T.C. Goh
Nanyang Technological University, Singapore

Wengang Zhang
Chongqing University, China

ABSTRACT

Squat walls are widely used in structural engineering. The empirical equations currently used to calculate the peak shear strength of squat walls do not correlate well with the experimental results. Another limitation is the reliance on the use of many assumed intermediate parameters. This chapter explores the use of MARS and BPNN approaches to build predictive peak shear strength models of squat walls based on an extensive experimental database from the literature. First the MARS methodology and its associated procedures will be explained in detail. Analyses of the database are then carried out to verify the MARS's capacity in modelling the non-linear interactions between variables without making any specific assumptions. The performances of the built MARS and BPNN models are compared in terms of predictive accuracy, parameter relative importance, parametric analysis and model interpretability. Design charts are also proposed based on parametric studies using the developed models.

DOI: 10.4018/978-1-5225-0588-4.ch010

INTRODUCTION

Reinforced concrete (RC) walls with a ratio of height to length of less than or equal to two, commonly described as squat or short walls, are widely used structural components in many commercial buildings and nearly all safety-related nuclear structures. Squat walls are generally grouped by plan geometry, namely, rectangular, barbell, and flanged. The performance of these short (squat) walls is most important because they are designed to provide most of the lateral stiffness and strength to resist earthquake shaking or wind loadings. Accurate modelling of the response of squat walls is essential considering that the conventional buildings are likely to experience multiple deformation cycles at or beyond yield in maximum earthquake shaking. Current design provisions in codes and standards for reinforced concrete walls, such as the ACI 318-08 (ACI, 2008) and ASCE 41-06 (ASCE, 2007), focus on tall (slender) walls and pay less attention to squat walls, although the latter is far more common in practice.

A desirable earthquake resistant design philosophy for RC walls is to prevent shear failure in earthquake shaking. Experimental studies have shown that well designed and detailed slender walls will yield in flexure and not fail in shear. In contrast, squat walls are prone to shear failure that is generally associated with rapid loss of strength and stiffness under cyclic loading. Three types of shear failure are commonly observed in squat walls, namely, diagonal tension, diagonal compression, and sliding shear (Paulay & Priestley, 1992). If a shear failure occurs after the wall achieves its flexural strength, the failure mode is considered mixed and termed a flexure-shear failure (e.g., flexure-diagonal tension). It is generally recognized that the design parameters such as horizontal and vertical web reinforcement ratios, wall geometry, and axial force also affect the behaviour of squat reinforced concrete walls.

A significant number of experiments on squat reinforced concrete walls have been completed in the past 60 years. The first experimental programs on squat walls were conducted at Massachusetts Institute of Technology and Stanford University (Galletly, 1952; Benjamin & Williams, 1956) on bar bell cross-section. These tests were performed using monotonic loading and focused mainly on evaluation of peak shear strength. In the United States, the first cyclic tests on squat walls were performed by Barda (1972). Almost in the same period, a substantial number of cyclic loading experiments on squat walls were conducted in Japan (Hirosawa, 1975).

Although there is substantial experimental data in the literature on squat walls, current design procedures have been essentially unchanged for more than 40 years. One such example is the prediction of peak shear strength, which is the key parameter in the design of squat walls. The empirical equations and procedures available to calculate the peakshear strength of squat walls do not correlate well with the experimental results with substantial scatter in the predictions. In addition, one limitation of using the commonly used five sets of predictive equations, i.e., Section 11.9 of ACI318-08 (ACI, 2008), Section 21.9 of ACI 318-08 (ACI, 2008), Barda et al. (1977), ASCE 43-05 (ASCE, 2007) and Wood (1990), is the reliance on the use of many assumed intermediate parameters and generally these intermediate parameters are functions of other introduced parameters. Some use of Artificial Neural Networks in structural engineering includes Soleimanbeigi and Hataf (2005, 2006), and Padmini et al. (2008). Recently, other soft computing techniques including the Support Vector Machine (e.g., Samui, 2012) and Relevant Vector Machine (e.g., Samui 2012), the Genetic Programming (e.g., Shahnazari & Tutunchian, 2012), and Fuzzy models (e.g., Padmini et al. 2008) have been applied to foundation engineering for estimation of the ultimate bearing capacity of footings on cohesionless soils. Despite reasonable accuracy obtained by these soft computing methods, the developed models are generally difficult to interpret.

Multivariate adaptive regression spline MARS (Friedman, 1991), which is a statistical method for fitting the relationship between a set of input variables and dependent variables, is a nonlinear and non-parametric regression method based on a divide and conquer strategy in which the training data sets are partitioned into separate piecewise linear segments (splines) of differing gradients (slope). No specific assumption about the underlying functional relationship between the input variables and the output is required. The end points of the segments are called knots. A knot marks the end of one region of data and the beginning of another. The resulting piecewise curves (known as basis functions), give greater flexibility to the model, allowing for bends, thresholds, and other departures from linear functions. It generates basis functions by searching in a stepwise manner. An adaptive regression algorithm is used for selecting the knot locations.

MARS models are constructed in a two-phase procedure. The forward phase adds functions and finds potential knots to improve the performance, resulting in an overfitted model. The backward pass prunes the model by removing the BFs with the lowest contribution to the model until it finds the best sub-model. The aim of the backward deletion procedure is to find a close to optimal model by removing extraneous variables. Thus, the BFs maintained in the final optimal model are selected from the set of all candidate BFs, used in the forward selection step. Model subsets are compared using the less computationally expensive method of Generalized Cross-Validation (GCV). After the optimal MARS model is determined, by grouping together all the BFs that involve one variable and another grouping of BFs that involve pairwise interactions (and even higher level interactions when applicable), the procedure known as analysis of variance (ANOVA) decomposition (Friedman, 1991) can be used to assess the contributions from the input variables and the BFs through comparing (testing) variables for statistical significance.

This chapter explores the use of MARS and BPNN (back propagation neural networks) approaches to build predictive peak shear strength models of squat walls based on an extensive experimental database compiled by Gulec (2009). First the MARS methodology and its associated procedures will be explained in detail. Analyses of the database are then carried out to verify the MARS's capacity in modeling the non-linear interactions between variables without making any specific assumptions. The performances of the built MARS and BPNN models are compared in terms of predictive accuracy, parameter relative importance, parametric analysis and model interpretability. Design charts are also proposed based on parametric studies using the developed models.

BPNN AND MARS ALGORITHMS

BPNN Algorithm

A three-layer, feed-forward neural network topology shown in Figure 1 is adopted in this study. As shown in Figure 1, the back-propagation algorithm involves two phases of data flow. In the first phase, the input data are presented forward from the input to output layer and produces an actual output. In the second phase, the error between the target values and actual values are propagated backwards from the output layer to the previous layers and the connection weights are updated to reduce the errors between the actual output values and the target output values. No effort is made to keep track of the characteristics of the input and output variables. The network is first trained using the training data set. The objective of the network training is to map the inputs to the output by determining the optimal connection weights and biases through the back-propagation procedure. The number of hidden neurons is typically

Figure 1. Back-propagation neural network architecture used in this study

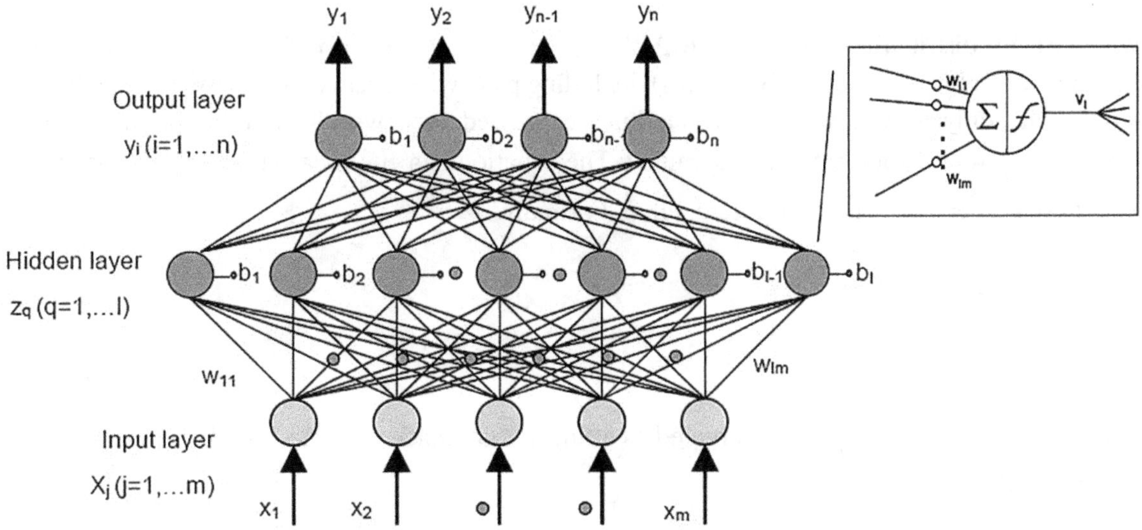

determined through a trial-and-error process; normally the smallest number of neurons that yields satisfactory results (judged by the network performance in terms of the coefficient of determination R^2 of the testing data set) is selected. In the present study, a Matlab-based back-propagation algorithm BPNN with the Levenberg-Marquardt (LM) algorithm (Demuth & Beale, 2003) was adopted for neural network modeling. The Levenberg-Marquardt back-propagation algorithm is a second-order nonlinear optimization technique that is usually faster and more reliable than other back-propagation techniques (Fistikoglu & Okkan, 2011).

MARS Algorithm

Friedman (1991) introduced MARS as a statistical method for fitting the relationship between a set of input variables and dependent variables. It is a nonlinear and nonparametric regression method based on a divide and conquer strategy in which the training data sets are partitioned into separate piecewise linear segments (splines) of differing gradients (slope). No specific assumption about the underlying functional relationship between the input variables and the output is required. The end points of the segments are called knots. A knot marks the end of one region of data and the beginning of another. The resulting piecewise curves (known as basis functions), give greater flexibility to the model, allowing for bends, thresholds, and other departures from linear functions.

MARS generates basis functions by searching in a stepwise manner. An adaptive regression algorithm is used for selecting the knot locations. MARS models are constructed in a two-phase procedure. The forward phase adds functions and finds potential knots to improve the performance, resulting in an overfit model. The backward phase involves pruning the least effective terms. An open source code on MARS from Jekabsons (2010) is used in carrying out the analyses presented in this paper.

Let $\boldsymbol{y}$ be the target output and $\boldsymbol{X} = (X_1, \ldots, X_P)$ be a matrix of P input variables. Then it is assumed that the data are generated from an unknown 'true' model. In case of a continuous response this would be

$$y = f\left(X_1, \ldots X_P\right) + e = f\left(X\right) + e \tag{1}$$

in which e is the distribution of the error. MARS approximates the function f by applying basis functions (BFs). BFs are splines (smooth polynomials), including piecewise linear and piecewise cubic functions. For simplicity, only the piecewise linear function is expressed. Piecewise linear functions are of the form $\max\left(0, x-t\right)$ with a knot occurring at value t. The equation $\max\left(.\right)$ means that only the positive part of $\left(.\right)$ is used otherwise it is given a zero value. Formally,

$$\max\left(0, x-t\right) = \begin{cases} x-t, & if\ x \geq t \\ 0, & otherwise \end{cases} \tag{2}$$

The MARS model $f\left(X\right)$, is constructed as a linear combination of BFs and their interactions, and is expressed as

$$f\left(X\right) = \beta_0 + \sum_{m=1}^{M} \beta_m \lambda_m \left(X\right) \tag{3}$$

where each $\lambda_m\left(X\right)$ is a basis function. It can be a spline function, or the product of two or more spline functions already contained in the model (higher orders can be used when the data warrants it; for simplicity, at most second-order is assumed in this chapter). The coefficients β are constants, estimated using the least-squares method.

Figure 2 shows a simple example of how MARS would use piecewise linear spline functions to attempt to fit data. The MARS mathematical equation is expressed as

$$y = -44.08 + 4.24 \times BF1 - 3.67 \times BF2 + 6.31 \times BF3 - 2.50 \times BF4 \tag{4}$$

where BF1 = max(0, 16–x), BF2 = max(0, x –10), BF3 = max(0, x – 5.5) and BF4 = max(0, 5.5–x). The knots are located at x = 5.5, 10 and 16. They delimit four intervals where different linear relationships are identified.

The MARS modeling is a data-driven process. To fit the model in Equation 3, first a forward selection procedure is performed on the training data. A model is constructed with only the intercept, β_0, and the basis pair that produces the largest decrease in the training error is added. Considering a current model with M basis functions, the next pair is added to the model in the form

$$\beta_{M+1}\lambda_m\left(X\right)\max\left(0, X_j - t\right) + \beta_{M+2}\lambda_m\left(X\right)\max\left(0, t - X_j\right) \tag{5}$$

with each β being estimated by the method of least squares. As a basis function is added to the model space, interactions between BFs that are already in the model are also considered. BFs are added until the model reaches some maximum specified number of terms leading to a purposely overfit model.

Figure 2. Knots and linear splines for a simple MARS example

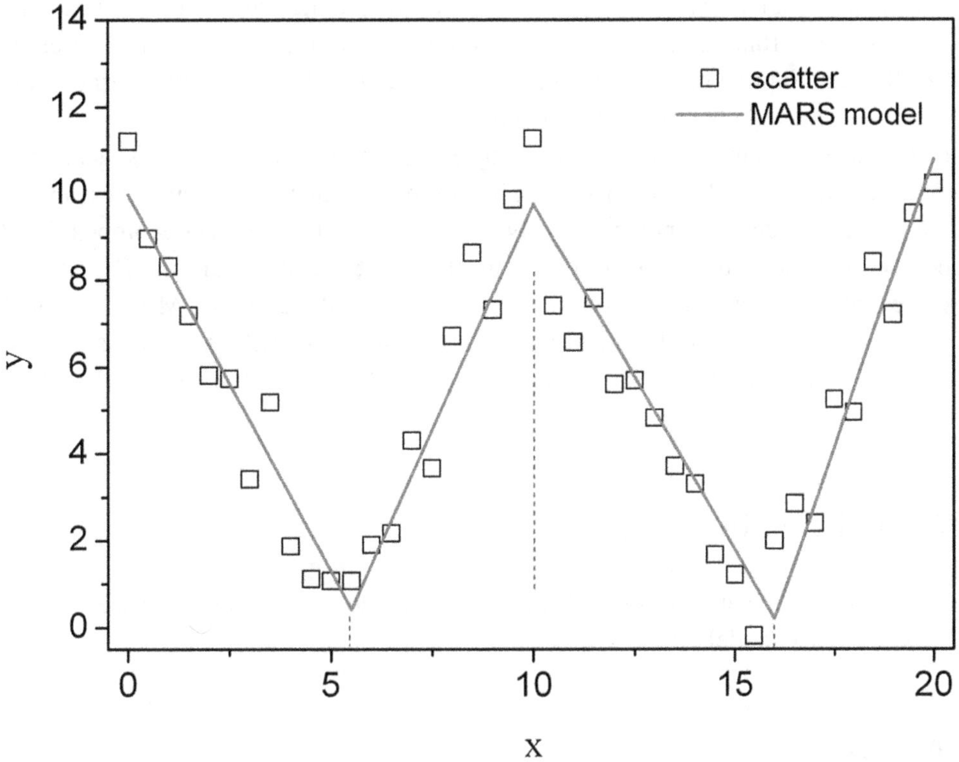

To reduce the number of terms, a backward deletion sequence follows. The aim of the backward deletion procedure is to find a close to optimal model by removing extraneous variables. The backward pass prunes the model by removing the BFs with the lowest contribution to the model until it finds the best sub-model. Thus, the BFs maintained in the final optimal model are selected from the set of all candidate BFs, used in the forward selection step. Model subsets are compared using the less computationally expensive method of Generalized Cross-Validation (GCV). The GCV equation is a goodness of fit test that penalizes large numbers of BFs and serves to reduce the chance of overfitting. For the training data with N observations, GCV for a model is calculated as follows (Hastie et al., 2009)

$$GCV = \frac{\frac{1}{N}\sum_{i=1}^{N}\left[y_i - f\left(x_i\right)\right]^2}{\left[1 - \frac{M + d \times \left(M-1\right)/2}{N}\right]^2} \tag{6}$$

in which M is the number of BFs, d is the penalizing parameter, N is the number of observations, and $f\left(x_i\right)$ denotes the predicted values of the MARS model. The numerator is the mean squared error of the evaluated model in the training data, penalized by the denominator. The denominator accounts for the increasing variance in the case of increasing model complexity. Note that $\left(M-1\right)/2$ is the number of hinge function knots. The GCV penalizes not only the number of the model's basis functions but also the number of knots. A default value of 3 is assigned to penalizing parameter *d* (Friedman, 1991). At each deletion step a basis function is removed to minimize Equation 6, until an adequately fitted model is found. MARS is an adaptive procedure because the selection of BFs and the variable knot locations are data-based and specific to the problem at hand.

After the optimal MARS model is determined, by grouping together all the BFs that involve one variable and another grouping of BFs that involve pairwise interactions (and even higher level interactions when applicable), the procedure known as analysis of variance (ANOVA) decomposition (Friedman, 1991) can be used to assess the contributions from the input variables and the BFs through comparing (testing) variables for statistical significance. Previous applications of MARS algorithm in civil engineering can be found in Attoh-Okine et al. (2009), Lashkari (2012), Mirzahosseinia et al. (2011), Zarnani et al. (2011), Samui (2011), Samui and Karup (2011), Zhang and Goh (2013, 2014), Goh and Zhang (2014), Zhang et al. (2015).

EVALUATION CRITERIA

A number of performance measures as listed and defined in Table 1 were used to assess the performance of the MARS and BPNN models.

THE DATABASE

Gulec (2005) reviewed and catalogued the results of tests of 352 reinforced concrete squat walls with three different cross-section types, namely, rectangular, barbell, and flanged. Later this database was expended to 434 walls (Gulec, 2009). Of the 434 walls in the database, 150 are of rectangular cross section and 284 have boundary elements (191 of barbell and 93 of flanges). Of the 434 data sets, seven cases were omitted due to incomplete data. The assembled data can be used to assess the performance of BPNN and MARS algorithms that are used to predict squat wall response and to develop new models. Of the 427 cases, 307 observations were randomly selected as the training patterns while 102 were used for testing purposes. The remaining 18 observations were adopted for validation use. A summary of the nine input parameters and single output are listed in Table 2. The basic parameters of the squat wall are shown in Figure 3. The data sets for the training and the testing patterns are listed in Table A and B, respectively, in the Appendix. The validation pattern is listed in Table 3.

THE DEVELOPED MARS AND BPNN MODELS

Model Interpretability

The optimal MARS model for squat wall peak shear strength prediction adopted 33 BFs of linear spline function, with 2nd order interaction. For the BPNN model, the optimum number of hidden neurons is six. The mathematical expression of this BPNN model is shown in Appendix C. It can be seen from Appendix C that the built BPNN model is difficult to interpret, as the weights and bias values give no

Table 1. Summary of performance measures

Measure	Calculation
Coefficient of determination (R^2)	$R^2 = 1 - \dfrac{\frac{1}{n}\sum_{i=1}^{n}\left(Y_i - \bar{Y}\right)^2}{\frac{1}{n}\sum_{i=1}^{n}\left(y_i - \bar{y}\right)^2}$
Coefficient of correlation (r)	$r = \dfrac{\sum_{i=1}^{N}\left(Y_i - \bar{Y}\right)\left(y_i - \bar{y}\right)}{\sqrt{\sum_{i=1}^{N}\left(Y_i - \bar{Y}\right)^2}\sqrt{\sum_{i=1}^{N}\left(y_i - \bar{y}\right)^2}}$
Root mean squared error (*RMSE*)	$RMSE = \sqrt{\frac{1}{N}\sum_{i=1}^{N}\left(Y_i - y_i\right)^2}$
Mean average error (*MAE*)	$MAE = \frac{1}{N}\sum_{i=1}^{N} abs\left(Y_i - y_i\right)$
Variance account factor (VAF)+	$VAF = \left(1 - \left(\mathrm{var}\left(Y_i - y_i\right)/\mathrm{var}\, y_i\right)\right)100$
Performance index (PI)+	$PI = Adj.R^2 + 0.01VAF - RRMSE$
Computational efficiency	CPU processing time (s)*

$\bar{y}$ is the mean of the target values of y_i; $\bar{Y}$ is the mean of the predicted Y_i; N denotes the number of data points in the used set, training set, testing set or the overall set; +VAF and PI definitions are based on Ceryan (2014).

* Using a PC with 2.40 GHz Intel(R) Core(TM) i7-4700 MQ processor, 16.0 GB RAM

Table 2. Statistical parameters of inputs and output for RC squat walls

Variable No.	Parameters	Physical Meaning	Ranges
1	t_w(m)	thickness of wall web	0.050-0.200
2	h_w(m)	height of wall	0.401-2.619
3	l_w (m)	length of wall	0.508-3.960
4	M/Vl_w	moment-to-shear ratio	0.06-2.13
5	ρ_v (%)	vertical web reinforcement ratio	0.00-2.87
6	ρ_{vall} (%)	ratio of total area of vertical reinforcement to wall area	0.30-5.91
7	ρ_h (%)	horizontal web reinforcement ratio	0.00-2.80
8	f_c' (kPa)	compressive strength of concrete	210-2187
9	T	Loading type: 1 for cyclic; 2 for monotonic; 3 for dynamic; 4 for repeated; 5 for blast; 6 for earthquake simulator	1-6
Output V_{peak} (kN)		peak shear strength of squat walls	85-7060

Figure 3. Typical geometry of RC squat wall: (a) wall sample; (b) configuration; and (c) front and cross-section view

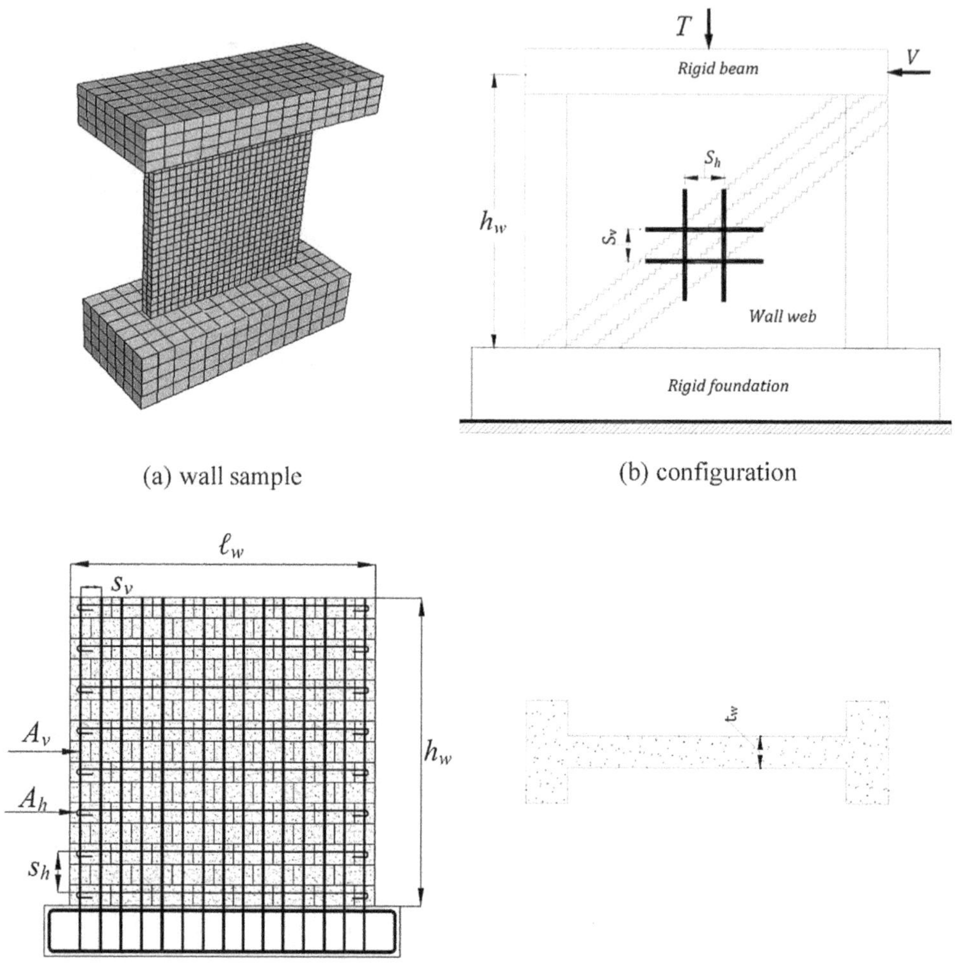

(a) wall sample

(b) configuration

(c) front and cross-section view

physical indication for engineers to have an insight and understanding of where significant changes in the data may occur.

The equation of MARS bearing capacity model is given by

$$\begin{aligned} V_{peak}\left(kN\right) &= 1945.6 + 122700 \times BF1 - 16578 \times BF2 - 4200 \times BF3 - 11454 \times BF4 + \\ & 5142 \times BF5 - 150090 \times BF6 - 256600 \times BF7 - 6424 \times BF8 + 15244 \times BF9 - \\ & 350 \times BF10 + 0.465 \times BF11 - 2.154 \times BF12 + 16.387 \times BF13 - 2373 \times BF14 + \\ & 3851 \times BF15 - 276 \times BF16 - 12583 \times BF17 + 11933 \times BF18 + 13085 \times BF19 - \\ & 37332 \times BF20 - 91507 \times BF21 - 1.62 \times BF22 + 0.7 \times BF23 - 6545 \times BF24 - \\ & 5022 \times BF25 - 49.6 \times BF26 + 1859 \times BF27 + 1396 \times BF28 + 139.4 \times BF29 - \\ & 8345 \times BF30 + 0.687 \times BF31 - 20.96 \times BF32 + 20.32 \times BF33 \end{aligned} \tag{7}$$

Table 3. Validation data sets for RC squat walls

No.	Geometric Properties				Reinforcement Measures			Material	Loading Type T	Output V_{peak} (kN)
	t_w (m)	h_w (m)	l_w (m)	M/Vl_w	ρ_v (%)	ρ_{vall} (%)	ρ_h (%)	f_c' (kPa)		
1	0.051	0.93	1.803	0.64	0.25	1.95	0.25	504	5	535
2	0.051	0.93	1.803	0.64	0.25	1.95	0.25	487	2	454
3	0.051	0.93	1.803	0.64	0.5	1.11	0.5	421	5	556
4	0.076	0.93	1.803	0.64	0.25	0.79	0.25	394	5	489
5	0.076	0.93	1.803	0.64	0.25	0.79	0.25	294	2	400
6	0.051	0.508	0.61	0.92	0	1.23	0	413	2	91
7	0.051	0.508	0.914	0.61	0	0.92	0	477	2	113
8	0.051	0.508	1.219	0.46	0.5	1.07	0.5	410	2	201
9	0.1	1.199	1.181	1.12	1.13	1.13	1.01	717	1	855
10	0.05	1.1	0.8	1.5	0.57	0.87	0.41	652	6	85
11	0.05	1.1	0.8	1.5	0.57	1.35	0.41	625	6	129
12	0.05	0.599	1.12	0.63	0.25	2.04	0.25	338	1	210
13	0.07	0.5	0.861	0.76	0.73	1.22	0.82	677	1	205
14	0.08	1.75	2.25	0.61	0.49	0.64	0.47	546	1	654
15	0.075	2.019	3.099	0.77	1.2	0.66	1.2	602	6	1636
16	0.075	1.42	3.076	0.59	1.2	1.2	1.2	664	1	1473
17	0.1	1.001	1.999	0.58	0.59	0.8	0.8	694	1	401
18	0.06	0.8	0.8	1.08	0.73	1.76	0.82	495	1	142

The mathematical expression of the BFs in Equation 7 can be found in Table 4. It is observed from Table 4 that interactions have occurred between BFs (22 of the 33 BFs are interaction terms). The interactions suggest that the built MARS model is not simply additive and that interactions play a significant role in building an accurate model for squat wall peak shear strength prediction. This indicates that MARS is capable of capturing the nonlinear and complex relationships between dependent response and a multitude of design parameters without making any specific assumption about the underlying functional relationship between them.

Predictive Accuracy

The predictions using MARS and BPNN models are shown in Figure 4 for the training, testing and the validation patterns. Table 5 lists the values of the performance measures for the developed MARS and BPNN models. Both r and R^2 express the degree of similarity between predicted and actual values, with r or R^2 values close to 1 indicating predicted and actual values are very similar. Low RMSE and MAE values indicate high confidence in model-predicted values. Error measures of VAF and PI in the last column of Table 5 indicate that the MARS model predictions are as accurate as BPNN. Comparisons of these measures indicate that the BPNN gives only marginally better predictions than MARS. For the training samples, the MARS model is more accurate than BPNN. Therefore, both MARS and BPNN models can give reasonable estimations of the peak shear strength.

Table 4. Expressions of BFs for MARS model

BF	Equation	BF	Equation
BF1	max(0, t_w -0.16)	BF18	max(0, 2.301 -l_w) × max(0, 0.65 - M/Vl_w)
BF2	max(0, 0.16 -t_w)	BF19	BF8 × max(0, 2.72 -l_w)
BF3	max(0, l_w -2.301)	BF20	BF3 × max(0, t_w -0.075)
BF4	BF2 × max(0, l_w -1.6)	BF21	BF3 × max(0, 0.075 -t_w)
BF5	BF2 × max(0, 1.6 -l_w)	BF22	BF12 × max(0, ρ_h -0.9)
BF6	BF1 × max(0, ρ_{vall} -0.84)	BF23	BF12 × max(0, 0.9 -ρ_h)
BF7	BF1 × max(0, 0.84 -ρ_{vall})	BF24	max(0, l_w -1.999)
BF8	max(0, M/Vl_w -0.35)	BF25	max(0, 1.999 -l_w)
BF9	max(0, 0.35 - M/Vl_w)	BF26	max(0, 4 - T)
BF10	max(0, 2.59 -ρ_{vall})	BF27	BF24 × max(0, h_w -1.4)
BF11	max(0,f_c' -725)	BF28	BF24 × max(0, 1.4 -h_w)
BF12	max(0, 725 -f_c')	BF29	BF10 × max(0, M/Vl_w -0.35)
BF13	BF2 × max(0, 905 -f_c')	BF30	BF10 × max(0, 0.35 - M/Vl_w)
BF14	BF10 × max(0, t_w -0.102)	BF31	BF10 ×max(0, 536 -f_c')
BF15	BF10 × max(0, 0.102 -t_w)	BF32	BF3 × max(0,f_c' -456)
BF16	max(0, 1.32 -ρ_v) × max(0, l_w -1.12)	BF33	BF3 × max(0,f_c' -435)
BF17	max(0, 2.301 -l_w) × max(0, M/Vl_w -0.65)		

Computational Efficiency

The second column of Table 5 lists the CPU processing time using MARS and BPNN. The advantage of the processing speed of MARS over BPNN is obvious.

Parameter Relative Importance

Table 6 displays the ANOVA decomposition of the developed MARS model for peak shear strength estimation. The first column lists the ANOVA function number. The second column gives an indication

Table 5. Performance measures of developed MARS and BPNN models

Model	CPU Time (s)	Patterns	Performance Measures					
			R^2	r	RMSE	MAE	VAF	PI
MARS	5.62	Training	0.976	0.988	142.0	93.8	97.6	1.779
		Testing	0.888	0.944	317.5	167.3	89.8	1.351
		Validation	0.828	0.910	180.7	133.6	82.9	1.280
BPNN	35.63	Training	0.974	0.987	146.7	94.3	97.4	1.769
		Testing	0.913	0.966	280.5	145.0	91.4	1.442
		Validation	0.836	0.932	176.5	129.6	86.7	1.335

Figure 4. Comparison between measured and predicted peak shear strength: (a) training, (b) testing, and (c) validation

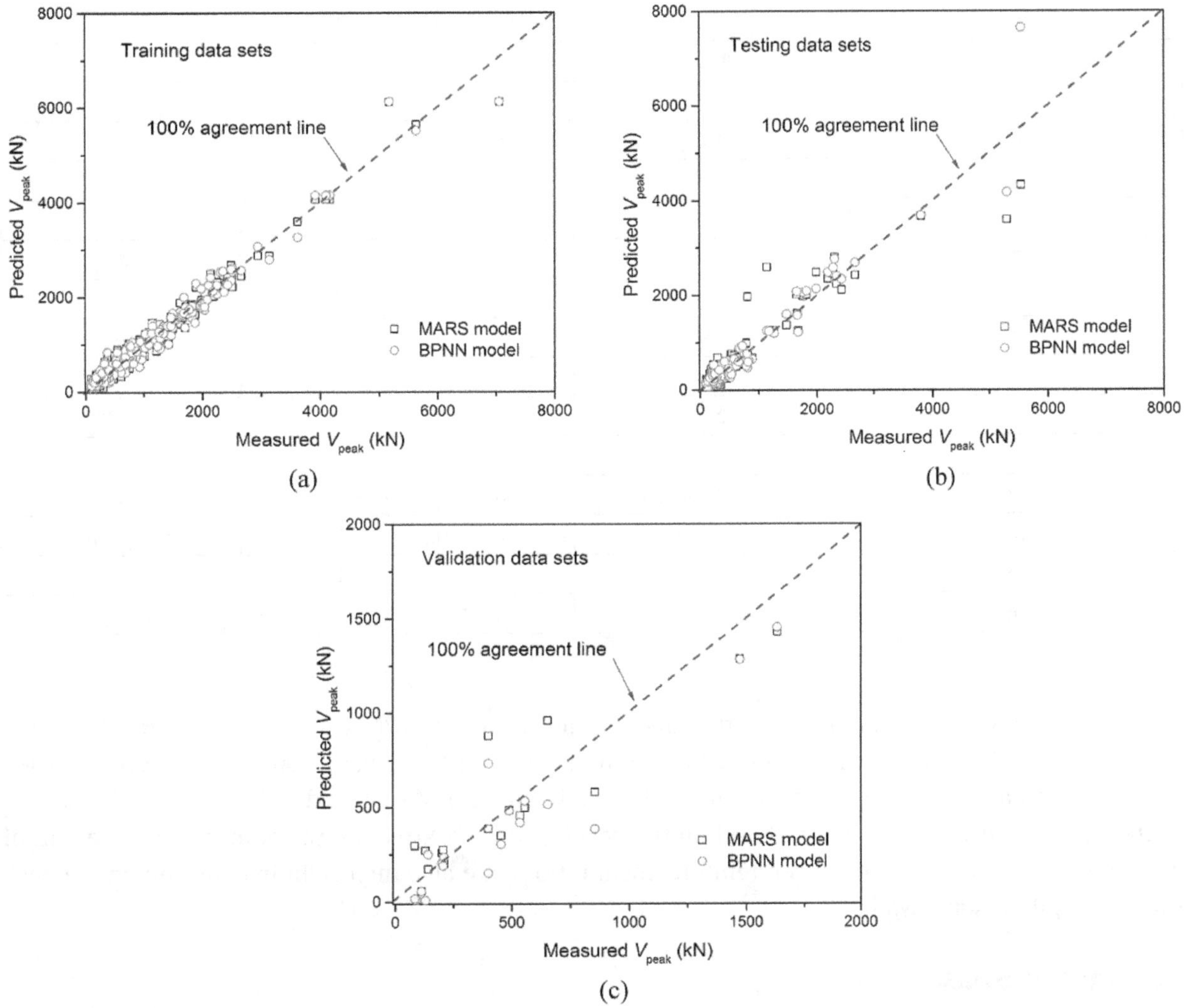

of the importance of the corresponding ANOVA function, by listing the GCV score for a model with all BFs corresponding to that particular ANOVA function removed. The third column provides the standard deviation of the function. It gives an indication of its relative importance to the overall model and can be interpreted in a manner similar to the standardized regression coefficient in a linear model. The fourth column gives the number of BFs comprising the ANOVA function. The last column gives the particular input variables associated with the ANOVA function.

Figure 5 gives the plots of the relative importance of the input variables for the developed MARS model, which is evaluated by the increase in the GCV value caused by removing the considered variables from the built MARS model. It can be observed that the MARS model identifies the length of wall l_w as the most influential parameter, followed by the moment-to-shear ratio M/Vl_w, the thickness of wall web t_w and the ratio of total area of vertical reinforcement to wall area ρ_{vall}. The horizontal web reinforcement ratio ρ_h and loading type T are the two least important parameters. For comparison, Figure

Table 6. ANOVA decomposition for the MARS bearing capacity model

Function	GCV (×10⁵)	STD	#Basis	Variable(s)
1	40.9	996	2	t_w
2	262.3	3388	3	l_w
3	266.6	2755	2	M/Vl_w
4	4.7	223	1	ρ_{vall}
5	4.1	338	2	$f_c^{'}$
6	0.7	47	1	T
7	16.8	937	4	t_w & l_w
8	1.7	275	4	t_w & ρ_{vall}
9	4.8	305	1	t_w & $f_c^{'}$
10	1.7	264	2	h_w & l_w
11	822.9	4310	3	l_w & M/Vl_w
12	1.0	146	1	l_w & ρ_v
13	0.4	60	2	l_w & $f_c^{'}$
14	4.7	489	2	M/Vl_w & ρ_{vall}
15	0.5	58	1	ρ_{vall} & $f_c^{'}$
16	0.5	66	2	ρ_h & $f_c^{'}$

5 also plots the relative importance of the input variables for the BPNN model, based on the method by Garson (1991). The percentage values of the relative importance based on both models are listed on top of the bar plots in Figure 5. It can be observed that both MARS and BPNN models identify l_w as the most significant parameter. Based on the developed BPNN model, the compressive strength of concrete $f_c^{'}$, ρ_{vall} and the vertical web reinforcement ratio ρ_v are also important in determining the peak shear strength of squat walls.

Parametric Study

In consideration of the model interpretability, for brevity this section focuses only on the developed MARS model. Figure 6 plots the required wall length l_w for different moment-to-shear ratios M/Vl_w with regard to the peak shear strength predicted by the MARS model, assuming the thickness of wall web t_w is 0.10 m, loading type is cyclic, and that the values of the other parameters correspond to their average values. It is obvious that the peak shear strength of squat wall increases with increase of wall length. This chart can be used as a preliminary tool in determining the required wall length for a given applied load, under similar parameter combinations.

Figure 7 gives the required thickness of wall web t_w for different wall length l_w with regard to the MARS peak shear strength model, assuming that the moment-to-shear ratio M/Vl_w is 0.36, and the loading type is cyclic. It is obvious that under the parameter combination as listed in Figure 7, for the same thickness of wall web, longer walls do not always provide the maximum shear strength.

Figure 8 plots the required moment-to-shear ratio M/Vl_w for different wall length l_w, assuming that the wall web t_w is 0.10 m, and the loading type is cyclic. Generally, M/Vl_w decreases with the increase of the required shear strength. This trend is more pronounced for walls of smaller length.

Figure 5. Relative importance of the input variables selected in the MARS and BPNN model

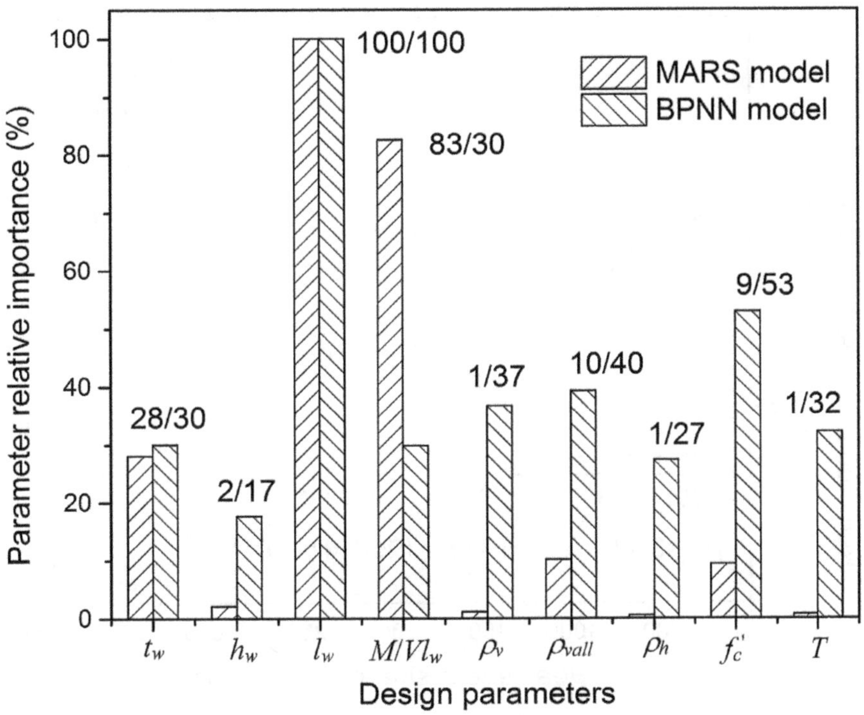

Figure 6. Plot of V_{peak} versus wall length l_w

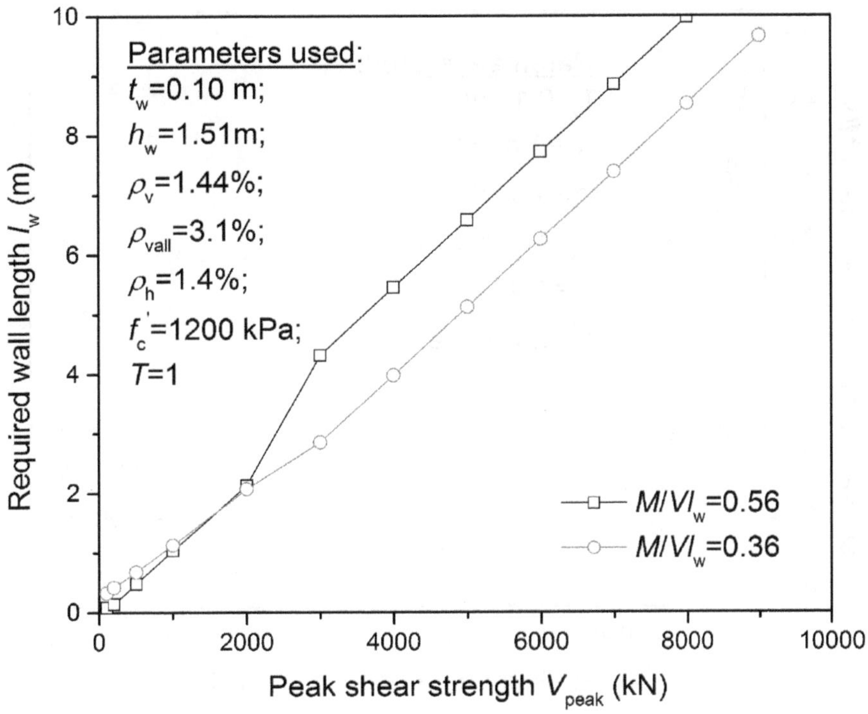

Figure 7. Plot of V_{peak} versus thickness of wall web t_w

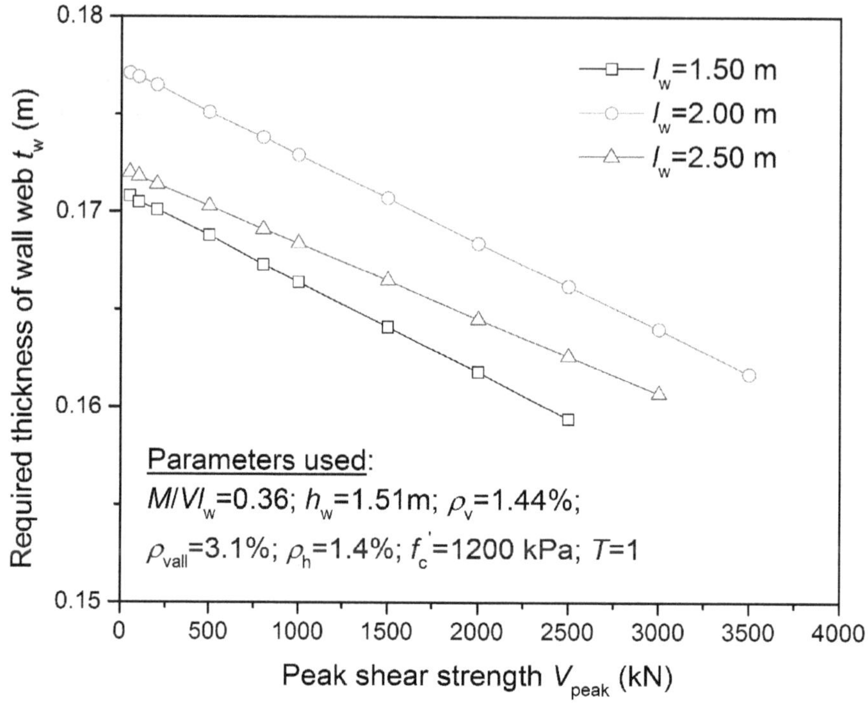

Figure 8. Plot of V_{peak} versus moment-to-shear ratio M/Vl_w

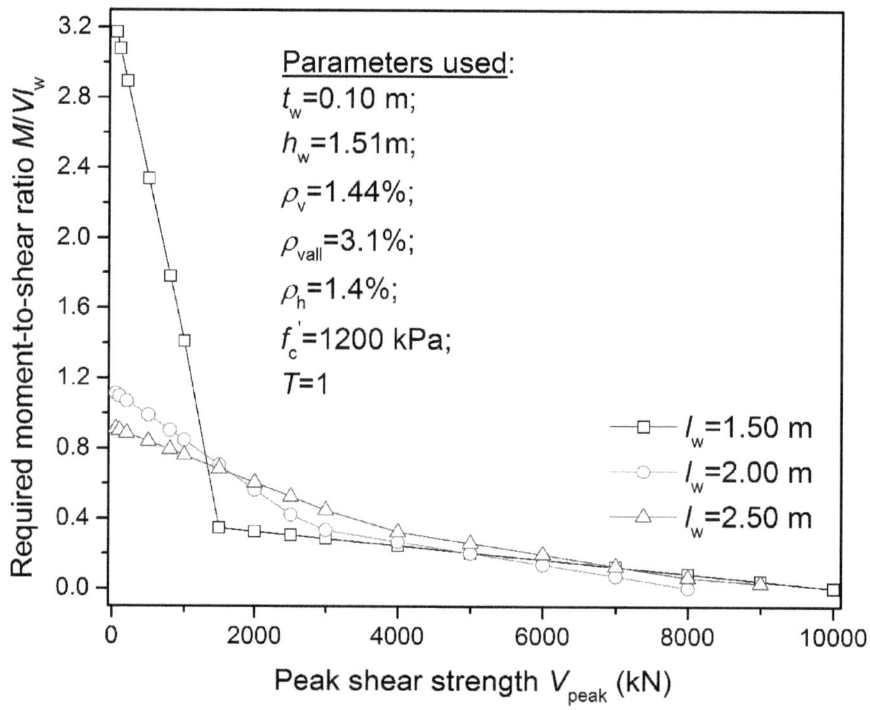

Figure 9 *a* to *c* gives the required moment-to-shear ratio M/Vl_w for different loading types include cyclic, monotonic and blast, for wall length l_w of 1.50 m, 2.00m and 2.50 m, respectively, assuming that the thickness of wall web t_w is 0.10 m. It is obvious that as T is identified as the least influential parameter by the MARS model, the impact of T on the required moment-to-shear ratio M/Vl_w is marginal.

SUMMARY AND CONCLUSION

This chapter demonstrates the viability of MARS and BPNN to predict the peak shear strength of squat walls with the influencing parameters, based on an extensive database comprising of load tests data. The performances of the developed MARS and BPNN models are compared in terms of predictive accuracy, computational efficiency and model interpretability. Based on the developed MARS model, some typi-

Figure 9. Plot of V_{peak} versus M/Vl_w for different loading type T: (a) l_w=1.50 m, (b) l_w=2.00 m, and (c) l_w=2.50 m

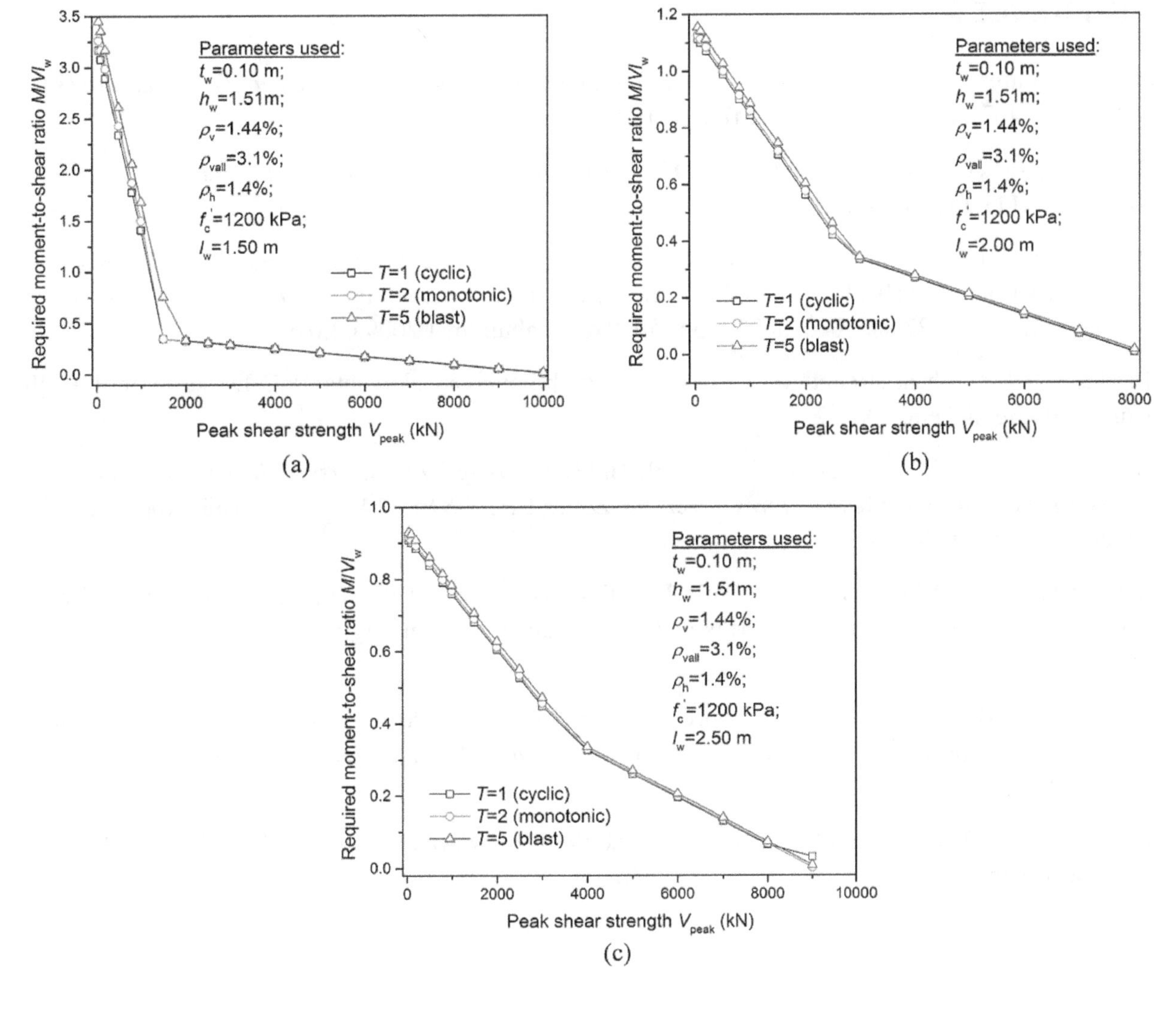

cal charts have been developed for estimating the peak shear strength for various squat wall geometries. These charts can be used as a preliminary design tool in determining the required wall sizes for a given applied load, under similar parameter combinations.

It should be noted that due to the data-driven nature of MARS and BPNN algorithms, interpolations within the ranges of design input variables are more accurate and reliable than extrapolations. Consequently, it is not recommended that the models be applied for values of input parameters beyond the specific ranges in this study. Additional new data sets are required to further evaluate and update the current MARS and BPNN squat wall shear strength models.

ACKNOWLEDGMENT

The authors would like to express their appreciation to Gulec (2009) for making his reinforced concrete squat wall database available for this work.

REFERENCES

ACI Committee 318. (2008). *Building code requirements for structural concrete and commentary (ACI 318-08)* (p. 467). Farmington Hills, MI: American Concrete Institute.

American Society of Civil Engineers. (2007). *Seismic rehabilitation of existing buildings (ASCE/SEI 41-06)* (p. 411). Reston, VA: ASCE.

Attoh-Okine, N. O., Cooger, K., & Mensah, S. (2009). Multivariate adaptive regression (MARS) and hinged hyperplanes (HHP) for doweled pavement performance modeling. *Journal of Construction and Building Materials*, *23*(9), 3020–3023. doi:10.1016/j.conbuildmat.2009.04.010

Barda, F. (1972). Shear strength of low-rise walls with boundary elements. PhD Dissertation, Lehigh University, Bethlehem, PA.

Barda, F., Hanson, J. M., & Corley, W. G. (1977). *Shear strength of low-rise walls with boundary elements. Reinforced Concrete Structures in Seismic Zones, SP 53-8* (pp. 149–202). Farmington Hills, MI: American Concrete Institute.

Benjamin, J. R., & Williams, H. A. (1956). Investigation of shear walls, part 12 – studies of reinforced concrete shear wall assemblies (Report No. 10). Department of Civil Engineering, Stanford University, Stanford.

Ceryan, N. (2014). Application of support vector machines and relevance vector machines in predicting uniaxial compressive strength of volcanic rocks. *Journal of African Earth Sciences*, *100*, 634–644. doi:10.1016/j.jafrearsci.2014.08.006

Demuth, H., & Beale, M. (2003). *Neural Network Toolbox for MATLAB-User Guide Version 4.1*. The Math Works Inc.

Fistikoglu, O., & Okkan, U. (2011). Statistical down scaling of monthly precipitation using NCEP/NCAR reanalysis data for Tahtali River Basin in Turkey. *Journal of Hydrologic Engineering*, *16*(2), 157–164. doi:10.1061/(ASCE)HE.1943-5584.0000300

Friedman, J. H. (1991). Multivariate adaptive regression splines. *Annals of Statistics*, *19*(1), 1–141. doi:10.1214/aos/1176347963

Galletly, G. D. (1952). *Behavior of reinforced concrete walls under static load*. Cambridge, MA: Department of Civil and Sanitary Engineering, Massachusetts Institute of Technology.

Garson, G. D. (1991). Interpreting neural-network connection weights. *AI Expert*, *6*(7), 47–51.

Goh, A. T. C., & Zhang, W. G. (2014). An improvement to MLR model for predicting liquefaction-induced lateral spread using Multivariate Adaptive Regression Splines. *Engineering Geology*, *170*, 1–10. doi:10.1016/j.enggeo.2013.12.003

Gulec, C. K. (2005). Ultimate shear strength of squat rectangular reinforced concrete walls [MS Thesis]. Department of Civil, Structural and Environmental Engineering, State University of New York at Buffalo, Buffalo, NY.

Gulec, C. K. (2009). Performance-based assessment and design of squat reinforced concrete shear walls [Ph.D. Thesis]. State University of New York at Buffalo.

Hastie, T., Tibshirani, R., & Friedman, J. (2009). *The elements of statistical learning: Data mining, inference and prediction* (2nd ed.). Springer. doi:10.1007/978-0-387-84858-7

Hirosawa, M. (1975). *Past experimental results on reinforced concrete shear walls and analysis on them. Kenchiku Kenkyu Shiryo, No. 6*. Tokyo, Japan: Building Research Institute, Ministry of Construction.

Jekabsons, G. (2010). VariReg: A software tool for regression modelling using various modeling methods. Riga Technical University. Retrieved from http://www.cs.rtu.lv/jekabsons/

Lashkari, A. (2012). Prediction of the shaft resistance of non-displacement piles in sand. *International Journal for Numerical and Analytical Methods in Geomechanics*, *37*(8), 904–931. doi:10.1002/nag.1129

Mirzahosseini, M., Aghaeifar, A., Alavi, A., Gandomi, A., & Seyednour, R. (2011). Permanent deformation analysis of asphalt mixtures using soft computing techniques. *Expert Systems with Applications*, *38*(5), 6081–6100. doi:10.1016/j.eswa.2010.11.002

Padmini, D., Ilamparuthi, K., & Sudheer, K. P. (2008). Ultimate bearing capacity prediction of shallow foundations on cohesionless soils using neurofuzzy models. *Computers and Geotechnics*, *35*(1), 33–46. doi:10.1016/j.compgeo.2007.03.001

Paulay, T., & Priestley, M. L. N. (1992). *Seismic Design of Reinforced Concrete and Masonry Buildings*. John Wiley & Sons. doi:10.1002/9780470172841

Samui, P. (2011). Determination of ultimate capacity of driven piles in cohesionless soil: A Multivariate Adaptive Regression Spline approach. *International Journal for Numerical and Analytical Methods in Geomechanics*, *36*(11), 1434–1439. doi:10.1002/nag.1076

Samui, P. (2012). Application of statistical learning algorithms to ultimate bearing capacity of shallow foundation on cohesionless soil. *Int. J. Numer. Anal. Met.*, *36*(1), 100–110. doi:10.1002/nag.997

Samui, P., & Karup, P. (2011). Multivariate adaptive regression splines and least square support vector machine for prediction of undrained shear strength of clay. *Applied Metaheuristic Computing*, *3*(2), 33–42.

Shahnazari, H., & Tutunchian, M. A. (2012). Prediction of ultimate bearing capacity of shallow foundations on cohesionless soils: An evolutionary approach. KSCE J. *Civil Engineering (New York, N.Y.)*, *16*(6), 950–957.

Soleimanbeigi, A., & Hataf, N. (2005). Predicting ultimate bearing capacity of shallow foundations on reinforced cohesionless soils using artificial neural networks. *Geosynthetics International*, *12*(6), 321–332. doi:10.1680/gein.2005.12.6.321

Soleimanbeigi, A., & Hataf, N. (2006). Prediction of settlement of shallow foundations on reinforced soils using neural networks. *Geosynthetics International*, *13*(4), 161–170. doi:10.1680/gein.2006.13.4.161

Wood, S. L. (1990). Shear strength of low-rise reinforced concrete walls. *ACI Structural Journal*, *87*(1), 99–107.

Zarnani, S., El-Emam, M., & Bathurst, R. J. (2011). Comparison of numerical and analytical solutions for reinforced soil wall shaking table tests. *Geotechnical Engineering*, *3*(4), 291–321.

Zhang, W. G., & Goh, A. T. C. (2013). Multivariate adaptive regression splines for analysis of geotechnical engineering systems. *Computers and Geotechnics*, *48*, 82–95. doi:10.1016/j.compgeo.2012.09.016

Zhang, W. G., & Goh, A. T. C. (2014). Multivariate adaptive regression splines model for reliability assessment of serviceability limit state of twin caverns. *Geotechnical Engineering*, *7*(4), 431–458.

Zhang, W. G., Goh, A. T. C., Zhang, Y. M., Chen, Y. M., & Xiao, Y. (2015). Assessment of soil liquefaction based on capacity energy concept and multivariate adaptive regression splines. *Engineering Geology*, *188*, 29–37. doi:10.1016/j.enggeo.2015.01.009

APPENDIX A

Table 7. Training data sets for RC squat walls

No.	Geometric Properties				Reinforcement Measures			Material	Loading Type T	Output V_{peak} (kN)
	t_w (m)	h_w (m)	l_w (m)	M/Vl_w	ρ_v (%)	ρ_{vall} (%)	ρ_h (%)	f_c' (kPa)		
1	0.051	0.930	1.803	0.64	0.25	0.95	0.25	419	5	423
2	0.051	0.930	1.803	0.64	0.25	0.95	0.25	461	5	378
3	0.051	0.930	1.803	0.64	0.25	0.95	0.25	458	2	360
4	0.051	0.930	1.803	0.64	0.25	1.95	0.25	548	5	511
5	0.051	0.930	1.803	0.64	0.25	1.95	0.25	531	5	483
6	0.051	0.930	1.803	0.64	0.25	0.95	0.25	743	5	636
7	0.051	0.930	1.803	0.64	0.25	0.95	0.25	662	5	523
8	0.051	0.930	1.803	0.64	0.5	1.11	0.5	387	2	414
9	0.051	0.930	1.803	0.64	0.5	1.11	0.5	447	5	489
10	0.051	0.930	1.803	0.64	1	1.42	1	426	5	830
11	0.051	0.930	1.803	0.64	1	1.42	1	421	5	756
12	0.051	0.930	1.803	0.64	0.5	1.11	0.5	866	2	409
13	0.051	0.930	1.803	0.64	0.5	1.11	0.5	562	5	578
14	0.051	0.930	1.803	0.64	0.5	1.11	0.5	562	2	445
15	0.051	0.930	1.803	0.64	0.5	1.11	0.5	510	5	476
16	0.051	0.930	1.803	0.64	0.5	1.11	0.5	490	5	516
17	0.051	0.930	1.803	0.64	0.5	1.11	0.5	536	2	405
18	0.051	0.930	1.803	0.64	0.5	1.11	0.5	518	5	454
19	0.051	0.930	1.803	0.64	0.5	1.11	0.5	594	2	360
20	0.051	0.930	1.803	0.64	0.5	2.1	0.5	509	5	641
21	0.051	0.930	1.803	0.64	0.5	2.1	0.5	714	5	676
22	0.051	0.930	1.803	0.64	0.5	2.1	0.5	480	2	471
23	0.051	0.930	1.803	0.64	0.5	2.1	0.5	725	5	560
24	0.051	0.930	1.803	0.64	0.5	2.1	0.5	663	5	592
25	0.076	0.930	1.803	0.64	0.25	0.79	0.25	428	2	409
26	0.076	0.930	1.803	0.64	0.25	0.79	0.25	249	5	387
27	0.076	0.930	1.803	0.64	0.25	0.79	0.25	377	5	489
28	0.076	0.930	1.803	0.64	0.5	0.96	0.5	383	5	498
29	0.076	0.930	1.803	0.64	0.5	0.96	0.5	433	5	520
30	0.076	0.930	1.803	0.64	0.5	0.96	0.5	303	2	427
31	0.051	0.930	3.327	0.34	0.5	0.88	0.5	479	2	494
32	0.051	0.930	3.327	0.34	0.5	0.88	0.5	483	5	792
33	0.051	0.930	3.327	0.34	0.5	0.88	0.5	395	5	676

continued on following page

Table 7. Continued

No.	Geometric Properties				Reinforcement Measures			Material	Loading Type T	Output V_{peak} (kN)
	t_w (m)	h_w (m)	l_w (m)	M/Vl_w	ρ_v (%)	ρ_{vall} (%)	ρ_h (%)	f_c' (kPa)		
34	0.051	0.930	3.327	0.34	0.5	0.88	0.5	412	2	600
35	0.080	1.199	2.720	0.5	0.71	1.28	0.76	414	1	931
36	0.080	1.199	2.720	0.5	0.71	1.28	0.76	544	1	1030
37	0.160	1.199	2.720	0.5	0.58	1.02	0.62	619	1	1554
38	0.080	1.199	2.720	0.5	0.71	3.89	0.76	501	1	1495
39	0.160	1.199	2.720	0.5	0.58	2.83	0.62	615	1	2309
40	0.102	0.876	1.905	0.5	0.5	1.06	0.48	609	2	1218
41	0.102	0.876	1.905	0.5	0.5	2.96	0.48	344	2	979
42	0.102	0.876	1.905	0.5	0.5	2	0.48	568	1	1108
43	0.102	0.876	1.905	0.5	0.5	2	0	400	1	1017
44	0.102	0.876	1.905	0.5	0	1.71	0.48	607	1	699
45	0.102	0.876	1.905	0.5	0.26	1.86	0.48	447	1	876
46	0.051	0.508	1.219	0.46	0	0.74	0	473	2	139
47	0.051	0.508	1.778	0.31	0	0.54	0	507	2	214
48	0.051	0.508	0.610	0.92	0.5	1.45	0.5	423	2	89
49	0.051	0.508	0.914	0.61	0.5	1.21	0.5	452	2	155
50	0.051	0.851	1.727	0.53	0.5	1.13	0.5	467	2	463
51	0.051	0.851	1.727	0.53	0.5	1.13	0.5	503	2	374
52	0.051	0.851	1.727	0.53	0.5	0.91	0.5	471	2	294
53	0.051	0.851	1.727	0.53	0	0.82	0	438	2	218
54	0.051	0.851	1.727	0.53	0	0.82	0	438	2	267
55	0.076	1.278	2.591	0.53	0	0.82	0	500	2	534
56	0.076	1.278	2.591	0.53	0.5	1.13	0.5	435	2	685
57	0.051	0.851	1.727	0.53	0.25	0.97	0.25	435	2	316
58	0.051	0.864	1.575	0.56	0.25	1.04	0.25	464	2	214
59	0.051	1.778	1.575	0.76	0.25	1.04	0.25	399	2	347
60	0.051	1.778	1.575	0.76	0.25	1.04	0.25	533	2	325
61	0.051	0.864	1.575	0.56	0.25	1.04	0.25	449	2	178
62	0.051	0.635	1.219	0.57	0.5	1.15	0.5	339	2	178
63	0.080	1.501	1.999	1	0.53	0.88	0.53	404	1	480
64	0.080	1.501	1.999	1	0.8	1.32	0.8	404	1	541
65	0.080	1.501	1.999	1	0.8	1.04	0.8	437	1	510
66	0.120	1.501	1.999	0.5	0.53	0.79	0.53	422	1	738
67	0.100	1.199	1.181	1.12	1.13	1.13	1.01	776	2	680
68	0.100	1.199	1.181	1.12	1.13	1.13	1.01	807	2	928
69	0.100	1.199	1.181	1.12	2.54	2.54	1.01	772	2	977

continued on following page

Table 7. Continued

No.	Geometric Properties				Reinforcement Measures			Material	Loading Type T	Output
	t_w (m)	h_w (m)	l_w (m)	M/Vl_w	ρ_v (%)	ρ_{vall} (%)	ρ_h (%)	f_c' (kPa)		V_{peak} (kN)
70	0.100	1.199	1.181	1.12	1.13	1.13	1.01	784	1	682
71	0.050	1.100	0.800	1.5	0.57	1.35	0.41	708	1	117
72	0.050	1.100	0.800	1.5	0.57	0.87	0.41	650	1	137
73	0.050	1.100	0.800	1.5	0.57	1.35	0.41	644	1	144
74	0.050	1.100	0.800	1.5	0.57	0.87	0.41	559	2	94
75	0.078	1.199	2.301	0.58	0.18	1.3	0.18	489	4	966
76	0.075	1.199	2.301	0.58	0.19	1.32	0.19	693	1	931
77	0.080	1.199	1.549	0.85	0.17	1.59	0.18	365	1	608
78	0.160	1.199	3.000	0.2	0.8	1.4	0.8	760	1	3620
79	0.200	1.199	3.000	0.2	0.4	0.84	0.6	846	1	5180
80	0.200	1.199	3.000	0.2	0.4	0.84	0.6	846	1	7060
81	0.200	1.199	3.000	0.2	0.4	0.67	0.4	846	1	3920
82	0.200	1.199	3.000	0.2	0.4	0.67	0.4	846	1	4170
83	0.200	1.199	3.000	0.2	0.6	0.98	0.6	846	1	5640
84	0.200	1.199	3.000	0.2	0.4	0.67	0.4	846	1	4100
85	0.050	0.599	1.120	0.63	0.25	2.04	0.25	338	2	226
86	0.050	0.599	1.120	0.63	0.25	2.04	0.25	311	1	334
87	0.050	0.599	1.120	0.63	0.5	2.18	0.5	311	1	235
88	0.050	0.599	1.120	0.63	0.5	2.18	0.5	322	1	304
89	0.050	0.599	1.120	0.63	0.5	2.18	0.5	322	1	282
90	0.050	0.599	1.120	0.63	0.25	2.04	0.25	357	2	240
91	0.050	0.599	1.120	0.63	0.25	2.04	0.25	357	1	238
92	0.050	0.599	1.120	0.63	0.25	2.04	0.25	349	1	277
93	0.050	0.599	1.120	0.63	0.25	2.04	0.25	349	1	332
94	0.050	0.599	1.120	0.63	0.25	2.04	0.25	357	1	216
95	0.050	0.599	1.120	0.63	0.25	2.04	0.25	357	1	203
96	0.050	0.599	1.120	0.63	0.25	2.04	0.25	245	1	129
97	0.050	0.599	1.120	0.63	0.25	2.04	0.25	245	1	134
98	0.050	0.599	1.120	0.63	0.25	2.04	0.25	210	1	193
99	0.074	1.199	2.301	0.58	0.18	1.33	0.18	507	1	834
100	0.083	1.199	2.301	0.58	0.07	1.2	0.07	530	1	804
101	0.120	1.440	3.960	0.18	0.66	1.11	0.66	433	2	2353
102	0.120	1.440	3.960	0.06	0.66	1.11	0.66	437	2	2942
103	0.120	1.440	3.960	0.18	0.66	1.11	0.66	447	2	3138
104	0.120	1.440	3.960	0.18	0.33	0.91	0.33	412	2	1814
105	0.120	1.440	3.960	0.18	0.33	0.91	0.33	437	2	1912

continued on following page

Table 7. Continued

No.	Geometric Properties				Reinforcement Measures			Material	Loading Type T	Output V_{peak} (kN)
	t_w (m)	h_w (m)	l_w (m)	M/Vl_w	ρ_v (%)	ρ_{vall} (%)	ρ_h (%)	f_c' (kPa)		
106	0.120	1.440	3.960	0.18	0.69	1.12	0.66	431	2	2138
107	0.067	0.813	0.508	1.77	1.97	5.68	1.89	660	4	162
108	0.067	0.813	0.508	1.77	2.53	3.37	2.58	629	4	109
109	0.067	0.813	0.508	1.77	2.53	5.91	2.58	676	4	174
110	0.067	0.813	0.508	0.8	1.97	3.14	1.89	625	4	195
111	0.067	0.813	0.508	0.8	1.97	5.68	1.89	602	4	184
112	0.070	0.500	0.861	0.76	0.73	1.22	0.82	620	1	204
113	0.070	0.500	0.861	0.76	0.73	1.22	0.82	652	1	205
114	0.070	0.500	0.861	0.76	0.73	1.22	0.82	788	1	223
115	0.070	0.500	0.861	0.76	0.73	1.22	0.82	788	1	231
116	0.070	0.500	0.861	0.76	0.73	1.22	0.82	839	1	250
117	0.070	0.500	0.861	0.76	0.58	1.12	0.82	379	1	193
118	0.070	0.500	0.861	0.76	0.58	1.12	0.82	379	1	217
119	0.070	0.500	0.861	0.76	0.58	1.12	0.82	625	1	203
120	0.070	0.500	0.861	0.76	0.58	1.12	0.82	646	1	246
121	0.070	0.500	0.861	0.76	0.58	1.12	0.82	635	1	200
122	0.070	0.500	0.861	0.76	0.58	1.12	0.82	635	1	210
123	0.070	0.500	0.861	0.76	0.58	1.12	0.82	826	1	219
124	0.070	0.500	0.861	0.76	0.58	1.12	0.82	778	1	205
125	0.070	0.500	0.861	0.76	0.58	1.12	0.82	725	1	210
126	0.070	0.500	0.861	0.76	0.58	1.12	0.82	1388	1	250
127	0.075	2.019	3.099	0.77	1.2	0.66	1.2	619	6	1618
128	0.075	2.019	3.076	0.76	0.82	0.49	0.76	456	1	1277
129	0.075	2.019	3.086	0.76	0.82	0.48	0.76	395	1	891
130	0.076	0.914	1.219	0.75	0.55	1.64	0.55	1041	1	606
131	0.076	0.914	1.219	0.75	0.55	1.64	0.55	1185	1	633
132	0.076	0.914	1.219	0.75	0.55	1.64	0.55	1046	1	453
133	0.076	0.914	1.219	0.75	1.09	1.96	1.09	1109	1	702
134	0.076	0.914	1.219	0.75	0.23	1.45	0.23	1015	1	553
135	0.076	0.914	1.219	0.75	1.09	1.96	1.09	1056	1	737
136	0.070	0.749	1.001	0.9	1.04	1.52	1.45	649	2	391
137	0.070	0.749	1.001	0.9	1.04	1.52	1.45	691	1	326
138	0.070	0.749	1.001	0.9	1.04	1.52	1.45	588	2	364
139	0.070	0.749	1.001	0.9	1.04	1.52	1.45	625	1	323
140	0.070	0.749	1.001	0.9	1.04	1.52	1.45	784	2	382
141	0.070	0.500	1.001	0.65	0.58	0.86	0.82	563	1	272

continued on following page

Table 7. Continued

No.	Geometric Properties				Reinforcement Measures			Material	Loading Type T	Output V_{peak} (kN)
	t_w (m)	h_w (m)	l_w (m)	M/Vl_w	ρ_v (%)	ρ_{vall} (%)	ρ_h (%)	f_c' (kPa)		
142	0.070	0.500	1.001	0.65	0.58	0.86	0.82	464	1	254
143	0.070	0.500	1.001	0.65	0.58	0.86	0.82	472	1	274
144	0.070	1.250	1.001	1.9	0.58	0.86	0.82	484	1	181
145	0.070	1.250	1.001	1.9	0.58	0.86	0.82	511	1	136
146	0.070	1.250	1.001	1.9	0.58	0.86	0.82	495	1	129
147	0.070	1.250	1.001	1.9	0.58	0.86	0.82	608	1	205
148	0.120	0.749	2.121	0.47	1.32	2.51	1.32	740	1	2657
149	0.120	0.749	2.121	0.47	0.9	2.24	0.9	802	1	2510
150	0.120	0.749	2.121	0.47	1.69	2.75	1.69	753	1	2510
151	0.120	0.749	2.121	0.47	1.32	2.51	1.32	753	1	2481
152	0.075	2.019	3.076	0.78	1.2	1.2	1.2	608	1	1464
153	0.075	2.619	3.076	0.98	1.2	1.2	1.2	633	1	1031
154	0.075	2.619	3.076	0.98	1.2	1.2	1.2	614	1	1170
155	0.150	2.200	2.149	1.12	0.45	1.39	0.45	906	1	2250
156	0.150	1.400	2.149	0.74	0.8	0.8	0.8	799	1	1680
157	0.150	1.400	2.149	0.74	0.6	0.6	0.6	753	1	1740
158	0.150	1.400	2.149	0.74	0.48	0.48	0.48	736	1	1740
159	0.150	1.001	2.149	0.56	0.8	0.8	0.8	843	1	2100
160	0.150	1.001	2.149	0.56	0.6	0.6	0.6	862	1	2190
161	0.150	2.200	2.149	1.12	1.16	3.57	1.16	763	1	2490
162	0.150	2.200	2.149	1.12	0.9	2.77	0.9	723	1	2490
163	0.150	2.200	2.149	1.12	0.72	2.22	0.72	782	1	2430
164	0.150	1.400	2.149	0.74	1.16	1.16	1.16	515	1	1800
165	0.150	1.400	2.149	0.74	0.9	0.9	0.9	585	1	1830
166	0.150	1.400	2.149	0.74	1.16	1.16	1.16	826	1	1890
167	0.150	1.400	2.149	0.74	0.9	0.9	0.9	816	1	2040
168	0.150	1.400	2.149	0.74	0.72	0.72	0.72	788	1	1980
169	0.150	1.001	2.149	0.56	1.16	1.16	1.16	702	1	2250
170	0.150	1.001	2.149	0.56	0.9	0.9	0.9	727	1	2370
171	0.150	1.400	2.149	0.74	1.6	1.6	1.6	576	1	1980
172	0.150	1.400	2.149	0.74	1.16	1.16	1.16	578	1	2040
173	0.150	1.400	2.149	0.74	0.96	0.96	0.96	589	1	2040
174	0.080	1.049	2.301	0.52	1.6	1.51	1.6	707	1	1677
175	0.080	1.049	2.301	0.52	2	1.86	2	728	1	1823
176	0.080	1.049	2.301	0.52	1.2	1.11	1.2	668	1	1515
177	0.080	0.650	2.301	0.35	1.2	1.11	1.2	695	1	1617

continued on following page

Table 7. Continued

No.	Geometric Properties				Reinforcement Measures			Material	Loading Type T	Output V_{peak} (kN)
	t_w (m)	h_w (m)	l_w (m)	M/Vl_w	ρ_v (%)	ρ_{vall} (%)	ρ_h (%)	f_c' (kPa)		
178	0.080	1.450	2.301	0.7	1.2	1.11	1.2	703	1	1343
179	0.080	1.049	2.301	0.52	0.6	0.85	0.6	612	1	1246
180	0.080	1.049	2.301	0.52	0.8	0.94	0.8	623	1	1307
181	0.080	1.049	2.301	0.52	1.2	1.11	1.2	604	1	1146
182	0.080	1.049	2.301	0.52	0	0.59	0	740	1	1193
183	0.080	1.049	2.301	0.52	0.3	0.72	0.3	740	1	1283
184	0.080	1.049	2.301	0.52	2.4	2.04	2.4	707	1	2003
185	0.080	1.049	2.301	0.52	2.8	2.21	2.8	666	1	1732
186	0.080	1.049	2.301	0.52	0	0.59	0	666	1	744
187	0.080	1.049	2.301	0.52	0	0.59	0	666	1	1421
188	0.080	1.049	2.301	0.52	0.6	0.85	0.6	740	1	1151
189	0.080	1.049	2.301	0.52	0.6	0.85	0.6	715	1	1698
190	0.080	0.650	2.301	0.35	0.6	0.85	0.6	707	1	1871
191	0.080	1.450	2.301	0.7	0.6	0.85	0.6	705	1	1275
192	0.080	0.650	2.301	0.35	2	1.86	2	715	1	2081
193	0.080	1.450	2.301	0.7	2	1.86	2	722	1	1656
194	0.080	1.049	2.301	0.52	2	1.86	2	709	1	1712
195	0.080	1.049	2.301	0.52	2	1.86	2	713	1	2035
196	0.080	1.049	2.301	0.52	2	1.45	2	746	1	1775
197	0.080	1.049	2.301	0.52	0.6	0.85	0.6	530	1	1276
198	0.080	1.049	2.301	0.52	1.2	1.11	1.2	544	1	1390
199	0.080	1.049	2.301	0.52	2	1.86	2	530	1	1491
200	0.080	1.049	2.179	0.55	1.2	1.72	1.2	571	1	1088
201	0.080	1.049	2.179	0.55	1.2	1.72	1.2	806	1	1334
202	0.080	1.049	2.179	0.55	1.2	1.72	1.2	1227	1	1461
203	0.080	1.049	2.179	0.55	0.6	1.3	0.6	767	1	1236
204	0.080	1.049	2.179	0.55	2	2.27	2	540	1	1137
205	0.080	1.049	2.179	0.55	2	2.27	2	782	1	1461
206	0.080	1.049	2.179	0.55	2	2.27	2	1223	1	1677
207	0.075	1.100	0.879	1.36	1.29	2.83	0.48	2187	1	845
208	0.075	1.100	0.879	1.36	1.29	2.83	0.48	1955	1	625
209	0.075	1.100	0.879	1.36	0.75	2.59	0.48	1808	1	866
210	0.075	1.100	0.879	1.36	0.75	2.59	0.48	1913	1	801
211	0.075	1.100	0.879	1.36	1.29	2.83	0.75	1766	1	745
212	0.075	1.100	0.879	1.36	1.29	2.83	0.75	1892	1	800
213	0.075	1.100	0.879	1.36	0.75	2.59	0.75	2145	1	845

continued on following page

Table 7. Continued

No.	Geometric Properties				Reinforcement Measures			Material	Loading Type T	Output V_{peak} (kN)
	t_w (m)	h_w (m)	l_w (m)	M/Vl_w	ρ_v (%)	ρ_{vall} (%)	ρ_h (%)	f_c' (kPa)		
214	0.102	1.372	2.743	0.5	0.3	0.3	0.3	435	1	329
215	0.076	1.905	1.905	1.08	0.85	2.32	0.27	905	2	519
216	0.076	1.905	1.905	1.08	2.87	2.87	0.93	905	2	679
217	0.076	1.905	1.905	1.08	0	1.64	0	848	2	306
218	0.076	1.905	1.905	1.08	0	2.36	0.93	808	2	658
219	0.076	1.905	1.905	1.08	2.87	2.87	0.93	914	1	632
220	0.06	1.199	0.599	2.13	0.5	2.83	0.31	812	1	108
221	0.06	1.199	0.599	2.13	0.59	3.02	0.39	673	1	129
222	0.06	1.199	0.599	2.13	0.5	2.93	0.28	963	1	95
223	0.1	0.61	1.001	0.69	0.34	0.34	0.37	1066	1	202
224	0.1	0.61	1.001	0.69	0.34	0.34	0	1072	1	203
225	0.08	0.61	0.899	0.69	0.39	0.39	0.26	423	1	174
226	0.08	0.61	0.899	0.69	0.39	0.39	0.26	513	1	135
227	0.12	1.999	1.001	1	0.25	1.9	0.13	423	1	198
228	0.12	1.999	1.001	1	0.25	1.9	0.25	429	1	270
229	0.12	1.801	1.3	0.69	0.13	1.41	0.25	397	1	364
230	0.12	1.801	1.3	0.69	0.26	1.51	0.25	343	1	374
231	0.1	1.801	1.3	0.69	0.26	1.6	0.26	385	1	258
232	0.08	1.801	1.3	0.69	0.25	1.66	0.25	358	1	187
233	0.1	1.399	1.399	0.5	0.13	1.24	0.26	372	1	304
234	0.1	1.399	1.399	0.5	0.26	1.35	0.26	397	1	289
235	0.08	1.199	1.699	0.35	0.13	0.98	0.26	416	1	368
236	0.08	1.199	1.699	0.35	0.25	1.08	0.25	412	1	362
237	0.1	1.801	1.3	0.69	0	0.92	0	530	1	258
238	0.1	1.801	1.3	0.69	0	0.92	0	376	1	222
239	0.1	1.399	1.399	0.5	0	0.86	0	387	1	262
240	0.1	1.399	1.399	0.5	0	1.3	0.25	522	1	491
241	0.1	1.399	1.399	0.5	0.25	1.06	0	509	1	258
242	0.08	1.049	1.501	0.35	0	1	0	507	1	400
243	0.08	1.049	1.501	0.35	0	1	0	391	1	356
244	0.08	1.049	1.501	0.35	0	1.33	0.25	507	1	391
245	0.16	1.6	1.699	1	0.51	1.54	0.26	363	1	772
246	0.16	1.6	1.699	1	0.51	1.54	0.26	437	1	775
247	0.16	1.6	1.699	1	0.51	1.54	0.57	437	1	790
248	0.16	1.6	1.699	1	0.51	1.54	1.08	386	1	875
249	0.16	1.6	1.699	1	0.51	0.91	0.61	437	1	662

continued on following page

Table 7. Continued

No.	Geometric Properties				Reinforcement Measures			Material	Loading Type T	Output V_{peak} (kN)
	t_w (m)	h_w (m)	l_w (m)	M/Vl_w	ρ_v (%)	ρ_{vall} (%)	ρ_h (%)	f_c' (kPa)		
250	0.16	1.6	1.699	1	0.51	0.91	0.61	289	1	591
251	0.16	1.6	0.851	2	0.4	2.31	0.57	375	1	304
252	0.16	1.6	0.851	2	0.4	2.01	1.08	375	1	319
253	0.16	1.6	0.851	2	0.4	2.01	1.08	437	1	336
254	0.07	0.749	0.749	1.1	2.14	2.49	1.17	1035	2	260
255	0.07	0.749	0.749	1.1	2.14	2.49	1.17	1063	2	340
256	0.07	0.749	0.749	1.1	2.14	2.49	1.17	844	2	320
257	0.07	0.749	0.749	1.1	2.14	2.49	1.17	1023	2	355
258	0.065	1.3	0.65	2.12	2.09	2.62	0.87	999	2	150
259	0.065	1.3	0.65	2.12	2.09	2.62	0.87	940	2	180
260	0.065	1.3	0.65	2.12	2.09	2.62	0.87	950	2	120
261	0.065	1.3	0.65	2.12	1.26	2.14	0.36	1063	1	111
262	0.065	1.3	0.65	2.12	1.26	2.14	0.36	969	1	112
263	0.1	1.199	1.181	1.12	1.02	1.02	1.01	692	2	392
264	0.08	1.1	0.8	1.5	0.71	0.97	0.51	511	3	107
265	0.08	1.1	0.8	1.5	0.71	0.97	0	606	3	90
266	0.08	1.1	0.8	1.5	0.71	0.97	0.51	706	1	89
267	0.08	1.1	0.8	1.5	0.71	0.97	0.51	565	1	129
268	0.1	1.001	1.999	0.58	0.59	0.8	0.26	526	1	532
269	0.1	0.5	1.999	0.33	0.59	0.8	0.26	463	1	685
270	0.1	1.001	1.999	0.58	1.07	1.2	1.2	568	1	544
271	0.1	1.501	1.999	0.82	0.59	0.8	0.8	946	1	375
272	0.1	1.801	1.199	1.6	0.28	0.69	0.28	507	1	173
273	0.1	1.801	1.199	1.6	0.57	1.02	0.57	578	1	187
274	0.1	0.749	1.001	0.9	0.43	0.43	0.57	526	2	252
275	0.1	0.5	1.001	0.65	0.43	0.43	0.57	561	1	299
276	0.1	0.749	1.001	0.9	0.43	0.43	0.57	590	1	245
277	0.1	0.5	1.001	0.65	0.71	0.71	0.71	712	1	324
278	0.1	0.5	1.001	0.65	0.43	0.43	0.57	567	4	189
279	0.1	0.5	1.001	0.65	0.43	0.43	0.57	547	1	179
280	0.1	0.5	1.001	0.65	0.43	0.43	0.57	573	1	189
281	0.1	0.5	1.001	0.65	0.77	0.77	1.03	557	2	279
282	0.1	0.5	1.001	0.65	0.77	0.77	1.03	561	4	272
283	0.1	0.5	1.001	0.65	0.77	0.77	1.03	594	1	276
284	0.1	0.5	1.001	0.65	0.79	0.79	0.57	548	2	248
285	0.1	0.5	1.001	0.65	0.76	0.76	0	681	2	243

continued on following page

Table 7. Continued

No.	Geometric Properties				Reinforcement Measures			Material	Loading Type T	Output V_{peak} (kN)
	t_w (m)	h_w (m)	l_w (m)	M/Vl_w	ρ_v (%)	ρ_{vall} (%)	ρ_h (%)	f_c' (kPa)		
286	0.1	0.5	1.001	0.65	0.76	0.76	1.14	660	2	313
287	0.1	0.749	1.001	0.9	0.76	0.76	0.57	557	2	221
288	0.1	0.749	1.001	0.9	0.79	0.79	0.57	547	1	180
289	0.1	0.749	1.001	0.9	0.78	0.78	0.57	557	2	219
290	0.1	0.749	1.001	0.9	0.76	0.76	0	516	2	157
291	0.1	0.749	1.001	0.9	0.76	0.76	0.57	433	2	211
292	0.1	1.501	3	0.57	0.81	0.98	1.61	572	1	773
293	0.152	1.219	2.438	0.59	0.37	0.37	0.31	616	1	1261
294	0.06	0.8	0.8	1.08	0.22	1.23	0.23	495	1	98
295	0.06	0.8	0.8	1.08	0.44	2.01	0.41	495	1	130
296	0.06	0.8	0.8	1.08	1.17	2.71	1.17	495	1	173
297	0.06	0.8	1.199	0.72	0.24	0.9	0.23	516	1	152
298	0.06	0.8	1.199	0.72	0.44	1.46	0.41	516	1	205
299	0.06	0.8	1.199	0.72	1.17	2.12	1.17	516	1	254
300	0.06	0.8	1.6	0.54	0.22	0.7	0.23	536	1	178
301	0.06	0.8	1.6	0.54	0.37	1.18	0.41	536	1	279
302	0.06	0.8	1.6	0.54	0.8	1.53	0.82	536	1	391
303	0.06	0.8	1.6	0.54	1.17	1.88	1.17	536	1	399
304	0.152	1.219	1.372	0.44	0.25	0.42	0.27	595	1	754
305	0.152	1.219	1.372	0.44	0.25	0.42	0.27	660	1	821
306	0.152	1.219	1.372	0.44	0.25	0.42	0.27	673	1	683
307	0.152	1.219	1.372	0.44	0.25	0.42	0.27	653	1	325

APPENDIX B

Table 8. Testing data sets for RC squat walls

No.	Geometric Properties				Reinforcement Measures			Material	Loading Type T	Output V_{peak} (kN)
	t_w (m)	h_w (m)	l_w (m)	M/Vl_w	ρ_v (%)	ρ_{vall} (%)	ρ_h (%)	f_c' (kPa)		
1	0.051	0.93	1.803	0.64	0.5	1.11	0.5	315	5	489
2	0.051	0.93	1.803	0.64	0.5	1.11	0.5	332	5	520
3	0.051	0.93	1.803	0.64	0.5	1.11	0.5	440	5	578
4	0.076	0.93	1.803	0.64	0.5	0.96	0.5	345	2	409
5	0.076	0.93	1.803	0.64	0.5	0.96	0.5	358	2	503
6	0.051	0.93	3.327	0.34	0.5	0.88	0.5	438	5	783
7	0.051	0.93	3.327	0.34	0.5	0.88	0.5	421	2	489
8	0.102	0.401	1.905	0.25	0.5	2	0.49	541	1	1140
9	0.102	1.829	1.905	1	0.5	2	0.5	493	1	886
10	0.051	0.508	1.778	0.31	0.5	0.92	0.5	555	2	294
11	0.051	0.851	1.727	0.53	0.5	1.4	0.5	444	2	187
12	0.051	0.851	1.727	0.53	0.25	0.97	0.25	423	2	249
13	0.051	0.635	1.219	0.57	0.5	1.15	0.5	339	2	178
14	0.051	0.635	1.219	0.57	0.5	1.15	0.5	339	2	160
15	0.08	1.501	1.999	0.75	0.27	0.44	0.27	404	1	441
16	0.1	1.199	1.181	1.12	1.13	1.13	0.57	748	2	667
17	0.16	1.199	3	0.2	0.8	1.4	0.8	846	1	5300
18	0.2	1.199	3	0.2	0.8	1.32	0.8	846	1	5540
19	0.2	1.199	3	0.2	0.11	0.65	0.11	846	1	3800
20	0.05	0.599	1.12	0.63	0.25	2.04	0.25	388	1	173
21	0.05	0.599	1.12	0.63	0.25	2.04	0.25	388	1	184
22	0.12	1.44	3.96	0.18	0.69	1.12	0.66	412	2	1981
23	0.12	1.44	3.96	0.18	0.77	1.17	0.74	439	2	2305
24	0.1	1.501	3	0.57	0.37	0.75	1.61	547	1	742
25	0.067	0.813	0.508	1.77	1.97	3.14	1.89	635	4	101
26	0.07	0.5	0.861	0.76	0.73	1.22	0.82	677	1	247
27	0.07	0.5	0.861	0.76	0.73	1.22	0.82	675	1	202
28	0.07	0.5	0.861	0.76	0.73	1.22	0.82	620	1	255
29	0.07	0.5	0.861	0.76	0.58	1.12	0.82	1388	1	227
30	0.1	0.749	1.501	0.25	0.8	1.17	0.8	601	1	806
31	0.08	1.75	2.25	0.61	0.16	0.47	0.14	518	1	522
32	0.08	1.75	2.25	0.83	0.16	0.47	0.14	546	1	524
33	0.05	1.75	2.25	0.61	0.26	0.58	0.22	518	1	467
34	0.05	0.749	2.25	0.39	0.79	0.8	0.76	546	1	783

continued on following page

Table 8. Continued

No.	Geometric Properties				Reinforcement Measures			Material	Loading Type T	Output V_{peak} (kN)
	t_w (m)	h_w (m)	l_w (m)	M/Vl_w	ρ_v (%)	ρ_{vall} (%)	ρ_h (%)	f_c' (kPa)		
35	0.076	0.914	1.219	0.75	1.09	1.96	1.09	1174	1	824
36	0.076	0.914	1.219	0.75	0.23	1.45	0.23	1200	1	676
37	0.076	0.914	1.219	0.75	0.23	1.45	0.23	1197	1	474
38	0.07	0.5	1.001	0.65	0.58	0.86	0.82	569	2	316
39	0.07	0.5	1.001	0.65	0.58	0.86	0.82	586	4	291
40	0.07	0.5	1.001	0.65	0.58	0.86	0.82	532	1	262
41	0.07	0.5	1.001	0.65	0.58	0.86	0.82	546	1	267
42	0.07	0.5	1.001	0.65	0.58	0.86	0.82	651	2	306
43	0.12	0.749	2.121	0.47	1.32	2.51	1.32	850	1	2667
44	0.12	1.75	2.121	0.94	1.32	2.51	1.32	697	1	1755
45	0.12	1.75	2.121	0.94	1.32	2.51	1.32	713	1	1648
46	0.15	0.749	2.121	0.47	1.06	2.16	1.06	522	1	2187
47	0.15	1.75	2.121	0.94	1.06	2.16	1.06	544	1	1814
48	0.075	1.42	3.076	0.59	1.2	1.2	1.2	621	1	1671
49	0.075	2.019	3.076	0.78	1.2	1.2	1.2	608	1	1264
50	0.15	1.4	2.149	0.74	1.6	1.6	1.6	879	1	2280
51	0.15	1.4	2.149	0.74	1.16	1.16	1.16	906	1	2340
52	0.15	1.4	2.149	0.74	0.96	0.96	0.96	938	1	2430
53	0.08	1.049	2.301	0.52	1.2	1.11	1.2	715	1	1658
54	0.08	1.049	2.301	0.52	0.6	0.85	0.6	621	1	1179
55	0.08	1.049	2.301	0.52	0.8	0.94	0.8	835	1	1475
56	0.076	1.905	1.905	1.08	2.87	2.87	0.27	893	2	570
57	0.076	1.905	1.905	1.08	0	2.36	0.69	803	2	609
58	0.06	1.199	0.599	2.13	0.5	2.83	0.39	776	1	105
59	0.06	1.199	0.599	2.13	0.59	3.02	0.31	669	1	112
60	0.06	1.199	0.599	2.13	0.5	2.93	0.56	818	1	98
61	0.12	1.999	1.001	1	0.25	2.32	0.38	427	1	324
62	0.12	1.801	1.3	0.69	0.26	1.51	0.13	385	1	309
63	0.1	1.399	1.399	0.5	0.26	1.35	0.13	355	1	235
64	0.08	1.199	1.699	0.35	0.25	1.08	0.13	374	1	255
65	0.1	1.801	1.3	0.69	0	1.71	0.25	530	1	333
66	0.1	1.801	1.3	0.69	0.25	1.12	0	524	1	232
67	0.1	1.399	1.399	0.5	0	0.86	0	524	1	352
68	0.08	1.049	1.501	0.35	0.25	1.2	0	509	1	344
69	0.16	1.6	1.699	1	0.51	1.54	0.57	289	1	812
70	0.16	1.6	1.699	1	0.51	1.54	1.08	309	1	794

continued on following page

Table 8. Continued

No.	Geometric Properties				Reinforcement Measures			Material	Loading Type T	Output V_{peak} (kN)
	t_w (m)	h_w (m)	l_w (m)	M/Vl_w	ρ_v (%)	ρ_{vall} (%)	ρ_h (%)	f_c' (kPa)		
71	0.16	1.6	1.699	1	0.51	0.91	1.08	309	1	657
72	0.16	1.6	1.699	1	0.51	0.91	1.08	386	1	713
73	0.16	1.6	0.851	2	0.4	2.31	0.57	437	1	318
74	0.07	0.749	0.749	1.1	2.14	2.49	1.17	786	2	330
75	0.07	0.749	0.749	1.1	2.14	2.49	1.17	818	2	265
76	0.07	0.749	0.749	1.1	2.14	2.49	0.37	950	2	247
77	0.065	1.3	0.65	2.12	2.09	2.62	0.87	833	2	127
78	0.065	1.3	0.65	2.12	2.09	2.62	0.4	563	2	123
79	0.065	1.3	0.65	2.12	1.26	2.14	0.36	563	2	118
80	0.065	1.3	0.65	2.12	1.26	2.14	0.36	671	1	116
81	0.1	1.199	1.181	1.12	1.02	1.02	0	614	2	342
82	0.08	1.1	0.8	1.5	0.71	1.38	0.51	509	3	127
83	0.1	1.501	1.501	1.09	0.51	0.8	0.8	946	1	225
84	0.1	1.801	1.199	1.6	0.57	1.02	0.57	549	1	198
85	0.1	1.199	1.199	1.1	0.57	1.02	0.57	467	1	260
86	0.1	1.199	1.199	1.1	0.28	0.69	0.28	454	1	185
87	0.1	1.199	1.199	1.1	0.28	0.69	0.28	503	1	251
88	0.1	0.5	1.001	0.65	0.43	0.43	0.57	516	2	341
89	0.1	0.5	1.001	0.65	0.71	0.71	0.71	524	1	315
90	0.1	0.749	1.001	0.9	1.27	1.27	1.18	606	1	336
91	0.1	0.5	1.001	0.65	0.43	0.43	0.57	547	2	210
92	0.1	0.5	1.001	0.65	0.77	0.77	1.03	573	1	242
93	0.1	0.5	1.001	0.65	0.76	0.76	0.57	567	2	270
94	0.1	0.5	1.001	0.65	0.79	0.79	0.57	548	1	222
95	0.1	0.5	1.001	0.65	0.78	0.78	0.57	557	2	248
96	0.1	0.749	1.001	0.9	0.79	0.79	0.57	547	2	222
97	0.06	0.8	0.8	1.08	0.73	2.36	0.82	495	1	161
98	0.06	0.8	1.199	0.72	0.78	1.41	0.82	516	1	210
99	0.06	0.8	1.199	0.72	0.78	1.81	0.82	516	1	243
100	0.06	0.8	1.6	0.54	0.8	1.23	0.82	536	1	322
101	0.152	1.219	1.372	0.44	0.25	0.42	0.27	671	1	649
102	0.152	1.219	1.372	0.44	0.25	0.42	0.27	629	1	405

APPENDIX C: CALCULATION OF BPNN OUTPUT V_{peak}

The transfer functions used for BP neural network output for V_{peak} are '*logsig*' transfer function for hidden layer to output layer and '*purelin*' transfer function for output layer to target. The calculation process of BPNN output for V_{peak} is elaborated in detail as follows:

From the connection weights for a trained neuron network, it is possible to develop a mathematical equation relating the input parameters and the single output parameter Y using

$$Y = f_{sig}\left\{b_0 + \sum_{k=1}^{h}\left[w_k f_{sig}\left(b_{hk} + \sum_{i=1}^{m} w_{ik} X_i\right)\right]\right\} \tag{C-1}$$

in which b_0 is the bias at the output layer, ω_k is the weight connection between neuron k of the hidden layer and the single output neuron, b_{hk} is the bias at neuron k of the hidden layer ($k=1, h$), ω_{ik} is the weight connection between input variable i ($i=1, m$) and neuron k of the hidden layer, X_i is the input parameter i, and f_{sig} is the sigmoid (*logsig* & *purelin*) transfer function.

Using the connection weights of the trained neural network, the following steps can be followed to mathematically express the BPNN model:

Step 1: Normalize the input values for t_w, h_w, l_w, M/Vl_w, ρ_v, ρ_{vall}, ρ_h, $f_c^{'}$ and T linearly using

$$X_{norm} = 2\left(x_{actual} - x_{\min}\right)/\left(x_{\max} - x_{\min}\right) - 1$$

Let the actual $t_w = X_{1a}$ and the normalized $t_w = X_1$

$$X_1 = -1 + 2\times\left(X_{1a} - 0.050\right)/\left(0.200 - 0.050\right) \tag{C-2}$$

Let the actual $h_w = X_{2a}$ and the normalized $h_w = X_2$

$$X_2 = -1 + 2\times\left(X_{2a} - 0.401\right)/\left(2.619 - 0.401\right) \tag{C-3}$$

Let the actual $l_w = X_{3a}$ and the normalized $l_w = X_3$

$$X_3 = -1 + 2\times\left(X_{3a} - 0.508\right)/\left(3.960 - 0.508\right) \tag{C-4}$$

Let the actual $M/Vl_w = X_{4a}$ and the normalized $M/Vl_w = X_4$

$$X_4 = -1 + 2\times\left(X_{4a} - 0.06\right)/\left(2.13 - 0.06\right) \tag{C-5}$$

Let the actual ρ_v= X_{5a} and the normalized ρ_v= X_5

$$X_5 = -1 + 2 \times X_{5a}/2.87 \tag{C-6}$$

Let the actual ρ_{vall} = X_{6a} and the normalized ρ_{vall} = X_6

$$X_6 = -1 + 2 \times (X_{6a} - 0.30)/(5.91 - 0.30) \tag{C-7}$$

Let the actual ρ_h = X_{7a} and the normalized ρ_h = X_7

$$X_7 = -1 + 2 \times X_{7a}/2.80 \tag{C-8}$$

Let the actual f_c'= X_{8a} and the normalized f_c'= X_8

$$X_8 = -1 + 2 \times (X_{8a} - 210)/(2187 - 210) \tag{C-9}$$

Let the actual T= X_{9a} and the normalized T= X_9

$$X_9 = -1 + 2 \times (X_{9a} - 1)/(6 - 1) \tag{C-10}$$

Step 2: Calculate the normalized value (Y_1) using the following expressions:

$$\begin{aligned} A_1 &= -1.425 + 2.560X_1 + 0.444X_2 + 7.885X_3 - 1.439X_4 \\ &-3.669X_5 + 3.943X_6 + 3.994X_7 + 73978X_8 + 3.071X_9 \end{aligned} \tag{C-11}$$

$$\begin{aligned} A_2 &= 30.358 + 23.776X_1 - 7.308X_2 + 82.770X_3 + 0.281X_4 \\ &-5.021X_5 - 1.357X_6 + 63968X_7 + 0.261X_8 + 18.548X_9 \end{aligned} \tag{C-12}$$

$$\begin{aligned} A_3 &= 5.148 - 1.625X_1 + 0.131X_2 + 1.582X_3 - 0.070X_4 \\ &+1.728X_5 + 4.959X_6 - 1.876X_7 + 1.794X_8 + 0.674X_9 \end{aligned} \tag{C-13}$$

$$\begin{aligned} A_4 &= 1.234 + 0.065X_1 - 0.822X_2 + 2.994X_3 + 1.912X_4 \\ &-1.065X_5 - 0.014X_6 + 0.124X_7 - 1.451X_8 + 1.010X_9 \end{aligned} \tag{C-14}$$

$$\begin{aligned} A_5 &= -8.609 - 2.539X_1 + 1.083X_2 - 2.438X_4 + 0.711X_4 \\ &-2.381X_5 - 3.583X_6 - 2.655X_7 - 3.983X_8 - 1.101X_9 \end{aligned} \tag{C-15}$$

$$A_6 = -0.914 + 0.116X_1 + 0.724X_2 - 2.752X_3 - 1.845X_4 + 1.031X_5 + 0.020X_6 - 0.121X_7 + 1.496X_8 - 0.957X_9 \quad \text{(C-16)}$$

$$B_1 = 16.892/\left(1 + \exp\left(-A_1\right)\right) \quad \text{(C-17)}$$

$$B_2 = 0.192/\left(1 + \exp\left(-A_2\right)\right) \quad \text{(C-18)}$$

$$B_3 = 0.345/\left(1 + \exp\left(-A_3\right)\right) \quad \text{(C-19)}$$

$$B_4 = 8.444/\left(1 + \exp\left(-A_4\right)\right) \quad \text{(C-20)}$$

$$B_5 = 0.344/\left(1 + \exp\left(-A_5\right)\right) \quad \text{(C-21)}$$

$$B_6 = 8.753/\left(1 + \exp\left(-A_6\right)\right) \quad \text{(C-22)}$$

$$C_1 = -9.922 + B_1 + B_2 + B_3 + B_4 + B_5 + B_6 \quad \text{(C-23)}$$

$$Y_1 = C_1 \quad \text{(C-24)}$$

Step 3: De-normalize the output to obtain the peak shear strength

$$V_{peak} = 85 + \left(7060 - 85\right) \times \left(Y_1 + 1\right)/2 \quad \text{(C-25)}$$

Chapter 11
Nonlinear System Identification of Smart Buildings

Soroush Mohammadzadeh
University of Oklahoma, USA

Yeesock Kim
Worcester Polytechnic Institute (WPI), USA

ABSTRACT

In this book chapter, a system identification method for modeling nonlinear behavior of smart buildings is discussed that has a significantly low computation time. To reduce the size of the training data used for the adaptive neuro-fuzzy inference system (ANFIS), principal component analysis (PCA) is used, i.e., PCA-based adaptive neuro-fuzzy inference system: PANFIS. The PANFIS model is evaluated on a seismically excited three-story building equipped with a magnetorheological (MR) damper. The PANFIS model is trained using an artificial earthquake that contains a variety of characteristics of earthquakes. The trained PANFIS model is tested using four different earthquakes. It was demonstrated that the proposed PANFIS model is effective in modeling nonlinear behavior of a smart building with significant reduction in computational loads.

INTRODUCTION

The use of smart control devices such as magnetorheological (MR), electrorheological (ER) damper, variable friction devices, and etc. promises a high potential of increase in the reliability of buildings under destructive environmental forces ranging from strong winds to earthquakes. These benefits come with the challenge of modeling the structure integrated with the nonlinear smart dampers. The integration of the building with smart dampers may introduce nonlinearities despite the building behaving linearly. (Kim *et al.* 2009). Thus, it is critically important to develop an effective model such that the nonlinear behavior of a smart building is captured. In this chapter, a method is studied that efficiently (in terms of computational load) identifies nonlinear behavior of seismically excited buildings equipped with smart dampers.

DOI: 10.4018/978-1-5225-0588-4.ch011

An essential part of modeling smart structures is system identification (SI) that develops a numerical model based on experimental data collected from the smart structure. The SI models are used to reliably estimate the complex behavior of the smart structure subject to a variety of dynamic loading scenarios using the inputs (in this case earthquake and control force) and outputs (in this case, displacements on the particular stories). There are two different SI approaches: parametric and nonparametric approaches (Bani-Hani *et al.,* 1999). The parametric SI approach identifies certain properties of the dynamic system to be modeled, e.g. stiffness and damping parameters of a building (Jalili-Kharaajoo, 2004). The nonparametric SI approach uses a black box model that is trained using the input-output relationship of the structural system (Filev, 1991). The nonparametric SI generally does not require accurate physical representation of the structure. Therefore, it is easily applicable to nonlinear modeling of the structural system, which has previously been successfully performed with neural networks and fuzzy logic systems.

One of the most commonly used nonparametric approaches of SI is fuzzy inference system which uses fuzzy set theory to build a set of rules. This can be effective in dealing with nonlinearities and uncertainties of dynamic systems (Gu & Oyadiji, 2008). Fuzzy logic has been used for many SI problems (Takagi & Sugeno, 1985; Yan & Langari, 1998; Kim *et al.* 2011) since the pioneering work of Zadeh (1965). Studies on Takagi-Sugeno (TS) fuzzy models conducted in recent years mostly deal with effective representation of nonlinear systems using fuzzy sets, fuzzy rules and a set of local liner models (Filev, 1991; Golparakrishnan *et al.* 2010; Johansen & Babuska, 2003). One of the main applications of fuzzy logic theory has been nonlinear fuzzy control system design in the field of large-scale infrastructure (Guo *et al.* 2011; Kim *et al.* 2009; Mitchell *et al.* 2012). Estimation of the parameters of a fuzzy inference system requires many trials and errors. Hence, neural networks have been used to determine these parameters.

Artificial neural networks (ANN) was developed inspired by the cognitive mechanism of the human brain (Wang *et al.* 2009). An ANN consists of linked nodes, where each nodes computes an output from its input. A link is created between each two nodes and the output from one node is fed in to other nodes. By adjusting the parameters of the network, the performance of the ANN improves, resulting in a more accurate model. Since ANN models are black box models, it is challenging to design them in a transparent way.

Adaptive neuro-fuzzy inference system (ANFIS) is the result of integrating features of ANN and fuzzy logic models to produce an effective SI model. Although the application of ANFIS for SI in civil engineering has been studied widely, it is still a relatively new research topic (Gu & Oyadiji, 2008; Golparakrishnan, 2010; Schurter *et al.* 2000; Ozbulut *et al.* 2007; Hakim& Abdul Razak, 2013). The main advantages of this modeling technique is to create nonlinear functions using adjustable parameters which include types and the number of membership functions (MF), learning process step size and number of epochs. A main disadvantage of the ANFIS model is that it can be computationally expensive or time-consuming (Wang *et al.* 2009). In real time situations or when dealing with large sets of data, this method can fall out of favor. To resolve this problem, principal component analysis (PCA) is integrated with ANFIS model to reduce the computation load.

PCA was first introduced in the context of data fitting by Pearson (1901). This method has been extensively used for dimensional reduction of measurement data in diverse fields (Jolliffe, 2002). This method can be used to decrease the amount of data needed for further use by discarding redundant data or variables that are less important. Examples of applications of PCA in the field of health monitoring and control of civil engineering includes the work of Sharifi *et al.* (2010) in sensor fault isolation and detection, Kuzniar and Waszczyszyn (2006) in identifying natural periods of buildings from measured data, Mujica *et al.* (2010) and Park *et al.* (2007) to assess and detect damages in civil infrastructure.

The use of PCA as a means of data compression for an efficient training of ANFIS model has been studied in the fields other than civil engineering. Examples include a hybrid PCA-ANFIS measurement system for product quality control by Wayne *et al.* (2004). An intelligent diagnosis system based on PCA and ANFIS has been proposed by Avci and Turkoglu (2009). Another application includes the use of PCA and ANFIS together for diagnosis of lymph disease by Polat and Gundes (2007). To date, however, there has been minimal research on use of PCA combined neuro-fuzzy modeling for estimation of nonlinear response of smart structures. In this chapter, PCA is implemented as a time series data compression method, where segments of data are effectively removed by compression while majority of the variation within the data is still available for modeling. The compressed data is then used for training of an ANFIS model and validated using four actual earthquakes.

BACKGROUND

PANFIS is an integration of PCA, ANN and fuzzy inference system. It can be used to model nonlinear behavior which uses a least-square method and well as back-propagation methods to train the fuzzy inference systems parameters using the PCA-based compressed input and output data sets.

Takagi-Sugeno Fuzzy Model

The backbone for the proposed PANFIS SI algorithm is the Takagi-Sugeno (TS) fuzzy model, which was proposed in 1985 as an effective way for modeling nonlinear dynamic systems. This was accomplished by introducing linear equations in consequent parts of a fuzzy model (Takagi & Sugeno, 1985). This method does not need any defuzzification, which leads to a significant reduction in computational costs. ANFIS is designed with TS fuzzy model, which takes the form

$$\begin{aligned} &R_j : IF\ u_{FZ}^1\ is\ P_{1,j}\ and\ u_{FZ}^2\ is\ P_{2,j} \ldots and\ u_{FZ}^i\ is\ P_{i,j} \\ &Then\ z_{FZ} = f_j\left(u_{FZ}^1, \ldots, u_{FZ}^i\right), j = 1, 2, \ldots, N_r \end{aligned} \tag{1}$$

where R_j is the jth fuzzy rule, N_r is the number of fuzzy rules, $P_{i,j}$ are fuzzy sets centered at the operating j^{th} point, and u_{FZ}^i are premise variables that can be either input or output values. The equation of the consequent part $z_{FZ} = f_j(u_{FZ}^1, \ldots, u_{FZ}^i)$ can have any linear form. . Note that the Equation 1 represents the jth local linear subsystem of a nonlinear system, i.e. a linear system model that is operating in only a limited region. All of the local subsystems are integrated by blending operating regions of each local subsystem using the fuzzy interpolation method as a global nonlinear system

$$y_{FZ} = \frac{\sum_{j=1}^{N_r} W_j\left(u_{FZ}^i\right)\left[f_j\left(u_{FZ}^1, \ldots, u_{FZ}^i\right)\right]}{\sum_{j=1}^{N_r} W_j\left(u_{FZ}^i\right)} \tag{2}$$

where $W_j\left(u_{FZ}^i\right) = \prod_{i=1}^{n} \mu_{P_{i,j}}\left(u_{FZ}^i\right)$ and $\mu_{P_{i,j}}\left(u_{FZ}^i\right)$ is the grade of membership of u_{FZ}^i in $P_{i,j}$. These parameters are optimized by the back-propagation algorithm. A typical architecture of fuzzy rules for a model with n membership functions for each input and n^2 rules are shown in Figure 1.

The parameters of the fuzzy need to be optimized which is another challenge in the application of a fuzzy model. The incorporation of neural networks creates an adaptive inference system that allows for optimization of fuzzy inference system parameters during computation as explained below.

Adaptive Neuro Fuzzy Inference System

The typical architecture of an AFNIS model is demonstrated in Figure 2. The architecture depicted represents one with two inputs and one output with n MFs for each input. The model used in the current study uses only two MFs for each input.

Each layer in the ANN has a particular task to perform before the data moves to the next layer. In the first layer (layer 1), the activation function of the node if represented by

$$F_{FZ}^{1,j} = \mu_{P_{i,j}}\left(u_{FZ}^i\right) \tag{3}$$

In the example of this chapter, Gaussian MFs are used which have the following form

$$\mu_{P_{i,j}}\left(u_{FZ}^i\right) = \exp\left(-\left(u - a_1\right)^2 / 2a_2^2\right) \tag{4}$$

Figure 1. Typical architecture of fuzzy rules for a model with n MFs
Mohammadzadeh et al. 2015.

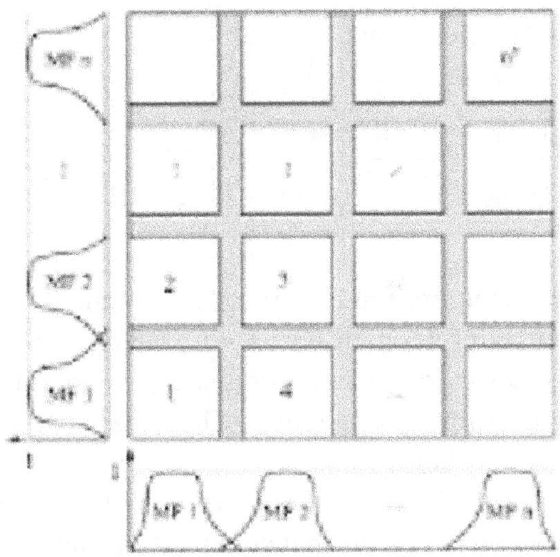

Figure 2. Typical ANFIS architecture with n MFs for each of the two inputs
Mohammadzadeh et al. 2015.

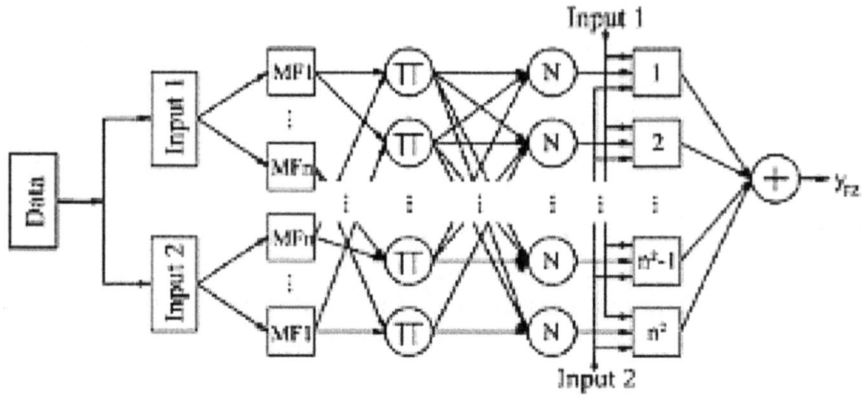

where a_1 and a_2 are parameters to be optimized by the neural network. The MFs are applied to each input in layer 1. The second layer (layer 2) produces the product of all inputs of layer 2, known as firing strengths

$$F_{FZ}^{2,j} = \mu_{P_{i,j}}\left(u_{FZ}^{1}\right) \times \mu_{P_{i,j}}\left(u_{FZ}^{2}\right) \ldots \mu_{P_{i,j}}\left(u_{FZ}^{i}\right) \tag{5}$$

The third layer (layer 3) normalizes the layer 2 outputs by taking a ratio of layer 2 firing strengths as follows

$$F_{FZ}^{3,j} = F_{FZ}^{2,j} \Big/ \prod_{i=1}^{n} \mu_{P_{i,j}}\left(u_{FZ}^{i}\right) \tag{6}$$

Finally, the last layer sums the layer inputs

$$F_{FZ}^{5} = \frac{\sum_j \prod_{i=1}^{n} \mu_{P_{i,j}}\left(u_{FZ}^{i}\right)\left[f_j\left(u_{FZ}^{1}, \ldots, u_{FZ}^{i}\right)\right]}{\sum_j \prod_{i=1}^{n} \mu_{P_{i,j}}\left(u_{FZ}^{i}\right)} \tag{7}$$

The output of this system is then used in a learning algorithm. The important parameters for the learning algorithm are the number of iterations or epochs, the type and number of MFs, and the step size of the algorithm. The MFs may have different forms such as sigmoid function, trapezoidal function, generalized bell function and other forms. Changing each of these parameters affects the learning and the model (Filev, 1991; Kim *et al.* 2011).

Although ANFIS is very effective in modeling nonlinear systems, it is computationally expensive, which can be resolved with integration of principal component analysis.

Principal Component Analysis (PCA)

Principal component analysis was introduced by Pearson (1901) in the context of data fitting and was independently developed by Hotelling (Jolliffe, 2002). The method of derivation used by Hotelling which was by using Lagrange multiplier method and eigenvalue/eigenvector analysis is explained in this section.

Let $\mathbf{V}_{obs.}$ be a matrix of N observations of p real random variables, organized in p rows and N columns. The mean of each random variable is subtracted from each element of the row i. The covariance matrix for this measurement data $\mathbf{V}_{obs.}$ can be constructed as follows

$$\mathbf{C}_{\mathbf{V}_{obs.}} = \frac{1}{N}\mathbf{V}_{obs.}\mathbf{V}_{obs.}^{T} \tag{8}$$

where each element $c_{i,j}$, $i, j = 1, \ldots, p$ of the covariance matrix, is the covariance between the ith and the jth variables. It is desirable to find a linear transformation of the observation matrix with the following form

$$\mathbf{V}_{tran.} = \mathbf{Q}\mathbf{V}_{obs.} \tag{9}$$

which leads to the linear combination of the p random variables having the maximum variance among all other linear combinations. The transformation matrix $\mathbf{Q}$ is desired to be a unit norm matrix, that is $\mathbf{Q}^T\mathbf{Q} = \mathbf{I}$. Therefore, the goal is to find the transformation $\mathbf{Q}$ with the constraint $\mathbf{Q}^T\mathbf{Q} = \mathbf{I}$ which leads to the elements of the covariance matrix being maximized

$$\underset{\mathbf{Q}}{Max}\left\{\mathbf{C}_{\mathbf{V}_{tran.}}\right\} \tag{10}$$

The transformed covariance matrix can be written as

$$\mathbf{C}_{\mathbf{V}_{tran.}} = \left(1/N\right)\mathbf{Q}\mathbf{C}_{\mathbf{V}_{obs.}}\mathbf{Q}^T \tag{11}$$

Using a Lagrange multiplier, the cost function is augmented as

$$\left(1/N\right)\mathbf{Q}\mathbf{C}_{\mathbf{V}_{obs.}}\mathbf{Q}^T - \lambda(\mathbf{Q}^T\mathbf{Q} - \mathbf{I}) \tag{12}$$

where λ is the Lagrange multiplier. To find the solution, Equation 12 is differentiated with respect to the transformation matrix $\mathbf{Q}$ which gives

$$\left(\left(1/N\right)\mathbf{C}_{\mathbf{V}_{obs.}} - \lambda\mathbf{I}\right)\mathbf{Q} = \mathbf{0} \tag{13}$$

$\mathbf{I}$ is an $p \times p$ identity matrix, λ is an eigenvalue of the covariance matrix of the original data, and $\mathbf{Q}$ is found to be the corresponding matrix of eigenvectors of all the eigenvalues λ. Therefore, the eigenvectors of $\mathbf{C}_{\mathbf{v}_{obs.}}$ (i.e. $\mathbf{Q}$) transforms $\mathbf{C}_{\mathbf{v}_{obs}}$ to a covariance matrix: the off-diagonal elements are zero and the diagonal elements have the maximum value given the data. The contribution of each random variable to each principal component is the corresponding element of the transformation matrix. Therefore, the variables with larger (in absolute value) corresponding elements in the transformation matrix contribute more to the variation of the principal components. So, it is possible to discard the variables that contribute less than a user-defined threshold and reduce the size or dimensions of the data needed for training. It should be noted that since in this study, the observations used are samples of an artificial earthquake ground acceleration, the units of the observations are the same and a drawback of the PCA method, that is the sensitivity of the PCA based on covariance to measurement units of observations (Jolliffe, 2002) is not an issue.

As mentioned before, to perform PCA, calculation of eigenvalues/eigenvectors of the covariance matrix of the observations is needed. For large data sets, direct computation of the eigenvalues/eigenvectors are computationally expensive. There are more efficient algorithms for calculation of the principal components such as the power method (Hotelling, 1933), an accelerated version of the power method (Hotelling, 1936), the QL algorithm (Wilkinson, 1965), the EM algorithm (Dempster *et al.*, 1977), and Singular Value Decomposition (SVD). Singular Value Decomposition (SVD) of the observation data matrix $\mathbf{V}_{obs}$ provides the same results, however, due to existence of efficient algorithms for calculation of SVD, it is favoured to direct calculation of principal components and other mentioned algorithms.

Singular Value Decomposition (SVD)

SVD is a factorization of real or complex matrices that has many applications, such as solving homogenous linear equations, least squares minimization, and low-rank approximation to name but a few (Golub & Van Loan, 1996). The process of SVD is as follows. Let $\mathbf{V}_{obs.}$ be the $p \times N$ matrix of observations. Then $\mathbf{V}_{obs.}$ can be factorized into the following form

$$\mathbf{V}_{obs.} = \mathbf{U}\mathbf{S}\mathbf{V}^T \tag{14}$$

where matrix $\mathbf{U}$ is an $p \times p$ orthogonal matrix with columns $\mathbf{u}_i, i = 1, \ldots, p$ that satisfy

$$\mathbf{u}_i^T \mathbf{u}_i = \delta_{ij} \tag{15}$$

where δ is the Kronecker delta function. Matrix $\mathbf{V}$ is an $N \times N$ orthogonal matrix with columns satisfying (15). Matrix $\mathbf{S}$ is a $p \times N$ matrix with singular values on the diagonal and zero off-diagonal elements. The matrix $\mathbf{U}$ diagonalizes the matrix $\mathbf{V}_{obs.}\mathbf{V}_{obs.}^T$ and the non-zero singular values (diagonal elements) of the matrix $\mathbf{S}$ are the square root of the non-zero eigenvalues of $\mathbf{V}_{obs.}\mathbf{V}_{obs.}^T$. The existence of the SVD algorithms is important for computation of principal components in two aspects. First, there is no need for direct computation of principal components, and second, SVD results in a scaled version of principal components where they have a variance of $1 / N$ (Jolliffe, 2002). Although SVD has been used to compute the principal components in this study, however, the discussion of the efficient algorithms for SVD are out of the scope of this chapter.

PCA-BASED ANFIS SYSTEM IDENTIFICATION

In system identification of structural systems, the nonlinear behaviour of the structures can be estimated with the use of black box modelling of measured data. Typically, the measured data include the inputs and outputs to the structure. The input contains earthquake signals and forces due to control devices that may include smart damper. The output data contains responses of the structure to the inputs which are the accelerations and displacements of different stories. In practice, due to the large scale of the structures and the need for more accuracy, the size of measured data for long periods of time can be very large. Therefore, it may be difficult to apply signal processing techniques to the lengthy data sets. Hence, it is of significant importance to reduce the number of data points in the measurement data set for use in modelling, which leads to noticeable reduction in training time of ANFIS models or other machine learning techniques.

An important input to the structural systems is the ground acceleration. By applying only a short duration of acceleration with a large variation, the computational efficiency of the training algorithm can be improved. These large variations can also produce a wide range of complex behavior of the structure which helps performing ac accurate and efficient system identification.

To find a short time-series of ground acceleration with maximum variation, it is proposed to divide the earthquake acceleration time-series $\ddot{\mathbf{x}}_g(t)$, $t \in [0, T]$ to N_t number of time-series

$$\mathbf{v}_i(t),\ t \in \left[(i-1)T/N_t, (i)T/N_t\right],\ i = 1, \ldots, N_t$$

with equal lengths called segments, keeping in mind that each segments should be small enough for a favourable training time and large enough for reasonable accuracy of the training of the ANFIS model. PCA can then be applied to the matrix of the segments

$$\mathbf{v} = \left[\mathbf{v}_1(0, T/N_t)\ \mathbf{v}_2(T/N_t, 2T/N_t) \ldots \mathbf{v}_N\left(\left(N_t - 1\right)T/N_t, T\right)\right] \tag{16}$$

to find the principal components of the time series that have the length T/N_t. The higher PCA coefficients (corresponding elements of the transformation matrix) represent the segments that have wider range of variation. Therefore, instead of training the ANFIS mode using the entire data set, the ANFIS model is trained using only the segments of the data with the highest PCA coefficients. In the example, PCA is applied to the earthquake signal and the PCA coefficients are found. Then the segments of other inputs and outputs corresponding to the selected earthquake segments are used for training. Figure 3 represents a conceptual example on the application of PCA to the input signal. Any signal (Figure 3(a)) can be divided into n segments with equal lengths using the window functions (applied point by point) as in Figure 3(b), and then the PCA is applied to the selected signals. The PCA coefficients for

Figure 3. Conceptual example on how the PCA method is applied to the input signal: (a) a typical input signal; (b) the window function used in this chapter

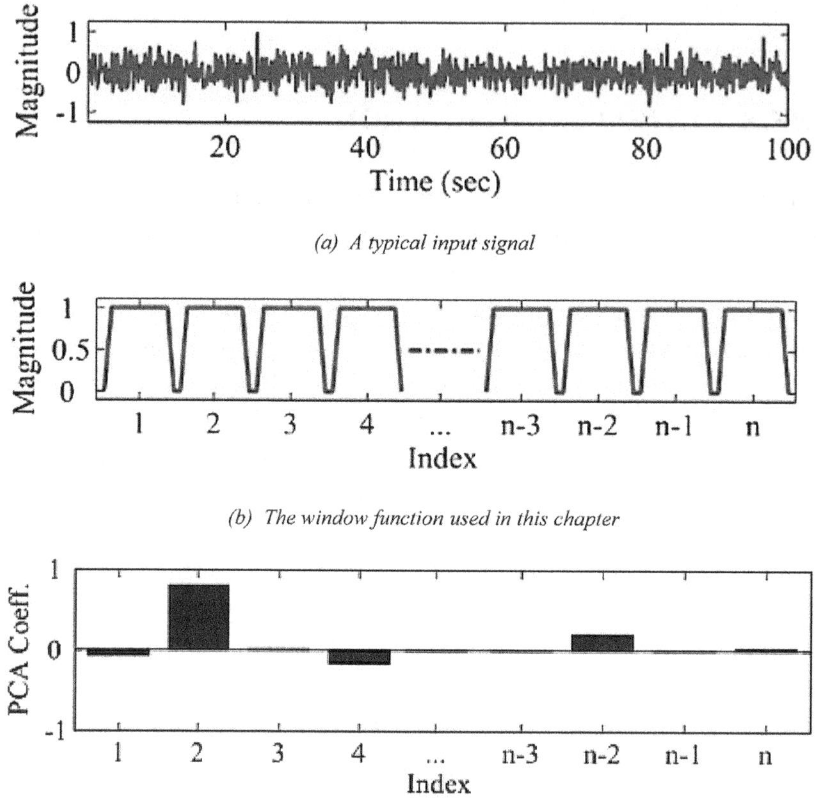

(a) A typical input signal

(b) The window function used in this chapter

the first component produce a signal with the maximum variance among the other linear combinations of the signal segments. Therefore, here, only the first component and its coefficients are considered. The PCA coefficients show where the important signal segments are located within the whole data set: for example, here, the segments number 2, 4, and *n-2* are the ones with the most variation and of importance. Using these three segments results in a training set that has almost the same variation as the original signal.

The higher PCA coefficients represent that the dynamic signals have the broader range of variations, which means that the signal includes the better information. Hence the signal with the highest PCA components is selected in this study. As shown in the concept example in Figure 3, the indices of 2, 4, and *n-2* have the higher PCA values than other indices. Thus, the application of the second index component to identifiers would produce the better identification results. It should be noted that PCA is applied to the earthquake time series to find the major contributing part to make the training data set smaller. Hence, instead of training the ANFIS model using the entire data sets, the ANFIS is developed using the selected time series with the high values of PCA coefficients and the corresponding input-output segments. In the following section, the effectiveness of the PANFIS modelling technique is demonstrated with examples using various earthquake signals.

EXAMPLE

The effectiveness of the proposed PCA-based adaptive neuro-fuzzy inference system (PANFIS) is demonstrated by identification of a three-story building equipped with a magnetorheological (MR) damper.

Magnetorheological (MR) Damper

Smart structures have emerged as a cost-effective structural system due to avoiding extra mass or high cost of control power to control their response. A smart structure is an integration of actuators, sensors, control units, and signal processing units with a structural system. Hence they may be called intelligent structures, adaptive structures, and the related terminologies adaptronics, structronics, etc. The materials commonly used in smart structures include piezoelectrics, shape memory alloys, electrostrictive/magnetostrictive materials, polymer gels, magnetorheological/electrorheological fluids, and etc. (Hurlebaus & Gaul, 2006).

Since semiactive controllers have favourable features of both active and passive control systems, they have gained wide-spread in large structures. These devices include but are not limited to variable-orifice dampers, variable-stiffness dampers, variable-friction dampers, controllable-fluid dampers, shape memory alloy actuators, piezoelectrics, and etc. (Hurlebaus & Gaul, 2006). Due to having many useful characteristics for structural system, controllable-fluid dampers, specifically magnetorheological (MR) dampers have attracted much attention in recent years.

MR dampers consist of a hydraulic cylinder, MR fluid, and magnetic coils. The MR fluid typically contains micron-sized magnetically polarizable particles floating within a fluid that is typically a form of oil. A schematic of a typical MR damper is shown in Figure 4. Without presence of a magnetic field within the fluid, the MR damper acts as a passive damper. However, in the presence of a magnetic field caused by the coils, the MR fluid changes into a semi-solid state in a few milliseconds. This is one of the most unique properties of the MR damper compared to active systems in that a failure in control feedback components, e.g., wires or sensors during a severe earthquake causes an active system to malfunction. However, the semiactive system can still be operated as a passive system in presence of malfunction of control feedback components (Kim *et al.* 2009).

A mathematical model is needed that represents the nonlinear behaviour of the MR damper in order to exploit the behaviour of the MR damper. Since the MR damper is a highly nonlinear hysteretic device, modelling of its behaviour poses a challenge. A modified Bouc-Wen model (Spencer *et al.* 1997) estimates the MR damper force $f_{MR}(t)$ governed by the following equations

$$f_{MR} = c_a \dot{u}_b + k_a \left(u_a - u_{a_0} \right) \tag{17}$$

$$\dot{z}_{BW} = -\gamma \left| \dot{u}_a - \dot{u}_b \right| z_{BW} \left| z_{BW} \right|^{n-1} - \beta \left(\dot{u}_a - \dot{u}_b \right) \left| z_{BW} \right|^{n} + A \left(\dot{u}_a - \dot{u}_b \right) \tag{18}$$

Figure 4. Schematic of the prototype 20-ton large-scale MR damper
Yang et al. 2002.

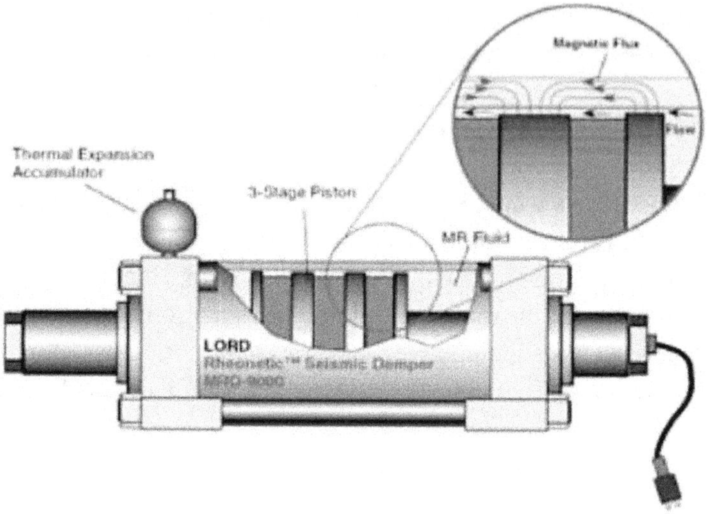

$$\dot{u}_b = \frac{1}{\left(c_a + c_b\right)}\left\{\alpha z_{BW} + c_b\dot{u}_a + k_b\left(u_a + u_b\right)\right\} \tag{19}$$

$$\alpha = \alpha_a + \alpha_b u_{MR} \tag{20}$$

$$c_a = c_{a_1} + c_{a_2} u_{MR} \tag{21}$$

$$c_b = c_{b_1} + c_{b_2} u_{MR} \tag{22}$$

$$\dot{u}_{MR} = -\eta\left(u_{MR} - v_{MR}\right) \tag{23}$$

where z_{BW} and α called evolutionary variables, describe the hysteretic behavior of the MR damper; c_b is the viscous damping parameter at high velocities; c_a is the viscous damping parameter for the force roll-off at low velocities; α_a, α_b, c_{b_1}, c_{b_2}, c_{a_1} and c_{a_2} are parameters that account for the dependence of the MR damper force on the voltage applied to the current driver; k_b controls the stiffness at large velocities; k_a represents the accumulator stiffness; u_{a_0} is the initial displacement of the spring stiffness k_a; γ, β and A are adjustable shape parameters of the hysteresis loops, i.e., the linearity in the unloading and the transition between pre-yielding and post-yielding regions; v_{MR} and u_{MR} are input and output voltages of a first-order filter, respectively and η is the time constant of the first-order filter. In this study, the structure itself is assumed to behave linearly. However, as mentioned before, the introduction of the MR damper to the structural system introduces nonlinearities which necessitate developing a mathematical model to portray this behavior which is usually the key part in the design of semiactive control systems.

Integrated Structure-MR Damper System

The motion of a structure equipped with an MR damper can be expressed with the following second order ODE

$$\mathbf{M}\ddot{\mathbf{x}} + \mathbf{C}\dot{\mathbf{x}} + \mathbf{K}\mathbf{x} = \text{“}\mathbf{f}_{MR}\left(t, \mathbf{x}_i, \dot{\mathbf{x}}_i, v_{MR_i}\right) - \mathbf{M}\text{›}\, \ddot{x}_g \tag{24}$$

where $\ddot{x}_g$ denotes the ground acceleration, $\mathbf{M}$ the mass matrix, $\mathbf{K}$ the stiffness matrix, $\mathbf{C}$ the damping matrix, and the vector $\mathbf{x}$ the displacement relative to the ground, $\dot{\mathbf{x}}$ the velocity, $\ddot{\mathbf{x}}$ the acceleration; $\mathbf{x}_i$ and $\dot{\mathbf{x}}_i$ are the displacement and the velocity at the ith floor level relative to the ground, respectively, v_{MR_i} is the voltage level to be applied, “ and › are location vectors of control forces and ground acceleration signal, respectively. The second order differential equation in Equation 24 can be converted into a state space model

$$\begin{aligned} \dot{\mathbf{z}} &= \mathbf{A}\mathbf{z} + \mathbf{B}\mathbf{f}_{MR}(t, \mathbf{x}_i, \dot{\mathbf{x}}_i, v_i) - \mathbf{E}\ddot{x}_g \\ \mathbf{y} &= \mathbf{C}\mathbf{z} + \mathbf{D}\mathbf{f}_{MR}(t, \mathbf{x}_i, \dot{\mathbf{x}}_i, v_i) + \mathbf{n} \end{aligned} \tag{25}$$

where the following parameters are used

$$\mathbf{A} = \begin{bmatrix} \mathbf{0} & \mathbf{I} \\ -\mathbf{M}^{-1}\mathbf{K} & -\mathbf{M}^{-1}\mathbf{C} \end{bmatrix} \tag{26}$$

$$\mathbf{B} = \begin{bmatrix} \mathbf{0} \\ \mathbf{M}^{-1}\mathbf{F} \end{bmatrix} \tag{27}$$

$$\mathbf{C} = \begin{bmatrix} \mathbf{I} & \mathbf{0} \\ \mathbf{0} & \mathbf{I} \\ -\mathbf{M}^{-1}\mathbf{K} & -\mathbf{M}^{-1}\mathbf{C} \end{bmatrix} \tag{28}$$

$$\mathbf{D} = \begin{bmatrix} \mathbf{0} \\ \mathbf{0} \\ \mathbf{M}^{-1}\mathbf{F} \end{bmatrix} \tag{29}$$

$$\mathbf{E} = \begin{bmatrix} \mathbf{0} \\ \mathbf{F} \end{bmatrix} \tag{30}$$

where $\mathbf{F}$ is the location matrix of Chevron braces within the building structure where the MR dampers are assumed to be installed, $\mathbf{n}$ is the measurement noise vector, $\mathbf{x}_i$ and $\dot{\mathbf{x}}_i$ are the displacement and the velocity at the ith floor level of the three-story building structure, respectively. Properties of the three-story building structure are adopted from Dyke *et al.* (1996). The mass $(\mathbf{M})$, stiffness $(\mathbf{K})$, and damping $(\mathbf{C})$ matrices of this example are as follows

$$\mathbf{M} = \begin{bmatrix} m_1 & 0 & 0 \\ 0 & m_2 & 0 \\ 0 & 0 & m_3 \end{bmatrix} \tag{31}$$

where $m_i, \mathrm{i} = 1,2,3$ are the mass of the 1st through 3rd floor, and

$$\mathbf{K} = \begin{bmatrix} k_1 + k_2 & -k_2 & 0 \\ -k_2 & k_2 + k_3 & -k_3 \\ 0 & -k_3 & k_3 \end{bmatrix} \tag{32}$$

where $k_i, \mathrm{i} = 1,2,3$ are the damping coefficient of the 1st through 3rd floor, and

$$\mathbf{C} = \begin{bmatrix} c_1 + c_2 & -c_2 & 0 \\ -c_2 & c_2 + c_3 & -c_3 \\ 0 & -c_3 & c_3 \end{bmatrix} \tag{33}$$

where $c_i, \mathrm{i} = 1,2,3$ are the damping coefficient of the 1st through 3rd floor. The $\mathbf{I}$ matrix here is an 3×3 identity matrix and $\mathbf{0}$ is a 3×3 zero matrix. The disturbance location matrix $\mathbf{F}$ is also as follows

$$\mathbf{F} = \begin{bmatrix} -1 & 1 & 0 \\ 0 & -1 & 1 \\ 0 & 0 & -1 \end{bmatrix} \tag{34}$$

SOLUTIONS AND RECOMMENDATIONS

A set of input-output data is produced from a structural system integrated with MR dampers to use for training and validation of the proposed PANFIS model. The input data are the MR damping force and the artificial earthquake while the output data consists of the structural systems response. To capture a wide range of nonlinear behavior of the structural system in the PANFIS model, it is critically important to use training data that includes a wide range of spectrums. Therefore, the MR damping force is generated using a random voltage generator so as much voltage information is included in the set of MR damper force as possible. The artificial earthquake (the second input) is generated with a spectral density defined by the Kanai-Tajimi spectrum (Yang *et al.* 1987). The proposed PANFIS modeling framework can be trained by any type of output data such as acceleration, velocities, displacements, or drifts of the storied of the structure to be modeled, the output selected in this study is the acceleration. Accelerometers are relatively cheap and easy to install into large-scale civil infrastructural system, hence the choice of acceleration as the output data. Furthermore, accelerometers provide absolute measurement value as opposed to relative values such as displacements measured by LVDT sensors. This acceleration-based estimation model will be useful in implementing output feedback controllers into large structural systems, i.e. acceleration feedback control systems. It should be noted that many structural control systems are implemented using acceleration responses (Schurter & Roschke, 2001).

The approach proposed in this section is implemented after the generation of the set of input and output data, i.e. PCA is used to compress the five-seconds-long artificial earthquake into a one-second-long signal using the window functions and the linear combinations described. Since the performance of the PANFIS model depends on the training data and the training data in turn depends on the width of the window used, it is reasonable to optimize the width of the window used. In this study, the width of the window was chosen as one-second-long, as it provided reasonable computation time and accuracy in training and validation. Furthermore, the use of adaptive window functions would improve the robustness of the proposed PANIFS model in detecting sudden peak in the structural responses during short periods where impulse-like loading occurs. This reduction in size of the input significantly decreases the computational cost of the modeling, yet provides reasonable accuracy in training and validation.

Generally, the architecture of the PANFIS model is determined via trial-and-error. The number of MFs chosen were selected to be two for both the earthquake signal and MR damper force. The type of the MFs were chosen to be Gaussian, the maximum number of epochs 150 and the step size was chosen as 0.001. The architecture of the PANFIS model can be optimized to best model the input-output relationship, however, this optimization is beyond the scope of this chapter.

The dynamic response of the original model and of the trained PANFIS model are shown as in Figure 5. It should be noted that the original model is an analytical model of the structure equipped with an MR damper and is subject to the artificial earthquake acceleration. Overall, there is a good match between the original and the trained model as seen in Figure 5. By increasing the number and type of modeling parameters, the performance of the model can be improved in training but not without an increase in the computational cost and/or decrease in reliability of the model in validation.

To assess the reliability of the PANFIS model, the trained model is validated using data that is not used for training the model to account for a range of possible earthquakes. These validation earthquakes include El-Centro, Kobe, Hachinohe, and Northridge. Since the original smart building model is a 1/5 scale model structure, the earthquakes used for training and validation are reproduced at a rate five times the recorded one (Chung *et al.* 1989, Dyke *et al.* 1998). Therefore, the width of the window function

Figure 5. Training results: original model response versus the trained model response to the artificial earthquake

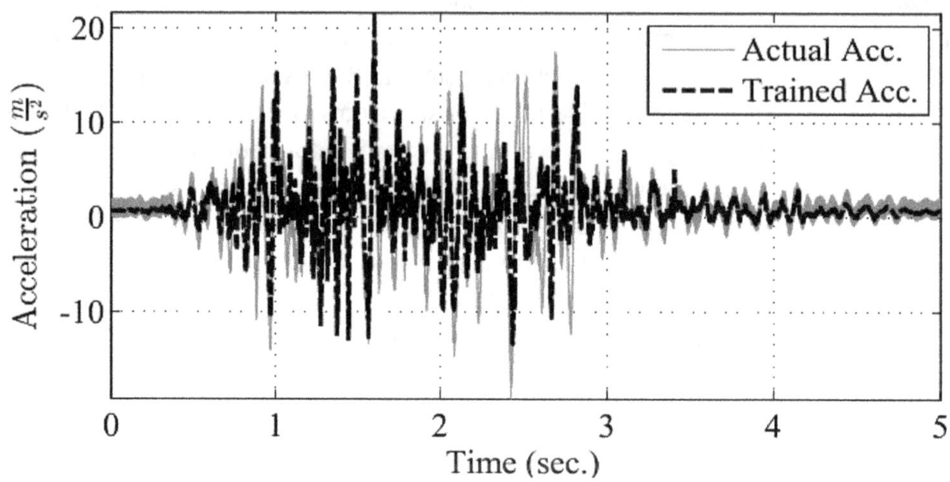

needs to be adjusted accordingly. Figure 6 represents a comparison of the original building third story level acceleration response with the estimated response obtained from the PANFIS model for the El-Centro validation earthquake. The trained and original response correlate well, meaning that the proposed PANFIS model is effective in modeling the complex nonlinear behavior of the structure. The accuracy of the identified model can be improved by increasing the number of MFs or the step size, however at the cost of losing the accuracy in the validation step.

FUTURE RESEARCH DIRECTION

The performance of the PANFIS model proposed in this chapter can be optimized over the type of the MFs used, the number of MFs used for each input, the maximum number of epochs and the step size of the ANFIS model used. Also, a dynamic width selection algorithm ca be devised to optimize the windowing of the input ground acceleration for use in a real-time modeling approach. The validation of the PANFIS model can also be extended to other ground accelerations to further assess the reliability of the model.

CONCLUSION

In this chapter, a computationally efficient PCA-based adaptive neuro-fuzzy inference system (PANFIS) is developed for fast system identification of highly nonlinear behavior of seismically excited structural systems equipped with magnetorheological (MR) dampers. The principal component analysis, artificial neural networks and Takagi-Sugeno fuzzy models are combined to create a powerful method called

Figure 6. Validation result: original model response versus the trained model response to the Kobe earthquake

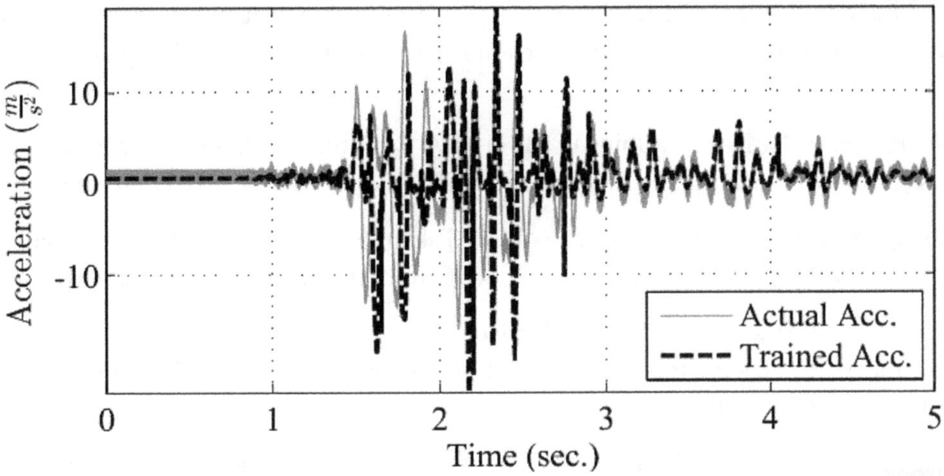

PANFIS. An artificial earthquake signal and a MR damper force signal generated from random voltage are used as a disturbance and control signal respectively, while the acceleration response of the structure is used as output data. To validate the trained model, a variety of earthquake records and MR damper forces and their associated response are used. Decoupling of the identification of the structure and the integrated nonlinear control device is not needed in this approach as the combined effects can be used to train and validate the model. The simulation results demonstrate the effectiveness of the proposed PANFIS model compared to the ANFIS model in identifying the nonlinear behavior of the seismically excited structure while using significantly less computational resources.

REFERENCES

Adeli, H., & Kim, H. (2004). Wavelet-Hybrid Feedback-Least Mean Square Algorithm for Robust Control of Structures. *Journal of Structural Engineering*, *130*(1), 128–137. doi:10.1061/(ASCE)0733-9445(2004)130:1(128)

Avci, E., & Turkoglu, I. (2009). An intelligent Diagnosis System Based on Principal Component Analysis and ANFIS for the Hearth Valve Diseases. *Expert Systems with Applications*, *36*(2), 2873–2878. doi:10.1016/j.eswa.2008.01.030

Bani-Hani, K., Ghaboussi, J., & Schneider, S. P. (1999). Experimental Study of Identification and Control of Structures using Neural Network Part 1: Identification. *Earthquake Engineering & Structural Dynamics*, *28*(9), 995–1018. doi:10.1002/(SICI)1096-9845(199909)28:9<995::AID-EQE851>3.0.CO;2-8

Chen, Y., Yang, B., Abraham, A., & Peng, L. (2007). Automatic Design of Hierarchical Takagi-Sugeno Type Fuzzy Systems using Evolutionary Algorithms. *IEEE Transactions on Fuzzy Systems*, *15*(3), 385–397. doi:10.1109/TFUZZ.2006.882472

Chung, L. L., Lin, R. C., Song, T. T., & Reinhorn, A. M. (1989). Experiments on active control for MDOF seismic structures. *Journal of Engineering Mechanics*, *115*(8), 1609–1627. doi:10.1061/(ASCE)0733-9399(1989)115:8(1609)

Dempster, A. P., Laird, N. M., & Rubin, D. B. (1977). Maximum Likelihood from Incomplete Data via the EM Algorithm. *Journal of the Royal Statistical Society. Series B. Methodological*, *39*(1), 1–38.

Du, H., & Zhang, N. (2008). Application of Evolving Takagi-Sugeno Fuzzy Model to Nonlinear System Identification. *Applied Soft Computing*, *8*(1), 676–686. doi:10.1016/j.asoc.2007.05.006

Dyke, S. J., Spencer, B. F. Jr, Sain, M. K., & Carlson, J. D. (1996). Modeling and Control of Magnetorheological Dampers for Seismic Response Reduction. *Smart Materials and Structures*, *5*(5), 565–575. doi:10.1088/0964-1726/5/5/006

Dyke, S. J., Spencer, B. F. Jr, Sain, M. K., & Carlson, J. D. (1998). An experimental study of MR dampers for seismic protection. *Smart Materials and Structures*, *7*(5), 693–703. doi:10.1088/0964-1726/7/5/012

Filev, D. P. (1991). Fuzzy Modeling of Complex Systems. *International Journal of Approximate Reasoning*, *5*(3), 281–290. doi:10.1016/0888-613X(91)90013-C

Golub, G., & Van Loan, C. F. (2013). Volume 3 of Johns Hopkins Studies in Mathematical Sciences Matrix Computations, Baltimore, MD: John Hopkins University Press.

Gopalakrishnan, K., & Khaitan, S. K., (2010). Finite Element Based Adaptive Neuro-Fuzzy Inference Technique for Parameter Identification of Multi-Layered Transportation Structures. *Transport,* 25(1), 58-65.

Gu, Z., & Oyadiji, S. (2008). Application of MR Damper in Structural Control Using ANFIS Method. *Computers & Structures*, *86*(3-5), 427–436. doi:10.1016/j.compstruc.2007.02.024

Hakim, S. J. S., & Abdul Razak, H. (2013). Adaptive neuro fuzzy inference system (ANFIS) and artificial neural networks (ANNs) for structural damage identification. *Structural Engineering & Mechanics*, *45*(6), 779–802. doi:10.12989/sem.2013.45.6.779

Hotteling, H. (1933). Analysis of a Complex of Statistical Variables Into Principal Components. *Journal of Educational Psychology*, *24*(7), 498–520. doi:10.1037/h0070888

Hotteling, H. (1936). Simplified Calculation of Principal Components. *Psychometrika*, *1*(1), 27–35. doi:10.1007/BF02287921

Housner, G., Bergman, L., Caughey, T., Chassiakos, A., Claus, R., Masri, S., & Yao, J. et al. (1997). Structural control: Past, present, and the future. *Journal of Engineering Mechanics*, *123*(9), 897–971. doi:10.1061/(ASCE)0733-9399(1997)123:9(897)

Hung, S. L., Huang, C. S., Wen, C. M., & Hsu, Y. C. (2003). Nonparametric Identification of a Building Structure from Experimental Data using Wavelet Neural Network. *Computer-Aided Civil and Infrastructure Engineering*, *18*(5), 356–368. doi:10.1111/1467-8667.t01-1-00313

Hurlebaus, S., & Gaul, L. (2006). Smart Structure Dynamics. *Mechanical Systems and Signal Processing*, *20*(2), 255–281. doi:10.1016/j.ymssp.2005.08.025

Jalili-Kharaajoo, M. (2004). Nonlinear system identification using ANFIS based on emotional learning. InAdvances in Artificial Intelligence,*LNCS* (Vol. *3315*, pp. 697–707). doi:10.1007/978-3-540-30498-2_70

Jang, J.-S. R. sd(1993). ANFIS: Adaptive-network-based Fuzzy Inference System. *IEEE Transactions on Systems, Man, and Cybernetics*, *23*(3), 665–685. doi:10.1109/21.256541

Jang, J.-S. R., Sun, C.-T., & Mizutani, E. (1997). *Neuro-Fuzzy and Soft Computing*. Upper Saddle River, New Jersey: Prentice Hall.

Johansen, T. A. (1994). Fuzzy Model Based Control: Stability, Robustness, and Performance Issues. *IEEE Transactions on Fuzzy Systems*, *2*(3), 221–234. doi:10.1109/91.298450

Johansen, T. A., & Babuškam, R. (2003). Multiobjective Identification of Takagi-Sugeno Fuzzy Models. *IEEE Transactions on Fuzzy Systems*, *11*(6), 847–860. doi:10.1109/TFUZZ.2003.819824

Jolliffe, I. T. (2002). *Principal component analysis*. New York: Springer.

Kim, Y., Hurlebaus, S., & Langari, R. (2011). MIMO Fuzzy Identification of Building-MR Damper System. *Journal of Intelligent and Fuzzy Systems*, *22*(4), 185–205.

Kim, Y., Hurlebaus, S., Sharifi, R., & Langari, R. (2009). Nonlinear Identification of MIMO Smart Structures.*ASME Dynamic Systems and Control Conference* (Vol. 1, pp. 33-40), Hollywood: California.

Kim, Y., & Langari, R. (2007). Nonlinear Identification and Control of a Building Structure with a Magnetorheological Damper System.*American Control Conference* (Vol. 1, pp. 3353-3358), New York: NY.

Kim, Y., Langari, R., & Hurlebaus, S. (2009). Semiactive Nonlinear Control of a Building with a Magnetorheological Damper System. *Mechanical Systems and Signal Processing*, *23*(2), 300–315. doi:10.1016/j.ymssp.2008.06.006

Kuzniar, K., & Waszczyszyn, Z. (2006). Neural networks and Principal Component Analysis for Identification of Building Natural Periods. *Journal of Computing in Civil Engineering*, *20*(6), 431–436. doi:10.1061/(ASCE)0887-3801(2006)20:6(431)

Langari, R. (1999). Past, Present and Future of Fuzzy Control: A Case for Application of Fuzzy Logic in Hierarchical Control.*Proceedings of 18th International Conference of the North American Fuzzy Information Processing Society-NAFIPS*, (pp. 760-765) New York City, New York, USA. doi:10.1109/NAFIPS.1999.781796

Lin, J. W., & Betti, R. (2004). On-line Identification and Damage Detection in Non-linear Structural Systems using a Variable Forgetting Factor Approach. *Earthquake Engineering & Structural Dynamics*, *33*(4), 419–444. doi:10.1002/eqe.350

Mohammadzadeh, S., Kim, Y., & Ahn, J. (2015). PCA-based Neuro-Fuzzy Model for System Identification of Smart Structures. *Smart Structures and Systems, 15*(4), 1139–1158. doi:10.12989/sss.2015.15.4.1139

Mujica, L. E., Rodellar, J., Fernandez, A., & Guemes, A. (2010). Q-statistic and T2-statistic PCA-based measures for damage assessment in structures. *Structural Health Monitoring, 10*(5), 539–553. doi:10.1177/1475921710388972

Ozbulut, O. E., Mir, C., Moroni, M. O., Sarrazin, M., & Roschke, P. N. (2007). A Fuzzy Model of Superelastic Shape Memory Alloys for Vibration Control in Civil Engineering Applications. *Smart Materials and Structures, 16*(3), 818–829. doi:10.1088/0964-1726/16/3/031

Palau, C. V., Arregui, F. J., & Carlos, M. (2012). Burst Detection in Water Networks Using Principal Component Analysis. *Journal of Water Resources Planning and Management, 138*(1), 47–54. doi:10.1061/(ASCE)WR.1943-5452.0000147

Park, S., Lee, J., Yun, C., & Inman, D. (2007). Electro-Mechanical Impedance-Based Wireless Structural Health Monitoring Using PCA-Data Compression and *k*-means Clustering Algorithms. *Journal of Intelligent Material Systems and Structures, 19*(4), 509–520. doi:10.1177/1045389X07077400

Pearson, K. (1901). On lines and planes of closest fit to systems of points in space. *Philosophical Magazine, 2*(6), 559–572. doi:10.1080/14786440109462720

Polat, K., & Gunes, S. (2007). Automatic Determination of Diseases Related to Lymph System from Lymphography Data Using Principal Component Analysis (PCA), Fuzzy Weighting Pre-Processing and ANFIS. *Expert Systems with Applications, 33*(3), 636–641. doi:10.1016/j.eswa.2006.06.004

Ramallo, J. C., Yoshioka, H., & Spencer, B. F. Jr. (2004). A Two-step Identification Technique for Semi-active Control Systems. *Structural Control and Health Monitoring, 11*(4), 273–289. doi:10.1002/stc.43

Schurter, K. C., & Roschke, P. (2000). Fuzzy Modeling of Magnetorheological Damper Using ANFIS. *Proceedings of the 9th IEEE International Conference on Fuzzy Systems*, San Antonio, TX, USA (Vol. 1, pp. 122-127).

Schurter, K. C., & Roschke, P. (2001). Neuro-fuzzy control of structures using acceleration feedback. *Smart Materials and Structures, 10*(4), 770–779. doi:10.1088/0964-1726/10/4/322

Sharifi, R., Kim, Y., & Langari, R. (2010). Sensor Fault Isolation and Detection of Smart Structures. *Smart Materials and Structures, 19*(10), 5001–5016. doi:10.1088/0964-1726/19/10/105001

Smyth, A. W., Masri, S. F., Chassiakos, A. G., & Caughey, T. K. (1999). On-line Parametric Identification of MDOF Nonlinear Hysteretic Systems. *Journal of Engineering Mechanics, 125*(2), 133–142. doi:10.1061/(ASCE)0733-9399(1999)125:2(133)

Spencer, B. F. Jr, Dyke, S. J., Sain, M. K., & Carlson, J. D. (1997). Phenomenological Model for Magnetorheological Dampers. *Journal of Engineering Mechanics, 123*(3), 230–238. doi:10.1061/(ASCE)0733-9399(1997)123:3(230)

Takagi, T., & Sugeno, M. (1985). Fuzzy Identification of Systems and Its Applications to Modeling and Control. *IEEE Transactions on Systems, Man, and Cybernetics, 15*(1), 116–132. doi:10.1109/TSMC.1985.6313399

Wang, H., & Hu, H. (2009). Hierarchical Fuzzy Identification of MR Damper.*Proceedings of SPIE 7493, Second International Conference on Smart Materials and Nanotechnology in Engineering*, Weihai, China. doi:10.1117/12.843077

Wang, L., & Fu, K. (2009). *Artificial Neural Networks*. Wiley Encyclopedia of Computer Science and Engineering. doi:10.1002/9780470050118.ecse021

Wang, L., & Langari, R. (1995). Decomposition Approach for Fuzzy Systems Identification.*Proceedings, the 34th IEEE Conference on Decision and Control*, New Orleans, LA, USA (pp. 261-265).

Warne, K., Prasad, G., Siddique, N. H., & Maguire, L. P. (2004). *Development of a Hybrid PCA-ANFIS Measurement System for Monitoring Product Quality in the Coating Industry. Systems, Man and Cybernetics, IEEE international Conference* (Vol. 4, pp. 3519–3524). UK: Deny.

Wilkinson, J. H. (1965). *The Algebraic Eigenvalue Problem*. Oxford: Oxford University Press.

Yager, R. R., & Filev, D. P. (1993). Unified Structure and Parameter Identification of Fuzzy Models. *IEEE Transactions on Systems, Man, and Cybernetics*, *23*(4), 1198–1205. doi:10.1109/21.247902

Yan, G., & Zhou, L. L. (2006). Integrated Fuzzy Logic and Genetic Algorithms for Multi-objective Control of Structures using MR Dampers. *Journal of Sound and Vibration*, *296*(1-2), 368–382. doi:10.1016/j.jsv.2006.03.011

Yan, J., & Langari, R. (1998). *Fuzzy Logic-Intelligence, Control, and Information, Upper Saddle River, New Jersey*. USA: Prentice Hall.

Yang, G., Spencer, B. F. Jr, Carlson, J. D., & Sain, M. K. (2002). Large-scale MR Fluid Dampers: Modeling and Dynamic Performance Considerations. *Engineering Structures*, *24*(3), 309–323. doi:10.1016/S0141-0296(01)00097-9

Yang, Y. N., Akbarpour, A., & Ghaemmaghami, P. (1987). New Optimal Control Algorithms for Structural Control. *Journal of Engineering Mechanics*, *113*(0), 1369–1386. doi:10.1061/(ASCE)0733-9399(1987)113:9(1369)

Yang, Y. N., & Lin, S. (2004). On-line Identification of Non-linear Hysteretic Structures using an Adaptive Tracking Technique. *International Journal of Non-linear Mechanics*, *39*(9), 1481–1491. doi:10.1016/j.ijnonlinmec.2004.02.010

Yang, Y. N., & Lin, S. (2005). Identification of Parametric Variations of Structures based on Least Squares Estimation and Adaptive Tracking Technique. *Journal of Engineering Mechanics*, *131*(3), 290–298. doi:10.1061/(ASCE)0733-9399(2005)131:3(290)

Zadeh, L. A. (1965). Fuzzy Sets. *Information and Control*, *8*(3), 338–353. doi:10.1016/S0019-9958(65)90241-X

Chapter 12
Polynomial Correlated Function Expansion

Souvik Chakraborty
Indian Institute of Technology Roorkee, India

Rajib Chowdhury
Indian Institute of Technology Roorkee, India

ABSTRACT

Uncertainty quantification of civil engineering structures are often time consuming. Hence, one has to rely on equivalent models (EM) that represent the original functions with a specific level of accuracy. In this chapter, a fully equivalent operational model, referred as, Polynomial Correlated Function Expansion (PCFE) has been discussed. PCFE facilitates a systematic mapping between the input and output by expressing the output as a ranked order of component functions, with higher order component functions representing higher-order cooperative effect. The component functions are expressed in terms of extended bases and the unknown coefficients associated with the bases are determined by employing homotopy algorithm (HA). HA determines the unknown coefficients by minimizing the least-squared error and imposing an additional criteria defined as an objective function. Implementation of PCFE has been illustrated with two large scale problems. Results obtained have been compared with other popular techniques. For both the problems, PCFE outperforms popular EMs available in literature.

INTRODUCTION

Uncertainties in load and system properties play a significant role in reliability analysis of structural/mechanical systems. The subject of uncertainty quantification has evolved over the last few decades to deal with various uncertainties. A well-developed body of literature (Ditlevsen & Madsen, 1996; Melchers, 1999) now exists which documents the status of this subject. However, most of the available methods for uncertainty quantification are suitable for simple academic problems involving very few random variables. Difficulties in the analysis of structures with parameter uncertainties primarily arise due to nonlinear variation of the response variables to the stochastic system parameters; this being true even

DOI: 10.4018/978-1-5225-0588-4.ch012

when structures are idealized to display linear material and deformation characteristics. The problem is further enhanced when nonlinear structural behavior is included in the analysis.

Unfortunately, stochastic methods that are utilized for uncertainty quantification are inaccurate and/or computationally inefficient when:

1. The input-output relationship is highly nonlinear;
2. The number of input random variables or fields is large; and
3. There are large statistical variations in input.

For example, the most common approach for uncertainty quantification based on Taylor's series expansion are not adequate for highly nonlinear problems. Simulation methods, which are usually employed for obtaining benchmark results, are not computationally efficient, and are not suitable to being embedded in optimization processes. Hence, new stochastic methods are needed to achieve greater accuracy and/or better efficiency than traditional methods.

In order to alleviate the computational issue associated with uncertainty quantification, various equivalent models (EM) has been developed. EM are utilized to generate surrogates for original function based on actual simulation at significantly reduces number of sample points. Parameters/unknown coefficients of the EM are determined by minimizing the difference between the two models (*i.e.,* original and equivalent) in some statistical sense. Once the unknown coefficients are determined, uncertainty of the response is quantified by performing simulations on the generated equivalent model. Application of EM can be found in design optimization (Dubourg, Sudret, & Bourinet, 2011), design space exploration (Adhikari, Chowdhury, & Friswell, 2011), sensitivity analysis (Sobol, 1993) *etc.*

The most popular method for constructing EM is the response surface method (RSM) based on least-square regression (LSR) (Wu, 2009). In RSM, an EM is constructed by utilizing the second order polynomials. The unknown coefficients are determined by minimizing the least squared error. Number of modifications to the conventional RSM (such as RSM based on least-angle regression, high-order stochastic response surface method (HO-SRSM) (Gavin & Yau, 2008) and RSM based on moving least-square regression (Goswami, Chakraborty, & Ghosh, 2013)), have been suggested by the researchers. However, all the variants of RSM yield satisfactory results only for problems having a lower degree of nonlinearity. Another method that is quite popular among researchers is the Kriging or Gaussian process regression (GPR) (Echard, Gayton, Lemaire, & Relun, 2013). In GPR, the interpolated values are modelled by Gaussian process governed by prior covariances, as opposed to a piece-polynomial spline to optimize smoothness of the predicted values. The basic idea of GPR is to predict the value of a function at a given point by computing a weighted average of the known values of the function in the neighbourhood of the point. Although GPR is mathematically closely related to the conventional regression analysis, the basic difference resides in the fact that while conventional regression is based on multiple observations of a multivariate data set, GPR is used for estimation of a single realization. However, GPR requires significant number of sample points (actual experiments/FE analysis), rendering the process inefficient for high dimensional problems.

A distinct approach to those discussed above is the polynomial chaos expansion (PCE) (Wiener, 1938). This method was developed as an extension of Volterra's theory of nonlinear functionals for stochastic systems and utilized Hermite polynomials for modelling stochastic process with Gaussian random variables. Xiu (Xiu & Karniadakis, 2002) in 2002 generalised the concept of PCE by coupling Askey-scheme for orthogonal polynomials with the PCE. The unknown coefficients associated with the bases are determined by utilizing the Galerkin projection. Application of PCE can be observed in various fields including fluid dynamics, flow through porous media, steady-state diffusion problem, uncertainty quantification, reliability based design optimization. However, PCE suffers from the *curse of dimensionality* and is thus unsuitable for problems that are truly high dimensional in nature. Other popular EM includes Bayesian linear regression (Hu, Zhao, & Lian, 2014; von Toussaint, 2015), Bayesian emulators (DiazDelaO, Adhikari, Saavedra Flores, & Friswell, 2013), high dimensional model representation (HDMR) (Chowdhury, Rao, & Prasad, 2007, 2009) and radial basis function (RBF) (Deng, Gu, Li, & Yue, 2005; Jamshidi & Kirby, 2010).

The primary concern regarding the EMs discussed above is mainly threefold. Firstly, application of all the methods discussed above is limited to small-scale problems only. This is because the number of sample points (actual experiments/FE analysis) required using conventional methods increases factorially with increase in number of input variables, which renders the procedure inefficient. Secondly, apart from a few exceptions (for e.g. HDMR), all the methods yield accurate results only for problems having a lower order of nonlinearity. For highly nonlinear problems, results obtained are unsatisfactory. Last but not the least, all the methods are applicable for problems involving uncorrelated variables only. If the system under consideration is having correlated variables, one has to rely on some *ad hoc* procedures for transforming the dependent variables into independent ones. This whole procedure compromises the accuracy of the methods.

The primary focus of this chapter is to present a fully operational equivalent model, referred to as Polynomial Correlated Function Expansion (PCFE) (Chakraborty & Chowdhury, 2014, 2015a, 2015b; Chakraborty, Mandal, Chowdhury, & Chakrabarti, 2016) for uncertainty quantification. PCFE can be viewed as an extension of Analysis of Variance (ANOVA) decomposition (Sobol', 2003) where the component functions are represented in terms of extended bases (Li et al., 2006). In literature, PCFE is often referred as generalized ANOVA (Hooker, 2007) or generalized HDMR (Li & Rabitz, 2012). The unknown coefficients associated with the bases are determined by employing Homotopy Algorithm (HA) (Chakraborty & Chowdhury, 2015c; Li & Rabitz, 2010; Li, Rey-de-Castro, & Rabitz, 2012).

Compared to popular EMs, PCFE has certain desirable features. Firstly, PCFE is convergent in mean-square sense. This is because, the unknown coefficients associated with the bases are determined by minimizing the L^2 error norm. Secondly, PCFE is a convergent series consisting of 2^N component functions. Thus, if the component functions are convergent, PCFE yields a convergent solution. Thirdly, PCFE is optimal in Fourier sense because apart from minimizing the L^2 error norm, it also ensures the orthogonality of the component functions. Finally, PCFE is capable of dealing with both correlated and uncorrelated random variables without the need of any *ad hoc* transformations.

In the chapter, detailed description of PCFE along with HA will be discussed. The formulation of objective function (to be used in conjunction with HA) by considering hierarchical orthogonality of the component functions of PCFE will also be presented. Performance of PCFE, as compared to other conventional techniques, shall be illustrated with two large scale problems. Finally, the chapter will be concluded by highlighting some possible future applications of PCFE.

PROBLEM DEFINITION

Suppose, $\mathbf{x} = \left(x_1, x_2, \ldots, x_N\right) : \Omega \to \mathbb{R}^N$ be a $N-$ dimensional random vector (denoting the input variables) with cumulative distribution function (CDF) $F_{\mathbf{x}}\left(\mathbf{X}\right) = \mathbb{P}\left(\mathbf{x} \le \mathbf{X}\right)$, where $\mathbb{P}$ denotes probability, Ω is the probability space and $\mathbf{x} \in \mathbb{R}^N$. Also assume Y denotes the output response such that,

$$Y = g\left(\mathbf{x}\right) \tag{1}$$

Naturally the uncertainty in the input variables $\mathbf{x}$ propagates and therefore, the output response Y is also uncertain. This uncertainty needs to be quantified. Application of uncertainty quantification can be observed in various disciplines including structural reliability, reliability based design optimization, robust design optimization and bound estimation. In this chapter, uncertainty quantification for structural reliability analysis will be discussed.

Traditionally, a reliability problem is defined by a performance function, often known as limit-state function $\mathcal{J}\left(\mathbf{x}\right)$, where $\mathcal{J}\left(\mathbf{x}\right) < 0$ denotes the failure domain Ω_F and $\mathcal{J}\left(\mathbf{x}\right) \ge 0$ denoting the safe region, *i.e.*,

$$\Omega_F \triangleq \left\{\mathbf{x} : \mathcal{J}\left(\mathbf{x}\right) < 0\right\} \tag{2}$$

The limit-state function $\mathcal{J}\left(\mathbf{x}\right)$ can be related to the output response as:

$$\mathcal{J}\left(\mathbf{x}\right) = y_{\text{thres}} - g\left(\mathbf{x}\right) \tag{3}$$

where y_{thres} denotes the threshold point. The failure probability P_f is defined as

$$P_f = \mathbb{P}\left(Y \in \Omega_F\right) = \int_{\Omega_F} dF_Y\left(Y\right) = \int_{\Omega} \xi_{\Omega_F}\left(Y\right) dF_Y\left(Y\right) \tag{4}$$

where $\xi_C\left(Y\right)$ is a characteristics function and satisfies

$$\xi_C\left(Y\right) = \begin{cases} 1 \ if \ Y \in C \\ 0 \ if \ Y \notin C \end{cases} \tag{5}$$

It is apparent that the limit-state function $\mathcal{J}\left(\mathbf{x}\right)$ plays an important role in determining the probability of failure. The most popular and widely accepted approach for evaluating failure probability is the Monte Carlo simulation (MCS) (Muscolino, Ricciardi, & Cacciola, 2003; Rubinstein, 1981; Shinozuka, 1972; Tamimi, Amadei, & Frangopol, 1989; Thakur & Misra, 1978). The foundation of this method is based on an algorithm for generating random numbers. Failure probability is obtained within some desired confidence bound by carrying out large number of simulation at randomly generated sample points and counting the number of failure points. MCS is easy to implement and robust because it gives the confidence interval for the estimate. However, MCS is computationally expensive specifically for problems that are already complicated in deterministic state.

As an alternative, an EM for the output/limit-state function belonging to a specific space (such as the space of random variables with finite variance) can be constructed. In this chapter, an efficient EM, referred to as PCFE has been utilized to generate an EM that represent the actual limit-state function in a hierarchical order of component functions. In the remainder of this chapter, procedure for constructing an expression based on the PCFE has been described. A homotopy algorithm for determining the unknown coefficients associated with the PCFE basis has also been presented. Lastly, two large scale problems are addressed to illustrate the performance of PCFE in structural reliability analysis.

POLYNOMIAL CORRELATED FUNCTION EXPANSION

From ANOVA to PCFE

ANOVA decomposition is a set of quantitative model assessment and analysis tool used which is used to capture high dimensional relationship between input and output model variables (Chowdhury et al., 2007, 2009). It is a well-known fact that most real-life problems only exhibits a lower order co-operative effect (Alis & Rabitz, 2001; Rabitz & Aliş, 1999). ANOVA decomposition utilizes this property to formulate an accurate hierarchical expression of the physical system. It is worth mentioning that ANOVA decomposition is also referred as HDMR (Alis & Rabitz, 2001; Li, Wang, Rosenthal, & Rabitz, 2001; Rabitz & Aliş, 1999) or dimensional decomposition (Rahman, 2006; Xu & Rahman, 2004) in some literatures.

Considering the $N-$dimensional random vector (denoting the input variables) defined previously, the response variable $Y=g(\mathbf{x})$ can be represented in a hierarchical order of component function as (Sobol, 1993)

$$g(\mathbf{x})=g_0+\sum_{k=1}^{N}\sum_{i_1\le i_2\le\cdots\le i_k}^{N} g_{i_1 i_2\ldots i_k}\left(x_{i_1},x_{i_2},\ldots,x_{i_k}\right) \tag{6}$$

Here, g_0 represents the mean response. Equation 6can also be represented as:

$$g(\mathbf{x})=g_0+\sum_{i_1} g_i\left(x_{i_1}\right)+\sum_{1\le i_1<i_2\le N} g_{ij}\left(x_{i_1},x_{i_2}\right)+\cdots+g_{12\ldots N}\left(x_1,x_2,\ldots,x_N\right) \tag{7}$$

Definition 1: The terms in Equation 7 reflecting independent effect only are termed as first order component function and denoted as $g_{i_1}\left(x_{i_1}\right)$. Similarly, the terms reflecting bivariate effect are termed as second order component function and denoted by $g_{i_1 i_2}\left(x_{i_1},x_{i_2}\right)$. The constant g_0 is termed as zeroth order component function.

Remark 1: First order component function does not indicate linear variation and consists of terms having high degree of non-linearity. The degree of non-linearity of a PCFE expression is controlled by number of basis.

Considering $x_1, x_2, \ldots, x_N$ to be independent, all the component functions, over the entire problem space of $\mathbf{x}$ $\left(\left[p_i, q_i\right], i = 1, 2, \ldots, N\right)$, can be determined by imposing a vanishing condition (Wang, Georgopoulos, Li, & Rabitz, 2003)

$$\int_{p_i}^{q_i} \varpi_k\left(x_k\right) g_{i_1 i_2 \ldots i_m}\left(x_{i_1}, x_{i_2}, \ldots, x_{i_m}\right) dx_k = 0, \forall k \in \left\{i_1, i_2, \ldots, i_m\right\} \tag{8}$$

where, ϖ_k denotes the pdf of x_k. Utilizing Equation 8, the resulting component function are given as

$$\begin{aligned} g_0 &= \left\langle g\left(\mathbf{x}\right)\right\rangle \\ g_i\left(x_i\right) &= \left\langle g\left(\mathbf{x}\right) \middle| x_i\right\rangle - \left\langle g\left(\mathbf{x}\right)\right\rangle \\ g_{ij}\left(x_i, x_j\right) &= \left\langle g\left(\mathbf{x}\right) \middle| x_i, x_j\right\rangle - \left\langle g\left(\mathbf{x}\right) \middle| x_i\right\rangle - \left\langle g\left(\mathbf{x}\right) \middle| x_j\right\rangle + \left\langle g\left(\mathbf{x}\right)\right\rangle \end{aligned} \tag{9}$$

where $\left\langle \bullet \right\rangle$ denotes expectation. However, Equation 9 does not stand when the variables are correlated and sequential determination of component function becomes impossible. Even if the variables are independent, practical determination of the component function comes down to an integral problem, solution of which requires huge number of sample points.

In order to address the issue of computational inefficiency, different variants of the ANOVA decomposition have been proposed by various researchers. The primary difference between these variants resides in the manner the component functions are determined. While classical ANOVA decomposition (Sobol, 1993) utilizes Equation 9 to determine the component function, anchored ANOVA (Tang, Congedo, & Abgrall, 2015) or cut-HDMR (Chowdhury et al., 2007, 2009) (also known as referential dimensional decomposition (Xu & Rahman, 2005)) determines the component functions by interpolating with respect to some planes passing through a reference point (mean in most cases). Other variants of the anchored ANOVA decomposition include factorized HDMR (formulated from anchored ANOVA), hybrid HDMR (formulated by combining anchored ANOVA with factorized ANOVA) and logarithmic HDMR (formulated in logarithmic scale). Recently, Polynomial based ANOVA decomposition or RS-HDMR (Li et al., 2006; Li, Rabitz, Hu, Chen, & Ju, 2007; Wang et al., 2003) (also known as polynomial dimensional decomposition (Rahman, 2008)) expresses the component function in terms of basis function and utilizes regression techniques to determine the unknown coefficients associated with the bases.

Remark 2: An essential condition associated with Equation 7 is the hierarchical orthogonality of the component functions. According to this criteria, a higher order component function must be orthogonal with all the lower order component functions. This criterion is essential to ensure the uniqueness of Equation 7 Interested reader may refer (Li et al., 2006) for detailed information regarding this criterion.

All the above mentioned variants of ANOVA decomposition are only suitable if the variables are independent. However, if the system under consideration involves correlated random variables, ANOVA decomposition yields erroneous results. One way-out of this problem is to adopt some suitable transformation to convert the dependent variables into independent random variables. But this introduces some error into the system. It is therefore essential to extend the concepts of ANOVA decomposition so that it can solve systems involving correlated random variables. This is achieved in PCFE by representing the unknown component functions in Equation 7 in terms of extended bases. In the next section, the concepts of extended and non-extended bases and their implementation into Equation 7 has been illustrated.

Extended vs. Non-Extended Bases

Definition 2: Assume, two subspace R and B in Hilbert space are spanned by basis $\{r_1, r_2, \ldots, r_l\}$ and $\{b_1, b_2, \ldots, b_m\}$ respectively. Now if (i) $B \supset R$ and (ii) $B = R + R^{\perp}$ where, $R^{\perp}$ is the orthogonal complement subspace of R in B, we term B as extended basis and R as non-extended basis (Li & Rabitz, 2012).

Assuming ψ to be a suitable basis (orthonormal basis in this case) for a given variable set $\mathbf{x} \subseteq \mathbf{X}$, where and using non-extended basis from *Definition 2*, the component functions described in Equation 7 can be represented as:

$$\begin{aligned} g_i\left(x_i\right) &= \sum_k \alpha_k^i \psi_k^i \\ g_{ij}\left(x_i, x_j\right) &= \sum_m \sum_n \alpha_{mn}^{(ij)(ij)} \psi_m^i \psi_n^j \\ &\vdots \end{aligned} \tag{10}$$

Substituting Equation 10 into Equation 7, one obtains:

$$g\left(\mathbf{x}\right) = g_0 + \sum_i \sum_k \alpha_k^{(i)i} \psi_k^i\left(x_i\right) + \sum_{1 \le i < j \le N} \left(\sum_m \sum_n \alpha_{mn}^{(ij)ij} \psi_m^i\left(x_i\right) \psi_n^j\left(x_j\right) \right) + \cdots\cdots \tag{11}$$

Equation 11 is the basic form for RS-HDMR proposed in (Wang et al., 2003).

The unknown coefficients are determined by employing the weighted orthonormality property of the basis function combined with Monte Carlo integration. Detailed description about this procedure can be found in (Li et al., 2006). However, this procedure introduces a small but significant error into the system and the component functions of the ANOVA decomposition are not exactly orthogonal. Moreover, the expression shown in Equation 11 is only valid for systems with only independent variables. In order to address this issue, the concept of extended basis has been developed.

Using the definition of extended bases, the component functions of Equation 7 can be represented as:

$$\begin{aligned} g_i\left(x_i\right) &= \sum_k \alpha_k^i \psi_k^i \\ g_{ij}\left(x_i, x_j\right) &= \sum_k \alpha_k^{(ij)i} \psi_k^i + \sum_k \alpha_k^{(ij)j} \psi_k^j + \sum_m \sum_n \alpha_{mn}^{(ij)(ij)} \psi_m^i \psi_n^j \\ &\vdots \end{aligned} \tag{12}$$

Substituting Equation 12 into Equation 7,

$$\begin{aligned} g\left(\mathbf{x}\right) &= g_0 + \sum_i \sum_k \alpha_k^{(i)i} \psi_k^i\left(x_i\right) \\ &+ \sum_{1 \le i < j \le N} \left(\sum_k \alpha_k^{(ij)i} \psi_k^i\left(x_i\right) + \sum_k \alpha_k^{(ij)j} \psi_k^j\left(x_j\right) + \sum_m \sum_n \alpha_{mn}^{(ij)ij} \psi_m^i\left(x_i\right) \psi_n^j\left(x_j\right) \right) \\ &+ \cdots\cdots \end{aligned} \tag{13}$$

Most real life problem exhibit only lower order cooperative effect and thus the higher order components of Equation 13 can be neglected. In-fact considering only the first and second order component function yields satisfactory result in most of the cases. Thus, neglecting the higher order component function of Equation 13, one obtains:

$$\begin{aligned} \tilde{g}\left(\mathbf{x}\right) &\approx g_0 + \sum_i \sum_k \alpha_k^{(i)i} \psi_k^i\left(x_i\right) \\ &+ \sum_{1 \le i < j \le N} \left(\sum_k \alpha_k^{(ij)i} \psi_k^i\left(x_i\right) + \sum_k \alpha_k^{(ij)j} \psi_k^j\left(x_j\right) + \sum_m \sum_n \alpha_{mn}^{(ij)ij} \psi_m^i\left(x_i\right) \psi_n^j\left(x_j\right) \right) \end{aligned} \tag{14}$$

Equation 14 is the basic functional form of 2nd order PCFE.

Remark 3: Unlike Equation 11, the hierarchical orthogonality of component functions is not automatically satisfied in Equation 14. Thus, it is essential to determine the unknown coefficients associated with the bases by satisfying the hierarchical orthogonality of the component functions.

Theorem 1: An nth order PCFE approximation always yields a higher-order approximation as compared with an nth order Taylor's series expansion as long as the number of the basis $k \ge n$.

Proof: Please refer the author's previous work (Chakraborty & Chowdhury, 2014) for this proof.

Remark 4: In practise for all approximation schemes including curtailed Taylor's and PCFE, the maximum value of n is often limited to 2 while that of k is limited to 3. Thus, Theorem I holds good and the approximation obtained using PCFE is of higher order.

Rewriting Equation 14 in matrix format:

$$\boldsymbol{\Psi}\boldsymbol{\alpha} = \boldsymbol{d} \tag{15}$$

where $\boldsymbol{\Psi}$ is a matrix consisting of the basis functions, $\boldsymbol{\alpha}$ consists all the unknown coefficient and $\boldsymbol{d}$ denotes the difference between response at the sample point and mean response. Premultiplying Equation 15 by $\boldsymbol{\Psi}^T$:

$$\boldsymbol{\Psi}^T\boldsymbol{\Psi}\alpha = \boldsymbol{\Psi}^T\boldsymbol{d} \tag{16}$$

or

$$\boldsymbol{B}\alpha = \boldsymbol{C} \tag{17}$$

Close inspection of Equation 17 reveals that $\boldsymbol{\Psi}$ has identical columns. Thus $\boldsymbol{B}$ has identical rows. These rows are redundants and therefore removed. Removing the identical rows yields:

$$\boldsymbol{B}'\alpha = \boldsymbol{C}' \tag{18}$$

where $\boldsymbol{B}'$ and $\boldsymbol{C}'$ are $\boldsymbol{B}$ and $\boldsymbol{C}$ after removing the redundants. Equation 18 represents a set of under-determined equations.

In PCFE, a novel regression technique, referred to as homotopy algorithm (HA) (Chakraborty & Chowdhury, 2015c; Li & Rabitz, 2010; Li et al., 2012) is implemented to determine the unknown coefficients. HA determines the unknown coefficients by minimizing the least squared error and satisfying the hierarchical orthogonality of the component function as an additional criterion. In the next section, the detailed procedure for determination of the unknown coefficients using HA is presented.

Solution Technique: HA

HA (Chakraborty & Chowdhury, 2015c; Li & Rabitz, 2010; Li et al., 2012) is a model exploration technique developed for solving an undetermined system of equation. Out of the infinite number of solutions available for an undetermined system, HA determines the best solution by minimizing the least squared error and satisfying an additional criteria defined in term of an objective function. In PCFE an objective function (Chakraborty & Chowdhury, 2014; Li & Rabitz, 2012) based on the hierarchical orthogonality of the component function is utilized to determine the unknown coefficient vector $\boldsymbol{\alpha}$.

Considering $\boldsymbol{B}'$ to be a $p \times q$, such that $p < q$, all the solutions of Equation 18 can be represented as:

$$\alpha(s) = \left(\boldsymbol{B}'\right)^{-1}\boldsymbol{C}' + \left(\boldsymbol{I} - \left(\boldsymbol{B}'\right)^{-1}\boldsymbol{B}'\right)v(s) \tag{19}$$

where $\left(\boldsymbol{B}'\right)^{-1}$ denotes the generalised inverse of $\boldsymbol{B}'$ satisfying either part or all four Penrose condition (Rao & Mitra, 1971). $\boldsymbol{I}$ represents an identity matrix of dimension $q \times q$. It is easily comprehendible that if $p = q$, the second part of Equation 19 vanishes and results in a unique solution.

Remark 5: It is to be noted that generalised inverse of a matrix is not unique. However, generalised inverse satisfying all four Penrose condition is unique. In PCFE, generalised inverse satisfying all four Penrose condition is utilized.

Considering $\boldsymbol{B}^{\dagger}$ to be the generalised inverse of $\boldsymbol{B}'$ satisfying all four Penrose condition and substituting into Equation 19, one obtains:

$$\boldsymbol{\alpha} = \left(\boldsymbol{B}'\right)^{\dagger} \boldsymbol{C}' + \boldsymbol{P}\boldsymbol{v}\left(s\right) \tag{20}$$

where

$$\boldsymbol{P} = \boldsymbol{I} - \left(\boldsymbol{B}'\right)^{\dagger} \boldsymbol{B}' \tag{21}$$

Remark 6: It is to be noted that $\boldsymbol{P}$ is an orthogonal projector and satisfies:

$$\boldsymbol{P}^2 = \boldsymbol{P}, \boldsymbol{P}^T = \boldsymbol{P} \tag{22}$$

HA searches for the best solution of $\boldsymbol{\alpha}$ which satisfies:

$$\frac{d\boldsymbol{\alpha}\left(s\right)}{ds} = \boldsymbol{P}\boldsymbol{v}'\left(s\right) \tag{23}$$

The free function vector $\boldsymbol{v}'\left(s\right)$ can be freely chosen. This not only enable broad choices for exploring $\boldsymbol{\alpha}\left(s\right)$ but also provide a scope for continuously reducing the predefined objective function along the exploration path. Choosing

$$\boldsymbol{v}'\left(s\right) = -\frac{\partial O\left(\boldsymbol{\alpha}\left(s\right)\right)}{\partial \boldsymbol{\alpha}} \tag{24}$$

where O is the objective function, one obtains:

$$\begin{aligned} \frac{dO\left(\boldsymbol{\alpha}\left(s\right)\right)}{ds} &= \left(\frac{dO\left(\boldsymbol{\alpha}\left(s\right)\right)}{d\boldsymbol{\alpha}\left(s\right)}\right)^T \frac{d\boldsymbol{\alpha}\left(s\right)}{ds} = \left(\frac{dO\left(\boldsymbol{\alpha}\left(s\right)\right)}{d\boldsymbol{\alpha}\left(s\right)}\right)^T \boldsymbol{P}\boldsymbol{v}'\left(s\right) \\ &= -\left(\boldsymbol{P}\frac{dO\left(\boldsymbol{\alpha}\left(s\right)\right)}{d\boldsymbol{\alpha}\left(s\right)}\right)^T \left(\boldsymbol{P}\frac{dO\left(\boldsymbol{\alpha}\left(s\right)\right)}{d\boldsymbol{\alpha}\left(s\right)}\right) \le 0 \end{aligned} \tag{25}$$

From Equation 25, it is evident that the objective function is continuously reduced in systematic manner over the course of traversing $s \ge 0$. Thus, the objective function is minimized at $s \to \infty$.

In order to determine the unknown coefficient $\boldsymbol{\alpha}$ at $s \to \infty$, an objective function of the form (Chakraborty & Chowdhury, 2015c; Li & Rabitz, 2010):

$$O = \frac{1}{2}\left[\boldsymbol{\alpha}(s)\right]^T \boldsymbol{W}\boldsymbol{\alpha}(s) \tag{26}$$

is considered where, $\boldsymbol{W}$ is a weight matrix which is symmetric and non-negative. Formulation of weight matrix $\boldsymbol{W}$ is essential for defining an appropriate objective function. A weight matrix, considering the hierarchical orthogonality of the component functions has been provided in later half of this chapter.

Substituting Equation 26 into Equation 24 yields

$$\boldsymbol{v}'(s) = -\boldsymbol{W}\boldsymbol{\alpha}(s) \tag{27}$$

The solution of Equation 23 after substituting Equation 27 is given as:

$$\boldsymbol{\alpha}(s) = \exp\left(-s\mathbf{P}\mathbf{W}\right)\boldsymbol{\alpha}(0) \tag{28}$$

where $\boldsymbol{\alpha}(0)$ is the least squared solution of Equation18. Using eigenvalue decomposition (EVD), $\boldsymbol{PW}$ can be decomposed as:

$$\boldsymbol{PW} = \boldsymbol{Q}\Lambda\boldsymbol{Q}^{-1} \tag{29}$$

where $\boldsymbol{Q}$ is the eigenvector matrix of $\boldsymbol{PW}$ and

$$\Lambda = \begin{pmatrix} \Lambda_r & 0 \\ 0 & 0 \end{pmatrix} \tag{30}$$

where

$$\Lambda_r = \begin{pmatrix} \Lambda_1 & & \\ & \ddots & \\ & & \Lambda_r \end{pmatrix} \tag{31}$$

Remark 7: Since the system under consideration is underdetermined, the rank of the matrix Λ will not be full. In Equation 30, the rank of Λ is assumed to be r.

Now partitioning $\boldsymbol{Q}$ and $\boldsymbol{Q}^{-1}$ as

$$\boldsymbol{Q} = \left[\boldsymbol{Q}_r \quad \boldsymbol{Q}_{q-r}\right] \tag{32}$$

and

$$\boldsymbol{Q}^{-1} = \begin{bmatrix} \boldsymbol{Q}_r^{-1} \\ \boldsymbol{Q}_{q-r}^{-1} \end{bmatrix} \tag{33}$$

$\boldsymbol{Q}_r$ and $\boldsymbol{Q}_{q-r}$ are the first r and last $q-r$ columns of $\boldsymbol{Q}$ respectively. Similarly, $\boldsymbol{Q}_r^{-1}$ and $\boldsymbol{Q}_{q-r}^{-1}$ are the first r and last $q-r$ rows of $\boldsymbol{Q}^{-1}$ respectively. Substituting Equation 32, Equation 33 and Equation 29 into Equation 28,

$$\begin{aligned} \boldsymbol{\alpha}(s) &= \boldsymbol{Q} \cdot \exp(-s\Lambda) \cdot \boldsymbol{Q}^{-1}\boldsymbol{\alpha}(0) \\ &= \left[\boldsymbol{Q}_r \quad \boldsymbol{Q}_{q-r}\right] \cdot \exp(-s\Lambda) \cdot \begin{bmatrix} \boldsymbol{Q}_r^{-1} \\ \boldsymbol{Q}_{q-r}^{-1} \end{bmatrix} \boldsymbol{\alpha}(0) \\ &= \left[\boldsymbol{Q}_r \quad \boldsymbol{Q}_{q-r}\right] \cdot \begin{bmatrix} \exp(-s\Lambda_r) & \\ & I_{q-r} \end{bmatrix} \cdot \begin{bmatrix} \boldsymbol{Q}_r^{-1} \\ \boldsymbol{Q}_{q-r}^{-1} \end{bmatrix} \boldsymbol{\alpha}(0) \\ &= \left[\boldsymbol{Q}_r \exp(-s\Lambda_r) \boldsymbol{Q}_r^{-1} + \boldsymbol{Q}_{q-r}\boldsymbol{Q}_{q-r}^{-1}\right] \boldsymbol{\alpha}(0) \end{aligned} \tag{34}$$

where

$$\exp(-s\Lambda_r) = \begin{bmatrix} \exp(-s\Lambda_1) & & \\ & \ddots & \\ & & \exp(-s\Lambda_r) \end{bmatrix} \tag{35}$$

Noting that all the eigenvalue of $\boldsymbol{PW}$ are non-negative, the first term on RHS of Equation 34 vanishes as $s \to \infty$. Thus,

$$\boldsymbol{\alpha}_\infty = \boldsymbol{Q}_{q-r}\boldsymbol{Q}_{q-r}^{-1}\boldsymbol{\alpha}(0) \tag{36}$$

Remark 8: The matrix $\boldsymbol{PW}$ is not symmetric and thus, direct implementation of EVD to determine the eigenvalues and eigenvectors may yield erroneous result. In PCFE, singular value decomposition (SVD) is employed to determine the eigenvalues and eigenvectors (to be specific $\boldsymbol{Q}_{q-r}$ and $\boldsymbol{Q}_{q-r}^{-1}$).

Using SVD,

$$\boldsymbol{PW} = \boldsymbol{U}\begin{bmatrix} \boldsymbol{J}_r & 0 \\ 0 & 0 \end{bmatrix}\boldsymbol{V}^T \tag{37}$$

where $\boldsymbol{U}$ and $\boldsymbol{V}$ are orthonormal matrices of dimension q and

$$\boldsymbol{J}_r = \begin{bmatrix} J_1 & & \\ & \ddots & \\ & & J_r \end{bmatrix} \tag{38}$$

Similar to the eigenvector matrix $\boldsymbol{U}$ and $\boldsymbol{V}$ are partitioned as

$$\begin{aligned} \boldsymbol{U} &= \begin{bmatrix} \boldsymbol{U}_r & \boldsymbol{U}_{q-r} \end{bmatrix}, \\ \boldsymbol{V} &= \begin{bmatrix} \boldsymbol{V}_r & \boldsymbol{V}_{q-r} \end{bmatrix} \end{aligned} \tag{39}$$

Comparing Equation 39 with Equation 32 and Equation 33, it is easy to comprehend that $\boldsymbol{V}_{q-r}$ and $\boldsymbol{U}_{q-r}^T$ are left and right eigenvectors of $\boldsymbol{PW}$ associated with eigenvalue 0. However, eigenvectors are not unique and thus, it is possible that

$$\boldsymbol{U}_{q-r}^T \boldsymbol{V}_{q-r} \neq \boldsymbol{I}_{q-r} \tag{40}$$

Setting

$$\boldsymbol{Q}_{q-r} = \boldsymbol{V}_{q-r} \tag{41}$$

and considering

$$\boldsymbol{U}_{q-r}^T \boldsymbol{V}_{q-r} = \boldsymbol{K} \tag{42}$$

where $\boldsymbol{K}$ is a non-singular square matrix of dimension $q-r$, one obtains

$$\boldsymbol{Q}_{q-r} \boldsymbol{Q}_{q-r}^T = \boldsymbol{V}_{q-r} \boldsymbol{K}^{-1} \boldsymbol{U}_{q-r}^T \tag{43}$$

Thus,

$$\alpha_\infty = \boldsymbol{V}_{q-r}\left(\boldsymbol{U}_{q-r}^T \boldsymbol{V}_{q-r}\right)^{-1} \boldsymbol{U}_{q-r} \alpha(0) \tag{44}$$

Equation 44 is the key formula for determining the unknown coefficients using HA. Next the construction of weight matrix based on the hierarchical orthogonality of the component functions has been discussed.

Construction of Weight Matrix: The Orthogonality Criteria

The objective function to be used with HA is constructed based on the orthogonality conditions of the component functions of PCFE (Chakraborty & Chowdhury, 2014; Li & Rabitz, 2012). First, the first order component function is considered. As per the hierarchical orthogonal condition, the first order function needs to be orthogonal with the constant g_0, i.e.

$$\int g_0 \left(\sum_k \alpha_k^{(i)i} \Psi_k^i \left(x_i \right) \right) \varpi_i dx_i = 0 \tag{45}$$

where, ϖ_i represent the probability distribution function of x_i. Noting that g_0 is the mean response and may not be zero, Equation 45 reduces to

$$\int \left(\sum_k \alpha_k^{(i)i} \Psi_k^i \left(x_i \right) \right) \varpi_i dx_i = 0 \tag{46}$$

Equation 46 can be represented as a summation series as:

$$\frac{1}{N} \sum_{n=1}^{N} \sum_k \alpha_k^{(i)i} \Psi_k^i \left(x_i^n \right) = 0 \tag{47}$$

Equation 47 can again be written in vectorial notation as:

$$\boldsymbol{F}_1 \left(x_i \right)^T \boldsymbol{\alpha}_1^i = 0, \forall i \tag{48}$$

Thus, the objective function for first order PCFE considering the orthogonality condition is

$$O_1^i = \frac{1}{2} \boldsymbol{T}_i^T \boldsymbol{T}_i \tag{49}$$

where

$$\boldsymbol{T}_i = \boldsymbol{F}_1 \left(x_i \right)^T \boldsymbol{\alpha}_1^i \tag{50}$$

Equation 50 can be written as:

$$O_1^i = \frac{1}{2}\left(\boldsymbol{\alpha}_1^i\right)^T \boldsymbol{W}_1^i \left(\boldsymbol{\alpha}_1^i\right) \tag{51}$$

where $\boldsymbol{W}_1^i$ is symmetric and non-negative matrix.

Similarly, the second order component function needs to be orthogonal to all the lower order component functions. The same can be achieved by setting the second order function orthogonal to all the basis used in lower order component functions, i.e., zeroth and first order component functions. Carrying out similar treatment as the first order component function, a set of $2k+1$ equations, where k is the number of basis. Similar to Equation 51, the objective function for second order component function is written as:

$$O_2^{ij} = \frac{1}{2}\left(\boldsymbol{\alpha}_2^{ij}\right)^T \boldsymbol{W}_2^{ij} \left(\boldsymbol{\alpha}_2^{ij}\right) \tag{52}$$

Combining Equation 51 and Equation 52, the combined objective function for a second order PCFE based on orthogonality condition of the cost function is given as:

$$\begin{aligned} O &= \sum_{i=1}^{N} O_1^i + \sum_{1\le i<j\le N} O_2^{ij} \\ &= \frac{1}{2}\boldsymbol{\alpha}^T \boldsymbol{W} \boldsymbol{\alpha} \end{aligned} \tag{53}$$

where

$$\boldsymbol{W} = \begin{pmatrix} \boldsymbol{W}_1^1 & 0 & \dots & 0 & 0 & \cdots & 0 \\ 0 & \boldsymbol{W}_1^2 & \cdots & 0 & 0 & \cdots & 0 \\ \vdots & \vdots & \ddots & \vdots & \vdots & \ddots & \vdots \\ 0 & 0 & \cdots & \boldsymbol{W}_1^N & 0 & \cdots & 0 \\ 0 & 0 & \cdots & 0 & \boldsymbol{W}_2^{12} & \cdots & 0 \\ \vdots & \vdots & \ddots & \vdots & \vdots & \ddots & \vdots \\ 0 & 0 & \cdots & 0 & 0 & \cdots & \boldsymbol{W}_2^{(N-1)N} \end{pmatrix} \tag{54}$$

Detailed derivation of Equation 52 – Equation 54 can be found in authors' previous work (Chakraborty & Chowdhury, 2015a).

Theorem 2: The extended basis used in PCFE reduces to non-extended basis if the variables are independent.

Proof: Proof of this theorem can be found in (Chakraborty & Chowdhury, 2014).

Remark 9: From theorem 2, it is evident that PCFE is applicable to systems having both correlated and uncorrelated random variables.

An algorithm depicting the step by step procedure of PCFE is shown in Algorithm 1.

Algorithm 1: Algorithm for PCFE

Initialize: Assume m to be the number of sample points, R_m to be the response at the sample points and k to be the number of basis. Also assume that the appropriate basis has been selected.

1. Scale the input variables in an appropriate manner. For example, if orthonormal basis for uniform distribution is used, all the input variables should be scaled in the range $\left[0,1\right]$.
2. Obtain the values of basis function for the scaled inputs and construct the $\boldsymbol{\Psi}$ matrix. Also obtain the response vector $\boldsymbol{d}$.
3. Choose the order of PCFE. If order=1; set $\boldsymbol{B}' = \boldsymbol{\Psi}$, $\boldsymbol{C}' = \boldsymbol{d}$ and go to step 6. Else continue.
4. Construct $\boldsymbol{B}$ matrix as $\boldsymbol{\Psi}^T\boldsymbol{\Psi}$ and $\boldsymbol{C}$ vector as $\boldsymbol{\Psi}^T\boldsymbol{d}$.
5. Remove the redundants to obtain $\boldsymbol{B}'$ and $\boldsymbol{C}'$.

Obtain least square solution $\left(\boldsymbol{b}_0\right)$ as

$$\boldsymbol{b}_0 = \left(\boldsymbol{B}'\right)^{\dagger}\boldsymbol{C}'$$

where, $\left(\boldsymbol{B}'\right)^{\dagger}$ denotes the generalised inverse of $\boldsymbol{B}'$.

7. Construct the weight matrix $\boldsymbol{W}$ and compute $\boldsymbol{P}$ as $\boldsymbol{P} = \boldsymbol{I} - \boldsymbol{B}^{\dagger}\boldsymbol{B}$ Thus obtain $\boldsymbol{PW}$.
8. Obtain $\boldsymbol{U}$ and $\boldsymbol{V}$ by SVD of matrix $\boldsymbol{PW}$
9. Obtain the unknown coefficients as

$$\boldsymbol{b}_{OA} = \left[\boldsymbol{V}_{p-r}\left(\boldsymbol{U}_{p-r}^T\boldsymbol{V}_{p-r}\right)^{-1}\boldsymbol{U}_{p-r}^T\right]\boldsymbol{b}_0$$

Finally, determine failure probability by performing MCS on the generate EM using PCFE.

APPLICATION TO RELIABILITY ANALYSIS

Five-Storey Building Subjected to Lateral Load

The problems consists of a three span five storey frame subjected to lateral loads as shown in Figure 1. The system involves total 21 random variables, including three applied loads, two Young's moduli, eight moments of inertias and eight cross-sectional areas. While Table 1 shows the grouping of the frame elements, Table 2 displays distribution type and parameters of the random variables. The correlation matrix for this problem is given as:

Figure 1. Frame structure considered in example 1

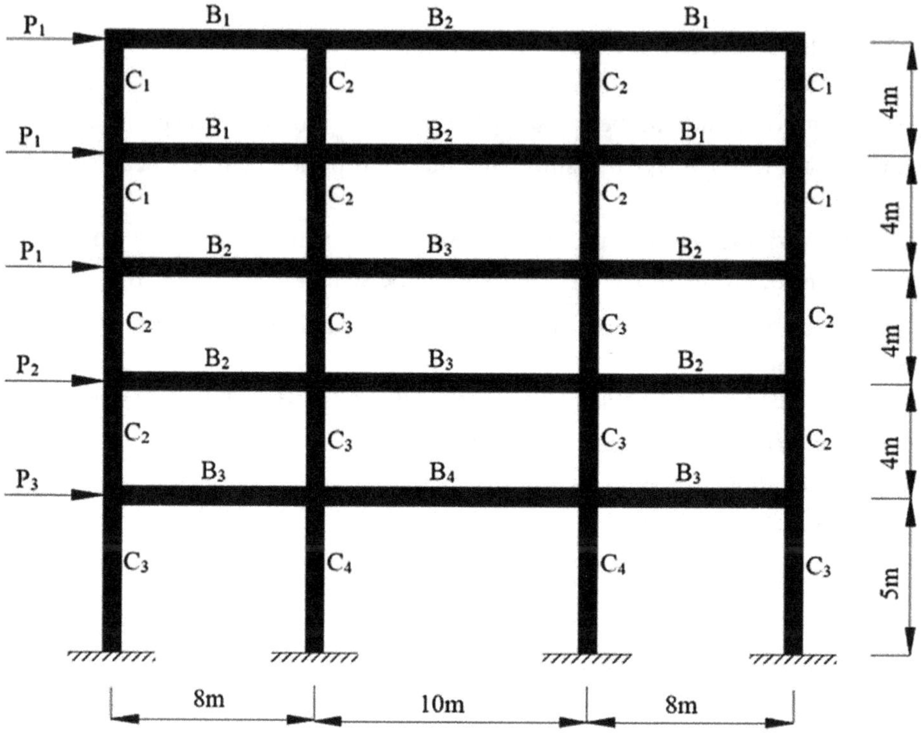

Table 1. Characteristics of the frame elements

Element	Young's Modulus	Moment of Inertia	Cross-Sectional Area
B_1	E_4	I_{10}	A_{18}
B_2	E_4	I_{11}	A_{19}
B_3	E_4	I_{12}	A_{20}
B_4	E_4	I_{13}	A_{21}
C_1	E_5	I_6	A_{14}
C_2	E_5	I_7	A_{15}
C_3	E_5	I_8	A_{16}
C_4	E_5	I_9	A_{17}

Table 2. Distribution type and parameters of random variables

Random Variable	Distribution	Mean	Standard Deviation
P_1	Rayleigh	$135\ kN$	$40\ kN$
P_2	Rayleigh	$90\ kN$	$35\ kN$
P_3	Rayleigh	$70\ kN$	$30\ kN$
E_4	Lognormal	$2.1\times10^{10}\ N/m^2$	$1.9\times10^{9}\ N/m^2$
E_5	Lognormal	$2.4\times10^{10}\ N/m^2$	$1.9\times10^{9}\ N/m^2$
I_6	Lognormal	$8.11\times10^{-3}\ m^4$	$1.04\times10^{-3}\ m^4$
I_7	Lognormal	$0.011\ m^4$	$1.30\times10^{-3}\ m^4$
I_8	Lognormal	$0.0213\ m^4$	$2.59\times10^{-3}\ m^4$
I_9	Lognormal	$0.0259\ m^4$	$3.02\times10^{-3}\ m^4$
I_{10}	Lognormal	$0.0108\ m^4$	$2.59\times10^{-3}\ m^4$
I_{11}	Lognormal	$0.0141\ m^4$	$3.45\times10^{-3}\ m^4$
I_{12}	Lognormal	$0.0232\ m^4$	$5.61\times10^{-3}\ m^4$
I_{13}	Lognormal	$0.0259\ m^4$	$6.47\times10^{-3}\ m^4$
A_{14}	Lognormal	$0.312\ m^2$	$0.056\ m^2$
A_{15}	Lognormal	$0.372\ m^2$	$0.074\ m^2$
A_{16}	Lognormal	$0.505\ m^2$	$0.093\ m^2$
A_{17}	Lognormal	$0.557\ m^2$	$0.111\ m^2$
A_{18}	Lognormal	$0.253\ m^2$	$0.093\ m^2$
A_{19}	Lognormal	$0.291\ m^2$	$0.102\ m^2$

continued on following page

Table 2. Continued

Random Variable	Distribution	Mean	Standard Deviation
A_{20}	Lognormal	$0.372\ m^2$	$0.121\ m^2$
A_{21}	Lognormal	$0.418\ m^2$	$0.14\ m^2$

$$\rho = \begin{matrix} P_1 \\ P_2 \\ P_3 \\ E_4 \\ E_5 \\ I_6 \\ \vdots \\ A_{21} \end{matrix} \begin{pmatrix} 1 & & & & & & & \\ 0.5 & 1 & & & & & & \\ 0.5 & 0.5 & 1 & & & & & \\ 0 & 0 & 0 & 1 & & & & \\ 0 & 0 & 0 & 0.9 & 1 & & & \\ 0 & 0 & 0 & 0 & 0.13 & 1 & & \\ \vdots & \vdots & \vdots & \vdots & \vdots & \ddots & \ddots & \\ 0 & 0 & 0 & 0 & 0.13 & \cdots & 0.13 & 1 \end{pmatrix} \quad (55)$$

The failure criteria of the frame have been defined in terms of horizontal component of the top-floor displacement $u(\mathbf{x})$ and given as

$$g(\mathbf{x}) = u_{\text{threshold}} - u(\mathbf{x}) \quad (56)$$

$u_{\text{threshold}}$ for this problem has been considered as $0.035\ m$. The structure is modelled using finite element package, ABAQUS CAE (version 6.13) ("Dassault Systmes Simulia Corp., ABAQUS documentation and theory manual," 2009). The actual failure probability of the system as obtained from full-scale MCS with target COV of 2% is 0.02084.

The failure probability of above mentioned system has been determined by employing both first and second order PCFE. For both first and second order PCFE, sample points are generated using the Sobol's sequence (Sobol, 1976). First order PCFE yields a failure probability of 0.01861 with just 82 actual function evaluation. In an effort to reduce the error, second-order PCFE has been employed. This yielded a failure probability of 0.02071. However, for second-order PCFE the increase in accuracy is accompanied by an increase in sample points to 2500. Table 3 shows the results obtained from various methods. Results obtained using PCFE has been compared with other popular techniques such as FORM, SORM and Kriging. It is observed that PCFE (second order) outperforms all the methods.

A Six Storey Building

This example examines the performance of proposed approach in predicting failure probability of a large scale system. The problem consists of a six-storey building (Chakraborty & Chowdhury, 2015a) subjected to lateral and gravity load as shown in Figure 2. The plan area of the building is 18(m)×18(m)

Table 3. Failure probability obtained using various methods

Method		Reliability Index	Failure Probability	N^*
FORM		2.0009	0.0227	162
SORM		2.0198	0.0217	602**
Kriging		2.0395	0.02070	5000
direct MCS		2.0367	0.02084	100000
PCFE	First order	2.0833	0.01861	82
	Second order	2.0393	0.02071	2500

* N indicates total number of actual function evaluation
** includes the initial function evaluations required for determining the MPP

Figure 2. Six-storey building considered in example 2

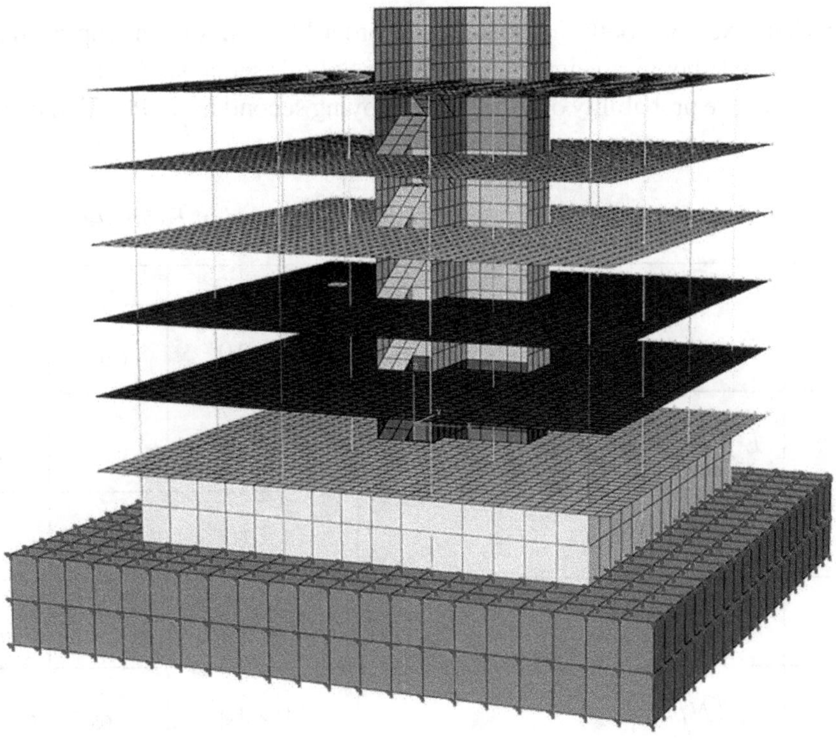

with a lift-wall and staircase located at its center. Each floor is having 16 columns at a spacing of 5(m). To increase the stiffness of the structure, the walls of the ground floor are designed as load-bearing wall. The floor height is considered as 3.6(m).

The building is modelled using *ABAQUS*^{TM} version 6.8 (Provide reference). While the floors and walls (lift wall and load bearing wall of ground floor) are modelled as 4-noded shell element (S4), the columns are modelled as 2-noded beam element (B31). The soil is modelled as 8-noded brick element (C3D8I).

The cross-section (both depth $\left(d_c\right)$ and width $\left(w_c\right)$) of columns, thickness of slabs (both floor $\left(t_{floor}\right)$ and staircase slab $\left(t_{stair}\right)$), thickness of walls $\left(t_{cellular}\right)$, density $\left(\rho\right)$ and elastic modulus $\left(E\right)$ of concrete, cellular material (used for load bearing wall) and soil, upper $\left(Y_U\right)$ and lower yield point $\left(Y_L\right)$ of concrete are considered to be independent random variables. The seven lateral loads, namely P_{f1}, P_{f2}, P_{f3}, P_{f4}, P_{f5}, P_{f6} and P_{surf}, acting at floor 1-6 and terrace lift wall respectively are also considered to be independent random variables. The system has 39 random variables. The details of random variables are shown in Table 4.

The limit-state function for the system described above is given as

$$g\left(\mathbf{x}\right) = \sigma_0 - \left(\sigma_{\text{mises}}\right)_{\max} \tag{57}$$

where $\left(\sigma_{\text{mises}}\right)_{\max}$ is the maximum von mises stress and σ_0 is the threshold value. In this example, $\sigma_0 = 1.1 \times 10^7 \; N/m^2$ is considered.

Similar to previous example, both first and second order PCFE has been employed to determine the failure probability. Sample points are generated by using Sobol's sequence (Bratley & Fox, 1988). First-order PCFE yields a failure probability of 0.1123. Employing second order PCFE, the failure probability

Table 4. Details of random variables for six-storey building defined in Example 3

Variable No.	Variable	Mean	COV	Distribution
1	$E_{Cellular}\left(N/m^2\right)$	2.5×10^5	0.05	Lognormal
2-4	$E_{col}, E_{floor}, E_{stair}\left(N/m^2\right)$	3.5×10^{10}	0.05	Lognormal
5	$G_{col}\left(N/m^2\right)$	1.4×10^{10}	0.05	Lognormal
6-8	$\rho_{col}, \rho_{floor}, \rho_{stairs}\left(kg/m^3\right)$	2500	0.1	Normal
9	$Y_L\left(N/m^2\right)$	2.0×10^7	0.05	Lognormal
10	$Y_U\left(N/m^2\right)$	2.2×10^7	0.05	Lognormal
11	$t_{stair}\left(m\right)$	0.1	0.2	Uniform
12	$t_{cellular}\left(m\right)$	0.5	0.2	Uniform

continued on following page

Table 4. Continued

Variable No.	Variable	Mean	COV	Distribution
13-18	$t_{floor1}\,(m) - t_{floor6}\,(m)$	0.125	0.2	Uniform
19	$E_{soil}\,(N/m^2)$	5×10^9	0.1	Lognormal
20	$\rho_{soil}\,(kg/m^3)$	1800	0.15	Normal
21-24	$d_{cf1}, w_{cf1}, d_{cf2}, w_{cf2}\,(m)$	0.8	0.2	Uniform
25-28	$d_{cf3}, w_{cf3}, d_{cf4}, w_{cf4}\,(m)$	0.6	0.2	Uniform
29-32	$d_{cf5}, w_{cf5}, d_{cf6}, w_{cf6}\,(m)$	0.4	0.2	Uniform
33	$P_{f1}\,(N/m)$	2722	0.15	Rayleigh
34	$P_{f2}\,(N/m)$	2528	0.15	Rayleigh
35	$P_{f3}\,(N/m)$	2917	0.15	Rayleigh
36	$P_{f4}\,(N/m)$	3111	0.15	Rayleigh
37	$P_{f5}\,(N/m)$	3306	0.15	Rayleigh
38	$P_{f6}\,(N/m)$	1750	0.15	Rayleigh
39	$P_{surf}\,(N/m^2)$	800	0.15	Uniform

is obtained as 0.1452. Table 5 shows the result obtained using various method. PCFE is found to yield the best result. Moreover, number of sample points for PCFE is considerably less as compared to other popular EMs.

CONCLUSION AND OUTLOOK

The method based on ANOVA decomposition has motivated many investigation over the last one decade. In this chapter an extension of the classical ANOVA decomposition, referred to as PCFE has

Table 5. Failure probability for six-storey building (Example 3) obtained using various method

Method		Reliability Index	Failure Probability	N_s*
FORM		Did not converge		
SORM				
Kriging		0.4563	0.3241	10000
RBF		1.0127	0.1556	7500
Full-scale MCS		1.0450	0.1480	20000
Proposed approach	First order PCFE	1.2144	0.1123	450
	Second order PCFE	1.0572	0.1452	5200

Ns* indicate total number of actual FE calculations

been discussed. The main advantage of PCFE, over the classical ANOVA decomposition, resides in the fact that it is capable of treating both correlated and uncorrelated random variables without any ad hoc transformations.

In this chapter application of PCFE in reliability analysis has been shown. The two examples presented demonstrate the advantage of PCFE over other popular methods such as Kriging, high-order stochastic response surface method, radial basis function etc. For both examples PCFE yields more accurate results as compared to the above mentioned methods. Moreover, the number of sample points required is also significantly less as compared to other popular techniques.

In addition to calculating probability of failure, PCFE can also be used in other fields of uncertainty quantification. In (Chakraborty & Chowdhury, 2014), response pdf of a nonlinear stochastic dynamic system having 1001 random variables have successfully predicted using PCFE. Thus, it is expected that PCFE will perform well in fields such as sensitivity analysis, robust design optimization, reliability based design optimization etc.

Last but not the least, PCFE is still under development. Advanced research on various variants of PCFE is being performed. Once completed, it is expected that the prediction accuracy will be further enhanced at a comparatively lower computational cost.

REFERENCES

Adhikari, S., Chowdhury, R., & Friswell, M. I. (2011). High dimensional model representation method for fuzzy structural dynamics. *Journal of Sound and Vibration*, *330*(7), 1516–1529. doi:10.1016/j.jsv.2010.10.010

Alis, O. F., & Rabitz, H. (2001). Efficient implementation of high dimensional model representations. *Journal of Mathematical Chemistry*, *29*(2), 127–142. doi:10.1023/A:1010979129659

Bratley, P., & Fox, B. L. (1988). Implementing Sobol's quasirandom sequence generator. *ACM Transactions on Mathematical Software*, *14*(1), 88–100. doi:10.1145/42288.214372

Chakraborty, S., & Chowdhury, R. (2014). Polynomial Correlated Function Expansion for Nonlinear Stochastic Dynamic Analysis. *Journal of Engineering Mechanics, 141*(3), 04014132:1–04014132:11.

Chakraborty, S., & Chowdhury, R. (2015a). A semi-analytical framework for structural reliability analysis. *Computer Methods in Applied Mechanics and Engineering*, *289*(1), 475–497. doi:10.1016/j.cma.2015.02.023

Chakraborty, S., & Chowdhury, R. (2015b). Assessment of polynomial correlated function expansion for high-fidelity structural reliability analysis. *Structural Safety, 59*(Accepted), 9–19.

Chakraborty, S., & Chowdhury, R. (2015c). Multivariate function approximations using D-MORPH algorithm. *Applied Mathematical Modelling*, *39*(23-24), 7155–7180. doi:10.1016/j.apm.2015.03.008

Chakraborty, S., Mandal, B., Chowdhury, R., & Chakrabarti, A. (2016). Stochastic free vibration analysis of laminated composite plates using polynomial correlated function expansion. *Composite Structures*, *135*, 236–249. doi:10.1016/j.compstruct.2015.09.044

Chowdhury, R., Rao, B. N., & Prasad, A. M. (2007). High dimensional model representation for piece-wise continuous function approximation. *Communications in Numerical Methods in Engineering*, *24*(12), 1587–1609. doi:10.1002/cnm.1053

Chowdhury, R., Rao, B. N., & Prasad, A. M. (2009). High-dimensional model representation for structural reliability analysis. *Communications in Numerical Methods in Engineering*, *25*(4), 301–337. Retrieved from http://doi.wiley.com/10.1002/cnm.1118 doi:10.1002/cnm.1118

Dassault Systmes Simulia Corp. (2009). ABAQUS documentation and theory manual.

Deng, J., Gu, D., Li, X., & Yue, Z. Q. (2005, June). Structural reliability analysis for implicit performance functions using artificial neural network. *Structural Safety*, *27*(1), 25–48. doi:10.1016/j.strusafe.2004.03.004

DiazDelaO. F. A., Adhikari, S., Saavedra Flores, E. I., & Friswell, M. I. (2013). Stochastic structural dynamic analysis using Bayesian emulators. *Computers & Structures, 120*, 24–32. Retrieved from http://apps.webofknowledge.com/full_record.do?product=UA&search_mode=GeneralSearch&qid=4&SID=X2RcsR1DGP199mjAKwr&page=1&doc=2

Ditlevsen, O., & Madsen, H. O. (1996). *Structural reliability methods* (1st ed.). New Jersey: Wiley-Blackwell.

Dubourg, V., Sudret, B., & Bourinet, M. (2011). Reliability-based design optimization using kriging surrogates and subset simulation. *Structural Multidisciplinary Optimization*, *44*(5), 673–690. doi:10.1007/s00158-011-0653-8

Echard, B., Gayton, N., Lemaire, M., & Relun, N. (2013). A combined importance sampling and Kriging reliability method for small failure probabilities with time-demanding numerical models. *Reliability Engineering & System Safety*, *111*, 232–240. doi:10.1016/j.ress.2012.10.008

Gavin, H. P., & Yau, S. C. (2008). High-order limit state functions in the response surface method for structural reliability analysis. *Structural Safety*, *30*(2), 162–179. doi:10.1016/j.strusafe.2006.10.003

Goswami, S., Chakraborty, S., & Ghosh, S. (2013). Adaptive response surface method in structural response approximation under uncertainty. *Proceedings of theInternational Conference on Structural Engineering and Mechanics*.

Hooker, G. (2007). Generalized Functional ANOVA Diagnostics for High-Dimensional Functions of Dependent Variables. *Journal of Computational and Graphical Statistics*, *16*(3), 709–732. doi:10.1198/106186007X237892

Hu, Y., Zhao, K., & Lian, H. (2014). Bayesian quantile regression for partially linear additive models. *Statistics and Computing*, *25*(3), 651–668. doi:10.1007/s11222-013-9446-9

Jamshidi, A. A., & Kirby, M. J. (2010). Skew-radial basis function expansions for empirical modeling. *SIAM Journal on Scientific Computing*, *31*(6), 4715–4743. doi:10.1137/08072293X

Li, G., Hu, J., Wang, S.-W., Georgopoulos, P. G., Schoendorf, J., & Rabitz, H. (2006). Random sampling-high dimensional model representation (RS-HDMR) and orthogonality of its different order component functions. *The Journal of Physical Chemistry A*, *110*(7), 2474–2485. doi:10.1021/jp054148m

Li, G., & Rabitz, H. (2010). D-MORPH regression: Application to modeling with unknown parameters more than observation data. *Journal of Mathematical Chemistry*, *48*(4), 1010–1035. doi:10.1007/s10910-010-9722-2

Li, G., & Rabitz, H. (2012). General formulation of HDMR component functions with independent and correlated variables. *Journal of Mathematical Chemistry*, *50*(1), 99–130. doi:10.1007/s10910-011-9898-0

Li, G., Rabitz, H., Hu, J., Chen, Z., & Ju, Y. (2007). Regularized random-sampling high dimensional model representation (RS-HDMR). *Journal of Mathematical Chemistry*, *43*(3), 1207–1232. doi:10.1007/s10910-007-9250-x

Li, G., Rey-de-Castro, R., & Rabitz, H. (2012). D-MORPH regression for modeling with fewer unknown parameters than observation data. *Journal of Mathematical Chemistry*, *50*(7), 1747–1764. doi:10.1007/s10910-012-0004-z

Li, G., Wang, S., Rosenthal, C., & Rabitz, H. (2001). High dimensional model representations generated from low dimensional data samples. I . mp-Cut-HDMR, *30*(1).

Melchers, R. E. (1999). *Structural reliability analysis and prediction* (2nd ed.). Balckwell publishers.

Muscolino, G., Ricciardi, G., & Cacciola, P. (2003). Monte Carlo simulation in the stochastic analysis of non-linear systems under external stationary poisson white noise input. *International Journal of Non-linear Mechanics*, *38*(8), 1269–1983. doi:10.1016/S0020-7462(02)00072-0

Rabitz, H., & Aliş, Ö. F. (1999). General foundations of high dimensional model representations. *Journal of Mathematical Chemistry*, *25*(2-3), 197–233. doi:10.1023/A:1019188517934

Rahman, S. (2006). A dimensional decomposition method for stochastic fracture mechanics. *Engineering Fracture Mechanics*, *73*(15), 2093–2109. doi:10.1016/j.engfracmech.2006.04.010

Rahman, S. (2008). A polynomial dimensional decomposition for stochastic computing. *International Journal for Numerical Methods in Engineering*, *76*(13), 2091–2116. doi:10.1002/nme.2394

Rao, C. R., & Mitra, S. K. (1971). *Generalized inverse of matrix and its applications*. New York, U.S.A.: Wiley.

Rubinstein, R. Y. (1981). *Simulation and the Monte Carlo method*. New York, U.S.A.: Wiley. doi:10.1002/9780470316511

Shinozuka, M. (1972). Monte Carlo solution of structural dynamics. *Composite Structures*, *2*(5-6), 855–874. doi:10.1016/0045-7949(72)90043-0

Sobol, I. M. (1976). Uniformly distributed sequences with an additional uniform property. *U.S.S.R. Computational Mathematics and Mathematical Physics*, *16*(5), 236–242. doi:10.1016/0041-5553(76)90154-3

Sobol, I. M. (1993). Sensitivity estimates for nonlinear mathematical models. *Mathematical Modeling and Computational Experiment*, *1*(4), 407–414.

Sobol', I. M. (2003). Theorems and examples on high dimensional model representation. *Reliability Engineering & System Safety*, *79*(2), 187–193. doi:10.1016/S0951-8320(02)00229-6

Tamimi, S., Amadei, B., & Frangopol, D. M. (1989). Monte-Carlo simulation of rock slope reliability. *Computers & Structures*, *33*(6), 1495–1505. doi:10.1016/0045-7949(89)90489-6

Tang, K., Congedo, P. M., & Abgrall, R. (2015). Sensitivity analysis using anchored ANOVA expansion and high-order moments computation. *International Journal for Numerical Methods in Engineering*, *102*(9), 1554–1584. doi:10.1002/nme.4856

Thakur, R., & Misra, K. B. (1978). Monte-Carlo simulation for reliability evaluation of complex systems. *International Journal of Systems Science*, *9*(11), 1303–1308. doi:10.1080/00207727808941777

von Toussaint, U. (2015). General Hyperplane Prior Distributions Based on Geometric Invariances for Bayesian Multivariate Linear Regression. *Entropy*, *17*(6), 3898–3912. doi:10.3390/e17063898

Wang, S. W., Georgopoulos, P. G., Li, G., & Rabitz, H. (2003). Random sampling–high dimensional model representation (RS–HDMR) with nonuniformly distributed variables: Application to an integrated multimedia/multipathway exposure and dose model for trichloroethylene. *The Journal of Physical Chemistry A*, *107*(23), 4707–4718. doi:10.1021/jp022500f

Wiener, N. (1938). The homogeneous chaos. *American Journal of Mathematics*, *60*(4), 897–936. doi:10.2307/2371268

Wu, L. (2009). Least square analysis of error-components models for the study of provisional economic performance. *Modern Physics Letters B*, *23*(03), 521–524. doi:10.1142/S0217984909018801

Xiu, D., & Karniadakis, G. E. (2002). The Wiener-Askey polynomial chaos for stochastic differential equations. *SIAM Journal on Scientific Computing*, *24*(2), 619–644. doi:10.1137/S1064827501387826

Xu, H., & Rahman, S. (2004). A generalized dimension-reduction method for multidimensional integration in stochastic mechanics. *International Journal for Numerical Methods in Engineering*, *61*(12), 1992–2019. doi:10.1002/nme.1135

Xu, H., & Rahman, S. (2005). Decomposition methods for structural reliability analysis. *Probabilistic Engineering Mechanics*, *20*(3), 239–250. doi:10.1016/j.probengmech.2005.05.005

Chapter 13
Prevention of Corrosion in Austenitic Stainless Steel through a Predictive Numerical Model Simulating Grain Boundary Chromium Depletion

M.K. Samal
Bhabha Atomic Research Centre, India

ABSTRACT

In this chapter, a mathematical model for rate of formation of chromium carbides near the grain boundary, which is a pre-cursor to chromium depletion and corresponding sensitization behavior in stainless steels, is presented. This model along with the diffusion equation for chromium in the grain has been used to obtain chromium depletion profiles at various time and temperature conditions. Finite difference method has been used to solve the above equations in the spherical co-ordinate system and the results of time-temperature-sensitization diagrams of four different types of alloys have been compared with those of experiment from literature. For the problem of low temperature sensitization and corresponding inter-granular corrosion in austenitic stainless steel, it is very difficult to carry out experiment at higher temperatures and justify its validity at lower operating temperatures by extrapolation. The development of predictive models is highly useful in order to design the structures for prevention of corrosion of the material in aggressive environments.

INTRODUCTION

In polycrystalline materials, corrosive attack of grain boundaries with little or no attack of the grains is called inter-granular corrosion. In practical application, the loss of cross section thickness and the introduction of cracks due to presence of inter-granular corrosion can have severe consequences for applications like pressure retaining containments. Extensive corrosion attack at the grain boundaries can

DOI: 10.4018/978-1-5225-0588-4.ch013

result in grains dropping or falling out of the metal surface resulting in the disintegration of the material (Verlinden et al., 2007). With austenitic stainless steels, inter-granular attack is usually the result of chromium carbide precipitation ($Cr_{23}C_6$ type) at grain boundaries, which produces a narrow zone of chromium depletion at the grain boundary.

Austenitic stainless steels are used in a wide range of applications due to their excellent corrosion resistance and good manufacturability property. However, during the process of welding or heat treatment (during solution annealing, stress-relieving etc.), there can be carbide precipitation at the grain boundaries leading to chromium depleted zones. This process is aggravated when the material is extensively heated or slowly cooled in the temperature range of 500 to 850 deg. C (Verlinden et al., 2007; Dayal et al., 2005). When chromium level near the grain boundaries falls below 12%, the material is said to be 'sensitized' as there is not enough chromium in that region to form a passive layer to prevent attack of corrosive media. Hence, as the sensitized material is exposed to corrosive media during service, inter-granular corrosion takes place.

Chromium is the primary alloying element that makes stainless steel corrosion resistant. The chromium-depleted regions are susceptible to preferential corrosion attack. It has been known over the years that inter-granular stress corrosion cracking in austenitic stainless steels occurs because the chromium content immediately adjacent to the carbide near the grain boundary may be below that required for the stainless steel alloy to actively prevent corrosion by forming an adherent oxide layer to the surface of the material. If the carbides form a continuous network on the grain boundary, then corrosion can produce a separation or gap at the boundary, possible grain dropping or loss of material, formation of grain boundary cracks etc.

The chromium carbides tend to precipitate at the grain boundaries of austenitic stainless steels in the 500 to 850°C temperature range. Any exposure or thermal excursion into this temperature range during metal manufacture, fabrication, or service could potentially sensitize the steel. Common manufacturing processes such as welding, stress relief, and hot working of the metal can expose the steel to the sensitizing temperature range. The formation of chromium carbides can be reversed by solution-anneal heat treatment. A test method has been described in ASTM A262 standard. This method can be used to detect susceptibility to inter-granular attack in austenitic stainless steels. The time and temperature required to produce susceptibility to inter-granular attack is highly dependent on alloy composition, particularly the carbon content.

To avoid inter-granular stress corrosion cracking, the carbon in the alloy should be kept in solution phase (preventing chromium carbide formation). This can be achieved by rapid cooling the alloy through the sensitizing temperature range from the solution-annealed stage. The usual solution-anneal temperature in order to avoid sensitization of the material depends on the alloy and is typically carried out in the range of 1050 to 1200°C followed by rapid cooling. Resistance to chromium depletion at the grain boundary can also be achieved by reducing the carbon content to below 0.030% level (Moura et al., 2009).

Over the years, various low carbon grades of stainless steels such as SS 304L, 316L, and 317L etc. have been developed so as to resist sensitization during typical welding operations performed on these alloys. These alloys however do not resist sensitization during long term exposure in the critical temperature range as occurs during service. The addition of stabilizing elements such as Ti, Nb and Ta improves the resistance to sensitization, especially for long-term exposures in the critical range in service. These elements are prone to chemically react with carbon in the alloy and tend to form carbides that are more stable than chromium carbide while cooling the material from solution-annealed state. Hence, chromium is saved from forming carbides and hence, it is available for corrosion protection in these alloys. These stabilized austenitic grades of stainless steel are SS 321, SS 347, 316Ti etc.

Inter-granular corrosion of ferritic stainless steels is similar to that of austenitic stainless steels. Ferritic grades contain nitrogen and it is known that the solubility of nitrogen is low in the ferritic crystal structure. In these alloys, nitrogen can also form nitrides with chromium and cause sensitization in ferritic grades of stainless steels. The compounds responsible for chromium depletion in ferritic stainless steels are usually chromium carbides ($Cr_{23}C_6$) and chromium nitrides (Cr_2N). For the ferritic grade stainless steel, sensitization occurs when the alloys is cooled from higher temperatures typically greater than 950°C.

The very high diffusion rates of carbon and nitrogen in the ferrite structure of ferritic grade stainless steel make it very difficult to design faster cooling rates in practice to prevent precipitation of carbides and nitrides at grain boundaries. Hence, the commercial ferritic grade manufacturers tend to prevent sensitization in these alloys by lowering the level of C and N. Other way is to add stabilizing elements such as Ti, Ta, or Nb in the matrix which are efficient carbide formers compared to chromium. One sensitization has occurred in a ferritic stainless steel, the chromium percentage at the grain boundaries may be partially restored by heating the material in the temperature range of 600 to 700°C for extended period of time. The standard method as outlined in ASTM A763 can also be used to detect susceptibility of the ferritic stainless steel alloy to inter-granular stress corrosion cracking.

Over the years, there has been substantial research on understanding of:

1. The mechanism of sensitization (Moura et al., 2009; Tsay et al., 2008),
2. The effect of different metallurgical variables including microstructure (e.g., grain size (Trillo et al., 1995; Beltran et al., 1997), nature of grain boundary (Parvathavarthini et al., 2009; Ortner & Randle, 1989; Lehockey et al., 1998; Zhou et al., 2001; Shimada et al., 2002; Wasnik et al., 2002), composition (% of C, N and other alloying elements) etc. on degree of sensitization (Parvathavarthini & Dayal, 2010; Watanabe et al., 2000; Trillo & Murr, 1998),
3. The effect of cooling rates during the process of heat treatment (Parvathavarthini & Dayal, 2010),
4. The effect of mechanical deformation (prior plastic strain) (Garcia et al., 2001; Fu et al., 2009),
5. The effect of environment (Tsay et al., 2008) and so on.

In Anderko et al. (2010), a new procedure has been developed for predicting the effect of thermal aging on the repassivation potential of austenitic alloys. A grain boundary microchemistry model for calculating the chromium and molybdenum depletion profiles in the vicinity of grain boundaries has been coupled to an electrochemical model that relates the repassivation potential to the microchemistry and environmental conditions, including temperature and solution chemistry. The model has been used to predict the repassivation potentials for thermally aged alloys 600 and 825 and their welded samples in chloride solutions. The influence of solute additions such as silicon and phosphorus on radiation-induced segregation at grain boundaries for high-purity Fe-15Cr-20Ni alloys was examined for austenitic model alloys in Sakaguchi et al. (2005). The alloy was exposed to irradiation by electrons. It was also observed in Sakaguchi et al. (2005) that the grain boundary depletion of chromium and segregation of nickel were significantly suppressed by addition of the above elements. A quantitative analysis of radiation-induced grain-boundary segregation was presented in Carter et al. (1994). A review on this subject has been recently presented by Dayal et al. (2005). On the other hand, there have been efforts in literature to estimate the size of chromium depletion zone using various approximations (Pragnell & Evans, 2006; Samal & Abhishek, 2011; Stawström & Hillert, 1969; Hall & Briant, 1984; Mozhi et al.,

1986, 1987; Was & Kruger, 1985; Grujisic & Tangrila, 1991; Bruemmer, 1990; Mayo, 1997; Sahaloui et al., 2002; Sourmail et al., 2003; Tedmon et al., 1971; Williams, 1982; Kohler, 1960; Natesan & Kassner, 1973; Richardson, 1953; Fullman, 1982; Wagner, 1952). These will be described here briefly. A two-dimensional finite-difference based model has been presented in Pragnell and Evans (2006) to predict depletion profiles arising from high temperature oxidation behavior of an alloy and a one dimensional finite-difference based model employing spherical co-ordinate system has been presented in Samal and Abhishek (2011) in order to predict chromium depletion in austenitic stainless steel.

Stawström and Hillert (1969) developed a model for estimating the chromium depletion zone near grain boundary in 18Cr-8Ni-C stainless steels. It was assumed that there exists a local equilibrium between the carbide and austenite at the interface. As carbon diffuses faster compared to chromium, it was assumed that the carbon activity in the whole material is constant. Later, these were modified by Hall and Briant (1984) and Mozhi et al. (1986, 1987) and others (Was & Kruger, 1985; Grujisic & Tangrila, 1991; Bruemmer, 1990; Mayo, 1997; Sahaloui et al., 2002; Sourmail et al., 2003) based on thermodynamic data from literature (Tedmon et al., 1971; Williams, 1982; Kohler, 1960; Natesan & Kassner, 1973; Richardson, 1953; Fullman, 1982; Wagner, 1952). Recently, Sourmail et al. (2003) have considered the diffusion of many different alloying elements to the grain boundary by satisfying their combined equilibrium and their participation in the carbide formation process in their model. However, a comprehensive model for prediction of time-temperature-sensitization behavior of these types of steels is lacking in literature. Moreover, the kinetics of chromium carbide precipitation is not explicitly considered while solving the diffusion equation for chromium in the matrix. The objective of this work is to develop a model for predicting the rate of carbide precipitation at the boundary which is a function of availability of chromium and carbon at the reaction site as well as their ability to diffuse to the site through the grain. This rate will in turn influence the chromium profile in the matrix as it will be coupled to the diffusion equation for chromium.

The chapter is divided in four sections. The mathematical model for carbide precipitation and its coupling to the diffusion equation for chromium is described in Section-2. The finite difference scheme implemented in the spherical co-ordinate system is discussed in this section. However, the methodology can be applied to the general 3-D Cartesian co-ordinates and other numerical methods such as finite element method can be used for the numerical solution. Section-3 discusses the application of the model for prediction of time-temperature-sensitization (TTS) behaviors of austenitic stainless steel (SS304) of four different chemical compositions and the results have been compared to that of experiment from literature. Section-4 concludes with a description of further modifications to this model to be carried out in future.

NUMERICAL MODEL TO SIMULATE CHROMIUM DEPLETION AT THE GRAIN BOUNDARY

For the austenitic stainless steel, we consider in this analysis, it is assumed that the type of carbide that precipitates mainly consists of $M_{23}C_6$ type where M stands for chromium (Cr) here and C stands for carbon. The chemical reaction for the carbide can be written as (Mozhi et al., 1986, 1987; Tedmon et al., 1971)

$$23Cr + 6C = Cr_{23}C_6 \tag{1}$$

Tedmon et al. (1971) expressed the chemical equilibrium constant k for the above reaction as

$$k = \frac{a_{Cr_{23}C_6}}{\left(\gamma_{Cr} x_{Cr}\right)^{23} \left(\gamma_C x_C\right)^{6}} \tag{2}$$

where $a_{Cr_{23}C_6}$ is the activity for chromium carbide which is assumed as unity by many authors including Mozhi et al. (1986). x_{Cr} and x_C are the equilibrium compositions of chromium and carbon respectively and these are expressed as mole fractions in calculation of the equilibrium constant according to Equation 2. The parameters γ_{Cr} and γ_C are the activity coefficients of chromium and carbon respectively.

The activity coefficient of carbon is assumed to be a constant in the whole material as it has very high mobility in the matrix (Tedmon et al., 1971). The activity coefficient of chromium and carbon are determined using free energy interaction coefficient of different components (e.g., Mo, Mn, Si, Cu, N, P and S etc.) present in the matrix according to the method outlined in Wagner (1952). However, there is no information in literature regarding relation of rate of carbide precipitation with the chemical equilibrium constant k of Equation 2 or concentration of the reactant species. The aim of this work is to develop a suitable expression for the rate of precipitation which depends upon the concentrations of reactant species in the Equation 1. This is described in the following paragraph.

The rate of precipitation reaction for formation of chromium carbide depends upon the available concentration of chromium and carbon at the reaction site and this is represented by the right hand side of Equation 3 as written below. The rate of diffusion of chromium at the interface will depend upon gradient of chromium concentration at the interface $\partial c_{cr} / \partial x$, diffusivity of chromium D_{cr}, the area of the surface A and volume V of the characteristics region containing the surface and it is represented by the left hand side of Equation 3. The equilibrium between the two expressions is written as

$$\frac{D_{cr} A}{V} \frac{\partial c_{cr}}{\partial x} = c_1 \frac{k_B T}{h} \left[c_{cr}\right]^{23} \left[c_C\right]^{6} \tag{3}$$

where c_1 is a constant which depends upon temperature and composition of the material, k_B is Boltzmann's constant, h is Plank's constant, c_{cr} and c_C stand for concentration of chromium and carbon respectively. In general, x can represent the 3-D Cartesian co-ordinate system components. It may be noted that the above Equation 3 serves as a boundary condition to the governing differential equation and it is nonlinear as the gradient of chromium concentration $\partial c_{cr} / \partial x$ at the interface depends upon the concentrations of chromium and carbon at this site. This model is solved along with the diffusion equation of chromium through the grain which is written as

$$\frac{\partial c_{cr}}{\partial t} = D_{cr} \nabla^2 c_{cr} \tag{4}$$

Finite Difference Scheme to Discretize the Chromium Diffusion Equation

We use the finite difference method to solve Equation 3 and 4. For simplicity and from spherical symmetry point of view (assumption), a spherical co-ordinate system is chosen with x as the dimension of the problem which is along the radial direction r of the sphere. Hence, the problem reduced to that of one dimension and the variable in space is denoted as r. Upon discretization, the coordinates of points in the finite element mesh are denoted as $r_i, where\ i = 1, n$. The grain is approximated as a sphere with carbide precipitation taking place at the outer surface. Discretizing Equation 4 using finite difference for both space and time, the equation is written as follows.

$$V_n \frac{x_n^i - x_n^{i-1}}{\Delta t} = -D_{\mathrm{Cr}} \left(A_n \frac{x_n^i - x_{n-1}^i}{\Delta r_{n-1}} - A_{n+1} \frac{x_{n+1}^i - x_n^i}{\Delta r_n} \right) \tag{5}$$

where x_n^i is the mole fraction of chromium at node 'n' at time 'i', Δr_n is the width of the nth spherical annulus, A_n is the surface area of the nth spherical shell, V_n is the volume of the nth spherical annulus, n is the node number in the finite difference grid, D_{Cr} is the diffusion coefficient of chromium in the matrix and Δt is the time increment. On rearrangement, Equation 5 can be written as

$$-\frac{DA_n \Delta t}{V_n \Delta r_{n-1}} x_{n-1}^i + \left\{ 1 + \frac{D\Delta t}{V_n} \left(\frac{A_n}{\Delta r_{n-1}} + \frac{A_{n+1}}{\Delta r_n} \right) \right\} x_n^i - \frac{DA_{n+1} \Delta t}{V_n \Delta r_n} x_{n+1}^i = x_n^{i-1} \tag{6}$$

The above equation is solved along with the discretized version of Equation 3 to obtain the distribution of chromium concentration at different points along the radius of the sphere as a function of time. As we know, the gradient of chromium concentration is high near the boundary and hence, we need to make a fine discretization near the boundary to catch the gradient accurately. For calculation of time for onset of sensitization, the region near the boundary up to a distance of 20 nm approximately is monitored to see if the chromium concentration falls below 12%. As this dimension is very small compared to the grain diameter (of the order of several microns), we need a mesh size, which is of the order of a nanometer at the boundary. This type of mesh over the whole domain will make the computation very expensive and moreover, we don't require such a fine mesh near the centre of the grain (where the concentration gradient is almost zero). Hence, a scheme for generation of non-uniform grid is presented in the following sub-section.

Non-Uniform Grid Generation Scheme for Minimizing Computation Time

The term $\nabla^2 c_{cr} \left(written\ as\ C_{xx} \right)$ in Equation 4 has been discretized considering non-uniform grid. We can approximate the second derivative C_{xx} using the four grid points $x_{i-2}, x_{i-1,} \mathrm{x}_i$ and x_{i+1} as

$$\left(C_{xx} \right)_i = \alpha_i^{II} C_{i-2} + \beta_i^{II} C_{i-1} - \left(a_i^{II} + \beta_i^{II} + \gamma_i^{II} \right) C_i + \gamma_i^{II} C_{i+1} + \varepsilon_i^{II} \tag{7}$$

where the coefficients $\alpha_i^{II}, \beta_i^{II}$ and γ_i^{II} are given by (Liu et al., 1995)

$$\alpha_i^{II} = \frac{2\left(d_{i+1} - d_{i-1}\right)}{\left[d_{i-2}\left(d_{i-2} + d_{i+1}\right)\left(d_{i-2} - d_{i-1}\right)\right]} \tag{8}$$

$$\beta_i^{II} = \frac{2\left(d_{i-2} - d_{i+1}\right)}{\left[d_{i-1}\left(d_{i-2} - d_{i-1}\right)\left(d_{i-1} + d_{i+1}\right)\right]} \tag{9}$$

$$\gamma_i^{II} = \frac{2\left(d_{i-2} + d_{i-1}\right)}{\left[d_{i+1}\left(d_{i-1} + d_{i+1}\right)\left(d_{i-2} + d_{i+1}\right)\right]} \tag{10}$$

This procedure helps in choosing a non-uniform grid depending upon the location of the steep concentration gradients. In the next section, this model has been applied to predict the chromium concentration profile in the grain and the grain boundary regions of austenitic stainless steels of different compositions.

Simulation of Time-Temperature-Sensitization Curves for Different Steels

The importance of this simulation lies in successful prediction of chromium concentration in the grain and the grain boundary as function of space and time. For prediction of time-temperature-sensitization (TTS) behavior of austenitic stainless steels (which is an important experimental data), the chromium concentration profile is used. The time at which the chromium concentration falls below 12% at a distance of 20 nanometer from the grain boundary is taken as the time for sensitization at that particular temperature. The chemical compositions of the four different materials considered in the study are shown in Table 1.

As can be seen from the Table 1, the carbon composition is different for the four materials and this can affect the rate of carbide formation and hence, the corresponding sensitization behavior. The material property like the diffusion co-efficient of chromium in the matrix as a function of temperature is taken from literature (Mozhi et al., 1986). The temperature and chemical composition dependent constant c_1 of Equation 3 is calibrated from some known experimental data points of TTS diagram for the material. The grain size considered in the analysis is of 100 micron diameter. Considering the symmetry of

Table 1. Compositions of different types of austenitic stainless steels

	C	Mn	P	S	Si	Ni	Cr	Mo	Cu	N
Material 1	0.053	1.77	0.031	0.008	0.41	8.49	19.27	0.36	0.16	0.04
Material 2	0.044	1.77	0.034	0.007	0.46	8.59	18.87	0.36	0.16	0.10
Material 3	0.043	1.78	0.028	0.008	0.41	8.63	18.80	0.38	0.16	0.16
Material 4	0.046	1.37	0.030	0.008	0.40	8.89	18.90	0.38	0.17	0.25

the spherical co-ordinate system, only 50 micron region along the radius has been discretized. The symmetric boundary condition (i.e., chromium concentration gradient as zero) has been applied at the centre of the sphere.

The predicted results from such a simulation is shown in Figure 1 where the variation of chromium concentration is shown as function of time at a distance of 20 nm from the grain boundary (calculation for material-2 at applied temperature of 942 K). It can be observed that the chromium concentration reaches a minimum and then increases. This phenomenon corresponds to the process of de-sensitization and hence, it is in-built in the model.

The de-sensitization phenomena are very important at higher temperatures and is often required to be predicted so that appropriate heat-treatment procedures may be applied to reclaim back a sensitized material zone. However, many other parameters seem to act during this regime and it will be discussed later. Similarly, Figure 2 shows the variation of chromium concentration as a function of distance from the centre of the grain as a function of time. The simulation is for material-2 at applied temperature of 942 K. This result can be used for construction of the TTS diagram as discussed later.

The effect of mesh size (i.e., size of discretization) and time step on the predicted results are very important in any numerical method and hence, a parametric study with respect to these variables is carried out. The effect of mesh size (at grain interior and grain boundary region as these are meshed with different sizes of the non-uniform grid discretization as discussed in Section-2) on the predicted chromium concentration at a distance of 20 nm from the grain boundary (for applied temperature of 1052 K in Material-2) is shown in Figure 3. It can be observed that the sizes of 10 nanometer at the

Figure 1. Predicted chromium profile at a distance of 20nm from boundary obtained for Material 2 at 942 K

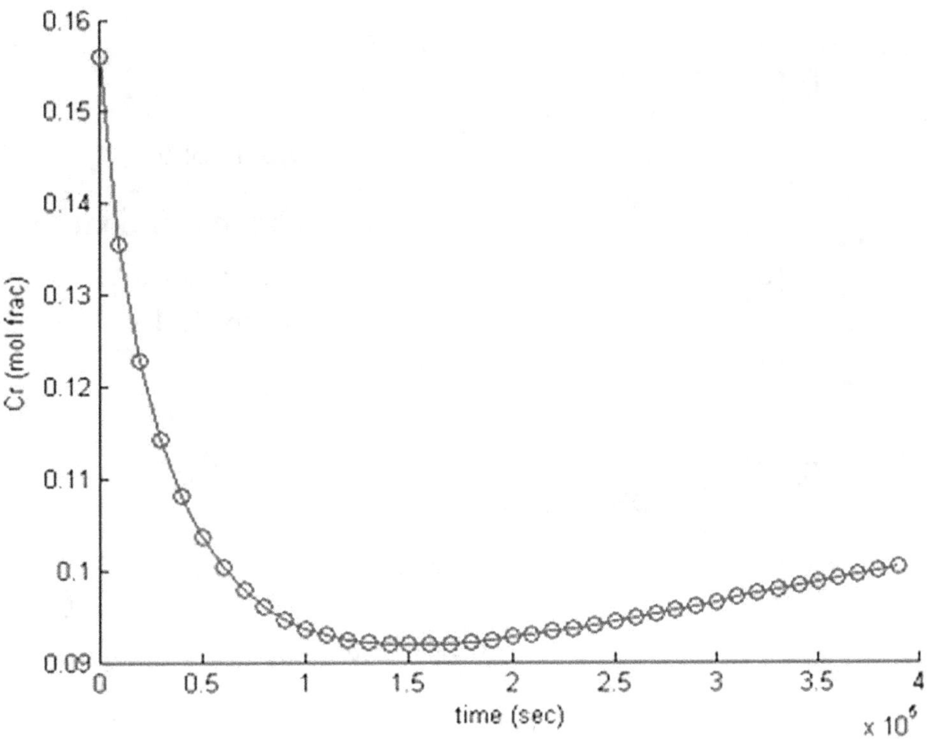

Figure 2. Variation of predicted chromium concentration profile with time for Material 2 at 942 K

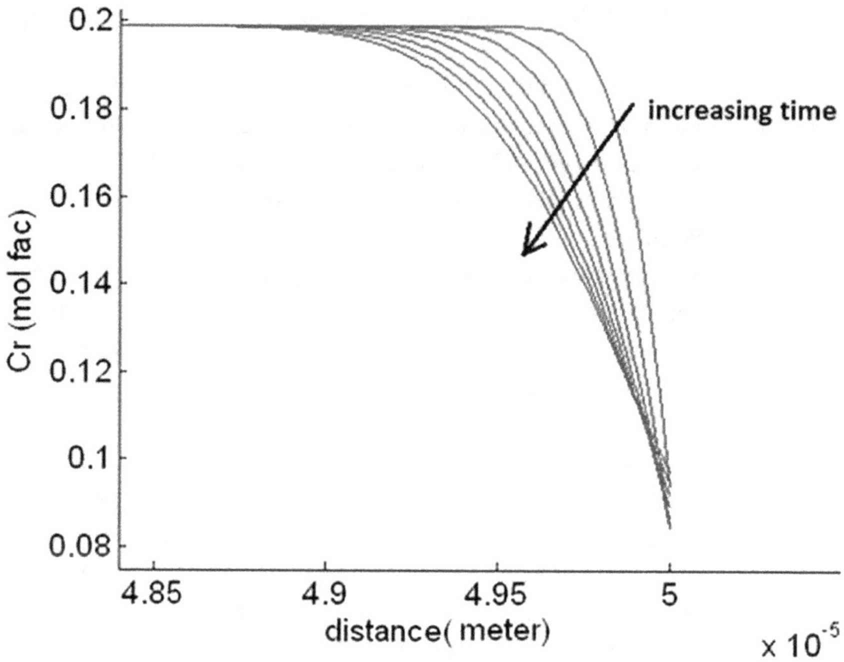

Figure 3. Effect of mesh size (at grain interior and grain boundary region) on the predicted chromium concentration at a distance of 20 nm from the grain boundary for temperature of 1052 K

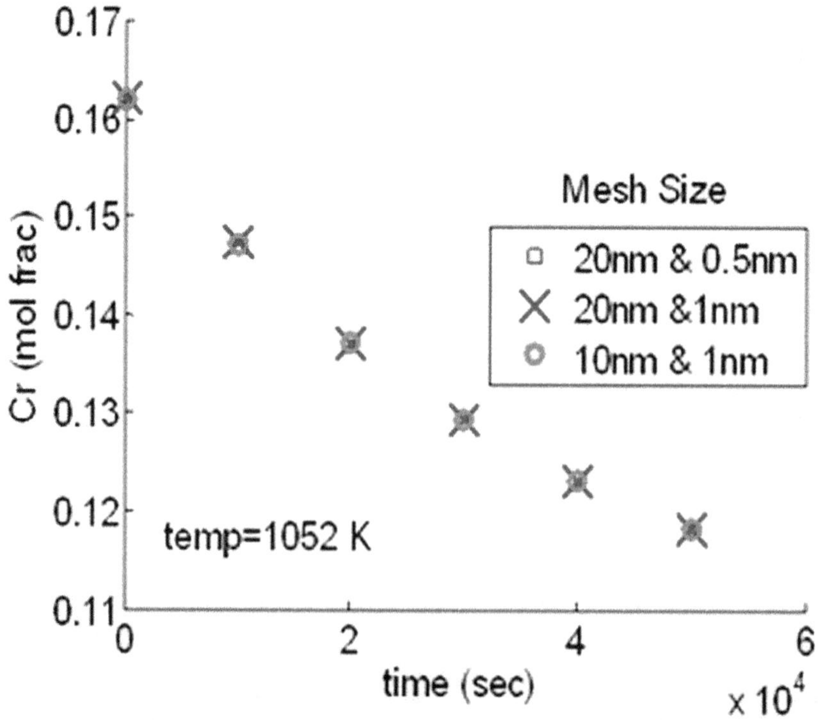

grain interior and 0.5 nanometer at the grain boundary region is sufficient for getting converged results. Similarly, the effect of time steps on the chromium profile for the same problem is shown in Fig. 4 and accordingly, the time step for subsequent simulations have been chosen as 50 seconds.

The time-temperature-sensitization (TTS) behaviour for the four different materials as obtained from experimental observation is shown in Figures 5 to 8 respectively. The experimental data is taken from Mozhi et al. (1986). The predicted TTS behaviour has been compared with those of simulation in Figures 5-8 for the four different materials. The material constant c_1 has been calibrated from the experimental data at lower temperatures (right end tail data of the TTS diagrams) and this has been used for simulation in the whole temperature range as shown in Figures 5-8.

It can be observed that the predicted results are in very good agreement with the experimental data especially in the sensitization region. As can be from Figures 5 to 8, the error between experiment and analysis is within ±5% in the sensitization part of the curves. In the desensitization region, the error is more than this limit. This is due to non-consideration of desensitization phenomena in the model. The top part of the TTS diagram where the sensitization time again increases (with temperature) after attaining a minimum represents the de-sensitization region. The kinetics of chromium diffusion in this region is affected by other parameters like the diffusion of nickel. As the chromium vacates the matrix due to carbide precipitation in the grain boundary, the nickel atom occupies the space by diffusion from the surrounding region. As diffusivities of chromium and nickel atoms are of the same order, there is a competition between the two atoms. Once chromium is used for carbide reaction at the boundary, more and more chromium migrates to the reaction site by the process of diffusion and these vacant sites are

Figure 4. Effect of time step size (seconds) on the predicted chromium concentration at a distance of 20 nm from the grain boundary for temperature of 1052 K

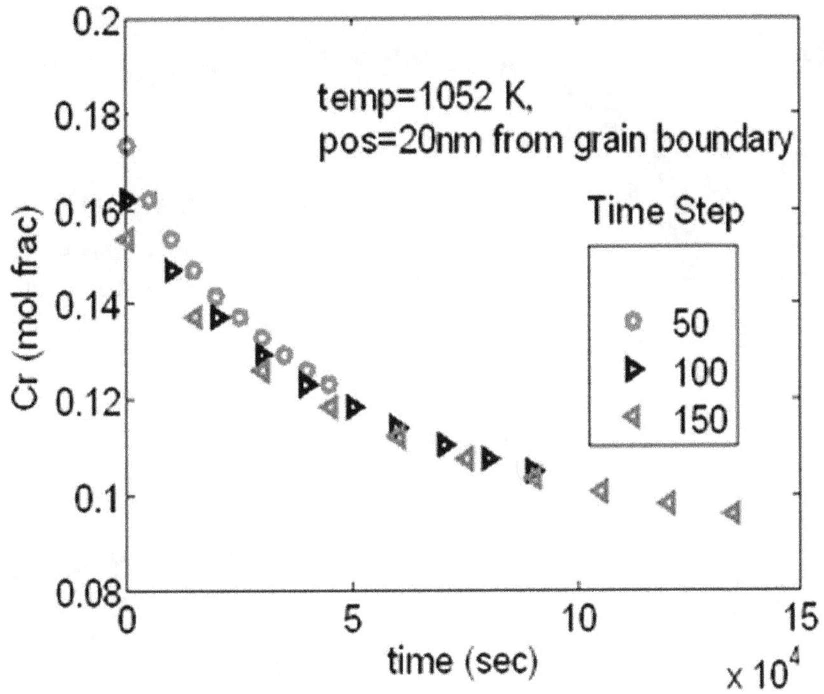

Figure 5. Comparison of predicted time-temperature-sensitization diagram for Material-1
Experimental data taken from Mozhi et al. (1986).

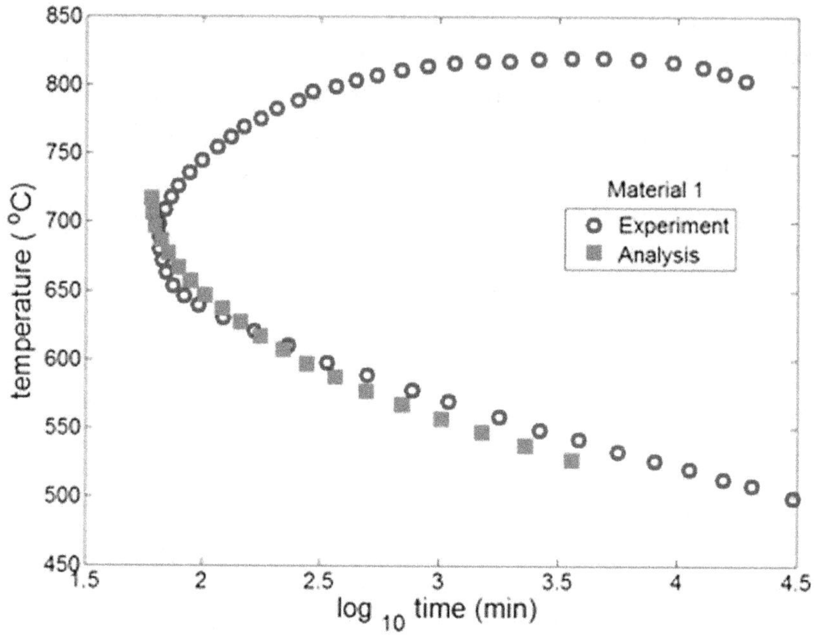

Figure 6. Comparison of predicted time-temperature-sensitization diagram for Material-2
Experimental data taken from Mozhi et al. (1986).

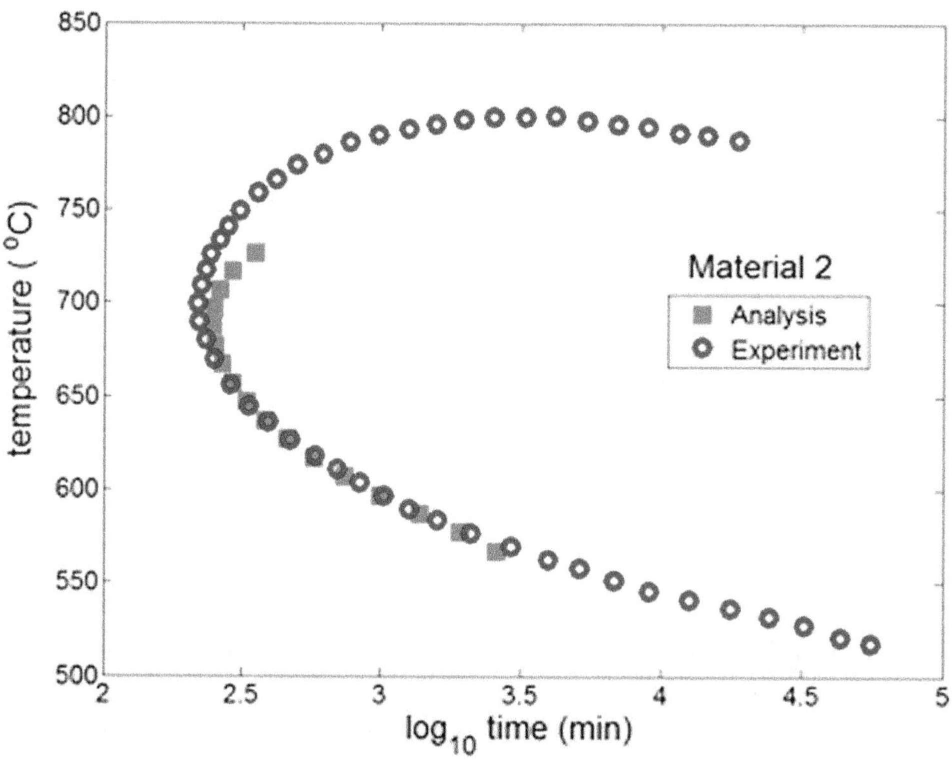

Figure 7. Comparison of predicted time-temperature-sensitization diagram for Material-3
Experimental data taken from Mozhi et al. (1986).

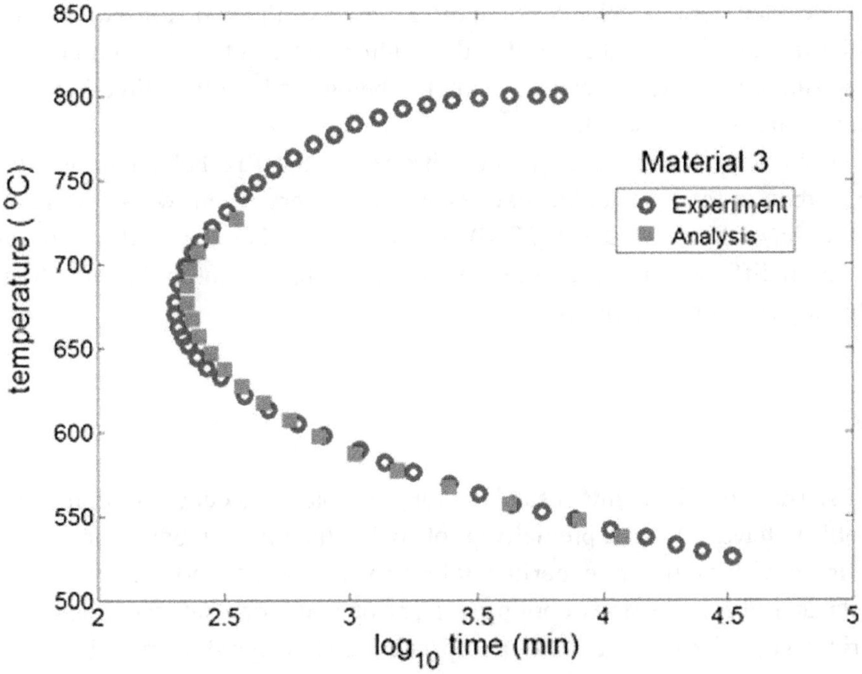

Figure 8. Comparison of predicted time-temperature-sensitization diagram for Material-4
Experimental data taken from Mozhi et al. (1986).

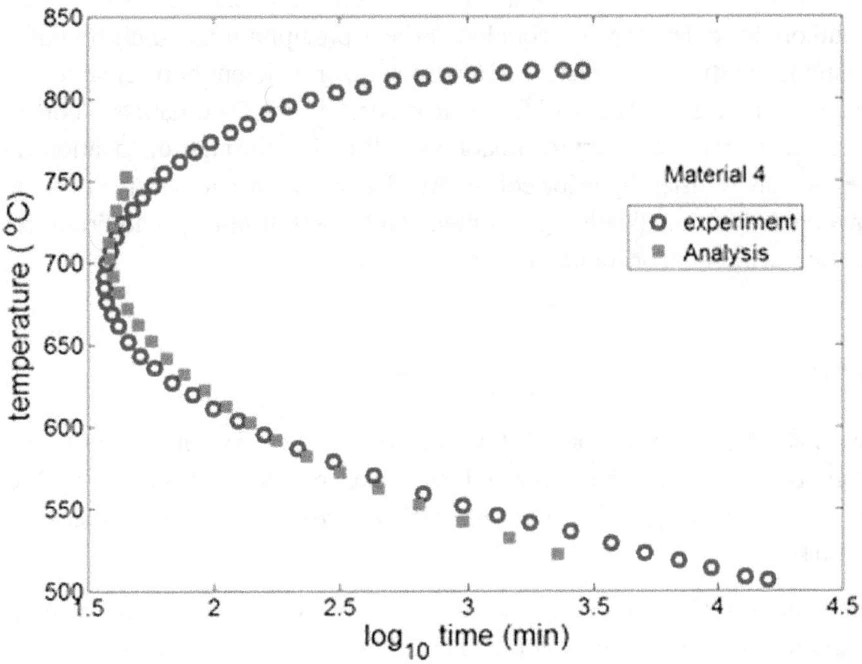

mostly occupied by nickel. After the rate of reaction decreases at the boundary and de-sensitization begins, migration of nickel and chromium atoms starts to the chromium depleted zone. However, as the site may be occupied with nickel and as the diffusivities of the two atoms are of similar order, one needs to consider the solutions of the coupled diffusion problem of nickel and chromium in order to predict the de-sensitization time correctly. Attempt is being made in this direction in order to include these effects in a future refined model.

There have been efforts to study the effect of nitrogen on the TTS behaviour as nitrogen is increasingly replacing carbon in these austenitic stainless steels. A very recent work towards this direction is presented in Parvathavarthini and Dayal (2010). The model needs to be modified in order to consider the effect of nitrogen diffusion. The present model has been an attempt in this direction and the results are quite satisfactory as evidenced in Figures 5-8.

CONCLUSION

The problem of sensitization in stainless steels is very complex and depends upon many parameters. It is always desirable to have a suitable predictive tool so that the material behavior can be predicted with reasonable accuracy with minimum experimental effort. It is not economically viable to consider all combinations of material compositions, cooling rates, prior cold-work and other micro-structural features etc. in the experiment in order to arrive at the appropriate decision regarding material selection. Moreover, there has been constant demand to design components for reliable performance for higher and higher life. Experimental data can be conducted for short times as planning and executing long-term experiments is expensive and cumbersome. Especially, for the problem of low temperature sensitization, it is very difficult to carry out experiment at higher temperatures and justify its validity at lower operating temperatures by extrapolation. The development of predictive models is an attempt in that direction. In this work, the authors have developed a coupled carbide-precipitation and chromium diffusion model and solved it using finite difference method. However, other efficient numerical techniques like finite element method etc. can be used. This will facilitate modeling in 3-D domain and consideration of other diffusion channels like grain boundary diffusion and diffusion through dislocation region etc. which are faster modes and are assisted by prior cold-work. The effect of competitive diffusion of chromium and nickel atoms on the de-sensitization phenomena and effect of nitrogen addition in the alloy and its corresponding effect will be incorporated in future work.

REFERENCES

Anderko, A., Sridhar, N., & Tormoenc, G. (2010). Localised corrosion of heat-treated alloys Part II -Predicting grain boundary microchemistry and its effect on repassivation potential. *Corrosion Engineering, Science and Technology. The International Journal of Corrosion Processes and Corrosion Control*, *45*(3), 204–223.

Beltran, R., Maldonado, J. G., Murr, L. E., & Fisher, W. W. (1997). Effects of strain and grain size on carbide precipitation and corrosion sensitization behavior in 304 stainless steel. *Acta Materialia*, *45*(10), 4351–4360. doi:10.1016/S1359-6454(97)00106-7

Bruemmer, S. M. (1990). Quantitative modeling of sensitization development in austenitic Stainless steels. *Corrosion*, *46*(9), 698–709. doi:10.5006/1.3585171

Carter, R. D., Damcott, D. L., Atzmon, M., Was, G. S., Bruemmer, S. M., & Kenik, E. A. (1994). Quantitative analysis of radiation-induced grain-boundary segregation measurements. *Journal of Nuclear Materials*, *211*(1), 70–84. doi:10.1016/0022-3115(94)90282-8

Dayal, R. K., Parvathavarthini, N., & Raj, B. (2005). Influence of metallurgical variables on sensitisation kinetics in austenitic stainless steels. *International Materials Reviews*, *50*(3), 129–155. doi:10.1179/174328005X14348

Fu, Y., Wu, X., Han, E.-H., Ke, W., Yang, K., & Jiang, Z. (2009). Effects of cold work and sensitization treatment on the corrosion resistance of high nitrogen stainless steel in chloride solutions. *Electrochimica Acta*, *54*(5), 1618–1629. doi:10.1016/j.electacta.2008.09.053

Fullman, R. L. (1982). A thermodynamic model of the effects of composition on the susceptibility of stainless steels to intergranular stress corrosion cracking. *Acta Metallurgica*, *30*(7), 1407–1415. doi:10.1016/0001-6160(82)90161-4

Garcia, C., Martín, F., Tiedra, P. D., Heredero, J. A., & Aparicio, M. L. (2001). Effects of prior cold work and sensitization heat treatment on chloride stress corrosion cracking in type 304 stainless steels. *Corrosion Science*, *43*(8), 1519–1539. doi:10.1016/S0010-938X(00)00165-7

Grujisic, M., & Tangrila, S. (1991). Thermodynamics and kinetic analysis of time-temperature-sensitization diagrams in austenitic stainless steels. *Materials Science and Engineering A*, *142*(2), 255–259. doi:10.1016/0921-5093(91)90664-9

Hall, E. L., & Briant, C. L. (1984). Chromium depletion in the vicinity of carbides in sensitized austenitic stainless steels. *Metallurgical Transactions*, *15A*(5), 793–810. doi:10.1007/BF02644554

Kohler, F. (1960). Zur berechung der thermodynamischen daten eines ternären systems ans den zugehörigen binären systemen. *Monatshefte fur Chemie*, *91*(4), 738–740. doi:10.1007/BF00899814

Lehockey, E. M., Palumbo, G., & Lin, P. (1998). Improving the weldability and service performance of nickel and iron based super alloys by grain boundary engineering. *Metallurgical Transactions*, *29A*(12), 3069–3079. doi:10.1007/s11661-998-0214-y

Liu, J., Pope, G. A., & Sepehrnoori, K. (1995). A high-resolution finite-difference scheme for nonuniform grids. *Applied Mathematical Modelling*, *19*(3), 162–172. doi:10.1016/0307-904X(94)00020-7

Mayo, W. E. (1997). Predicting IGSCC/IGA susceptibility of Ni-Cr-Fe alloys by modeling of grain boundary chromium depletion. *Materials Science and Engineering A*, *232*(1-2), 129–139. doi:10.1016/S0921-5093(97)00110-X

Moura, V., Kina, Y. A., Tavares, S. S. M., de Faria, M. M. S. G., & Mainier, F. B. (2009). Investigation of cracks and sensitization in an AISI 304L stainless steel exposed to 500-600 °C. *Engineering Failure Analysis*, *16*(1), 545–551. doi:10.1016/j.engfailanal.2008.02.002

Mozhi, T. A., Betrabet, H. S., Jagannathan, V., Wilde, B. E., & Clark, W. A. T. (1986). Thermodynamic modeling of sensitization of AISI 304 stainless steels containing nitrogen. *Scripta Metallurgica, 20*(5), 723–728. doi:10.1016/0036-9748(86)90499-0

Mozhi, T. A., Juhas, M. C., & Wilde, B. E. (1987). Modeling low temperature sensitization of austenitic stainless steels. *Scripta Metallurgica, 21*(11), 1547–1552. doi:10.1016/0036-9748(87)90298-5

Natesan, K., & Kassner, T. F. (1973). Thermodynamics of carbon in nickel, iron-nickel and iron-chromium-nickel alloys. *Metallurgical Transactions, 4*(11), 2557–2566. doi:10.1007/BF02644258

Ortner, S. R., & Randle, V. (1989). A study of the relation between grain boundary type and sensitization in a partially sensitized AISI 304 stainless steel using electron back-scattering patterns. *Scripta Metallurgica, 23*(11), 1903–1908. doi:10.1016/0036-9748(89)90480-8

Parvathavarthini, N., & Dayal, R. K. (2010). Time-temperature-sensitization diagrams and critical cooling rates of different nitrogen containing austenitic stainless steels. *Journal of Nuclear Materials, 399*(1), 62–67. doi:10.1016/j.jnucmat.2010.01.003

Parvathavarthini, N., Mulki, S., Dayal, R. K., Samajdar, I., Mani, K. V., & Raj, B. (2009). Sensitization control in AISI 316L(N) austenitic stainless steel: Defining the role of the nature of grain boundary. *Corrosion Science, 51*(9), 2144–2150. doi:10.1016/j.corsci.2009.05.045

Pragnell, W. M., & Evans, H. E. (2006). A finite-difference model to predict 2D depletion profiles arising from high temperature oxidation of alloy. *Modelling and Simulation in Materials Science and Engineering, 14*(4), 733–740. doi:10.1088/0965-0393/14/4/014

Richardson, F. D. (1953). The thermodynamics of metallurgical carbides and of carbon in iron. *Journal of the Iron and Steel Institute, 167*, 33–45.

Sahaloui, H., Sidhom, H., & Philibert, J. (2002). Prediction of chromium depleted zone evaluation during aging of Ni-Cr-Fe alloys. *Acta Materialia, 50*(6), 1383–1392. doi:10.1016/S1359-6454(01)00444-X

Sakaguchi, N., Takahashi, H., & Ichinose, H. (2005). Influence of Silicon and Phosphorus on Radiation-Induced Segregation of Chromium and Nickel in Austenitic Model Alloys. *Materials Transactions, 46*(3), 440–444. doi:10.2320/matertrans.46.440

Samal, M. K., & Abhishek, A. (2011). A new model for prediction of chromium depletion near grain boundaries and corresponding sensitization in austenitic stainless steels. *IMechE Part C Journal of Mechanical Engineering Science, 225*(4), 809–815. doi:10.1243/09544062JMES2327

Shimada, M., Kokawa, H., Wang, Z. J., Sato, Y. S., & Karibe, I. (2002). Optimization of grain boundary character distribution for intergranular corrosion resistant 304 stainless steel by twin-induced grain boundary engineering. *Acta Materialia, 50*(9), 2331–2341. doi:10.1016/S1359-6454(02)00064-2

Sourmail, T., Too, C. H., & Bhadeshia, H. K. D. H. (2003). Sensitization and evolution of chromium-depleted zones in Fe-Cr-Ni-C systems. *ISIJ International, 43*(11), 1814–1820. doi:10.2355/isijinternational.43.1814

Stawström, C., & Hillert, M. (1969). An improved depletion zone theory of intergranular corrosion of 18-8 stainless steel. *Journal of the Iron and Steel Institute, 207*, 77–90.

Tedmon, C. S. Jr, Vermilyea, D. A., & Rosolowski, T. H. (1971). Intergranular corrosion of austenitic stainless steel. *Journal of the Electrochemical Society*, *118*(2), 192–201. doi:10.1149/1.2407966

Trillo, E. A., Beltran, R., Maldonado, J. G., Romero, R. J., Murr, L. E., Fisher, W. W., & Advani, A. H. (1995). Combined effects of deformation (strain and strain state), grain size, and carbon content on carbide precipitation and corrosion sensitization in 304 stainless steel. *Materials Characterization*, *35*(2), 99–112. doi:10.1016/1044-5803(95)00072-0

Trillo, E. A., & Murr, L. E. (1998). Effects of carbon content, deformation, and interfacial energetics on carbide precipitation and corrosion sensitization in 304 stainless steel. *Acta Materialia*, *47*(1), 235–245. doi:10.1016/S1359-6454(98)00322-X

Tsay, L. W., Liua, Y.-F., Huang, R.-T., & Kuo, R.-C. (2008). The effect of sensitization on the hydrogen-enhanced fatigue crack growth of two austenitic stainless steels. *Corrosion Science*, *50*(5), 1360–1367. doi:10.1016/j.corsci.2008.01.005

Verlinden, B., Driver, J., Samajdar, I., & Doherty, R. D. (2007). *Thermo-mechanical processing of metallic materials, Pergamon materials series*. Amsterdam: Elsevier.

Wagner, C. (1952). *Thermodynamics of alloys* (pp. 51–52). Cambridge, Massachusetts: Addison-Wesley.

Was, G. S., & Kruger, R. M. (1985). Chromium depletion in Ni-Cr-Fe alloys. *Acta Metallurgica*, *33*(5), 841–854. doi:10.1016/0001-6160(85)90108-7

Wasnik, D. N., Kain, V., Samajdar, I., Verlinden, B., & De, P. K. (2002). Resistance to sensitization and intergranular corrosion through extreme randomization of grain boundaries. *Acta Materialia*, *50*(18), 4587–4601. doi:10.1016/S1359-6454(02)00306-3

Watanabe, Y., Kain, V., Tonozuka, T., Shoji, T., Kondo, T., & Masuyama, F. (2000). Effect of Ce addition on the sensitization properties of stainless steels. *Scripta Materialia*, *42*(3), 307–312. doi:10.1016/S1359-6462(99)00352-8

Williams, R. O. (1982). Thermodynamics of the superalloys. *Metallurgical Transactions*, *13A*(6), 959–965. doi:10.1007/BF02643391

Zhou, Y., Aust, K. T., Erb, U., & Palumbo, G. (2001). Effects of grain boundary structure on carbide precipitation in 304L stainless steel. *Scripta Materialia*, *45*(1), 49–54. doi:10.1016/S1359-6462(01)00990-3

Chapter 14
Risk Analysis of Structural Engineering Systems Using Bayesian Inference

Sharvil Alex Faroz
Indian Institute of Technology Bombay, India

Nikil Narayan Pujari
Indian Institute of Technology Bombay, India

Rohit Rastogi
Bhabha Atomic Research Centre, India

Siddhartha Ghosh
Indian Institute of Technology Bombay, India

ABSTRACT

Engineering phenomena described through models come with uncertainties arising out of the modelling assumptions and observational statistical discrepancies. A probabilistic framework helps in quantifying the uncertainties involved. Conventional statistics are not very useful for any inferring from a limited amount of data which is often the case for many real life engineering problem. Bayesian inference tackles this issue by integrating past experience with current information. This chapter introduces the basic concepts of Bayesian inference and illustrates through three problems in structural engineering, how it can be useful. The emphasis is in showing how Markov chain Monte Carlo (MCMC) simulation is used in Bayesian inference.

INTRODUCTION

As far as the laws of mathematics refer to reality, they are not certain; and as far as they are certain they do not refer to reality. — Albert Einstein

Structural engineering problems are usually analysed with the aid of models to predict the behaviour of structures under consideration. These models can be defined either in terms of derived physics based approaches or probabilistic approaches (Der Kiureghian & Ditlevsen, 2009). Subjected to environmental loads, a structure's performance is usually measured in terms of stresses induced, deformations,

DOI: 10.4018/978-1-5225-0588-4.ch014

deterioration levels, and other damage measures. Prediction of these quantities, by adopting either of these modelling approaches, results in an improper estimate as the models themselves may contain some uncertainty. There are various sources of uncertainty which need to be taken into account for a better prediction of response quantities. Based on the available literature (Pate-Cornell, 1996; Der Kiureghian & Ditlevsen, 2009; Kennedy *et al.*, 1980; Kennedy & Ravindra, 1984) uncertainties are characterised in two basic categories *viz.* aleatory and epistemic. Aleatory uncertainty includes intrinsic randomness in predicting the outcome of a system. This type of uncertainty always exists as there is inherent natural randomness in any system. This randomness can very well be explained by an example of a concrete cube compressive test. No two test results of a concrete cube test (with the same target f_{ck}) are the same, because of the inherent randomness that exists in the material and other properties. Epistemic uncertainty is due to a lack of knowledge of the behaviour of the structure being studied/analysed. These may be due to the differences in modelling approach for a single system. Inclusion of these uncertain quantities in prediction model will result in an accurate probabilistic estimation of structural responses.

Usually, we adopt simplified models based on our (prior) beliefs and assumptions within the accepted principles of mechanics. These assumptions include over simplification of mathematical models, conversion of mathematical expressions into computational models (e.g., Finite Element (FE) models) and knowledge acquired from limited experiments. Such models limit our capability of predicting the actual behaviour of a structure. Additionally, observations are typically replete with errors due to imperfections in measurements (especially non-destructive measurements), difficulty in acquiring data (Oden *et al.*, 2010) and lack of understanding about the behaviour of the structure. Moreover, due to limited availability of these data, a proper inference of the system may lack clarity.

Probabilistic assessment of a structure requires the model to be defined in the probabilistic domain. Simulation methods, for example Monte Carlo simulation (MCS), Latin hypercube sampling (LHS), importance sampling and subset simulation (SS), rely on the existing models to accurately predict the desired response. This prediction does not take into account the information/data available through newly performed testing/measurements. Bayesian inference is one such method which can incorporate a newly acquired information to update the existing model. Conventional statistics relies on a large amount of data to take safety related decisions. In the presence of a limited number of real observations available for a structure, pursuing the Bayesian inference is apt for the treatment of uncertainties (Ang & Tang, 2007). The Bayesian approach facilitates the integration of multiple prediction models, various sources of uncertainty, errors, and inspection data, towards overall uncertainty quantification for the engineering system/phenomenon under consideration. In the present work, Bayesian updating is used for incorporating uncertainties occurring in structural engineering systems.

In this chapter, we start with an introduction to the basic concepts for Bayesian inference. To illustrate the theory of Bayesian inference, three examples in the structural engineering domain are presented. The first application deals with updating the corrosion model of reinforced concrete (RC) structures using non-destructive measurements of steel loss. The second application illustrates the seismic fragility estimation of a nuclear containment structure and updating its parameters based on structural analysis results. The third example deals with predicting the fatigue crack growth of a piping component based on in-service inspection. We finally conclude with a brief summary of few caveats in adopting Bayesian inferences.

BAYESIAN INFERENCE

In statistics, probability plays an important role in drawing inferences about the nature of a model. Bayes' theorem originates from the concept of conditional probability. The probability of occurrence of an event A given that the event E has already occurred is,

$$\mathbb{P}(A \mid E) = \frac{\mathbb{P}(A \cap E)}{\mathbb{P}(E)} \tag{1}$$

Similarly, the probability of an event E occurring given the event A has already occurred is

$$\mathbb{P}(E \mid A) = \frac{\mathbb{P}(E \cap A)}{\mathbb{P}(A)} \tag{2}$$

Using Eq. (1) and (2) and recognising that $\mathbb{P}(A \cap E) = \mathbb{P}(E \cap A)$, we get Bayes' theorem

$$\mathbb{P}(A \mid E) = \frac{\mathbb{P}(E \mid A)\mathbb{P}(A)}{\mathbb{P}(E)} \tag{3}$$

Suppose A is subdivided into n mutually exclusive and collectively exhaustive events A_1, A_2,..., A_n, then by total using the probability theorem, $\mathbb{P}(E)$ is given by

$$\mathbb{P}(E) = \mathbb{P}(E \mid A_1)\mathbb{P}(A_1) + \mathbb{P}(E \mid A_2)\mathbb{P}(A_2) + \ldots + \mathbb{P}(E \mid A_n)\mathbb{P}(A_n) = \sum_{i=1}^{n} \mathbb{P}(E \mid A_i)\mathbb{P}(A_i) \tag{4}$$

Therefore, with the inclusion of n discrete events, $\mathbb{P}(A_i \mid E)$ can be written using Eq. (3) and Eq. (4) as,

$$\mathbb{P}(A_i \mid E) = \frac{\mathbb{P}(E \mid A_i)\mathbb{P}(A_i)}{\sum_{i=1}^{n} \mathbb{P}(E \mid A_i)\mathbb{P}(A_i)} \tag{5}$$

Eq. (5) is a standard form of Bayes' theorem for discrete events. For continuous random variables X and Y, Bayes' theorem can be formulated in terms of their respective probability density functions:

$$f_{X|Y}(X \mid Y = y) = \frac{f_{Y|X}(y \mid X = x) f_X(x)}{f_Y(y)} = \frac{f_{Y|X}(y \mid x) f_X(x)}{\int_{\mathcal{D}_X} f_{Y|X}(y \mid x) f_X(x) dx} \tag{6}$$

This concept of Bayes' theorem is used in various engineering applications to update an existing numerical model containing continuous random variables as its parameters. Updating of these models are subjected to the availability of observed/measured values either through experimental investigation or numerical simulations. For example, assessing the degradation of the concrete compressive strength of a beam over time. Here the strength of concrete can be modelled as a function of parameters which are uncertain in nature, such as the water-cement ratio, aggregate size and grade of cement. We check the influence of the measured compressive strength value, obtained from say, the rebound hammer test, on the existing beliefs (in the form of a probability density function) of these three parameters. Bayes' theorem provides an inference for updating of available or 'prior' models based on measured data. The updated model is characterised by the "posterior" density in Bayesian inference. Considering Ω as a vector of the observed/measured data of a system, the updated/posterior density of the model parameter Θ is given by,

$$\begin{aligned} f_{\Theta}(\boldsymbol{\theta} \mid \Omega) &= \frac{f_{\Omega}(\Omega \mid \boldsymbol{\theta}) f_{\Theta}(\boldsymbol{\theta})}{\int_{\mathcal{D}_{\Theta}} f_{\Omega}(\Omega \mid \boldsymbol{\theta}) f_{\Theta}(\boldsymbol{\theta}) d\boldsymbol{\theta}} \\ &= \mathcal{C}_0 f_{\Omega}(\Omega \mid \boldsymbol{\theta}) f_{\Theta}(\boldsymbol{\theta}) \\ &\propto f_{\Omega}(\Omega \mid \boldsymbol{\theta}) f_{\Theta}(\boldsymbol{\theta}) \\ &\text{Posterior} \propto \text{Likelihood} \times \text{Prior} \end{aligned} \tag{7}$$

where $f_{\Theta}(\boldsymbol{\theta})$ denotes the prior joint probability density function of the model parameters, $f_{\Omega}(\Omega \mid \boldsymbol{\theta})$ is the likelihood function which represents the contribution of measured data in formation of posterior density, $f_{\Theta}(\boldsymbol{\theta} \mid \Omega)$, $\mathcal{C}_0 = 1 / f_{\Omega}(\Omega)$ is the normalising constant expressed in the form of total probability theorem in Eq. (7), such that integrating $f_{\Theta}(\boldsymbol{\theta} \mid \Omega)$ over the parameter space of $\mathcal{D}_{\Theta}$ yields 1, hence making the posterior distribution a perfect PDF.

$$\int_{\mathcal{D}_{\Theta}} f_{\Theta}(\boldsymbol{\theta} \mid \Omega) d\boldsymbol{\theta} = \mathcal{C}_0 \int_{\mathcal{D}_{\Theta}} f_{\Omega}(\Omega \mid \boldsymbol{\theta}) f_{\Theta}(\boldsymbol{\theta}) d\boldsymbol{\theta} = 1 \tag{8}$$

Prior Distribution

Prior distribution $f_{\Theta}(\boldsymbol{\theta})$ contains the initial information of the parameters which may be based on prior beliefs, previous knowledge or expert opinion. While performing Bayesian analysis, it is always recommended to probe the data having different prior beliefs. This gives the knowledge, of up to what extent the data impose inferences on the prior about uncertainties involved in the properties of parameter Θ. This prior density, when combined with actual measurements of the system, gives us an updated version of our belief known as posterior density. Decision on type of priors to be selected is discussed in detail by Kass and Wasserman (1996). Priors should be ideally selected based on the available database or from previously performed experiments of similar nature. Based on a thorough literature investigation, one should be able to decide and justify the use of priors. Priors are mainly classified into two main categories: 1) informative priors and 2) non-informative priors. Informative priors are based on the past history of the parameter. These priors may be posterior distribution of previously performed Bayesian updating or a collection of statistical database available in literature. Non-informative priors, as the name suggests, is used when there is no information available of the parameter. The simplest adaptation of a non-informative prior is the use of a uniform distribution by assigning equal probabilities to all the possible outcomes of a parameter. Choosing a non-informative prior allows one to determine the posterior based on present information/data available only through the likelihood function. Non-informative priors give us the liberty of choosing a distribution which is mathematically easy to compute, the most popular choice being the use of conjugate and improper priors.

Priors which are chosen to be from the same family of distribution as that of the posterior distribution are called the conjugate priors (Raiffa & Schlaifer, 1961). This posterior-prior paring is known as a conjugate distribution. The prior is selected in a way such that its combination with the likelihood function yields the same type of distribution whose standard solution is available. This closed-form solution allows an easy numerical integration of the denominator in Eq. (7), which is generally difficult to evaluate otherwise. Table 1 gives a list of conjugate distributions which can be used in the absence of any informative prior. The main idea of this section is just to highlight the different types of prior distribution which can be used for Bayesian inference for an overall understanding.

Similarly, the use of improper priors is debatable as it should be used only as a last alternative when the use of conjugate or uniform prior are not feasible. The selection of these type of priors is based on Jeffrey's rule (Kass & Wasserman, 1996). Basically, a chosen prior is said to be improper prior if,

$$\int_{\mathcal{D}_\Theta} f_\Theta(\boldsymbol{\theta})d\boldsymbol{\theta} \neq 1 \tag{9}$$

Likelihood Distribution

The likelihood function $f_\Omega(\Omega \mid \boldsymbol{\theta})$ supplies the prior distribution with data needed to update an existing model. Generally, there are two types of observation that can be incorporated into a likelihood function (Straub & Papaioannou, 2014): 1) observations providing equality information, and 2) observation providing inequality information. The equality type observation is incorporated in the form $\Omega_i = g_i(\boldsymbol{\theta}) + \varepsilon_i$.

Table 1. Conjugate distributions for Bayesian updating

Posterior $\left(f_\Theta(\boldsymbol{\theta} \mid \Omega)\right)$	Prior $\left(f_\Theta(\boldsymbol{\theta})\right)$	Likelihood $\left(f_\Omega(\Omega \mid \boldsymbol{\theta})\right)$
Normal	Normal	Normal
Student's t	Inv. Gamma	Normal
Student's t	Inv. Chi-square	Normal
Negative binomial	Gamma	Poisson
Beta-binomial	Beta	Binomial

Here, Ω_i is the observed data, $g_i(\boldsymbol{\theta})$ is the predicted data based on the model available/adopted with Θ as the vector of its model parameters, and ε_i is the error in predicting the measured value. The likelihood function for such observations is written as

$$L_i(\boldsymbol{\theta}) = f_\varepsilon[\Omega_i - g_i(\boldsymbol{\theta})] = f_{\varepsilon|\Theta}(\varepsilon_i \mid \Theta = \boldsymbol{\theta}) \tag{10}$$

This type of likelihood is usually defined in terms of a PDF of the error in observation $\left(\varepsilon_i\right)$ which is often expressed as zero mean Gaussian distribution. The inequality type information are usually obtained from discrete outcomes when the measured value exceeds a threshold/limit state.

$$Z_i = \boldsymbol{I}[g_i(\boldsymbol{\theta}) \leq 0] \tag{11}$$

$g_i(\boldsymbol{\theta})$ is defined as limit state function (LSF) in structural reliability problems and $\boldsymbol{I}$ is the indicator function which gives an output of 0 or 1 when the $g_i(\boldsymbol{\theta}) \geq 0$ or $g_i(\boldsymbol{\theta}) \leq 0$, respectively. This LSF corresponds to a likelihood function of

$$L_i(\boldsymbol{\theta}) = \mathbb{P}(Z_i \mid \boldsymbol{\Theta} = \boldsymbol{\theta}_i) = \boldsymbol{I}[g_i(\boldsymbol{\theta}) \leq 0] \tag{12}$$

In general, when there are n such observed/measured statistically independent data, the likelihood function is represented as

$$L(\boldsymbol{\theta}) = \prod_{i=1}^{n} L_i(\boldsymbol{\theta}) \tag{13}$$

Posterior Distribution

The main aim of Bayesian updating is to compute the posterior distribution in Eq. (7). There are several limitations in computing this posterior distribution. This requires calculating the integral in the denominator which might be intractable and difficult to compute especially in cases where there are greater number (more than two) of model parameters to be updated. Even if it is possible to compute the denominator, the posterior may not be of standard form to draw any inference. To make any inference from the posterior distribution, one needs to draw random samples from this derived distribution. Using Monte Carlo simulation (MCS) is one of the easiest yet computationally intensive methods, which can be adopted in Bayesian updating given that it requires the posterior distribution is in some standard form. Markov chain Monte Carlo (MCMC) is one technique which is adopted here for sampling from any desired probability distribution which are otherwise difficult to sample from, by creating Markov chains until it reaches the target/stationary distribution (posterior distribution). In the following section we describe in detail the MCMC technique using Metropolis-Hastings (MH) algorithm to sample from complex distributions which is often encountered while using Bayesian inference. Bayesian inference can thus also be referred to as an optimisation problem where the optimum model parameters are obtained based on the available information.

Markov Chain Monte Carlo

Markov chain Monte Carlo (MCMC) method, is adopted for drawing random samples from a distribution of interest which may be otherwise difficult to sample from. Markov chain is a sequence of state of random variables such that the next state of the sequence depends only on the previous one. Markov process is a memory-less process which depends only on the current state and not the sequence of event that precedes it (Gamerman & Lopes, 2006). The basic difference between a MCS and a MCMC simulation is that, in the former, the random sampling is identical and independently distributed (i.i.d.), whereas those in the latter are correlated. Consider a sequence of random variable Θ_t on the state space $\mathcal{D}_\Theta$ for $t = 0, 1, 2, \ldots$ then Θ_t is called a Markov Chain if it satisfies the following property:

$$\mathbb{P}(\Theta_{t+1} = \theta_{t+1} \mid \Theta_t = \theta_t, \Theta_{t-1} = \theta_{t-1}, \ldots, \Theta_0 = \theta_0) = \mathbb{P}(\Theta_{t+1} = \theta_{t+1} \mid \Theta_t = \theta_t) \tag{14}$$

where the next state at $t+1$ is distributed according to $\mathbb{P}(\Theta_{t+1} = \theta_{t+1} \mid \Theta_t = \theta_t)$, which is also called the transition kernel. One of the most fundamental properties of MCMC is that no matter what the starting state $\left(\Theta_0 = \theta_0\right)$ is, after few iterations it will converge to a stationary/target/posterior distribution. The number of iterations to be performed depends on the condition that the number of samples generated from the Markov chain fully represents the target distribution. Suppose the chain has run for m iterations and we assume that the sample points θ_t, $t = m+1, \ldots, n$ are distributed according to the posterior distribution, then we can discard the first m samples and use the remaining $n - m$ samples to get an estimate of the expected value just like in the conventional MC simulation. The number of samples m that are discarded is called as the *burn-in*. The size of burn-in is not defined but Geyer (1992) suggested that the burn-in should be anywhere in between 1% to 2% of the total number of samples to be generated (n). Now, we must decide on how large should n be to get a precise estimate of the expected value. One way to determine the size of n through simulations is to run multiple chains in parallel, each starting with a different start point (Martinez & Martinez, 2002). These random samples are dependent samples and are typically auto-correlated. To reduce the auto-correlation, the so-called "thinning" technique is employed, where only the samples from every kth ($k > 1$) iteration, after the burn-in are stored for the output analysis (Congdon, 2006; Link & Eaton, 2012). The samples generated after the burn-in period using a suitable thinning interval can be used to evaluate the probabilistic characteristics of the joint posterior distribution or the marginal distribution of a given parameter. The major advantage of MCMC is that it allows us to sample from almost any target distribution. There are many tools that allow sampling from complex distributions. One such method which is adopted in the present study is Metropolis-Hastings algorithm. Details discussion about this algorithm is provided in the next section.

Metropolis-Hastings Algorithm

The Metropolis-Hastings (MH) algorithm is an MCMC method for obtaining samples from a stationary distribution by performing a random walk using the proposal density of a known nature. In the current study, the MH algorithm is adopted for Bayesian inference because of its ability to sample from posterior distribution. Here, instead of sampling from the posterior distribution, which may be of a complex form (not a common density function), a known density function is taken as proposal distribution from which random samples are generated. These samples are then subjected to a screening algorithm known as the acceptance-rejection (AR) algorithm where only those samples lying in the posterior region get accepted. The steps to be followed while adopting the MH algorithm is as follows (Chib & Greenberg, 1995):

1. Assume an initial start point $\boldsymbol{\theta}_{t=0}$ (i.e. $\boldsymbol{\theta}_0$)
2. Consider a proposal distribution of a known form $q(\boldsymbol{\theta}_{t+1} \mid \boldsymbol{\theta}_t)$ (say, normal distribution), where $\boldsymbol{\theta}_t$ is the current state of Markov chain. Our final objective is to generate candidate $\boldsymbol{\theta}^*$ which lies within the target distribution $f_\Theta(\boldsymbol{\theta} \mid \Omega)$.
3. Sample $\boldsymbol{\theta}_{t+1}$ from proposal distribution $q(\boldsymbol{\theta}_{t+1} \mid \boldsymbol{\theta}_t)$ considering $\boldsymbol{\theta}_t$ as its parameter.

$$\boldsymbol{\theta}_{t+1} \mid \theta_t \sim \mathcal{N}(\mu_{\boldsymbol{\theta}_t}, \sigma) \tag{15}$$

Here, $\mu_{\boldsymbol{\theta}_t} = \boldsymbol{\theta}_t$ and $\boldsymbol{\sigma}$ is an appropriate standard deviation chosen to ensure sufficient mixing of the Markov chains (Chib & Greenberg, 1995).

4. Check whether the sampled candidate $\boldsymbol{\theta}_{t+1}$ lies within the posterior distribution by making use of AR algorithm:

   Find acceptance probability:

$$A(\cdot \mid \theta_t) = \min\left\{1, \frac{f_\Theta(\boldsymbol{\theta}_{t+1} \mid \Omega) q(\boldsymbol{\theta}_t \mid \boldsymbol{\theta}_{t+1})}{f_\Theta(\boldsymbol{\theta}_t \mid \Omega) q(\boldsymbol{\theta}_{t+1} \mid \boldsymbol{\theta}_t)}\right\} \tag{16}$$

   Generate $u \sim \mathcal{U}(0,1)$ from a uniform distribution. If $u \leq A(\cdot \mid \boldsymbol{\theta}_t)$, then accept

$$\boldsymbol{\theta}^* = \boldsymbol{\theta}_{t+1}$$

else

$\boldsymbol{\theta}^* = \boldsymbol{\theta}_t$ i.e. the chain does not move forward

5. Set $t = t+1$ and repeat steps 3 and 4 until sufficient samples are generated which fully represent the target distribution. This generates a Markov chain ($\boldsymbol{\theta}_i^*$), because the transition probabilities from $\boldsymbol{\theta}_t$ to $\boldsymbol{\theta}_{t+1}$ depends only on $\boldsymbol{\theta}_t$ and not $\boldsymbol{\theta}_0,\ldots,\boldsymbol{\theta}_{t-1}$. Following a sufficient burn-in period (of say, m steps), the chain approaches its stationary distribution and the remaining samples ($\boldsymbol{\theta}_{m+1}^*,\ldots,\boldsymbol{\theta}_N^*$) are regarded as samples from $f_\Theta(\boldsymbol{\theta} \mid \Omega)$.

Two things that are to be noted are that, firstly in Eq. (16) the target or posterior distribution is independent of the normalising constant as given in the denominator of Eq. (7) and hence the ratio $f_\Theta(\boldsymbol{\theta}_{t+1} \mid \Omega) \,/\, f_\Theta(\boldsymbol{\theta}_t \mid \Omega)$ is easy to calculate. Secondly, if the proposal distribution is taken to be symmetric (such as a normal distribution), then the ratio $q(\boldsymbol{\theta}_t \mid \boldsymbol{\theta}_{t+1}) \,/\, q(\boldsymbol{\theta}_{t+1} \mid \boldsymbol{\theta}_t)$ in Eq. (16) eventually becomes 1. This means that the acceptance probability is now totally dependent on the target distribution. Such class of algorithms which chooses a symmetric proposal distribution is known as the Metropolis algorithm which is a special case of the Metropolis-Hastings algorithm.

EXAMPLE 1: CORROSION MODEL UPDATING OF AN RC BEAM

Corrosion of rebars in concrete is one of the primary causes of degradation of infrastructural systems/structures and it leads to a reduction in the capacity of structural members over time (Faroz *et al.*, 2014). Experimental evidences suggest that one parameter which can be associated with various damage levels of a reinforced concrete (RC) structure, is the amount of steel lost due to corrosion (Alonso *et al.*, 1996; Almusallam *et al.*, 1996). The process of corrosion in RC is highly uncertain and thus it is essential to formulate a probabilistic model to describe the randomness (Engelund *et al.*, 1999). Considering the random nature of the corrosion process, limited amount of data and lack of confidence in prediction, Faroz *et al.* (2014) developed a methodology for updating the corrosion model using the measurements available on steel loss. In this example, the likelihood function used for the Bayesian updating incorporates a model for the random measurement error (along with the model/prediction error).

Corrosion Model

Faraday's law has been traditionally adopted to model the loss of steel due to corrosion. However, the average mass loss predicted by Faraday's law has seen to be differing from the gravimetric mass loss (Malumbela *et al.*, 2010) and this law usually overestimates the loss of steel (Malumbela *et al.*, 2012). This is because Faraday's law does not consider the effect of thickening of the rust layer over the corroding rebar surface. As corrosion progresses, the thickness of rust goes on increasing and the rate of rust formation depends on the thickness of rust already produced (Liu and Weyers, 1998). In the present example, the corrosion model developed by Liu and Weyers, 1998. and later improved by Bhargava *et al.* (2006) is used. The mass loss of steel due to corrosion is expressed as (Bhargava *et al.* 2006):

$$W_s(t) = \sqrt{2\pi D_0 i_{corr} A_w \alpha_{sr}^2 t} \tag{17}$$

where, D_0 is the original diameter of the rebar and i_{corr} is the mean annual rate of corrosion. In this equation, the parameters A_w and α_{sr} are related to the corrosion process and the amount of rust generated, respectively. Therefore, these parameters cannot be measured directly and are treated in the present work as un-observable parameters. Bhargava *et al.* (2006) gave estimates for these quantities based on the experimental evaluations by Liu (1996). Performing a regression analysis on these laboratory-based data for D_0 measured in mm, i_{corr} in $\mu A / cm^2$, W_s in mg/mm and t in years, they found: $A_w = 2.48614$ and $\alpha_{sr} = 0.61309$. Their time-varying steel loss model for uniform corrosion, thus, can be expressed as

$$W_s(t) = 2.423\sqrt{D_0 i_{corr} t} \tag{18}$$

Bayesian Framework

For rebar corrosion in RC we have scientific or empirical models, but they are not very good predictors. These models rely on some (un-observable) parameters whose values come from limited statistical data available to the researchers. Considering that the monitoring of corrosion through non-destructive technique (NDT) is a fairly routine event for important structures these days, a Bayesian updating of the available corrosion model is presented here based on limited but actual evaluations of the in-situ condition.

In the steel loss model of Eq. (17), A_w and α_{sr} are the un-observable parameters that are usually obtained by a regression analysis of data obtained under a specific set of conditions. Therefore, there exists significant uncertainty associated with these parameters due to the difference between the test and field conditions. As a consequence, these parameters need to be treated as random variables in the corrosion model. The Bayesian updating in this work considers $\boldsymbol{\theta} = \left[A_w \quad \alpha_{sr}\right]^T$, whose probability models require to be updated based on steel loss measurements.

Prior

Through a compilation of data reported in literature, Jamali *et al.* (2013) gave the values of α_{sr} in the range of 0.348 to 0.778 corresponding to various possible corrosion products (Table 2). In the absence of any further information on the composition of rust products, we conservatively adopt a uniform distribution for the prior probability model of α_{sr}:

$$\alpha_{sr} \sim \mathcal{U}(0.348, 0.778) \tag{19}$$

Table 2. α_{sr} of corrosion products

FeO	Fe_3O_4	Fe_2O_3	FeOOH	$Fe(OH)_2$	$Fe(OH)_3$	$Fe(OH)_3 \cdot 3H_2O$
0.778	0.723	0.700	0.629	0.622	0.523	0.348

The experimental results of Liu (1996) (Table 3) were used by Bhargava *et al.* (2006) to estimate the coefficients of the steel loss model. In order to derive the prior distribution of A_w from (Table 3) available data, the following procedure based on the steps proposed by Bhargava *et al.* (2006) is adopted to obtain samples of A_w.

- Compute the product $\pi D_0 i_{corr}$ for each specimen using the data from Table 3.
- Compute $k_w = W_s^2 / (2\alpha_{sr}^2 t)$ for all specimens and all values of α_{sr} from Table 2.
- Calculate $A_w = k_w / (\pi D_0 i_{corr})$.

Following the above mentioned steps, 28 values of A_w are generated. The prior probability model for A_w is obtained by fitting this statistic using the Kolmogorov-Smirnov goodness-of-fit test. A lognormal distribution is found suitable to model A_w:

$$\ln A_w \sim \mathcal{N}(0.9073, 0.6752) \tag{20}$$

Figure 1(a) shows the empirical and the fitted CDFs of A_w. The prior PDF used in the updating is presented in Figure 1(b). The use of a lognormal distribution is also justified based on Eq. (17), which requires A_w to be non-negative.

The joint prior distribution is taken as the product of the densities of individual parameters, by assuming each parameter to be independent of each other:

$$f'(\Theta) = f'(A_w) f'(\alpha_{sr}) \tag{21}$$

Likelihood: Errors in Modelling and Measurement

The likelihood function serves as a very useful means of combining the observed/measured data with the prior knowledge. It also accounts for the discrepancy between the prediction model and the inspection results. Engineering models are imperfect, however in a probabilistic domain they can be assumed

Table 3. Steel loss data

Specimen Designation	D_0 (mm)	i_{corr} (μ A/cm²)	W_s (mg/mm)	t_{insp} (Years)
SL1	16.0	2.41	19.75	1.84
SL2	16.0	1.79	30.21	3.54
SL3	16.0	3.75	14.98	0.72
SL4	12.7	1.80	15.64	2.38

Figure 1. (a) Empirical and fitted CDFs of A_w; (b) prior PDF of A_w

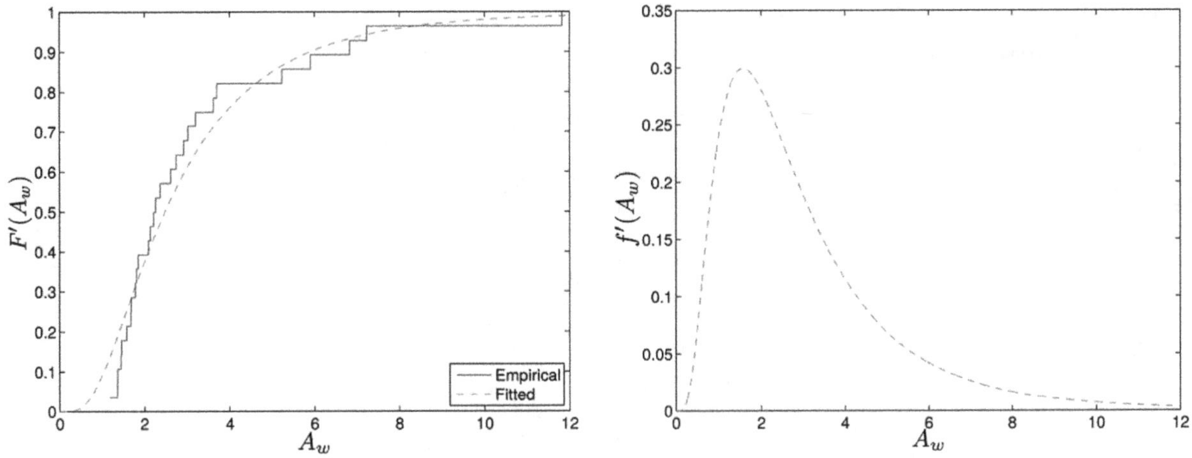

"correct" by incorporation of correcting random variables (Ditlevsen, 1982). A correct corrosion model may be assumed as

$$W_s = W_f + \varepsilon_M \tag{22}$$

where W_f is the "true" or field steel loss and $\varepsilon_M \sim \mathcal{N}\left(0, \sigma_{\varepsilon_M}\right)$ is the randomness in the corrosion model prediction. Relating the readings from an NDT to the "true value" of the observed parameter is possible through a linear measurement error model (Zheng & Ellingwood, 1998):

$$W_f = \beta_0 + \beta_1 W_m + \varepsilon_m \tag{23}$$

where, W_m denotes the measured steel loss, $\beta_0 \in \mathbb{R}$ is the constant bias (typically expressed in % of W_0), $\beta_1 \in \mathbb{R}$ is the scaling bias of the NDT instrument, and $\varepsilon_m \sim \mathcal{N}(0, \sigma_{\varepsilon_m})$ denotes the random error in measurements. $\beta_0 = 0$ and $\beta_1 = 1$ signifies an unbiased NDT tool. Determining the value of these biases may be termed as the "calibration" of an NDT tool, which Al-Amin *et al.* (2012) demonstrated using Bayesian inference. The present example does not focus on obtaining biases for a specific NDT instrument, rather it shows how the biases for a calibrated instrument can be incorporated in the updated prediction and its importance in the prediction of steel loss over time. Fitting a straight line on the squared steel loss data reported by Torres-Acosta and Sagues (2004), β_0 is obtained as 1.42% of W_0 (as presented later in Figure 2(b)). The non-dimensional scaling bias β_1 is assumed to take values as 0.75 and 1.25, to which the sensitivity of the results is discussed.

The difference between the predicted value and the true value ($W_s - W_f$) has to be zero for a correct prediction by the model. However due to the presence of randomness, a global random error $\varepsilon \sim \mathcal{N}\left(0, \sigma_\varepsilon\right)$

Figure 2. (a) Prediction following the existing model; (b) calculation of the constant bias β_0

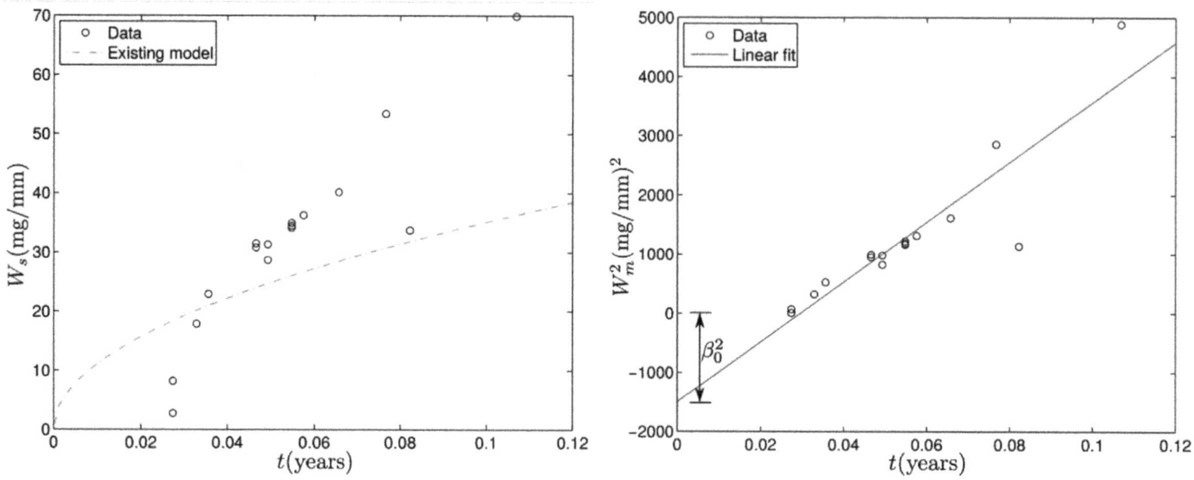

is defined to incorporate the deficiencies of the model and the measurement techniques. The global error consists of the measurement error (ε_m) and the model error (ε_M), with $\sigma_\varepsilon^2 = \sigma_{\varepsilon_M}^2 + \sigma_{\varepsilon_m}^2$. The quantity $\left[W_s - \left(\beta_0 + \beta_1 W_m\right)\right]$ represents each realisation of the zero mean Gaussian random variable ε. Typically, σ_ε is estimated by the user based on the beliefs about the model uncertainty and the presence of noise and bias in the measurements (Karandikar *et al.*, 2012).

The likelihood function is assumed to follow a normal distribution, with σ_ε as the standard deviation of the quantity $\varepsilon = W_s - \left(\beta_0 + \beta_1 W_m\right)$. The likelihood for the ith observation ε_i can be expressed as

$$f_\varepsilon(\varepsilon_i \mid \Theta) = \frac{1}{\sqrt{2\pi}\sigma_\varepsilon} \exp\left[-\frac{\{W_{s_i} - (\beta_0 + \beta_1 W_{m_i})\}^2}{2\sigma_\varepsilon^2}\right] \tag{24}$$

For n measured data, either a recursive scheme can be adopted (where the previously updated density will be used as the prior for the next sequence of updating) or, if the data become available for all past measurements together, the likelihood function can be constructed under the assumption that the observations are statistically independent:

$$f_\varepsilon(\Omega \mid \Theta) = \prod_{i=1}^{n} f_\varepsilon(\varepsilon_i \mid \Theta) \tag{25}$$

The choice of the likelihood function does not affect the updating procedure. However, as suggested by Karandikar *et al.* (2012), one should consider the aspect that, since the likelihood is based on the user's beliefs, the effect of the likelihood spread on the updated estimations should be evaluated.

MCMC Simulation for the Posterior Model

With the availability of the prior density function and the likelihood, incorporating the measured data, the posterior or updated density of the desired parameter(s) can now be computed using Bayesian updating. The proposal density chosen in this paper is a bi-variate normal distribution for the Metropolis-Hastings (MH) algorithm:

$$q(\Theta'' \mid \Theta') \sim \mathcal{N}\left(\Theta', \begin{bmatrix} \sigma_1^2 & 0 \\ 0 & \sigma_2^2 \end{bmatrix}\right) \tag{26}$$

where, Θ' and Θ'' are the subsequent samples on the Markov chain. The values of σ_1^2 and σ_2^2 are selected such that the acceptance rate lies between 0.23 to 0.45 to ensure sufficient mixing of a Markov chain (Chib and Greenberg, 1995), and are kept constant throughout the simulation. To reduce the autocorrelation between these 'dependent' samples, the 'thinning' technique is employed in this work, where – post burn-in – samples are stored only after a 'lag' of 500 iterations. A total of 5000 samples are used after burn-in.

Samples from the posterior density function, $f''(\Theta \mid \Omega)$ are then used to estimate various functions of the variables A_w and α_{sr}, such as $W_s(t)$. This probabilistic characterisation not only provides the expectation of W_s over time, but also its variance, which can be used to estimate confidence intervals on the steel loss prediction.

Bayesian Updating for Steel Loss over Time

For the present work, the steel loss data are adopted from the work of Torres-Acosta and Sagues (2004). Where in the results of an experimental investigation are presented on a corroding 21 mm diameter rebar in concrete over an anodic region, as shown in Table 4. The experiment was conducted using a constant current density with different measured values of i_{corr} for each specimen. For the present study, all observations are scaled to an i_{corr} =100.0 μ A/cm². From the same work, additional data are used for the validation of the updated model. These data, including the ones used for updating, are shown in Figure 2(a) along with the estimates based on the model of Eq. (18). Based on these data points, Figure 2(b) illustrates the calculation of the constant bias β_0 in measurements.

The updating is performed for a set of values of σ_ε (2.5% and 20.0% of W_f). Figure 3 shows the effect of the adopted σ_ε value on $W_s(t)$ for an instrument with $\beta_0 = 1.42\%$ and $\beta_1 = 1.0$. The enhancement in prediction of steel loss over time is clearly visible in the plots of the updated model when compared to the prediction based on the existing model (Eq. (18)), against the measured data set. In addition to the expected value of the probabilistic steel loss model used here, these plots also show the 95% confidence interval on the updated prediction of $W_s(t)$. This confidence interval typically envelops the measured data points. As expected, an increase in σ_ε widens the 95% confidence interval. Besides, this increase changes the mean (updated) estimation as well. Further, effects of the measurement bias are

Table 4. Steel loss data

Specimen	t (Days)	W_m (mg/mm)	Specimen	t (Days)	W_m (mg/mm)
CPA1	13	22.89	CPE1	39	69.88
CPA2[†]	17	31.47	CPE2	20	34.94
CPB1	24	40.15	CPF1	18	31.32
CPB2	20	34.46	CPF2[†]	17	30.79
CPC1	21	36.24	CPG1	10	8.240
CPC2	28	53.44	CPG2	10	2.726
CPD1	18	28.64	CP1	12	17.89
CPD2	20	34.13	CP2	30	33.73

[†]Data used for updating.

Figure 3. Prediction using the updated corrosion model (β_0 = 1.42%, β_1 = 1.0) for (a) $\sigma_\varepsilon = 2.5\%$, *and (b)* $\sigma_\varepsilon = 20.0\%$

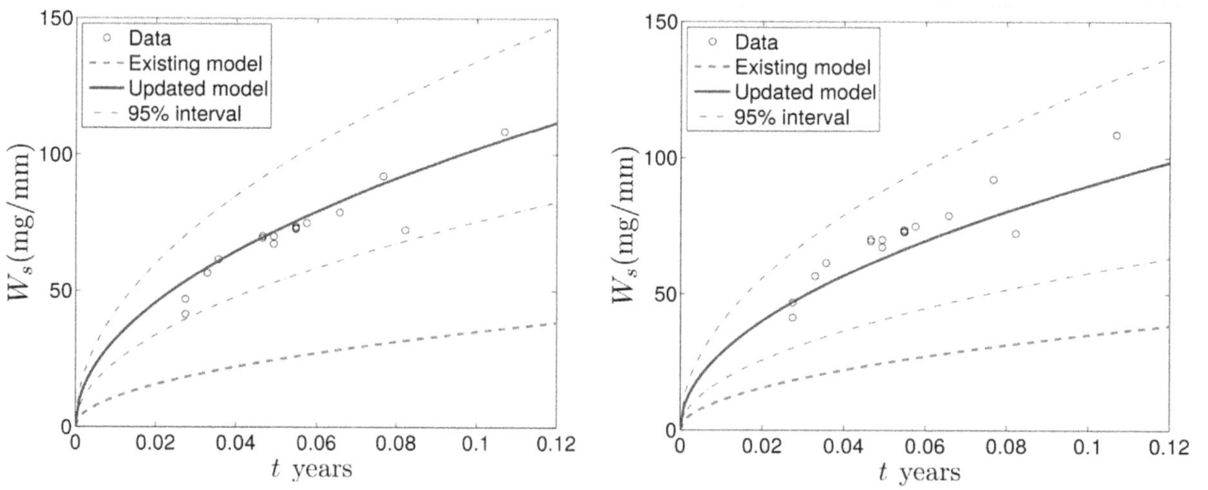

studied by considering different β_1 values as noted in Table 5. Table 5 presents the change in the parameters for updated distributions of A_w and α_{sr}. Figure 4 shows how β_1 changes the results for an assumed $\sigma_\varepsilon = 2.5\%$. These plots show that a change in the scaling bias changes the expected value of W_s more than it changes its confidence interval.

The results of this Bayesian updating of the time-varying steel loss model makes a definite case for using such an approach in corrosion studies. Figure 3 and 4 clearly show the improvements that can be achieved by using Bayesian updating over the existing corrosion estimation models.

Table 5. Measurement biases considered and their effects on updated A_w and α_{sr} ($\sigma_\varepsilon = 2.5\%$)

β_0 (% of W_0)	β_1	Updated A_w		Updated α_{sr}	
		Sample Mean	Sample SD	Sample Mean	Sample SD
1.42	0.75	12.76	3.109	0.6991	0.06568
1.42	1.25	19.19	3.917	0.7125	0.05711

Figure 4. Prediction using the updated corrosion model ($\sigma_\varepsilon = 2.5\%$) for $\beta_0 = 1.42\%$ and (a) $\beta_1 = 0.75$ and (b) $\beta_1 = 1.25$

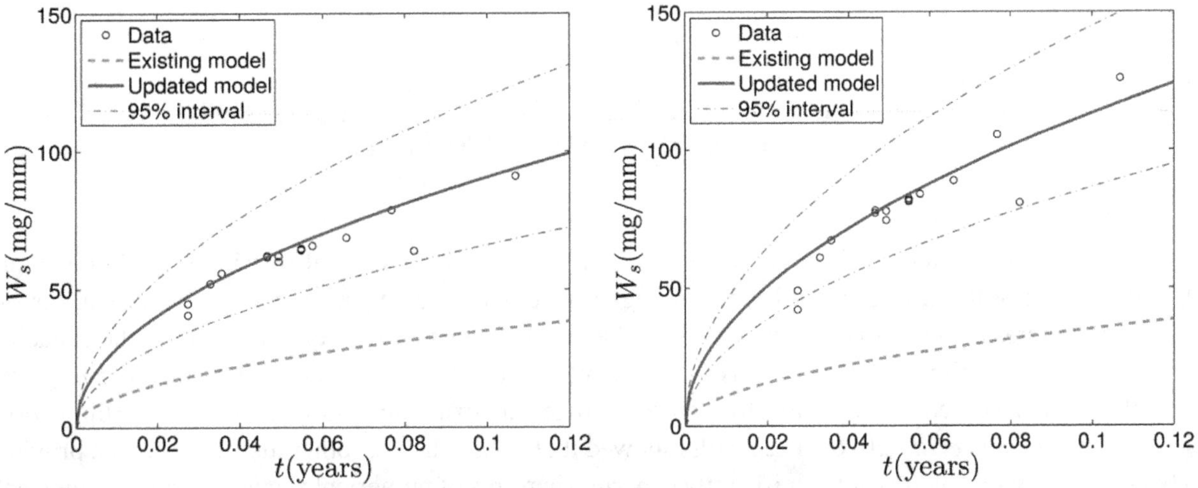

EXAMPLE 2: SEISMIC FRAGILITY ESTIMATION OF A CONTAINMENT STRUCTURE USING BAYESIAN INFERENCE

The seismic fragility (F_r) of a structure is the conditional probability of a damage measure (DM) exceeding a threshold damage measure (DM_l) given a particular intensity measure (IM) level:

$$F_r = \mathbb{P}(\mathrm{DM} > \mathrm{DM}_l \mid \mathrm{IM}) \tag{27}$$

DMs for a civil structure can be defined in terms of its material parameters like strength of concrete, physical deformations like interstorey drift ratio, plastic rotations, deflection of beam/slabs, strain measurements, etc. IMs can be defined in terms of peak ground acceleration/velocity (PGA/PGV), spectral acceleration/velocity (S_a / S_v), ARIAS intensity (I_A), etc. The seismic fragility of a structure can be defined at multiple limit states to indicate the performance level of the structure. A sample fragility curve is shown in Figure 5.

Figure 5. Sample fragility curve

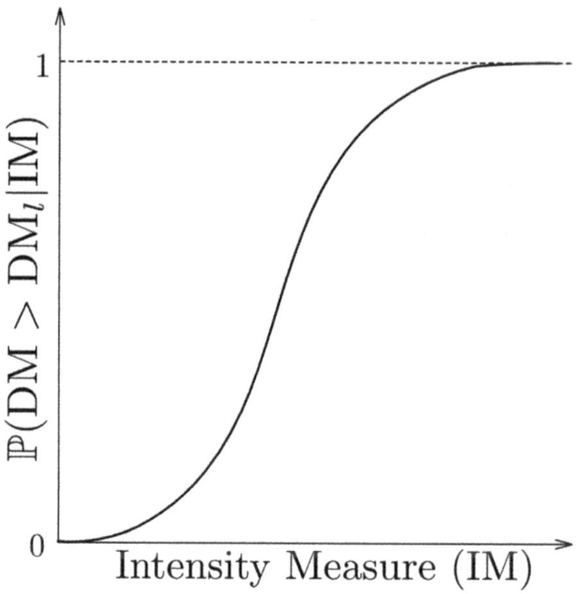

Traditionally, seismic fragility curves are assumed to follow lognormal distribution (Ellingwood, 1990). In most of the cases, the DM is a non-negative number and hence choosing a lognormal distribution eliminates the use of negative numbers. Also it gives more flexibility in evaluating the capacity of components which may be defined as a product of several random variables. Use of other fragility models have been tested and compared by many researchers for example by Goodman (1985); Ellingwood (1990); and Mandal *et al.* (2016). The results showed that although a few other distributions can provide a better fragility model, it is not significantly superior (in terms of numerical accuracy) to the lognormal fragility model used in practice and the lognormal model allows a lot of ease in its implementation. In one of the pioneering paper by Kennedy *et al.* (1980), lognormal distribution was used to estimate the seismic fragility:

$$F_r = \Phi\left[\frac{\ln(x / m_a) + \beta_U \Phi^{-1}(Q)}{\beta_R}\right] \tag{28}$$

where, Φ (.) is the standard normal CDF. m_a is the median of ground acceleration capacity of the structure, β_R the variability (randomness) in m_a, which is an aleatory uncertainty; β_U is the variability (uncertainty) due to our lack of knowledge in predicting m_a, which is the epistemic uncertainty; and Q represents the non-exceedance probability level. Many of the researchers, such as Shinozuka *et al.* (2000); Ellingwood *et al.* (2007) have suggested the use of a two parameter (m_a and β_R) lognormal curve instead of the three parameters used in Eq. (28). This is because of the fact that the parameter β_R which is the randomness associated with the ground motion far supersedes the uncertainty $\left(\beta_U\right)$ in the estimation of

m_a. This assumption has been validated and used in probabilistic seismic analyses of buildings and bridges by many researchers (Kwon and Elnashai (2006); Choi *et al.* (2008), for example). Thus Eq. (28) can be simplified as:

$$F_r = \Phi\left[\frac{\ln(x / m_a)}{\beta_R}\right] \tag{29}$$

All the fragility models considered in this section makes use of the two-parameter lognormal fragility curve. A proper estimation of these parameters is required for obtaining the seismic fragility of a structure. The concept of seismic fragility was first introduced in the probabilistic safety assessment (PSA) of the safety of critical nuclear structures. In the nuclear industry, seismic fragility analysis is generally carried out with PGA as the IM. The primary reason is that the seismic fragility analysis of different components need to be compared at selected seismic hazard intensities. Most of the fragility analyses which deal with individual bridges and building use S_a as its IM. However, for fragility analysis of many of the components of the NPP, their fundamental periods of vibration $\left(S_{a_{T1}}\right)$ are in wide range of each other. Moreover, the chosen IM for components should be independent of structural properties. Hence choosing PGA as IM is justified for NPP's.

IDA Based Fragility

There are two basic ways of evaluating the seismic fragility parameters of a structure. First one being the use of factor of safety (FoS) method proposed by Kennedy and Ravindra (1984). In FoS method, the median ground acceleration capacity $\left(m_a\right)$ is expressed in terms of FoS and safe shutdown ground acceleration capacity of the containment. Different FoS were applied to take into account the different uncertainties involved in estimation of ground acceleration capacity. Second being selection of optimised fragility parameters based on the actual available structure damage data at particular IM level. In absence of such data one can perform nonlinear static push over analysis (NSPA) up to its failure or perform nonlinear response history analyses (NLRHA) at multiple IM levels for different earthquakes on a mathematical model which adequately represents the real structure. This series of NLRHA performed for different ground motion (GM) records is known as the incremental dynamic analysis (IDA) (Vamvatsikos and Cornell, 2002, 2004). IDA is basically a plot of IM vs. DM based on a series of NLRHA at different scales of GM records. IDA is useful in probabilistic seismic demand evaluation of structures. Sample IDA curves for 25 GMs are shown in Figure 6. Here, maximum interstorey drift ratio $\left(\theta_{\max}\right)$ response are plotted against the scaled levels of IM for different ground motions. In this figure, LS1, LS2 and LS3 represent the three limit states that give different structural failure criteria. IDA provides a number of advantages in estimating the seismic demand, which include (1) flexibility in choosing any of the site-specific/artificial/synthetic GM records and (3) direct demand estimation at any intensity of the hazard (no scaling of demand parameters). Through detailed IDA (measuring response at minute IM intervals, computationally demanding) one can obtain a large number of fragility data by simply counting the number of curves exceeding the limit state (LS1, LS2 and LS3, in this case) one at a time to the total

Figure 6. Sample IDA curve with PGA as IM for 25 GM records

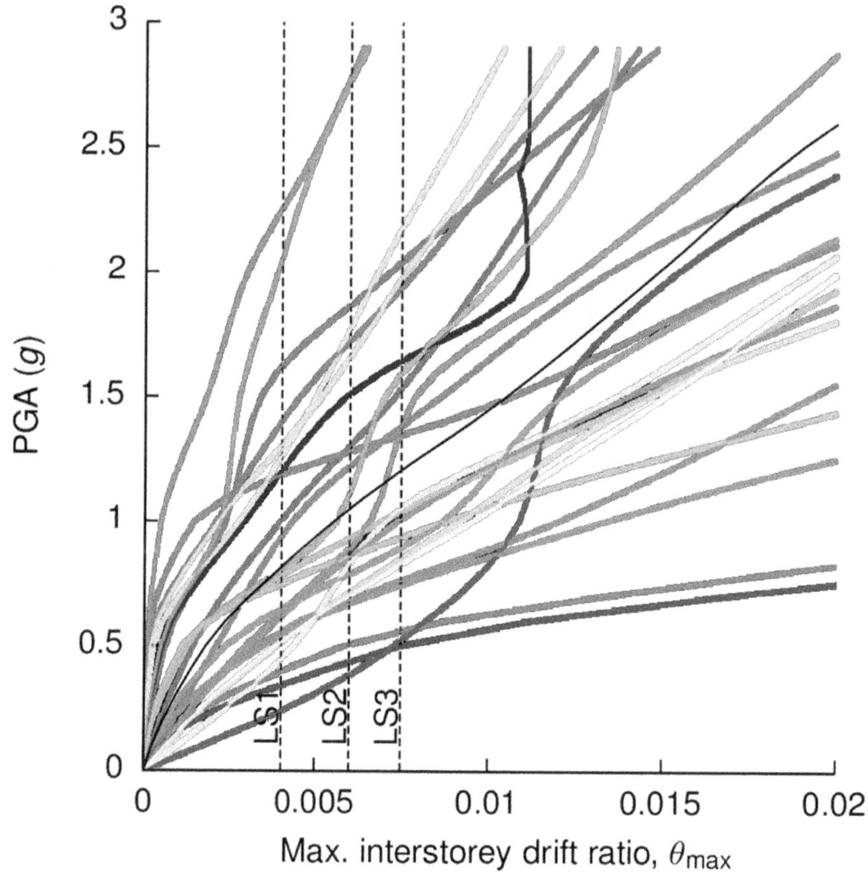

For a more accurate representation of this figure, please see the electronic version of this chapter.

number of IDA curves. For example, to find the fragility at 0.5g, we count the number of curves exceeding $\theta_{\max,LS1}$ at 0.5g. From Eq. (27) we get:

$$F_r = \mathbb{P}(\theta_{\max} > \theta_{\max,LS1} \mid IM = 0.5g) = 5 / 25 = 0.2 \tag{30}$$

This data now represents a single point (of many points) on a fragility plot which is helpful in determining the optimum fragility parameters (m_a and β_R) using techniques like likelihood method (Shinozuka *et al.*, 2000), regression method (Ellingwood *et al.*,2007) and Bayesian inference (Marin & Robert, 2007). In this particular section, we will demonstrate the use of Bayesian inference to give a reasonably better fragility estimate.

Bayesian Updating of Seismic Fragility

With the advent of better computational capabilities and with the ease in availability of recorded accelerograms, seismic fragility analysis techniques have evolved over the past few decades. Of the many available evaluation techniques, methods proposed by Kennedy and Ravindra (1984) ('Conventional'), Shinozuka *et al.* (2000) ('Likelihood') and Shinozuka *et al.* (2000) ('Regressed') have significantly influenced researchers towards PSA. To highlight the differences in adopting these techniques, a comparative study was performed by Mandal *et al.* (2016). A new method was also proposed for better estimation of seismic fragility parameters along with a modified variant which provided a single value of parameters for all limit states considered simultaneously to reduce the computation.

In this section, we present an illustrative example of Bayesian updating on the seismic fragility parameters of a nuclear containment structure. The primary containment (PC) considered here is of a 700 MWe typical Indian PHWR as shown in Figure 7. The containment structure consists of a prestressed concrete cylindrical wall capped by a segmental prestressed concrete dome over a massive ring beam. The PC structure is modelled in Simulia (2013) with detailed concrete damage plasticity (CDP) model along with the effect of prestressing tendons. The PC structure is modelled using four noded shell elements (S4R) with embedded reinforcement and prestressing details. The meshed structure (Figure 8b) has 16403 shell elements and 16472 nodes. The base nodes are constrained in all direction (that is, boundary conditions for the model are adopted to be fixed in all direction). Further details of containment modelling (material and geometry) are presented in Pujari *et al.* (2015). The PC structure is analysed for through crack formation anywhere within the containment. Concrete damage plasticity (CDP) model (Lubliner *et al.*, 1989; Lee & Fenves, 1998) is used to represent the behaviour of concrete under seismic action. In this model tensile cracking in concrete is defined when the equivalent plastic tensile strain ($\tilde{\varepsilon}_t^{pl}$) takes a positive value

Figure 7. Damage plasticity model for concrete in tension

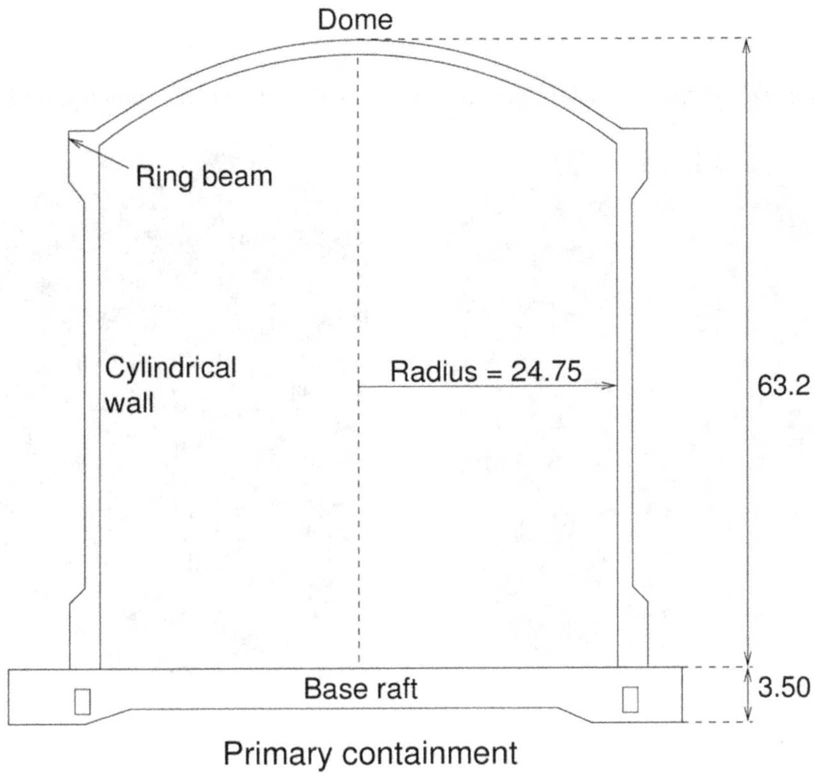

$$\tilde{\varepsilon}_t^{pl} > 0 \tag{31}$$

M45 grade of concrete is used for modelling the PC. Details of CDP model parameters are taken from Pujari *et al.* (2015).

In-order to evaluate the actual seismic fragility of containment, material parameters should be evaluated at their mean value rather than at their characteristics values. A series of NLRHA is performed at specific scaled ground motion levels of PGA based on the level of accuracy to be achieved. A total of 30 spectrum matched GMs are used for NLRHA. The whole matching procedure with the target spectrum along the list of 30 spectrally matched GMs are presented in Pujari *et al.* (2015). NLRHA is computation intensive, moreover performing a nonlinear analysis on 3D shell model of a containment will be severely demanding. A sample NLRHA requires 1:36 h on 8-core 2.4 GHz processors. Bayesian updating technique provides a reasonable fragility estimates even with less number of data. A sample NLRHA capturing the $\tilde{\varepsilon}_t^{pl}$ at PGA = 2.0 g is shown in Figure 9. From this figure it can be observed that through-wall cracks occur mainly around at the air-lock openings. These cracks are mainly due to shearing effects. Results at this higher PGA is shown only to illustrate the strain contour more prominently. For finding the fragility curve of the PC, $\tilde{\varepsilon}_t^{pl}$ values are noted at PGA levels not higher than 0.8 g (discussed later).

The two parameters of the fragility model in Eq. (29) are treated as Bayesian parameters for updating: $\boldsymbol{\theta} = \left[m_a \quad \beta_R\right]^T$.

$$F_r(x;\boldsymbol{\theta}) = \Phi\left[\frac{\ln(x / m_a)}{\beta_R}\right] \tag{32}$$

Figure 8. 3D model of the primary containment in Abaqus: (a) partitioned geometry, and (b) meshed model

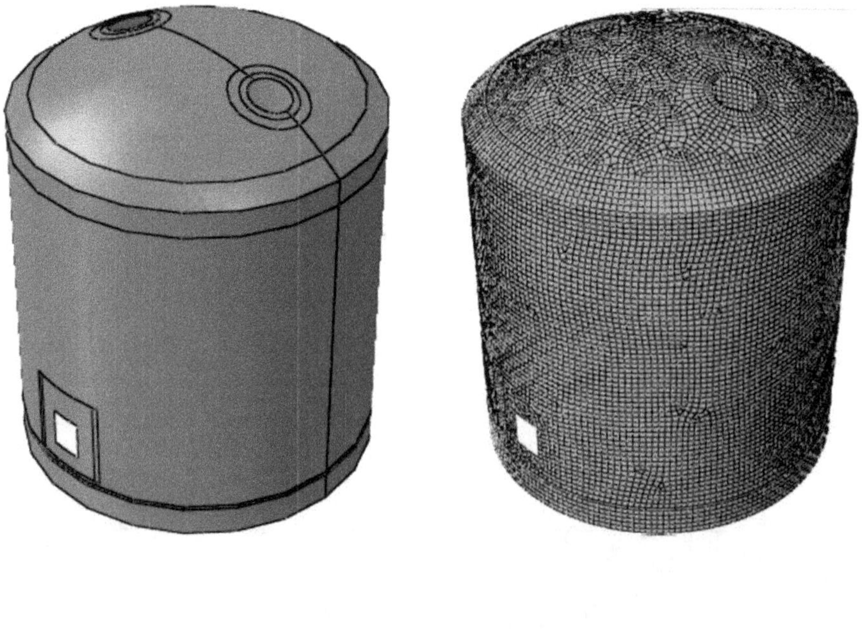

Figure 9. Overall and detailed views of $\tilde{\varepsilon}_t^{pl}$ ('PEEQT' in Abaqus) contours obtained from NLRHA

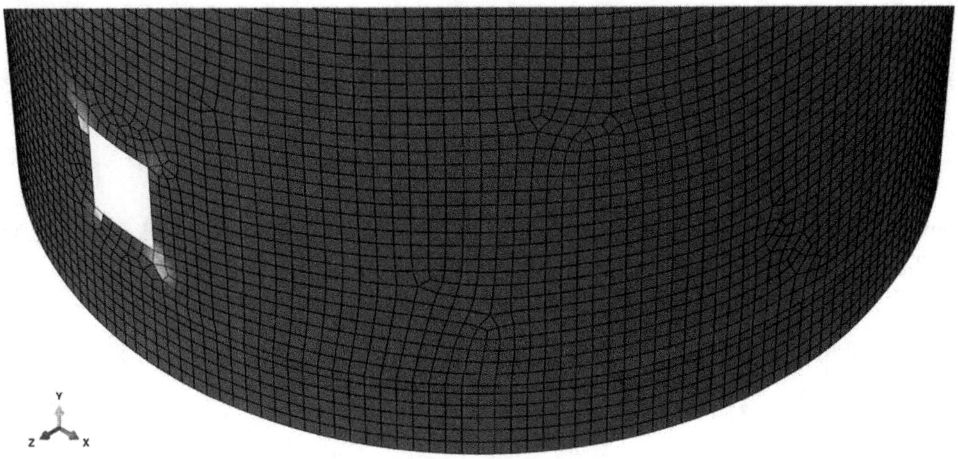

The prior distributions $p(\boldsymbol{\theta})$ are taken as joint distribution of two independent parameters m_a and β_R. Due to absence of any standard model for fragility based on through-wall cracking of a containment, $p(m_a)$ and $p(\beta_R)$ are taken as somewhat unbiased prior based on maximum interstorey drift based limits for a 2D model of the same containment (Mandal *et al.*, 2016). Although later we find that the results are not very sensitive to this prior. With an assumption that crack based fragility will be more critical than drift based fragility, the prior is modelled with m_a having mean lower than the value considered by Mandal et al. ($m_a = 0.6875$) and β_R having mean larger than their value for β_R ($= 0.2347$).

$$\begin{aligned} m_a &\sim \mathcal{U}(0.1, 0.7)\,(g) \\ \beta_{RR} &\sim \mathcal{U}(0.1, 0.8) \end{aligned} \tag{33}$$

The fragility data (d) obtained from the NLRHA is used in likelihood function $p(d \mid \boldsymbol{\theta})$ for updating the fragility parameters ($\boldsymbol{\theta}$) in Eq. (32). For NLRHA obtained fragility data point, we make use of likelihood function in the form binomial distribution.

$$\begin{aligned} p(\boldsymbol{d} \mid \boldsymbol{\theta}) &= \prod_{i=1}^{N_s} \left[1 - F(x_i; \boldsymbol{\theta})\right]^{N_r - f_i} F(x_i; \boldsymbol{\theta})^{f_i} \\ &= L(\boldsymbol{\theta} \mid \boldsymbol{d}) \end{aligned} \tag{34}$$

This type of likelihood function falls in the category of observations providing inequality information as described earlier. Here N_s is the total number of intensity scales and N_r is the total number of records used at each intensity scale to perform NLRHA. f_i is the total number of records indicating failure at the ith intensity:

$$
\begin{aligned}
f_i &= \sum_{j=1}^{N_r} \boldsymbol{I}_{ji} \\
&\text{where, for the jth record scaled to the ith intensity} \qquad (35) \\
\boldsymbol{I}_{ji} &= 1,\ if\ \tilde{\varepsilon}_t^{pl} > 0 \\
&= 0, \text{ otherwise}
\end{aligned}
$$

The updated fragility parameters are obtained based on Eq. (7) using a Markov chain Monte Carlo simulation. The posterior are sampled from the a bi-normal proposal density,

$$q(\boldsymbol{\theta}_{t+1} \mid \boldsymbol{\theta}_t) \sim \mathcal{N}\left(\boldsymbol{\theta}_t, \begin{bmatrix} \sigma_1^2 & 0 \\ 0 & \sigma_2^2 \end{bmatrix}\right) \qquad (36)$$

where σ_1 and σ_2 are iteratively selected such the acceptance ratio lies between 0.23 to 0.5 to ensure sufficient mixing of Markov chains. For the present study a total of 15,000 MCMC simulations where performed with a burn-in of 1000 and a lag 10.

To obtain the fragility of the PC using Bayesian updating, NLRHA were performed at 10 intensity scales. Figure 10 shows the histograms of the posterior samples along with the priors of m_a and β_R. It is observed that the updated m_a posterior histogram follows has a sample mean of 0.2385g and with a standard deviation of 0.008413. The obtained sample mean of β_R is 0.3431. These updated values of fragility parameters are used to construct the updated fragility curve. The standard deviation in m_a is the epistemic uncertainty (β_U) in using a Bayesian updated model for the discrete NLRHA data points, which can be used to draw credible intervals to the fragility plot. However, this epistemic uncertainty ($\beta_U = 0.008413$) is negligible compared to the other (aleatory, β_R) uncertainties and hence ignored.

Figure 10. Histogram of prior and posterior samples of (a) m_a and (b) β_{RR}

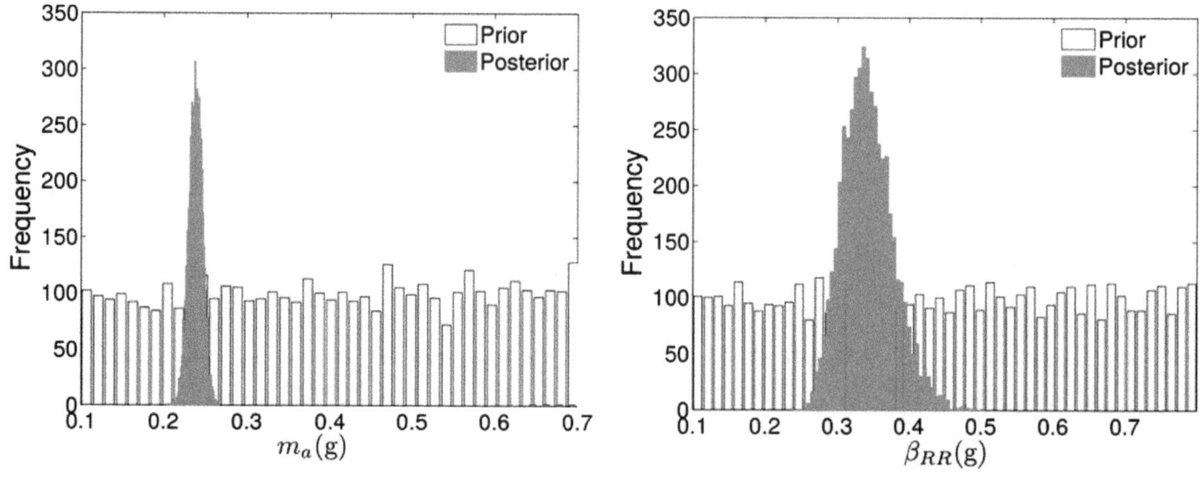

Similarly, the value of standard deviation in β_R, which is 0.03649, is also ignored in the fragility estimation which has little significance in current scenario. Figure 11 shows the prior and the updated fragility curves, along with the discrete fragility values based on f_i / N_r at 10 intensity points. The fragility curve can also be obtained using these NLRHA data with a MLE method, however, the MLE method does not provide any quick estimation of β_U.

EXAMPLE 3: PREDICTING FATIGUE CRACK GROWTH USING BAYESIAN UPDATING

Piping components in a process plant are subjected to in-service inspections to check for flaws. If flaws like crack are detected during the inspection, these are analysed for fitness for service. Fatigue is generally the dominant reason for crack growth. The fatigue crack growth (FCG) is predicted by using the Paris' law given as

Figure 11. Comparison of updated fragility curves based on 10 data points and 3 data points

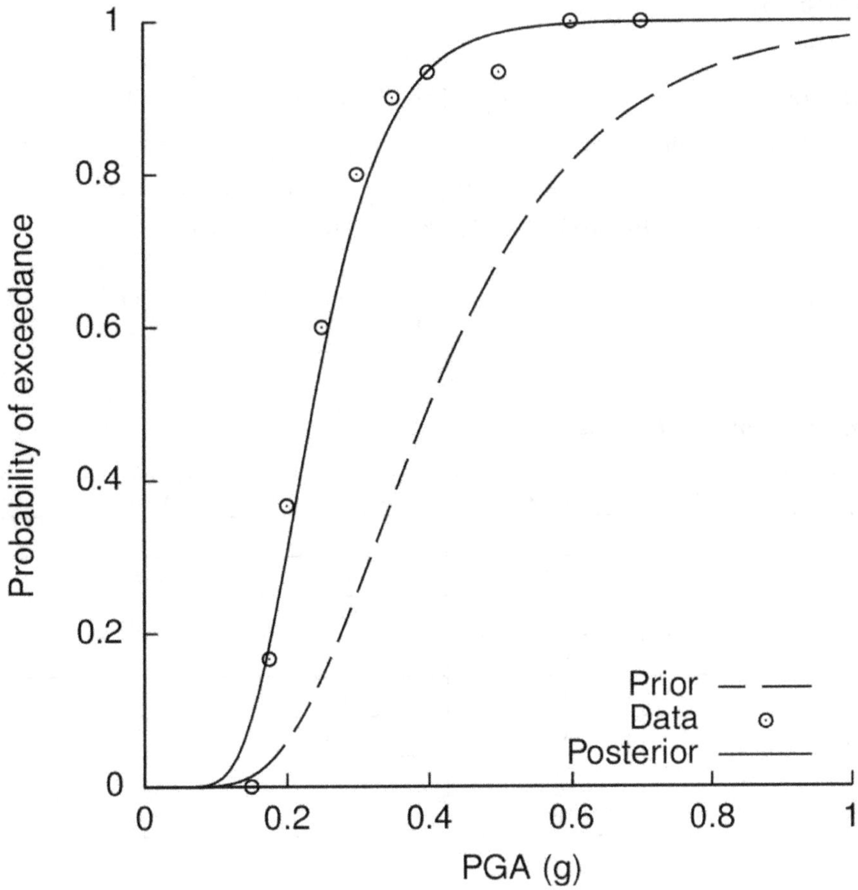

$$\frac{da}{dN} = C(\Delta K)^m \tag{37}$$

where a is the crack length in mm, N is the number of load cycles, ΔK is the range of stress intensity factor in MPa$\sqrt{\text{m}}$, and C and m are the crack growth parameters (Paris' coefficient).

The values of the Paris' coefficients are not known precisely and are uncertain (Karandikar *et al.*, 2012). These are modelled as random variables and the fatigue crack growth is evaluated typically at the mean value. The crack size measured from the in-service inspections can be used to reduce the uncertainty in the estimate of these coefficients and the fatigue life can be estimated with lesser ambiguity. This reduction in uncertainty can be achieved by applying the Bayes' theorem.

A methodology based on Bayesian updating is demonstrated in this example using the fatigue crack growth data from an experiment performed on a cracked pipe by Singh *et al.* (2008). The experiment on fatigue crack growth was conducted on a pipe with a pre-machined notch in the circumferential direction on the outer surface. The outer diameter of the pipe was 324 mm and the thickness was 21.5 mm. The stress ratio was kept as 0.1. The notch length was kept as 100 mm and the notch depth was kept at 5 mm. The stress range (difference between the maximum and the minimum values) applied for fatigue loading was 156 MPa (Table 6).

Bayesian Updating

The vector of random variable is $\boldsymbol{\theta} = \begin{bmatrix} C & m \end{bmatrix}^T$. The distribution of C is taken as lognormal and that of m as normal. The mean value of the parameters, C is 2.143×10^{-9} and of m is 3.21. The coefficient of variation in $\ln(C)$ and m are 4% and 6% respectively. For a mathematical model $\mathcal{P}$ such as the Paris law. To account for modelling discrepancy, the true value is given as

$$\mathcal{P}(N) = a_t(N) + \varepsilon_M \tag{38}$$

Table 6. Fatigue crack growth data

Cycles	Crack Depth (mm)	Cycles	Crack Depth (mm)	Cycles	Crack Depth (mm)
0	5	26500†	9.9	39706	14.8
2100	5	28500	10.4	40701	14.9
5000	5.4	29500	10.4	42100	15.4
10560†	6.2	30500†	10.6	42000	15.7
14036	6.9	31800	11.1	42000	16.2
16536	7.6	33000	11.4	43000	16.6
18350†	8	34000†	11.9	44000	16.9
21000	8.7	36632	12.6	45000	17.2
24350	9.6	38695	13.7		

†Data used for updating.

where a_t is the "true" crack size and $\varepsilon_M \sim \mathcal{N}\left(0, \sigma_{\varepsilon_M}\right)$ is the error in prediction. The crack size is examined during the in-service inspection. Let this measurement be represented by $a_d(N)$. This measurement of the crack size will be related to the true value as given by

$$a_t(N) = a_d(N) + \varepsilon_m \tag{39}$$

where $\varepsilon_m \sim \mathcal{N}\left(0, \sigma_{\varepsilon_m}\right)$ is the randomness in measurement. Using Eq. (38) and (39) the relationship between the crack size predicted using the model and the one detected during a measurement can be written as:

$$a_d(N) = \mathcal{P}(N) + \varepsilon \tag{40}$$

The likelihood is assumed to be Gaussian for the zero mean quantity $\varepsilon = a_d(N) - \mathcal{P}(N)$, called the global error with standard deviation σ_ε,

$$f_\varepsilon(\varepsilon_i \mid \boldsymbol{\theta}) = \frac{1}{\sqrt{2\pi}\sigma_\varepsilon} \exp\left[-\frac{\{a_d(N) - \mathcal{P}(N)\}^2}{2\sigma_\varepsilon^2}\right] \tag{41}$$

For multiple observations, the likelihood becomes

$$f_\varepsilon(\Omega \mid \boldsymbol{\theta}) = \prod_{i=1}^{n} f_\varepsilon(\varepsilon_i \mid \boldsymbol{\theta}) \tag{42}$$

MCMC Simulation

In the present work, the MH algorithm is initiated from four different starting points. The starting point of the four chains is taken as the combination of 10 and 90 percentile values of m and $\ln C$. All the chains explore the same domain of the random variables. Four Markov chains of length 2000 each are simulated, the first 1000 values of each are considered as burn-in. Thus, a total of 4000 random samples are available. The total numbers of simulations required for generating 8000 Markov chain elements are 52124.

Results

For the purpose of illustration the prior and posterior histograms of the variables m and $\ln C$ are shown in Figure 12. The global error factor is assumed $\sigma_\varepsilon = 3$. Five observations from the experiment conducted by Singh *et al.* (2008) are used to derive this set of results. The coefficient of variation for m is estimated to be 4%, while that for $\ln C$ is estimated equal to 2%, based on these histograms. The statistics estimated from the four chains are listed in Table 7. It is seen that the statistics (mean, standard

Figure 12. Relative frequency histogram (a) m *; (b)* $\ln C$

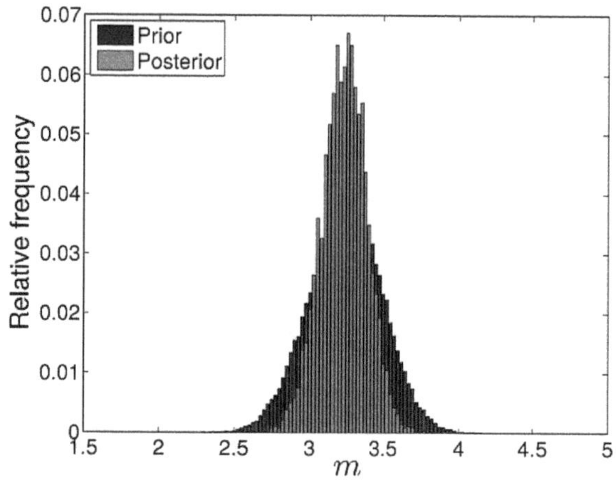

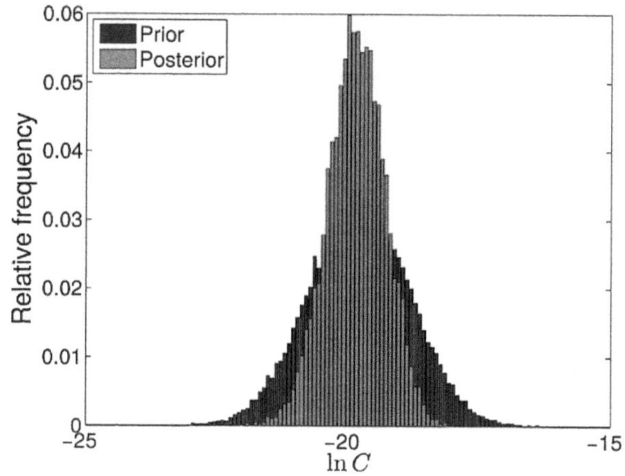

Table 7. Statistics from four chains of m *and* $\ln C$

Starting Point		Mean		95 Percentile		5 Percentile		Standard Deviation		CoV	
m	ln C	m	ln C	m	ln C	m	ln C	m	ln C	m	ln C
2.96	-20.98	3.21	-19.88	3.44	-19.04	2.98	-20.75	0.14	0.52	0.05	0.03
2.96	-18.94	3.22	-19.87	3.45	-18.98	2.97	-20.76	0.15	0.55	0.05	0.03
3.46	-20.98	3.21	-19.88	3.44	-19.05	2.99	-20.74	0.14	0.51	0.04	0.03
3.46	-18.94	3.22	-19.92	3.46	-19.08	2.96	-20.86	0.15	0.54	0.05	0.03
All chain		3.22	-19.89	3.45	-19.03	2.98	-20.76	0.15	0.53	0.05	0.03

deviation, CoV, 5 percentile and 95 percentile values) of the four chains are comparable. A good mixing of chains is visible in the time series plot for m and $\ln C$ as shown in Figure 13.

Figure 14 shows the FCG curve using the posterior information obtained using Bayesian updating of the random variables. The posterior curve is drawn using the mean values of m and $\ln C$, obtained from the converged Markov chains. This curve is compared with that obtained using the prior information and the measured data. The graph shows that Bayesian updating has improved the prediction as the updated curve is closer to the experimental observations.

Figure 15 gives the confidence bounds on the FCG curve on using the posterior information from Bayesian updating. 4000 values of the Paris' constants obtained from the four chains are used to estimate 4000 fatigue crack growth curves. Thus, 4000 points are available at each value of cycle (N). The bounds on the crack size for different cycles are obtained from this data set. The median of the FCG curve is near to the measured data.

The σ_ε parameter is varied to study the sensitivity of the results. The number of sampled points which are accepted in the MH algorithm as part of the Markov chain, is depended on the value of σ_ε.

Figure 13. Mixing of the four chains (a) m *; (b)* $\ln C$

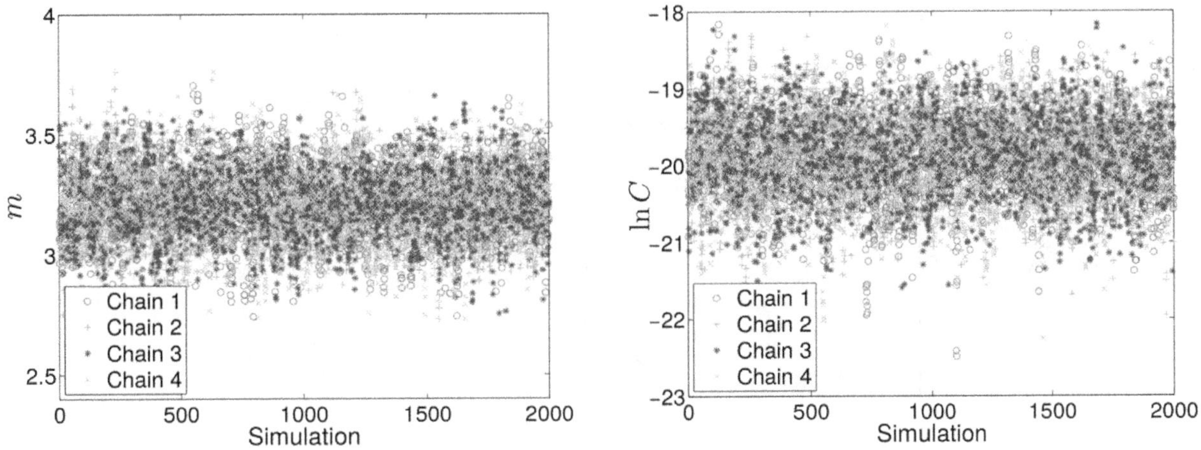

Figure 14. Comparison of the updated FCG Curve with experimental observations and prior curve

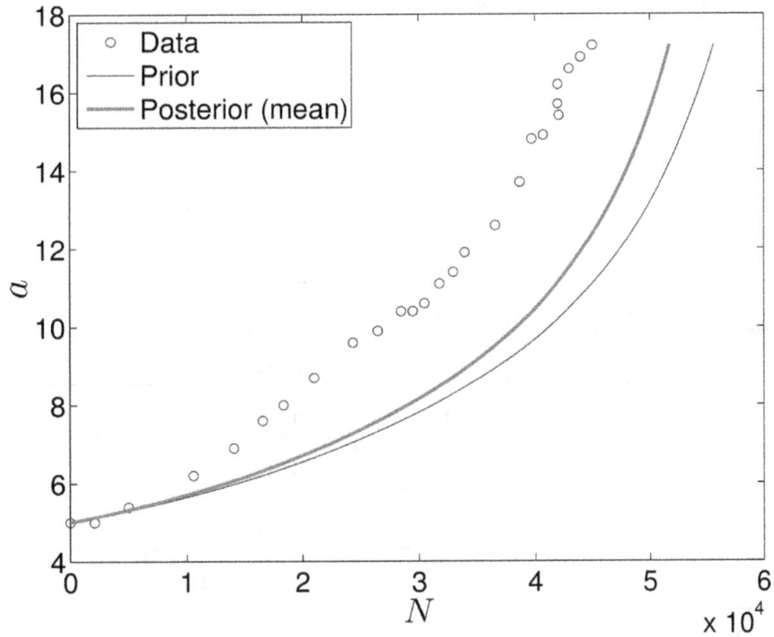

The acceptance ratio is found to be inversely proportional to the value of σ_ε. The statistics of the results is shown in Table 8. The posterior samples required for the convergence are more when is σ_ε higher. Thus, if a low value of σ_ε is adopted, the acceptance ratio reduces. The samples generated from a low σ_ε have a lower variance and as a result it requires a shorter length of the Markov chain for convergence.

The confidence bounds shown by Figure 15 are obtained for $\sigma_\varepsilon = 3$. Narrower bounds are obtained when $\sigma_\varepsilon = 1$ (Figure 16). Some of the experimental points however are outside these narrow bounds. This indicates that the actual uncertainty present is more than that given by $\sigma_\varepsilon = 1$.

Figure 15. Confidence bounds on the posterior estimates with $\sigma_{err} = 3$

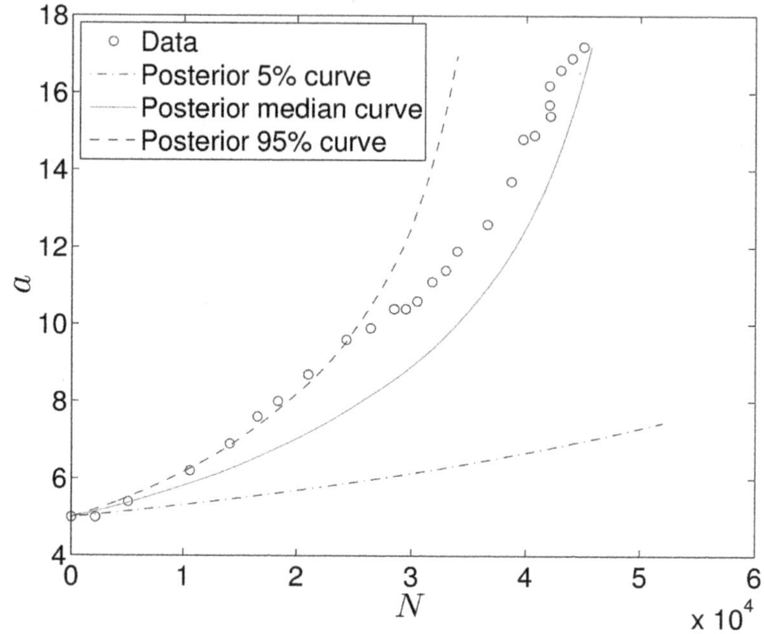

Table 8. Statistical dependence on σ_ε

σ_ε	0.5	1	2	3	4
Acceptance ratio	0.016	0.033	0.076	0.152	0.25
Mean of m	3.179	3.221	3.228	3.213	3.189
Standard Deviation of m	0.141	0.144	0.148	0.153	0.163
Coefficient of variation on m	0.044	0.045	0.046	0.048	0.051
Mean of $\ln C$	-19.509	-19.664	-19.75	-19.876	-20.013
Standard Deviation of $\ln C$	0.483	0.491	0.51	0.549	0.617
Coefficient of variation on $\ln C$	0.025	0.025	0.026	0.028	0.031

In the results presented so far, five experimental points are considered for Bayesian updating (marked with superscript † in Table 6). The effect of the number of experimental observations on the accuracy of the posterior FCG curve is also studied here. Alternate updating schemes are adopted which use 1, 2, 3, 4 and 5 observations. For the case with a single observation, the first observation in the Table 6 marked with the superscript † is considered. The case with updating using two observations, the initial two observations with superscript † are considered and so on.

Figure 16. Confidence bounds on the posterior estimates with $\sigma_{err} = 1$

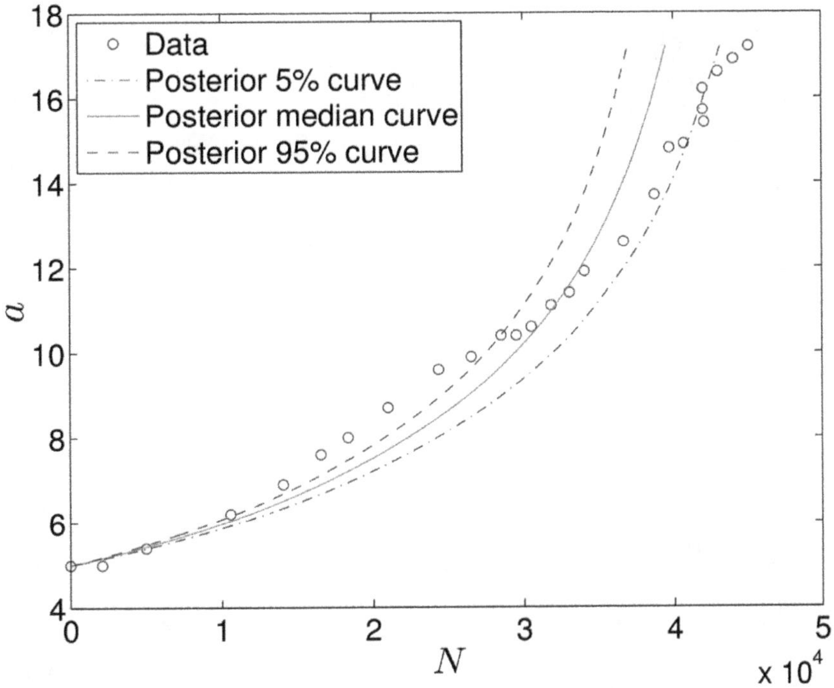

Figure 17 gives the results obtained after updating using five different number of experimental observations. These results are for $\sigma_{\varepsilon} = 3$. When the number of observations used for updating is less than five, the predictions from updated FCG curve deviate more from the experimental curve when compared with the specimen curve. For updating using five data points, predicted FCG curve is nearer to the experimental curve. Figure 18 gives similar curves when $\sigma_{\varepsilon} = 1$. In this case, updating is effective even with two experimental observations. Table 9 gives the measure of the average error for different cases with respect to the experimental curve. The average error is calculated by summing the squares of the residuals and taking their average.

It can be seen from Table 9 that with a lower value of σ_{ε} the average error is less. As the number of observations increase, the difference in the average error decreases. These results indicate that a lower value of σ_{ε} would help in arriving at the realistic solution. However, in probabilistic study, the bounds on the predictions are also important. These bounds help in tracking the conservative as well as the best-fit results for a prediction. It can be seen from the earlier results (Figures 15 and 16) that the smaller value of σ_{ε} gives tighter but non-ocnservative bounds for the problem solved in this example. Thus, for a few number of observations, a higher value of σ_{ε} should be used. If the number of observations are significant (5 or more), a lower value of σ_{ε} can be used.

This example has presented a methodology based on Markov chain Monte Carlo to improve the prediction of the life of deteriorating structures. This method is validated using the results from the measurements conducted on cracked pipes. The controlling parameter for simulation is the σ_{ε} term

Figure 17. Confidence bounds on the posterior estimates with $\sigma_{err} = 3$

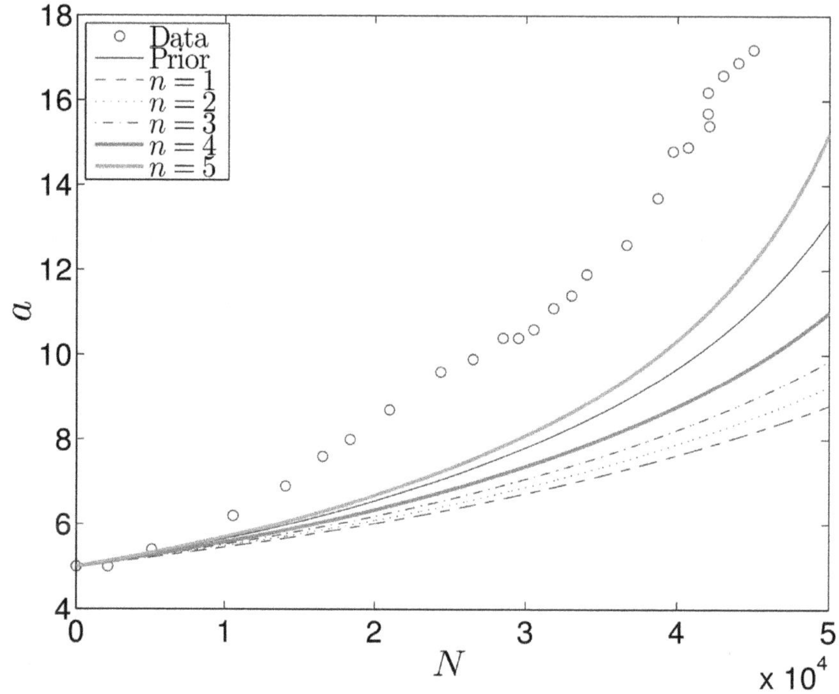

Figure 18. Comparison of the updated fatigue crack growth curve with experimental observations and prior curve, $\sigma_{err} = 1$

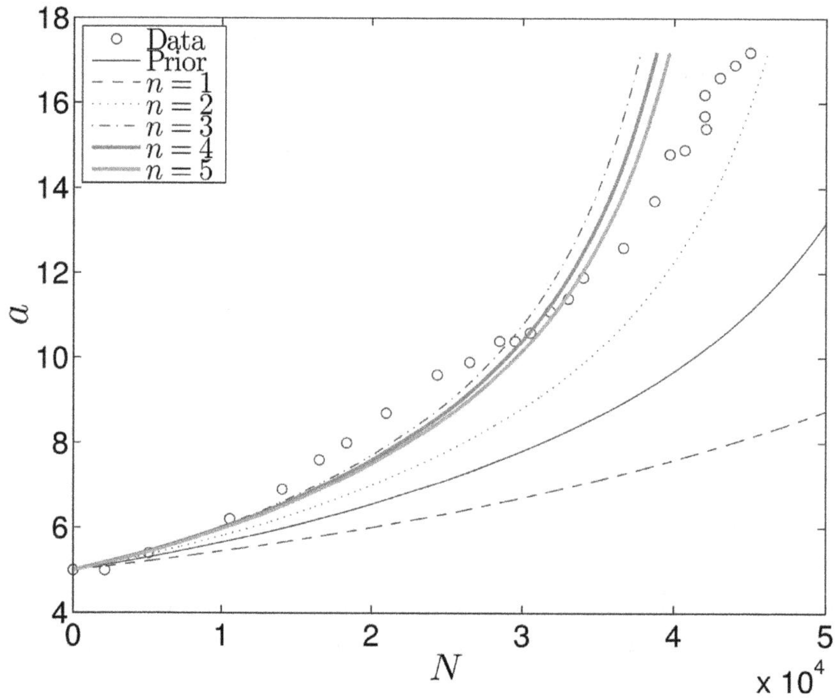

Table 9. The average error in updating with different number of observations

Number of Observations Used for Updating	Average Error	
	$\sigma_\varepsilon = 1$	$\sigma_\varepsilon = 3$
1	1.09	1.085
2	0.358	1.045
3	0.244	0.993
4	0.259	0.907
5	0.204	0.661

which represents the uncertainty in the FCG model and the uncertainty in the measurement. This example also presents studies related to the estimation of confidence bounds on the predicted results. It is shown that a smaller value of σ_ε, results in narrower bounds on the predicted values. The simulation however is expensive if a low value of σ_ε is used. The effect of number of observations used for updating is also studied. The methodology presented in this example can be applied to other degradation mechanisms active in nuclear plants.

CONCLUSION

Bayesian inference is an effective technique to handle scenarios which are plagued with limited amount of information, and to quantify uncertainties in the adopted mathematical models of prediction and in the measurements. It also offers a flexible framework to address different kinds of uncertainty. A growing number of researchers in the area of structural engineering is realising the potential of using the Bayesian framework, and has started to effectively adopt it in both theoretical research and practical implementation.

With the availability of mathematical algorithms and the processing ability of today's computers, Bayesian inference can be easily adopted for uncertainty quantification in structural engineering. However, a user should not treat these tools as black-boxes which may result in unrealistic estimation of structural responses. For example, choosing an effective variance of the proposal density function (in the M-H algorithm) requires significant experience. Similarly, a suitable 'lag' needs to be selected for reducing the auto-correlation among MCMC samples. However, careful handling of such issues results in an efficient use of Bayesian inference for most of the practical scenarios.

REFERENCES

Al-Amin, M., Zhou, W., Zhang, S., Kariyawasam, S., & Wang, H. (2012). Bayesian model for calibration of ILI tools.*Proceedings of the Biennial International Pipeline Conference, IPC* (Vol. 2, pp. 201-208). doi:10.1115/IPC2012-90491

Almusallam, A. A., Al-Gahtani, A. S., Aziz, A. R., & Rasheeduzzafar, . (1996). Effect of reinforcement corrosion on bond strength. *Construction & Building Materials*, *10*(2), 123–129. doi:10.1016/0950-0618(95)00077-1

Alonso, C., Andrade, C., Rodriguez, J., & Diez, J. M. (1996). Factors controlling cracking of concrete affected by reinforcement corrosion. *Materials and Structures*, *31*(211), 435–441.

Ang, A. H.-S., & Tang, W. H. (2007). *Probability Concepts in Engineering Emphasis on Application in Civil & Environmental Engineering* (2nd ed.). New York, USA: John Wiley & Sons.

Bhargava, K., Ghosh, A. K., Mori, Y., & Ramanujam, S. (2006). Analytical model for time to cover cracking in rc structures due to rebar corrosion. *Nuclear Engineering and Design*, *236*(11), 1123–1139. doi:10.1016/j.nucengdes.2005.10.011

Chib, S., & Greenberg, E. (1995). Understanding metropolis-hastings algorithm. *The American Statistician*, *49*(4), 327–335.

Choi, I.-K., Choun, Y.-S., Ahn, S.-M., & Seo, J.-M. (2008). Probabilistic seismic risk analysis of candu containment structure for near-fault earthquakes. *Nuclear Engineering and Design*, *238*(6), 1382–1391. doi:10.1016/j.nucengdes.2007.11.001

Congdon, P. (2006). *Bayesian Statistical Modelling*. John Wiley & Sons. doi:10.1002/9780470035948

Der Kiureghian, A., & Ditlevsen, O. (2009). Aleatory or epistemic? does it matter? *Structural Safety*, *31*(2), 105–112. doi:10.1016/j.strusafe.2008.06.020

Ditlevsen, O. (1982). Model uncertainty in structural reliability. *Structural Safety*, *1*(1), 73–86. doi:10.1016/0167-4730(82)90016-9

Ellingwood, B. (1990). Validation studies of seismic PRAs. *Nuclear Engineering and Design*, *123*(2-3), 189–196. doi:10.1016/0029-5493(90)90237-R

Ellingwood, B. R., Celik, O. C., & Kinali, K. (2007). Fragility assessment of building structural systems in Mid-America. *Earthquake Engineering & Structural Dynamics*, *36*(13), 1935–1952. doi:10.1002/eqe.693

Engelund, S., Sϕrensen, J. D., & Sϕrensen, B. (1999). Evaluation of repair and maintenance strategies for concrete coastal bridges on a probabilistic basis. *ACI Materials Journal*, *96*(2), 160–166.

Faroz, S. A., Pujari, N. N., & Ghosh, S. (2014). A Bayesian Markov Chain Monte Carlo approach for the estimation of corrosion in reinforced concrete structures. In B. H. V. Topping & P. Ivanyi (Eds.), *Proceedings of the Twelfth International Conference on Computational Structures Technology*, Stirlingshire, UK. doi:10.4203/ccp.106.150

Gamerman, D., & Lopes, H. (2006). *Markov Chain Monte Carlo: Stochastic Simulation for Bayesian Inference* (2nd ed.). Boca Raton, USA: Chapman & Hall/CRC.

Geyer, C. J. (1992). Practical markov chain Monte Carlo. *Statistical Science*.

Goodman, J. (1985). Structural fragility and principle of maximum entropy. *Structural Safety*, *3*(1), 37–46. doi:10.1016/0167-4730(85)90006-2

Jamali, A., Angst, U., Adey, B., & Elsener, B. (2013). Modeling of corrosion-induced concrete cover cracking: A critical analysis. *Construction & Building Materials*, *42*, 225–237. doi:10.1016/j.conbuildmat.2013.01.019

Karandikar, J. M., Kim, N. H., & Schmitz, T. L. (2012). Prediction of remaining useful life for fatigue-damaged structures using Bayesian inference. *Engineering Fracture Mechanics*, *96*, 588–605. doi:10.1016/j.engfracmech.2012.09.013

Kass, R. E., & Wasserman, L. (1996). The selection of prior distributions by formal rules. *Journal of the American Statistical Association*, *91*(435), 1343–1370. doi:10.1080/01621459.1996.10477003

Kennedy, R. P., Cornell, C. A., Campbell, R. D., Kaplan, S., & Perla, H. F. (1980). Probabilistic seismic safety study of an existing nuclear power plant. *Nuclear Engineering and Design*, *59*(2), 315–338. doi:10.1016/0029-5493(80)90203-4

Kennedy, R. P., & Ravindra, M. K. (1984). Seismic fragilities for nuclear power plant risk studies. *Nuclear Engineering and Design*, *79*(1), 47–68. doi:10.1016/0029-5493(84)90188-2

Kwon, O. S., & Elnashai, A. (2006). The effect of material and ground motion uncertainty on the seismic vulnerability curves of RC structure. *Engineering Structures*, *28*(2), 289–303. doi:10.1016/j.engstruct.2005.07.010

Lee, J., & Fenves, G. L. (1998). Plastic-damage model for cyclic loading of concrete structures. *Journal of Engineering Mechanics*, *124*(8), 892–900. doi:10.1061/(ASCE)0733-9399(1998)124:8(892)

Link, W. A., & Eaton, M. J. (2012). On thinning of chains in MCMC. *Methods in Ecology and Evolution*, *3*(1), 112–115. doi:10.1111/j.2041-210X.2011.00131.x

Liu, Y. (1996). *Modeling the time to corrosion cracking of the cover concrete in chloride contaminated reinforced concrete structures* (Ph.D. thesis). Virginia Polytechnic Institute and State University, Blacksburg, USA.

Liu, Y., & Weyers, R. E. (1998). Modeling the time-to-corrosion cracking in chloride contaminated reinforced concrete structures. *ACI Materials Journal*, *95*(6), 675–681.

Lubliner, J., Oliver, J., Oller, S., & Onate, E. (1989). A plastic-damage model for concrete. *International Journal of Solids and Structures*, *25*(3), 299–329. doi:10.1016/0020-7683(89)90050-4

Malumbela, G., Alexander, M., & Moyo, P. (2010). Variation of steel loss and its effect on the ultimate fexural capacity of rc beams corroded and repaired under load. *Construction & Building Materials*, *24*(6), 1051–1059. doi:10.1016/j.conbuildmat.2009.11.012

Malumbela, G., Moyo, P., & Alexander, M. (2012). A step towards standardizing accelerated corrosion tests on laboratory reinforced concrete specimens. *Journal of the South African Institution of Civil Engineering*, *54*(2), 78–85.

Mandal, T. K., Ghosh, S., & Pujari, N. N. (2016). Seismic fragility analysis of a typical Indian PHWR containment: Comparison of fragility models. *Structural Safety*, *58*, 11–19. doi:10.1016/j.strusafe.2015.08.003

Marin, J. M., & Robert, C. P. (2007). *Bayesian core: a practical approach to computational Bayesian statistics*. New York, USA: Springer Texts in Statistics.

Martinez, W. L., & Martinez, A. R. (2002). *Computational Statistics Handbook with MATLAB*. Boca Raton, USA: Chapman & Hall/CRC.

Oden, T., Moser, R., & Ghattas, O. (2010). Computer Predictions with Quantified Uncertainty, Part I. *SIAM News*, 43(9).

Pate-Cornell, M. E. (1996). Uncertainties in risk analysis: Six levels of treatment. *Reliability Engineering & System Safety*, *54*(2-3), 95–111. doi:10.1016/S0951-8320(96)00067-1

Pujari, N. N., Ghosh, S., & Lala, S. (2015). Seismic fragility estimation of a containment shell based on the formation of through-wall cracks. *ASCE-ASME Journal of Risk and Uncertainty in Engineering Systems, Part A. Civil Engineering*.

Raiffa, H., & Schlaifer, R. (1961). *Applied statistical decision theory*. Applied Statistical Decision Theory.

Shinozuka, M., Feng, M. Q., Lee, J., & Naganuma, T. (2000). Statistical analysis of fragility curves. *Journal of Engineering Mechanics*, *126*(12), 1224–1231. doi:10.1061/(ASCE)0733-9399(2000)126:12(1224)

Simulia. (2013). *Abaqus FEA User's Manual, Version 6.13*. France: Velizy-Villacoublay.

Singh, P. K., Bhasin, V., Vaze, K. K., Ghosh, A. K., Kushwaha, H. S., Murthy, D. S. R., & Sivaprasad, S. et al. (2008). Fatigue studies on carbon steel piping materials and components: Indian phwrs. *Nuclear Engineering and Design*, *238*(4), 801–813. doi:10.1016/j.nucengdes.2007.09.002

Straub, D., & Papaioannou, I. (2014). Bayesian updating with structural reliability methods. *Journal of Engineering Mechanics*, *141*(3).

Torres-Acosta, A. A., & Sagues, A. A. (2004). Concrete cracking by localized steel corrosion - Geometric effects. *ACI Materials Journal*, *101*(6), 501–507.

Vamvatsikos, D., & Cornell, C. A. (2002). Incremental dynamic analysis. *Earthquake Engineering & Structural Dynamics*, *31*(3), 491–514. doi:10.1002/eqe.141

Vamvatsikos, D., & Cornell, C. A. (2004). Applied incremental dynamic analysis. *Earthquake Spectra*, *20*(2), 523–553. doi:10.1193/1.1737737

Zheng, R., & Ellingwood, B. R. (1998). Role of non-destructive evaluation in time-dependent reliability analysis. *Structural Safety*, *20*(4), 325–339. doi:10.1016/S0167-4730(98)00021-6

Chapter 15
Simulating Post-Earthquake Fire Loading in Conventional RC Structures

Behrouz Behnam
Amirkabir University of Technology (Tehran Polytechnic), Iran

ABSTRACT

Post-earthquake fire (PEF) is one of the most complicated problems resulting from earthquake, presenting a serious risk to urban buildings. As most standards and codes ignore the possibility of PEF, buildings are too weak under PEF loads. This chapter is to investigate the effects of PEF loads on partially damaged RC buildings located in urban regions. To do that, a methodology named sequential analysis is introduced here via which the structural performance at various performance levels is evaluated under fire and PEF scenarios. Numerically, in order to simulate the earthquake loads, conventional pushover analysis is employed, with an explanation presented in the chapter to introduce the pushover analysis, its advantages and its limitations. To simulate the fire loads, standard fire curve (ISO 834) is used for simplicity.

INTRODUCTION

Strong earthquakes may bring about significant damage to all kinds of infrastructure, including buildings in urban areas. In addition to the damage caused by the earthquake itself, post-earthquake events resulting from the shock, such as fire, can potentially create even more significant damage than the earthquake itself. Past events have proven that lack of adequate attention paid to post-earthquake fire (PEF) can result in significantly increased fatalities and casualties. The 1906 San Francisco PEF is considered one of the largest urban disasters in human history. The fire swept over nearly a quarter of the city, including the entire downtown area. Over 3,000 people died as a result of the disaster. It is estimated that the economic value of the damage was more than US$524 million (in 1906 dollars) (Fitzpatrick, 1914). The fire accounted for 80% of the total damage. The 1923 Kanto PEF in Japan is also said to be one of the largest urban disasters, resulting from a strong earthquake. More than 140,000 people lost their lives

DOI: 10.4018/978-1-5225-0588-4.ch015

and around 447,000 houses were destroyed. It is estimated that 77% of the total damage was due to the fire. The fire following the 1989 California earthquake (known as the Loma Prieta Earthquake) caused more than US$6 billion damage to the Bay area of California. The fire displaced more than 12000 people and caused 3700 casualties. The strong 1995 Kobe (Japan) earthquake resulted in significant damage to the city. It is recorded that more than 100 fires broke out within a few minutes of the earthquake, in both residential and commercial buildings. Two hours after the earthquake, several conflagrations had developed. It is estimated that over 6,500 buildings burnt, in an area of nearly 700,000 square meters. The most recent serious earthquake (with a magnitude of 8.9 on the Richter scale) happened in the north of Japan in 2011. This earthquake caused a tsunami, which created one of the worst disasters in the history of Japan. Thousands of people were killed or injured, and more than 150,000 people became homeless, many being forced to evacuate their homes and leave their possessions. The city of Kesen-numa in the northern part of Japan was completely engulfed in flames following this earthquake, the biggest on record in the country.

The devastating effect of PEF on buildings arises mainly because the majority of structural elements are not designed for extreme loading events combined with gravity loads, lateral loads and fire loads. Consequently, even those structures only moderately damaged by an earthquake can be rapidly destroyed in a subsequent fire. PEF is a high-risk load, which needs to be scrutinized further and allowed for in the codes of design. This scrutiny and modification to codes should include providing sufficient time under PEF, during which the building remains gravity load bearing, until the building's inhabitants can safely evacuate. The time would naturally not only depend on the building's structural properties, the severity of earthquake, and the extent of fire load, but also on the availability and response time of fire brigades. The first priority is thus to simulate the PEF loading in a building a structure. This priority is further highlighted with this reality that there are currently limited user-friendly computer packages capable of performing a coupled analysis, i.e. considering seismic analysis and its subsequent fire, allowing for the variation of material properties at elevated temperature. Sequential analysis would be an alternative to this limitation. Sequential analysis is a functional tool that considers the loads and corresponding changes in geometry in a number of steps. It allows the effects of residual deformations, as well as degradation in stiffness and strength from the earthquake, to be used in the fire analysis. Utilizing a commonly used structural analysis software package, and following the simple application of gravity loads, for which the structure mostly remains in the elastic region, the seismic analysis is performed and the equivalent seismic loads are derived. These loads are transferred into the second software package, which is capable of performing the structural fire analysis. There is often a degree of programming involved in connecting the two software packages together so that the process becomes automatic.

SEISMIC ANALYSIS

Structural seismic analysis can be performed through static or dynamic methods, which can be linear or non-linear. Although most static and linear methods are simple to use, they cannot entirely demonstrate the inelastic deformation or damage in structures, which is an important deficiency of these methods. On the other hand, earthquake forces have a dynamic nature and therefore, only a dynamic analysis (as the most rigorous approach) can meet a realistic assessment. Nevertheless, dynamic analyses are not routinely employed in everyday designs, mostly because of the complex nature of the analysis, in particular with respect to inelastic behavior. In addition, different ground motions may bring about different non-linear responses, so choosing the right excitation is an important and sometimes determining consideration.

In view of the abovementioned difficulties, there is a more practical method, which is used to determine the design forces. This method is known as static pushover analysis and was developed over the past two decades in accordance with the concept of design based on performance. It is becoming a popular tool for seismic performance evaluation of existing and new buildings, because the procedure is not only simple to use but also covers the non-linear behavior. Nevertheless, as certain simplifications are made during the analysis, the method becomes approximate only. Pushover method is based on the fundamental assumption that the structural response can be approximated by the response of an equivalent single degree-of-freedom (SDOF) system taken as a pre-assumed mode shape that remains constant throughout the analysis. This is obviously not a correct assumption, yet there exists a consensus on the use of the method as a rational tool to estimate the maximum seismic response of multi degree-of-freedom (MDOF) structures, if the response of the structure is mostly governed by a single mode. For those structures that cannot meet these mentioned limitations, advanced pushover-based methods can be employed, such as adaptive pushover analysis and modal pushover analysis, in order to consider the effects of higher modes in the process. Conventional pushover analysis is the non-linear incremental-iterative solution of the equilibrium equation $KU = P$ in a finite element formulation, in which K, U and P stand for the non-linear stiffness matrix, the displacement vector and a predefined load vector (which is laterally applied to the structure), respectively.

The analysis is carried out in two steps: the load-controlled step (controlled by gravity loads), and the displacement-controlled step (controlled by earthquake loads). While, the former step normally happens prior to notable yielding, the latter step occurs after significant yielding (Papanikolaou & Elnashai, 2005). It is also worth noting that the lateral load may be a set of displacements or forces, but should have a constant ratio and a constant shape during the analysis. In this way, at the end of each iteration, the reaction vector (P_e) of the structure is assembled from the contribution of all finite elements. Then, the out-of-balance forces are iteratively reapplied to the structure until the specific predefined tolerance is reached, as represented by Equation 1.

$$\Delta U = \left[K_T\right]^{-1}\left(\lambda P_0 - P_e\right) \tag{1}$$

where ΔU is the calculated displacement increment within an iteration, K_T is the current non-linear stiffness matrix, λ is the load factor within the corresponding load increment, P_0 is the initial load, and P_e is the reaction of the previous iteration. The reaction load vector (P_e) is then accounted for using Equation 2.

$$P_e = \sum \int_v B_T \cdot \sigma_{NL} \cdot dV \tag{2}$$

where B is the strain-displacement matrix of each element and σ_{NL} is the vector of the element non-linear stress, which is determined by its material constitutive law.

The process terminates when either a predefined limit state is reached or the structural collapse is identified. The plotted roof displacement versus base shear is then interpreted as the capacity curve. In conventional pushover analysis, it is assumed that the target displacement of a SDOF structure corresponds to a MDOF structure, provided that the first mode of the structure is the predominant mode. In addition, studies have shown that for structures with a fundamental period smaller than 2 sec., this assumption is relatively accurate, specifically for elastic systems. For inelastic systems, however, the results are mostly conservative (Krawinkler & Seneviratna, 1998). On the other hand, and for the sake of simplification, the non-linear base shear versus displacement curve can be replaced with an idealized bilinear diagram, incorporating the effective lateral stiffness and the effective yield strength as shown in Figure 1. In the figure, K_i, K_e and V_y stand for elastic lateral stiffness, effective lateral stiffness and effective yield strength of the structure in the direction under consideration, respectively. The post-yield slope of the figure, $K_s = \alpha K_e$, then, may be positive or negative. In addition, V_y and V_T are the yield base-shear and the total base-shear, corresponding to the yield displacement (Δ_y) and the target displacement (Δ_T). For the pushover curve, the lateral load shape (force or displacement) and the way it is distributed over the height of the structure (Kalkan & Kunnath, 2007; Tirca & Chen, 2012), the number of load steps and the convergence criteria are required, which play important roles in the level of accuracy of the analysis.

Several methodologies for estimation of the target displacement have been proposed, some of the more well-known of which are the N2 method (P. Fajfar, 2000), ATC40 method (P. Fajfar, 1999), FEMA356 method (FEMA356, 2000) and NERHP (Council, Agency, & Council, 2003) method. These methods are mostly based on statistical relationships of the lateral displacement estimated by linear and non-linear analyses.

PLASTIC HINGES AND ACCEPTANCE CRITERIA

Performing the pushover analysis results in computation of inertia forces generated at places of significant masses, and in a structure, which is displaced to the target displacement. The analysis may lead to some of the structural elements exhibiting yield in their sections, resulting in the formation of plastic hinges. It is worth mentioning that appropriately defining the non-linear (plastic) characterization of the members is important. When an inelastic deformation is experienced by an RC member, cracks tend to spread, resulting in the distribution of the curvature, as schematically shown in Figure 2 (Roh, Reinhorn, & Lee,

Figure 1. Idealized force-displacement curve

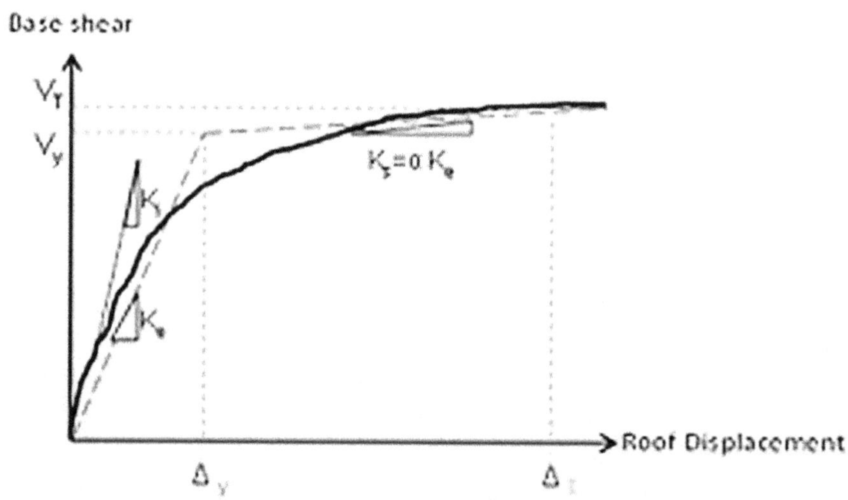

Figure 2. Distribution of curvature in a member
Roh et al., 2012.

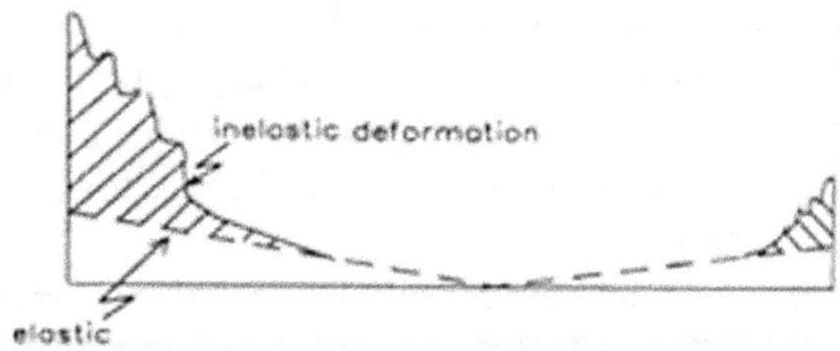

2012). It is mentionable that rebar is customarily modelled as an elastic perfectly plastic material with parabolic strain hardening (Karthik & Mander, 2011). In order to simulate the inelastic behavior of the members, two main models have been developed – concentrated plasticity (lumped plasticity) models, and spread plasticity (distributed plasticity) models, the parameters of which are quantified using the members' strength and deformation capacity (Firat Alemdar, 2010). The distributed plasticity models provide a skeleton for non-linear structural analysis by defining the moment curvature at different locations of each member, which require more computational effort than the lumped plasticity models (Roh et al., 2012).

The lumped plasticity models, which are based on defining concentrated plastic hinges, however, are usually assigned at the expected locations of plasticity spread. This is shown schematically using idealized curvature distribution along a cantilever in Figure 3. As is seen in the figure, assuming a ductile behavior for the RC element, a typical moment-curvature curve can be idealized to a tri-linear curve, with three stages – elastic stage, cracking stage and steel yielding stage (Kwak & Kim, 2002).

Before the cracking moment (M_{cr}), the distribution of the stress across the cross-section is linear and elastic. Assuming complete bond between concrete and rebar in the cross-section, the concrete and the

Figure 3. Actual and idealized curvature distribution in a cantilever
Park. R & Paulay.T, 1975.

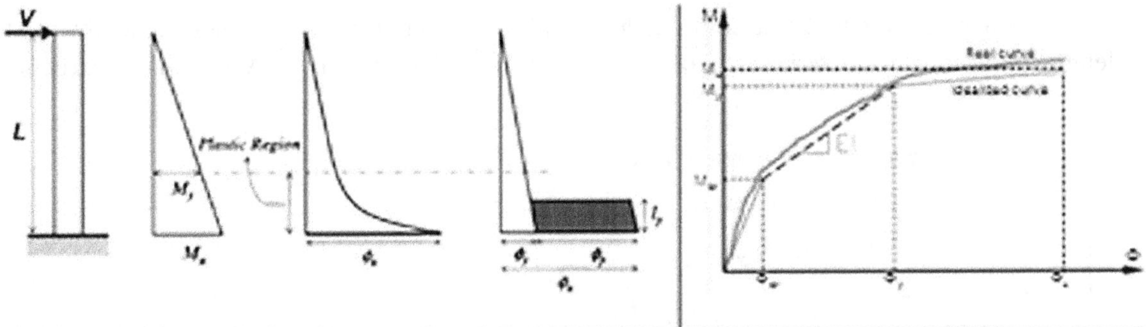

M_{cr} = cracking moment ; M_y = yield moment ; M_u = ultimate moment ; l_p = plastic hinge length ; Φ_{cr} = cracking curvature ; Φ_y = yield curvature ; Φ_p = plastic curvature ; Φ_u = ultimate curvature

rebar experience similar strain and similar modular stress ratios (Warner, 1998). The modular stress ratio is defined as the ratio of the modulus of elasticity of the rebar to that of the concrete. However, when the maximum tensile stress in the cross-section increases to the modulus of rupture (defined as the cracking stress of concrete), the concrete cracks. Assuming that plane cross-section remains plane even after cracking, a linear strain distribution throughout the depth will be developed. Then, alongside the increasing of the bending moment to the yield moment (M_y), the neutral axis shifts towards the compressive face, as shown in Figure 4.

At the last stage when the rebar yields, the curvature increases significantly, while the moment ascends gradually to its maximum value. At that stage, the bending moment can still increase until the ultimate moment (M_u) reaches the ultimate condition (corresponding to the ultimate curvature (Φ_u). Several ultimate conditions can then be considered: a) when the moment capacity of the member drops by 20% (Bae & Bayrak, 2008b); b) when the rebar's tensile strain reaches ultimate (Park. R & Paulay.T, 1975); and c) when the ultimate compressive strain in concrete reaches a limiting value. In most design procedures, limiting values of 0.003 or 0.0035 are conservatively assumed for unconfined reinforcement, after which concrete crushes in compression (Park. R & Paulay.T, 1975). The stress-strain relation of both unconfined and confined concrete has been investigated in numerous studies (Kent & Park, 1971; Scott BD, Park R, & Priestley MJN, 1982; Teng-Hooi & Ngoc-Ba, 2005), some of the more well-known of which are those of Scott et al. (Scott BD et al., 1982), Mander et al. (Mander, Priestley, & Park, 1988) and Priestley et al. (Priestley, Calvi, & Seible, 1996). As an example, Figure 5 shows the concrete stress-strain relation proposed by Scott et al. (Scott BD et al., 1982).

The member will then experience plastic rotation as a result of plasticity developing over a certain length, called the plastic hinge length. This plastic rotation is defined as the difference between the ultimate curvature and the yield curvature multiplied by the plastic hinge length. Several empirical and numerical investigations have been performed to evaluate the equivalent length of the plastic hinge (l_p) (Bae & Bayrak, 2008a; Bayrak & Sheikh, 2001). Of these, Park and Paulay's (Park. R & Paulay.T, 1975) formula is the simplest, as represented by Equation 3, in which H is the section height.

$$l_p = 0.5H \tag{3}$$

After deciding on the method of accounting for the ultimate condition of the concrete and the plastic hinge length, the acceptance criteria can be defined. For example, based on the FEMA356 code, structures

Figure 4. Behavior of reinforced concrete member
Warner, 1998.

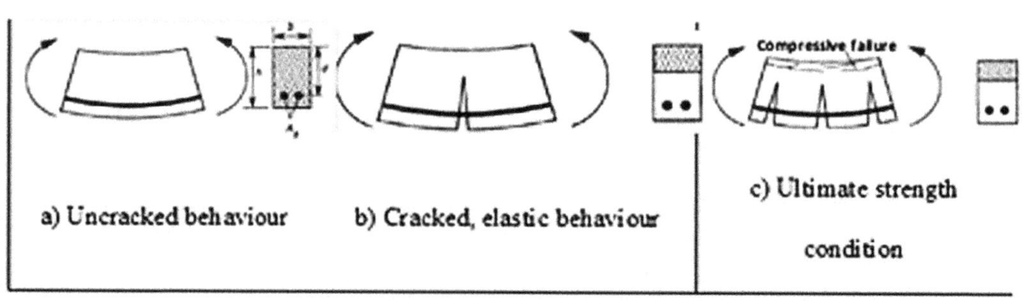

Figure 5. Stress-strain relation for monotonic loading of confined and unconfined concrete
Scott BD et al., 1982.

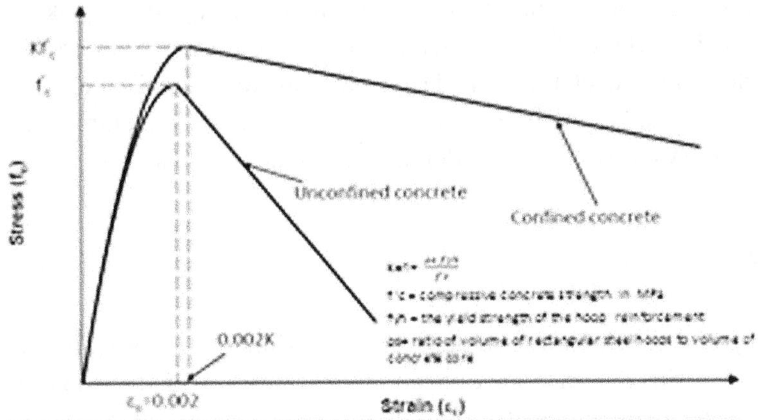

are designed to meet different performance levels such as Operational (O), Immediate Occupancy (IO), Life Safety (LS), and Collapse Prevention (CP). As is seen in Figure 6, at the IO, LS, and CP levels of performance, drift value is limited to 1%, 2% and 4%, respectively.

According to the performance design criteria, the expected performance of structures shall be controlled by assignment of structures to one of several Seismic Use Groups. For example, based on the type of occupancy, there are three "Seismic Use Groups" introduced in FEMA450. Those structures that contain substances which, if released into the environment, are deemed to be hazardous to the public are classified as Group I or special structures. The structures in this special class are supposed to remain in the Immediate Occupancy category when subjected to the design earthquake, and remain operational under all frequent earthquakes. One class lower in performance requirement lie the Group II structures, which contain a large number of occupants and those where the occupants' ability to exit is restrained. Schools, day care centres, and suburban medical facilities are examples of this category.

Figure 6. Drift values at various performance levels in RC structures

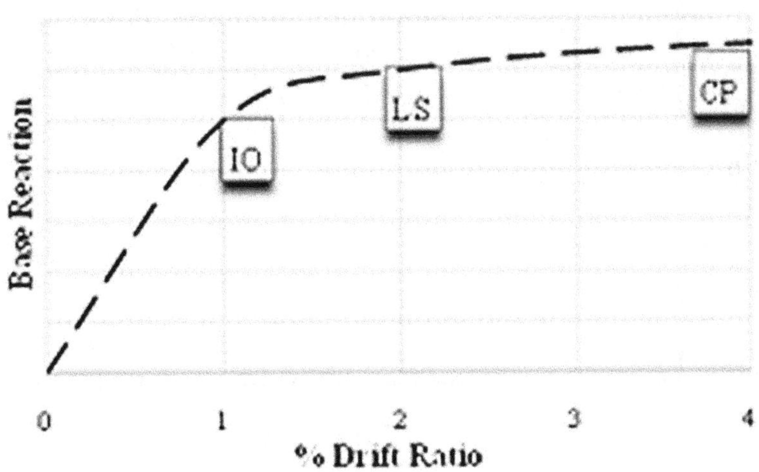

Group III, then, are those ordinary structures such as residential buildings, which shall remain in the Life Safety level if subjected to the design earthquake. The various performance levels required for buildings of different categories can implicitly be met by increasing the design earthquake by a factor called the importance factor. The importance factor basically adjusts the intensity of earthquake in the design so that the required performance level under the design earthquake is met. Therefore, if these performance targets were known, it would be possible to control whether the defined target is met. Figure 7 shows the idealized moment-curvature curve for different performance levels.

FIRE ANALYSIS

A structural fire analysis should include two steps: a sectional analysis in order to account for the distribution of the temperature inside the cross-section, and a structural fire analysis to account for the stresses, displacements, and forces in the elements. To do that, a special non-linear finite element program that allows simulation of the structural behavior under ambient and elevated temperature is needed. It is worth noting that in Eurocode 2, the total strain ε_{tot} is implicitly considered as the sum of free thermal strain (FTS) ε_{th}, mechanical strain $\varepsilon_{m,}$ and possible basic creep strain ε_{cr}, as shown in Equation 4.

$$\varepsilon_{tot} = \varepsilon_{th} + \varepsilon_{m} \left(+\varepsilon_{cr}\right) \tag{4}$$

The total strain shall however be explicitly considered by splitting into FTS, instantaneous stress-related strain, ε_{σ}, and transient creep strain, ε_{tr}, (and possibly basic creep strain, ε_{cr}), as shown in Equation 5.

$$\varepsilon_{tot} = \varepsilon_{th} + \varepsilon_{\sigma} + \varepsilon_{tr} \left(+\varepsilon_{\sigma}\right) \tag{5}$$

Figure 7. Idealized moment-curvature curve for various performance levels
IO *immediate occupancy,* **LS** *life safety,* **CP** *collapse prevention.*

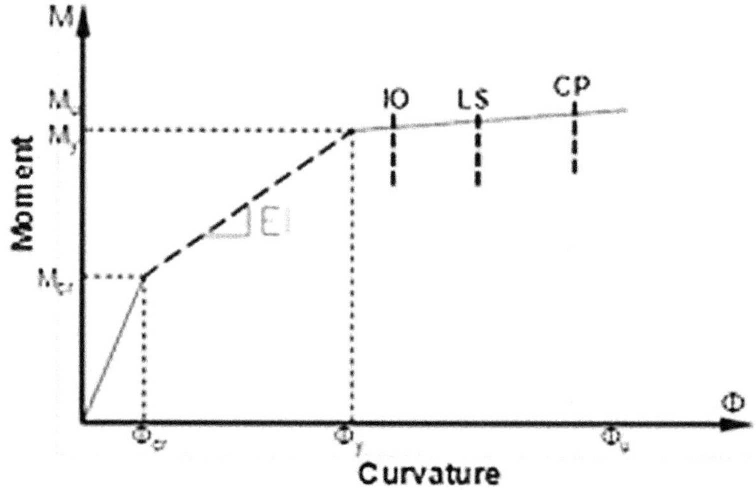

Overall, the structural fire analysis is performed in several steps, which are cross-sectional analysis, torsional analysis (if required) and structural analysis. The first step is thermal analysis, which is performed in order to predict the temperature distribution inside the cross-section. To do so, the thermal characteristics of materials are either manually introduced or chosen from default values already available within the program. The cross-sections are then discretized into several fibres as shown in Figure 8. The fibres are then simulated as two-dimensional solid elements.

In order to reduce the computational effort, fibres can be modelled using iso-perimetric triangle or quadrilateral solids. For two-dimensional models, it might be assumed that no heat is transferred along the longitudinal axis of the beam. Conduction can be considered the main mode of heat transfer in plane and solid sections. However, for those sections with internal cavities, radiation shall also be considered. This is particularly important to consider when hollow-core concrete slabs are to be modelled. The elastic torsional analysis may then be performed for three-dimensional structures with beam elements subjected to warping and at the ambient temperature. The analysis is performed based on the geometry of the members, while cracking in concrete caused by the structural loading is ignored. Hence, the torsional results are more conservative than reality. On the other hand, the elastic torsional stiffness is related to the modulus of elasticity and Poisson's ratio. As both mentioned characteristics decrease as the temperature increases, the elastic torsional stiffness also decreases. Nevertheless, the reduction is normally ignored, mostly to reduce the complexity of the analysis. The results are consequently used for the three-dimensional structural analysis. Finally, the structural analysis is performed in either two or three-dimensional fashion, in order to determine the structural performance subjected to the applied loads (gravity, lateral, and thermal loads). For this, the generated time-temperature profile resulting from the thermal and/or torsional analysis with consideration of the non-linear temperature-dependent mechanical properties is utilized in the structural analysis. Then, using an iterative technique based on the Newton-Raphson method, the equilibrium between external loads and internal stress at each time step are computed, as shown in Figure 9. The iteration will run as long as convergence is achieved. Consequently, the required information, such as displacements at the nodes, axial and bending moment in the elements, stress, strain and tangent modulus of the fibres, can be generated at every time step.

The following assumptions are made during the modelling of the structure:

Figure 8. Element and section discretization
Lee & Mosalam, 2004.

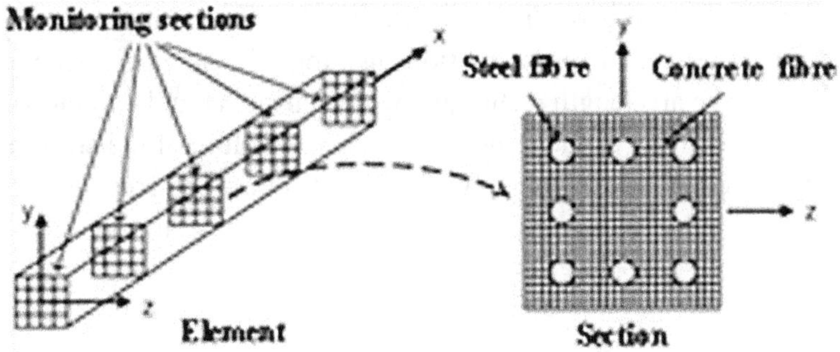

Figure 9. Convergence iterations for the structural analysis
J. M. Franssen, Kodur, & Mason, 2002,

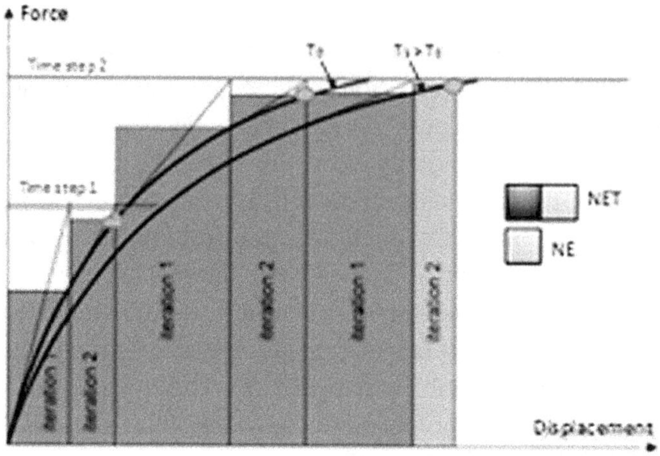

NE: The norm of the energy computed at each iteration

NET: The norm of the total energy computed as the summation of all the NEs

- Based on Bernoulli theory, plane sections remain plane, even after bending. Therefore, a linear distribution of strain throughout the cross-section of the reinforced concrete sections can be assumed (Irschik, 1991).
- For the elements that have experienced plastic deformations under compression, the unloading path is linear, from the maximum compressive strain in the loading curve parallel to the tangent at the origin (Gernay & Franssen, 2011).
- The plastification is considered only in the longitudinal direction of beam elements.
- In the beam elements and three-dimensional structures, non-uniform torsion is considered.
- In the beam elements with steel fibres, local buckling is not considered.

The above-mentioned steps can be done using non-linear software called SAFIR. The software package is capable of performing these steps and has been developed by Franssen at the University of Liege, Belgium (J.-M. Franssen, 2005). Since the software employs a fibre model, spread of plasticity can be considered appropriately. Indeed, unlike the lumped plasticity model, in the fibre element model, the plasticity is spatially distributed both in cross-section and along the member. Another important point to mention is that SAFIR uses the arc-length technique, which ensures that local failure of a single element does not result in the overall failure of the structure. Thus, the failure of the structure in fire becomes a global failure, which is more realistic (Chang, 2007). In addition, as co-rotational procedure (Crisfield, 1981) is used to solve for the large displacements, the swift increase in the displacement can then be tolerated.

SEQUENTIAL ANALYSIS

Modelling of PEF response is a complex process, which involves modelling the response under gravity loads, seismic loads, and thermal loads. The states of the structure under (displacements and stress states) gravity and seismic loads have to be transferred to the fire modelling. Nonetheless, in the absence of suitable simulation techniques, such as computer tools, to perform all the processes successively, substitute solutions may be suggested in a rational manner. Most of the conventional finite elements analysis software packages can perform a structural seismic analysis adequately. The advantage of using these software packages is their conformity with the established seismic codes and standards. Therefore, the results of the analysis can be utilized for interpretation of the structural seismic behavior of buildings that are code-approved, at least for conventional loading situations. When it comes to fire, however, almost none of the software products are capable of performing the analysis. On the other hand, there are some more advanced but complex finite element programs such as ANSYS and ABAQUS with the ability of performing structural analysis subjected to fire and earthquake, albeit with high complexity. This inadequacy is the result of complexity of the phenomenon occurring in fire. For example, in a reinforced concrete building designed to meet the LS level of performance, it is expected that, if the building is subjected to the design earthquake, the cover of the elements will be spalled-off in the vicinity of the created plastic hinges. This, in turn, speeds up the degradation of strength and stiffness of the elements in the subsequent PEF, because it directly exposes the rebars to fire. Sequential analysis is introduced here as a functional method for considering the effects of earthquake and fire on a structure. Figure 10 schematically shows stages of the non-linear sequential analysis.

Figure 10. Stages of the sequential analysis
Behnam, 2014.

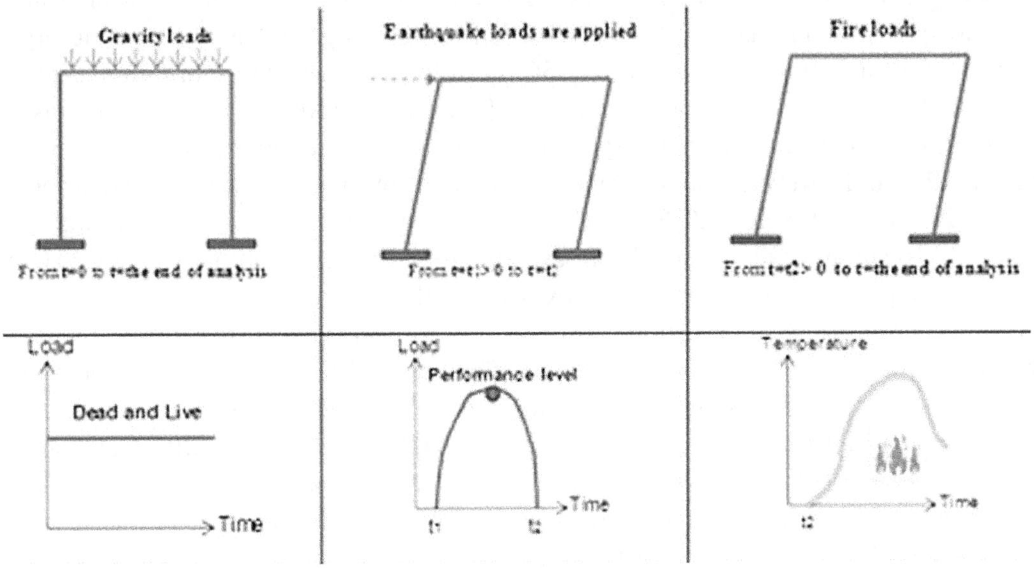

The first stage of loading is the application of gravity loads, consisting of dead and live loads, which are assumed to be static and uniform. Indeed, using a time-load fashion, the gravity loads are applied to the structure from t=0, when the analysis initiates, until the analysis terminates. Simulation of the seismic loads is then performed using a pseudo earthquake load, which can be in a pushover style. The merits of the pushover analysis were mentioned in the previous sections. For the purpose of the study undertaken here, the earthquake loads are applied similarly to the pushover analysis (thus no dynamic effects are considered), with the difference that the structure is unloaded after reaching a certain level of load corresponding to the defined performance level, such as IO, LS or CP. This assumption is in line with the seismic design philosophy in which the performance level of structures shall not exceed the assumed level when subjected to the design earthquake. Therefore, the structures are pushed to these levels and are then unloaded. Clearly, during this time, the application of the gravity loads is continued. Given that long-term effects such as creep and shrinkage are not included, load duration is not important for either gravity or earthquake loads. Finally, the fire loads are applied to the structure, using either standard fire curves or natural fire curves. Prior to fire loading, the properties of the structures have to be set to the reference temperature, but evidently, during fire, the mechanical properties will vary with temperature.

For the structural analysis, including gravity and earthquake loads, the well-accepted finite element program, SAP2000, is employed. Although the program has limited potential for performing the thermal analysis, it is good at handling linear and non-linear seismic analysis, including pushover analysis. For the acceptance criteria, moreover, the program is compatible with most of well-established seismic codes, such as FEMA356, in a way that the formation and state of hinges, in the pushover analysis, can be found at every load step. It is also worth mentioning that lumped (concentrated) plasticity models are used in SAP2000. The hinges are thus defined at the expected locations of plasticity spread and based on the idealized curvature distribution along the structural member. The results of the pushover analysis, including the lateral forces and the state of the created plastic hinges, can therefore be stored in a file.

On the other hand, although SAFIR is a finite element tool specifically written for fire analysis, and considering some of its limitations, it can potentially be employed for performing an analysis including gravity, lateral and thermal loads. Nevertheless, the program has no compatibility with the well-established seismic codes such as FEMA356. Therefore, no acceptance or failure criteria can be defined for the structure, if a seismic analysis is to be performed. Moreover, as cross-sections in SAFIR are discretized into fibres, the distributed type plasticity is considered in the analysis, which makes it different to SAP2000. However, it does not mean that the results are necessarily different too. Behnam and Ronagh (Behnam & Ronagh, 2013) have already compared the two software packages in this respect. Their investigation, which was made on several reinforced concrete frames subjected to pushover analysis with both programs, revealed that in SAFIR only those parts of the elements located adjacent to the joints sustained plastic deformations, while the rest remained elastic throughout the loading history. The locations matched those of the lumped plasticity under SAP2000. Here, in order to perform a sequential analysis, results of the stored pushover analysis by SAP2000 were imported into SAFIR. To do so, a function had to be developed as a linkage between SAP2000 and SAFIR. Clearly, using SAP2000, the state of the created plastic hinges can also be known. While for steel structures, only extracting the pushover forces resulting from SAP2000 and casting in SAFIR seems adequate for the subsequent fire analysis, the states of the potential hinges becomes important in reinforced concrete structures. This is because there is no difference between the fire frontiers of cross-sections of bare steel structures exposed to fire before or after earthquake, while for RC structures, the fire frontiers of the cross-sections may vary before or after earthquake, which, in turn, may speed-up the degradation of strength and stiffness of the members. Indeed, an important distinction to be made for the PEF analysis of RC structures is that the cross-sectional fire frontiers vary depending on the performance criterion used in the design against earthquake. For example, earthquake design for the IO level of performance only allows minor residual displacement, meaning that at the IO level of performance only minor cracking is expected to occur. Minor cracking would impose no significant effect on the position of the fire frontiers (Ervine, Gillie, Stratford, & Pankaj, 2012). However, at LS level of performance, along with some residual deformation and degradation in strength and stiffness, the removal of cover in regions around plastic hinges should also be considered. There are a number of studies to find out the damage state at different performance levels. Apart from FEMA356 prescriptive explanation as pointed out above, there are also some tests. Based on the Japan Building Disaster Prevention Association (JBDPA), Meada et al. (Meada & Kang, 2009) and Nakano et al. (Nakano, Maeda, Kuramoto, & Murakami., 2004) showed in several studies that when a structure sustains severe damage (corresponding to the CP performance level in FEMA356) crushing and spalling of concrete cover with exposed reinforcement is observed. The author has also carefully monitored the results of a test conducted at the laboratory of the University of Queensland on an RC joint in order to find the extent of damage after it is subjected to cyclic loading (Behnam, Ronagh, & Lim, 2015). The test setup is shown in Figure 11 and the results are shown in Figure 12. The monitored observations of the test showed a great similarity between the joint's responses at different performance levels, i.e. IO, LS and CP, with those mentioned by Meada et al. and the perspective definition of damage in FEMA356. The application of the sequential analysis to steel or RC structures at various performance levels is shown in Figure 13.

Figure 11. Test conducted on an RC specimen at the University of Queensland
Behnam et al., 2015.
Note: Arrows show fire frontiers.

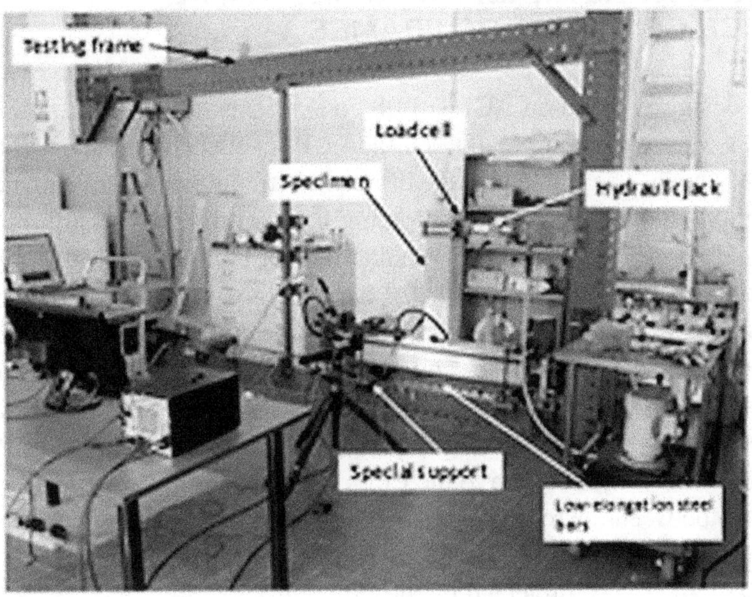

Figure 12. Damage state of the specimen at different performance levels and corresponding the applied fire frontiers
Behnam et al., 2015.

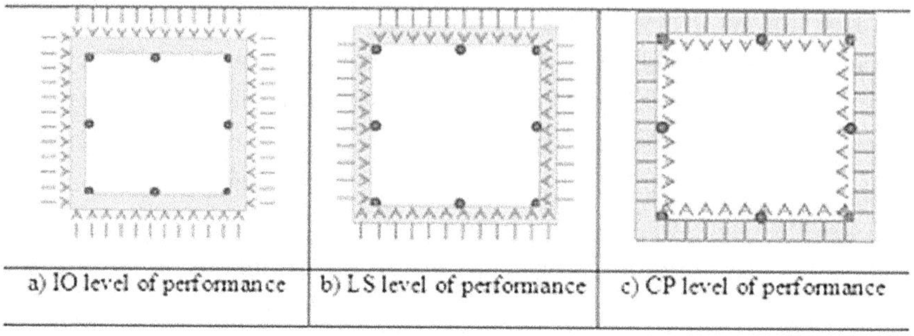

Figure 13. The sequential analysis for performing PEF analysis

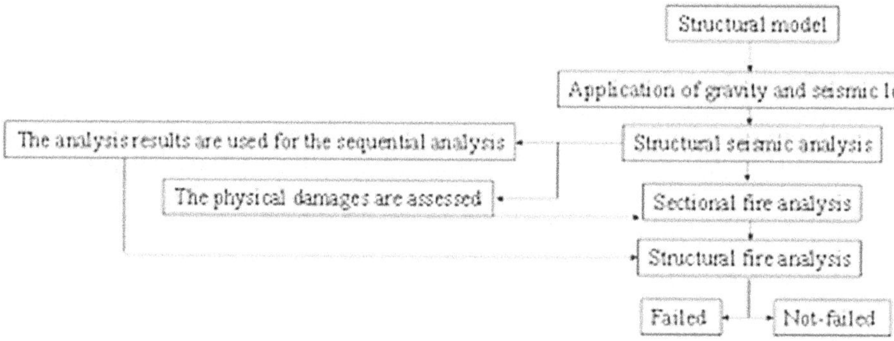

USING THE SEQUENTIAL ANALYSIS TO DETERMINE PEF RESISTANCE OF PORTAL RC FRAMES

A portal RC frame designed based on ACI 318-08 code as shown in Figure 14 is pushed to arrive at different lateral drifts, corresponding to IO, LS, and CP levels of performance, and using pushover analysis as explained above. The frame is made of normal-strength concrete with compressive strength of 25MPa and longitudinal and transverse reinforcing bars with yield stress of 400MPa. The frame is dimensioned for a height of 3.0 m and load combinations of 8.0kPa for dead load and 2.5kPa for live load. The combination of 100% dead load and 20% live load is used to find the required mass for calculating the earthquake load. Furthermore, the frame is exposed to ISO834 fire curve and three different situations of fire: only beams, only columns and both beam and columns. For the thermal analysis, it is assumed that the concrete moisture content is 20 kg/m3. Moreover, the thermal expansion coefficient of rebar and concrete are assumed to be 12×10^{-6} /°C and 10×10^{-6} /°C, respectively. Poisson's ratio of 0.2 is considered for the concrete. In addition to the PEF analysis, the fire-alone analysis is also performed to make a comparison between the results. The exterior side of the columns and top side of the beam is not exposed to fire. This is because; it is assumed no external fire occurs during the analysis. The top side of the beam is assumed to be protected by the concrete slab.

Figure 14. Geometric properties of the case study

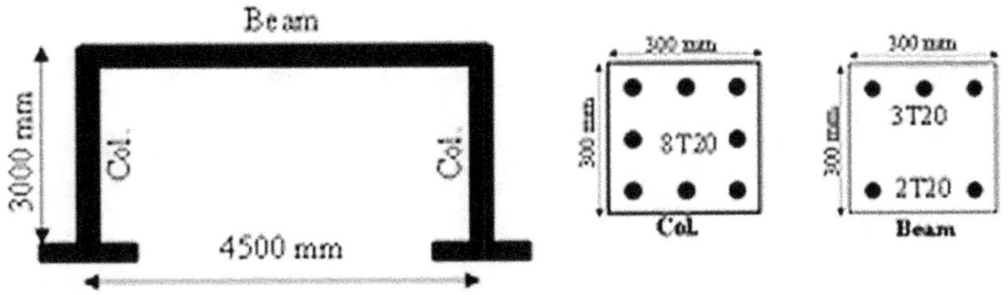

The sequential analysis comprises three main stages, which are gravity loading, followed by seismic pushover analysis, and finally PEF. In the seismic analysis, the structure is subjected to a monotonically increasing lateral load to meet the specified performance levels. Accordingly, three different levels of performance, i.e. IO, LS, and CP, are met after the pushover analysis. In this respect, the SAP2000 program is employed to perform the nonlinear pushover analysis. Furthermore, FEMA356 procedure is used to define the hinges for the nonlinear pushover analysis. The lateral forces corresponding to the target displacement at every performance level are extracted from the SAP2000 program, and are then input to the SAFIR program for performing the sequential analysis. Final stage of the sequential analysis is to apply PEF to the frame. As mentioned, several scenarios are used for the fire analysis; in one case, the undamaged frame is subjected to fire loading, while in the second case, the damaged frame is exposed to fire load. In this way, in the first case, the fire load follows the gravity loads, but in the second case, the fire load follows gravity and earthquake loads. Figure 15 shows the temperature distribution in the column at different performance levels.

Figures 16 and 17 show displacement against time for the case study, which implies the fire resistance of the frames in seconds for both scenarios (fire and PEF). The fire resistance is defined as the time at which the displacements, either globally (i.e. the drift of a certain point) or locally (i.e. the deformations at the middle of a beam), go beyond chosen thresholds. The thresholds have been identified by the curve for displacements versus time step merging towards a vertical asymptote by 1% error. These thresholds implicitly represent the definition of fire resistance of a member as described earlier, where

Figure 15. Distribution of temperature in the column at various performance levels
Ronagh & Behnam, 2012.

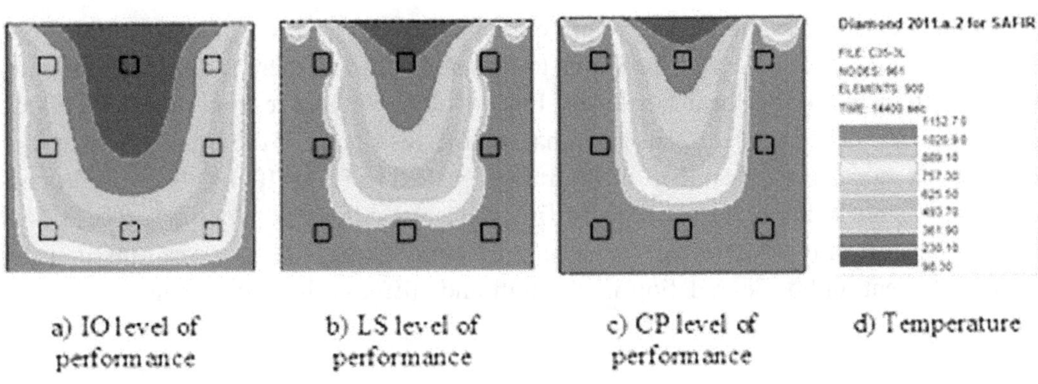

Figure 16. Fire resistance of the case study when members are separately exposed to fire
Ronagh & Behnam, 2012.

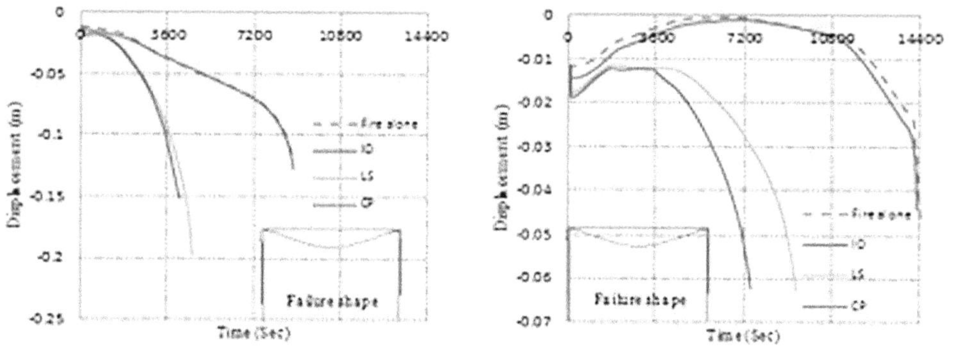

a) Only the **beam** exposed to fire b) Only the **columns** exposed to fire

IO: Immediate Occupancy , LS: Life Safety , CP: Collapse Prevention

Figure 17. Fire resistance of the case study when all the members exposed to fire
Ronagh & Behnam, 2012.

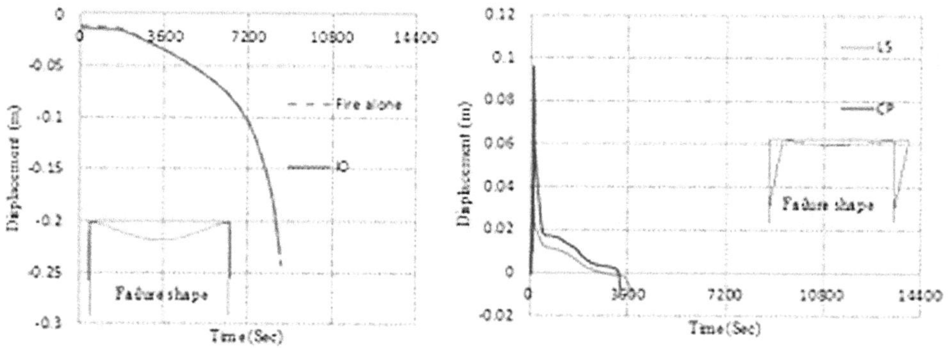

a) Fire alone and IO level b) LS and CP level

IO: Immediate Occupancy , LS: Life Safety , CP: Collapse Prevention

the member is not able to resist the initially applied gravity loads (Kodur & Dwaikat, 2007). As is seen in the figures, regardless of subjecting the frame to fire alone or PEF, there is a correlation between the fire-resistance rating and the performance levels. Indeed, along with increasing the lateral displacement in the frame, the fire resistance decreases such that the fire resistance of the frames pushed to CP level of performance is much lower than that of the frames pushed to LS or IO levels of performance. The figures also show a minor difference between the fire resistance at IO level of performance and fire alone. That is mostly because at IO level of performance, only minor damage occurs, resulting in insignificant residual displacement and/or degradation in strength and stiffness. It is also seen that fire resistance declines considerably when only the beam is exposed to fire, compared to exposing the columns to fire. In other words, it seems that the beam is more sensitive to fire than the columns; the reason of which can be correlated to the strong column-weak beam concept, which is highly advised to meet for seismically designed structures. Interestingly, Figures 16 and 17 show that there is similarity in fire resistance when all members are exposed to fire and when only the beam is exposed to fire. This implies that the fire resistance of a frame is mostly dependent on the fire resistance of the beams.

In the figures, the shape of failure is also shown. As is seen, two types of failure are observed; local and global. While local collapse depends largely on the collapse of beams (Figures 16a-b and Figure 17-a), global collapse is mainly governed by considerable lateral displacement of the columns (Figure 17-b). It is evident that the frame fails locally in case of fire alone and the IO level of performance. However, it fails globally when pushed to LS or CP levels of performance. The sharp increase and then decrease in Figure 17-b represents the displacement resulted from lateral pushover.

SUMMARY

Simulation of PEF in a structure is performed in three steps: i) application of the gravity loads; followed by ii) application of the seismic loads; and then iii) application of the fire loads. The simulation of seismic loads can be made through static and dynamic methods, which may be linear or non-linear. While static linear methods are simpler to use, their results cannot be used entirely to demonstrate the inelastic deformation or damage in structures. Dynamic non-linear methods, by contrast, are much more complex but provide more information on the inelastic deformations. Pushover method, being one such method, allows a good understanding of the state of plasticity in the structure while it is still simple enough to be used for everyday designs. The use of pushover analysis has been made here for the seismic evaluation portion of the sequential analysis. Following this, it is necessary to transfer the end geometrical configuration and the resulting damage to the beginning of the fire analysis and continue with the analysis. It is important to note that the materials' mechanical characteristics vary as temperature starts to rise. Therefore, the computer tool has to have the ability of altering the materials' characteristics when the structure is exposed to thermal loads. The geometry of fire frontiers of cross-sections of the members may also change in an earthquake-damaged structure. While usually no change is made between the steel cross-sections before or after an earthquake, the same assumption cannot be made for the cross-sections of reinforced concrete members. As a specific case, at the LS and the CP levels of performance, it is expected that cover of the members will be removed in the vicinity of the plastic hinges. Hence, the computer tool should also have the ability to change the fire frontiers of the cross-sections, if needed. Sequential analysis is a functional method for consideration of the effects of earthquake and fire on a structure. For the purpose of this project, the structural analysis (including gravity and seismic loads) is firstly performed by SAP2000. The results of the pushover analysis are then stored in a file to be used for the next stage, PEF analysis. For the fire part, SAFIR, a well-accepted finite element-based program, is employed. Results of the pushover analysis are cast in SAFIR (by a written function) in order to perform the sequential analysis. Meanwhile, fire frontiers of the cross-sections of the structural members are also defined. The fire resistance of the structure is consequently determined.

REFERENCES

Bae, S., & Bayrak, O. (2008a). Plastic Hinge Length of Reinforced Concrete Columns. *ACI Structural Journal, 105*(3), 290–300.

Bae, S., & Bayrak, O. (2008b). Seismic Performance of Full-Scale Reinforced Concrete Columns. *ACI Structural Journal, 105*(2), 123–133.

Bayrak, O., & Sheikh, S. A. (2001). Plastic hinge analysis. *Journal of structural engineering (New York, N.Y.), 127*(9), 1092–1100.

Behnam, B. (2014). Post-Earthquake Fire Analysis and Risk Mitigation Strategies. (PhD), the University of Queensland, Brisbane, Australia. doi:10.14264/uql.2014.122

Behnam, B., & Ronagh, H. (2014). Performance-Based Vulnerability Assessment of Multi-Story Reinforced Concrete Structures Exposed to Pre- and Post-Earthquake Fire. *Journal of Earthquake Engineering, 18*, 853–875.

Behnam, B., Ronagh, H. R., & Lim, P. J. (2015). Numerical evaluation of the post-earthquake fire resistance of CFRP-strengthened reinforced concrete joints based on experimental observations. *European Journal of Environmental and Civil Engineering*, 1-19. doi:10.1080/19648189.2015.1018448

Chang, J. J. (2007). *Computer simulation of hollowcore concrete flooring systems exposed to fire. (Doctor of Philosophy research)*. Christchurch, New Zealand: University of Canterbury.

Council, B. S. S., Agency, U. S. F. E. M., & Council, A. T. (2003). *NEHRP guidelines for the seismic rehabilitation of buildings*. Washington, D.C.: Federal Emergency Management Agency.

Crisfield, M. A. (1981). A fast incremental/iterative solution procedure that handles "snap-through". *Computers & Structures, 13*(1-3), 55–62. doi:10.1016/0045-7949(81)90108-5

Ervine, A., Gillie, M., Stratford, T. J., & Pankaj, P. (2012). Thermal propagation through tensile cracks in reinforced concrete. *Journal of Materials in Civil Engineering, 24*(5), 516–522. doi:10.1061/(ASCE)MT.1943-5533.0000417

Fajfar, P. (1999). Capacity spectrum method based on inelastic demand spectra. *Earthquake Engineering & Structural Dynamics, 28*(9), 979–993. doi:10.1002/(SICI)1096-9845(199909)28:9<979::AID-EQE850>3.0.CO;2-1

Fajfar, P. (2000). A Nonlinear Analysis Method for Performance Based Seismic Design. *Earthquake Spectra, 16*(3), 573–592. doi:10.1193/1.1586128

FEMA356. (2000). *Prestandard and commentary for the seismic rehabilitation of buildings Rehabilitation Requirements*. Washington, DC: American Society of Civil Engineers.

Firat Alemdar, Z. (2010). *Plastic hinging behavior of reinforced concrete bridge columns* [Ph.D. Thesis]. University of Kansas, Kansas, USA. Retrieved from http://search.proquest.com/docview/305210088?accountid=14723

Fitzpatrick, F. (1914). *Fireproof construction*. Chicago: American School of Correspondence.

Franssen, J.-M. (2005). SAFIR: A thermal/structural program for modeling structures under fire. *English Journal, 42*(3), 143–150.

Franssen, J. M., Kodur, V. K. R., & Mason, J. (2002). *Elements of Theory for SAFIR2001 Free: A Computer Program for Analysis of Structures at Elevated Temperature Conditions*. Belgium: University of Liege.

Gernay, T., & Franssen, J. M. (2011). *SAFIR MANUAL Materials SILCON_ETC and CALCON_ETC. University of Liege*. Belgium: Environnement & Constructions Structural Engineering.

Irschik, H. (1991). Analogy between refined beam theories and the Bernoulli-Euler theory. *International Journal of Solids and Structures, 28*(9), 1105–1112. doi:10.1016/0020-7683(91)90105-O

Kalkan, E., & Kunnath, S. K. (2007). Assessment of current nonlinear static procedures for seismic evaluation of buildings. *Engineering Structures, 29*(3), 305–316. doi:10.1016/j.engstruct.2006.04.012

Karthik, M. M., & Mander, J. B. (2011). Stress-block parameters for unconfined and confined concrete based on a unified stress-strain model (Technical Notes, Author abstract). *Journal of structural engineering (New York, N.Y.), 137*(2), 270.

Kent, D. C., & Park, R. (1971). Flexural members with confined concrete. *Journal of the Structural Division, 97*(ST7), 1969–1990.

Kodur, V. K. R., & Dwaikat, M. (2007). Performance-based Fire Safety Design of Reinforced Concrete Beams. *Journal of Fire Protection Engineering, 17*(4), 293–320. doi:10.1177/1042391507077198

Krawinkler, H., & Seneviratna, G. D. P. K. (1998). Pros and cons of a pushover analysis of seismic performance evaluation. *Engineering Structures, 20*(4–6), 452–464. doi:10.1016/S0141-0296(97)00092-8

Kwak, H.-G., & Kim, S.-P. (2002). Nonlinear analysis of RC beams based on moment–curvature relation. *Computers & Structures, 80*(7-8), 615–628. doi:10.1016/S0045-7949(02)00030-5

Lee, T.-H., & Mosalam, K. M. (2004). Probabilistic fiber element modeling of reinforced concrete structures. *Computers & Structures, 82*(27), 2285–2299. doi:10.1016/j.compstruc.2004.05.013

Mander, J. B., Priestley, M. J. N., & Park, R. (1988). Theoretical Stress-Strain Model for Confined Concrete. *Journal of structural engineering (New York, N.Y.), 114*(8), 1804–1826.

Meada, M., & Kang, D. (2009). Post-earthquake damage evaluation of RC buildings. *Journal of Advanced Concrete Technology, 7*(3), 327–335. doi:10.3151/jact.7.327

Nakano, Y., Maeda, M., Kuramoto, H., & Murakami, M. (2004). Guidelines for post-earthquake damage eveluation and rehabilitation of RC buildings in Japan. *Paper presented at the13th World Conference on Earthquake Engineering*, Vancouver, Canada.

Papanikolaou, V. K., & Elnashai, A. S. (2005). Evaluation of conventional and adaptive pushover analysis i: methodology. *Journal of Earthquake Engineering, 9*(6), 923–941. doi:10.1080/13632460509350572

Park, R., & Paulay, T. (1975). *Reinforced concrete structures*. New York: John Wiley & Sons. doi:10.1002/9780470172834

Priestley, M. J. N., Calvi, G. M., & Seible, F. (1996). *Seismic design and retrofit of bridges*. New York: Wiley. doi:10.1002/9780470172858

Roh, H., Reinhorn, A. M., & Lee, J. S. (2012). Power spread plasticity model for inelastic analysis of reinforced concrete structures. *Engineering Structures*, *39*(0), 148–161. doi:10.1016/j.engstruct.2012.01.019

Ronagh, H. R., & Behnam, B. (2012). Investigating the Effect of Prior Damage on the Post-earthquake Fire Resistance of Reinforced Concrete Portal Frames. *International Journal of Concrete Structures and Materials*, *6*(4), 209–220. doi:10.1007/s40069-012-0025-9

Scott, B. D., Park, R., & Priestley, M. J. N. (1982). Stress-Strain Behavior of Concrete Confined by Overlapping Hoops at Low and High Strain Rates. *Journal of the American Concrete Institute*, *79*, 13–27.

Teng-Hooi, T., & Ngoc-Ba, N. (2005). Flexural Behavior of Confined High-Strength Concrete Columns. *ACI Structural Journal*, *102*(2), 198–205.

Tirca, L., & Chen, L. (2012). The influence of lateral load patterns on the seismic design of zipper braced frames. *Engineering Structures*, *40*, 536–555. doi:10.1016/j.engstruct.2012.03.017

Warner, R. F. (1998). *Concrete structures*. South Melbourne, Vic.: Longman.

Chapter 16
Structural Integrity Assessment and Control of Ageing Onshore and Offshore Structures

R.M. Chandima Ratnayake
University of Stavanger, Norway

S.M. Samindi Samarakoon
University of Stavanger, Norway

ABSTRACT

Structural integrity assessment and control (SIA & C) are vital for existing ageing as well as newly built offshore and onshore structures. The SIA & C becomes highly sensitive to interventions under a potential loss of structural integrity (SI) especially when there are inherent constraints present in carrying out engineering work in hazardous environments such as petroleum production and process facilities (P&PFs). The challenges have been further exacerbated by the constantly ageing onshore and offshore structures whilst it is the necessity of carrying out life extension at the verge of their design service lives. Local and international regulations demand the implementation of appropriate strengthening, modification and repair (SMR) plans when significant changes in the SI have been revealed. This chapter presents potential simulation approaches that are possible to utilize for SIA&C in relation to prioritization as well as evaluation of current status of a SIA&C organization.

INTRODUCTION

Structural integrity assessment and control (SIA&C) is an increasingly important element of a plant engineer's role (Bea, 1998; Samarakoon, Lee & Miles, 2013; Ratnayake & Siriwardane, 2013) and becomes vital when a structure has been undergone different loading circumstances for a significant amount of time and/or when it has been in operation beyond its designed service life (COMAH, 2010, Bea, 1998, Engineerlive, 2013, Val and Stewart, 2005). In this context, the SI is defined as "all of the structural aspects necessary to enable an industrial installation to function in accordance with stated duty via protecting health, safety, environment and quality (HSE&Q) performance requirements" (HSE,

DOI: 10.4018/978-1-5225-0588-4.ch016

2009). SI inherently deteriorates due to 'ageing' of the existing physical assets (or plant). However, the term 'ageing plant' misleads up to a certain extent as it is not related to the age and design life of a plant, system, part of the structure/structural element, or piece of equipment (HSE UK, 2010). The term 'ageing' or 'ageing plant' is described as (HSE UK, 2006): Ageing is not about how old the structure or structural element under the consideration is, where, it is about its present condition, and how it has been changing over time from the date of the installation. Consequently, ageing is the effect whereby a part of the structure or structural element under the consideration suffers some form of material deterioration and damage (usually, but not necessarily, associated with time in service) with an increasing likelihood of failure over the lifetime (HSE UK, 2006). In this context, the ageing structural element is an element "for which there is evidence or likelihood of significant deterioration and damage taking place since new, or for which there is insufficient information and knowledge available to know the extent to which this possibility exists" (HSE UK, 2006). Moreover, "the significance of deterioration and damage relates to the potential effect" on the "structural element's functionality, availability, reliability and safety". For instance, "just because an item of equipment is old does not necessarily mean that it is significantly deteriorating and damaged". All types of structural elements are susceptible to ageing mechanisms. Hence, "ageing plant is a plant which is, or may be, no longer considered fully fit for purpose due to deterioration or obsolescence in its integrity or functional performance. 'Ageing' is not directly related to chronological age. There are many examples of very old plant remaining fully fit for purpose, and of recent plant showing evidence of accelerated or early ageing, e.g. due to corrosion, fatigue or erosion failures" (HSE UK, 2006). Hence, the ageing has direct correlation between integrity assessment and control approaches. Figure 1 illustrates inherent structural integrity declination vs. the result of *ad hoc* improvements in an isolated fashion (Ratnayake, 2012c).

Figure 1. Inherent structural integrity declination vs. improvements in an isolated fashion

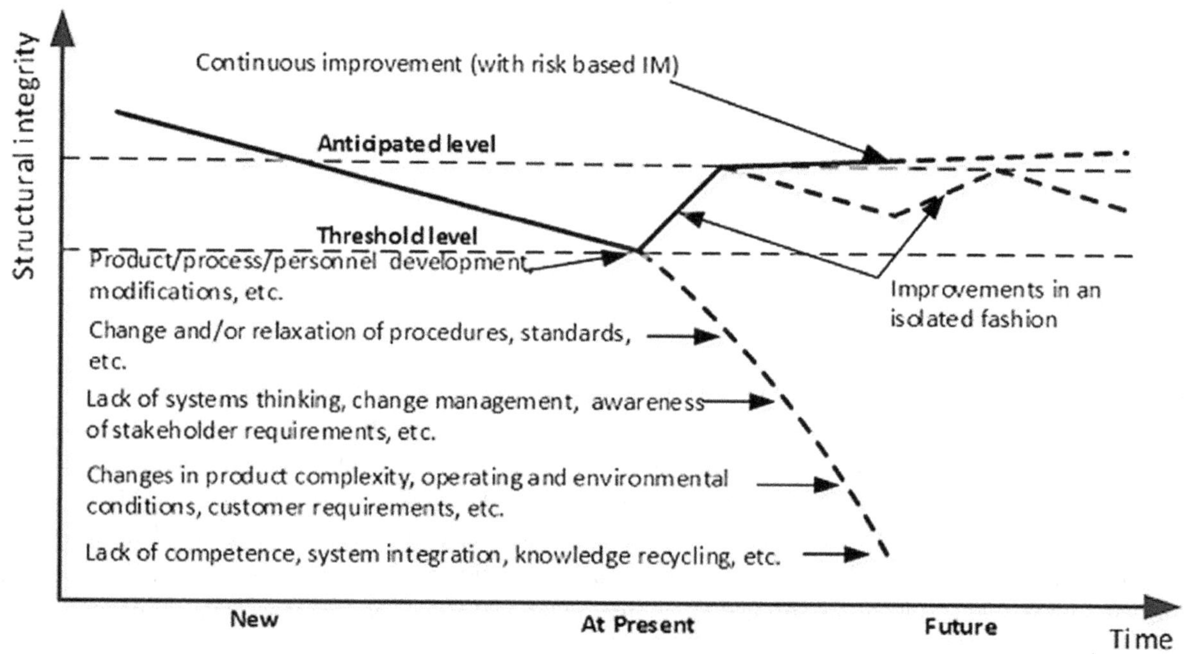

In order to cater the SIA&C, it is vital to understand the interfaces of a physical assets in a plant to other critical kinds of assets such as human, information, intangible and financial assets. The interfaces of physical assets to other critical kinds of assets are illustrated in Figure 2 (PAS 55-1, 2004).

The role of SIA&C is to ensure that the human, information, and intangible assets (i.e. people, systems, processes and resources) that deliver SI (i.e. “the ability of a structure to perform its required function effectively and efficiently over a defined time period whilst protecting health, safety, environment” and quality) are in place, in use and will perform when required over the whole lifecycle of the structure (HSE, 2009).

The SIA&C of ageing assets is conventionally a challenging task, which requires various factors to be dealt with (i.e. damage, consequences, repair methods, safety, confidence, people, etc.) (Bea, 1998; Sorensen, 2002). Moreover, confirmation of the provision of scientific facts that ageing assets are fit for service results in a positive impact on health, safety and the environment (Samarakoon & Ratnayake, 2013). In other words, assessment of the SI of ageing structures helps to identify the potential restoration (i.e. strengthening, modification and repair) of elements within those structures (HSE UK, 2006; Dier, 1996). In this context, strengthening, modification and repair (SMR) provides a backbone, ensuring SI and continued operation whilst assuring the HSE&Q of industrial installations. However, SMR requires rather different skills (e.g. welding, grouting, structural analysis, etc.) and knowledge, which are considered as specialists’ work (Dier, 1996; MSL, 2004; Etterdal, Askheim & Gringorian, 2001). For instance, in the case of structures in offshore P&PPs, it is vital to plan SMR to optimize the high expenditure associated with offshore work. To date, deployed SMR schemes are inappropriate, unnecessary or expensive, mainly due to the lack of skills/knowledge (Dier, 1996; Scherf, Etterdal & Monshaugen,

Figure 2. Physical assets vs. other critical kinds of assets

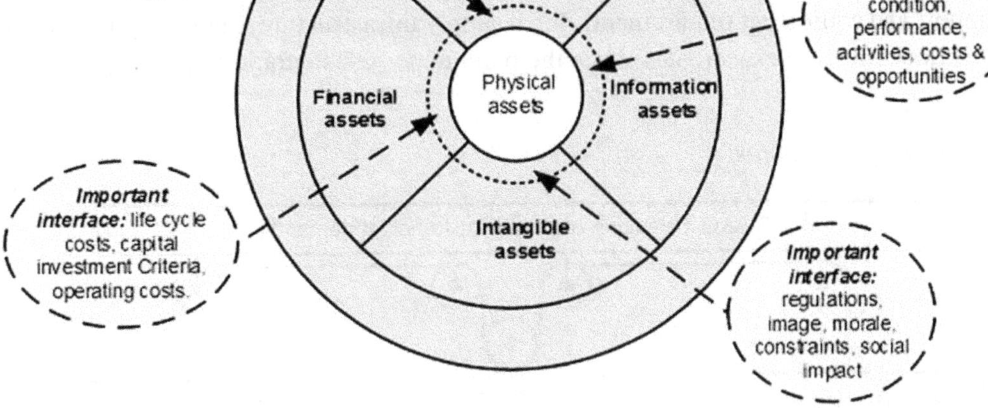

1999). Hence, the selection of optimal SMR schemes and individual SMR techniques is important for reducing in-service inspection.

Normally, a structural assessment is carried out to ascertain the requirement of SMR schemes. The structural assessment includes both platform analysis (to establish member loads or the resistance of the whole platform to design events) and code or other checks on the (intact or damaged) component capacity (Ersdal, 2005; Dong, Moan & Gao, 2012). Once the necessity for an SMR scheme/individual techniques is confirmed, then a decision on optimal SMR techniques could be based on criteria such as technical performance, reliability, costs, depth limitations, offshore support requirements, existing applications, extent of background knowledge, timescales for design/fabrication/installation, tolerance acceptability, post-installation inspection requirements, potential problem areas, remaining life of installation, environmental and other legislative requirements, and operator preferences (MSL, 2004; Val & Stewart, 2005; Este & Frangopol, 2001). However, the selection of an inappropriate method results in wastage of time and adverse effects on environmental, economic, health and safety aspects (Ratnayake & Markeset, 2010; Sorensen, 2002). Therefore, it is essential to make the right decision on SMR techniques in a selection process with the aid of expert judgments.

First part of this chapter illustrates the use of analytic hierarchy process (AHP) approach for the selection of the optimal SMR techniques to cut down wastage in terms of cost, time and adverse effects on environmental, economic, health and safety aspects (Bevilacqua & Braglia, 2000; Zio, 1996). The AHP enables simulating SIA&C related criteria along a hierarchy for developing a consensus for making balanced decisions considering different stakeholders' demands (Saaty, 1990; Ratnayake, 2012a, 2012b). Figure 3 illustrates notion of balanced physical assets' performance in relation to financial and HSE&Q performance (Ratnayake, 2012d).

Second part of this chapter illustrate the evaluation of the current status of a SIA&C organization via fuzzy logic based simulation.

BACKGROUND: SIA&C CHALLENGE

For onshore and offshore structures, the selection of the optimal SMR is an important aspect in ensuring the safe continuation of the operation of production and process plants. However, the selection process remains a challenge as it requires knowledge of the available range of techniques, including variants, their strengths and weaknesses, and an appreciation of other factors such as ease of design, buildability, offshore support and equipment requirements, local supply infrastructure, regulatory requirements, etc. The aforementioned further exacerbates when the plant under consideration operating in geographical

Figure 3. Balanced performance

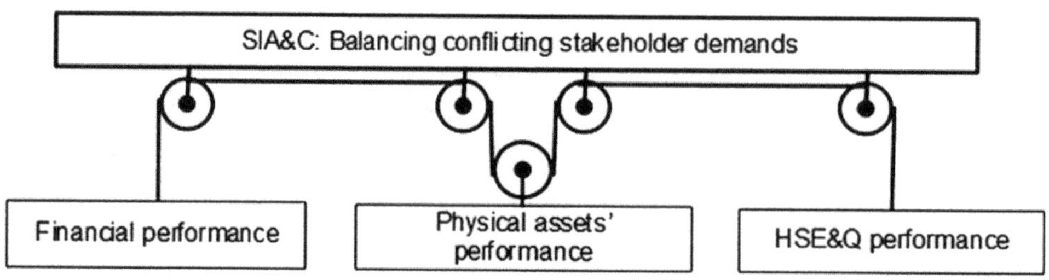

areas (e.g. Norwegian Continental Shelf) which requires satisfying several stringent requirements during the commissioning, maintenance-modification-and-operation (MMO) and life extension (i.e. defined in this context as "the demonstration of safe operation of the structure to the end of the extended anticipated operating life" (HSE, 2009) of a plant-structure which is at the verge of pre-determined design life (Ratnayake & Markeset, 2010c). In this context, the assessment, restoration and control of SI for an extended lifetime of a plant-structure is also an essential task. To restore the SI, structural assessments of structures and/or structural elements are performed to identify the requirement for SMR (HSE UK, 2006; Dier, 1996). Hence, SMR plays a vital role in ensuring safe operations for the extended duration of the life time of overall plant or selected structural elements. Figure 4 illustrates the process for meeting the SI requirements for life extension of ageing plants (HSE, 2009).

In the context of life extension of a structure or structural element, the corresponding structure needs to be reanalyzed in relation to the SMR scheme/technique (i.e. when a local or global SMR scheme/technique has been selected) to ascertain whether the required level of SI would be reached or not. In essence, SMR techniques are classified into welding (e.g. dry or wet welding), use of clamp technology (e.g. use of mechanical clamp, grouted clamp/sleeve), grout filling of members or joints, weld improvements (e.g. toe grinding) or other techniques (e.g. members' removal) (Dier, 1996; MSL, 2004). Figure 5 illustrates the currently used SMR techniques for ageing plant-structures.

Figure 4. Process for meeting the SI requirements for life extension
HSE, 2009.

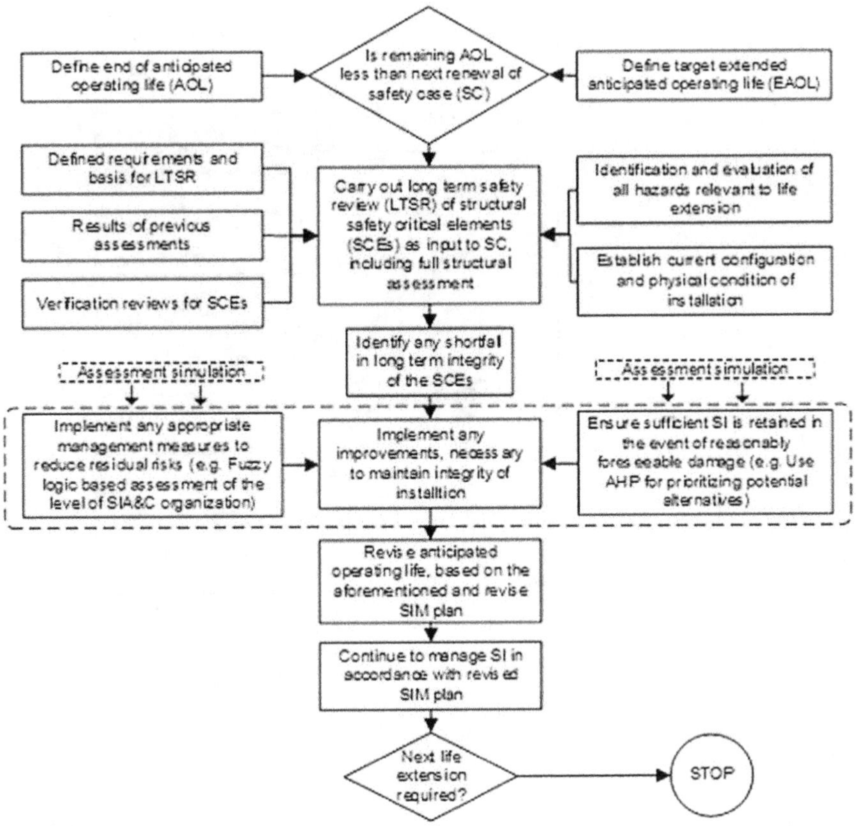

Figure 5. Available SMR techniques

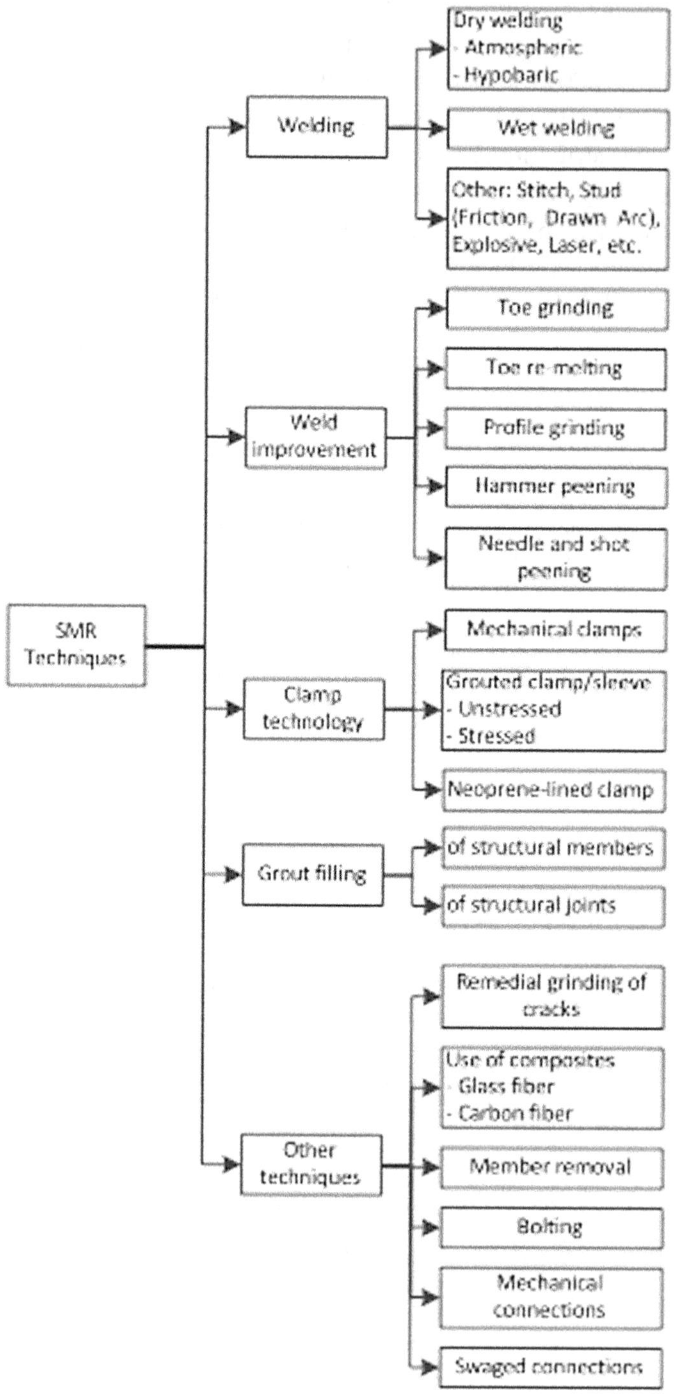

It is also important to take into account various criteria such as technical performance, reliability, costs, depth limitations, offshore support requirements, existing applications, extent of background knowledge, timescales for design/fabrication/installation, tolerance acceptability, post-installation inspection requirements, potential problem areas, remaining life of installation, environmental and other legislative requirements, and operator preferences, etc. However, this poses a significant challenge for practicing engineers to make such decisions by taking both quantitative and qualitative information into consideration. For instance, Table 1 compares the different SMR techniques and their installation cost, time taken for installations, etc. Table 2 indicates the potential applicability of SMR techniques for the selected defect scenario.

Interrelationship of SMR Techniques

The evaluation of the SI and fitness-for-purpose of a structure is vital to identify the extent of SMR required. Essentially, it is necessary to identify whether a 'local' SMR will suffice, or whether a more elaborate 'global' solution is required. Basically, a local SMR offers a cheaper solution, provided that it meets the technical and nontechnical requirements. In this context, there are five basic approaches to SMR:

1. Remove damage;
2. Reduce loadings;
3. Local strengthening/modification/repair;
4. Global strengthening/modification/repair by provision of new members;
5. Total strengthening/modification/repair by tying into a new adjacent structure.

However, "It is prudent to consider a number of SMR techniques to ensure that the chosen solution is the best from a technical, operational and economic standpoint, even though a particular technique may appear an obvious choice" (MSL, 2004). The possible techniques were listed in Figure 1. Table 1 illustrated salient information on each technique, whilst Table 2 illustrated the potential applicability of SMR techniques for selected defect scenarios. Figure 6 illustrates the interrelationship between commonly occurring scenarios and the SMR techniques which it is possible to employ on offshore structures.

Among the SMR approaches (see Tables 1 and 2), welding repairs, mechanical clamping, grouted repairs, and member removal or reduction of loading plays a significant role when it comes to ageing structures under routine operations. The aforementioned SMR approaches have to be prioritized in relation to: applicability; cost of design; fabrication and installation, timescales for design, fabrication and installation; technical performance; reliability; as well as legislative requirements. Hence, AHP approach is proposed to make the prioritization. Figure 7 illustrates the SIC approach suggested in this chapter and the role of AHP.

Aside from the level of reliability, it is also important to take into consideration technical and non-technical factors (such as applicability, cost of design/fabrication/installation, timescales for design/fabrication/installation, technical performance, legislative requirements, static strength/fatigue strength, and tolerance applicability) in the evaluation of structural integrity (SI) and fitness-for-purpose (Wang, Rabiei, Hurtado, Modarres & Hoffman, 2009; UN, 2014). To fulfill the aforementioned, it is necessary to employ AHP.

Table 1. Comparison of SMR techniques

Technique	Used Offshore	Data Available for		Equipment Needs	Offshore Installation Timescales	Onshore Fabrication Costs	Load Penalties		Relative Post-Installation Inspection Requirements	Design Guidance Available
		Static Strength	Fatigue Strength				Weight	Wave Load		
Dry welding	Yes	Yes	Yes	Heavy	Very slow	High for habitat	None	None	Low	Yes
Wet welding	Yes	Yes	No	Moderate	Quick	None	None	None	Low	Yes
Toe grinding	Yes	N/A	Yes	Low	Moderate	None	None	None	Moderate	Yes
Remedial grinding	Yes	Yes	Yes	Low	Moderate	None	None	None	Moderate	Yes
Hammer peening	Yes	N/A	Yes	Low	Quick	None	None	None	Moderate	Yes
Stressed mechanical clamps	Yes	Yes	Yes	Moderate	Moderate	High	Moderate	High	High	Yes
Unstressed grouted connections	Yes	Yes	Yes	Moderate	Moderate	Low	Low	Low	Low	Yes
Unstressed grouted clamps without shear keys	Yes	Yes	Yes	Moderate	Moderate	Moderate	Moderate	Moderate	Moderate	Yes
Unstressed grouted clamps with shear keys	Yes	Yes	Yes	Heavy	Slow	Moderate	Moderate	Moderate	Moderate	Yes
Stressed grouted clamps	Yes	Yes	Yes	Moderate	Slow	High	Moderate	High	High	Yes
Elastomer-lined clamps	Yes	Yes	Yes	Moderate	Moderate	High	Moderate	High	High	Yes
Pressurized connections	No	Yes	No	Light	Slow	Moderate	Low	Low	Low	No
Grout filling members	Yes	Yes	No	Light	Quick	Low	High	None	Low	No
Grout filling joints	Yes	Yes	No	Light	Quick	Low	High	None	Low	No
Bolting	Yes	Yes	Yes	Light	Moderate	Low	Low	Low	Moderate	Yes
Member removal	Yes	N/A	N/A	Moderate	Quick	None	None	None	None	N/A
Adhesives	Yes	Yes	Yes	Light	Quick	Low	Low	Low	Low	No
Composites	Yes	Yes	Yes	Light	Quick	Moderate	Low	Low	Low	Yes
Swaging	Yes	Yes	Yes	Moderate	Quick	Moderate	Low	None	Low	Yes

MSL, 2004.
Note: N/A – not applicable.

Table 2. SMR techniques vs. possible scenarios

Technique	Defect								
	Fatigue Crack	Non-fatigue Crack	Dent	Corrosion	Inadequate Static Strength		Inadequate Fatigue Strength		Understrength Topsides Plating
					Member	Joint	High Loads	Fabrication Fault	
Dry welding	Yes (1)	Yes	Yes (3)	Yes (3)	Yes (1)	Yes (1)	No	Yes	Yes
Wet welding	No (2)	Yes	Yes (3)	Yes (3)	Yes (1)	Yes (1)	No	Yes	No
Toe grinding	No	No	No	No	No	No	Yes	No	Yes
Remedial grinding	Yes	Yes (1)	No	No	No	No	No	No	No
Hammer peening	No	No	No	No	No	No	Yes	No	Yes
Stressed mechanical clamp	Yes	Yes	No	Yes	Yes	No	Yes	Yes	No
Unstressed grouted clamp	Yes	Yes	Yes	Yes	Yes	Yes	Yes	Yes	No
Stressed grouted clamp	Yes	Yes	Yes	Yes	Yes	Yes	Yes	Yes	No
Stressed elastomer-lined clamp	No	Yes	No	Yes	No	No	No	No	No
Grout filling members	No	No	Yes	No	Yes	Yes (4)	Yes (4)	No	No
Grout filling of joints	No	No	Yes	No	Yes	Yes (4)	Yes (4)	No	No
Bolting	No	Yes	No	No	No	No	No	No	Yes
Member removal	Yes (5)	Yes (5)	Yes (5)	Yes (5)	No	No	Yes (5)	Yes (5)	No
Adhesives	Yes (6)	Yes (6)	Yes (3 or 6)	Yes (3 or 6)	Yes (3 or 6)	Yes (6)	Yes (3 or 6)	Yes (3 or 6)	Yes (3)
Composites	Yes	Yes	No	Yes	Yes	Yes	Yes	Yes	Yes

MSL, 2004.
Notes:
(1) Usually in conjunction with additional strengthening measures.
(2) Except to apply weld beads in unstressed grouted connection/clamp repairs.
(3) To apply patch plates.
(4) Applicability depends on type and sense of loading.
(5) If member is redundant.
(6) Used as epoxy grout in clamps.

USE OF AHP IN SELECTING THE OPTIMAL SMR TECHNIQUE

The selection of SMR techniques is a challenging task with the presence of limited quantitative data and a simultaneous comparison of different criteria/alternatives. In this context, the AHP approach helps to lay out all the relevant factors of concern, tangible and intangible, for a practicing engineer making an optimal decision (Saaty, 1990; Zio, 1996). In essence, this approach enables the decision maker to

Figure 6. The interrelationship between commonly occurring scenarios and SMR techniques
Adopted from MSL (2004).

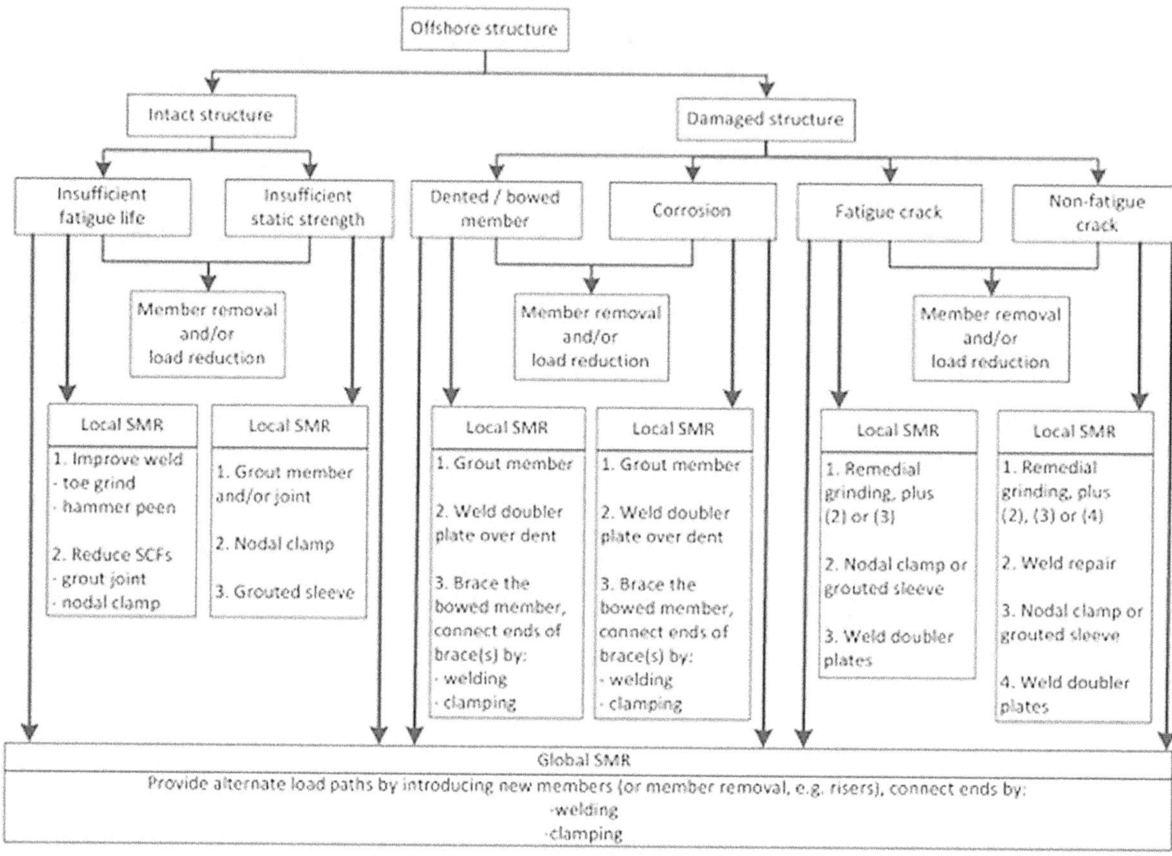

Figure 7. Phases of an SIC cycle
Adapted from ISO 19902 (2007).

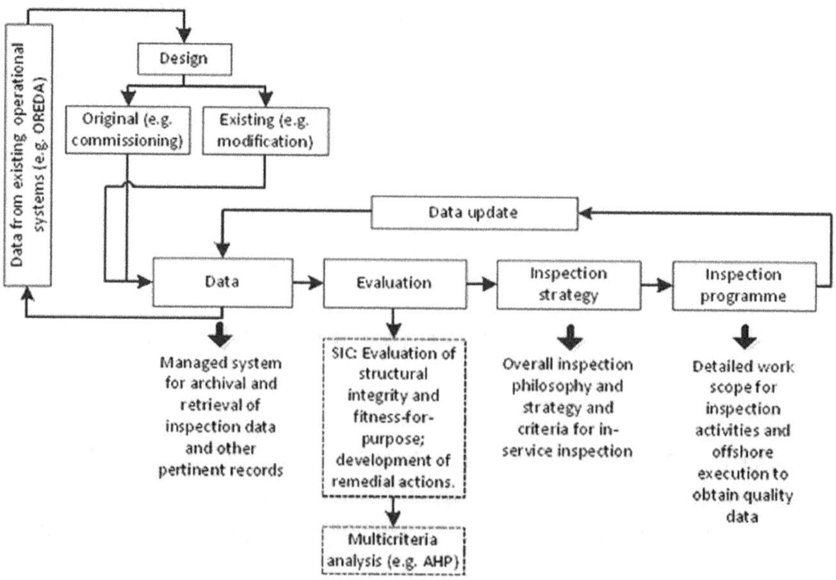

develop priorities by converting human judgments (i.e. data, experiences, the intuitions and intentions of experts in different disciplines) into numerical values. Hence, the AHP has been chosen as an appropriate methodology for selecting the optimal SMR technique due to its ability to handle qualitative and quantitative data simultaneously. Moreover, the AHP also offers a convenient way to decompose a problem by seeking cause-effect explanations in steps, particularly by using the incomplete design process (Saaty, 1990; Samarakoon & Ratnayake, 2013; Ratnayake & Markeset, 2010). Primarily, AHP is built on the following three underlying concepts:

1. Structuring the decision problem as a hierarchy of goals, criteria, and alternatives.
2. Pairwise comparison of elements at each level of the hierarchy with respect to each criterion on the preceding level.
3. Vertically synthesizing the judgments over the different levels of the hierarchy.

The AHP was pioneered by Saaty (1990). The methodology has also been accepted by the international scientific community as a robust and flexible multi-criteria analysis (MCA) tool (Daim, Yates, Peng & Jimenez, 2009; Samarakoon &Ratnayake, 2013; Ratnayake & Markeset, 2010). The AHP method is becoming quite popular in the research domain as its effectiveness is more significant than that of other rating methods (Scholl, Manthey, Helm & Steiner, 2005). It has been adopted successfully in many areas including resource allocation (Cheng & Li, 2001; Ratnayake & Markeset, 2010), quality management (Samarakoon & Ratnayake, 2013; Ratnayake & Markeset, 2010; Chin, Pun, Xu & Chan, 2002), technology selection (Daim, Yates, Peng & Jimenez, 2009), maintenance strategy selection (Bevilacqua & Braglia, 2000), safety assessment (Ha & Seong, 2004), etc. In addition, ExpertChoice.com references over 1000 articles and doctoral dissertations using AHP (Daim, Yates, Peng & Jimenez, 2009). Hence, ExpertChoice has been utilized for estimating priorities and carrying out sensitivity analysis.

In this approach, each criterion/alternative existing in the hierarchy is assumed to be independent of one another (Saaty, 1990). In addition, in a decision-making problem, it may be useful to know about the consistency of the decision makers' responses. The AHP measures the consistency of judgments using a consistency ration (CR). Throughout an evaluation process, the CR should be less than the threshold value 0.1 (see (Daim, Yates, Peng & Jimenez, 2009)) to ensure that the decision maker was consistent in assigning paired comparisons. For instance, if the CR is greater than 0.1 then the judgments are considered to be untrustworthy and the exercise is valueless or must be repeated.

ILLUSTRATIVE CASE FOR AN OPTIMAL SMR TECHNIQUE SELECTION

In the selection of SMR techniques particular to the strengthening of an existing structure or making an evaluation for a newly built structure, it is vital to take into consideration factors indicated in Tables 1 and 2 along with other technical and/or nontechnical requirements. In this context, it is essential to have a formal mechanism to consider all the factors systematically so that a team of experts with different expertise can independently input their experience (along with insight, and intuition) together with existing data and information (e.g. OREDA - Offshore REliability DAta) in a logical and thorough way (ISO 31000, 2010, ISO 19901-3, 2010). For instance, OREDA has established a comprehensive databank of reliability and maintenance data for exploration and production equipment from a wide variety of geographic areas, installations, equipment types and operating conditions. Primarily, offshore subsea and topside equipment are covered (note: it also includes onshore equipment). Moreover, international standards like ISO 31000:2009 (as well as ISO standards regulating reliability analysis) specify a lower bond for risk and reliability management in terms of accountability, establishing internal/external communication, applicability, cost, technical performance, etc. In order to include all the aforementioned in an evaluation, the SMR selection process is modeled in a hierarchical structure, indicating the relationships among the goal, criteria, sub-criteria and alternatives. The hierarchical structure has been incorporated along with technical, nontechnical and reliability requirements in the SMR techniques' evaluation process (see Figure 8).

To construct the hierarchical structure, the prevailing practices (see Figure 8), regulations, experiences, and literature on existing SMR techniques have been used. The hierarchical structure has been developed using technical and nontechnical factors; applicability, cost of design/fabrication/installation, timescale for design/fabrication/installation, technical performance, reliability, and legislative requirements make up the elements in the first level of the hierarchy (i.e. criteria), and static strength/fatigue strength & tolerance acceptability comprise the elements in the second level of the hierarchy (i.e. sub-criteria). Finally, at the third level, the alternative SMR techniques have been introduced to the hierarchy.

Data Collection

In the AHP approach, expert judgments provide the necessary information for the ranking of criteria. For instance, in the current SMR techniques' selection study there are four alternatives: grout filling, member removal, clamping and welding (see Figure 8). It is necessary to perform a prioritization of the four alternatives (i.e. SMR techniques) in this illustrative case in relation to six criteria and two sub-

Figure 8. AHP hierarchical structure for selection of SMR techniques
Samarakoon & Ratnayake, 2015.

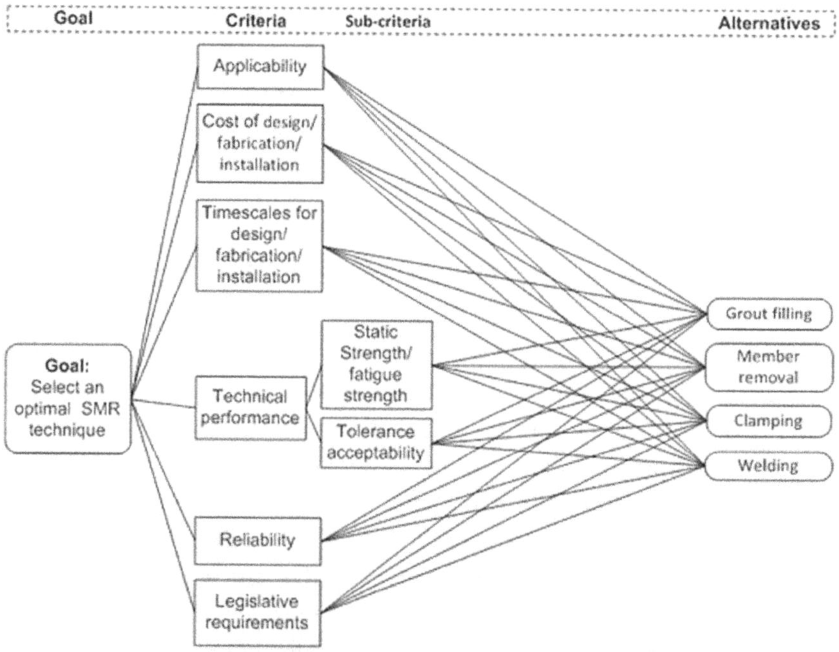

criteria (see Figure 9. However, depending on the particular case, the number of criteria, sub-criteria and alternatives can change. Figure 9 illustrates the model of the hierarchical structure developed for the prioritization of SMR techniques.

In this context, both qualitative and quantitative criteria are compared using informed judgments [with the help of particular discipline expert(s)] to derive weights and priorities (Ratnayake, 2012b). Using the scale illustrated in Table 3, pairwise comparison matrices are developed for the analysis (Saaty, 1990).

The evaluations are conducted with experts using a questionnaire similar to a sample (i.e. a part of the questionnaire) given in Figure 10 (Samarakoon & Ratnayake, 2015).

The judgment value is a full number if the comparison is regarded as favorable and is a fraction if the comparison is regarded less favorably. When the evaluations are made at different levels along the hierarchy, if no consensus is reached, then a geometric mean $\left(\overline{w}\right)$ is used to average the judgments (i.e. weights, w_i). The geometric mean is given by Equation 1.

$$\overline{w} = \left(\prod_{i=1}^{n} w_i\right)^{1/n} \quad (1)$$

Figure 9. Hierarchical structure for SMR techniques' prioritization (G: goal; C_i = i^{th} criteria; SB_i = i^{th} sub-criteria; A_i = i^{th} alternative)
Samarakoon & Ratnayake, 2015.

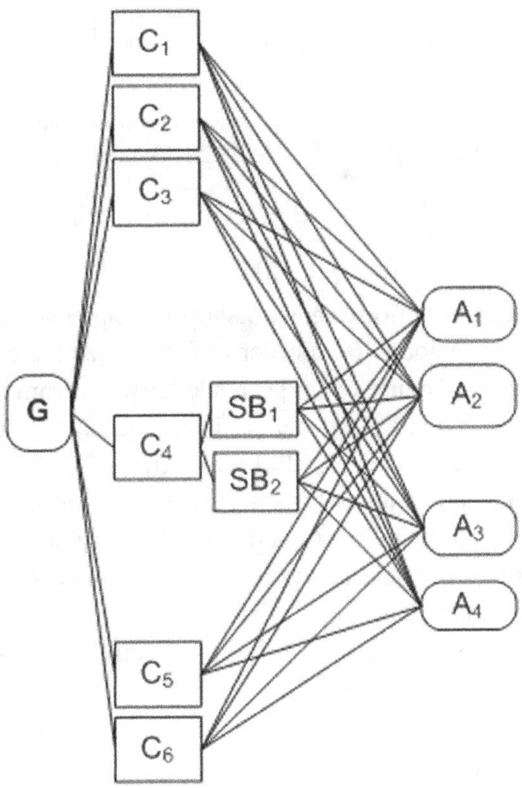

Table 3. Fundamental scale for assessing the importance of activities (pairwise comparison of activities)

Intensity of Importance	Definition	Explanation
1	Equal importance	Two activities contribute equally to the objective
3	Moderate importance one over another	Experience and judgment favor one activity over another
5	Essential or strong importance	Experience and judgment strongly favor one activity over another
7	Very strong importance	An activity is strongly favored and its dominance is demonstrated in practice
9	Extreme importance	The evidence favoring one activity over another is of the highest possible order of affirmation
2,4,6,8	Intermediate values between the two adjacent judgments	When compromise is needed

Saaty, 1990.

Figure 10. Excerpted sample questions of the AHP interviews

Question (2): Please select a question to answer, as per the relevance (i.e. at C_2).

2.1 How much more strongly does **"Alternative A_1"** contribute to, dominate, influence, satisfy or benefit, than does **"Alternative A_1"** in relation to the **"Criteria – C_2"**?

1	2	3	4	5	6	7	8	9

2.2 How much more strongly does **"Alternative A_1"** contribute to, dominate, influence, satisfy or benefit, than does **"Alternative A_2"** in relation to the **"Criteria – C_2"**?

1	2	3	4	5	6	7	8	9

2.3 How much more strongly does **"Alternative A_1"** contribute to, dominate, influence, satisfy or benefit, than does **"Alternative A_3"** in relation to the **"Criteria – C_2"**?

1	2	3	4	5	6	7	8	9

2.4 How much more strongly does **"Alternative A_1"** contribute to, dominate, influence, satisfy or benefit, than does **"Alternative A_4"** in relation to the **"Criteria – C_2"**?

1	2	3	4	5	6	7	8	9

It is important to provide an introduction (e.g. workshop, examples) regarding the evaluation approach prior to the distribution of the questionnaire in order to familiarize the evaluation team with the AHP. In addition, an evaluation team shall consist of experienced experts from different fields (e.g. lead engineer, senior engineer mechanical/structural, specialists in engineering control, senior engineer in design, senior engineer in testing, etc.) (Ratnayake & Markeset, 2010a, 2010b). The AHP approach provides the possibility to carry out interviews and analysis independently for different layers based on different expertise. Hence, it is possible, for instance, to address reliability, legislative requirements, technical performance, cost, etc. based on a particular expert's relevance. The aforementioned enables MCA to be performed in an effective and efficient manner. The authors have utilized the relative importance of two factors for making comparisons, using the facility provided by the ExpertChoice software.

Computational Steps in the AHP Approach

It is necessary to establish priorities, after modeling the SMR techniques' prioritization process in a hierarchy. Each node is evaluated against each of its peers in relation to its parent node. These evaluations are referred to as "pairwise comparisons". The pairwise comparisons of the elements at each level of the SMR prioritization hierarchical model are made in terms of either: *Importance* – when comparing goal, criteria, sub-criteria or alternatives in relation to their relative importance; *Preference* – when comparing the preference alternatives in relation to a goal, criteria, sub-criteria; *Likelihood* – when comparing uncertain events or scenarios in relation to the probability of their occurrence (Forman & Sally, 2001). When comparing a pair of factors (i.e. criteria, sub-criteria or alternatives), a ratio of relative importance, preference, or likelihood of the factors is established (Forman & Sally, 2001). In the pairwise comparison process, it is also possible to take the probability and consequence of a possible event (i.e. risk aspect) into the evaluation.

The numerical values (i.e. weights) that represent the judgments of the comparisons are arranged in a matrix for further calculations. The weights in the pairwise comparison matrices (PCMs) are composed using the numerical values in Table 3, whilst comparing all elements at each level of the SMR technique prioritization hierarchical model. The PCM "$\underline{A}$" for comparing "*n*" elements are given by:

$$\underline{A} = \left[a_{ij}\right] \text{where, } a_{ij} = \left[\frac{1}{a_{ij}}\right],\ a_{ij} = 1,\ (1 \le i \le n) \text{and } (1 \le j \le n) \tag{2}$$

The PCM (i.e. $\underline{A}$) has a form as follows:

$$\underline{A} = \begin{pmatrix} \frac{w_1}{w_1} & \frac{w_1}{w_2} \cdots & \frac{w_1}{w_n} \\ \vdots & \ddots & \vdots \\ \frac{w_n}{w_1} & \frac{w_n}{w_2} \cdots & \frac{w_n}{w_n} \end{pmatrix}$$

Once the weights of the PCM have been established then the matrix product of the matrix $\underline{A}$ with vector $\underline{w} = \left[w_1 \quad w_2 \quad w_3 \cdots w_n\right]$ enables calculation of the weights of $\underline{w}$:

$$\begin{pmatrix} \frac{w_1}{w_1} & \frac{w_1}{w_2} \cdots & \frac{w_1}{w_n} \\ \vdots & \ddots & \vdots \\ \frac{w_n}{w_1} & \frac{w_n}{w_2} \cdots & \frac{w_n}{w_n} \end{pmatrix} * \begin{pmatrix} w_1 \\ w_2 \\ \vdots \\ w_n \end{pmatrix} = \begin{pmatrix} nw_1 \\ nw_2 \\ \vdots \\ nw_n \end{pmatrix}$$

$$\underline{Aw} = n\underline{w} \tag{3}$$

The problem of solving Equation 3 for a nonzero solution is an eigenvalue problem which is given by:

$$\underline{A}\underline{w} = \lambda \underline{w}$$

As the matrix $\underline{A}$ has a special form (i.e. each row is a constant multiple of the first row), the rank of the matrix is one, and except for one eigenvalue, all the other eigenvalues of the matrix $\underline{A}$ are zero. However, as the sum of the eigenvalues of a positive matrix is equal to the trace of the matrix, the non-zero eigenvalue has a value of n (i.e. the size of matrix). This is referred to as λ_{max} and used for consistency analysis.

The consistency analysis is carried out to determine the consistency of experts' judgments. For instance, if the matrix $\underline{w}$ is not known, only the estimates of $a_{ij's}$ in matrix $\underline{A}$ are known, and no strong consistency property (i.e. small errors and inconsistencies in judgments) exists, then the eigenvalue problem for the inconsistent case is:

$$\underline{A}\underline{w} = \lambda_{max} \underline{w} \tag{4}$$

where $\lambda_{max} \geq n$ and other $\lambda's \sim 0$ the estimates of weights for the activities are calculated by normalizing the eigenvector corresponding to the largest eigenvalue in Equation 4. The normalization is performed by summing each column and then dividing each column entry by its respective column sum. The normalization process is given by:

$$a' = \left[a'_{ij}\right] \text{where } a'_{ij} = \frac{a_{ij}}{\sum_{k=1}^{n} a_{ik}} \text{ for } (1 \leq i \leq n) \text{ and } (1 \leq j \leq n) \tag{5}$$

Then, the average value in each row of the normalized matrix is calculated to obtain the relative weights or eigenvector, which is determined by:

$$\underline{w} = \left[w_k\right] \text{where } w_k = \sum_{i=1}^{n} \frac{a'_{ij}}{n} \text{ for } (1 \leq i \leq n),\ (1 \leq j \leq n) \text{ and } (1 \leq k \leq n) \tag{6}$$

The notion here is that the closer the λ_{max} is to *n*, the more consistent the judgments. Hence, it is possible to use the difference $(\lambda_{max} - n)$ as a measure of inconsistency (i.e. the perfect consistency refers to the zero difference). Hence, the consistency index (*CI*) is defined (see (Saaty, 1982) as Equation 7, as it represents the average of the remaining eigenvalues.

$$CI = \frac{(\lambda_{max} - n)}{(n-1)} \tag{7}$$

To derive a meaningful interpretation of the *CI*, random pairwise comparisons have been simulated at the inception of the AHP for different sizes of matrices. The calculations have been performed for calculating *CIs* and arriving at an average *CI* for random judgments for each size of matrix [i.e. referred to as random consistency index (RCI)]. The size of the matrix *vs.* RCI is illustrated in Table 4 (Saaty, 1982).

Hence, the consistency ratio (*CR*) is defined as the ratio of the CI for a particular set of judgments, to the average CI (or RCI) for random comparisons for a matrix of the same size (see Equation 8).

$$CR = \frac{CI}{RCI} \tag{8}$$

where if CR<0.1(or 10%) then the judgments are considered to be consistent.

Analysis Results Using the AHP Approach

The judgments using the pairwise comparisons have been performed (and alternative PCMs have been established) starting with the goal and working down to the alternatives (refer to Figure 4). For instance, the PCM for the goal *vs.* criteria is given in Figure 11 (a) and the sub-criterion reliability *vs.* alternatives (i.e. grouted repair, member removal or reduction of loading, mechanically clamped repair, and welded repair) is given in Figure 11 (b).

The 'pairwise graphical comparison' option, which is available in the 'Expertchoice' software, has been used to insert exact values in PCMs. The judgments about the relative importance of criteria have been made with respect to the parent node in the hierarchy (either the goal or higher level criteria). Finally, the judgments about the relative preference of alternatives are made with respect to each covering criteria (i.e. the lowest level criteria in the hierarchy). The final selection is considered to be consistent as the overall inconsistency is 0.01. The 'sensitivity analyses' support the identification of how the alternatives' priorities change when the objectives' priorities increase or decrease. Figure 12 demonstrates a performance sensitivity analysis that has been performed for the illustrative case study (Samarakoon and Ratnayake, 2015).

Table 4. Size of the matrix vs. RCI

Size of the Matrix	1	2	3	4	5	6	7	8	9	10
RCI	0.00	0.00	0.58	0.90	1.12	1.24	1.32	1.41	1.45	1.49

Figure 11. (a). PCM for the goal vs. criteria; (b). PCM for the sub-criterion reliability vs. alternatives

$$\begin{bmatrix} 1.000 & 0.625 & 0.699 & 0.641 & 0.518 & 0.571 \\ 1.600 & 1.000 & 1.000 & 0.741 & 0.613 & 0.730 \\ 1.430 & 1.000 & 1.000 & 0.676 & 1.000 & 0.847 \\ 1.560 & 1.350 & 1.480 & 1.000 & 1 & 1 \\ 1.930 & 1.630 & 1.000 & 1 & 1.000 & 1 \\ 1.750 & 1.370 & 1.180 & 1 & 1 & 1.000 \end{bmatrix} \quad \begin{bmatrix} 1.000 & 1.120 & 1.250 & 1.330 \\ 0.893 & 1.000 & 1.000 & 1.000 \\ 0.800 & 1.000 & 1.000 & 0.840 \\ 0.752 & 1.000 & 1.190 & 1.000 \end{bmatrix}$$

(a) (b)

Figure 12. Performance sensitivity (goal: selection of an optimal SMR technique)

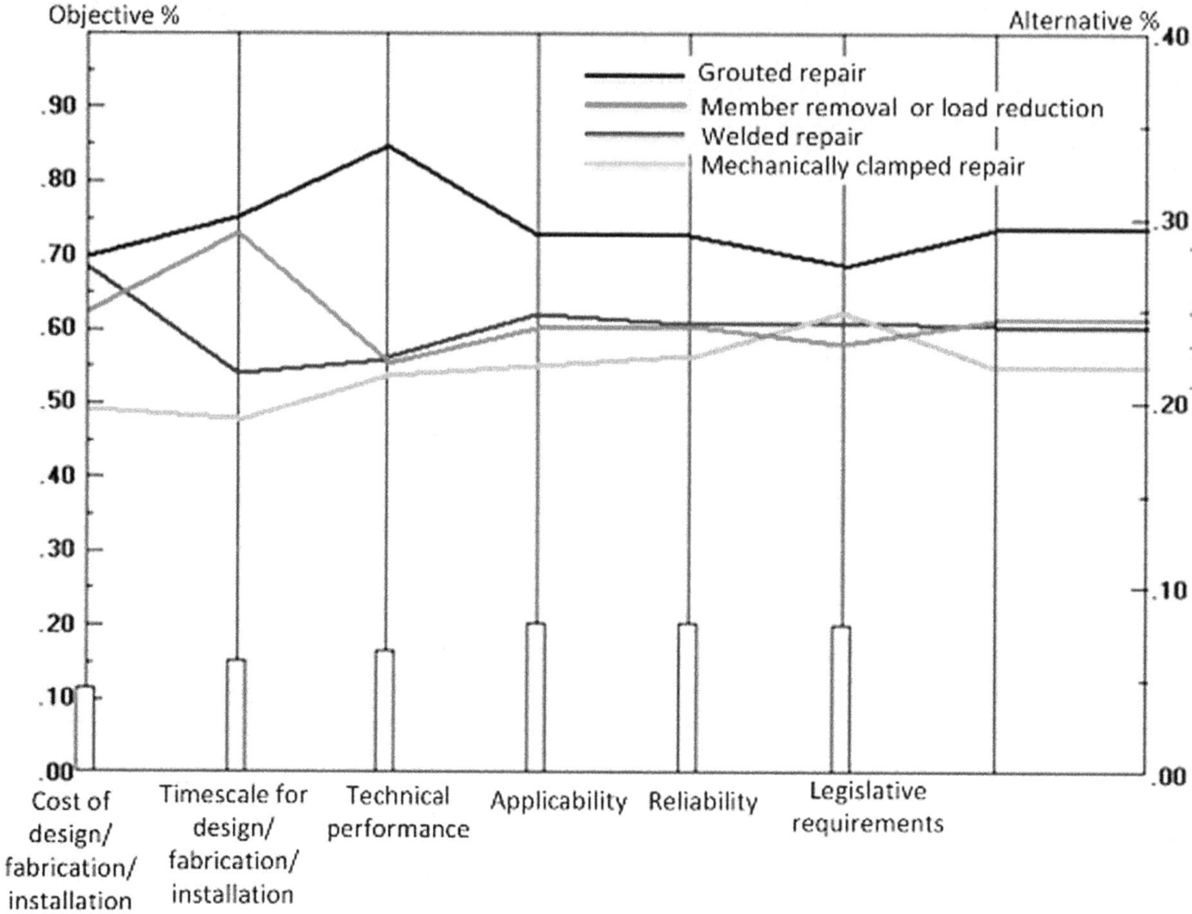

Figure 12 also indicates the synthesis with respect to the goal – the selection of optimal SMR technique – on the right 'y-axis'. Furthermore, it illustrates how the alternatives have been prioritized relative to other alternatives with respect to each criterion as well as overall. The best SMR technique compared to the others has been illustrated by the overall priority from the intersection of 'right y-axis'. It presents the overall priority for each alternative. For instance, in this case, grouted repair is approximately 0.294 (i.e. considered to be the optimal selection with the highest weight), member removal or reduction of loading is approximately 0.245, welded repair is approximately, 0.241 and mechanically clamped repair is approximately 0.220 (note that the priorities for the alternatives sum to one) (see right 'y-axis' in Figure 8). The left 'y-axis' in Figure 8 illustrates each criterion priority (based on the engineering expert's paired comparisons). For instance, cost of design/installation/fabrication is about 10.8%, timescale for design/fabrication/installation is about 14.7%, technical performance is about 15.8%, applicability is 19.7%, reliability 19.8%, and legislative requirements is 19.3%.

The gradient sensitivity analysis results are illustrated in Figure 13. It shows the alternatives' priorities with respect to one criterion at a time. For instance, Figure 13 demonstrates the cost of design/fabrication/installation (Samarakoon & Ratnayake, 2015).

Figure 13. Gradient sensitivity with respect to cost of design/fabrication/installation

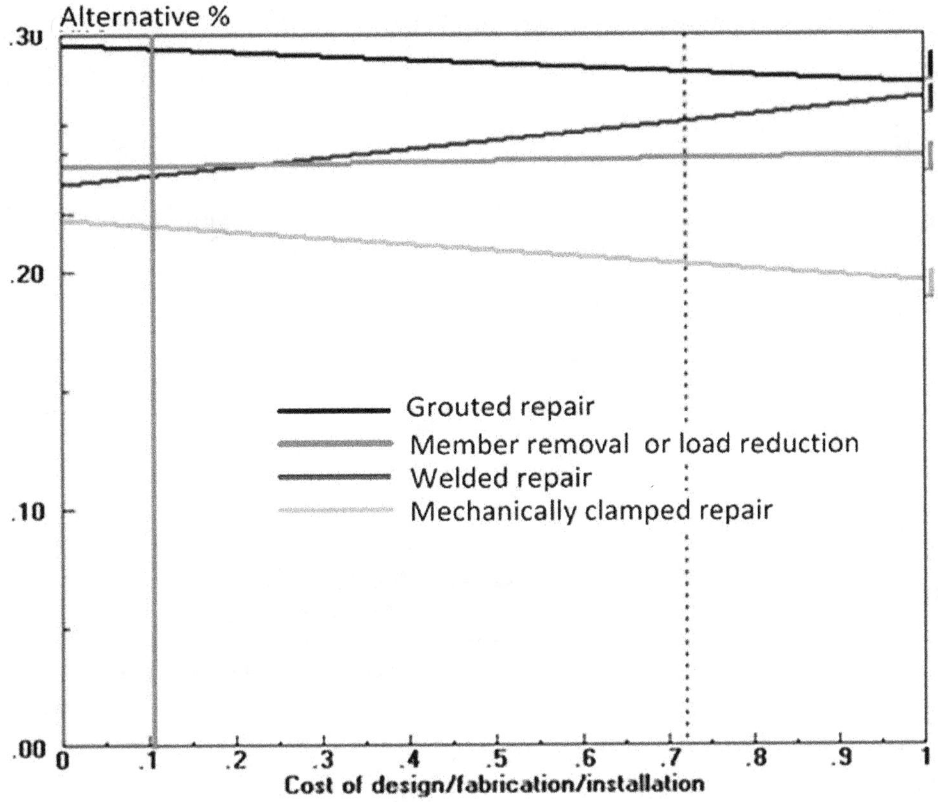

The red vertical line indicates the criterion's priority (based on the decision maker's paired comparisons). Dragging the red bar to the left or right, it is possible to indicate how a criterion's priority changes. For instance, this is indicated as a blue dashed vertical line. Such analysis supports making a logical trade-off between different alternatives, instead of moving to *ad hoc* conclusions.

Two-dimensional sensitivity analyses for nodes have been carried out to illustrate the alternatives' priorities with respect to two criteria at a time. For instance, Figure 14 illustrates two-dimensional sensitivities for cost of design/fabrication/installation vs. timescale for design/fabrication/installation (Samarakoon & Ratnayake, 2015).

Essentially, here the area of the two-dimensional plot is divided into quadrants. The most favorable alternatives with respect to the criteria on the two axes are those in the upper right quadrant (i.e. the closer to the upper right corner, the better the alternative). The least favorable alternatives are shown in the lower left quadrant (i.e. the closer to the lower left corner, the less favorable the alternative). If alternatives are located in the upper left and lower right quadrants, then they indicate key trade-offs where there is a conflict between the two selected criteria.

Discussion

In the SIC&A, the AHP approach provides a backbone to carry out such analysis in a formal manner with a defined mechanism to incorporate criteria and sub-criteria for evaluating the alternative SMR

Figure 14. Two-dimensional sensitivity (goal: selection of an optimal individual SMR technique)

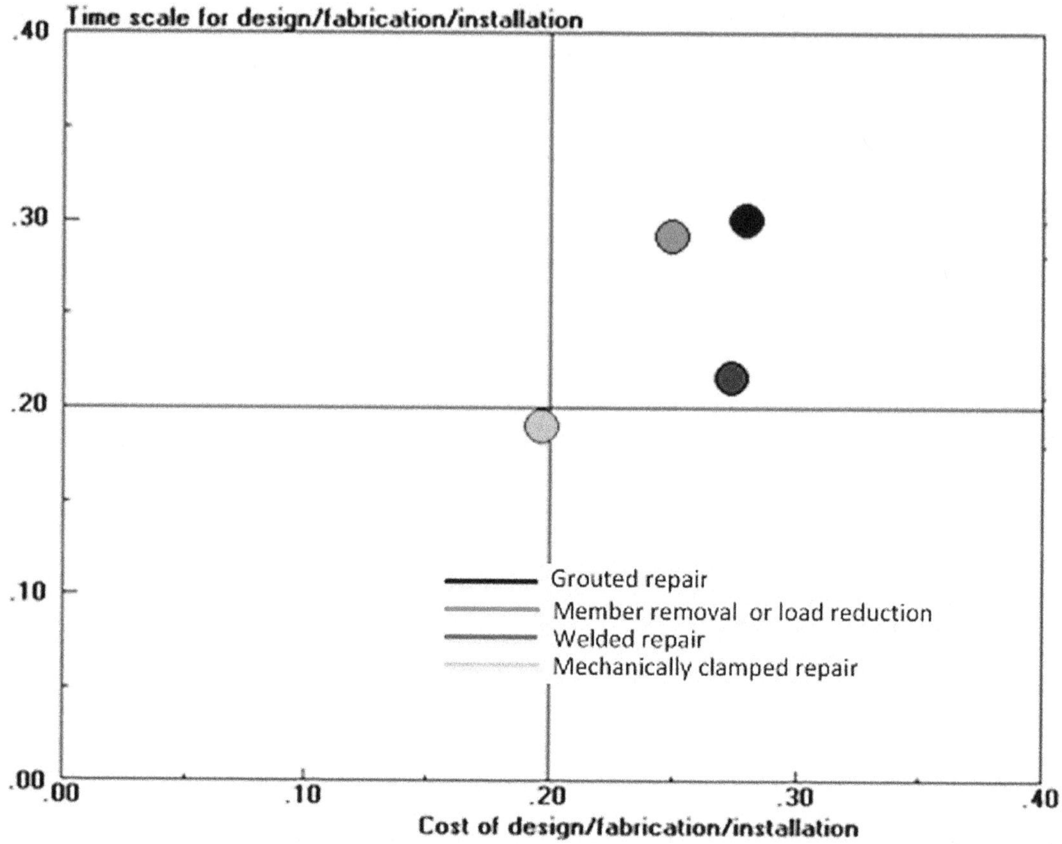

techniques. In addition, the suggested approach enabled visualizing the SMR decision challenge along a hierarchy, facilitating the interpretation of the whole set of criteria and sub-criteria with the potential alternatives. Also, it has been illustrated how to use AHP for making an optimal SMR technique selection and possible sensitivity analysis, evaluating possible trade-offs. Overall synthesis provides the possible priorities based on the pairwise comparisons. Aside from that, performance, gradient and two-dimensional sensitivity analyses provide alternative means to study how the final selection has been made. This enables one to see how different criteria and sub-criteria contributed to the final priorities. Hence, the suggested approach provides transparency in selecting the optimal SMR technique in relation to criteria and sub-criteria. However, the drawbacks, such as 'ranking irregularities', 'need of decomposing MCA problem into a number of subsystems', and 'requirement of performing substantial number of comparisons within which and between which', limit the use of AHP in SIC-related prioritizations. Also, the aforementioned approach has the limitation regarding the number of pairwise comparisons to be made (i.e. the number of comparisons shall become very large [n(n−1)/2], whilst modelling and analysis shall become a lengthy task).

ILLUSTRATIVE CASE: FUZZY LOGIC BASED APPROACH FOR ASSESSMENT OF CURRENT STATUS OF A SIA&C ORGANIZATION

The purpose of this section is to assist with the implementation of the SIA&C framework and reveal the current status of the SIA&C organization (see Figure 4). This is achieved by providing an illustrative case for identifying the current status in SIA&C practice (i.e. the status of SIA&C organization). This enables to prioritize the needs, assess resource needs and costs, and managing the implementation. The identification of the current status should be performed in line with the processes, information and documentation, and personnel competence. Identification of current practice of the SIA&C organization should be carried out in a structured manner and be documented. A rating system that shall be used to assess the current level of the SIA&C organization (i.e. maturity of the current practice with respect to individual activities) is illustrated in Table 5.

As the SIA&C organization's level assessment measures; processes (PROC), information and documentation (INFO&DOC), personnel (PER) and organization (ORG) have ratings, it is vital to develop

Table 5. Example rating system for assessment of current level of the SIA&C organization

Rating	Processes	Information and Documentation	Personnel	Organization
1	**Not in place (NIP)** – The need has not been recognized or the need has been recognized but no action has been taken.	**Poor (P)** – quality and quantify are below the level required to undertake basic SIM activities.	**Unsatisfactory (U)** – competence below minimum requirements in many areas and/or sever staff shortages.	**Poor (P)** – structure not suitable for adequate SIM. Roles and responsibilities not appropriately defined.
2	**Implementation (I) –** the need has been recognized and a plan for implementation is currently being developed or is already being implemented	**Basic (B)** – inventory is adequate but quality is poor.	**Needs improvement (NI)** – staff competence below minimum requirements in some areas and/or some staff shortage.	**Needs improvement (NI)** – Roles and responsibilities appropriately defined but not clearly communicated.
3	**Recently implemented (RI)** – has recently been implemented and the user is in the early stages of training and usage.	**Fair (F)** – inventory is complete and quality sufficient to support basic SIM activities.	**Satisfactory (S)** – meet minimum competence and resource requirements, but skills and capacity for innovation may be insufficient.	**Fair (F)** – Appropriate organizational structure in place and all responsibilities clearly allocated and communicated.
4	**Established (E)** – has been in place for a number of years and as such is documented and associated training is in place. It may not be fully integrated with other processes.	**Good (G)** – inventory is complete and quality and quantity is improving.	**Competent (CPT)** – competence and/or resource above minimum requirements, and skills and capacity available to promote innovation and improve efficiency.	**Good (G)** – All responsibilities clearly allocated and communicated. Some competency management and emergency response arrangements in place.
5	**Fully embedded (FE)** – mature and fully documented, associated training is in palace and it is integrated with other processes.	**Comprehensive (CPH)** – inventory is complete, accurate, up-to-date and sufficient to support all SIM activities.	**Excellent (E)** – competence and resources above minimum requirements with suitable skills and capacity to actively innovate and improve efficiency.	**Excellent (E)** – All responsibilities clearly allocated and communicated. Full competency management and emergency response systems in place and periodically reviewed.

HSE, 2009.

set of rules (see Table 6) to take all the possible combinations. The current SIA&C organization level is categorized as; Poor (P); Needs Improvement (NI); Fair (F); Good (G); Excellent (E). In order to demonstrate the approach, the combination; PROC and INFO&DOC vs. ORG has been used (i.e. to assess the current status of a SIA&C organization).

The selection of the aforementioned combination shall be based on the assessors' focus. For instance, it is possible to have following combinations; PROC and INFO&DOC vs. ORG, PROC and PER vs. ORG, or PER and INFO&DOC vs. ORG.

FUZZY LOGIC BASED SIA&C ORGANIZATION ASSESSMENT APPROACH

In order to cater for rapid changes at the boundaries of each SIA&C organization level (see Table 3), it is possible to introduce fuzzy membership functions (MFs) for; PROC, INFO&DOC, PER and ORG. The introduction of fuzzy MFs enables the assessing the performance of SIA&C organization more realistic. Furthermore, a fuzzy inference system (FIS) enables the mitigation of discrepancies that may occur during the SIA&C organization level assessment process with different combinations (i.e. PROC and INFO&DOC vs. ORG, PROC and PER vs. ORG, or PER and INFO&DOC vs. ORG). Moreover, it is possible to mitigate the inherent challenges related to SIA&C organization level assessments such as: uncertainty that has been caused by: fuzziness (i.e. "lack of definite or sharp distinctions" due to "vagueness, cloudiness, haziness, unclearness, indistinctness, and sharpness"); ambiguity (i.e. "one to many relationships"); discord (i.e. "disagreement in choosing among several alternatives" due to "dissonance, incongruity, discrepancy, and conflict") and nonspecificity (i.e. "two or more alternatives are left unspecified" due to "variety, generality, diversity, equivocation, and imprecision") via fuzzification as well as the development of MFs and a rule base with the help of experts' knowledge (Klir & Yuan, 1995)

Methodology

Figure 15 illustrates the methodology suggested for the assessment of fuzzy rank of the level of a SIA&C organization (Ratnayake, 2014).

It is possible to perform the assessments in relation to the current status of the performance measures which are used for assessing the SIA&C organization. The input and output variables shall consist of quantitative, qualitative and judgmental (i.e. linguistic) data. Using an appropriate MF, the user has

Table 6. Matrix for assessing the level of SIA&C organization

Level of SIA&C Organization		INFO&DOC				
		1	2	3	4	5
PROC	1	P	P	NI	NI	NI
	2	P	NI	F	F	G
	3	P	NI	F	G	G
	4	P	NI	F	G	E
	5	P	NI	G	E	E

Poor (P); Needs Improvement (NI); Fair (F); Good (G); Excellent (E).

Figure 15. Estimation of fuzzy rank of the level of a SIA&C organization

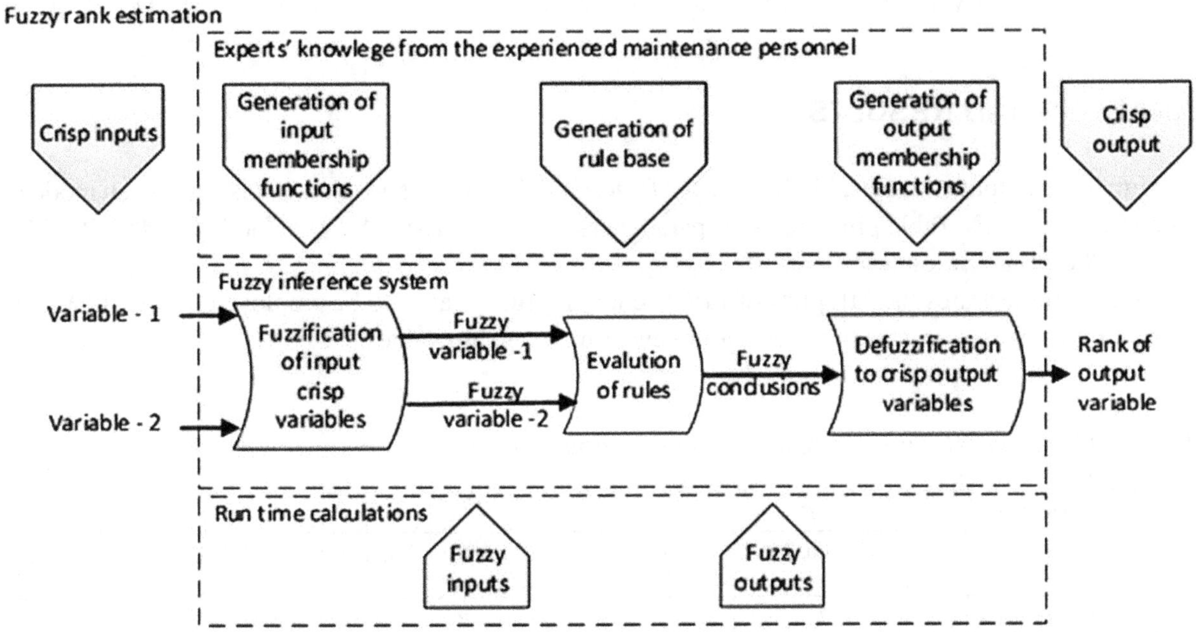

'more confidence' that the input parameter lies in the center of the interval than at the edges. In this study, the author has incorporated Gaussian membership functions (Tay & Lim, 2008) which are defined by Equation 1:

$$Gaussian(x;c;\sigma) = e^{\frac{-(x-c)^2}{2\sigma^2}} \tag{1}$$

where, c represents the center and σ determines the width of the MFs. To model the MFs, the Gaussian combination MF (GCMF) (i.e. 'gauss2mf'), which is available in MATLAB (R2013a), has been utilized (Mathworks, 2014). The function 'gauss2mf' is a combination of two parameters [i.e. (c, σ)]] indicated in Equation (2). It follows the following syntax (Mathworks, 2014):

$$y = gauss2mf\left\{x, \left[\sigma_1 \quad c_1 \quad \sigma_2 \quad c_2\right]\right\} \tag{2}$$

The first part of the function of the GCMF is specified by σ_1 and c_1, which determine the shape of the left-most curve. The second part of the GCMF, specified by σ_2 and c_2, determines the shape of the right-most curve. Whenever $c_1 \leq c_2$, the 'gauss2mf' function reaches a maximum value of 1. Otherwise, the maximum value is less than one. The order of the parameters is as follows: [σ_1, c_1, σ_2, c_2] (Mathworks, 2014). Moreover, the other parameters of the fuzzy logic based expert system that have been selected for the current analysis are as follows: 'AND' operator with 'minimum', 'OR' operator with 'maximum', 'Implication' with 'minimum', 'Aggregation' with 'maximum' and 'Defuzzification' with 'centroid' algorithm.

A fuzzy rule base has been developed using the table-look-up approach (see Table 3). The toolbox simulator tool of MATLAB (R2013a) has been utilized to execute the suggested FES (MATLAB, 2013).

ANALYSIS AND RESULTS

The input and output variables; PROC, INFO&DOC and SIA&C organization levels, have been modeled using Gaussian MFs. Table 7 illustrates the parameters of each Gaussian MF (i.e. for PROC, INFO&DOC and SIA&C organization levels).

Figure 16 illustrates the MF plots of PROC enabling fuzzy ranks to be employed in the SIA&C organization level assessments together with relevant membership values.

Table 7. Gaussian MF parameters for input and output variables

Input Variable 1	***NIP***	***I***	***RI***	***E***	***FE***
PROC	[0.45, 1.0, 0.45, 1.0]	[0.4, 2.0, 0.4, 2.0]	[0.4, 3.0, 0.4, 3.0]	[0.4, 4.0, 0.4, 4.0]	[0.4, 0.5, 0.4, 5.0]
Input Variable 2	***P***	***I***	***RI***	***E***	***FE***
INFO&DOC	[0.45, 1.0, 0.45, 1.0]	[0.4, 2.0, 0.4, 2.0]	[0.4, 3.0, 0.4, 3.0]	[0.4, 4.0, 0.4, 4.0]	[0.4, 5.0, 0.4, 5.0]
Output Variable	***P***	***NI***	***F***	***G***	***E***
SIA&C organization levels	[0.4 1.0 0.4 1.0]	[0.4 2.0 0.4 2.0]	[0.4 3.0 0.4 3.0]	[0.4 4.0 0.4 4.0]	[0.4 5.0 0.4 5.0]

Figure 16. MF plots of PROC

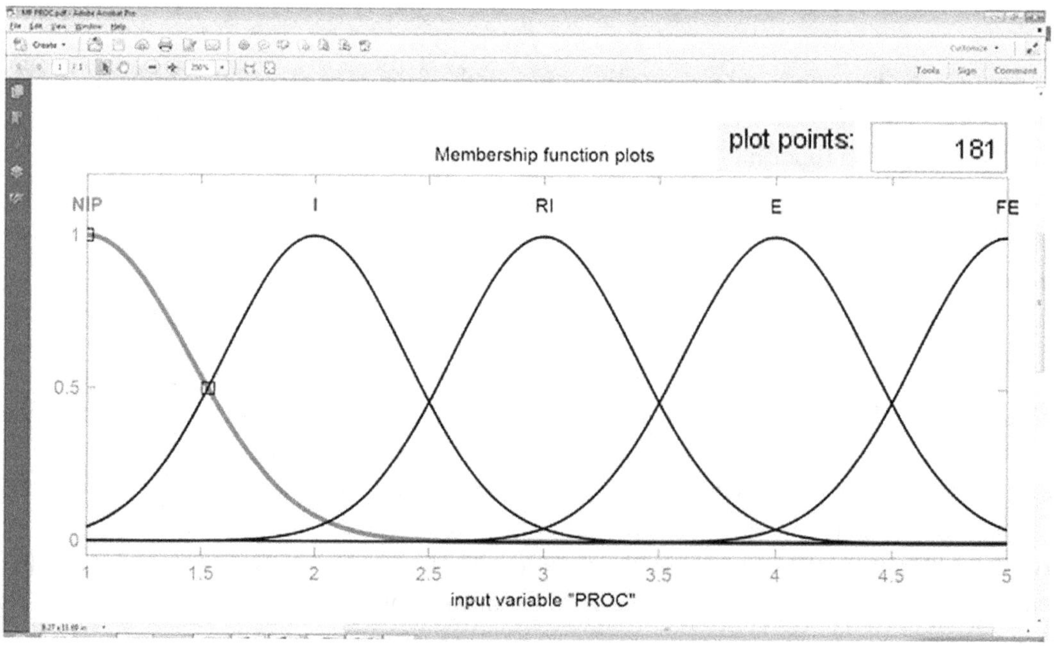

Figure 17 illustrates the MF plots of INFO&DOC enabling fuzzy ranks to be employed in the SIA&C organization level assessments together with relevant membership values.

Figure 18 illustrates the MF plots of SIA&C organization levels enabling fuzzy ranks to be employed in the SIA&C organization level assessments together with relevant membership values.

Figure 17. MF plots of INFO&DOC

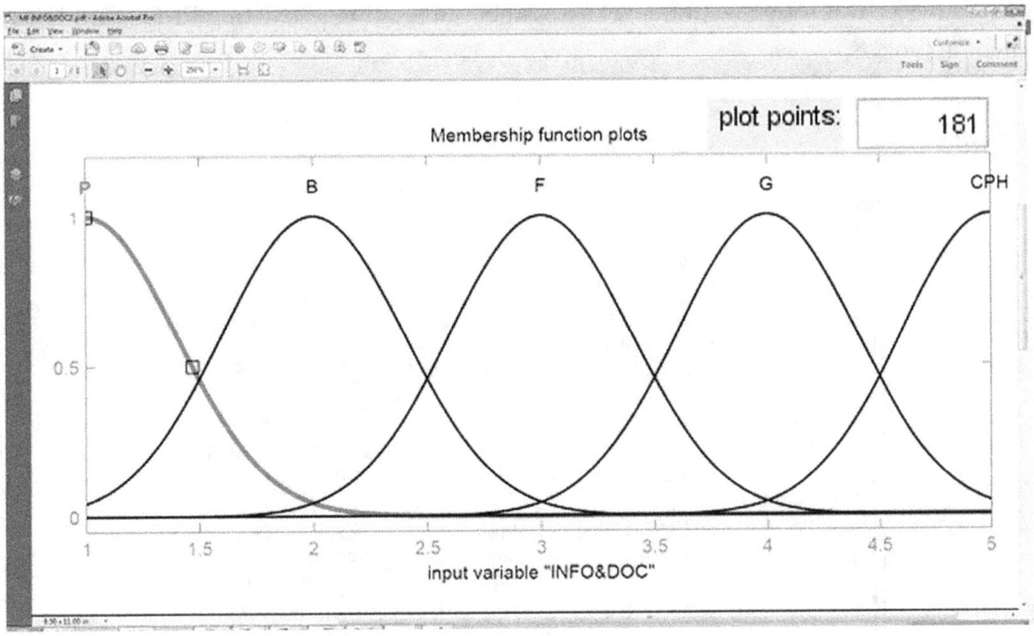

Figure 18. MF plots of SIA&C organization levels

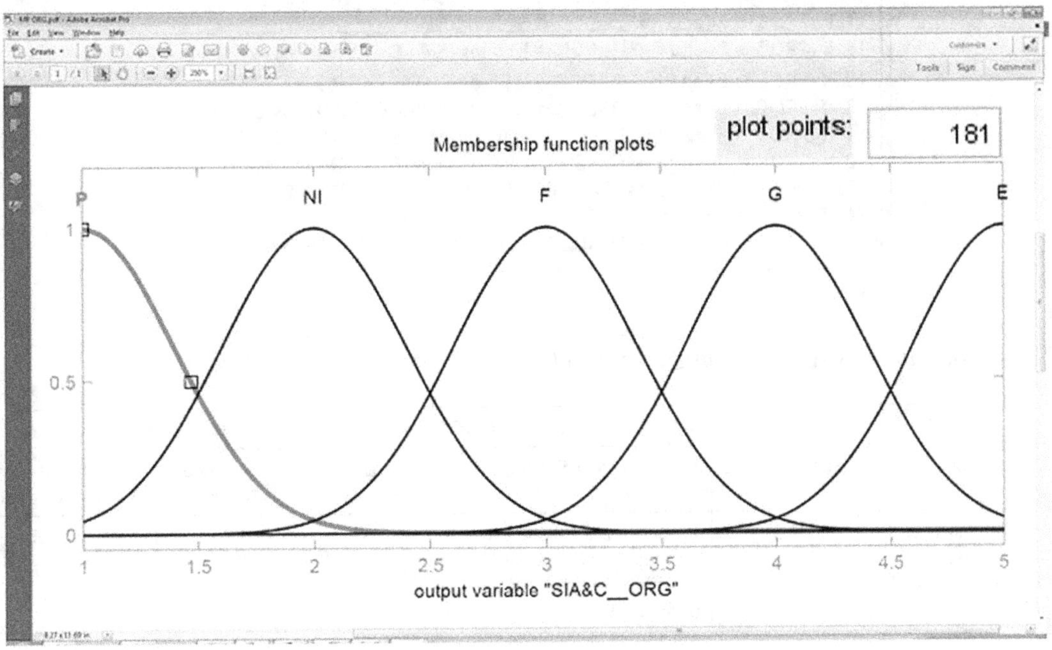

Figure 19 illustrates the rule base that has been developed using the table-look-up method. Table 3 has been utilized for developing the rule base.

The selected MFs and defined rule base provide the SIA&C organization levels' surface as illustrated in Figure 20. The SIA&C organization levels' surface evidences that, within the ranges of PROC vs. INFO&DOC ranks, the values of the SIA&C organization level are distributed smoothly.

Figure 21 shows an example calculation [i.e. for PROC rating = 3.2 and INFO&DOC rating = 3.4 (see Table 5&6 for explanations)].

An illustrative SIA&C organization level assessment in relation to existing PROC and INFO&DOC is presented in Table 8.

Figure 19. Rule base

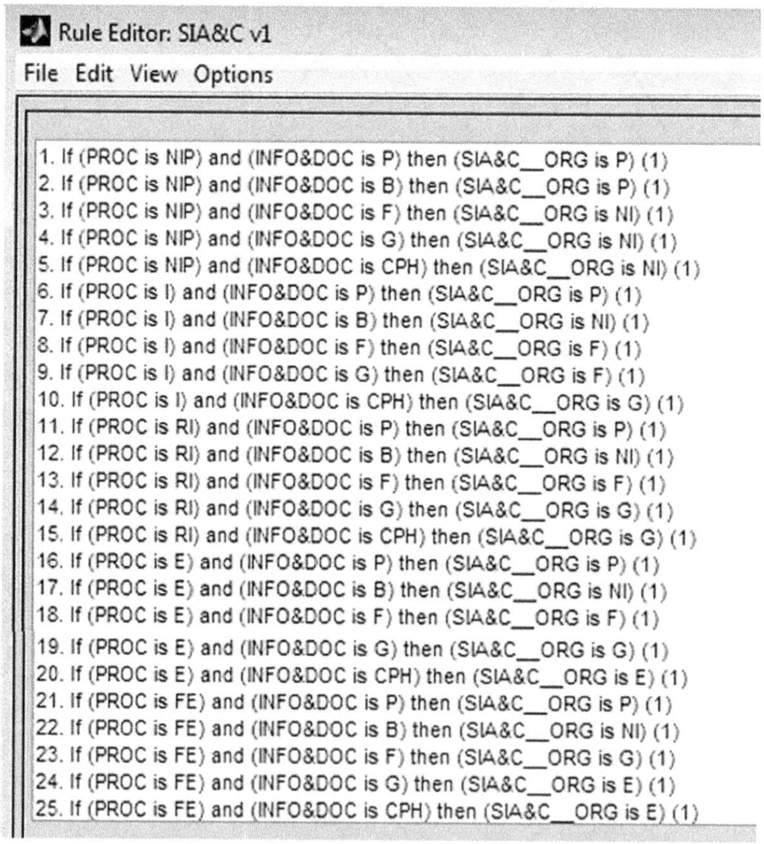

Table 8. An illustrative SIA&C organization level assessment

PROC		INFO&DOC		SIA&C Organization Level	
Current Status	**Fuzzy Rating**	**Current Status**	**Fuzzy Rating**	**Calculated Value**	**Assessed Level of SIA&C Organization**
Implementation has been started however user training has not yet started. (refer Table 5)	2.3	Inventory is adequate and some quality improvement initiatives have been taken (refer Table 5).	3.4	3.26	Fair

Figure 20. SIA&C organization levels' surface

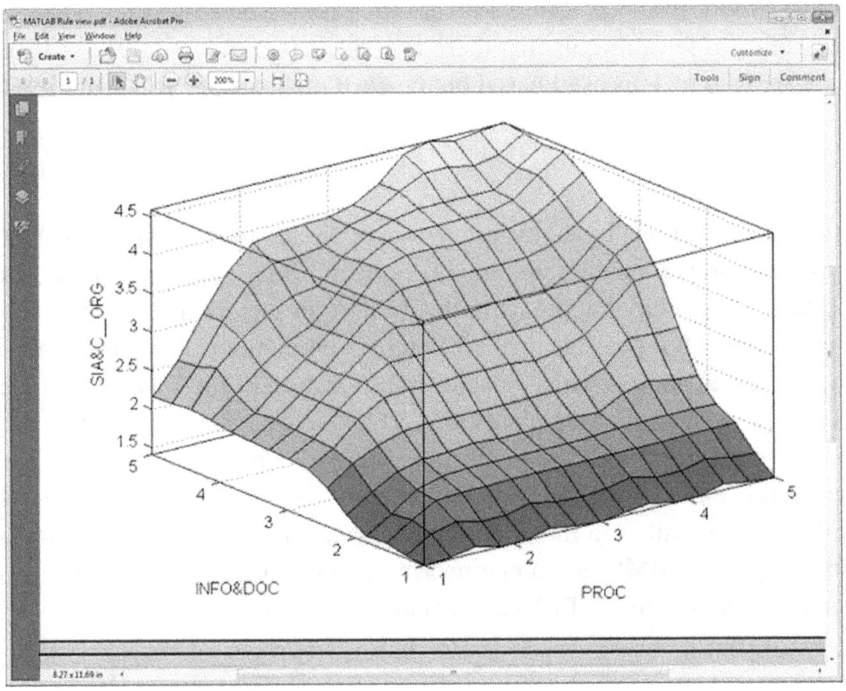

Figure 21. Example calculation of SIA&C organization level

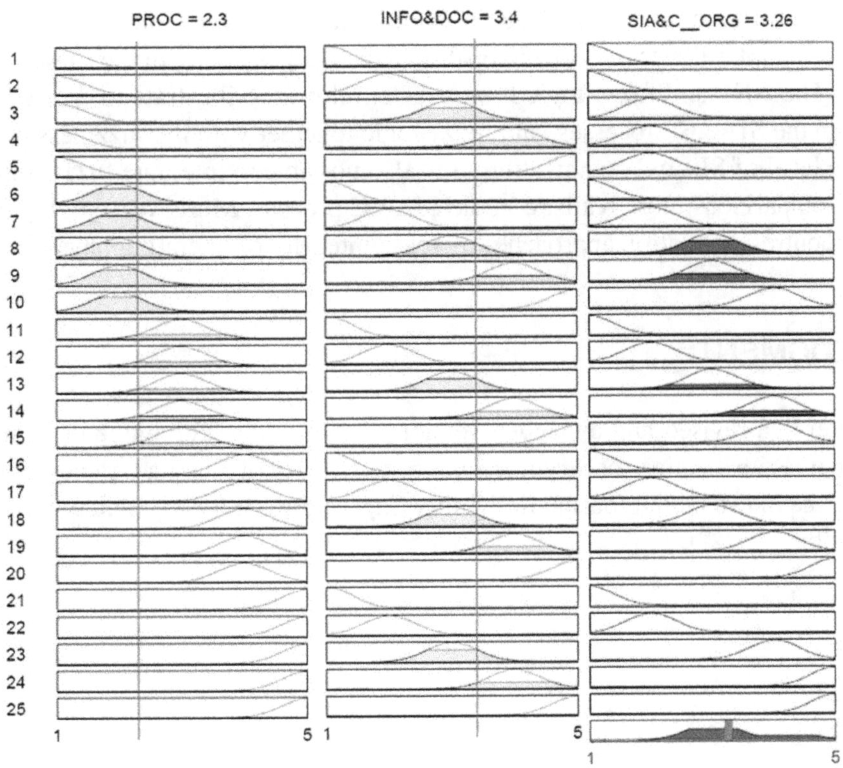

It is possible to recommend a necessary improvement actions based on the assessed SIA&C organization level (see Table 6). For instance, Table 8 reveals that the assessed level of SIA&C organization is 'Fair'. The aforementioned level reveals that there is enormous room to develop SIA&C organization. Also, it reveals the areas to be improved based on the current status of PROC and INFO&DOC.

Discussion

This study illustrates a fuzzy logic based expert system approach for assessing a SIA&C organization's level. The suggested approach enables minimizing uncertainties that may cause assessments to be performed as it provides structured approach to take different factors influencing the level of SIA&C organization's performance. It is possible to use different combinations of inputs as appropriately depend on the focus. The suggested approach also enables minimization of the waste (i.e. in terms of time and other resources) in onshore and offshore plants' SIA&C. It also enables minimizing ad hoc approaches that have been used for making SIA&C organizations' levels assessments, whilst allowing the 'human touch' for further improvements with relevant experts via dynamic re-evaluation to MFs and the rule base. Moreover, it is also possible to integrate the suggested approach in an existing structures information management system (SIMS) in an engineering contractor organization or in a computer aided maintenance management system (CMMS) in an operator's (or assets owner's) industrial organization.

CONCLUSION

The SIA&C of physical assets depends on the other kinds of critical assets such as human, information, intangible and financial assets involved in the integrity assessment and control process as well as organization. This chapter illustrated how to deal with challenges facing in the SIA&C process as well as organization. The AHP and fuzzy logic have been employed to illustrate to deal with multi-criteria nature presents in the SIA&C processes, and fuzzy logic together with the fuzziness presents in the assessments of the level of SIA&C organization. It is also possible to integrate fuzzy logic into the AHP for making the comparisons more realistic in the SIA&C process. Moreover, employing multi-criteria assessment and control simulation approaches increase integrity of operating physical assets.

ACKNOWLEDGMENT

This book chapter is a revised and expanded version of the article "Strengthening, Modification and Repair Techniques' Prioritization for Structural Integrity Control of Ageing Offshore Structures" which was published in the International Journal Reliability Engineering & System Safety, Volume 135, March 2015, Pages 15–26.

REFERENCES

ABS. (2013), Guidance notes on: accidental load analysis and design for offshore structures, American Bureau of Shipping. http://www.eagle.org/eagleExternalPortalWEB/ShowProperty/BEA%20Repository/Rules&Guides/Current/197_accload/Guide

Bea, R. G. (1998). Human and organization factors: Engineering operating safety into offshore structures. *Reliability Engineering & System Safety*, *61*(1–2), 109–126. doi:10.1016/S0951-8320(97)00058-6

Bevilacqua, M., & Braglia, M. (2000). The analytic hierarchy process applied to maintenance strategy selection. *Reliability Engineering & System Safety*, *70*(1), 71–83. doi:10.1016/S0951-8320(00)00047-8

British Standards Institution. (2004). Asset management, Part 1: Specification for the optimized management of physical infrastructure assets BSI PAS 55-1.

British Standards Institution. (2010). Risk management – principles and guidelines BS ISO 31000:2009.

Cheng, E. W. L., & Li, H. (2001). Information priority-setting for better resource allocation using Analytic Hierarchy Process (AHP). *Information Management & Computer Security*, *9*(2), 61–70. doi:10.1108/09685220110388827

Chin, K. S., Pun, K. F., Xu, Y., & Chan, J. S. F. (2002). An AHP based study of critical factors for TQM implementation in Shanghai manufacturing industries. *Technovation*, *22*(11), 707–715. doi:10.1016/S0166-4972(01)00065-7

COMAH. (2010). Ageing Plant Operational Delivery Guide. Control of Major Accident Hazards-COMAH Competent Authority. Retrieved from http://www.hse.gov.uk/comah/guidance/ageing-plant-core.pdf

Daim, T., Yates, D., Peng, Y., & Jimenez, B. (2009). Technology assessment for clean energy technologies: The case of the Pacific Northwest. *Technology in Society*, *31*(3), 232–243. doi:10.1016/j.techsoc.2009.03.009

Dier, A.F. (1996, May 6-9). Background to new design manual for platform strengthening, modification and repair. *Proceedings of OTC Offshore Conference*, Houston, Texas.

Dong, W., Moan, T., & Gao, Z. (2012). Fatigue reliability analysis of the jacket support structure for offshore wind turbine considering the effect of corrosion and inspection. *Reliability Engineering & System Safety*, *106*(0), 11–27. doi:10.1016/j.ress.2012.06.011

Engineerlive. (2013). Repairing and strengthening of ageing offshore structures. Oil and Gas Engineer - Exploration/Drilling. Retrieved from http://www.engineerlive.com/content/21403

Ersdal, G. (2005). *Assessment of existing offshore structures for life extension* [Doctoral Thesis].

Estes, A. C., & Frangopol, D. M. (2001). Minimum expected cost-oriented optimal maintenance planning for deteriorating structures: Application to concrete bridge decks. *Reliability Engineering & System Safety*, *73*(3), 281–291. doi:10.1016/S0951-8320(01)00044-8

Etterdal, B., Askheim, D. & Grigorian, H. (2001, 30 April - 3 May). Strengthening of offshore components using high-strength grout: component testing and analytical methods. *Proceedings of OCT Offshore Technology Conference*, Houston, Texas.

Forman, E., & Sally, M. (2001). Decision by Objectives. World Scientific.

Ha, J. S., & Seong, P. H. (2004). A method for risk-informed safety significance categorization using the analytic hierarchy process and Bayesian belief networks. *Reliability Engineering & System Safety*, *83*(1), 1–15. doi:10.1016/j.ress.2003.08.002

HSE. (2009). Structural integrity management framework for fixed jacket structures. RR684 Research Report. Retrieved from http://www.hse.gov.uk/research/rrpdf/rr684.pdf

HSE UK. (2006). Plant ageing: Management of equipment containing hazardous fluids or pressure. Retrieved from http://www.hse.gov.uk/research/rrpdf/rr509.pdf

HSE UK. (2010). Managing Ageing Plant: A Summary Guide. Retrieved from http://www.hse.gov.uk/offshore/ ageing/ ageing-plant-summary-guide.pdf

ISO 19901-3. (2010). Petroleum and natural gas industries -- Specific requirements for offshore structures (Part 3: Topsides structure). Retrieved from http://www.iso.org/iso/home/store/catalogue_tc/catalogue_detail.htm?csnumber=43812

Klir, G. J., & Yuan, B. (1995). *Fuzzy Sets and Fuzzy Logic: Theory and Applications*. Upper Saddle River, New Jersey: Prentice Hall PTR.

Lee, H.W. & Miles, R. (2013). Protecting your assets against corrosion and environmental attack. Offshore Pipeline Integrity Services & Solutions.

Mathworks, (2014). Fuzzy Inference System Modeling: Gaussian Combination Membership Function, http://www.mathworks.se/help/fuzzy/gauss2mf.html (accessed on 20.06.2014).

MATLAB. (2013). MATLAB (R2013a): "Fuzzy Logic Toolbox", 1984-2013 The MathWorks, Inc.

MSL. (2004). Assessment of Repair Techniques for Ageing or Damaged Structures. Project #502 - DOC REF C357R001 Rev 1. Retrieved from http://www.bsee.gov/Research-and-Training/Technology-Assessment-and-Research/tarprojects/500-599/502AA/

Ratnayake, R. M. C. (2012a). Translating sustainability concerns at plant level asset operations: Industrial performance assessment. *International Journal of Sustainable Strategic Management*, *3*(4), 314–339. doi:10.1504/IJSSM.2012.052655

Ratnayake, R. M. C. (2012b). A decision model for executing plant strategy: Maintaining the technical integrity of petroleum flowlines. *International Journal of Decision Sciences. Risk Management*, *4*(1/2), 1–24.

Ratnayake, R. M. C. (2012c). Modelling of asset integrity management process: A case study for computing operational integrity preference weights. *International Journal of Computational Systems Engineering*, *1*(1), 3–12. doi:10.1504/IJCSYSE.2012.044738

Ratnayake, R. M. C. (2014). Application of a fuzzy inference system for functional failure risk rank estimation: RBM of rotating equipment and instrumentation. *Journal of Loss Prevention in the Process Industries, 29*, 216–224. doi:10.1016/j.jlp.2014.03.002

Ratnayake, R. M. C., & Markeset, T. (2010a). Implementing company policies in plant level asset operations: Measuring organizational alignment. *European Journal of Industrial Engineering, 4*(3), 355–371. doi:10.1504/EJIE.2010.033335

Ratnayake, R. M. C., & Markeset, T. (2010b). Technical integrity management: Measuring HSE awareness using AHP in selecting a maintenance strategy. *Journal of Quality in Maintenance Engineering, 16*(1), 44–63. doi:10.1108/13552511011030327

Ratnayake, R. M. C., & Markeset, T. (2012d). Asset integrity management for sustainable industrial operations: Measuring the performance. *International Journal of Sustainable Engineering, 5*(2), 145–158. doi:10.1080/19397038.2011.581391

Saaty, T. L. (1982). *Decision making for leaders: the analytical hierarchy process for decisions in a complex world*. Belmont, CA: Lifetime Learning Publications.

Saaty, T. L. (1990). How to make a decision: The analytical hierarchy process. *European Journal of Operational Research, 48*(1), 9–26. doi:10.1016/0377-2217(90)90057-I

Samarakoon, S. M. S. M. K., & Ratnayake, R. M. C. (2013). Improving efficiency of technology qualification process in industrial sectors. *International Journal of Continuing Engineering Education and Lifelong Learning, 24*(3/4), 362–376. doi:10.1504/IJCEELL.2014.063084

Samarakoon, S. M. S. M. K., & Ratnayake, R. M. C. (2015). Strengthening, modification and repair techniques' prioritization for structural integrity control of ageing offshore structures. *International Journal Reliability Engineering & System Safety, 135*, 15–26. doi:10.1016/j.ress.2014.10.023

Samarakoon, S. M. S. M. K., Ratnayake, R. M. C., & Siriwardane, S. A. S. C. (2013). Structural integrity control of ageing offshore structures: Repairing and strengthening with grouted connections. *Proceedings of the 32ndInternational Conference on Ocean, Offshore and Arctic Engineering OMAE '13*. doi:10.1115/OMAE2013-10925

Scherf, I., Etterdal, B., & Monshaugen, T. (1999). Cost-efficient structural upgrade and life extension of North Sea jacket platforms with use of modern reassessment techniques. *Paper presented at the 31st Offshore Technology Conference*. doi:10.4043/10851-MS

Scholl, A., Manthey, L., Helm, R., & Steiner, M. (2005). Solving multi-attribute design problems with analytic hierarchy process and conjoint analysis: *An empirical comparison. European Journal of Operational Research, 164*(3), 760–777. doi:10.1016/j.ejor.2004.01.026

Sorensen, J. N. (2002). Safety culture: A survey of the state-of-the-art. *Reliability Engineering & System Safety, 76*(2), 189–204. doi:10.1016/S0951-8320(02)00005-4

Tay, K. M., & Lim, C. P. (2008). On the Use of Fuzzy Inference Techniques in Assessment Models: Part II: Industrial Applications. *Fuzzy Optimization and Decision Making, 7*(3), 283–302. doi:10.1007/s10700-008-9037-y

Val, D. V., & Stewart, M. G. (2005). Decision analysis for deteriorating structures. *Reliability Engineering & System Safety*, *87*(3), 377–385. doi:10.1016/j.ress.2004.06.006

Wang, X., Rabiei, M., Hurtado, J., Modarres, M., & Hoffman, P. A. (2009). Probabilistic-based airframe integrity management model. *Reliability Engineering & System Safety*, *94*(5), 932–941. doi:10.1016/j.ress.2008.10.010

Zio, E. (1996). On the use of the analytic hierarchy process in the aggregation of expert judgments. *Reliability Engineering & System Safety*, *53*(2), 127–138. doi:10.1016/0951-8320(96)00060-9

Compilation of References

ABS. (2013), Guidance notes on: accidental load analysis and design for offshore structures, American Bureau of Shipping. http://www.eagle.org/eagleExternalPortalWEB/ShowProperty/BEA%20Repository/Rules&Guides/Current/197_accload/Guide

ACI Committee 318. (2008). *Building code requirements for structural concrete and commentary (ACI 318-08)* (p. 467). Farmington Hills, MI: American Concrete Institute.

Adduri, P. R., & Penmetsa, R. C. (2008). Confidence bounds on component reliability in the presence of mixed uncertain variables. *International Journal of Mechanical Sciences*, *50*(3), 481–489. doi:10.1016/j.ijmecsci.2007.09.015

Adeli, H., & Kim, H. (2004). Wavelet-Hybrid Feedback-Least Mean Square Algorithm for Robust Control of Structures. *Journal of Structural Engineering*, *130*(1), 128–137. doi:10.1061/(ASCE)0733-9445(2004)130:1(128)

Adhikari, S., Chowdhury, R., & Friswell, M. I. (2011). High dimensional model representation method for fuzzy structural dynamics. *Journal of Sound and Vibration*, *330*(7), 1516–1529. doi:10.1016/j.jsv.2010.10.010

Akpan, U. O., Koko, T. S., Orisamolu, I. R., & Gallant, B. K. (2001a). Practical fuzzy finite element analysis of structures. *Finite Elements in Analysis and Design*, *38*(2), 93–111. doi:10.1016/S0168-874X(01)00052-X

Akpan, U. O., Koko, T. S., Orisamolu, I. R., & Gallant, B. K. (2001b). Fuzzy finite element analysis of smart structures. *Smart Materials and Structures*, *10*(2), 273–284. doi:10.1088/0964-1726/10/2/312

Al-Amin, M., Zhou, W., Zhang, S., Kariyawasam, S., & Wang, H. (2012). Bayesian model for calibration of ILI tools. *Proceedings of the Biennial International Pipeline Conference, IPC* (Vol. 2, pp. 201-208). doi:10.1115/IPC2012-90491

Alefeld, G., & Herzberger, J. (1983). *Introduction to Interval Computations*. New York, USA: Academic Press.

Alis, O. F., & Rabitz, H. (2001). Efficient implementation of high dimensional model representations. *Journal of Mathematical Chemistry*, *29*(2), 127–142. doi:10.1023/A:1010979129659

Almusallam, A. A., Al-Gahtani, A. S., Aziz, A. R., & Rasheeduzzafar, . (1996). Effect of reinforcement corrosion on bond strength. *Construction & Building Materials*, *10*(2), 123–129. doi:10.1016/0950-0618(95)00077-1

Alonso, C., Andrade, C., Rodriguez, J., & Diez, J. M. (1996). Factors controlling cracking of concrete affected by reinforcement corrosion. *Materials and Structures*, *31*(211), 435–441.

Al-Qureshi, H. A. (1999). Elastic-plastic analysis of tube bending. *International Journal of Machine Tools & Manufacture*, *39*(1), 87–104. doi:10.1016/S0890-6955(98)00012-1

American Bureau of Shipping (ABS). (2014). *Bottom founded offshore wind turbine installation* (2014th ed.). New York.

American Petroleum Institute (API). (2014). *Recommended Practice for Planning*. Dallas: Design and Constructing Fixed Offshore Platform.

American Society of Civil Engineers. (2007). *Seismic rehabilitation of existing buildings (ASCE/SEI 41-06)* (p. 411). Reston, VA: ASCE.

American Society of Mechanical Engineers Boiler and Pressure Vessel Code. (2000). *Sec. III, Nuclear Power Plant Components*.

Anderko, A., Sridhar, N., & Tormoenc, G. (2010). Localised corrosion of heat-treated alloys Part II -Predicting grain boundary microchemistry and its effect on repassivation potential. *Corrosion Engineering, Science and Technology. The International Journal of Corrosion Processes and Corrosion Control*, *45*(3), 204–223.

Anderson, G. L., Crowson, A., & Chandra, J. (1992). Introduction to smart structures. In *Intelligent Structural Systems* (pp. 1–8). Springer Netherlands. doi:10.1007/978-94-017-1903-2_1

Anderson, J. D. (2003). *Fundamentals of Aerodynamics* (4th ed.). U.S.A.: McGraw-Hill.

Ang, A. H.-S., & Tang, W. H. (2007). *Probability Concepts in Engineering Emphasis on Application in Civil & Environmental Engineering* (2nd ed.). New York, USA: John Wiley & Sons.

ANSYS APDL 14, ANSYS Inc. (n. d.). Basic Analysis Procedures Guide Release 14. Retrieved from http://www.ansys.com

Arsene, S., & Bai, J. (1996). A New Approach to measuring transverse properties of structural tubing by a ring test. *Journal of Testing and Evaluation*, *24*(6), 386–391. doi:10.1520/JTE11461J

Arsène, S., & Bai, J. (1998). A new approach to measuring transverse properties of structural tubing by a ring test–experimental investigation. *ASTM Journal of Testing and Evaluation*, *26*(1), 26–30. doi:10.1520/JTE11966J

Ashley, H., & Rodden, W. P. (1972). Wing-Body Aerodynamic Interaction. *Annual Review of Fluid Mechanics*, *4*(1), 431–472. doi:10.1146/annurev.fl.04.010172.002243

Attoh-Okine, N. O., Cooger, K., & Mensah, S. (2009). Multivariate adaptive regression (MARS) and hinged hyperplanes (HHP) for doweled pavement performance modeling. *Journal of Construction and Building Materials*, *23*(9), 3020–3023. doi:10.1016/j.conbuildmat.2009.04.010

Au, S. K., & Beck, J. L. (2003). Subset Simulation & its application to seismic risk based on dynamic analysis. *Journal of Engineering Mechanics*, *129*(8), 901–917. doi:10.1061/(ASCE)0733-9399(2003)129:8(901)

Avci, E., & Turkoglu, I. (2009). An intelligent Diagnosis System Based on Principal Component Analysis and ANFIS for the Hearth Valve Diseases. *Expert Systems with Applications*, *36*(2), 2873–2878. doi:10.1016/j.eswa.2008.01.030

Aydinoglu, M. N. (2004, June 28 – July 1). An Improved Pushover Procedure for Engineering Practice: Incremental Response Spectrum Analysis (IRSA) in Performance-based seismic design. In Krawinkler & Fajfar (Eds.), Concepts and implementation PEER report, Bled, Slovenia (pp. 345-356).

Azarpeyvand, M. (2005). Active Noise Cancellation of a Spherical Multipole Source. *Acoustical Physics*, *51*(6), 6. doi:10.1134/1.2130891

Bae, B.-K., Cho, S.-K., & Seok, C.-S. (2008). A study on ring tensile specimens. *Materials Science and Engineering A*, *483-484*, 248–250. doi:10.1016/j.msea.2006.10.203

Bae, S., & Bayrak, O. (2008a). Plastic Hinge Length of Reinforced Concrete Columns. *ACI Structural Journal*, *105*(3), 290–300.

Bae, S., & Bayrak, O. (2008b). Seismic Performance of Full-Scale Reinforced Concrete Columns. *ACI Structural Journal, 105*(2), 123–133.

Balakrishnan, K. S., Samal, M. K., Parashar, J., Tiwari, G. P., & Anantharaman, S. (2014). Suitability of miniature tensile specimens for estimating the mechanical property data of pressure tubes- An assessment. *Transactions of the Indian Institute of Metals, 67*(1), 47–55. doi:10.1007/s12666-013-0316-0

Balu, A. S., & Rao, B. N. (2012a). High dimensional model representation based formulations for fuzzy finite element analysis of structures. *Finite Elements in Analysis and Design, 50*, 217–230. doi:10.1016/j.finel.2011.09.012

Balu, A. S., & Rao, B. N. (2012b). Multicut-HDMR for structural reliability bounds estimation under mixed uncertainties. *Computer-Aided Civil and Infrastructure Engineering, 27*(6), 419–438. doi:10.1111/j.1467-8667.2012.00763.x

Balu, A. S., & Rao, B. N. (2012c). Inverse structural reliability analysis under mixed uncertainties using high dimensional model representation and fast Fourier transform. *Engineering Structures, 37*, 224–234. doi:10.1016/j.engstruct.2011.12.043

Balu, A. S., & Rao, B. N. (2013). Confidence bounds on design variables using HDMR based inverse reliability analysis. *Journal of Structural Engineering, 139*(6), 985–996. doi:10.1061/(ASCE)ST.1943-541X.0000709

Balu, A. S., & Rao, B. N. (2014). Efficient assessment of structural reliability in presence of random and fuzzy uncertainties. *Journal of Mechanical Design, 136*(5), 051008. doi:10.1115/1.4026650

Bani-Hani, K., Ghaboussi, J., & Schneider, S. P. (1999). Experimental Study of Identification and Control of Structures using Neural Network Part 1: Identification. *Earthquake Engineering & Structural Dynamics, 28*(9), 995–1018. doi:10.1002/(SICI)1096-9845(199909)28:9<995::AID-EQE851>3.0.CO;2-8

Barda, F. (1972). Shear strength of low-rise walls with boundary elements. PhD Dissertation, Lehigh University, Bethlehem, PA.

Barda, F., Hanson, J. M., & Corley, W. G. (1977). *Shear strength of low-rise walls with boundary elements. Reinforced Concrete Structures in Seismic Zones, SP 53-8* (pp. 149–202). Farmington Hills, MI: American Concrete Institute.

Bayrak, O., & Sheikh, S. A. (2001). Plastic hinge analysis. *Journal of structural engineering (New York, N.Y.), 127*(9), 1092–1100.

Bazant, Z. P., & Pijaudier-Cabot, G. (1988). Nonlocal continuum damage, localization instability and convergence. *Journal of Applied Mechanics, 55*(2), 287–293. doi:10.1115/1.3173674

Bea, R. G. (1998). Human and organization factors: Engineering operating safety into offshore structures. *Reliability Engineering & System Safety, 61*(1–2), 109–126. doi:10.1016/S0951-8320(97)00058-6

Beauvilain, T. A., & Bolton, J. S. (1993). Cancellation of radiated sound fields by the use of multipole secondary sources. *Proceedings of the second Conference on Recent Advances in Active Control of Sound and Vibration* (Technomic, Lancaster, PA) (pp. 957-968).

Beer, G., & Watson, J. O. (1992). Introduction to Finite and Boundary Element Methods for Engineers. Chichester, England: John Wiley & Sons Ltd.

Behnam, B. (2014). Post-Earthquake Fire Analysis and Risk Mitigation Strategies. (PhD), the University of Queensland, Brisbane, Australia. doi:10.14264/uql.2014.122

Behnam, B., Ronagh, H. R., & Lim, P. J. (2015). Numerical evaluation of the post-earthquake fire resistance of CFRP-strengthened reinforced concrete joints based on experimental observations. *European Journal of Environmental and Civil Engineering*, 1-19. doi:10.1080/19648189.2015.1018448

Behnam, B., & Ronagh, H. (2014). Performance-Based Vulnerability Assessment of Multi-Story Reinforced Concrete Structures Exposed to Pre- and Post-Earthquake Fire. *Journal of Earthquake Engineering*, *18*, 853–875.

Belgasmia, M., & Mossaoui, S. (2015, February). Comparison between static nonlinear and time history analysis using Flexibility-based model for an existing structure and effect of taking into account soil using domain reduction method for single media. *KSCE Journal of Civil Engineering*.

Belgasmia, M., Spacone. E., Urbanski, A., & Zimmermann, Th. (2007, August). Seismic evaluation of construction: static pushover procedure and time history analysis for nonlinearframes. Swiss Federal Institute of Technology in Lausanne, Switzerland, LSC Internal Report.

Belgasmia, M., & Moussaoui, S. (2013). Comparison of Static Pushover Analysis in the case of Small and Large Deformation with Time History Analysis Using Flexibility-Based Model for an existing Structure. *International Journal of Current Engineering and Technologie*, *3*(2), 655–665.

Beltran, R., Maldonado, J. G., Murr, L. E., & Fisher, W. W. (1997). Effects of strain and grain size on carbide precipitation and corrosion sensitization behavior in 304 stainless steel. *Acta Materialia*, *45*(10), 4351–4360. doi:10.1016/S1359-6454(97)00106-7

Benallal, A., Billardon, R., & Geymonat, G. (1993). Bifurcation and localization in rate-independent materials. Some general considerations. In Q. S. Nguyen (Ed.), *Bifurcation and Stability of dissipative Systems 1993* (pp. 1–44). Berlin: Springer Verlag. doi:10.1007/978-3-7091-2712-4_1

Benjamin, J. R., & Williams, H. A. (1956). Investigation of shear walls, part 12 – studies of reinforced concrete shear wall assemblies (Report No. 10). Department of Civil Engineering, Stanford University, Stanford.

Berryman, M., & Angus, S. D. (2010). *Tutorials on Agent-based modelling with NetLogo and Network Analysis with Pajek* [Doctoral dissertation]. World Scientific Publishing Company.

Bertolino, G., Meyer, G., & Perez Ipiña, J. (2003). In-situ crack growth observation and fracture toughness measurement of hydrogen charged zircaloy-4. *Journal of Nuclear Materials*, *322*(1), 57–65. doi:10.1016/S0022-3115(03)00305-2

Bertolino, G., Meyer, G., & Perez, I. J. (2003). Effects of hydrogen content and temperature on fracture toughness of Zircaloy-4. *Journal of Nuclear Materials*, *320*(3), 272–279. doi:10.1016/S0022-3115(03)00193-4

Bertolotti, T. (2014). Generative and Demonstrative Experiments. In *Model-Based Reasoning in Science and Technology* (pp. 479–498). Springer Berlin Heidelberg. doi:10.1007/978-3-642-37428-9_27

Bertsch, J., & Hoffelner, W. (2006). Crack resistance curve determination of tube cladding material. *Journal of Nuclear Materials*, *352*(1-3), 116–125. doi:10.1016/j.jnucmat.2006.02.045

Bevilacqua, M., & Braglia, M. (2000). The analytic hierarchy process applied to maintenance strategy selection. *Reliability Engineering & System Safety*, *70*(1), 71–83. doi:10.1016/S0951-8320(00)00047-8

Bhargava, K., Ghosh, A. K., Mori, Y., & Ramanujam, S. (2006). Analytical model for time to cover cracking in rc structures due to rebar corrosion. *Nuclear Engineering and Design*, *236*(11), 1123–1139. doi:10.1016/j.nucengdes.2005.10.011

Bhattacharjya, S., & Chakraborty, S. (2011). Robust optimization of Structures subjected to stochastic earthquake excitation under limited information on system parameters uncertainty. *Engineering Optimization*, *43*(12), 1311–1330. doi:10.1080/0305215X.2011.554545

Bisplinghoff, R. L., Ashley, H., & Halfman, R. (1955). *Aeroelasticity*. Reading, Massachusetts: Addison Wesley.

Bolt, S.E., & Greenstreet, W.L. (1971). Experimental determination of plastic collapse loads for pipe elbows. ASME Paper 71-PVP-37.

Bolton, J. S., & Beauvilain, T. A. (1992). Multipole sources for cancellation of radiated sound fields. *The Journal of the Acoustical Society of America*, *91*(4), 2349. doi:10.1121/1.403449

Bolton, J. S., Gardner, B. K., & Beauvilain, T. A. (1995). Cancellation by the use of secondary multipoles. *The Journal of the Acoustical Society of America*, 1995, 98.

Bond, M. P. (1971). *A theoretical and experimental investigation of multi mitred pipe bends subjected to various types of loading* [PhD thesis]. Institute of Science & Technology, Victoria University of Manchester.

Bond, M. P., & Kitching, R. (1972). Stress and flexibility factors for multi-mitred pipe bends subjected to internal pressure combined with external loadings. *J Strain Anal*, *7*(2), 97–108. doi:10.1243/03093247V072097

Boore, D. M. (2003). Simulation of ground motion using the stochastic method. *Pure and Applied Geophysics*, *160*(3), 635–676. doi:10.1007/PL00012553

Bordini, R. H., Braubach, L., Dastani, M., Seghrouchni, A. E. F., Gomez-Sanz, J. J., Leite, J., ... & Ricci, A. (2006). A survey of programming languages and platforms for multi-agent systems. *Informatica, 30*(1).

Boyle, J. T., & Spence, J. (1981). A simple stress analysis for out-of-round pressurised pipe bends. *International Journal of Pressure Vessels and Piping*, *9*(4), 251–261. doi:10.1016/0308-0161(81)90027-2

Brabin, T. A., Christopher, T., & Rao, B. N. (2011). Bursting pressure of mild steel cylindrical vessels. *International Journal of Pressure Vessels and Piping, Vol*, *88*, 11–22.

Branimir Cace. (2006). ENTORAMA multi-agent modeling and simulation tool. Retrieved from http://www.entorama.com/

Bratley, P., & Fox, B. L. (1988). Implementing Sobol's quasirandom sequence generator. *ACM Transactions on Mathematical Software*, *14*(1), 88–100. doi:10.1145/42288.214372

Bratman, M. E., Israel, D., & Pollack, M. E. (1988). Plans and resource-bounded practical reasoning. *Computational Intelligence*, *4*(4), 349–355. doi:10.1111/j.1467-8640.1988.tb00284.x

Brazier, L. G. (1927). On The Flexure of Thin Cylindrical Shells and Other Thin Sections. Proceedings of Royal Society, London, U.K.

British Standard Europaische Norm. (2004). (*BS*). U.K.: Wind Turbines-Protective Measures-Requirements for Design Operation and Maintenance.

British Standards Institution. (2004). Asset management, Part 1: Specification for the optimized management of physical infrastructure assets BSI PAS 55-1.

British Standards Institution. (2010). Risk management – principles and guidelines BS ISO 31000:2009.

Brooks, R. (1986). A robust layered control system for a mobile robot. *IEEE Journal of* Robotics and Automation, *2*(1), 14–23.

Bruemmer, S. M. (1990). Quantitative modeling of sensitization development in austenitic Stainless steels. *Corrosion*, *46*(9), 698–709. doi:10.5006/1.3585171

Bucher, C., & Most, T. (2008). A comparison of approximate response functions in structural reliability analysis. *Probabilistic Engineering Mechanics*, *23*(2-3), 154–163. doi:10.1016/j.probengmech.2007.12.022

Bucher, G. C., & Bourgund, U. (1990). A fast and efficient response surface approach for structural reliability problems. *Structural Safety*, *7*(1), 57–66. doi:10.1016/0167-4730(90)90012-E

Burton, T., Sharpe, D., Jenkins, N., & Bossanyi, E. (2001). *Wind Energy Handbook*. New York: Wiley. doi:10.1002/0470846062

Calladine, C. R. (1974). Limit analysis of curved tubes. *Journal of Mechanical Engineering Science*, *16*(2), 85–87. doi:10.1243/JMES_JOUR_1974_016_016_02

Carter, R. D., Damcott, D. L., Atzmon, M., Was, G. S., Bruemmer, S. M., & Kenik, E. A. (1994). Quantitative analysis of radiation-induced grain-boundary segregation measurements. *Journal of Nuclear Materials*, *211*(1), 70–84. doi:10.1016/0022-3115(94)90282-8

Cerovsek, T. (2011). A review and outlook for a 'Building Information Model'(BIM): A multi-standpoint framework for technological development. *Advanced Engineering Informatics*, *25*(2), 224–244. doi:10.1016/j.aei.2010.06.003

Ceryan, N. (2014). Application of support vector machines and relevance vector machines in predicting uniaxial compressive strength of volcanic rocks. *Journal of African Earth Sciences*, *100*, 634–644. doi:10.1016/j.jafrearsci.2014.08.006

Chakraborty, S., & Chowdhury, R. (2014). Polynomial Correlated Function Expansion for Nonlinear Stochastic Dynamic Analysis. *Journal of Engineering Mechanics*, *141*(3), 04014132:1–04014132:11.

Chakraborty, S., & Chowdhury, R. (2015b). Assessment of polynomial correlated function expansion for high-fidelity structural reliability analysis. *Structural Safety*, *59*(Accepted), 9–19.

Chakraborty, S., & Chowdhury, R. (2015a). A semi-analytical framework for structural reliability analysis. *Computer Methods in Applied Mechanics and Engineering*, *289*(1), 475–497. doi:10.1016/j.cma.2015.02.023

Chakraborty, S., & Chowdhury, R. (2015c). Multivariate function approximations using D-MORPH algorithm. *Applied Mathematical Modelling*, *39*(23-24), 7155–7180. doi:10.1016/j.apm.2015.03.008

Chakraborty, S., Mandal, B., Chowdhury, R., & Chakrabarti, A. (2016). Stochastic free vibration analysis of laminated composite plates using polynomial correlated function expansion. *Composite Structures*, *135*, 236–249. doi:10.1016/j.compstruct.2015.09.044

Chakraborty, S., & Sen, A. (2014). Adaptive response surface based efficient finite element model updating. *Finite Elements in Analysis and Design*, *80*, 33–40. doi:10.1016/j.finel.2013.11.002

Chan, A. S. (2012). Control and Suppression of Laminar Vortex Shedding off Two-Dimensional Bluff Bodies [PhD thesis]. Stanford.

Chan, A. S. (2012). Control And Suppression Of Laminar Vortex Shedding Off Two-Dimensional Bluff Bodies [PhD thesis]. Stanford.

Chandler-Wilde, S., & Langson, S. (2007). Boundary element methods for acoustics. Retrieved from http://www.personal.reading.ac.uk/~sms03snc/fe_bem_notes_sncw.pdf

Chandrasekaran, S. (2015). *Dynamic Analysis and Design of Offshore Structures*. India: Springer. doi:10.1007/978-81-322-2277-4

Chang, J. J. (2007). *Computer simulation of hollowcore concrete flooring systems exposed to fire. (Doctor of Philosophy research)*. Christchurch, New Zealand: University of Canterbury.

Chattopadhyay, J., Natahani, D. K., Dutta, B. K., & Kushwaha, H. S. (2000). Closed-form collapse moment quations of elbows under internal pressure and in-plane bending moment. *J. Pressure Vessel Technol.*, *6*(4), 431–436. doi:10.1115/1.1285988

Chen, S. S. (1985). Flow-Induced Vibration Of Circular Cylindrical Structures. Argonne National Laboratory, The University of Chicago.

Chen, C. (1996). *Fuzzy Logic and Neural Network Handbook*. New York: McGraw-Hill.

Cheng, E. W. L., & Li, H. (2001). Information priority-setting for better resource allocation using Analytic Hierarchy Process (AHP). *Information Management & Computer Security*, *9*(2), 61–70. doi:10.1108/09685220110388827

Chen, L., & Rao, S. S. (1997). Fuzzy finite element approach for the vibration analysis of imprecisely defined systems. *Finite Elements in Analysis and Design*, *27*(1), 69–83. doi:10.1016/S0168-874X(97)00005-X

Chen, W. F., & El-Metwally, S.E.D.E. (2011). *Understanding Structural Engineering: From Theory to Practice*. CRC Press.

Chen, Y., Yang, B., Abraham, A., & Peng, L. (2007). Automatic Design of Hierarchical Takagi-Sugeno Type Fuzzy Systems using Evolutionary Algorithms. *IEEE Transactions on Fuzzy Systems*, *15*(3), 385–397. doi:10.1109/TFUZZ.2006.882472

Chib, S., & Greenberg, E. (1995). Understanding metropolis-hastings algorithm. *The American Statistician*, *49*(4), 327–335.

Chin, A. L., Worth, A. D., Greenough, A. C., Coakley, S., Holcombe, M., & Kiran, M. (2012). FLAME: An approach to the parallelisation of agent-based applications. *Work (Reading, Mass.)*, *501*, 63259.

Chin, K. S., Pun, K. F., Xu, Y., & Chan, J. S. F. (2002). An AHP based study of critical factors for TQM implementation in Shanghai manufacturing industries. *Technovation*, *22*(11), 707–715. doi:10.1016/S0166-4972(01)00065-7

Choi, I.-K., Choun, Y.-S., Ahn, S.-M., & Seo, J.-M. (2008). Probabilistic seismic risk analysis of candu containment structure for near-fault earthquakes. *Nuclear Engineering and Design*, *238*(6), 1382–1391. doi:10.1016/j.nucengdes.2007.11.001

Chojaczyk, A. A., Teixeira, A. P., Neves, L. C., Cardoso, J. B., & Guedes Soares, C. (2015). Review and application of Artificial Neural Networks models in reliability analysis of steel structures. *Structural Safety*, *52*, 78–89. doi:10.1016/j.strusafe.2014.09.002

Chowdhury, R., Rao, B. N., & Prasad, A. M. (2008). High dimensional model representation for piece wise continuous function approximation. *Communications in Numerical Methods in Engineering*, *24*(12), 1587–1609. doi:10.1002/cnm.1053

Chowdhury, R., Rao, B. N., & Prasad, A. M. (2009). High-dimensional model representation for structural reliability analysis. *Communications in Numerical Methods in Engineering*, *25*(4), 301–337. Retrieved from http://doi.wiley.com/10.1002/cnm.1118 doi:10.1002/cnm.1118

Christo Michael, T., Veerappan, A. R., & Shanmugam, S. (2011). Effect of cross section on collapse load in pipe bends subjected to in-plane closing moment. *International Journal of Engineering Science and Technology*, *3*(6), 247–256.

Christo Michael, T., Veerappan, A. R., & Shanmugam, S. (2012). Comparison of plastic limit and collapse loads in pipe bends with shape imperfections under in-plane bending and an internal pressure. *International Journal of Pressure Vessels and Piping*, *99-100*, 23–33. doi:10.1016/j.ijpvp.2012.07.013

Chung, L. L., Lin, R. C., Song, T. T., & Reinhorn, A. M. (1989). Experiments on active control for MDOF seismic structures. *Journal of Engineering Mechanics*, *115*(8), 1609–1627. doi:10.1061/(ASCE)0733-9399(1989)115:8(1609)

Colwell, S., & Basu, B. (2009). Tuned Liquid Column Dampers in Offshore Wind Turbines for Structural Control. *Engineering Structures*, *31*(2), 358–368. doi:10.1016/j.engstruct.2008.09.001

COMAH. (2010). Ageing Plant Operational Delivery Guide. Control of Major Accident Hazards-COMAH Competent Authority. Retrieved from http://www.hse.gov.uk/comah/guidance/ageing-plant-core.pdf

Command, S. (2013). Pushover Analysis with Zsoil Taking Soil into Account. *Proceedings of the Numerics in Geotechnics and Structures Symposium, 28 years of Zsoil and structure PC*, Switzerland.

Congdon, P. (2006). *Bayesian Statistical Modelling*. John Wiley & Sons. doi:10.1002/9780470035948

Cornell, C. A., Jalayer, F., Hamburger, R. O., & Foutch, D. A. (2002). The probabilistic basis for the 2000 SAC Federal Emergency Management Agency steel moment frame guidelines. *Journal of Structural Engineering*, *128*(4), 526–533. doi:10.1061/(ASCE)0733-9445(2002)128:4(526)

Council, B. S. S., Agency, U. S. F. E. M., & Council, A. T. (2003). *NEHRP guidelines for the seismic rehabilitation of buildings*. Washington, D.C.: Federal Emergency Management Agency.

Craig, R. R., & Kurdila, A. J. (2006). *Fundamentals of Structural Dynamics* (2nd ed.). U.S.A.: John Wiley and Sons.

Crisfield, M. A. (1981). A fast incremental/iterative solution procedure that handles "snap-through". *Computers & Structures*, *13*(1-3), 55–62. doi:10.1016/0045-7949(81)90108-5

Daim, T., Yates, D., Peng, Y., & Jimenez, B. (2009). Technology assessment for clean energy technologies: The case of the Pacific Northwest. *Technology in Society*, *31*(3), 232–243. doi:10.1016/j.techsoc.2009.03.009

Danish Wind Turbine Manufacturers Association. (2000). *Wind Energy*. Pioneer Poul la Cour.

Dan, V. (2009). The Influence Of The Initial Ovality Tolerance On The Nonlinear Cycling analysis Of Piping Bends. *Journal of Pressure Vessel Technology*, *131*(4).

Dassault Systmes Simulia Corp. (2009). ABAQUS documentation and theory manual.

Daum, R. S., Majumdar, S., Tsai, H., Bray, T. S., Koss, D. A., Motta, A. T., & Billone, M. C. (2002). Mechanical property testing of irradiated Zircaloy cladding under reactor transient conditions. In M. Sokolov, J. D. Landes, & G. E. Lucas (Eds.), *ASTM STP 1418. Small Specimen Test Techniques: fourth volume*. West Conshohocken, PA, USA: ASTM International. doi:10.1520/STP10821S

Davenport, A. G. (1961). The Spectrum of Horizontal Gustiness Near The Ground in High Winds. *Journal of Royal Metrological Society.*, *87*(372), 194–211. doi:10.1002/qj.49708737208

Davidsson, P., & Boman, M. (2000). A multi-agent system for controlling intelligent buildings. *Proceedings of the Fourth International Conference on MultiAgent Systems '00* (pp. 377-378). IEEE. doi:10.1109/ICMAS.2000.858483

Dayal, R. K., Parvathavarthini, N., & Raj, B. (2005). Influence of metallurgical variables on sensitisation kinetics in austenitic stainless steels. *International Materials Reviews*, *50*(3), 129–155. doi:10.1179/174328005X14348

DeGersem, H., Moens, D., Desmet, W., & Vandepitte, D. (2007). Interval and fuzzy dynamic analysis of finite element models with superelements. *Computers & Structures*, *85*(5–6), 304–319. doi:10.1016/j.compstruc.2006.10.011

Dempster, A. P., Laird, N. M., & Rubin, D. B. (1977). Maximum Likelihood from Incomplete Data via the EM Algorithm. *Journal of the Royal Statistical Society. Series B. Methodological*, *39*(1), 1–38.

Demuth, H., & Beale, M. (2003). *Neural Network Toolbox for MATLAB-User Guide Version 4.1*. The Math Works Inc.

Deng, J., Gu, D., Li, X., & Yue, Z. Q. (2005, June). Structural reliability analysis for implicit performance functions using artificial neural network. *Structural Safety*, *27*(1), 25–48. doi:10.1016/j.strusafe.2004.03.004

Der Kiureghian, A., & Ditlevsen, O. (2009). Aleatory or epistemic? does it matter? *Structural Safety*, *31*(2), 105–112. doi:10.1016/j.strusafe.2008.06.020

Deshmukh, V., Singha, U., Tonpe, S., & Saibaba, N. (2010). Effect of process variables on texture and creep of pressurized heavy water reactor pressure tubes. *Transactions of the Indian Institute of Metals*, *63*(2-3), 397–402. doi:10.1007/s12666-010-0054-5

Desquines, J., Koss, D. A., Motta, A. T., Cazalis, B., & Petit, M. (2011). The issue of stress state during mechanical tests to assess cladding performance during a reactivity-initiated accident (RIA). *Journal of Nuclear Materials*, *412*(2), 250–267. doi:10.1016/j.jnucmat.2011.03.015

Dessombz, O., Thouverez, F., Laîné, J. P., & Jézéquel, L. (2001). Analysis of mechanical systems using interval computations applied to finite element methods. *Journal of Sound and Vibration*, *239*(5), 949–968. doi:10.1006/jsvi.2000.3191

Det Norske Veritas. (2004). Design of Offshore Wind Turbine Structures. Offshore Standard DNV-OS-J101. Norway.

Det Norske Veritas. (2009). Offshore Substations for Wind Farms. Offshore Standard DNV-OS-J201. Norway.

Dhingra, A. K., Rao, S. S., & Kumar, V. (1992). Nonlinear membership function in multi-objective fuzzy optimization of mechanical and structural systems. *AIAA Journal*, *30*(1), 251–260. doi:10.2514/3.10906

DiazDelaO. F. A., Adhikari, S., Saavedra Flores, E. I., & Friswell, M. I. (2013). Stochastic structural dynamic analysis using Bayesian emulators. *Computers & Structures, 120*, 24–32. Retrieved from http://apps.webofknowledge.com/full_record.do?product=UA&search_mode=GeneralSearch&qid=4&SID=X2RcsR1DGP199mjAKwr&page=1&doc=2

Dier, A.F. (1996, May 6-9). Background to new design manual for platform strengthening, modification and repair. *Proceedings of OTC Offshore Conference*, Houston, Texas.

Ditlevsen, O. (1982). Model uncertainty in structural reliability. *Structural Safety*, *1*(1), 73–86. doi:10.1016/0167-4730(82)90016-9

Ditlevsen, O., & Madsen, H. O. (1996). *Structural reliability methods* (1st ed.). New Jersey: Wiley-Blackwell.

Djojodihardjo, H & Tendean, E. (2004, October). A Computational Technique for the Dynamics of Structure Subject to Acoustic Excitation. *Proceedings of ICAS '04*, Vancouver.

Djojodihardjo, H. (2007). BEM-FEM Acoustic-Structural Coupling for Spacecraft Structure Incorporating Treatment of Irregular Frequencies. *Proceedings of the58th International Astronautical Congress*, Hyderabad.

Djojodihardjo, H. (2008). Unified BE-FE Aerodynamic-Acoustic-Structure Coupling Scheme for Acoustic Effects on Aeroelastic Structures. *Proceedings of the 26th International Congress Of the Aeronautical Sciences*, Anchorage.

Djojodihardjo, H. (2009). Application of Fast Multipole Boundary Element Method for Unified BEM-FEM Acoustic-Structural Coupling. *Proceedings International Astronautical Congress.*

Djojodihardjo, H., & Safari, I. (2005). Further Development of the Computational Technique for Dynamics of Structure Subjected to Acoustic Excitation, paper IAC-05-C2.2.08. *Presented at the56th International Astronautical Congress*, Fukuoka Japan.

Djojodihardjo, H., & Safari, I. (2006). Unified Computational Scheme For Acoustic Aeroelastomechanic Interaction, paperIAC-06-C2.3.09. *Presented at the57th International Astronautical Congress*, Valencia, Spain.

Djojodihardjo, H., & Tendean, F. (2001). Structure-Acoustic Analysis Using BEM/FEM; Implementation [Master's Thesis]. Structural Mechanics & Engineering Acoustics, LTH, Sweden.

Djojodihardjo, H. (2013). BEM-FEM Coupling for Acoustic Effects On Aeroelastic Stability Of Structures. *CMES*, *91*(3), 205–234.

Djojodihardjo, H. (2013). *Unified Aerodynamic-Acoustic Formulation for Aero-Acoustic Structure Coupling. Journal of Mechanics Engineering and Automation, 3*, 209–220.

Djojodihardjo, H. (2015). Vibro-Acoustic Analysis of the Acoustic-Structure Interaction of Flexible Structure Due to Acoustic Excitation. *Acta Astronautica, 108*(March–April).

Djojodihardjo, R. H., & Widnall, S. E. (1969). A Numerical Method for the Calculation of Nonlinear, Unsteady Lifting Potential Flow Problems. *AIAA Journal, 7*(10), 2001–2009. doi:10.2514/3.5494

Donders, S., Vandepitte, D., Van de Peer, J., & Desmet, W. (2005). Assessment of uncertainty on structural dynamic responses with the short transformation method. *Journal of Sound and Vibration, 288*(3), 523–549. doi:10.1016/j.jsv.2005.07.003

Dong, W. M., & Shah, H. C. (1987). Vertex method for computing functions of fuzzy variables. *Fuzzy Sets and Systems, 24*(1), 65–78. doi:10.1016/0165-0114(87)90114-X

Dong, W. M., & Wong, F. S. (1987). Fuzzy weighted averages and implementation of the extension principle. *Fuzzy Sets and Systems, 21*(2), 183–199. doi:10.1016/0165-0114(87)90163-1

Dong, W., Moan, T., & Gao, Z. (2012). Fatigue reliability analysis of the jacket support structure for offshore wind turbine considering the effect of corrosion and inspection. *Reliability Engineering & System Safety, 106*(0), 11–27. doi:10.1016/j.ress.2012.06.011

Dowling, A. P., & Ffowcs Williams, J. E. (1983). Sound and Sources of Sound. Chichester: John Wiley & Sons.

Drucker, H., Burges, C. J. C., Kaufman, L., Smola, A., & Vapnik, V. (1997). Support vector regression machines. *Advances in Neural Information Processing Systems, 9*, 155.

Dubourg, V., Sudret, B., & Bourinet, M. (2011). Reliability-based design optimization using kriging surrogates and subset simulation. *Structural Multidisciplinary Optimization, 44*(5), 673–690. doi:10.1007/s00158-011-0653-8

Du, H., & Zhang, N. (2008). Application of Evolving Takagi-Sugeno Fuzzy Model to Nonlinear System Identification. *Applied Soft Computing, 8*(1), 676–686. doi:10.1016/j.asoc.2007.05.006

Duraiswami, R. (2006). The Wave and Helmholtz Equations. *MD Medical Newsmagazine*.

Dyke, S. J., Spencer, B. F. Jr, Sain, M. K., & Carlson, J. D. (1996). Modeling and Control of Magnetorheological Dampers for Seismic Response Reduction. *Smart Materials and Structures, 5*(5), 565–575. doi:10.1088/0964-1726/5/5/006

Dyke, S. J., Spencer, B. F. Jr, Sain, M. K., & Carlson, J. D. (1998). An experimental study of MR dampers for seismic protection. *Smart Materials and Structures, 7*(5), 693–703. doi:10.1088/0964-1726/7/5/012

Dymiotis, C., Kappos, A. J., & Chryssanthopoulos, M. K. (1999). Seismic reliability assessment of RC frames with uncertain drift and member capacity. *Journal of Structural Engineering, 125*(9), 1038–1047. doi:10.1061/(ASCE)0733-9445(1999)125:9(1038)

Echard, B., Gayton, N., Lemaire, M., & Relun, N. (2013). A combined importance sampling and Kriging reliability method for small failure probabilities with time-demanding numerical models. *Reliability Engineering & System Safety, 111*, 232–240. doi:10.1016/j.ress.2012.10.008

Eclipse. (2015). Agent Modeling Platform from the Eclipse Foundation. Retrieved from http://www.eclipse.org/amp/

Edinger, R., & Kaul, S. (2007). *Renewable Resources for Electric Power: Prospects and challenges*. Westport: Quorum.

Edsinger, K., Davies, J. H., & Adamson, R. B. (2000). Degraded fuel cladding fractography and fracture behaviour. In *Ziconium in nuclear industry* (pp. 316-339). ASTM.

Elhewy, A. H., Mesbahi, E., & Pu, Y. (2006). Reliability analysis of structures using neural network method. *Probabilistic Engineering Mechanics*, *21*(1), 44–53. doi:10.1016/j.probengmech.2005.07.002

Elishakoff, I. (1995). Essay on uncertainties in elastic and viscoelastic structures: From A.M. Freudenthal's criticism to modern convex modelling. *Computers & Structures*, *56*(2), 871–895. doi:10.1016/0045-7949(94)00499-S

Ellingwood, B. (1990). Validation studies of seismic PRAs. *Nuclear Engineering and Design*, *123*(2-3), 189–196. doi:10.1016/0029-5493(90)90237-R

Ellingwood, B. R., Celik, O. C., & Kinali, K. (2007). Fragility assessment of building structural systems in Mid-America. *Earthquake Engineering & Structural Dynamics*, *36*(13), 1935–1952. doi:10.1002/eqe.693

Engelund, S., Sϕrensen, J. D., & Sϕrensen, B. (1999). Evaluation of repair and maintenance strategies for concrete coastal bridges on a probabilistic basis. *ACI Materials Journal*, *96*(2), 160–166.

Engineerlive. (2013). Repairing and strengthening of ageing offshore structures. Oil and Gas Engineer - Exploration/ Drilling. Retrieved from http://www.engineerlive.com/content/21403

Ersdal, G. (2005). *Assessment of existing offshore structures for life extension* [Doctoral Thesis].

Ervine, A., Gillie, M., Stratford, T. J., & Pankaj, P. (2012). Thermal propagation through tensile cracks in reinforced concrete. *Journal of Materials in Civil Engineering*, *24*(5), 516–522. doi:10.1061/(ASCE)MT.1943-5533.0000417

Estes, A. C., & Frangopol, D. M. (2001). Minimum expected cost-oriented optimal maintenance planning for deteriorating structures: Application to concrete bridge decks. *Reliability Engineering & System Safety*, *73*(3), 281–291. doi:10.1016/ S0951-8320(01)00044-8

Etterdal, B., Askheim, D. & Grigorian, H. (2001, 30 April - 3 May). Strengthening of offshore components using high-strength grout: component testing and analytical methods. *Proceedings of OCT Offshore Technology Conference*, Houston, Texas.

Eurocode 8. (2003). Design of Structures for Earthquake Resistance. European Committee for Standardization.

Fajfar, P. (1999). Capacity Spectrum Method Based on Inelastic Demand Spectra. *Earthquake Engineering & Structural Dynamics*, *28*(9), 979–993. doi:10.1002/(SICI)1096-9845(199909)28:9<979::AID-EQE850>3.0.CO;2-1

Fajfar, P. (2000). A Nonlinear Analysis Method for Performance Based Seismic Design. *Earthquake Spectra*, *16*(3), 573–592. doi:10.1193/1.1586128

Fajfar, P. (2002). Structural Analysis in Earthquake Engineering – A Breakthrough of Simplified Non-Linear Methods. *Proc. 12th European Conference on Earthquake Engineering.*

Faltinsen, O. M. (1999). *Sea Loads on Ships and Offshore Structures.* U.S.A.: Cambridge University Press.

Fang, K. T. (1980). Experimental design by uniform distribution. *Acta Mathematice Applicatae Sinica*, *3*, 363–372.

Faravelli, L. (1989). Response surface approach for reliability analyses. *Journal of Engg. Mech. ASCE*, *115*(2), 2763–2781. doi:10.1061/(ASCE)0733-9399(1989)115:12(2763)

Farkas, L., Moens, D., Vandepitte, D., & Desmet, W. (2008). Application of fuzzy numerical techniques for product performance analysis in the conceptual and preliminary design stage. *Computers & Structures*, *86*(10), 1061–1079. doi:10.1016/j.compstruc.2007.07.012

Faroz, S. A., Pujari, N. N., & Ghosh, S. (2014). A Bayesian Markov Chain Monte Carlo approach for the estimation of corrosion in reinforced concrete structures. In B. H. V. Topping & P. Ivanyi (Eds.), *Proceedings of the Twelfth International Conference on Computational Structures Technology*, Stirlingshire, UK. doi:10.4203/ccp.106.150

FEMA356. (2000). *Prestandard and commentary for the seismic rehabilitation of buildings Rehabilitation Requirements*. Washington, DC: American Society of Civil Engineers.

Ferguson, I. A. (1992). *Touring Machines: An architecture for dynamic, rational, mobile agents* [Doctoral dissertation]. University of Cambridge.

Filev, D. P. (1991). Fuzzy Modeling of Complex Systems. *International Journal of Approximate Reasoning*, *5*(3), 281–290. doi:10.1016/0888-613X(91)90013-C

Firat Alemdar, Z. (2010). *Plastic hinging behavior of reinforced concrete bridge columns* [Ph.D. Thesis]. University of Kansas, Kansas, USA. Retrieved from http://search.proquest.com/docview/305210088?accountid=14723

Fistikoglu, O., & Okkan, U. (2011). Statistical down scaling of monthly precipitation using NCEP/NCAR reanalysis data for Tahtali River Basin in Turkey. *Journal of Hydrologic Engineering*, *16*(2), 157–164. doi:10.1061/(ASCE)HE.1943-5584.0000300

Fitzpatrick, F. (1914). *Fireproof construction*. Chicago: American School of Correspondence.

Forman, E., & Sally, M. (2001). Decision by Objectives. World Scientific.

Franssen, J. M., Kodur, V. K. R., & Mason, J. (2002). *Elements of Theory for SAFIR2001 Free: A Computer Program for Analysis of Structures at Elevated Temperature Conditions*. Belgium: University of Liege.

Franssen, J.-M. (2005). SAFIR: A thermal/structural program for modeling structures under fire. *English Journal*, *42*(3), 143–150.

Friedman, J. H. (1991). Multivariate adaptive regression splines. *Annals of Statistics*, *19*(1), 1–141. doi:10.1214/aos/1176347963

Fukuda, T., Itagaki, N., Kamimura, J., Mozumi, Y., & Furuya, T. (2003, March 16-19). Fracture toughness of hydrogen-absorbed zircaloy-2 cladding tube. Proceedings of Topfuel 2003, Wurzburg, Germany.

Fullman, R. L. (1982). A thermodynamic model of the effects of composition on the susceptibility of stainless steels to intergranular stress corrosion cracking. *Acta Metallurgica*, *30*(7), 1407–1415. doi:10.1016/0001-6160(82)90161-4

Fu, Y., Wu, X., Han, E.-H., Ke, W., Yang, K., & Jiang, Z. (2009). Effects of cold work and sensitization treatment on the corrosion resistance of high nitrogen stainless steel in chloride solutions. *Electrochimica Acta*, *54*(5), 1618–1629. doi:10.1016/j.electacta.2008.09.053

Galletly, G. D. (1952). *Behavior of reinforced concrete walls under static load*. Cambridge, MA: Department of Civil and Sanitary Engineering, Massachusetts Institute of Technology.

Gamerman, D., & Lopes, H. (2006). *Markov Chain Monte Carlo: Stochastic Simulation for Bayesian Inference* (2nd ed.). Boca Raton, USA: Chapman & Hall/CRC.

Garcia, C., Martín, F., Tiedra, P. D., Heredero, J. A., & Aparicio, M. L. (2001). Effects of prior cold work and sensitization heat treatment on chloride stress corrosion cracking in type 304 stainless steels. *Corrosion Science*, *43*(8), 1519–1539. doi:10.1016/S0010-938X(00)00165-7

Garson, G. D. (1991). Interpreting neural-network connection weights. *AI Expert*, *6*(7), 47–51.

Gasch, R., & Twele, J. (2007). *Windkraftanlagen: Grundlagen, Entwurf, Planung und Betrieb*. Teubner Verlag.

Gasparini, D. A., & Vanmarcke, E. H. (1976). *SIMQKE, a program for artificial motion generation, user's manual and documentation. Publication R76-4*. Cambridge, Massachusetts: MIT Press.

Gauger, U., Turrin, S., Hanss, M., & Gaul, L. (2008). A new uncertainty analysis for the transformation method. *Fuzzy Sets and Systems*, *159*(11), 1273–1291. doi:10.1016/j.fss.2007.12.027

Gavin, H. P., & Yau, S. C. (2007). High-order limit state functions in the response surface method for structural reliability analysis. *Structural Safety*, *30*(2), 162–179. doi:10.1016/j.strusafe.2006.10.003

Geers, M. G. D., De Borst, R., Brekelmans, W. A. M., & Peerlings, R. H. J. (1998). Strain-based transient-gradient damage model for failure analyses. *Computer Methods in Applied Mechanics and Engineering*, *60*(1-2), 133–153. doi:10.1016/S0045-7825(98)80011-X

Georgeff, M. P., & Lansky, A. L. (1987, July). *Reactive reasoning and planning* (Vol. 87). AAAI.

Germain, P. (2000, May). *Active Control of Run-up Noise from Propeller Aircraft* [M.App.Sc. Thesis]. Uni. British Columbia.

Gernay, T., & Franssen, J. M. (2011). *SAFIR MANUAL Materials SILCON_ETC and CALCON_ETC. University of Liege*. Belgium: Environnement & Constructions Structural Engineering.

Geyer, C. J. (1992). Practical markov chain Monte Carlo. *Statistical Science*.

Global Wind Report. (2014). Annual Market Update. Retrieved from http://www.gwec.net/publications/global-wind-report-2/global-wind-report-2014-annual-market-update/

Goh, A. T. C., & Zhang, W. G. (2014). An improvement to MLR model for predicting liquefaction-induced lateral spread using Multivariate Adaptive Regression Splines. *Engineering Geology*, *170*, 1–10. doi:10.1016/j.enggeo.2013.12.003

Golub, G., & Van Loan, C. F. (2013). Volume 3 of Johns Hopkins Studies in Mathematical Sciences Matrix Computations, Baltimore, MD: John Hopkins University Press.

Goodall, I. W. (1978). Large deformations in plastically deforming curved tubes subjected to in-plane bending. *Research Division Report RD/B/N4312, Central Electricity Generating Board*, England.

Goodall, I. W. (1978). *Lower bound limit analysis of curved tubes loaded by combined internal pressure and in-plane bending moment, Research Division report RD/B/N4360*. England: Central Electricity Generating Board.

Goodman, J. (1985). Structural fragility and principle of maximum entropy. *Structural Safety*, *3*(1), 37–46. doi:10.1016/0167-4730(85)90006-2

Gopalakrishnan, K., & Khaitan, S. K., (2010). Finite Element Based Adaptive Neuro-Fuzzy Inference Technique for Parameter Identification of Multi-Layered Transportation Structures. *Transport,* 25(1), 58-65.

Goswami, S., Chakraborty, S., & Ghosh, S. (2013). Adaptive response surface method in structural response approximation under uncertainty. *Proceedings of theInternational Conference on Structural Engineering and Mechanics*.

Gresnigt, A. M. (2002). Elastic and plastic design of mitred bends.*Proceedings of the 12th international offshore and polar engineering conference*, Kitakyushu, Japan (pp. 26–31).

Griffin, J. R. (2006). The Control Of Interior Cabin Noise Due To A Turbulent [MSc Thesis].

Griffiths, J. E. (1978). The effect of cracks on the limit load of pipe bends under in-plane bending experimental study. *International Journal of Mechanical Sciences*, *21*(2), 119–130. doi:10.1016/0020-7403(79)90038-9

Grigoriev, V., & Jakobsson, R. (2000, November). Application of the Pin Loading Tension test to measurements of Delayed Hydride Cracking Velocity in Zircaloy cladding.

Grigoriev, V., Josefsson, B., & Rosborg, B. (1996). Fracture Toughness of Zircaloy Cladding Tubes. Zirconium In nuclear Industry. *Proc. of 11th Int. Symposium ASTM STP* (pp. 431-447). doi:10.1520/STP16184S

Grigoriev, V., & Jakobsson, R. (2005). Delayed Hydrogen Cracking Velocity and J-Integral Measurements on Irradiated BWR Cladding. *Journal of ASTM International*, *2*(8), 1–16. doi:10.1520/JAI12434

Grigoriev, V., Josefsson, B., Lind, A., & Rosborg, B. (1995). A pin-loading tension test for evaluation of thin-walled tubular materials. *Scripta Metallurgica et Materialia*, *33*(1), 109–114. doi:10.1016/0956-716X(95)00110-H

Grigoriev, V., Josefsson, B., Rosborg, B., & Bai, J. (1997). A novel fracture toughness testing method for irradiated tubing-experimental results and 3D numerical evaluation.*Transactions of the 14th Int. Conference on Structural Mechanics in Reactor Technology*, Lyon, France (pp. 57-64).

Grujisic, M., & Tangrila, S. (1991). Thermodynamics and kinetic analysis of time-temperature-sensitization diagrams in austenitic stainless steels. *Materials Science and Engineering A*, *142*(2), 255–259. doi:10.1016/0921-5093(91)90664-9

Gulec, C. K. (2005). Ultimate shear strength of squat rectangular reinforced concrete walls [MS Thesis]. Department of Civil, Structural and Environmental Engineering, State University of New York at Buffalo, Buffalo, NY.

Gulec, C. K. (2009). Performance-based assessment and design of squat reinforced concrete shear walls [Ph.D. Thesis]. State University of New York at Buffalo.

Gurson, A. L. (1977). Continuum theory of ductile rupture by void nucleation and growth: Part-I: Yield criteria and flow rules for porous ductile media.*Transactions of ASME Journal Engineering Materials and Technology (Vol. 99*, pp. 2-15). doi:10.1115/1.3443401

Gu, Z., & Oyadiji, S. (2008). Application of MR Damper in Structural Control Using ANFIS Method. *Computers & Structures*, *86*(3-5), 427–436. doi:10.1016/j.compstruc.2007.02.024

Ha, J. S., & Seong, P. H. (2004). A method for risk-informed safety significance categorization using the analytic hierarchy process and Bayesian belief networks. *Reliability Engineering & System Safety*, *83*(1), 1–15. doi:10.1016/j.ress.2003.08.002

Hakim, S. J. S., & Abdul Razak, H. (2013). Adaptive neuro fuzzy inference system (ANFIS) and artificial neural networks (ANNs) for structural damage identification. *Structural Engineering & Mechanics*, *45*(6), 779–802. doi:10.12989/sem.2013.45.6.779

Haldar, A., & Mahadevan, S. (2000). *Reliability Assessment Using Stochastic Finite Element Analysis*. New York: John Wiley & Sons, Inc.

Hall, E. L., & Briant, C. L. (1984). Chromium depletion in the vicinity of carbides in sensitized austenitic stainless steels. *Metallurgical Transactions*, *15A*(5), 793–810. doi:10.1007/BF02644554

Han, J. J., Lee, K. H., Kim, N. H., Kim, Y. J., Jerng, D. W., & Budden, P. J. (2012). Comparison of existing plastic collapse load solutions with experimental data for 90° elbows. *International Journal of Pressure Vessels and Piping*, *89*, 19–27. doi:10.1016/j.ijpvp.2011.09.001

Hansen, C. H., Qiu, X.-J., Barrault, G., Howard, C. Q., Petersen, C. D., & Singh, S. (2007). Optimization Of Active And Semi-Active Noise and Vibration Control Systems. *Proceedings of the14th International Congress of Sound and Vibration*, Cairns, Australia.

Hansen, E. (1992). *Global Optimization Using Interval Analysis*. New York, USA: Marcel Dekker, Inc.

Hansen, M. (2008). *Aerodynamics of Wind Turbines* (2nd ed.). U.K.: Earthscan.

Hanss, M. (2002). The transformation method for the simulation and analysis of systems with uncertain parameters. *Fuzzy Sets and Systems*, *130*(3), 277–289. doi:10.1016/S0165-0114(02)00045-3

Harvey, R. L. (1994). *Neural network principles*. NJ, USA: Prentice-Hall Inc.

Hassan, T., Modlin, M., & Matzen, V. C. (2000). *Monotonic in-plane tests on 2-inch, schedule 10* elbows. In *Data developed at the center for nuclear power plant structures, equipment and piping*. NC: North Carolina State University.

Hasselman, K., Barnett, T. P., Bouws, E., et al. (1973) Measurement of Wind Wave Growth and Swell Decay During the Joint North Sea Wave Project (JONSWAP).

Hastie, T., Tibshirani, R., & Friedman, J. (2009). *The elements of statistical learning: Data mining, inference and prediction* (2nd ed.). Springer. doi:10.1007/978-0-387-84858-7

Heng, L., He, Y., Mei, Z., & Jie, G. R. (2006). Forming characteristics of thin-walled tube bending process with small bending radius. *Transactions of Nonferrous Metals Society of China*, *16*, 613–623. doi:10.1016/S1003-6326(06)60266-5

Hess, J. L., & Smith, A. M. O. (1967). Calculation of Potential Flow about Arbitrary Bodies. *Progress in Aeronautical Sciences*, *8*, 1–138. doi:10.1016/0376-0421(67)90003-6

Hilsenkopf, P., Boneh, B., & Sollogoub, P. (1998). Experimental study of behavior and functional capability of ferric steel elbows and austenitic stainless steel thin walled elbows. *International Journal of Pressure Vessels and Piping*, *33*(2), 111–128. doi:10.1016/0308-0161(88)90065-8

Hirosawa, M. (1975). *Past experimental results on reinforced concrete shear walls and analysis on them. Kenchiku Kenkyu Shiryo, No. 6*. Tokyo, Japan: Building Research Institute, Ministry of Construction.

Holmes, J. D. (2010). *Wind Loading of Structures* (2nd ed.). U.S.A.: W.I.T. Press.

Hooker, G. (2007). Generalized Functional ANOVA Diagnostics for High-Dimensional Functions of Dependent Variables. *Journal of Computational and Graphical Statistics*, *16*(3), 709–732. doi:10.1198/106186007X237892

Hose, D.R. (1988). *Glass reinforced plastic pipework of mixed wall construction* [PhD thesis]. University of Manchester.

Hotteling, H. (1933). Analysis of a Complex of Statistical Variables Into Principal Components. *Journal of Educational Psychology*, *24*(7), 498–520. doi:10.1037/h0070888

Hotteling, H. (1936). Simplified Calculation of Principal Components. *Psychometrika*, *1*(1), 27–35. doi:10.1007/BF02287921

Houbolt, J.C. (1969). Some New Concepts in Oscillatory Lifting Surface Theory.

Housner, G., Bergman, L., Caughey, T., Chassiakos, A., Claus, R., Masri, S., & Yao, J. et al. (1997). Structural control: Past, present, and the future. *Journal of Engineering Mechanics*, *123*(9), 897–971. doi:10.1061/(ASCE)0733-9399(1997)123:9(897)

HSE UK. (2006). Plant ageing: Management of equipment containing hazardous fluids or pressure. Retrieved from http://www.hse.gov.uk/research/rrpdf/rr509.pdf

HSE UK. (2010). Managing Ageing Plant: A Summary Guide. Retrieved from http://www.hse.gov.uk/ offshore/ ageing/ageing-plant-summary-guide.pdf

HSE. (2009). Structural integrity management framework for fixed jacket structures. RR684 Research Report. Retrieved from http://www.hse.gov.uk/research/rrpdf/rr684.pdf

Hsu, H. H., Chien, K. F., Chu, H. C., & Liaw, P. K. (2001). In R. Chona (Ed.), *An X-specimen test for determination of thin walled tube fracture toughness. Fatigue and Fracture Mechanics: 32nd Volume, ASTM STP 1406* (pp. 214–226). West Conshohocken, PA: Americal Society for Testing and Materials.

Hung, S. L., Huang, C. S., Wen, C. M., & Hsu, Y. C. (2003). Nonparametric Identification of a Building Structure from Experimental Data using Wavelet Neural Network. *Computer-Aided Civil and Infrastructure Engineering*, *18*(5), 356–368. doi:10.1111/1467-8667.t01-1-00313

Hurlebaus, S., & Gaul, L. (2006). Smart Structure Dynamics. *Mechanical Systems and Signal Processing*, *20*(2), 255–281. doi:10.1016/j.ymssp.2005.08.025

Hu, Y., Zhao, K., & Lian, H. (2014). Bayesian quantile regression for partially linear additive models. *Statistics and Computing*, *25*(3), 651–668. doi:10.1007/s11222-013-9446-9

Hu, Z., & Li, J. Q. (1999). Computer simulation of pipe-bending processes with small bending radius using local induction heating. *Journal of Materials Processing Technology*, *91*(1-3), 75–79. doi:10.1016/S0924-0136(98)00425-7

Hu, Z., Zhang, J., & Deng, Z. (2008). Construction process simulation and safety analysis based on building information model and 4D technology. *Tsinghua Science and Technology*, *13*(Suppl. 1), 266–272. doi:10.1016/S1007-0214(08)70160-3

Iervolino, I., Fabbrocino, G., Manferedi, G. (2004). Seismic vulnerability of standardized industrial components: application to oil storage tanks. *Proceedings of the 13th WCEE*, Vancouver.

International Electro-technical Commission (IEC). (2005). Wind Turbine: Part-1. Switzerland, Edition 2005.

International Electro-technical Commission (IEC). (2009). Wind Turbine: Part-3. Switzerland, Edition 2009.

International Electro-technical Commission (IEC). (2010). Wind Turbine: Part-22. Switzerland, Edition 2010.

International Electro-technical Commission (IEC). (2010). Wind Turbine: Part-24. Switzerland, Edition 2010.

International Organization for Standardization (ISO). (2014). *Condition Monitoring and Diagnostics of Wind Turbines-Part 1: General Guidelines*. Switzerland: Edition.

Irschik, H. (1991). Analogy between refined beam theories and the Bernoulli-Euler theory. *International Journal of Solids and Structures*, *28*(9), 1105–1112. doi:10.1016/0020-7683(91)90105-O

IS1893. (2002). *Criteria for Earthquake Resistant Design of Structures, Part 1: General Provisions and Buildings*. Fifth Revision.

Ishida, T. (1984). Ring test of 304 stainless steel tube oxidized in high temperature steam. *Journal of Nuclear Materials*, *125*(1), 33–51. doi:10.1016/0022-3115(84)90513-0

Islam, A. S., Hussain, R. R., Jameel, M., & Jumaat, M. Z. (2012). Non-linear time domain analysis of base isolated multi-storey building under site specific bi-directional seismic loading. *Automation in Construction*, *22*, 554–566. doi:10.1016/j.autcon.2011.11.017

ISO 19901-3. (2010). Petroleum and natural gas industries -- Specific requirements for offshore structures (Part 3: Topsides structure). Retrieved from http://www.iso.org/iso/home/store/catalogue_tc/catalogue_detail.htm?csnumber=43812

Jacobsen, F., & Juhl, P. (2011). *Sound Radiation, U*. Southern Denmark.

Jalili-Kharaajoo, M. (2004). Nonlinear system identification using ANFIS based on emotional learning. InAdvances in Artificial Intelligence,*LNCS* (Vol. *3315*, pp. 697–707). doi:10.1007/978-3-540-30498-2_70

Jamali, A., Angst, U., Adey, B., & Elsener, B. (2013). Modeling of corrosion-induced concrete cover cracking: A critical analysis. *Construction & Building Materials*, *42*, 225–237. doi:10.1016/j.conbuildmat.2013.01.019

Jamshidi, A. A., & Kirby, M. J. (2010). Skew-radial basis function expansions for empirical modeling. *SIAM Journal on Scientific Computing*, *31*(6), 4715–4743. doi:10.1137/08072293X

Jang, J.-S. R. sd(1993). ANFIS: Adaptive-network-based Fuzzy Inference System. *IEEE Transactions on Systems, Man, and Cybernetics*, *23*(3), 665–685. doi:10.1109/21.256541

Jang, J.-S. R., Sun, C.-T., & Mizutani, E. (1997). *Neuro-Fuzzy and Soft Computing*. Upper Saddle River, New Jersey: Prentice Hall.

Janiszewski, J. (2012). Ductility of selected metals under electromagnetic ring test loading conditions. *International Journal of Solids and Structures*, *49*(7-8), 1001–1008. doi:10.1016/j.ijsolstr.2012.01.005

Jekabsons, G. (2010). VariReg: A software tool for regression modelling using various modeling methods. Riga Technical University. Retrieved from http://www.cs.rtu.lv/jekabsons/

Ji, J., Zhang, W., Zhao, W., Yuan, C., & Liu, Y. (2012). Research of Seismic Testing and Dynamic Character of High-rise Building Structure Based on ANSYS. *JDCTA: International Journal of Digital Content Technology and its Applications*, 6(8), 63-71.

Jiang, L., Jonas, J. J., Boyle, K., & Martin, P. (2008). Deformation behavior of two Mg alloys during ring hoop tension testing. *Materials Science and Engineering A*, *492*(1–2), 68–73. doi:10.1016/j.msea.2008.04.028

Jin, Z., Luo, S., & Fang, X. D. (2001). KBS-aided design of tube bending processes. *Engineering Applications of Artificial Intelligence*, *14*(5), 599–606. doi:10.1016/S0952-1976(01)00016-1

Johansen, T. A. (1994). Fuzzy Model Based Control: Stability, Robustness, and Performance Issues. *IEEE Transactions on Fuzzy Systems*, *2*(3), 221–234. doi:10.1109/91.298450

Johansen, T. A., & Babuškam, R. (2003). Multiobjective Identification of Takagi-Sugeno Fuzzy Models. *IEEE Transactions on Fuzzy Systems*, *11*(6), 847–860. doi:10.1109/TFUZZ.2003.819824

Johansson S., Claesson I., Nordebo S. & Sjösten P. (1997, June). Evaluation of Multiple Reference Active Noise Control Algorithms on Dornier 328 Aircraft Data. *IEEE Trans. on Speech and Audio Processing*.

Johansson, S. (2000, December). Active Control of Propeller Induced Noise in Aircraft - Algorithms & Methods [PhD Thesis]. Department of Telecommunications and Signal Processing, Blekinge Institute of Technology, Sweden.

Johansson, S. (2000). Active Control of Propeller Induced Noise in Aircraft[Ronneby, Sweden]. *Department of Telecommunications and Signal Processing Blekinge Institute of Technology*, *SE–372*, 25.

Jolliffe, I. T. (2002). *Principal component analysis*. New York: Springer.

Jones, D. R., Schonlau, M., & Welch, W. J. (1998). Efficient global optimization of expensive black-box functions. *Journal of Global Optimization*, *13*(4), 455–492. doi:10.1023/A:1008306431147

Jonkman, J. (2007). *Dynamics Modeling and Loads Analysis of an Offshore Floating Wind Turbine. NREL/TP-500 - 41958*. Colden, Colorado: National Renewable Energy Laboratory. doi:10.2172/921803

Josefsson, B., & Grigoriev, V. (1996, May 22). Modified Ring Tensile Testing and a New Method for Fracture Toughness Testing of Irradiated Cladding. Studsvik Material, Nykoping, Sweden.

Juang, B. H. (1993). *Fundamentals of speech Recognition*. Prentice-Hall, Inc.

Kado, S. Z., Ibrahim, R., Khaidzir, M., & Anwar, K. (2014, May). Agent parameters for building consultant teamwork in the agent-based simulation model development. *Proceedings of the 2014 International Symposium on Technology Management and Emerging Technologies (ISTMET)* (pp. 212-217). IEEE.

Kale, A. V., & Thorat, H. T. (2009). Effect of precompression on thickness of pipe during bending. *Journal of Pressure Vessel Technology*, *131*(031201), 1–8.

Kalkan, E., & Kunnath, S. K. (2007). Assessment of current nonlinear static procedures for seismic evaluation of buildings. *Engineering Structures*, *29*(3), 305–316. doi:10.1016/j.engstruct.2006.04.012

Kang, S.-C., Koh, H.-M., & Choo, J. F. (2010). An efficient response surface method using moving least squares approximation for structural reliability analysis. *Probabilistic Engineering Mechanics*, *25*(4), 365–371. doi:10.1016/j.probengmech.2010.04.002

Karandikar, J. M., Kim, N. H., & Schmitz, T. L. (2012). Prediction of remaining useful life for fatigue-damaged structures using Bayesian inference. *Engineering Fracture Mechanics*, *96*, 588–605. doi:10.1016/j.engfracmech.2012.09.013

Karthik, M. M., & Mander, J. B. (2011). Stress-block parameters for unconfined and confined concrete based on a unified stress-strain model (Technical Notes, Author abstract). *Journal of structural engineering (New York, N.Y.)*, *137*(2), 270.

Kass, R. E., & Wasserman, L. (1996). The selection of prior distributions by formal rules. *Journal of the American Statistical Association*, *91*(435), 1343–1370. doi:10.1080/01621459.1996.10477003

Kaul, M. K. (1978). Stochastic characterization of earthquakes through their response spectrum. *Earthquake Engineering & Structural Dynamics*, *6*(5), 497–509. doi:10.1002/eqe.4290060506

Kaynak, C., Erdiller, E. S., Parnas, L., & Senel, F. (2005). Use of split-disk tests for the process parameters of filament wound epoxy composite tubes. *Polymer Testing*, *24*(5), 648–655. doi:10.1016/j.polymertesting.2005.03.012

Kazantzi, A. K., Righiniotis, T. D., & Chryssanthopoulos, M. K. (2008). Fragility and hazard analysis of a welded steel moment resisting frame. *Journal of Earthquake Engineering*, *12*(4), 596–615. doi:10.1080/13632460701512993

Kelly, K. (1995). *Out of control: the new biology of machines, social systems and the economic world.* Basic Books.

Kempton, J. (1976). The ambiguity of acoustic sources: A possibility for active noise control? *Journal of Sound and Vibration*, *48*(4), 475–483. doi:10.1016/0022-460X(76)90551-4

Kennedy, R. P., Cornell, C. A., Campbell, R. D., Kaplan, S., & Perla, H. F. (1980). Probabilistic seismic safety study of an existing nuclear power plant. *Nuclear Engineering and Design*, *59*(2), 315–338. doi:10.1016/0029-5493(80)90203-4

Kennedy, R. P., & Ravindra, M. K. (1984). Seismic fragilities for nuclear power plant risk studies. *Nuclear Engineering and Design*, *79*(1), 47–68. doi:10.1016/0029-5493(84)90188-2

Kent, D. C., & Park, R. (1971). Flexural members with confined concrete. *Journal of the Structural Division*, *97*(ST7), 1969–1990.

Khanal, B. (2010). A Numerical Investigation of the Aerodynamic Noise Generation Mechanism in Transonic Cavity Flows [PhD Thesis]. Cranfield.

Kima, J. W., Leeb, S. H., & Parkb, C. Y. (2009). *Experimental Evaluation Of The Effect of Local Wall Thinning On The Failure Pressure Of Elbows. Nuclear Engineering and Design, 239*, 2737–2746.

Kim, C., Wang, S., & Choi, K. K. (2005). Efficient response surface modeling by using moving least-squares method and sensitivity. *AIAA Journal*, *43*(1), 2404–2411. doi:10.2514/1.12366

Kim, H.-G., Yang, J.-H., Kim, W.-J., & Koo, Y.-H. (2016). Development status of accident-tolerant fuel for light water reactors in Korea. *Nuclear Engineering and Technology*, *48*(1), 1–15. doi:10.1016/j.net.2015.11.011

Kim, J. H., Lee, M. H., Choi, B. K., & Jeong, Y. H. (2006). Effects of oxide and hydrogen on the circumferential mechanical properties of Zircaloy-4 cladding. *Nuclear Engineering and Design*, *236*(18), 1867–1873. doi:10.1016/j.nucengdes.2006.02.010

Kim, J. H., Lee, M. H., Choi, B. K., & Jeong, Y. H. (2007). Effect of the hydrogen contents on the circumferential mechanical properties of zirconium alloy claddings. *Journal of Alloys and Compounds*, *431*(1-2), 155–161. doi:10.1016/j.jallcom.2006.05.074

Kim, J. H., Lee, M. H., Jeong, Y. H., & Lim, J. G. (2008). Behavior of zirconium fuel cladding under fast pressurization rates. *Nuclear Engineering and Design*, *238*(6), 1441–1447. doi:10.1016/j.nucengdes.2007.10.020

Kim, W.-J., Kim, D.-J., & Park, Y. (2013). Fabrication and material issues for the application of SiC composites to LWR fuel cladding. *Nuclear Engineering and Technology*, *45*(4), 565–572. doi:10.5516/NET.07.2012.084

Kim, Y. J., Kim, N. H., Oh, C. Y., & Oh, C. S. (2008). A method to estimate plastic loads for elbows with non-uniform thicknesses. *Fatigue Fract. Eng. Mater. Struct.*, *31*, 822–837.

Kim, Y. J., Lee, K.-H., Oh, C.-S., Yoo, B., & Park, C.-Y. (2007). Effect of bend angle on plastic loads of pipe bends under internal pressure and in-plane bending. *International Journal of Mechanical Sciences*, *49*(12), 1413–1424. doi:10.1016/j.ijmecsci.2007.03.006

Kim, Y. J., & Oh, C. S. (2006). Closed-form plastic collapse loads of pipe bends under combined pressure and in-plane bending. *Engineering Fracture Mechanics*, *73*(11), 1437–1754. doi:10.1016/j.engfracmech.2006.02.001

Kim, Y. J., & Oh, C. S. (2006). Limit loads for pipe bends under combined pressure and in-plane bending based on finite element limit analysis. *International Journal of Pressure Vessels and Piping*, *83*(2), 148–153. doi:10.1016/j.ijpvp.2005.11.001

Kim, Y. J., & Oh, C.-S. (2007). Effect of attached straight pipes on finite element limit analysis for pipe bends. *International Journal of Pressure Vessels and Piping*, *84*(3), 177–184. doi:10.1016/j.ijpvp.2006.09.017

Kim, Y., Hurlebaus, S., & Langari, R. (2011). MIMO Fuzzy Identification of Building-MR Damper System. *Journal of Intelligent and Fuzzy Systems*, *22*(4), 185–205.

Kim, Y., Hurlebaus, S., Sharifi, R., & Langari, R. (2009). Nonlinear Identification of MIMO Smart Structures.*ASME Dynamic Systems and Control Conference* (Vol. 1, pp. 33-40), Hollywood: California.

Kim, Y., & Langari, R. (2007). Nonlinear Identification and Control of a Building Structure with a Magnetorheological Damper System.*American Control Conference* (Vol. 1, pp. 3353-3358), New York: NY.

Kim, Y., Langari, R., & Hurlebaus, S. (2009). Semiactive Nonlinear Control of a Building with a Magnetorheological Damper System. *Mechanical Systems and Signal Processing*, *23*(2), 300–315. doi:10.1016/j.ymssp.2008.06.006

Kitching, R., & Bond, M. P. (1972). *Out-of-circularity effects in a pipe bend subjected to pressure and bending load.* J Strain Anal.

Kitching, R., Zarrabi, K., & Moore, M. A. (1979). Limit load for a smooth pipe bend under in-plane bending. *International Journal of Mechanical Sciences*, *21*(12), 731–738. doi:10.1016/0020-7403(79)90053-5

Klein, J. (2008). Breve: a 3d Simulation Environment for Multi-Agent Simulations and Artificial Life. Retrieved from http://www.spiderland.org/

Klein, L., Kwak, J. Y., Kavulya, G., Jazizadeh, F., Becerik-Gerber, B., Varakantham, P., & Tambe, M. (2012). Coordinating occupant behavior for building energy and comfort management using multi-agent systems. *Automation in Construction*, *22*, 525–536. doi:10.1016/j.autcon.2011.11.012

Klir, G. J., & Yuan, B. (1995). *Fuzzy Sets and Fuzzy Logic: Theory and Applications*. Upper Saddle River, New Jersey: Prentice Hall PTR.

Klopfer, E., Scheintaub, H., Huang, W., Wendel, D., & Roque, R. (2009). The simulation cycle: Combining games, simulations, engineering and science using StarLogo TNG. *E-Learning and Digital Media*, *6*(1), 71–96.

Kochekseraii, S. B. (1994). *Mitred pipe bends subject to internal pressure and in-plane bending moment* [PhD thesis]. University of Manchester.

Kodur, V. K. R., & Dwaikat, M. (2007). Performance-based Fire Safety Design of Reinforced Concrete Beams. *Journal of Fire Protection Engineering*, *17*(4), 293–320. doi:10.1177/1042391507077198

Kohler, F. (1960). Zur berechung der thermodynamischen daten eines ternären systems ans den zugehörigen binären systemen. *Monatshefte fur Chemie*, *91*(4), 738–740. doi:10.1007/BF00899814

Koopman, G. H., Song, L., & Fahnline, J. B. (1989). A method for computing acoustic fields based on the principle of wave superposition. *The Journal of the Acoustical Society of America*, *86*(6), 2433–2438. doi:10.1121/1.398450

Kravari, K., & Bassiliades, N. (2015). A Survey of Agent Platforms. *Journal of Artificial Societies and Social Simulation*, *18*(1), 11. doi:10.18564/jasss.2661

Krawinkler, H., & Seneviratna, G. D. P. K. (1998). Pros and cons of a pushover analysis of seismic performance evaluation. *Engineering Structures*, *20*(4–6), 452–464. doi:10.1016/S0141-0296(97)00092-8

Kuo, M., & Morgan, D. R. (1996). *Active Noise Control Systems. John Wiley & Sons, Inc.*

Kusni, M, & Soenarko, B. (2001). The Active Noise Control of Free Sound Field Using A multipole Secondary Source. *J. Science and Math*, 2(2).

Kusni, M. (2002). The Active Noise Control Of Propeller Noise Using a Multipole Secondary Source.

Kusni, M., & Soenarko, B. (2001). Dipole Wave Cancellation Using a Multipole Secondary Source.*Proc. of the Symposium on Wave Propagation, Institut Teknologi Bandung*.

Kusni, M., Soenarko, B., & Djojodihardjo, H. (2014). Application of a Multipole Secondary Source for Propeller Active Noise Control, Applied. *Mechanics of Materials*, *564*, 135–142. doi:10.4028/www.scientific.net/AMM.564.135

Kussmaul, K., Eisele, U., & Seidenfuss, M. (1995). On the applicability of local approach models for the determination of the failure behaviour of steels with different toughness. In Mehta et al (Eds.), Fatigue and fracture mechanics in pressure vessel and piping (Vol. 304, pp. 17-25). New York: ASME.

Kuzniar, K., & Waszczyszyn, Z. (2006). Neural networks and Principal Component Analysis for Identification of Building Natural Periods. *Journal of Computing in Civil Engineering*, *20*(6), 431–436. doi:10.1061/(ASCE)0887-3801(2006)20:6(431)

Kwak, H.-G., & Kim, S.-P. (2002). Nonlinear analysis of RC beams based on moment–curvature relation. *Computers & Structures*, *80*(7-8), 615–628. doi:10.1016/S0045-7949(02)00030-5

Kwon, O. S., & Elnashai, A. S. (2006). The effect of material and ground motion uncertainty on the seismic vulnerability curves of RC structure. *Engineering Structures*, *28*(2), 289–303. doi:10.1016/j.engstruct.2005.07.010

Lagö, Th. L. (1996). *Frequency Analysis of Helicopter Sound in the AS323 Super Puma* (Technical Report).

Lagö, Th. L., Johansson, S., & Hellström, P. A. (1997, February). Analysis of Helicopter Sound for the Development of a New Generation Active Headset. *Proceedings of the IMAC-XV Conference*, Orlando, Florida.

Laiarinandrasana, L., Devilliers, C., Oberti, S., Gaudichet, E., Fayolle, B., & Lucatelli, J. M. (2011). Ring tests on high density polyethylene: Full investigation assisted by finite element modeling. *International Journal of Pressure Vessels and Piping*, *88*(1), 1–10. doi:10.1016/j.ijpvp.2010.12.002

Lane, P.H.R. & Rose, T.T. RT. (1960). Experiments on fabricated pipe bends. B.W.R.A, report no. D5. 12-60.

Langari, R. (1999). Past, Present and Future of Fuzzy Control: A Case for Application of Fuzzy Logic in Hierarchical Control.*Proceedings of 18th International Conference of the North American Fuzzy Information Processing Society-NAFIPS*, (pp. 760-765) New York City, New York, USA. doi:10.1109/NAFIPS.1999.781796

Lashkari, A. (2012). Prediction of the shaft resistance of non-displacement piles in sand. *International Journal for Numerical and Analytical Methods in Geomechanics*, *37*(8), 904–931. doi:10.1002/nag.1129

Lättilä, L., Hilletofth, P., & Lin, B. (2010). Hybrid simulation models–when, why, how? *Expert Systems with Applications*, *37*(12), 7969–7975. doi:10.1016/j.eswa.2010.04.039

Law, M & Bowie, G. (2006). Failure strain in high yield-to-tensile ratio linepipes. *Journal of Pipeline Integrity*,5, 25-36.

Law, M., & Bowie, G. (2007). Prediction of failure strain and burst pressure in high yield to tensile strength ratio line pipe. *International Journal of Pressure Vessels Piping, 84*(8), 487–492. doi:10.1016/j.ijpvp.2007.04.002

Leclercq, S., Parrot, A., & Leroy, M. (2008). Failure characteristics of cladding tubes under RIA conditions. *Nuclear Engineering and Design*, *238*(9), 2206–2218. doi:10.1016/j.nucengdes.2008.03.017

Lee, H.W. & Miles, R. (2013). Protecting your assets against corrosion and environmental attack. Offshore Pipeline Integrity Services & Solutions.

Lee, J., & Fenves, G. L. (1998). Plastic-damage model for cyclic loading of concrete structures. *Journal of Engineering Mechanics*, *124*(8), 892–900. doi:10.1061/(ASCE)0733-9399(1998)124:8(892)

Lee, K. W., & Hong, S. I. (2002). Zirconium hydrides and their effect on the circumferential mechanical properties of Zr-Sn-Fe-Nb tubes. *Journal of Alloys and Compounds*, *346*(1-2), 302–307. doi:10.1016/S0925-8388(02)00527-3

Lee, K. W., Kim, S. K., Kim, K. T., & Hong, S. I. (2001). Ductility and strain rate sensitivity of Zircaloy-4 nuclear fuel claddings. *Journal of Nuclear Materials*, *295*(1), 21–26. doi:10.1016/S0022-3115(01)00509-8

Lee, T.-H., & Mosalam, K. M. (2004). Probabilistic fiber element modeling of reinforced concrete structures. *Computers & Structures*, *82*(27), 2285–2299. doi:10.1016/j.compstruc.2004.05.013

Legendre, C. (2014). On the interactions of sound waves and vortices [PhD Thesis]. Université Libre de Bruxelles.

Lehockey, E. M., Palumbo, G., & Lin, P. (1998). Improving the weldability and service performance of nickel and iron based super alloys by grain boundary engineering. *Metallurgical Transactions*, *29A*(12), 3069–3079. doi:10.1007/s11661-998-0214-y

Lernout, E.C., Pellissetti, M., Pradlwarter, H., Schueller, G.I., & Soize, C. (2006). Data and model uncertainties in complex aerospace engineering systems. *Journal of Sound and Vibration,* 295(3–5), 923–938.

Li, G., Wang, S., Rosenthal, C., & Rabitz, H. (2001). High dimensional model representations generated from low dimensional data samples. I . mp-Cut-HDMR, *30*(1).

Li, G., Hu, J., Wang, S.-W., Georgopoulos, P. G., Schoendorf, J., & Rabitz, H. (2006). Random sampling-high dimensional model representation (RS-HDMR) and orthogonality of its different order component functions. *The Journal of Physical Chemistry A*, *110*(7), 2474–2485. doi:10.1021/jp054148m

Li, G., & Rabitz, H. (2010). D-MORPH regression: Application to modeling with unknown parameters more than observation data. *Journal of Mathematical Chemistry*, *48*(4), 1010–1035. doi:10.1007/s10910-010-9722-2

Li, G., & Rabitz, H. (2012). General formulation of HDMR component functions with independent and correlated variables. *Journal of Mathematical Chemistry*, *50*(1), 99–130. doi:10.1007/s10910-011-9898-0

Li, G., Rabitz, H., Hu, J., Chen, Z., & Ju, Y. (2007). Regularized random-sampling high dimensional model representation (RS-HDMR). *Journal of Mathematical Chemistry*, *43*(3), 1207–1232. doi:10.1007/s10910-007-9250-x

Li, G., Rey-de-Castro, R., & Rabitz, H. (2012). D-MORPH regression for modeling with fewer unknown parameters than observation data. *Journal of Mathematical Chemistry*, *50*(7), 1747–1764. doi:10.1007/s10910-012-0004-z

Li, G., Rosenthal, C., & Rabitz, H. (2001a). High dimensional model representations. *The Journal of Physical Chemistry A*, *105*(33), 7765–7777. doi:10.1021/jp010450t

Li, G., Wang, S. W., & Rabitz, H. (2001b). High dimensional model representations generated from low dimensional data samples–I. mp–Cut–HDMR. *Journal of Mathematical Chemistry*, *30*(1), 1–30. doi:10.1023/A:1013172329778

Lighthill, M. J. (1952). On sound generated aerodynamically, General Theory, Proceedings of the Royal Society of London. *Series A, Mathematical and Physical Sciences*, 211(1107), 564-587.

Lin, D. K. J., & Tu, W. (1995). Dual Response Surface Optimization. *Journal of Quality Technology*, *21*(1), 34–39.

Lin, J. W., & Betti, R. (2004). On-line Identification and Damage Detection in Non-linear Structural Systems using a Variable Forgetting Factor Approach. *Earthquake Engineering & Structural Dynamics*, *33*(4), 419–444. doi:10.1002/eqe.350

Link, T. M., Koss, D. A., & Motta, A. T. (1998). Failure of Zircaloy cladding under transverse plane-strain deformation. *Nuclear Engineering and Design*, *186*(3), 379–394. doi:10.1016/S0029-5493(98)00284-2

Link, W. A., & Eaton, M. J. (2012). On thinning of chains in MCMC. *Methods in Ecology and Evolution*, *3*(1), 112–115. doi:10.1111/j.2041-210X.2011.00131.x

Liu, Y. (1996). *Modeling the time to corrosion cracking of the cover concrete in chloride contaminated reinforced concrete structures* (Ph.D. thesis). Virginia Polytechnic Institute and State University, Blacksburg, USA.

Liu, J., Pope, G. A., & Sepehrnoori, K. (1995). A high-resolution finite-difference scheme for nonuniform grids. *Applied Mathematical Modelling*, *19*(3), 162–172. doi:10.1016/0307-904X(94)00020-7

Liu, Y., & Weyers, R. E. (1998). Modeling the time-to-corrosion cracking in chloride contaminated reinforced concrete structures. *ACI Materials Journal*, *95*(6), 675–681.

Li, Y. W., & Yao, L. (2001). Laser bending of tubes: Mechanism, analysis, and prediction. *Journal of Manufacturing Science and Engineering*, *123*(4), 674–681. doi:10.1115/1.1392992

Lloyd, G. (2005). Rules and Guidelines for the Certification of Offshore Wind Turbines (2005th ed.). Germany.

Lubliner, J., Oliver, J., Oller, S., & Onate, E. (1989). A plastic-damage model for concrete. *International Journal of Solids and Structures*, *25*(3), 299–329. doi:10.1016/0020-7683(89)90050-4

Luke, S., Cioffi-Revilla, C., Panait, L., Sullivan, K., & Balan, G. (2005). Mason: A multiagent simulation environment. *Simulation*, *81*(7), 517–527. doi:10.1177/0037549705058073

Luo, D. Z. (1992). Effect of outside diameter, wall thickness and bend radius on ovality of nylon tube during bending operation. *International Journal of Pressure Vessels and Piping*, *52*(1), 145–148. doi:10.1016/0308-0161(92)90065-N

Maalal, S., & Addou, M. (2012). A new approach of designing Multi-Agent Systems. With a practical sample. *International Journal of Advanced Computer Science and Applications*, *2*(11), 148–157.

Macal, C. M., & North, M. J. (2010). Tutorial on agent-based modelling and simulation. *Journal of simulation*, 4(3), 151-162.

Mackenzie, D. (2004). Plastic collapse of pipe bends under combined internal pressure and in-plane bending. *International Journal of Pressure Vessels and Piping*, *82*, 407–416.

Malumbela, G., Alexander, M., & Moyo, P. (2010). Variation of steel loss and its effect on the ultimate fexural capacity of rc beams corroded and repaired under load. *Construction & Building Materials*, *24*(6), 1051–1059. doi:10.1016/j.conbuildmat.2009.11.012

Malumbela, G., Moyo, P., & Alexander, M. (2012). A step towards standardizing accelerated corrosion tests on laboratory reinforced concrete specimens. *Journal of the South African Institution of Civil Engineering*, *54*(2), 78–85.

Mandal, T. K., Ghosh, S., & Pujari, N. N. (2016). Seismic fragility analysis of a typical Indian PHWR containment: Comparison of fragility models. *Structural Safety*, *58*, 11–19. doi:10.1016/j.strusafe.2015.08.003

Mander, J. B., Priestley, M. J. N., & Park, R. (1988). Theoretical Stress-Strain Model for Confined Concrete. *Journal of structural engineering (New York, N.Y.)*, *114*(8), 1804–1826.

Manwell, J., McGowan, J., & Rogers, A. (2002). *Wind Energy Explained*. New York: Wiley. doi:10.1002/0470846127

Marin, J. M., & Robert, C. P. (2007). *Bayesian core: a practical approach to computational Bayesian statistics*. New York, USA: Springer Texts in Statistics.

Markl, A. R. C. (1947). *Fatigue tests of welding elbows and comparable double-mitre bends* (Vol. 69). Trans ASME.

Martinez, W. L., & Martinez, A. R. (2002). *Computational Statistics Handbook with MATLAB*. Boca Raton, USA: Chapman & Hall/CRC.

Martín-Rengel, M. A., Gómez Sánchez, F. J., Ruiz-Hervías, J., Caballero, L., & Valiente, A. (2009, September 6-10). New Method to Calculate the Mechanical Properties of Unirradiated Hydrogen-Charged Fuel Cladding from Ring Tensile Tests. *Proceedings of the Water Reactor Fuel Performance Meeting (WRFPM Top Fuel 2009)*, Paris, France.

Martín-Rengel, M. A., Gómez Sánchez, F. J., Ruiz-Hervías, J., Caballero, L., & Valiente, A. (2012). Revisiting the method to obtain the mechanical properties of hydride fuel cladding in the hoop direction. *Journal of Nuclear Materials*, *429*(1-3), 276–283. doi:10.1016/j.jnucmat.2012.06.003

Mathworks, (2014). Fuzzy Inference System Modeling: Gaussian Combination Membership Function, http://www.mathworks.se/help/fuzzy/gauss2mf.html (accessed on 20.06.2014).

MATLAB. (2010). Mathworks Inc., Basic Guidelines. Retrieved from http://www.matlab.com

MATLAB. (2013). MATLAB (R2013a): "Fuzzy Logic Toolbox", 1984-2013 The MathWorks, Inc.

Mayo, W. E. (1997). Predicting IGSCC/IGA susceptibility of Ni-Cr-Fe alloys by modeling of grain boundary chromium depletion. *Materials Science and Engineering A*, *232*(1-2), 129–139. doi:10.1016/S0921-5093(97)00110-X

Meada, M., & Kang, D. (2009). Post-earthquake damage evaluation of RC buildings. *Journal of Advanced Concrete Technology*, *7*(3), 327–335. doi:10.3151/jact.7.327

Melchers, R. E. (1999). *Structural reliability analysis and prediction* (2nd ed.). Balckwell publishers.

Michael, T. C., Veerappan, A. R., & Shanmugam, S. (2012). Effect of ovality and variable wall thickness on collapse loads in pipe bends subjected to in-plane bending closing moment. *Engineering Fracture Mechanics*, *79*, 138–148. doi:10.1016/j.engfracmech.2011.10.009

Michael, T. C., Veerappan, A. R., & Shanmugam, S. (2013). Effect Of Internal Pressure And Shape Imperfections On Plastic Loads Of Pipe Bends Under In-Plane Closing Moment. *Engineering Fracture Mechanics*, *105*, 1–15. doi:10.1016/j.engfracmech.2013.03.031

Mirzahosseini, M., Aghaeifar, A., Alavi, A., Gandomi, A., & Seyednour, R. (2011). Permanent deformation analysis of asphalt mixtures using soft computing techniques. *Expert Systems with Applications*, *38*(5), 6081–6100. doi:10.1016/j.eswa.2010.11.002

Moens, D., & Vandepitte, D. (2004). Non-probabilistic approaches for non-deterministic dynamic FE analysis of imprecisely defined structures.*Proceedings of the International Conference on Noise and Vibration Engineering*, Leuven (pp. 3095–3119).

Mohammadzadeh, S., Kim, Y., & Ahn, J. (2015). PCA-based Neuro-Fuzzy Model for System Identification of Smart Structures. *Smart Structures and Systems*, *15*(4), 1139–1158. doi:10.12989/sss.2015.15.4.1139

Möller, B., Graf, W., & Beer, M. (2000). Fuzzy structural analysis using α-level optimization. *Computational Mechanics*, *26*(6), 547–565. doi:10.1007/s004660000204

Möller, O., Foschi, R. O., Rubinstein, M., & Quiroz, L. (2009). Seismic structural reliability using different nonlinear dynamic response surface approximations. *Structural Safety*, *31*(5), 432–442. doi:10.1016/j.strusafe.2008.12.001

Moore, R. E. (1979). *Methods and Applications of Interval Analysis*. Philadelphia, USA: SAIM Publication. doi:10.1137/1.9781611970906

Morison, J. R., O'Brien, M. P., Johnson, J. W., & Schaaf, S. A. (1950). *The Force Exerted by Surface Waves on Piles. Petroleum Transactions*. American Institute of Mining Engineers.

Mourad, H. M. (1999). *Elastic-plastic behaviour and limit load analysis of pipe bends under out-of-plane moment loading and internal pressure* [Thesis, Master of Science in Engineering]. The American University in Cairo, School of Sciences and Engineering.

Moura, V., Kina, Y. A., Tavares, S. S. M., de Faria, M. M. S. G., & Mainier, F. B. (2009). Investigation of cracks and sensitization in an AISI 304L stainless steel exposed to 500-600 °C. *Engineering Failure Analysis*, *16*(1), 545–551. doi:10.1016/j.engfailanal.2008.02.002

Mozhi, T. A., Betrabet, H. S., Jagannathan, V., Wilde, B. E., & Clark, W. A. T. (1986). Thermodynamic modeling of sensitization of AISI 304 stainless steels containing nitrogen. *Scripta Metallurgica*, *20*(5), 723–728. doi:10.1016/0036-9748(86)90499-0

Mozhi, T. A., Juhas, M. C., & Wilde, B. E. (1987). Modeling low temperature sensitization of austenitic stainless steels. *Scripta Metallurgica*, *21*(11), 1547–1552. doi:10.1016/0036-9748(87)90298-5

Mr. Brush's Windmill Dynamo. (1890). *Scientific American*, 63(25), 54.

MSL. (2004). Assessment of Repair Techniques for Ageing or Damaged Structures. Project #502 - DOC REF C357R001 Rev 1. Retrieved from http://www.bsee.gov/Research-and-Training/Technology-Assessment-and-Research/tarprojects/500-599/502AA/

Muhanna, R. L., & Mullen, R. L. (2001). Uncertainty in mechanics problems interval-based approach. *Journal of Engineering Mechanics*, *127*(6), 557–566. doi:10.1061/(ASCE)0733-9399(2001)127:6(557)

Mujica, L. E., Rodellar, J., Fernandez, A., & Guemes, A. (2010). Q-statistic and T2-statistic PCA-based measures for damage assessment in structures. *Structural Health Monitoring*, *10*(5), 539–553. doi:10.1177/1475921710388972

Müller, J. P., & Pischel, M. (2011). The agent architecture inteRRaP: Concept and application.

Munck, M. D., Moens, D., Desmet, W., & Vandepitte, D. (2008). A response surface based optimisation algorithm for the calculation of fuzzy envelope FRFs of models with uncertain properties. *Computers & Structures*, *86*(10), 1080–1092. doi:10.1016/j.compstruc.2007.07.006

Murray, , N.W. (1979). Bursting Tests On Welded Mitred Bends For Gas Pipelines. *Aust. Weld Res.*, *7*, 1–26.

Murty, K. L. (1999). *Texture based physical, mechanical and corrosion characteristics of Zirconium alloys. Textures in Materials Research* (pp. 113–126). New Delhi: Oxford and IBH co. Ltd.

Murty, K. L. (2002, September 11-13). Deformation, texture and microstructure- property correlation and in-reactor performance of Zircaloy cladding in LWRs.*Proceedings of the international symposium on Advances in Zirconium*, Mumbai, India (pp. 28-45).

Muscolino, G., Ricciardi, G., & Cacciola, P. (2003). Monte Carlo simulation in the stochastic analysis of non-linear systems under external stationary poisson white noise input. *International Journal of Non-linear Mechanics*, *38*(8), 1269–1983. doi:10.1016/S0020-7462(02)00072-0

Nakano, Y., Maeda, M., Kuramoto, H., & Murakami, M. (2004). Guidelines for post-earthquake damage eveluation and rehabilitation of RC buildings in Japan. *Paper presented at the13th World Conference on Earthquake Engineering*, Vancouver, Canada.

Nakatsuka, M., Kubo, T., & Hayashi, Y. (1991). Fatigue Behavior of Neutron Irradiated Zircaloy-2 Fuel Cladding Tubes. In *Ziconium in the Nuclear Industry* (pp. 230-245). American Society for Testing and Materials. doi:10.1520/STP25510S

NASA. (2013). Development of Vibroacoustic and Shock Design and Test Criteria. *NASA, MSFC-STD, 3676-A*(January), 10.

Natesan, K., & Kassner, T. F. (1973). Thermodynamics of carbon in nickel, iron-nickel and iron-chromium-nickel alloys. *Metallurgical Transactions*, *4*(11), 2557–2566. doi:10.1007/BF02644258

Needleman, A., & Tvergaard, V. (1984). An analysis of ductile rupture in notched bars. *Journal of the Mechanics and Physics of Solids*, *32*(6), 461–490. doi:10.1016/0022-5096(84)90031-0

Neilson, R., Wood, J., Hamilton, R., & Li, H. (2010). A Comparison Of Plastic Collapse And Limit Loads For Single Mitred Pipe Bends Under In-Plane Bending. *International Journal of Pressure Vessels and Piping*, *87*(10), 550–558. doi:10.1016/j.ijpvp.2010.08.015

Nelson, P.A., Curtis, A.R.D. & Elliott, S.J. (1986). Optimal multipole source distribution for the active suppression and absorption of acoustic radiation. *Proc. Euromech Colloc.*

Nemat-Alla, M. (2003). Reproducing hoop stress–.strain behavior for tubular material using lateral compression test. *International Journal of Mechanical Sciences*, *45*(4), 605–621. doi:10.1016/S0020-7403(03)00115-2

Nigam, N. C., & Narayanan, S. (1994). *Applications of Random Vibrations*. India: Springer.

Nikolai, C., & Madey, G. (2009). Tools of the trade: A survey of various agent based modeling platforms. *Journal of Artificial Societies and Social Simulation*, *12*(2), 2.

Nikulina, A. V. (2003). Zirconium-Niobium Alloys for Core Elements of Pressurized Water Reactors. *Metal Science and Heat Treatment*, *45*(7-8), 287–292. doi:10.1023/A:1027388503837

Nikulina, A. V. (2004). Zirconium alloys in nuclear power engineering. *Metal Science and Heat Treatment*, *46*(11-12), 458–462. doi:10.1007/s11041-005-0002-x

North, M. J., Collier, N. T., Ozik, J., Tatara, E. R., Macal, C. M., Bragen, M., & Sydelko, P. (2013). Complex adaptive systems modeling with repast simphony. *Complex adaptive systems modeling, 1*(1), 1-26.

Norton, M.P. & Karzcub, D.G. (2003, October 16). Fundamentals of Noise and Vibration Analysis for Engineers. Cambridge University Press.

Nwana, H. S. (1996). Software agents: An overview. *The Knowledge Engineering Review*, *11*(03), 205–244. doi:10.1017/S026988890000789X

O' Brien, D. M., & Ferguson, W. G. (1983). Modified manjoine WOL specimen for J_{IC} and δ_i tests. *Engineering Fracture Mechanics*, *18*(2), 453–465. doi:10.1016/0013-7944(83)90153-4

Oden, T., Moser, R., & Ghattas, O. (2010). Computer Predictions with Quantified Uncertainty, Part I. *SIAM News*, 43(9).

Ortner, S. R., & Randle, V. (1989). A study of the relation between grain boundary type and sensitization in a partially sensitized AISI 304 stainless steel using electron back-scattering patterns. *Scripta Metallurgica*, *23*(11), 1903–1908. doi:10.1016/0036-9748(89)90480-8

Orynyak, V., & Radchenko, S. A. (2004). A Mixed-approach analysis of deformations in pipe bends. *Strength of Materials*, *36*(4), 365–376. doi:10.1023/B:STOM.0000041537.43590.f9

Ostemin, A. A., & Dil'man, V. L. (2002). Calculating wall thickness for major oil product pipelines: Analysis of regulation documents. *Chemical and Petroleum Engineering*, *38*(1/2), 58–63. doi:10.1023/A:1015278306138

Ott, L. J., Robb, K. R., & Wang, D. (2014). Preliminary assessment of accident-tolerant fuels on LWR performance during normal operation and under DB and BDB accident conditions. *Journal of Nuclear Materials*, *448*(1-3), 520–533. doi:10.1016/j.jnucmat.2013.09.052

Ozbulut, O. E., Mir, C., Moroni, M. O., Sarrazin, M., & Roschke, P. N. (2007). A Fuzzy Model of Superelastic Shape Memory Alloys for Vibration Control in Civil Engineering Applications. *Smart Materials and Structures*, *16*(3), 818–829. doi:10.1088/0964-1726/16/3/031

Pabst, O., Kletschkowski, Th., & Sachau, D. (2009, July 5-9). Combined Active Noise Control and Audio In A Light Jet. *Proceedings of the 16th Int.Congress on Sound and Vibration*, Krakow.

Padmini, D., Ilamparuthi, K., & Sudheer, K. P. (2008). Ultimate bearing capacity prediction of shallow foundations on cohesionless soils using neurofuzzy models. *Computers and Geotechnics*, *35*(1), 33–46. doi:10.1016/j.compgeo.2007.03.001

Palau, C. V., Arregui, F. J., & Carlos, M. (2012). Burst Detection in Water Networks Using Principal Component Analysis. *Journal of Water Resources Planning and Management*, *138*(1), 47–54. doi:10.1061/(ASCE)WR.1943-5452.0000147

Papanikolaou, V. K., & Elnashai, A. S. (2005). Evaluation of conventional and adaptive pushover analysis i: methodology. *Journal of Earthquake Engineering*, *9*(6), 923–941. doi:10.1080/13632460509350572

Parker, M. T. (2001). What is Ascape and why should you care. *Journal of Artificial Societies and Social Simulation*, *4*(1), 5.

Park, J., & Towashiraporn, P. (2014). Rapid seismic damage assessment of railway bridges using the response-surface statistical model. *Structural Safety*, *47*, 1–12. doi:10.1016/j.strusafe.2013.10.001

Park, R., & Paulay, T. (1975). *Reinforced concrete structures*. New York: John Wiley & Sons. doi:10.1002/9780470172834

Park, S., Lee, J., Yun, C., & Inman, D. (2007). Electro-Mechanical Impedance-Based Wireless Structural Health Monitoring Using PCA-Data Compression and *k*-means Clustering Algorithms. *Journal of Intelligent Material Systems and Structures*, *19*(4), 509–520. doi:10.1177/1045389X07077400

Parvathavarthini, N., & Dayal, R. K. (2010). Time-temperature-sensitization diagrams and critical cooling rates of different nitrogen containing austenitic stainless steels. *Journal of Nuclear Materials*, *399*(1), 62–67. doi:10.1016/j.jnucmat.2010.01.003

Parvathavarthini, N., Mulki, S., Dayal, R. K., Samajdar, I., Mani, K. V., & Raj, B. (2009). Sensitization control in AISI 316L(N) austenitic stainless steel: Defining the role of the nature of grain boundary. *Corrosion Science*, *51*(9), 2144–2150. doi:10.1016/j.corsci.2009.05.045

Pate-Cornell, M. E. (1996). Uncertainties in risk analysis: Six levels of treatment. *Reliability Engineering & System Safety*, *54*(2-3), 95–111. doi:10.1016/S0951-8320(96)00067-1

Patrascu, M., Dragoicea, M., & Ion, A. (2014, October). Emergent intelligence in agents: A scalable architecture for smart cities. *Proceedings of the 2014 18th International Conference on System Theory, Control and Computing (ICSTCC)* (pp. 181-186). IEEE. doi:10.1109/ICSTCC.2014.6982412

Pătraşcu, M., & Drăgoicea, M. (2014). Integrating agents and services for control and monitoring: managing emergencies in smart buildings. In *Service Orientation in Holonic and Multi-Agent Manufacturing and Robotics* (pp. 209–224). Springer International Publishing. doi:10.1007/978-3-319-04735-5_14

Patrascu, M., Dumitrache, I., & Patrut, P. (2012). A comparative study for advanced seismic vibration control algorithms. *UPB Sci. Bull., Series C*, *74*(4), 3–16.

Patrascu, M., & Ion, A. (2015). *Seismic Vibration Control with Semi-Active Dampers*. Bucharest: MatrixRom.

Paulay, T., & Priestley, M. L. N. (1992). *Seismic Design of Reinforced Concrete and Masonry Buildings*. John Wiley & Sons. doi:10.1002/9780470172841

Pearson, K. (1901). On lines and planes of closest fit to systems of points in space. *Philosophical Magazine*, *2*(6), 559–572. doi:10.1080/14786440109462720

Peerlings, R. H. J., De Borst, R., Brekelmans, W. A. M., & Geers, M. G. D. (2002). Localisation issues in local and nonlocal continuum approaches to fracture. *European Journal of Mechanics. A, Solids*, *21*(2), 175–189. doi:10.1016/S0997-7538(02)01211-1

Pellissetti, M. F., & Schueller, G. I. (2007). The effects of uncertainties in structural analysis. *Journal of Structural Engineering and Mechanics*, *25*(3), 311–330. doi:10.12989/sem.2007.25.3.311

Pierson, W. J. Jr, & Moskowitz, L. (1964). A Proposed Spectral Filter For Fully-Developed Wind Seas Based on The Similarity Theory of S.A. Kitaigorodski. *Journal of Geophysics*, *69*(24), 181–203. doi:10.1029/JZ069i024p05181

Pint, B. A., Terrani, K. A., Brady, M. P., Cheng, T., & Keiser, J. R. (2013). High temperature oxidation of fuel cladding candidate materials in steam-hydrogen environments. *Journal of Nuclear Materials*, *440*(1-3), 420–427. doi:10.1016/j.jnucmat.2013.05.047

Pitard-Bouet, J.-M., Seidenfuss, M., Bethmont, M., & Kussmaul, K. (1999). Experimental investigations on the shallow crack effect on the 10MnMoNi55 steel and computational analysis in the upper shelf by means of the global and local approaches. *Nuclear Engineering and Design*, *190*(1-2), 171–190. doi:10.1016/S0029-5493(98)00322-7

Polat, K., & Gunes, S. (2007). Automatic Determination of Diseases Related to Lymph System from Lymphography Data Using Principal Component Analysis (PCA), Fuzzy Weighting Pre-Processing and ANFIS. *Expert Systems with Applications*, *33*(3), 636–641. doi:10.1016/j.eswa.2006.06.004

Pollack, M. E., & Ringuette, M. (1990). *Introducing the Tileworld: Experimentally evaluating agent architectures* (pp. 183–189). Defense Technical Information Center.

Pragnell, W. M., & Evans, H. E. (2006). A finite-difference model to predict 2D depletion profiles arising from high temperature oxidation of alloy. *Modelling and Simulation in Materials Science and Engineering*, *14*(4), 733–740. doi:10.1088/0965-0393/14/4/014

Price, E. G. (1972). Hydride orientation and tensile properties of Zr-2.5 wt % Nb pressure tubing hydrided while internally pressurized. *Canadian Metallurgical Quarterly*, *11*(1), 129–136. doi:10.1179/cmq.1972.11.1.129

Priestley, M. J. N., Calvi, G. M., & Seible, F. (1996). *Seismic design and retrofit of bridges*. New York: Wiley. doi:10.1002/9780470172858

Privara, S., Široký, J., Ferkl, L., & Cigler, J. (2011). Model predictive control of a building heating system: The first experience. *Energy and Building*, *43*(2), 564–572. doi:10.1016/j.enbuild.2010.10.022

Pujari, N. N., Ghosh, S., & Lala, S. (2015). Seismic fragility estimation of a containment shell based on the formation of through-wall cracks. *ASCE-ASME Journal of Risk and Uncertainty in Engineering Systems, Part A. Civil Engineering*.

Qingquan, D., Hong, Z., Feng, Y., & Changyi, D. (2009). Hydrostatic Burst Test of X80 Grade steel pipe. *Journal of Loss Prevention in the Process Industries*, *22*(6), 897–900. doi:10.1016/j.jlp.2009.09.003

Qiu, Z., & Elishakoff, I. (1998). Antioptimization of structures with large uncertain-but-non-random parameters via interval analysis. *Computer Methods in Applied Mechanics and Engineering*, *152*(3–4), 361–372. doi:10.1016/S0045-7825(96)01211-X

Rabitz, H., & Aliş, Ö. F. (1999). General foundations of high dimensional model representations. *Journal of Mathematical Chemistry*, *25*(2-3), 197–233. doi:10.1023/A:1019188517934

Rahman, S. (2006). A dimensional decomposition method for stochastic fracture mechanics. *Engineering Fracture Mechanics*, *73*(15), 2093–2109. doi:10.1016/j.engfracmech.2006.04.010

Rahman, S. (2008). A polynomial dimensional decomposition for stochastic computing. *International Journal for Numerical Methods in Engineering*, *76*(13), 2091–2116. doi:10.1002/nme.2394

Raiffa, H., & Schlaifer, R. (1961). *Applied statistical decision theory*. Applied Statistical Decision Theory.

Rajashekhar, M. R., & Ellingwood, B. R. (1993). A new look at the response surface approach for reliability analysis. *Structural Safety*, *12*(3), 205–220. doi:10.1016/0167-4730(93)90003-J

Ramallo, J. C., Yoshioka, H., & Spencer, B. F. Jr. (2004). A Two-step Identification Technique for Semiactive Control Systems. *Structural Control and Health Monitoring*, *11*(4), 273–289. doi:10.1002/stc.43

Rao, C. R., & Mitra, S. K. (1971). *Generalized inverse of matrix and its applications*. New York, U.S.A.: Wiley.

Rao, S. S., & Berke, L. (1997). Analysis of uncertain structural systems using interval analysis. *AIAA Journal*, *34*(4), 727–735. doi:10.2514/2.164

Rashid, Y. R., Lemaignan, C., & Strasser, A. (2000, April). Evaluation of fracture toughness and extension in fuel cladding. American Nuclear Society, Park City, UT.

Ratnayake, R. M. C. (2012a). Translating sustainability concerns at plant level asset operations: Industrial performance assessment. *International Journal of Sustainable Strategic Management*, *3*(4), 314–339. doi:10.1504/IJSSM.2012.052655

Ratnayake, R. M. C. (2012b). A decision model for executing plant strategy: Maintaining the technical integrity of petroleum flowlines. *International Journal of Decision Sciences. Risk Management*, *4*(1/2), 1–24.

Ratnayake, R. M. C. (2012c). Modelling of asset integrity management process: A case study for computing operational integrity preference weights. *International Journal of Computational Systems Engineering*, *1*(1), 3–12. doi:10.1504/IJCSYSE.2012.044738

Ratnayake, R. M. C. (2014). Application of a fuzzy inference system for functional failure risk rank estimation: RBM of rotating equipment and instrumentation. *Journal of Loss Prevention in the Process Industries*, *29*, 216–224. doi:10.1016/j.jlp.2014.03.002

Ratnayake, R. M. C., & Markeset, T. (2010a). Implementing company policies in plant level asset operations: Measuring organizational alignment. *European Journal of Industrial Engineering*, *4*(3), 355–371. doi:10.1504/EJIE.2010.033335

Ratnayake, R. M. C., & Markeset, T. (2010b). Technical integrity management: Measuring HSE awareness using AHP in selecting a maintenance strategy. *Journal of Quality in Maintenance Engineering*, *16*(1), 44–63. doi:10.1108/13552511011030327

Ratnayake, R. M. C., & Markeset, T. (2012d). Asset integrity management for sustainable industrial operations: Measuring the performance. *International Journal of Sustainable Engineering*, *5*(2), 145–158. doi:10.1080/19397038.2011.581391

Reddy, T. Y., & Reid, S. R. (1979). On obtaining material properties from the ring compression test. *Nuclear Engineering and Design*, *52*(2), 257–263. doi:10.1016/0029-5493(79)90055-4

Reichl, B. P. (2013) *Procedural modeling of buildings* [Dissertation]. Masarikova University, Brno.

Reusch, F., Svendsen, B., & Klingbeil, D. (2003a). A non-local extension of Gurson-based ductile damage modelling. *Computational Materials Science*, *26*, 219–229. doi:10.1016/S0927-0256(02)00402-0

Reusch, F., Svendsen, B., & Klingbeil, D. (2003b). Local and non-local Gurson-based ductile damage and failure modelling at large deformation. *European Journal of Mechanics. A, Solids*, *22*(6), 779–792. doi:10.1016/S0997-7538(03)00070-6

Richardson, F. D. (1953). The thermodynamics of metallurgical carbides and of carbon in iron. *Journal of the Iron and Steel Institute*, *167*, 33–45.

Robertson, A., Li, H., & Mackenzie, D. (2005). Plastic collapse of pipe bends under combined internal pressure and in-plane bending. *International Journal of Pressure Vessels and Piping*, *82*(5), 407–416. doi:10.1016/j.ijpvp.2004.09.005

Robert, W. (1996). *Wind Energy in America: A History*. U.S.A.: University of Oklahoma Press.

RoboCup Rescue League. (2006). RoboCup Rescue. Retrieved from http://www.robocuprescue.org/

Rodabaugh, E.C. & Moore, S.E. (1978). Evaluation of the plastic characteristics of piping products in relations to ASME Code *Criteria. NUREG Report* CR-0261.

Rodabaugh, E. C. (1975). *Review of data on mitre joints in piping to establish maximum angularity for fabrication of girth butt welds* (Vol. 208). WRC Bulletin.

Rodabaugh, E. C., & George, H. H. (1957). *Effect of internal pressure on flexibility and stress intensification factors of curved pipe or welding elbows* (Vol. 79). Trans ASME.

Rodden, W. P., & Johnson, E. H. (1994). MSC/NASTRAN Aeroelastic Analysis User's Manual, Version 68, pp. 237-257.

Roh, H., Reinhorn, A. M., & Lee, J. S. (2012). Power spread plasticity model for inelastic analysis of reinforced concrete structures. *Engineering Structures*, *39*(0), 148–161. doi:10.1016/j.engstruct.2012.01.019

Ronagh, H. R., & Behnam, B. (2012). Investigating the Effect of Prior Damage on the Post-earthquake Fire Resistance of Reinforced Concrete Portal Frames. *International Journal of Concrete Structures and Materials*, *6*(4), 209–220. doi:10.1007/s40069-012-0025-9

Rousselier, G. (1987). Ductile fracture models and their potential in local approach of fracture. *Nuclear Engineering and Design*, *105*(1), 97–111. doi:10.1016/0029-5493(87)90234-2

Rubinstein, R. Y. (1981). *Simulation and the Monte Carlo method*. New York, U.S.A.: Wiley. doi:10.1002/9780470316511

Rudnicki, J. W., & Rice, J. R. (1975). Conditions for the localization of deformation in pressure-sensitive dilatant materials. *Journal of the Mechanics and Physics of Solids*, *23*(6), 371–394. doi:10.1016/0022-5096(75)90001-0

Ruijgrok, G. J. J. (1993). *Elements of Aviation Acoustics*. Delft University Press.

Saaty, T. L. (1982). *Decision making for leaders: the analytical hierarchy process for decisions in a complex world*. Belmont, CA: Lifetime Learning Publications.

Saaty, T. L. (1990). How to make a decision: The analytical hierarchy process. *European Journal of Operational Research*, *48*(1), 9–26. doi:10.1016/0377-2217(90)90057-I

Sahaloui, H., Sidhom, H., & Philibert, J. (2002). Prediction of chromium depleted zone evaluation during aging of Ni-Cr-Fe alloys. *Acta Materialia*, *50*(6), 1383–1392. doi:10.1016/S1359-6454(01)00444-X

Sainte Catherine, C., Le Boulch, D., Carassou, S., Ramasubramanian, N., & Lemaignan, C. (2006). An internal conical mandrel technique for fracture toughness measurements on nuclear fuel cladding. *ASTM Journal of Testing and Evaluation*, *34*(5), 373–382.

Sakaguchi, N., Takahashi, H., & Ichinose, H. (2005). Influence of Silicon and Phosphorus on Radiation-Induced Segregation of Chromium and Nickel in Austenitic Model Alloys. *Materials Transactions*, *46*(3), 440–444. doi:10.2320/matertrans.46.440

Samal, M. K., Balakrishnan, K. S., Bhasin, V., Vaze, K. K., & Anantharaman, S. (2012, December 10-12). A combined experimental and FE analysis procedure to evaluate tensile behavior of Zircaloy pressure tubes. *Proceedings of theInternational Congress on Computational Mechanics and Simulation (ICCMS)*, IIT Hyderabad, India.

Samal, M. K., & Abhishek, A. (2011). A new model for prediction of chromium depletion near grain boundaries and corresponding sensitization in austenitic stainless steels. *IMechE Part C Journal of Mechanical Engineering Science*, *225*(4), 809–815. doi:10.1243/09544062JMES2327

Samal, M. K., Balakrishnan, K. S., & Balakrishnan, S. (2015). A practical approach to evaluate stress-strain behavior of remotely handled pressure tubes of nuclear reactors using ring tension test. *Transactions of the Indian Institute of Metals*, *68*(2), 299–310. doi:10.1007/s12666-014-0461-0

Samal, M. K., Balakrishnan, K. S., Parashar, J., & Tiwari, G. P. (2014). Investigation of deformation behaviour of ring-tensile specimens machined from pressure tubes of Indian PHWR. *Transactions of the Indian Institute of Metals*, *67*(2), 167–176. doi:10.1007/s12666-013-0314-2

Samal, M. K., Balakrishnan, K. S., Parashar, J., Tiwari, G. P., & Anantharaman, S. (2013). Estimation of tensile behaviour of Zirconium alloy pressure tubes using ring tensile test and finite element analysis. *Part C. Journal of Mechanical Engineering Science*, *227*(6), 1177–1186. doi:10.1177/0954406212460474

Samal, M. K., Sanyal, G., & Chakravartty, J. K. (2010a). Experimental and numerical study of fracture behaviour of tubular specimens in a Pin-Loading-Tension setup. *IMechE Part C Journal of Mechanical Engineering Science*, *224*(1), 1–12. doi:10.1243/09544062JMES1781

Samal, M. K., Sanyal, G., & Chakravartty, J. K. (2010b). Estimation of fracture behaviour of thin walled nuclear reactor fuel pins using pin-loading tension (PLT) test. *Nuclear Engineering and Design*, *240*(12), 4043–4050. doi:10.1016/j.nucengdes.2010.09.030

Samal, M. K., Sanyal, G., & Chakravartty, J. K. (2011). Investigation of failure behavior of two different types of Zircaloy clad tubes used as nuclear reactor fuel pins. *Engineering Failure Analysis*, *18*(8), 2042–2053. doi:10.1016/j.engfailanal.2011.06.009

Samal, M. K., Seidenfuss, M., Roos, E., Dutta, B. K., & Kushwaha, H. S. (2008). Finite element formulation of a new nonlocal damage model. *Finite Elements in Analysis and Design*, *44*(6-7), 358–371. doi:10.1016/j.finel.2007.12.002

Samarakoon, S. M. S. M. K., & Ratnayake, R. M. C. (2013). Improving efficiency of technology qualification process in industrial sectors. *International Journal of Continuing Engineering Education and Lifelong Learning*, *24*(3/4), 362–376. doi:10.1504/IJCEELL.2014.063084

Samarakoon, S. M. S. M. K., & Ratnayake, R. M. C. (2015). Strengthening, modification and repair techniques' prioritization for structural integrity control of ageing offshore structures. *International Journal Reliability Engineering & System Safety*, *135*, 15–26. doi:10.1016/j.ress.2014.10.023

Samarakoon, S. M. S. M. K., Ratnayake, R. M. C., & Siriwardane, S. A. S. C. (2013). Structural integrity control of ageing offshore structures: Repairing and strengthening with grouted connections. *Proceedings of the 32ndInternational Conference on Ocean, Offshore and Arctic Engineering OMAE '13*. doi:10.1115/OMAE2013-10925

Samui, P. (2011). Determination of ultimate capacity of driven piles in cohesionless soil: A Multivariate Adaptive Regression Spline approach. *International Journal for Numerical and Analytical Methods in Geomechanics*, *36*(11), 1434–1439. doi:10.1002/nag.1076

Samui, P. (2012). Application of statistical learning algorithms to ultimate bearing capacity of shallow foundation on cohesionless soil. *Int. J. Numer. Anal. Met.*, *36*(1), 100–110. doi:10.1002/nag.997

Samui, P., & Karup, P. (2011). Multivariate adaptive regression splines and least square support vector machine for prediction of undrained shear strength of clay. *Applied Metaheuristic Computing*, *3*(2), 33–42.

Samyn, P., De Baets, P., Schoukens, G., & Van Peteghem, A. P. (2006). Large-scale tests on friction and wear of engineering polymers for material selection in highly loaded sliding systems. *Materials & Design*, *27*(7), 535–555. doi:10.1016/j.matdes.2004.12.021

Sanyal, G., & Samal, M. K. (2012). Fracture behavior of thin-walled Zircaloy fuel clad tubes of Indian pressurized heavy water reactor. *International Journal of Fracture*, *173*(2), 175–188. doi:10.1007/s10704-012-9678-2

Saragoni, G. R., & Hart, G. C. (1973). Simulation of artificial earthquakes. *Earthquake Engineering & Structural Dynamics*, *2*(3), 249–267. doi:10.1002/eqe.4290020305

Sarpkaya, T., & Isaacson, M. (1981). *Mechanics of Wave Forces on Offshore Structures*. New York: Van Nostrand Reinhold.

Scherf, I., Etterdal, B., & Monshaugen, T. (1999). Cost-efficient structural upgrade and life extension of North Sea jacket platforms with use of modern reassessment techniques. *Paper presented at the 31st Offshore Technology Conference*. doi:10.4043/10851-MS

Scholl, A., Manthey, L., Helm, R., & Steiner, M. (2005). Solving multi-attribute design problems with analytic hierarchy process and conjoint analysis: *An empirical comparison. European Journal of Operational Research, 164*(3), 760–777. doi:10.1016/j.ejor.2004.01.026

Schueller, G. I., Pradlwarter, H. J., & Koutsourelakis, P. (2004). A critical appraisal of reliability estimation procedures for high dimensions. *Probabilistic Engineering Mechanics, 19*(4), 463–474. doi:10.1016/j.probengmech.2004.05.004

Schurter, K. C., & Roschke, P. (2000). Fuzzy Modeling of Magnetorheological Damper Using ANFIS. *Proceedings of the 9th IEEE International Conference on Fuzzy Systems*, San Antonio, TX, USA (Vol. 1, pp. 122-127).

Schurter, K. C., & Roschke, P. (2001). Neuro-fuzzy control of structures using acceleration feedback. *Smart Materials and Structures, 10*(4), 770–779. doi:10.1088/0964-1726/10/4/322

Schwartz, L. W. (1974). Computer Extension and Analytic Continuation of Stokes' Expansion for Gravity Waves. *Journal of Fluid Mechanics, 62*(3), 553–578. doi:10.1017/S0022112074000802

Scott, B. D., Park, R., & Priestley, M. J. N. (1982). Stress-Strain Behavior of Concrete Confined by Overlapping Hoops at Low and High Strain Rates. *Journal of the American Concrete Institute, 79*, 13–27.

Seok, C. S., Bae, B. K., Koo, J. M., & Murty, K. L. (2006). The properties of the ring and burst creep of ZIRLO cladding. *Engineering Failure Analysis, 13*(3), 389–397. doi:10.1016/j.engfailanal.2005.02.009

SESAM. (2012). SeSAm Multiagent Simulation. Retrieved from http://130.243.124.21/sesam/index.php

Seybert, A. F., Soenarko, B., Rizzo, F. J., & Shippy, D. J. (1984). Application of the BIE Method to Sound Radiation Problems Using an Isoparametric Element. *J. of Vib., Acoust. Stress and Reliab., 106*(3), 414–420. doi:10.1115/1.3269211

Shahnazari, H., & Tutunchian, M. A. (2012). Prediction of ultimate bearing capacity of shallow foundations on cohesionless soils: An evolutionary approach. KSCE J. *Civil Engineering (New York, N.Y.), 16*(6), 950–957.

Shah, P. K., Samal, M. K., Dubey, J. S., Shriwastaw, R. S., Balakrishnan, K. S., Anantharaman, S., & Chakravartty, J. K. (2011, November6-11). Finite element study on mechanical behaviour of Zircaloy fuel clad tubes.*Proceedings of the 21st International conference on Structural Mechanics in Reactor Technology (SMiRT21)*, New Delhi, India.

Shalaby, M. S., & Younan, M. Y. A. (1998). Limit loads for pipe elbows with internal pressure under in- plane closing moments. *J Press Vessel Technol., 120*(1), 35–42. doi:10.1115/1.2841882

Sharifi, R., Kim, Y., & Langari, R. (2010). Sensor Fault Isolation and Detection of Smart Structures. *Smart Materials and Structures, 19*(10), 5001–5016. doi:10.1088/0964-1726/19/10/105001

Sharobeam, M. H., & Landes, J. D. (1991). The load separation criterion and methodology in ductile fracture mechanics. *International Journal of Fracture, 47*(2), 81–104. doi:10.1007/BF00032571

Sharobeam, M. H., & Landes, J. D. (1993). The load separation and η_{pl} development in precracked specimen test records. *International Journal of Fracture, 59*(3), 213–226. doi:10.1007/BF02555184

Shimada, M., Kokawa, H., Wang, Z. J., Sato, Y. S., & Karibe, I. (2002). Optimization of grain boundary character distribution for intergranular corrosion resistant 304 stainless steel by twin-induced grain boundary engineering. *Acta Materialia, 50*(9), 2331–2341. doi:10.1016/S1359-6454(02)00064-2

Shinozuka, M. (1972). Monte Carlo Solution of Structural Dynamics. *International Journal of Computers and Structures., 2*(5-6), 855–874. doi:10.1016/0045-7949(72)90043-0

Shinozuka, M., Feng, M. Q., Lee, J., & Naganuma, T. (2000). Statistical analysis of fragility curves. *Journal of Engineering Mechanics, 126*(12), 1224–1231. doi:10.1061/(ASCE)0733-9399(2000)126:12(1224)

Simpson, T. W., Poplinski, J. D., Koch, P. N., & Allen, J. K. (2001). Metamodels for computer-based engineering design: Survey and recommendations. *Engineering with Computers*, *17*(2), 129–150. doi:10.1007/PL00007198

Simulation and Modelling of Terrestrial Systems. (2015, June 22). *TerraME*. Retrieved from http://www.terrame.org/doku.php?id=start&rev=1434979167

Simulia. (2013). *Abaqus FEA User's Manual, Version 6.13*. France: Velizy-Villacoublay.

Singh, P. K., Bhasin, V., Vaze, K. K., Ghosh, A. K., Kushwaha, H. S., Murthy, D. S. R., & Sivaprasad, S. et al. (2008). Fatigue studies on carbon steel piping materials and components: Indian phwrs. *Nuclear Engineering and Design*, *238*(4), 801–813. doi:10.1016/j.nucengdes.2007.09.002

Smyth, A. W., Masri, S. F., Chassiakos, A. G., & Caughey, T. K. (1999). On-line Parametric Identification of MDOF Nonlinear Hysteretic Systems. *Journal of Engineering Mechanics*, *125*(2), 133–142. doi:10.1061/(ASCE)0733-9399(1999)125:2(133)

So, H. S. (1971). *Single mitred pipe bends under bending and internal pressure loads* [MSc thesis]. Victoria University of Manchester.

Sobol, I. M. (1976). Uniformly distributed sequences with an additional uniform property. *U.S.S.R. Computational Mathematics and Mathematical Physics*, *16*(5), 236–242. doi:10.1016/0041-5553(76)90154-3

Sobol', I. M. (1993). Sensitivity estimates for nonlinear mathematical models. *Mathematical Modeling and Computational Experiment*, *1*, 407–414.

Sobol', I. M. (2003). Theorems and examples on high dimensional model representations. *Chemical Reliability Engineering and System Safety*, *79*(2), 187–193. doi:10.1016/S0951-8320(02)00229-6

Soize, C. (2009). Nonparametric probabilistic approach of uncertainties for elliptic boundary value problem. *International Journal for Numerical Methods in Engineering*, *80*(6&7), 673–688. doi:10.1002/nme.2563

Soleimanbeigi, A., & Hataf, N. (2005). Predicting ultimate bearing capacity of shallow foundations on reinforced cohesionless soils using artificial neural networks. *Geosynthetics International*, *12*(6), 321–332. doi:10.1680/gein.2005.12.6.321

Soleimanbeigi, A., & Hataf, N. (2006). Prediction of settlement of shallow foundations on reinforced soils using neural networks. *Geosynthetics International*, *13*(4), 161–170. doi:10.1680/gein.2006.13.4.161

Solovyev, A., Mikheev, M., Zhou, L., Dutta-Moscato, J., Ziraldo, C., An, G., & Mi, Q. et al. (2010, April). SPARK: a framework for multi-scale agent-based biomedical modeling.*Proceedings of the 2010 Spring Simulation Multiconference* (p. 3). Society for Computer Simulation International. doi:10.1145/1878537.1878541

Sorensen, J. N. (2002). Safety culture: A survey of the state-of-the-art. *Reliability Engineering & System Safety*, *76*(2), 189–204. doi:10.1016/S0951-8320(02)00005-4

Sound Generation-Compendium, Sources Of Sound. (2013). Retrieved from www.akustik.tu-berlin.de

Sourmail, T., Too, C. H., & Bhadeshia, H. K. D. H. (2003). Sensitization and evolution of chromium-depleted zones in Fe-Cr-Ni-C systems. *ISIJ International*, *43*(11), 1814–1820. doi:10.2355/isijinternational.43.1814

Spacone, E. (1994). Flexibility-Based Finite Element Models for the Nonlinear Static and Dynamic Analysis of Concrete Frame Structures [Ph.D. Dissertation]. Department of Civil Engineering, University of California, Berkeley.

Spacone, E., Filippou, F.C., & Taucer F.F. (1996). Fiber Beam-Column Model for Nonlinear Analysis of R/C Frames. Part I: Formulation, . *Earthquake Engineering & Structural Dynamics*, *25*(7), 711–742.

Spence, J., & Findlay, G. E. (1973). Limit load for elbows under in-plane bending. *Proceedings of second international conference on pressure vessel technology*, San Antonio (Vol. 28, pp. 393–399).

Spencer, B. F. Jr, Dyke, S. J., Sain, M. K., & Carlson, J. D. (1997). Phenomenological Model for Magnetorheological Dampers. *Journal of Engineering Mechanics*, *123*(3), 230–238. doi:10.1061/(ASCE)0733-9399(1997)123:3(230)

Stawström, C., & Hillert, M. (1969). An improved depletion zone theory of intergranular corrosion of 18-8 stainless steel. *Journal of the Iron and Steel Institute*, *207*, 77–90.

Stempien, J. D., Carpenter, D. M., Kohse, G., & Kazimi, M. S. (2013). Characteristics of composite silicon carbide fuel cladding after irradiation under simulated PWR conditions. *Nuclear Technology*, *183*, 13–29.

Stone, J. G., Schleicher, R., Deck, C. P., Jacobsen, G. M., Khalifa, H. E., & Back, C. A. (2015). Stress analysis and probabilistic assessment of multi-layer SiC-based accident tolerant nuclear fuel cladding. *Journal of Nuclear Materials*, *466*, 682–697. doi:10.1016/j.jnucmat.2015.08.001

Straub, D., & Papaioannou, I. (2014). Bayesian updating with structural reliability methods. *Journal of Engineering Mechanics*, *141*(3).

Suneel, K. G., Vivek, B., Vaze, K. K., Ghosh, A. K., & Kushwaha, H. S. (2007). General limit load and B2 stress index equation for pipe bends under in-plane bending. Trans, SMiRT (Vol. 19).

Taflanidis, A. A., & Cheung, S. H. (2012). Stochastic sampling using moving least squares response surface approximations. *Probabilistic Engineering Mechanics*, *28*, 216–224. doi:10.1016/j.probengmech.2011.07.003

Takagi, T., & Sugeno, M. (1985). Fuzzy Identification of Systems and Its Applications to Modeling and Control. *IEEE Transactions on Systems, Man, and Cybernetics*, *15*(1), 116–132. doi:10.1109/TSMC.1985.6313399

Tamimi, S., Amadei, B., & Frangopol, D. M. (1989). Monte-Carlo simulation of rock slope reliability. *Computers & Structures*, *33*(6), 1495–1505. doi:10.1016/0045-7949(89)90489-6

Tang, K., Congedo, P. M., & Abgrall, R. (2015). Sensitivity analysis using anchored ANOVA expansion and high-order moments computation. *International Journal for Numerical Methods in Engineering*, *102*(9), 1554–1584. doi:10.1002/nme.4856

Tan, W., Chai, Y., & Liu, Y. (2011). A message-driving formalism for modeling and simulation of multi-agent supply chain systems. *Journal of Systems Science and Systems Engineering*, *20*(4), 385–399. doi:10.1007/s11518-011-5182-x

Tay, K. M., & Lim, C. P. (2008). On the Use of Fuzzy Inference Techniques in Assessment Models: Part II: Industrial Applications. *Fuzzy Optimization and Decision Making*, *7*(3), 283–302. doi:10.1007/s10700-008-9037-y

Tedmon, C. S. Jr, Vermilyea, D. A., & Rosolowski, T. H. (1971). Intergranular corrosion of austenitic stainless steel. *Journal of the Electrochemical Society*, *118*(2), 192–201. doi:10.1149/1.2407966

Teng-Hooi, T., & Ngoc-Ba, N. (2005). Flexural Behavior of Confined High-Strength Concrete Columns. *ACI Structural Journal*, *102*(2), 198–205.

Terrani, K. A., Wang, D., Ott, L. J., & Montgomery, R. O. (2014). The effect of fuel thermal conductivity on the behavior of LWR cores during loss-of-coolant accidents. *Journal of Nuclear Materials*, *448*(1-3), 512–519. doi:10.1016/j.jnucmat.2013.09.051

Terrani, K. A., Zinkel, S. J., & Snead, L. L. (2013). Advanced oxidation-resistance iron-based alloys for LWR fuel cladding. *Journal of Nuclear Materials*, *448*(1-3), 420–435. doi:10.1016/j.jnucmat.2013.06.041

Thakur, R., & Misra, K. B. (1978). Monte-Carlo simulation for reliability evaluation of complex systems. *International Journal of Systems Science*, *9*(11), 1303–1308. doi:10.1080/00207727808941777

Tirca, L., & Chen, L. (2012). The influence of lateral load patterns on the seismic design of zipper braced frames. *Engineering Structures*, *40*, 536–555. doi:10.1016/j.engstruct.2012.03.017

Torres-Acosta, A. A., & Sagues, A. A. (2004). Concrete cracking by localized steel corrosion - Geometric effects. *ACI Materials Journal*, *101*(6), 501–507.

Touboul, F., Ben, D. M., & Acker, D. (1988) Design criteria for piping components against plastic collapse: application to pipe bend experiments, *Pressure vessel technologyProceedings of the 6th international conference*, Beijing.

Towashiraporn, P. (2004). Building seismic fragility using response surface metamodel [PhD Thesis]. Georgia Inst. of Tech.

Trillo, E. A., Beltran, R., Maldonado, J. G., Romero, R. J., Murr, L. E., Fisher, W. W., & Advani, A. H. (1995). Combined effects of deformation (strain and strain state), grain size, and carbon content on carbide precipitation and corrosion sensitization in 304 stainless steel. *Materials Characterization*, *35*(2), 99–112. doi:10.1016/1044-5803(95)00072-0

Trillo, E. A., & Murr, L. E. (1998). Effects of carbon content, deformation, and interfacial energetics on carbide precipitation and corrosion sensitization in 304 stainless steel. *Acta Materialia*, *47*(1), 235–245. doi:10.1016/S1359-6454(98)00322-X

Tsay, L. W., Liua, Y.-F., Huang, R.-T., & Kuo, R.-C. (2008). The effect of sensitization on the hydrogen-enhanced fatigue crack growth of two austenitic stainless steels. *Corrosion Science*, *50*(5), 1360–1367. doi:10.1016/j.corsci.2008.01.005

Tsuji, M., & Meshii, T. (2013). Influence Of Circumferential Flaw Length On Internal Burst Pressure Of A Wall-Thinned Pipe. *Nuclear Engineering and Design, 255*, 59–67. doi:10.1016/j.nucengdes.2012.10.002

Tvergaard, V., & Needleman, A. (1984). Analysis of cup-cone fracture in a round tensile bar. *Acta Metallurgica*, *32*(1), 157–169. doi:10.1016/0001-6160(84)90213-X

Ueda, T., & Dowell, E. H. (1982). A New Subsonic Method for Lifting Surfaces in Subsonic Flow. *AIAA Journal*, *20*(3), 348–355. doi:10.2514/3.7916

Unnikrishnan, U., Prasad, A. M., & Rao, B. N. (2013). Development of fragility curves using high-dimensional model representation. *Earthquake Engineering & Structural Dynamics*, *42*(3), 419–430. doi:10.1002/eqe.2214

Val, D. V., & Stewart, M. G. (2005). Decision analysis for deteriorating structures. *Reliability Engineering & System Safety*, *87*(3), 377–385. doi:10.1016/j.ress.2004.06.006

Valipour, H. R., & Foster, S. J. (2010). A Total Secant Flexibility-Based Formulation for Frame Elements with Physical and Geometrical Nonlinearities. *Finite Elements in Analysis and Design*, *46*(3), 288–297. doi:10.1016/j.finel.2009.11.002

Valliappan, S., & Pham, T. D. (1995). Fuzzy logic applied to numerical modeling of engineering problems. *Computational Mechanics Advances*, *2*, 213–281.

Vamvatsikos, D., & Cornell, C. A. (2002). Incremental dynamic analysis. *Earthquake Engineering & Structural Dynamics*, *31*(3), 491–514. doi:10.1002/eqe.141

Vamvatsikos, D., & Cornell, C. A. (2004). Applied incremental dynamic analysis. *Earthquake Spectra*, *20*(2), 523–553. doi:10.1193/1.1737737

Vaughan, R. (2008). Massively multi-robot simulation in stage. *Swarm Intelligence*, *2*(2-4), 189–208. doi:10.1007/s11721-008-0014-4

Veerappan, A. R., & Shanmugam, S. (2006). Stress analysis of pipe bends subjected to internal fluid pressure using the finite element technique. *Journal of Strain Analysis*, *41*(8), 561–573. doi:10.1243/03093247JSA213

Veerappan, A. R., & Shanmugam, S. (2008). Analysis for flexibility in the ovality and thinning limits of pipe bends. *ARPN J Eng Appl Sci.*, *1*(1), 31–41.

Vere, S., & Bickmore, T. (1990). A basic agent. *Computational Intelligence*, *6*(1), 41–60. doi:10.1111/j.1467-8640.1990.tb00128.x

Verlinden, B., Driver, J., Samajdar, I., & Doherty, R. D. (2007). *Thermo-mechanical processing of metallic materials, Pergamon materials series*. Amsterdam: Elsevier.

Videm, K., & Lunde, L. (1979). Stress corrosion cracking and growth formation of pellet-clad interaction defects. In *Ziconium in nuclear industry* (pp. 230-237). Americal Society for Testing and Materials.

Vidic, T., Fajfar, P., & Fischinger, M. (1994). Consistent Inelastic Design Spectra: Strength and Displacement. *Earthquake Engineering & Structural Dynamics*, *23*(5), 502–521. doi:10.1002/eqe.4290230504

von Toussaint, U. (2015). General Hyperplane Prior Distributions Based on Geometric Invariances for Bayesian Multivariate Linear Regression. *Entropy*, *17*(6), 3898–3912. doi:10.3390/e17063898

Wagner, C. (1952). *Thermodynamics of alloys* (pp. 51–52). Cambridge, Massachusetts: Addison-Wesley.

Wang, H., Bouchard, R., Eagleson, R., Martin, P., & Tyson, W. R. (2002). Ring hoop tension test (RHTT): A test for transverse tensile properties of tubular materials. *Journal of Testing and Evaluation*, *30*(5), 381–391.

Wang, H., & Hu, H. (2009). Hierarchical Fuzzy Identification of MR Damper.*Proceedings of SPIE 7493, Second International Conference on Smart Materials and Nanotechnology in Engineering*, Weihai, China. doi:10.1117/12.843077

Wang, L., & Fu, K. (2009). *Artificial Neural Networks*. Wiley Encyclopedia of Computer Science and Engineering. doi:10.1002/9780470050118.ecse021

Wang, L., & Langari, R. (1995). Decomposition Approach for Fuzzy Systems Identification.*Proceedings, the 34th IEEE Conference on Decision and Control*, New Orleans, LA, USA (pp. 261-265).

Wang, L., & Zhang, Y. (2011). Plastic collapse analysis of thin-walled pipes based on uniform yield criterion. *International Journal Mechanical Sciences, 53*(5), 348–354. doi:10.1016/j.ijmecsci.2011.02.004

Wang, S. W., Georgopoulos, P. G., Li, G., & Rabitz, H. (2003). Random sampling–high dimensional model representation (RS–HDMR) with nonuniformly distributed variables: Application to an integrated multimedia/multipathway exposure and dose model for trichloroethylene. *The Journal of Physical Chemistry A*, *107*(23), 4707–4718. doi:10.1021/jp022500f

Wang, X., Rabiei, M., Hurtado, J., Modarres, M., & Hoffman, P. A. (2009). Probabilistic-based airframe integrity management model. *Reliability Engineering & System Safety*, *94*(5), 932–941. doi:10.1016/j.ress.2008.10.010

Wang, Y., & Murty, K. L. (1998, June 21-25). Effect of temperature on mechanical anisotropy of re-crystallized Zircaloy-4 sheet.*Proc. 4th Asia-Pacific symposium on Advances in Engineering plasticity and its applications, AEPA 98*, Seoul, South Korea.

Warne, K., Prasad, G., Siddique, N. H., & Maguire, L. P. (2004). *Development of a Hybrid PCA-ANFIS Measurement System for Monitoring Product Quality in the Coating Industry. Systems, Man and Cybernetics, IEEE international Conference* (Vol. 4, pp. 3519–3524). UK: Deny.

Warner, R. F. (1998). *Concrete structures*. South Melbourne, Vic.: Longman.

Was, G. S., & Kruger, R. M. (1985). Chromium depletion in Ni-Cr-Fe alloys. *Acta Metallurgica*, *33*(5), 841–854. doi:10.1016/0001-6160(85)90108-7

Wasnik, D. N., Kain, V., Samajdar, I., Verlinden, B., & De, P. K. (2002). Resistance to sensitization and intergranular corrosion through extreme randomization of grain boundaries. *Acta Materialia*, *50*(18), 4587–4601. doi:10.1016/S1359-6454(02)00306-3

Watanabe, Y., Kain, V., Tonozuka, T., Shoji, T., Kondo, T., & Masuyama, F. (2000). Effect of Ce addition on the sensitization properties of stainless steels. *Scripta Materialia*, *42*(3), 307–312. doi:10.1016/S1359-6462(99)00352-8

Weaver, W., & Johnston, P. R. (1987). *Structural Dynamics by Finite Elements*. Englewood Cliffs, New Jersey: Prentice-Hall, Inc.

Wiener, N. (1938). The homogeneous chaos. *American Journal of Mathematics*, *60*(4), 897–936. doi:10.2307/2371268

Wilensky, U. (1999). NetLogo. Retrieved from http://ccl.northwestern.edu/netlogo/

Wiley, A. (1956). Design of Piping System (2nd ed.). Interscience Publication.

Wilkins, K., Tan, Y., & Matzen, V.C. (2001). Monotonic Out-Of Plane Tests On 2-Inch, Schedule 10, Stainless Steel, 90° Elbows. *Proceedings of the 2003 ASME pressure vessels and piping conference.*

Wilkinson, J. H. (1965). *The Algebraic Eigenvalue Problem*. Oxford: Oxford University Press.

Williams, R. O. (1982). Thermodynamics of the superalloys. *Metallurgical Transactions*, *13A*(6), 959–965. doi:10.1007/BF02643391

Wood, J. (2008). A review of literature for the structural assessment of mitred bends. *International Journal of Pressure Vessels and Piping*, *85*(5), 275–294. doi:10.1016/j.ijpvp.2007.11.003

Wood, S. L. (1990). Shear strength of low-rise reinforced concrete walls. *ACI Structural Journal*, *87*(1), 99–107.

Wooldridge, M. (2009). *An introduction to multiagent systems*. John Wiley & Sons.

Wooldridge, M., & Jennings, N. R. (1995). Agent theories, architectures, and languages: a survey. In *Intelligent agents* (pp. 1–39). Springer Berlin Heidelberg. doi:10.1007/3-540-58855-8_1

Wrobel, L. C. (2002). The Boundary Element Methods, Vol.1, Applications in Thermo-Fluids and Acoustics. Chichester, West Sussex England: John Wiley & Sons.

Wu, L. (2009). Least square analysis of error-components models for the study of provisional economic performance. *Modern Physics Letters B*, *23*(03), 521–524. doi:10.1142/S0217984909018801

Xian-Kui, Z., & Leis, B. N. (2012). Evaluation Of Burst Pressure Prediction Models For Line pipes. *International Journal of Pressure Vessels and Piping, 89*, 85–97. doi:10.1016/j.ijpvp.2011.09.007

Xiao-bo, C., Jing, L., & Jian-yun, C. (2009). Wind-Induced Response Analysis of a Wind Turbine tower including the blade-tower coupling effect. *Journal of Zhejiang University Applied Physics & Engineering*, *10*(11), 1573–1580.

Xiu, D., & Karniadakis, G. E. (2002). The Wiener-Askey polynomial chaos for stochastic differential equations. *SIAM Journal on Scientific Computing*, *24*(2), 619–644. doi:10.1137/S1064827501387826

Xu, H., & Rahman, S. (2004). A generalized dimension-reduction method for multidimensional integration in stochastic mechanics. *International Journal for Numerical Methods in Engineering*, *61*(12), 1992–2019. doi:10.1002/nme.1135

Xu, H., & Rahman, S. (2005). Decomposition methods for structural reliability analysis. *Probabilistic Engineering Mechanics*, *20*(3), 239–250. doi:10.1016/j.probengmech.2005.05.005

Yager, R. R., & Filev, D. P. (1993). Unified Structure and Parameter Identification of Fuzzy Models. *IEEE Transactions on Systems, Man, and Cybernetics*, *23*(4), 1198–1205. doi:10.1109/21.247902

Yagnik, S. K., Ramasubramanian, N., Grigoriev, V., Sainte-Catherine, C., Bertsch, J., Adamson, R., & Oberländer, B. C. (2008). Round-Robin Testing of Fracture Toughness Characteristics of Thin-Walled Tubing. *Journal of ASTM International*, *5*(2), 1–21. doi:10.1520/JAI101140

Yamada, R., Suzuki, M., & Harayama, Y. (1977). Application of finite element method to ring compression test. *Nuclear Engineering and Design*, *44*(1), 75–85. doi:10.1016/0029-5493(77)90124-8

Yan, G., & Zhou, L. L. (2006). Integrated Fuzzy Logic and Genetic Algorithms for Multi-objective Control of Structures using MR Dampers. *Journal of Sound and Vibration*, *296*(1-2), 368–382. doi:10.1016/j.jsv.2006.03.011

Yang, G., Spencer, B. F. Jr, Carlson, J. D., & Sain, M. K. (2002). Large-scale MR Fluid Dampers: Modeling and Dynamic Performance Considerations. *Engineering Structures*, *24*(3), 309–323. doi:10.1016/S0141-0296(01)00097-9

Yang, Y. N., Akbarpour, A., & Ghaemmaghami, P. (1987). New Optimal Control Algorithms for Structural Control. *Journal of Engineering Mechanics*, *113*(0), 1369–1386. doi:10.1061/(ASCE)0733-9399(1987)113:9(1369)

Yang, Y. N., & Lin, S. (2004). On-line Identification of Non-linear Hysteretic Structures using an Adaptive Tracking Technique. *International Journal of Non-linear Mechanics*, *39*(9), 1481–1491. doi:10.1016/j.ijnonlinmec.2004.02.010

Yang, Y. N., & Lin, S. (2005). Identification of Parametric Variations of Structures based on Least Squares Estimation and Adaptive Tracking Technique. *Journal of Engineering Mechanics*, *131*(3), 290–298. doi:10.1061/(ASCE)0733-9399(2005)131:3(290)

Yan, J., & Langari, R. (1998). *Fuzzy Logic-Intelligence, Control, and Information, Upper Saddle River, New Jersey.* USA: Prentice Hall.

Yau, J. F., Wang, S. S., & Corten, H. T. (1980). A mixed-mode crack analysis of isotropic solids using conservation laws of elasticity. *Journal of Applied Mechanics*, *47*(2), 335–341. doi:10.1115/1.3153665

Yoshitake, T., Abe, Y., Akasaka, N., Ohtsuka, S., Ukai, S., & Kimura, A. (2004). Ring-tensile properties of irradiated oxide dispersion strengthened ferritic/martensitic steel claddings. *Journal of Nuclear Materials*, *329-333*, 342–346. doi:10.1016/j.jnucmat.2004.04.084

Youinou, G. J., & Sen, R. S. (2014). Impact of accident-tolerant fuels and claddings on the overall fuel cycle: A preliminary systems analysis. *Nuclear Technology*, *188*, 123–128.

Yuan, S. J., Xu, Z., Wang, Z. R., & Hai, W. (1998). The integral hydro-forming process of pipe elbows. *International Journal of Pressure Vessels and Piping*, *75*(1), 7–9. doi:10.1016/S0308-0161(97)00107-5

Yu, K. H., Kadarman, A. H., & Djojodihardjo, H. (2010). Development and implementation of some BEM variants - A critical review. *Engineering Analysis with Boundary Elements*, *34*(10), 884–899. doi:10.1016/j.enganabound.2010.05.001

Zadeh, L. (1965). Fuzzy sets. *Information and Control*, *8*(3), 338–353. doi:10.1016/S0019-9958(65)90241-X

Zahoor, A. (1991). *Ductile fracture handbook* (Vol. 3). USA: Novotech Corp.

Zarnani, S., El-Emam, M., & Bathurst, R. J. (2011). Comparison of numerical and analytical solutions for reinforced soil wall shaking table tests. *Geotechnical Engineering*, *3*(4), 291–321.

Zhang, J., & Foschi, R. O. (2004). Performance-based design and seismic reliability analysis using designed experiments and neural networks. *Probabilistic Engineering Mechanics*, *19*(3), 259–267. doi:10.1016/j.probengmech.2004.02.009

Zhang, W. G., & Goh, A. T. C. (2013). Multivariate adaptive regression splines for analysis of geotechnical engineering systems. *Computers and Geotechnics*, *48*, 82–95. doi:10.1016/j.compgeo.2012.09.016

Zhang, W. G., & Goh, A. T. C. (2014). Multivariate adaptive regression splines model for reliability assessment of serviceability limit state of twin caverns. *Geotechnical Engineering*, *7*(4), 431–458.

Zhang, W. G., Goh, A. T. C., Zhang, Y. M., Chen, Y. M., & Xiao, Y. (2015). Assessment of soil liquefaction based on capacity energy concept and multivariate adaptive regression splines. *Engineering Geology*, *188*, 29–37. doi:10.1016/j.enggeo.2015.01.009

Zheng, R., & Ellingwood, B. R. (1998). Role of non-destructive evaluation in time-dependent reliability analysis. *Structural Safety*, *20*(4), 325–339. doi:10.1016/S0167-4730(98)00021-6

Zhou, Y., Aust, K. T., Erb, U., & Palumbo, G. (2001). Effects of grain boundary structure on carbide precipitation in 304L stainless steel. *Scripta Materialia*, *45*(1), 49–54. doi:10.1016/S1359-6462(01)00990-3

Zhu, X. K., & Leis, B. N. (2004). Accurate prediction of burst pressure for linepipes. *Journal of Pipeline Integrity*, *4*, 195–206.

Zhu, X. K., & Leis, B. N. (2007). Deterministic and probabilistic of pipeline burst failure.*Proceedings of ASME Pressure Vessels and Piping Conference*, San Antonio, Texas, USA (pp. 22-26). doi:10.1115/PVP2007-26272

Zimmermann, Th., Truty, A., Urbanski, A., & Podles, K. (2008). *Z-Soil User Manual*. Switzerland: Zace Services.

Zinkle, S. J., Terrani, K. A., Gehin, J. C., Ott, L. J., & Snead, L. L. (2014). Accident tolerant fuels for LWRs: A perspective. *Journal of Nuclear Materials*, *448*(1-3), 374–379. doi:10.1016/j.jnucmat.2013.12.005

Zio, E. (1996). On the use of the analytic hierarchy process in the aggregation of expert judgments. *Reliability Engineering & System Safety*, *53*(2), 127–138. doi:10.1016/0951-8320(96)00060-9

Zona Technology, Inc. (2004). ZAERO Theoretical Manual (13th ed.).

About the Contributors

Pijush Samui is working as an associate professor in civil engineering department at NIT Patna, India. He graduated in 2000, with a B.Tech. in Civil Engineering from Indian Institute of Engineering Science and Technology, Shibpur, India. He received his M.Sc. in Geotechnical Earthquake Engineering from Indian Institute of Science, Bangalore, India (2004). He holds a Ph.D. in Geotechnical Earthquake Engineering (2008) from Indian Institute of Science, Bangalore, India. He was a postdoctoral fellow at University of Pittsburgh (USA) (2008-2009) and Tampere University of Technology (Finland) (2009-2010). At University of Pittsburgh, he worked on design of efficient tool for rock cutting and application of Support Vector Machine (SVM) in designing of geostructure. At Tampere University of Technology, he worked on design of railway embankment, slope reliability and site characterization. In 2010, Dr. Pijush joined in the Center for Disaster Mitigation and Management at VIT University as an Associate Professor. He was promoted to a Professor in 2012. Dr. Pijush's research focuses on the application of Artificial Intelligence for designing civil engineering structure, design of foundation, stability of railway embankment, reliability analysis, site characterization, and earthquake engineering. Dr. Pijush is the recipient of the prestigious CIMO fellowship (2009) from Finland, for his integrated research on the design of railway embankment. He was awarded Shamsher Prakash Research Award (2011) by IIT Roorkee for his innovative research on the application of Artificial Intelligence in designing civil engineering structure. He was selected as the recipient of IGS Sardar Resham Singh Memorial Award – 2013 for his innovative research on infrastructure project. He was elected Fellow of International Congress of Disaster Management in 2010. He served as a guest in disaster advance journal. He also serves as an editorial board member in several international journals. Dr. Pijush is active in a variety of professional organizations including the Indian Geotechnical Society, Indian Science Congress, Institution of Engineers, World federation of Soft Computing, and Geotechnical Engineering for Disaster Mitigation and Rehabilitation. He has organized numerous workshops and conferences on the applications of artificial intelligence in civil engineering design.

Subrata Chakraborty is currently a Professor at the Indian Institute of Engineering Science and Technology, Shibpur. He is a fellow of the Indian National Academy Engineering and the Institution of Engineers (India). Prof. Chakraborty did his Bachelor in Civil Engineering from Bengal Engineering College, Shibpur, M. Tech. and Ph.D. from IIT Kharagpur. He was a postdoctoral researcher at University of Cambridge, UK and University of Arizona, USA and Technical University of Aachen, Germany. In general, Prof. Chakraborty's research interest lies in the field of computational mechanics under uncertainty, Seismic Safety Assessment, vibration control, structural health monitoring, composite mechanics etc.

Dookie Kim received his Bachelor degree in Civil and Environmental Engineering from Korea University in 1993, and he earned his Master and Ph.D. degrees in Civil Engineering from Korea Advanced Institute of Science and Technology (KAIST) in 1995 and 1999 respectively. He is as a post-doctoral researcher and at Korea Atomic Energy Research Institute (KAERI), University of California, and Irvine (UCI) and as a senior researcher at UNISON Co. Ltd in September 1999 - February 2003. Since March 2003, Dr. Kim has been teaching, researching, and consulting at the Department of Civil Engineering, Kunsan National University (KNU), Jeonbuk, Korea.

* * *

Arundhuti Banerjee is an Assistant Professor, Department of Civil Engineering.

Behrouz Behnam finished his PhD at the University of Queensland (UQ), Australia, in Risk Management of Post-earthquake Fire. He holds a Master of Civil Engineering in Construction Management and a Bachelor of Civil Engineering from Amirkabir University of Technology (AUT), Iran. He is currently working as Assistant Professor at AUT and research associate in Structural Retrofit Research Group (SRRG) at UQ. Focusing on PEF, he has written 3 books, more than 40 peer-reviewed journal papers, and 9 conference papers.

Souvik Chakraborty is a research scholar in Department of Civil Engineering, Indian Institute of Technology Roorkee. His research interest includes uncertainty quantification, structural reliability, robust design optimization, reliability based design optimization, multiscale analysis, polymer composites and wind engineering.

Tanusree Chakraborty holds a Bachelor of Engineering (B.E.) degree in Civil Engineering from Bengal Engineering Collage, Shibpur; Master of Technology (M.Tech.) degree in Rock Mechanics from IIT Delhi; PhD degree in Geotechnical Engineering from Purdue University, USA. She is an Assistant Professor in the Department of Civil Engineering at IIT Delhi. Her areas of research interest are foundation engineering, soil plasticity, constitutive modelling, soil-structure interaction, underground construction in soil and rock, energy geotechnical engineering, offshore geotechnical engineering, blast loading in soil and rock, and blast resistant design of structures.

Rajib Chowdhury is currently working as an Assistant Professor of IIT Roorkee. Prior this position, he worked as a Lecturer of College of Engineering in Swansea University. He received the prestigious Newton International Fellowship (2008) and received his PhD in 2008 from Indian Institute of Technology Madras. He was also working as a research associate at the Indian Institute of Technology Madras (February 04-November 08). His research areas are multidisciplinary in nature and include: Nano-Structural System, Atomistic Simulation, High Dimensional Model Representation and Probabilistic Methods in Science & Engineering. He has published 52 peer reviewed journal papers and 27 conference papers in these areas and attracted £100,080 research funding. He is also a member of professional bodies such as IEEE, American Nano Society, The Institute of Nanotechnology (IoN), and a reviewer of 6 international journals. In 2011, he received an outstanding paper award on stochastic sensitivity analysis from Emerald Literati Network. He was the finalist for the best student paper competition in the ASME-PVP 2005 Conference at Denver, Colorado. Specialties: Probabilistic modelling.

Harijono Djojodihardjo (Sarjana Teknik Mesin, Institut Teknologi Bandung, 1962; M.Sc. in Mechanical Engineering, University of Kentucky, 1964; Mech.E. 1965, S.M. in Naval Architecture and Marine Engineering, 1966 and Sc.D. in Aerodynamics and Gas Dynamics, 1968, all from Massachusetts Institute of Technology). Insinyur Professional Utama (IPU, Premier Professional Engineer) certification was conferred by the Indonesian Institution of Engineers (PII) in 2009 and ASEAN Chartered Professional Engineer in September 2010 by the ASEAN Chartered Professional Engineer. He is an Academician of the International Academy of Astronautics (since 2008). His latest assignment was AIROD and Visiting Professor at the Aerospace Engineering Department, Universiti Putra Malaysia, Serdang, Selangor, Malaysia, 2009-2015. Formerly he was Professor of Aerospace Engineering at Universiti Sains Malaysia (2004-2008), Universitas Al-Azhar Indonesia (2005-2010) and Institut Teknologi Bandung (1989-2005), and Visiting Professor at Technische Universität München (2002) and Toyohashi University of Technology (2002 - 2003), Chairman, Indonesian National Institute of Aeronautics and Space (Lembaga Penerbangan dan Antariksa Nasional – LAPAN), and Secretary, Indonesian National Council for Aeronautics and Space (DEPANRI - Feb.1999-June.2000), Adviser for Aerospace Affairs to the Vice President of the Republic of Indonesia, April-June 1998, Inspector General for Development, Technology and Strategic Industries, at the Office of the President of the Republic of Indonesia (June 1998 – January 1999), Deputy Chairman for Technology Development, The Agency for the Assessment and Application of Technology/ Badan Pengkajian dan Penerapan Teknologi (1991-1998). He has authored or co-authored 12 books, 13 book chapters, more than 110 journal publications and some 50 conference papers, and is reviewer of some 15 International Journals. He is the recipient of The International Academy of Astronautics Engineering Award in 2015. He has devoted a combined teaching, research, industrial and management experience totaling of 53 years, which is a manifestation of his career objectives, strong interest, dedication and commitments in education, science and technology. He is presently the President of The Institute for the Advancement of Aerospace Science and Technology.

Sharvil Alex Faroz is currently pursuing PhD at the Indian Institute of Technology Bombay under Prof. Siddhartha Ghosh. His research focuses on quantifying uncertainty in reinforced concrete corrosion degradation model through Bayesian updating for service life prediction.

Shyamal Ghosh is a PhD Research Scholar at Indian Institute of Engineering Science and Technology, Shibpur.

Siddhartha Ghosh is a Professor in the Department of Civil Engineering, Indian Institute of Technology Bombay (IIT Bombay), India. His research interests are primarily in the application of probabilistic methodologies in earthquake risk reduction. He also heads the Structural Safety, Risk and Reliability (SSRR) Lab at IIT Bombay.

Swarup Ghosh is a PhD Research Scholar at Indian Institute of Engineering Science and Technology, Shibpur.

Anthony T.C. Goh is an Associate Professor in the School of Civil and Environmental Engineering, Nanyang Technological University. He received his B.Eng (First Class Honours) and PhD from Monash University. Dr Goh is a registered Professional Engineer in Singapore. His teaching, research and professional practice have covered many aspects of civil engineering including soft computing, finite element analysis, earth retaining structures, pile foundations, slope stability, and soil-structure interaction.

Andreea Ion received her MSc (2014) in Intelligent Control Systems and a BSc (2011) in Control Engineering from the Faculty of Automatic Control and Computer Science, University Politehnica of Bucharest, Romania. She is currently working on her PhD in multi-agent systems and artificial intelligence at University Politehnica of Bucharest, Romania. She is a Teaching Assistant at the Department of Automatic Control and Systems Engineering, University Politehnica of Bucharest. Her research interests include artificial intelligence, multi-agent systems applied in the fields of intelligent control systems like fuzzy techniques and genetic algorithms.

Soham Mitra received his B.E degree in Civil Engineering from IIEST, Shibpur in 2015. He also worked as a research associate in the University of Windsor, Canada from Jun-Jul,14. His research interest includes Probabilistic reliability analysis, Concrete Technology.

Monica Patrascu received a PhD in Systems Engineering from University Politehnica of Bucharest in 2012, an MSc in Intelligent Control Systems and a BSc in Control Engineering from the Faculty of Automatic Control and Computer Science, University Politehnica of Bucharest, Romania. She is an Assistant Professor at the Department of Automatic Control and Systems Engineering, University Politehnica of Bucharest. Research interests include intelligent control systems, evolutionary computing, large scale systems, heuristic optimization, emergent systems, multi-agent systems, and complex systems. She is the co-founder of the EMAS (Emergent Multiscale Agents and Services) for a Smarter World Research Group.

Nikil Narayan Pujari is currently pursuing PhD in Department of Civil Engineering, Indian Institute of technology Bombay. His research focuses on evaluation of the seismic fragility estimation of the nuclear containment considering its structural integrity and functional requirement. This involves quantification of material, ground motion and pressure uncertainties.

Rohit Rastogi has been contributing in the domain of Structural Reliability, Fitness for Service Assessment and Software Development for past 20 years. His current research interests are Development of Design Codes based on Probabilistic Methods, Modeling of Flow Accelerated Corrosion and System Identification based on Markov Chain Monte Carlo Methods.

R.M. Chandima Ratnayake is a Professor (100%) in the Department of Mechanical & Structural Engineering and Materials Science (IKM), University of Stavanger (UiS) – Norway, where he has been since 2006. He has a B.Sc. in Production Engineering, M.Sc. in Manufacturing Engineering and PhD in Offshore Engineering (Asset Integrity Management). Professor Ratnayake is also a 'Senior Integrity Management - Technical Advisor, Wood Group Integrity Management, Wood Group Kenny Norway AS. He has also served as a 'Maintenance Engineering Specialist' (20%) in ApplySørco, AS (2012.11-2014.02), Norway and as a 'Senior Engineer' in AkerSolutions Offshore Partner (2009-2011), Stavanger, Norway. He also served in the capacities of Associate Professor, visiting Associate Professor and Assistant Professor from August 2007 to July 2010 in the University of Stavanger, Norway and as a lecturer to a number of universities & educational institutions in Sri Lanka. Presently he lectures: Manufacturing & Production Engineering and Product Development for B.Sc. students, Computer Aided Engineering for M.Sc. students. He also offers two PhD courses (Industrial asset integrity assessment & control and Multi-criteria Decision Analysis). Moreover, he conducts courses for practicing engineers in Oil & Gas industry and manufacturing sector in national and international level. He has been invited for numerous international level conferences as a speaker within the contexts of Asset integrity control and management, Health, Safety & Environmental issues, Sustainability, Maintenance, Criticality Analysis, Risk based inspection and Reliability centered maintenance, Offshore Data Management, Human factors, etc. Professor Ratnayake supervises B.Sc., M.Sc., and PhD projects. He has published number of articles in peer reviewed journals & conference proceedings and book chapters. He is an Editor of the Central European Journal of Engineering (Industrial Engineering/Offshore Engineering) and a member of editorial board of International Journal of Computational Systems Engineering. He has also been served as an ad hoc reviewer and member of technical committee for several international conferences and journals.

Mahendra Kumar Samal is a senior scientist at Reactor Safety Division of Bhabha Atomic Research Centre (BARC) and associate professor in the divison of engineering sciences at Homi Bhabha National Istitute (HBNI), Mumbai, India. He has more than 20 years of research and teaching experience. He received his Ph.D. from University of Stuttgart, Germany and has carried out post-doctoral research work at Ohio State University, USA. His teaching and research interests include finite element analysis, multi-scale material modeling, smart structures, creep, fatigue, modeling of nolinear behavior of piezoelectric material, fracture and damage mechanics. He has implemented the online creep and fatigue damage monitoring system at various chemical process and thermal power plants of India. He has developed finite element based packages such as DEFINE for damage mechanics and CRYSP for crystal plasticity which are widely used by his students and colleagues for carrying out advanced nonlinear finite element analysis. Under his guidance, 9 students have competed master's thesis. Currently, 4 students are carrying out master's theses work and 4 students are pursuing Ph.D. work. He has 220 publications including 65 papers in various international journals.

Subramaniam Shanmugam has earned Honours degree in bachelor of engineering in the discipline of Mechanical Engineering from Coimbatore Institute of Technology, Coimbatore which was affiliated to erstwhile Madras University. He has got master's degree in Production Engineering from PSG College of Technology, Coimbatore affiliated to Bharathiar University, Coimbatore and obtained Doctor of Philosophy in Mechanical Engineering from Indian Institute of Technology Madras, Chennai. He has been working at National Institute of Technology Tiruchirappalli since 1991. He has rich teaching and research experience of over quarter century. He has published more than seventy research articles in reputed international and national journals. He has presented more than forty articles in conferences.

A. R. Veerappan completed his B.E. degree in Mechanical Engg from Madurai Kamaraj University in 1992 and M.E.degree in Design & Production of Thermal Power Equipment from R.E.C. Trichy in 1994. He acquired a Ph.D. (Gold medal) in Mechanical Engineering from National Institute of Technology Tiruchirappalli in 2008. Currently, he is working as an Associate Professor in the Department of Mechanical Engineering, NIT, Trichy. He has more than thirty-five papers published in journals of good repute and presented around nineteen articles in national and international conferences.

Wengang Zhang was a Research Fellow in the School of Civil and Environmental Engineering, Nanyang Technological University, Singapore. He received his B.Eng from Hohai University, China and PhD from Nanyang Technological University. Now he is working in Chongqing University, China, School of Civil Engineering, as a professor. Dr. Zhang's research interests include: slope stability, reliability based design, braced excavation and soil liquefaction.

www.ingramcontent.com/pod-product-compliance
Lightning Source LLC
LaVergne TN
LVHW060349300926
849162LV00006B/1229